200 Years of British Hydrogeology

Special Publication reviewing procedures

The Society makes every effort to ensure that the scientific and production quality of its books matches that of its journals. Since 1997, all book proposals have been refereed by specialist reviewers as well as by the Society's Books Editorial Committee. If the referees identify weaknesses in the proposal, these must be addressed before the proposal is accepted.

Once the book is accepted, the Society has a team of Book Editors (listed above) who ensure that the volume editors follow strict guidelines on refereeing and quality control. We insist that individual papers can only be accepted after satisfactory review by two independent referees. The questions on the review forms are similar to those for *Journal of the Geological Society*. The referees' forms and comments must be available to the Society's Book Editors on request.

Although many of the books result from meetings, the editors are expected to commission papers that were not presented at the meeting to ensure that the book provides a balanced coverage of the subject. Being accepted for presentation at the meeting does not guarantee inclusion in the book.

Geological Society Special Publications are included in the ISI Index of Scientific Book Contents, but they do not have an impact factor, the latter being applicable only to journals.

More information about submitting a proposal and producing a Special Publication can be found on the Society's web site: www.geolsoc.org.uk.

It is recommended that reference to all or part of this book should be made in one of the following ways:

MATHER, J. D. (ed.) 2004. *200 Years of British Hydrogeology*. Geological Society, London, Special Publications, **225.**

ROSE, E.P.F., MATHER, J.D. & PEREZ, M. 2004. British attempts to develop groundwater and water supply on Gibraltar 1800-1985. *In*: MATHER, J. D. (ed.) 2004. *200 Years of British Hydrogeology*. Geological Society, London, Special Publications, **225,** 239-262.

GEOLOGICAL SOCIETY SPECIAL PUBLICATION NO. 225

200 Years of British Hydrogeology

EDITED BY

J. D. MATHER

University of London, UK

2004

Published by

The Geological Society

London

THE GEOLOGICAL SOCIETY

The Geological Society of London (GSL) was founded in 1807. It is the oldest national geological society in the world and the largest in Europe. It was incorporated under Royal Charter in 1825 and is Registered Charity 210161.

The Society is the UK national learned and professional society for geology with a worldwide Fellowship (FGS) of 9000. The Society has the power to confer Chartered status on suitably qualified Fellows, and about 2000 of the Fellowship carry the title (CGeol). Chartered Geologists may also obtain the equivalent European title, European Geologist (EurGeol). One fifth of the Society's fellowship resides outside the UK. To find out more about the Society, log on to www.geolsoc.org.uk.

The Geological Society Publishing House (Bath, UK) produces the Society's international journals and books, and acts as European distributor for selected publications of the American Association of Petroleum Geologists (AAPG), the American Geological Institute (AGI), the Indonesian Petroleum Association (IPA), the Geological Society of America (GSA), the Society for Sedimentary Geology (SEPM) and the Geologists' Association (GA). Joint marketing agreements ensure that GSL Fellows may purchase these societies' publications at a discount. The Society's online bookshop (accessible from www.geolsoc.org.uk) offers secure book purchasing with your credit or debit card.

To find out about joining the Society and benefiting from substantial discounts on publications of GSL and other societies worldwide, consult www.geolsoc.org.uk, or contact the Fellowship Department at: The Geological Society, Burlington House, Piccadilly, London W1J 0BG: Tel. +44 (0)20 7434 9944; Fax +44 (0)20 7439 8975; E-mail: enquiries@geolsoc.org.uk.

For information about the Society's meetings, consult *Events* on www.geolsoc.org.uk. To find out more about the Society's Corporate Affiliates Scheme, write to enquiries@geolsoc.org.uk.

Published by The Geological Society from:
The Geological Society Publishing House
Unit 7, Brassmill Enterprise Centre
Brassmill Lane
Bath BA1 3JN, UK
(*Orders*: Tel. +44 (0)1225 445046
Fax +44 (0)1225 442836)

Online bookshop: http://bookshop.geolsoc.org.uk

British Library Cataloguing in Publication Data
A catalogue record for this book is available from the British Library.

ISBN 1-86239-155-6

Typeset by Servis Filmsetting Limited, Manchester, UK

Printed by Antony Rowe Ltd, Chippenham, UK

Distributors

USA
AAPG Bookstore
PO Box 979
Tulsa
OK 74101-0979
USA
Orders: Tel. + 1 918 584-2555
Fax + 1 918 560-2652
E-mail bookstore@aapg.org

India
Affiliated East-West Press PVT Ltd
G-1/16 Ansari Road, Daryaganj,
New Delhi 110 002
India
Orders: Tel. +91 11 327-9113
Fax +91 11 326-0538
E-mail affiliat@nda.vsnl.net.in

Japan
Kanda Book Trading Company
Cityhouse Tama 204
Tsurumaki 1-3-10
Tama-shi, Tokyo 206-0034
Japan
Orders: Tel. +81 (0)423 57-7650
Fax +81 (0)423 57-7651
E-mail geokanda@ma.kcom.ne.jp

Contents

Preface

The idea for this book arose when I was asked to prepare a paper for an international conference held in London in July/August 1997 to mark the bicentenary of the birth of Sir Charles Lyell. The brief which I was given was to review how the science of hydrogeology had developed in the UK during the 19th century (Mather 1998).

Searching the literature it was soon clear that, although there was a wealth of published information, this was hidden away in obscure journals and forgotton pamphlets and had been ignored by historians of the science. Thus the distinguished American hydrologist Oscar Meinzer documented the contributions made by French, German and Italian engineers, geologists and drillers but amongst British scientists found only the work of William Smith worthy of note (Meinzer 1934). Recording the early years of the 20th century he wrote '. . . British hydrologists have been active in developing groundwater supplies but have perhaps contributed less notably to the science of ground-water hydrology' (Meinzer 1934 p. 20).

The papers in this book represent an attempt to document the achievements of British hydrogeologists during the last 200 years and to show that not all of their work has been descriptive. 25 contributions are included, the majority of which deal with the period between 1800 and 1975. 15 of these were presented at a joint meeting of the History of Geology and Hydrogeological Groups of the Geological Society held in London on 12th December 2002.

Following an introductory paper by **Mather**, papers by **Torrens**, **Price**, **George**, **Mather** *et al.*, **Tellam** and **Preene** record the careers and achievements of William Smith, John Snow, William Whitaker, Joseph Lucas, Robert Stephenson and the members of the Geological Societies of Liverpool and Manchester, all of whom made major impacts on hydrogeological thinking during the 19th century. Papers by **Younger**, **Rose** and **Robins** *et al.* concentrate on groundwater work in Ireland and Scotland and the contribution of mining and military scientists and engineers to groundwater development.

The history of some of the British spas and hydropathic establishments is reviewed in **Edmunds**, **Fuller** and **Spence & Robins** and the work of British hydrogeologists overseas in **Lloyd**, **Hazell**, **Rose** *et al.* and **Barker**. From 1935 the Geological Survey began to have a significant role. Together with some of the personel involved, this is reviewed in **Downing**, **Downing & Gray**, **Gray & Mather** and **Gray**. The final chapters by **Downing** *et al.*, **Downing**, **Headworth** and **Brassington** review the innovative work of Norman Boulton and the more recent developments which have taken place since 1965. There is some minor overlap between the later chapters but this has been allowed to remain because of the different perspective provided by their authors.

In any book of this kind it is inevitable that some omissions will occur and that the contribution of some workers will not receive the recognition which it perhaps deserves. However, this does not set out to be a complete record of the development of hydrogeology in Britain over the last 200 years, rather a series of snapshots and reviews of particular periods and the people involved. Despite this it is hoped that most of the important events and individuals have been covered.

I am especially grateful to the following friends and colleagues who refereed the papers contained in this book and whose help and support have improved it immensely: R. P. Ashley, S. A. Baldwin, D. Banks, A. H. Bath, J. H. Black, F. C. Brassington, M. A. C. Browne, W. G. Burgess, P. J. Chilton, L. J. Clark, L. R. M. Cocks, J. Davies, N. J. Dottridge, R. A. Downing, K. J. Edworthy, J. Flude, J. G. C. M. Fuller, D. A. Gray, J. M. Hancock, R. C. Harris, I. B. Harrison, R. J. Howarth, J. A. Heathcote, M. P. Henton, R. A. Herbert, K. M. Hiscock, G. P. Jones, J. B. Joseph, C. V. Knipe, C. L. E. Lewis, D. N. Lerner, J. W. Lloyd, G. I. Lumsden, A. J. Martin, C. P. Nathanail, D. Norbury, M. O. Rivett, N. S. Robins, M. S. Rosenbaum, G. M. Reeves, A. C. Skinner, S. I. Snow, H. S. Torrens, W. B. Wilkinson, G. Wright and P. L. Younger.

References

MATHER, J. D. 1998. From William Smith to William Whitaker: the development of British hydrogeology in the nineteenth century. *In*: BLUNDELL, D. J. & SCOTT, A. C. (eds) *Lyell: the past is the Key to the Present.* Geological Society, London, Special Publications, **143**, 183–196.

MEINZER, O. E. 1934. The history and development of ground-water hydrology. *Journal Washington Academy of Science*, **24**, 6–32.

John Mather,
Emeritus Professor of Geology,
Royal Holloway University of London,
Egham, Surrey, TW20 0EX, UK.

200 years of British hydrogeology – an introduction and overview

J. D. MATHER

Department of Geology, Royal Holloway University of London, Egham, Surrey, TW20 0EX, UK

Abstract: In the early years of the 19th century William Smith and his pupil John Farey began to apply stratigraphic principles to the sinking of water wells. Between 1840 and 1870 engineers, such as Robert Stephenson, and geologists, such as James Clutterbuck, Joseph Prestwich and David Ansted began to make systematic observations. After 1870 Geological Survey officers, particularly William Whitaker, Joseph Lucas and Charles de Rance, became involved in groundwater work. Lucas introduced the term hydrogeology and produced the first British hydrogeological maps. The first half of the 20th century was a period of missed opportunities with the significant advances in hydrogeology taking place in mainland Europe and North America. The Water Act of 1945 marked the start of a new era in which the Geological Survey and, after 1965, the Water Resources Board led the way. Hydrogeology is now a mainstream branch of geology in Britain and interest in the subject is such that the Hydrogeological Group of the Geological Society has a membership of around 1050.

Until relatively recent times our forebears had rather primitive ideas about the origins of groundwater. In the 17th century most scholars believed that groundwater had flowed underground from the sea as they could not believe that rainfall and snowmelt alone were sufficient to support the flow of springs and rivers throughout the year. The challenge of explaining how seawater raised itself to great heights above the sea, losing its salinity during transit, was taken up with enthusiasm and many fanciful theories were advanced (Adams 1938).

The Frenchman Pierre Perrault (1611–1680) was the first, in 1674, to demonstrate experimentally that rainfall is more than adequate to account for the flow of rivers and springs (Perrault 1674). The concept of the hydrological cycle was further developed by the British astronomer Edmond Halley (1656–1742) who showed that water evaporated from the oceans and returned as rainfall (Halley 1687). Halley was the first British investigator to use quantitative results in a hydrological context. Just over a century later John Dalton (1766–1844), working in NW England, was the first to express the basic components of the evaporation process from a free water surface in quantitative terms (Dalton 1802). Dalton's law, although never expressed by the author in mathematical terms, has formed the starting point of much of the subsequent work on evaporation (Ward & Robinson 1990). Also towards the close of the 18th century the Derbyshire physician Erasmus Darwin (1731–1802), grandfather of Charles, used his knowledge of inclined strata to improve the water quality in an old well in his garden. His paper, read to the Royal Society in November 1784, concisely summarises the principles of the artesian well (Darwin 1785). Thus, as in many other branches of science, British investigators were at the forefront in the early development of hydrological concepts and the role of groundwater in the hydrological cycle.

The distinguished American hydrogeologist Stanley Davis has given a number of reasons why todays hydrogeologists should look back at the accomplishments of their predecessors (Davis 2000). The first lesson of history is that hydrogeologists need to avoid professional tunnel vision as many advances have been made by involving scientists, and embracing ideas, from other areas of science and technology. For example it was the cooperation with physicists which led to the recognition of the important role of matrix flow within the British Chalk aquifer (Smith *et al.* 1970).

A careful look at history should prevent continuous reinvention of the wheel. A recent paper contained in its conclusion the statement that 'An increasing understanding of pollution has helped identify many previous unthought-of potentially polluting activities, one of which is cemeteries.' (Lelliott 2002). The idea that cemeteries have been recognized, only recently, as a source of pollution would have amused his Victorian precedessors, one of whom provided guidence on their location to avoid the pollution of underground water supplies (Woodward 1897).

Davis pointed out that careful and accurate measurements do not decrease in value with time. Old borehole records contain irreplaceable data on groundwater levels and chemistry. The maintenance of such records has, for example, enabled the fluctuating groundwater levels beneath many British cities to be monitored since Victorian times and the present rising levels to be put in context (e.g. Environment Agency 2002).

Finally history proves repeatedly that great men can be wrong. The great engineer, Robert Stephenson, considered that on the Chalk 'The rapidity with which water finds its way into the bowels of the earth also in great measure prevents evaporation, and we are therefore justified in assuming that the quantity which

From: MATHER, J. D. (ed.) 2004. *200 Years of British Hydrogeology*. Geological Society, London, Special Publications, **225**, 1-13. 0305-8719/04/$15 © The Geological Society of London.

descends upon the surface of the chalk finds its way, with very slight diminution, into the fissures below.' (Stephenson 1840). Hydrogeologists have known for many years that such an assumption is not justified.

In Britain, the past 200 years have witnessed enormous changes in our use of groundwater. The impetus in the 19th and early 20th centuries came from industry and the needs of an expanding population. Development took place in a piecemeal fashion with the emphasis on local problems and solutions. The Water Act of 1945 saw the beginnings of a quantitative approach to hydrogeology and the recognition that sectional interests needed to be subordinate to the national interest. From 1945 onwards there has been a continual increase in the number of practising hydrogeologists with a gradual change in the emphasis of their work from resource assessment and development to quality issues. The objective of this introductory paper is to summarise the changes and developments which have taken place over the last 200 years and to identify some of the principal scientists who have contributed.

William Smith, artesian wells and the period to 1840

In British geology many practical applications of the subject originate with William Smith (1769–1839) and hydrogeology is no exception. His first known order of strata, drawn up in 1797, shows a keen awareness of the occurrence of spring lines and his better known 1799 version shows springs tabulated against five separate formations (Torrens 2004). He had a good understanding of the stratigraphic control of spring lines, as well as the significance of hydraulic head, using this knowledge to advise landowners and canal companies on water supplies (Phillips1844; Sheppard 1920).

William Smith was one of the three prominent authorities on practical or applied geology practising at the beginning of the 19th century, none of whom were ever elected to membership of the Geological Society of London. Of the others Robert Bakewell (1767–1843) produced one of the best early textbooks on geology but it contains almost nothing on groundwater (Bakewell 1813). John Farey (1766–1826) recognized the importance of springs in mining, agriculture and engineering and noted that few subjects had been encumbered with more wild and fanciful theories (Farey 1811 p. 500). He recognized the significance of Smith's work to the origin and course of springs and applied it to the sinking of deep wells, particularly in the Thames Basin (Farey 1807).

The gentlemen of the Geological Society of London, although not interested in the practical applications of geology, were involved in the collection of facts by experiment or observation in order to provide a basis for geological theories (Rudwick 1963). In 1818 one of their members, William Phillips (1775–1828), summarized published strata descriptions to provide an outline of the geology of England and Wales (Phillips 1818). Although this contained little concerning groundwater, the second edition, to which William Daniel Conybeare (1787–1857) added much original material, contains some hydrogeological information (Conybeare & Phillips 1822). Within each strata description, following discussion of the character, extent, thickness, etc., is a section entitled 'phenomena of water and springs'. Although some of these sections are only 2 or 3 lines in length others, such as those referring to the London Clay and Chalk, run to several pages. Terms, such as 'porous', 'porosity' and 'impervious to water' are used and the poor quality of water in the London Clay is ascribed to the decomposition of pyrite. Water levels in four wells over a distance of 15 miles from the River Thames to Epping are plotted on a diagram which is probably the earliest British hydrogeological cross-section. A brave attempt is made by Conybeare to explain the variations in the water levels observed and he concluded that 'The only general rule that can be deduced is, – that the water of wells can in no case rise to a higher level, than the highest point of the strata collecting them . . .' (Conybeare & Phillips 1822 p. 36). In contrast the 'Principles of Geology', the influential book by Charles Lyell (1797–1855), contains almost nothing on groundwater. Chapter 12 of Volume 1 is concerned with springs but only in their role as a means of transferring material in solution from depth and not in their role as a source of potable water (Lyell 1830–1833).

In the early part of the 19th century significant advances in the study of groundwater were being made in France largely as a result of the drilling of artesian wells. The term artesian was not used by Conybeare & Phillips (1822) although they described the phenomenon of overflowing wells. The term seems to have first appeared in British journals in 1823 the year in which John Farey called for the translation of relevant material published abroad (Farey 1823). Over the next decade the practical memoir of Garnier (1822) and the more theoretical work of Héricart de Thury (1829) were used widely in the absence of any comparable text in English the first of which did not appear until 1849 (Swindell 1849). The theory behind the occurrence of artesian flow was well understood by William Buckland (1784–1856) who produced excellent explanatory sections in his Bridgewater Treatise (Buckland 1836) based largely on French texts available in English in the 1830s (Héricart de Thury 1830; Arago 1835).

From the 1820s onwards papers related to groundwater began to appear in the Transactions and Proceedings of the Geological Society and from

1836 in the Transactions of the Institution of Civil Engineers. Many of the published items are merely descriptions of wells or boreholes together with the strata intersected (e.g. Yeats 1826; Donkin 1836). However, in 1831 the Proceedings of the Geological Society record that a letter was read 'On the Influence of Season over the Depth of Water in Wells' (Bland 1831). The work was later published in the Philosophical Magazine (Bland 1832) and described a series of monthly observations made in a well near Sittingbourne in Kent from January 1819 to June 1831. These observations represented the first British systematic record of fluctuations in groundwater level and demonstrated that levels varied seasonally with the greatest depth of water 'at and about the longest day' and the least depth 'at or about the shortest day'. Bland also measured the water levels along three traverses across the Chalk and was able to demonstate that the height to which water rose in wells correlated with the rise and fall of the hills (Bland 1832). At about the same time, the mining geologist William Jory Henwood (1805–1875) began to relate the amount of water pumped out of the Cornish Mines with rainfall, mine depth and whether the mine was in granite or slate (Henwood 1831; 1843).

By 1840 the construction of artesian wells and boreholes was generally understood and many had been sunk or drilled (Mylne 1840). However, the concept did not meet with universal approval amongst engineers and there were some who felt that boring should be the last method of resort for the purpose of supplying a large town (Seaward 1836). Many boreholes failed because they were drilled by engineers with no geological background who assumed that you could drill almost anywhere and find artesian water (see Farey 1822). Although Britain lagged behind in the development of drilling technology, pumping technology was well advanced largely because of equipment inherited from the mining industry (Younger 2004).

Stephenson, Clutterbuck, Prestwich and the years from 1840 to 1870

By 1840 the provision of an adequate water supply for London had become a major issue. Most parts of London were supplied either from shallow wells subject to contamination or from the highly polluted River Thames. A number of Committees had been set up to try to tackle the water supply problem and the distinguished engineer Thomas Telford was engaged to suggest a practicable plan. However nothing came of this and in 1840 another distinguished engineer Robert Stephenson (1803–1859) was commissioned to produce a report. Stephenson proposed to supply NW London from a well into the unconfined Chalk at Bushey Meads near Watford in Hertfordshire (Preene 2004). Stephenson's proposals were based on the premise that recharge was so rapid that there was little evaporation and most of the rainfall accumulated in the lower part of the Chalk forming an enormous natural reservoir from which water could be abstracted without affecting surface springs and streams.

The scheme was vigorously opposed by a number of local landowners and mill-owners and in particular by the Reverend James Charles Clutterbuck (1801–1885), the vicar of Long Wittenham in Oxfordshire but a native of Watford. Pamphlets and counter-pamphlets were issued and the debate became extremely acrimonious (Preene 2004). The debate is important because it demonstrates how significantly knowledge of groundwater had advanced since 1800. Stephenson (1841) accurately described the shape of the cone of depression around a pumping well. Clutterbuck began a series of detailed systematic observations of groundwater levels the conclusions from which were reported in a series of presentations to the Institution of Civil Engineers (Clutterbuck 1842, 1843, 1850). His observations enabled him to recognize the intimate relationship between surface water and groundwater, the depressed water levels beneath London caused by pumping, and the possibility of saline intrusion into the Chalk from the River Thames. He was the first British worker to make systematic observations of groundwater levels and apply these in a practical and innovative way to the study of groundwater flow.

Joseph Prestwich (1812–1896), one of the gentleman geologists of the Geological Society, also became interested in the problems of London's water supply. Later to become Professor of Geology at the University of Oxford, in the 1840s he was a wine merchant with offices in London. In 1850 he read a paper to the Royal Institute of British Architects suggesting that the Upper and Lower Greensands beneath London might provide a suitable auxiliary source of supply (Prestwich 1850). This paper was later expanded into a book (Prestwich 1851) which became widely quoted and used, so much so that it was reissued shortly before Prestwich's death (Prestwich 1895). Although the Lower Greensand had a more limited range beneath London than Prestwich envisaged, and hence never yielded the volume of water anticipated, the book is significant for a number of reasons. Prestwich states that he was drawn to the subject by the lack of geological information bearing upon the question of artesian wells beneath London and his book presents a review of the geology of the country around London followed by an appraisal of their hydrogeological properties and likely safe yields; it is in effect the first British hydrogeological memoir. Prestwich uses the term permeability in its modern context and provides a map and sections which divide the strata according to their permeability (Prestwich 1851). The map is the first British geological map to show hydrogeological information. Later Prestwich was a member of the Royal Commission on Water Supply, which reported in 1869, and became a widely respected authority on the application of geology to water supplies.

Working at about the same time as Prestwich, David Thomas Ansted (1814–1880) began experiments on the absorption of water by chalk (Ansted 1850) demonstrating that at least one third of the bulk of fully saturated chalk consisted of water. Data on the absorbent power of various rock types had been collected previously by the Commissioners appointed to select the building stone for the new Houses of Parliament, one of whom was William Smith (Royal Commission 1839).

It was not only in SE England that geologists and engineers became interested in groundwater. Whilst working in the coalfields of Lancashire and Cheshire, Edward Hull (1829–1917) recognized the value of the Permo-Triassic sandstones as a source of water to supply local towns. He recommended the line of a fault in these rocks as the best site for a well as it was certain to draw in water from a long distance (Hull 1865). Although the Geological Survey had probably been answering enquiries in relation to water supply since its foundation in 1835, Hull seems to have been the first to have an active interest in publication. The Permo-Triassic rocks of the Liverpool and Manchester area were also the subject of considerable research activity from the mid-1800s. Most workers were amateur members of the local geological societies (Tellam 2004) but Robert Stephenson also contributed (Preene 2004).

Water quality became an issue early in the century but it was not until 1854 that John Snow (1813–1858), working in Westminster, demonstrated beyond doubt that cholera was spread by contamination of drinking water (Price 2004). Subsequent research showed that sewage effluent derived from a local cesspool was to blame. Snow's study represents one of the first, if not the first, study of an incident of groundwater contamination in Britain.

Analysts became more confident in their determination of the constituents of water. Mineral springs such as those of Bath (Edmunds 2004), Tunbridge Wells (Fuller 2004) and Bridge of Allan (Spence & Robins 2004) became field sites for testing developments in analytical chemistry, but doubts about the true composition of such waters remained (see discussion in Hamlin 1990).

As early as 1850 Lyon Playfair (1819–1898), at that time the chemist at the School of Mines and later a noted administrator and educational reformer, provided an excellent description of cation exchange in Chalk groundwaters (Playfair in discussion of Clutterbuck 1850 p. 160). However, most of the early papers published by chemists were reports on the species dissolved in groundwater although some speculated on the source of groundwater and how it achieved its composition (Campbell 1857).

The Report of the Royal Commission on Metropolitan Water Supply (1869) shows that by 1869 a number of large conurbations were supplied by groundwater. Nottingham and parts of Liverpool and Birkenhead relied on wells in the Permo-Triassic sandstones, Sunderland and South Shields on the Permian Magnesian Limestone and Croydon and parts of SE London on the Chalk. Government offices around Westminster together with the fountains in Trafalgar Square were supplied by wells sunk to the Chalk in 1844 (Amos 1860).

The Geological Survey and the years from 1870 to 1900

Prior to 1870 the Geological Survey of Great Britain made little contribution to the application of geology to water supply. As early as 1850, some 15 years after its formation and under its first Director Henry Thomas de la Beche (1796–1855), complaints were already being made that mapping work should be transferred from North Wales, where no need existed for early geological information, to the metropolitan districts to investigate the deep water-bearing strata (Clutterbuck 1850). The second Director of the Survey, Roderick Impey Murchison (1792–1871), was a distinguished gentleman geologist and Fellow of the Royal Society, who had little interest in the economic applications of geology (Flett 1937). However, three surveyors whom he recruited were to make an impact in the last 30 years of the 19th century.

According to the bibliography prepared by Whitaker (1888) Hull's memoir on the geology of the country around Bolton-le-Moors in Lancashire was the first to include well sections (Hull 1862). However, after Murchison's death in 1871 the number of memoir pages devoted to well sections increased enormously covering some 141 pages in the London Basin Memoir (Whitaker 1872). Subsequently all memoirs relating to SE England had lists of well sections and information concerning water supply. Eventually these lists came to dominate the geological memoirs such that in 1899 the first Water Supply Memoir, on Sussex, was produced (Whitaker & Reid 1899).

Many of the memoirs were authored by William Whitaker (1836–1925) who had joined the Survey in 1857 (George 2004). Whitaker was an assiduous collector of records of well sections and temporary exposures and he also compiled numerous bibliographies. However, he made little use of the data he collected, preparing lists of information rather than using this information to understand hydrogeological processes. According to Wilson (1985 p. 125) 'There is no doubt . . . that Whitaker was the father of English hydrogeology' although Mather (1998) has characterized his contribution as merely 'worthy'.

Joseph Lucas (1846–1926) joined the Geological Survey in 1867 and spent 9 years mapping in Yorkshire before being forced to resign for a disciplinary offence (Mather *et al.* 2004). In 1874, whilst still with the Survey, he was the first person to use the term 'hydrogeology' in its modern context (Lucas 1874; Mather 2001) and defined this new subject in a series of papers in the 1870s (Lucas 1877*a*). He used water level data to draw the first British maps showing groundwater contours and described how to carry out a hydrogeological survey (Lucas 1874, 1877*b*, 1878). His innovative work was immediately taken up by other workers and the term hydrogeology was widely adopted (e.g. Prestwich 1876: Inglis 1877). For the next ten years of his life Lucas lobbied for hydrogeological surveys to be carried out over the whole country but his ideas were ignored.

The third geologist in the Survey to make an impact was Charles Eugene de Rance (1847–1906) who joined in 1868, the year after Lucas. On behalf of the Survey he assisted the Rivers Pollution Commission in the preparation of its sixth report on domestic water supply (Rivers Pollution Commission 1875). The Commissioners concluded that 'spring waters' and 'deep well waters' were the best sources of supply providing wholesome and palatable water for drinking and cooking. This testimonial provided a boost to the development of groundwater. In 1874 he was appointed the secretary of a Committee appointed by the British Association for the Advancement of Science for the purpose of investigating the circulation of underground waters. Edward Hull was the Chairman of this Committee and although it sat for 20 years it achieved little of substance. De Rance's work on these two bodies together with his own research enabled him to produce a 600 page volume on the water supply of England and Wales (de Rance 1882). The book is an account of the water supplied to each town and urban sanitary authority in England and Wales and includes a small hydrogeological map, on which the strata are divided in terms of their permeability.

In the latter part of the 19th century the involvement of the Geological Survey was but one part of a trend which saw the study of the geological aspects of water supply becoming respectable. Joseph Prestwich was elected President of the Geological Society in 1870 and in 1872 chose as the subject of his Presidential Address 'Our Springs and Water-supply and Our Coal-measures and Coal-supply' (Prestwich 1872). In 1876 his successor John Evans (1823–1908) also chose 'Water Supply' as one of the topics of his Presidential Adddress (Evans 1876) and in 1898 Whitaker was elected President chosing 'Water-supply and sanitation' as his first topic (Whitaker 1899).

Throughout the country groundwater was now being exploited for both urban and rural supplies and geologists, chemists and engineers were involved in exploration and development. Numerous papers and reports appeared, many emphazising the vastness of the underground water supplies available. The debate about supplying London from the Chalk continued with geologists emphazising the interdependence of groundwater and surface water and the dangers of overexploitation (e.g. Evans 1876; Hopkinson 1891) whilst most engineers still considered that rivers took their supply 'from the skin of the chalk formation' with a much larger quantity of water available for extraction from 'the great body of the Chalk below' (Harrison 1891 p. 21). Ansted wrote a text book on 'Water and Water-Supply' but his death in 1880 meant that only the first part on 'Surface Waters' was ever published (Ansted 1878).

Many papers were descriptive but some contributed towards an understanding of hydrogeological processes. In 1884 an earthquake caused much damage to buildings in Essex and had a significant impact on groundwater levels causing rises of up to 7 feet in some wells. Numerous reasons for the rise were advanced the most widely accepted being put forward by de Rance who suggested that the shock caused a widening of the fissures in the Chalk increasing the flow which led in turn to a rise in water level (de Rance 1884). Whitaker was involved as a consultant in an early study of groundwater pollution (Whitaker 1886). A brewery well in Brentford, west London, was found to be polluted from another well 297ft (90.5 m) away which had been turned into the drainage for a privy belonging to a printing works. The connection between the wells was proved by the use of lithium chloride as a tracer. The case, Ballard v. Tomlinson established the principle that no owner had the right to pollute a source of water common to his own and other wells.

The influence of barometric pressure on the discharge of springs was described (Latham 1882) and the dual permeability of the Chalk was recognized. Adits or headings were dug laterally from shafts to intersect as many fissures as possible and these reached a length of over two miles at the Ramsgate Waterworks in Kent. It was recognized that vast amounts of water were available in the body of the Chalk but that '. . . the failure of wells, sunk in chalk free from fissures and cavities, proves that capillary water does not travel with sufficient swiftness to be available' (Dawkins 1898, p. 262). Nearly all the work was concentrated in England largely because of the abundance of good quality surface waters in Wales, Scotland and Ireland (Robins *et al.* 2004).

Missed opportunities and the years from 1900 to 1945

At the beginning of the 20th century attention focussed on the ownership of groundwater and the 'Kent Water Preservation Association' was formed to conserve the waters of Kent for its own inhabitants and to prevent them being exploited to supply London. The association was later extended to include other counties, the name altered to 'The Underground Water Preservation Association' and a pamphlet issued (Beadle 1902). It was now generally admitted that groundwater levels beneath London were falling (Beadle 1903) and this was demonstrated conclusively by Barrow and Wills (1913).

In the years between 1900 and 1938 the Geological Survey published 27 Water Supply Memoirs. These consisted mostly of records of wells and boreholes many of which could be characterized as 'unchecked records of wells that have been sunk through ill-defined strata' (Bailey 1952 p. 202). Only in two of these memoirs, those for Essex (Whitaker and Thresh 1916) and the County of London (Buchan 1938) was any attempt made to produce maps of standing water levels and the Essex map was produced by the consultants H. Rofe and Sons. Prior to 1940 the approach of the Survey was extremely conservative and progress compared very unfavourably with that made by the United States Geological Survey during this period. The general descriptive approach of the British Survey is well illustrated by the textbook on the 'Geology of Water Supply' written by Horace Bolingbroke Woodward (1848–1914) a former senior member of the Directorate (Woodward 1910).

Some workers knew of the pioneering work going on in the USA and Europe and were able to use it to examine groundwater flow in British aquifers (e.g. Baldwin-Wiseman 1907). Later Norman Savage Boulton (1899–1984), in work which predated that of Theis in the USA, examined the time-variant flow to a pumped well in a confined aquifer (Downing *et al.* 2004). However, to his great disappointment his manuscript was rejected for publication, a sad reflection of the conservative views of his peers in the UK in the early 1930s.

The innovative regional work, such as it was, was carried out by consultants such as Herbert Lapworth (1875–1933) and Partners. One of Lapworth's assistants, Rupert Cavendish Skyring Walters (1888–1980) summarized the hydrogeology of the Chalk compiling groundwater contour maps for the Chalk throughout England (Walters 1929). Later he completed a comparable study of the Jurassic Oolitic Limestone (Walters 1936). Arthur Beeby-Thompson (1873–1968) was another well known consultant who was still active in his 80s. He worked extensively overseas (Beeby-Thompson 1969) including West Africa (Hazell 2004) and Gibraltar (Rose *et al.* 2004).

Many geologists worked for the Geological Surveys of Britain's overseas colonies and protectorates where they became involved in the provision of water supplies (Hazell 2004; Lloyd 2004). The use of geophysical techniques in borehole siting was pioneered in West Africa (Hazell 2004) and Southern Rhodesia, now Zimbabwe (Barker 2004). The British colonial geologist Frank Dixey (1892–1982) wrote a practical textbook which was widely used overseas (Dixey 1931).

In 1901 the origin of alkaline waters in the chalk was explained by the gradual dissolution of sodium carbonate (Fisher 1901). Later John Clough Thresh (1850–1932) showed that the rocks themselves possessed 'the power of softening hard water by substituting sodium salts for those of calcium and magnesium' (Thresh 1912 p. 43).

Groundwater pollution, particularly outbreaks of typhoid fever resulting from sewage leaking directly into wells, received considerable publicity. However the inherent safety of well waters was recognized and experience showed that 'water in slowly percolating through a few feet of compact soil cannot carry with it the microbe causing typhoid fever' (Thresh 1908 p. 109). Sections on contamination and the risk involved were included in some of the longer Geological Survey Water Supply Memoirs (e.g. Whitaker 1912). Parliament empowered many water boards to make bye-laws for protecting land around wells and in the Margate Act of 1902 the water board was given the power to control drains, closets, cesspools etc. over an area of 1500 yards (1372 m) from any well or adit (Thresh & Beale 1925).

During 1932–1934 there was a severe drought in southern and central England. This led to the appointment of an Inland Water Survey Committee in1935 and the eventual formation of a Water Unit within the Geological Survey in 1937 (Downing 2004). The first geologist assigned to this Unit, Francis Hereward Edmunds (1893–1960), began to introduce order to the Survey's water records. The state of knowledge in the Survey at that time was summarized by the then Director, Bernard Smith (1881–1936), in his Cantor lectures to the Royal Society of Arts (Smith 1935).

The outbreak of war in 1939 diverted many staff to water supply work and contributed to a build up of expertise within the Survey which was to prove valuable in postwar Britain. The urgent need for additional and emergency supplies both at home and overseas also involved academics such as William Bernard Robinson King (1889–1963), then Professor of Geology at University College, London, and Frederick William Shotton (1906–1990), a lecturer at Cambridge, who advised the army (Rose 2004) and Percy George Hamnall Boswell (1886–1960),

Professor of Geology at Imperial College London, who advised the Metropolitan Water Board (Boswell 1949). In addition consultants continued to make a major input (e.g. Herbert Lapworth Partners 1946).

Expansion, central control and the period from 1945 to 1975

The Water Act of 1945 was the first piece of major legislation affecting water supply in the UK for almost 100 years and was part of the social revolution which followed the end of the Second World War (Downing 1993). The period between 1945 and 1963 has been described as an era of resource assessment by Downing and Headworth (1990) and the 1945 Act provided the framework within which these assessments were made. Information on the areal distribution of rainfall was already well established in the UK and the work of Penman (1948) enabled evaporation to be estimated. Thus there was the opportunity to assess groundwater resources with far greater accuracy than in the past.

The Geological Survey took the lead in this work as, following the 1945 Act, they became advisors to the Government on technical aspects of groundwater development (Downing 2004*a*). Under the leadership of Stevenson Buchan (1907–1996), the Survey recruited a group of staff who came to consider themselves as hydrogeologists rather than geologists with a peripheral interest in groundwater (Gray & Mather 2004). They were primarily engaged on regional hydrogeological surveys to assess groundwater resources but they were also responsible for testing all new major public supply and industrial wells in order to assess appropriate sustainable yields before abstraction licences were issued by the Government.

Using data from many of the wells tested by the Survey, Jack Ineson (1917–1970) began to apply methods developed in the USA to heterogeneous British aquifers such as the Chalk (Downing & Gray 2004). Other staff pioneered the use of down-hole logging techniques adapted from the oil industry, analysed the river/groundwater interface and synthesized the well records using the data to compile hydrogeological maps (Downing 2004*a*; Gray & Mather 2004).

Outside the Survey, Boulton was working on the delayed yield observed when unconfined aquifers are pumped under non-steady state conditions (Boulton 1954; Downing *et al.* 2004). The potential of artificial recharge began to be taken seriously in the London Basin (Boniface 1959) and natural radioactivity in groundwaters became an issue (Turner *et al.* 1961).

In the late 1950s and early 1960s, the Ministry of Housing and Local Government, in cooperation with the Survey, made a series of hydrological surveys to identify the availability and use of water resources in England and Wales. Unfortunately the 1945 Water Act did not recognize the close links between groundwater and surface water which led to difficulties, as aquifer development and river management were the responsibility of different organisations. The reduction in stream flows as a consequence of increasing groundwater development, particularly in the Chalk, became a major factor leading to the Water Resources Act of 1963 (Downing 2004*b*). This created the Water Resources Board (WRB), to plan water resources development on a national scale, and 29 catchment-based river authorities to control abstraction, prevent pollution, drain land and protect fisheries. The period which ensued from 1963 to 1974 has been called an era of groundwater management by Downing & Headworth (1990).

The WRB made three major regional studies of water resources and, in 1973, proposed a national water strategy (Downing 2004*b*). The core of its Geology Division was made up of seven staff transferred from the Geological Survey, led by Ineson. It was enthusiastic in its support for the development of groundwater within integrated water resource systems, in particular the use of groundwater for river regulation, its use in conjunction with surface water and artificial recharge. In association with Ken Rushton, of Birmingham University, it built some of the first electrical analogue and then mathematical models used in Britain (Downing 2004*b*).Work carried out by Wantage Research Laboratory, in cooperation with WRB, on the movement of thermonuclear tritium through the unsaturated zone demonstrated that only 10 to 15% of infiltration to the Chalk flows in fissures with the remainder moving through the matrix by a form of piston flow (Smith *et al.* 1970) This important conclusion had major implications for the movement of solutes and contaminant transport.

The 1963 Act removed both experienced staff and the groundwater advisory service from the Geological Survey. However, with the support of successive Directors, the Survey increased its research role, undertaking both fundamental and applied research projects in the UK and overseas (Gray 2004). Laboratories were commissioned for research into groundwater modelling, hydrogeochemistry and the measurement of aquifer physical properties. Production of hydrogeological maps was continued using data held by the Survey and groundwater contamination, particularly from landfill leachates and intensive agriculture received significant support from central Government (Gray 2004).

Many of the studies carried out by WRB relied on the cooperation of the water supply and river authorities some of which began to recruit their own geological staff. One of the first of these was

Howard Headworth who joined the Hampshire River Authority in 1965 and subsequently became involved in a range of schemes (Headworth 2004). Another was Rodney Aspinwall who joined the Essex River Authority in the late 1960s. He soon became involved in the burgeoning problems of groundwater pollution and left the Authority to establish his own successful consulting practice – one of the first set up specifically to provide geological and hydrogeological expertise to the water and waste management industries.

At the University of Bath, John Napier Andrews (1930–1994) studied the release of radon from rock matrices and its entry into groundwater (Andrews & Wood 1972). He subsequently became a pioneer in the application of noble gases to problems in hydrogeology (e.g. Andrews & Lee 1979).

The years between 1963 and 1975 were probably the most significant in the history of British hydrogeology. The number of individuals involved in groundwater work, excluding those within consulting engineering companies, rose from less than 20 to around 150. This increasing demand resulted in the setting up of Masters Courses in Hydrogeology at University College London in 1965 and Birmingham University in 1971. The need for a discussion forum led to the formation of the Hydrogeological Group of the Geological Society in 1974 and the Sub-Committee for Hydrogeology of the British National Committee for Geology, serviced by the Royal Society, in 1975. During this period hydrogeology changed from a fringe subject to a mainstream branch of geology in the UK.

Regional development, groundwater quality and the period from 1975 to the present

Since the mid-1970s the development of hydrogeology in Britain has been intimately connected with changes in legislation. A weakness of the Water Resources Act was that there was inadequate provision for the coordination of water resources development and water quality control which were in the hands of separate organisations (Downing 1993). By the mid-1970s there was also a move to devolve power to the regions rather than concentrate it centrally and the Water Act of 1973 was part of the Government's reorganization of local government to achieve this aim. The WRB was disbanded and ten regional water authorities created whose areas of operation were defined by river catchment boundaries so that the whole of the water cycle in a particular area, including water supply and sewage treatment, was under the control of a single body (Brassington 2004). A further restructuring took place in 1989 when the ten water authorities were privatized to become water supply and sewage utility companies with their regulatory function transfered to a new body, the National Rivers Authority. In 1996 the latter became part of the new Environment Agency. The greater emphasis on quality from 1975 onwards prompted Downing & Headworth (1990) to define this period as an era of groundwater quality.

In Scotland and Northern Ireland groundwater received little attention before the mid-1970s when the Geological Survey appointed dedicated hydrogeologists in Edinburgh and Belfast (Gray 2004; Robins *et al.* 2004). The water industry in Scotland was not privatized in the same way as that in England and Wales and is now the responsibility of one multi-functional authority – Scottish Water. Since 1996 environmental regulation, including groundwater protection has been the responsibility of the Scottish Environmental Protection Agency. In Northern Ireland responsibilities for water supply and regulation are vested in the Department of the Environment for Northern Ireland.

The need of regional authorities for personnel to staff water resources and regulatory sections meant an increasing demand for hydrogeologists. Since the changes of 1989 in England and Wales some reduction in numbers has taken place as both the utilities and regulators increased their reliance on consultants. However, with the increase in contract work, the consultants themselves have a requirement for hydrogeologists which many find difficult to satisfy. Both the Geological Survey and the Water Research Centre (formed in 1974 from the research arm of WRB and the Water Research Association, the research organisation funded by the water supply industry) have continued with research and contract work The interest in hydrogeology is now such that the Hydrogeological Group of the Geological Society has a membership of around 1050. The work carried out by this large cohort is described by Brassington (2004).

One major change over the last 30 years has been the diversification of the scope of the work now carried out by the hydrogeological community in Britain. The traditional fields of water supply and water quality continue to dominate but hydrogeologists have also made major contributions to studies on the disposal of radioactive wastes (Chaplow 1996); geothermal energy (Barker *et al.* 2000); the problem of rising groundwater levels (Brassington 1990); mineralization (Barker *et al.* 1999) and climate change (Edmunds & Milne 2001).

Discussion

This paper has identified a number of scientists and engineers who have made a significant contribution to hydrogeological thinking in the UK. William

Smith had an excellent understanding of the stratigraphic control of spring lines and his pupil, John Farey, recognized how Smith's work could be applied in the sinking of wells. Robert Stephenson understood the principles of groundwater flow but his ideas on recharge were in error. James Clutterbuck was the first British worker to make systematic observations on groundwater levels and apply these in a practical and innovative way. Joseph Prestwich wrote the first hydrogeological memoir and John Snow demonstrated that cholera was spread by contaminated well water.

William Whitaker was an avid collector of records of well sections and, although most of his work was descriptive, its sheer volume enhanced the profile of groundwater as a source for water supply. His colleague Joseph Lucas introduced the term hydrogeology, drew the first British maps to show groundwater contours and was the first person to call himself a hydrogeologist. Charles de Rance produced the first hydrogeological map of the whole of England and Wales.

In the early 20th century John Thresh became an authority on groundwater chemistry and pollution. Consultants, such as Herbert Lapworth and Arthur Beeby-Thompson, were prominent but their contributions are difficult to assess because they were not recorded in formal publications. Norman Boulton made significant contributions to the theory of groundwater hydraulics and Jack Ineson first introduced quantitative methods to the study of water resources in the UK.

Aside from those who made scientific contributions there are those who provided the encouragement and financial environment in which such work could proceed. Geological Survey Directors Sir Andrew Crombie Ramsey (1814–1891) and Sir Edward Battersby Bailey (1881–1965), in 1871 and 1937 respectively, both presided over changes which were of benefit to the growth of hydrogeology in the UK. Since the Second World War others have abandoned their own research careers and taken on the role of 'enablers' to oversee the tremendous advances which have taken place.

Out of the individuals mentioned above it is William Whitaker who has been described as the 'Father of English Hydrogeology' (Wilson 1985). However from the viewpoint of a 21st century hydrogeologist, Whitaker made little innovative contribution to the science. His contemporary Joseph Lucas has a much stronger claim to such a title. Whitaker himself said of Lucas that '. . . the important subject of Hydrogeology . . . you have made quite your own' (Lucas 1888). Shortly after the death of both Lucas and Whitaker, Walters (1929) drew special attention to the work of Lucas feeling that '. . . the importance of [his work] has received but scant recognition.' (Walters 1929 p. 86).

Lucas, writing in 1877, commented on the few engineers or geologists qualified to undertake hydrogeological work. He wrote that 'William Smith knew more than any geologist who has followed him; and . . . the science has found in the Rev. J. C. Clutterbuck an able master and one whom all succeeding labourers in the same field must venerate as the Father of the subject . . .' (Lucas 1877*c* p. 3). It is difficult to disagree with Lucas and, if the title of the 'Father of English Hydrogeology' is worth having, it must surely go to Clutterbuck who first made systematic observations on groundwater levels using these to plot hydrogeological cross sections and demonstrate the relationship between surface and groundwaters.

Clutterbuck (1850) was also the first to suggest a survey of the deep water-bearing strata of the London area, a subject later taken up by Lucas (1874). Lucas (1877*c*) records that the direct suggestion that the Geological Survey should take up this work elicited the reply that it should be left to the engineer as the cost was not covered by the Survey's grant and the subject was an engineering not a geological one. Despite lobbying no Government staff were appointed to work on underground water until the mid-1930s. This may be the reason why, for the 60 years following 1880, the important advances in hydrogeology took place in mainland Europe and North America rather than in Britain.

The Water Act of 1945 provided, for the first time, a framework within which groundwater resources could be assessed and heralded a new era for hydrogeology in which the Geological Survey, joined after 1965 by the Water Resources Board, led the way. The period between 1963 and 1975 was probably the most significant in the development of British hydrogeology and saw the subject develop into a mainstream branch of geology, taught at University level and represented within the Learned Societies. Subsequent years have seen the consolidation of this position so that, at the beginning of the 21st century, legislation to control groundwater abstraction and contamination, and the expertise necessary to investigate quantity and quality, are in place to meet the challenges which the future will undoubtedly bring.

Many thanks are due to R. A. Downing, D. A. Gray, N. S. Robins and H. S. Torrens who read and commented on the first draft of this paper. However, the views expressed remain my own.

References

ADAMS, F. S. 1938. *The birth and development of the geological sciences*. The Williams and Wilkins Co., Baltimore.

AMOS, C. E. 1860. On the government waterworks in Trafalgar Square. *Proceedings of the Institution of Civil Engineers*, **19**, 21–52.

ANDREWS, J. N. & WOOD. D. F. 1972. Mechanism of radon release into rock matrices and entry into groundwaters. *Transactions of the Institute of Mining and Metallurgy*, **B81**, 198–209

ANDREWS, J. N. & LEE, D. J. 1979. Inert gases in groundwater from the Bunter Sandstone of England as indicators of age and palaeoclimatic trends. *Journal of Hydrology*, **41**, 233–252.

ANSTED, D. T. 1850. On the absorbent power of Chalk and its water contents under different conditions. *Proceedings of the Institution of Civil Engineers*, **9**, 360–375.

ANSTED, D. T. 1878. *Water and Water-Supply, chiefly in reference to the British Islands. Surface Waters.* William H. Allen and Co., London.

ARAGO, D. F. J. 1835. On springs,artesian wells, and spouting fountains. *Edinburgh New Philosophical Journal*, **18**, 205–246.

BAILEY, E. B. 1952. *Geological Survey of Great Britain.* Thomas Murby and Co., London. 278p.

BAKEWELL, R. 1813. *Introduction to Geology*. J. Harding, London.

BALDWIN-WISEMAN, W. R. 1907. The influence of pressure and porosity on the motion of sub-surface water. *Quarterly Journal of the Geological Society of London*, **63**, 80–105.

BARKER, J. A., DOWNING, R. A., GRAY, D. A., FINDLEY, J., KELLAWAY, G. A., PARKER, R. H. & ROLLIN, K. E. 2000. Hydrogeothermal studies in the United Kingdom. *Quarterly Journal of Engineering Geology and Hydrogeology,* **33**, 41–58.

BARKER, J. A., DOWNING, R. A., HOLLIDAY, D. W. & KITCHING, R. 1999. Hydrogeology. *In:* PLANT, J. A. & JONES, D. G. (eds) *Development of regional exploration criteria for buried carbonate-hosted mineral deposits: a multidisciplinary study in Northern England.* British Geological Survey, Technical Report, **WP/91/1**, 119–126.

BARKER, R. D. 2004. The first use of geophysics in borehole siting in hardrock areas of Africa. *In:* MATHER, J. D. (ed.) *200 Years of British Hydrogeology.* Geological Society, London, Special Publications **225**, 000–000.

BARROW, G. & WILLS, L. J. 1913. *Records of London Wells.* Memoirs of the Geological Survey of England and Wales HMSO, London.

BEADLE, C. 1902. The abstraction of underground water and its local effects. *Journal of the Sanitary Institute*, **23**, 467–474.

BEADLE, C. 1903. Evidence as to the cause and effect of the lowering of the permanent water levels in the London water basin. *Journal of the Sanitary Institute*, **24**, 400–403.

BEEBY-THOMPSON, A. 1969. *Exploring for water*. Villiers Publications, London.

BLAND, W. 1831. On the influence of season over the depth of water in wells. *Proceedings of the Geological Society of London*, **21**, 339–340.

BLAND, W. 1832. Letter from William Bland, Jun. Esq., of New Place in the Parish of Hartlip, near Sittingbourne, Kent, to Dr Buckland; recording a series of observations made by himself, on the rise and fall of water in wells in the County of Kent. *Philosophical Magazine and Annals of Philosophy, New Series*, **11**, 88–96.

BONIFACE, E. S. 1959. Some experiments in artificial recharge in the lower Lee Valley. *Proceedings of the Institution of Civil Engineers*, **14**, 325–338.

BOSWELL, P. G. H. 1949. *A review of the resources and consumption of water in the Greater London area.* Metropolitan Water Board, London.

BOULTON, N. S. 1954. The drawdown of the water table under non-steady conditions near a pumped well in an unconsolidated formation. *Proceedings of the Institution of Civil Engineers,* **3**, 564–579.

BRASSINGTON, F. C. 1990. A review of rising groundwater levels in the United Kingdom. *Proceedings of the Institution of Civil Engineers, Part 1*, **88**, 1037–1057.

BRASSINGTON, F. C. 2004. Developments in UK hydrogeology since 1974. *In:* MATHER, J. D. (ed.) *200 Years of British Hydrogeology*. Geological Society, London, Special Publications, **225**, 000–000.

BUCHAN, S. 1938. *Water supply of the County of London from underground sources.* Memoirs of the Geological Survey of Great Britain, HMSO, London.

BUCKLAND, Rev. W. 1836. *Geology and mineralogy considered with reference to natural theology.* Bridgewater Treatise 6, William Pickering, London.

CAMPBELL, D. 1857. On the source of the water of the deep wells in the Chalk under London. *Quarterly Journal of the Chemical Society,* **9**, 21–27.

CHAPLOW, R. 1996. The geology and hydrogeology of Sellafield: an overview. *Quarterly Journal of Engineering Geology*, **29**, S1–S12.

CLUTTERBUCK, J. C. 1842. Observations on the periodical drainage and replenishment of the subterraneous reservoir in the Chalk Basin of London. *Proceedings of the Institution of Civil Engineers*, **2**, 155–160.

CLUTTERBUCK, J. C. 1843. Observations on the periodical drainage and replenishment of the subterraneous reservoir in the Chalk basin of London – continuation of the paper read at the Institution, May 31st 1842. *Proceedings of the Institution of Civil Engineers*, **3**, 156–165.

CLUTTERBUCK, J. C. 1850. On the periodical alternations, and progressive permanent depression, of the Chalk water level under London. *Proceedings of the Institution of Civil Engineers*, **9**, 151–180.

CONYBEARE, W. D. & PHILLIPS, W. 1822. *Outlines of the geology of England and Wales, with an introductory compendium of the general principles of that science, and comparative views of the structure of foreign countries. Part 1.* William Phillips, London.

DALTON, J. 1802. Experiments and observations to determine whether the quantity of rain and dew is equal to the quantity of water carried off by the rivers and raised by evaporation with an enquiry into the origin of springs. *Memoirs and Proceedings of the Manchester Literary and Philosophical Society*, **5**, 346–372.

DARWIN, E. 1785 An account of an artificial spring of water. *Philosophical Transactions of the Royal Society of London*, **75**, 1–7.

DAVIS, S. N. 2000. *Heroes of hydrogeology and their messages for today.* Chester C. Kisiel 19th Memorial Lecture, April 5, 2000, Department of Hydrology and Water Resources, University of Arizona.

DAWKINS, W. B. 1898. On the relation of geology to engineering (The James Forrest Lecture). *Proceedings of the Institution of Civil Engineers*, **134**, 254–277.

DIXEY, F. 1931. *A practical handbook of water supply*. Thomas Murby, London.

DONKIN, J. 1836. Some accounts of borings for water in London and its vicinity. *Transactions of the Institution of Civil Engineers*, **1**, 155–156.

DOWNING, R. A. 1993. Groundwater resources, their development and management in the UK: an historical perspective. *Quarterly Journal of Engineering Geology*, **26**, 335–358.

DOWNING, R. A. 2004*a*. The development of groundwater in the UK between 1935 and 1965 – the role of the Geological Survey of Great Britain. *In:* MATHER, J. D. (ed.) *200 Years of British Hydrogeology*. Geological Society, London, Special Publications, **225**, 000–000.

DOWNING, R. A. 2004*b*. Groundwater in a national water strategy, 1964–79. *In:* MATHER, J. D. (ed.) *200 Years of British Hydrogeology*. Geological Society, London, Special Publications, **225**, 323–338.

DOWNING, R. A. & GRAY, D. A. 2004. Jack Ineson (1917–1970). The instigator of quantitative hydrogeology in Britain. *In:* MATHER, J. D. (ed.) *200 Years of British Hydrogeology*. Geological Society, London, Special Publications, **225**, 283–286.

DOWNING, R. A. & HEADWORTH, H. G. 1990. The hydrogeology of the Chalk in the UK; the evolution of our understanding. *In: Chalk*. Thomas Telford, London, 555–570.

DOWNING, R. A., EASTWOOD, W. & RUSHTON, K. R. 2004. Norman Savage Boulton (1899–1984): civil engineer and groundwater hydrologist. *In:* MATHER, J. D. (ed.) *200 Years of British Hydrogeology*. Geological Society, London, Special Publications, **225**, 000–000.

EDMUNDS, W. M. 2004. Bath thermal waters: 400 years in the history of geochemistry and hydrogeology. *In:* MATHER, J. D. (ed.) *200 Years of British Hydrogeology*. Geological Society, London, Special Publications, **225**, 000–000.

EDMUNDS, W. M. & MILNE, C. J. (eds) 2001. Palaeowaters in Coastal Europe: evolution of groundwater since the late Pleistocene. Geological Society, London, Special Publications, **189**, 344pp.

ENVIRONMENT AGENCY. 2002. *Rising groundwater levels in the chalk-basal sands aquifer of the Central London Basin, May 2002*. Environment Agency, Reading.

EVANS, J. 1876. Anniversary address of the President Geological Society. Deep-sea deposits. Artic researches. Climatal changes. Geological progress. Water-supply. *Quarterly Journal of the Geological Society of London*, **32**, 91–121.

FAREY, J. 1807. On the means of obtaining water. *Monthly Magazine and British Register*, **23**, 211–212.

FAREY, J. 1811. *General view of the agriculture and minerals of Derbyshire with observations on the means of their improvement, volume 1*. Board of Agriculture, London.

FAREY, J. 1822. On overflowing wells and boreholes. *Monthly Magazine and British Register*, **54**, 35– 37.

FAREY, J. 1823. On artesian wells and boreholes. *Monthly Magazine and British Register*, **56**, 309.

FISHER, W. W. 1901. On alkaline waters from the chalk. *Analyst*, **26**, 202–213.

FLETT, J. S. 1937. *The first hundred years of the Geological Survey of Great Britain*. HMSO, London.

FULLER, J. G. C. M. 2004. Chalybeate springs at Tunbridge Wells: site of a 17th-century New Town. *In:* MATHER, J. D. (ed.) *200 Years of British Hydrogeology*. Geological Society, London, Special Publications, **225**, 201–211.

GARNIER, F. A. J. 1822. *De l'art du fontenier sondeur, et des puits artésiens*. Paris.

GEORGE, W. H. 2004. William Whitaker (1836–1925) – geologist, bibliographer and a pioneer of British hydrogeology. *In:* MATHER, J. D. (ed.) *200 Years of British Hydrogeology*. Geological Society, London, Special Publications, **225**, 51–66.

GRAY, D. A. 2004. Groundwater studies in the Institute of Geological Sciences between 1965 and 1977. *In:* MATHER, J. D. (ed.) *200 Years of British Hydrogeology*. Geological Society, London, Special Publications, **225**, 295–318.

GRAY, D. A. & MATHER, J. D. 2004. Stevenson Buchan (1907–1996): field geologist, hydrogeologist and administrator. *In:* MATHER, J. D. (ed.) *200 Years of British Hydrogeology*. Geological Society, London, Special Publications, **225**, 287–294.

HALLEY, E. 1687. An estimate of the quantity of vapour raised out of the sea by the warmth of the sun; derived from an experiment shown before the Royal Society, at one of their late meetings. *Philosophical Transactions of the Royal Society of London,* **16** (for 1686–1692), 366–370.

HAMLIN, C. 1990. *A science of impurity, water analysis in nineteenth century Britain*. Adam Hilger, Bristol.

HARRISON, J. H. 1891. On the subterranean water in the chalk formation of the Upper Thames, and its relation to the supply of London. *Proceedings of the Institution of Civil Engineers,* **105**, 2–25.

HAZELL, R. 2004. British hydrogeologists in West Africa – an historical evaluation of their role and contribution.

HEADWORTH, H. G. 2004. Recollections of a golden age: the groundwater schemes of Southern Water 1970 to 1990. *In:* MATHER, J. D. (ed.) *200 Years of British Hydrogeology*. Geological Society, London, Special Publications, **225**, 339–362.

HENWOOD, W. J. 1831. Facts bearing on the theory of the formation of springs, and their intensity at various periods of the year. *Philosophical Magazine and Annals of Philosophy, New Series,* **9**, 170–177.

HENWOOD, W. J. 1843. On the quantities of water which enter the Cornish mines. *Transactions of the Royal Geological Society of Cornwall*, **5**, 411–444.

HÉRICART DE THURY, L. É. F. 1829. *Considérations géologiques et physiques sur la cause du jaillissement des eaux des puits forés ou fontaines artificielles et récherches sur l'origine ou l'invention de la sonde, l'état de l'art du fontenier-sondeur, et le degré de probabilité du succès des puits forés*. Bachelier, Paris.

HÉRICART DE THURY, L. É. F. 1830. Observations on the cause of the spouting of overflowing wells or artesian fountains. *Edinburgh New Philososphical Journal*, **9**, 157–165.

HOPKINSON, J. 1891. Water and water supply with special reference to the supply of London from the chalk of Hertfordshire. *Transactions of the Hertfordshire Natural History Society*, **6**, 129–161.

HULL, E. 1862. *The geology of the country around Bolton-le-Moors, Lancashire*. Geological Survey of England and Wales, Memoirs, HMSO, London.

HULL, E. 1865. On the New Red Sandstone as a source of water supply for the central towns of England. *Quarterly Journal of Science*, **2**, 418–429.

INGLIS, J. C. 1877. On the hydrogeology of the Plymouth District. *Annual Report and Transactions of the Plymouth Institution and Devon and Cornwall Natural History Society*, **6**, 105–121.

LAPWORTH PARTNERS, HERBERT, 1946. *A hydro-geological survey of Kent*. The Advisory Committee on Water Supplies for Kent, London.

LATHAM, B. 1882. On the influence of barometric pressure on the discharge of water from springs. *In: Report of the 51st Meeting of the British Association for the Advancement of Science, York, August/September 1881*. John Murray, London, 614.

LELLIOTT, M. 2002. Hydrogeology, pollution and cemeteries. *Teaching Earth Science*, **27**, 68–73.

LLOYD, J. W. 2004. British hydrogeologists in North Africa and the Middle East – an historical perspective. *In:* MATHER, J. D. (ed.) *200 Years of British Hydrogeology*. Geological Society, London, Special Publications, **225**, 219–228.

LUCAS, J. 1874. *Horizontal Wells. A new application of geological principles to effect the solution of the problem of supplying London with pure water*. Edward Stanford, London.

LUCAS, J. 1877*a*. Hydrogeology: one of the developments of modern practical geology. *Transactions of the Institution of Surveyors*, **9**, 153–184.

LUCAS, J. 1877*b*. *Hydrogeological Survey, Sheet 1*. Edward Stanford, London.

LUCAS, J. 1877*c*. *Hydrogeological Survey, Explanation accompanying Sheet 1*. Edward Stanford, London.

LUCAS, J. 1878. *Hydrogeological Survey, Sheet 2*. Edward Stanford, London.

LUCAS, J. 1888. *Testimonials with Memorandum and Appendix [submitted by J. Lucas as a candidate for the Professorship of Geology in the University of Oxford]*. Privately printed.

LYELL, C. 1830–1833. *Principles of Geology, 3 vols*. John Murray, London.

MATHER, J. D. 1998. From William Smith to William Whitaker: the development of British hydrogeology in the nineteenth century. *In:* BLUNDELL, D. J. & SCOTT, A. C. (eds) *Lyell: the Past is the Key to the Present*. Geological Society, London, Special Publications, **143**, 183–196.

MATHER, J. D. 2001. Joseph Lucas and the term 'hydrogeology'. *Hydrogeology Journal*, **9**, 413–415.

MATHER, J. D., TORRENS, H. S. & LUCAS, K. J. 2004. Joseph Lucas (1846–1926) – Victorian polymath and a key figure in the development of British hydrogeology. *In:* MATHER, J. D. (ed.) *200 Years of British Hydrogeology*. Geological Society, London, Special Publications, **225**, 67–88.

MYLNE, R. W. 1840. On the supply of water from artesian wells in the London Basin, with an account [by W. C. Mylne] of the sinking of the well at the reservoir of the New River Company, in the Hampstead Road. *Transactions of the Institution of Civil Engineers*, **3**, 229–244.

PENMAN, H. L. 1948. Natural evaporation from open water, bare soil and grass. *Proceedings of the Royal Society*, **193**, 120–146.

PERRAULT, P. 1674. *De l'origine des fontaines. (On the origin of springs, trans. A. la Roque, 1967)* Hafner Publishing Co., New York.

PHILLIPS, J. 1844. *Memoirs of William Smith, LL.D., author of the 'Map of the strata of England and Wales'*. John Murray, London.

PHILLIPS, W. 1818. *Selection of facts from the best authorities, arranged so as to form an outline of the geology of England and Wales*. William Phillips, London.

PREENE, M. 2004. Robert Stephenson (1803–1859) – The first groundwater engineer. *In:* MATHER, J. D. (ed.) *200 Years of British Hydrogeology*. Geological Society, London, Special Publications, **225**, 107–120.

PRESTWICH, J. 1850. On the geological conditions which determine the relative value of the water-bearing strata of the Tertiary and Cretaceous Series, and on the probability of finding in the lower members of the latter beneath London fresh and large sources of water supply. *Proceedings Royal Institute of British Architects*, **8th July 1850**, 14p.

PRESTWICH, J. 1851. *A geological inquiry respecting the water-bearing strata of the country around London, with reference especially to the water-supply of the metropolis;and including some remarks on springs*. Van Vorst, London.

PRESTWICH, J. 1872. Anniversary address of the President. Our Springs and Water-supply. Our Coal measures and Coal-supply. *Quarterly Journal of the Geological Society*, **28**, iii–xc.

PRESTWICH, J. 1876. *On the geological conditions affecting the water supply to houses and towns, with special reference to the modes of supplying Oxford: being a lecture given on October 22,1875, with additional notes*. James Parker, Oxford and London.

PRESTWICH, J. 1895. *A geological inquiry respecting the water-bearing strata of the country around London, with reference especially to the water-supply of the metropolis; and including some remarks on springs. Re-issue, with additions by the author*. Gurney and Jackson, London.

PRICE, M. 2004. Dr John Snow and an early investigation of groundwater contamination. *In:* MATHER, J. D. (ed.) *200 Years of British Hydrogeology*. Geological Society, London, Special Publications, **225**, 31–50.

DE RANCE, C. E. 1882. *The water supply of England and Wales; its geology, underground circulation,surface distribution, and statistics*. Edward Stanford, London.

DE RANCE, C. E. 1884. The recent earthquake. *Nature*, **May 8th 1884**, 31.

RIVERS POLLUTION COMMISSION. 1875. *Sixth Report of the Commissioners appointed in 1868 to inquire into the best means of preventing the pollution of rivers. The domestic water supply of Great Britain*. HMSO, London.

ROBINS, N. S, BENNETT, J. R. P. & CULLEN, K. T. 2004. Groundwater versus surface water in Scotland and Ireland – the formative years. *In:* MATHER, J. D. (ed.) *200 Years of British Hydrogeology*. Geological Society, London, Special Publications, **225**, 183–192.

ROSE, E. P. F. 2004. The contribution of geologists to the development of emergency groundwater supplies by the British army. *In:* MATHER, J. D. (ed.) *200 Years of British Hydrogeology*. Geological Society, London, Special Publications, **225**, 159–182.

ROSE, E. P. F., MATHER, J. D. & PEREZ, M. 2004. British attempts to develop groundwater and water supply on Gibraltar 1800–1985. *In:* MATHER, J. D. (ed.) *200 Years of British Hydrogeology*. Geological Society, London, Special Publications, **225**, 239–262.

ROYAL COMMISSION ON THE STONE TO BE USED FOR THE HOUSES OF PARLIAMENT. 1839. *Report of the Commissioners*. HMSO, London.

ROYAL COMMISSION ON METROPOLITAN WATER SUPPLY. 1869. *Report of the Commissioners*. HMSO, London.

RUDWICK, M. J. S. 1963. The foundation of the Geological Society of London: its scheme for co-operative research and its struggle for independence. *British Journal for the History of Science*, **1**, 325–355.

SEAWARD, J. 1836. On procuring supplies of water for cities and towns, by boring. *Transactions of the Institution of Civil Engineers*, **1**, 145–150.

SHEPPARD, T. 1920. *William Smith: his maps and memoirs*. A. Brown and Sons, Hull.

SMITH, B. 1935. *Geological aspects of underground water supplies*. Cantor Lectures November/December, 1935, Royal Society of Arts, London, 55p.

SMITH, D. B., WEARN, P. L., RICHARDS, H. J. & ROWE, P. C. 1970. Water movement in the unsaturated zone of high and low permeability strata using natural tritium. *In: Isotope Hydrology 1970*. International Atomic Energy Authority, Vienna, 73–87.

SPENCE, I. M. & ROBINS, N. S. 2004. The Scottish hydropathic establishments and their use of groundwater. *In:* MATHER, J. D. (ed.) *200 Years of British Hydrogeology*. Geological Society, London, Special Publications, **225**, 212–218.

STEPHENSON, R. 1840. *Report to the Provisional Committee of the London and Westminster Water-Works, Etc., Etc.* Reproduced in the Morning Advertiser of December 29th, 1840.

STEPHENSON, R. 1841. *London Westminster and Metropolitan Water Company. Second Report to the Directors*. London.

SWINDELL, J. G. 1849. *Rudimentary treatise on well-digging, boring, and pump work with illustrations*. John Weale, London.

TELLAM, J. H. 2004. 19th century studies of the hydrogeology of the Permo-Triassic Sandstones of the northern Cheshire Basin, England. *In:* MATHER, J. D. (ed.) *200 Years of British Hydrogeology*. Geological Society, London, Special Publications, **225**, 89–106.

THRESH, J. C. 1908. The detection of pollution in underground waters, and methods of tracing the source thereof. *Transactions British Association Waterworks Engineers*, **12**, 108–137.

THRESH, J. C. 1912. The alkaline waters of the London Basin. *Chemical News*, **106**, 25–27 and 40–44.

THRESH, J. C. & BEALE, J. F. 1925. *The Examination of Waters and Water Supplies. 3rd Edition.* J & A Churchill, London.

TORRENS, H. S. 2004. The water-related work of William Smith (1769–1839). *In:* MATHER, J. D. (ed.) *200 Years of British Hydrogeology*. Geological Society, London, Special Publications, **225**, 15–30.

TURNER, R. C., RADLEY, J. M. & MAYNEORD, W. V. 1961. Naturally occuring alpha-activity of drinking waters. *Nature*, **189**, 348–352.

WALTERS, R. C. S. 1929. The hydro-geology of the Chalk of England. *Transactions of the Institution of Water Engineers*, **34**, 79–110.

WALTERS, R. C. S. 1936. The hydro-geology of the Lower Oolite rocks of England. *Transactions of the Institution Water Engineers*, **41**, 134–158.

WARD, R. C. & ROBINSON, M. *Principles of Hydrology*, 3rd edition, McGraw-Hill, London.

WHITAKER, W. 1872. *The geology of the London Basin. Part 1 – The Chalk and the Eocene Beds of the Southern and Western tracts*. Geological Survey of England and Wales, Memoirs, **4**, HMSO, London.

WHITAKER, W. 1886. On a recent legal decision of importance in connection with water supply from wells. *Geological Magazine*, **3**, 111–114.

WHITAKER, W. 1888. Cronological list of works referring to underground water, England and Wales. Appendix in 13th Report of the British Association Committee appointed for the purpose of investigating the circulation of underground waters in the permeable formations of England and Wales and the quantity and character of water supplied to various towns and districts from these formations. *In; Report of the 57th Meeting of the British Association, Manchester, August/September 1887*, John Murray, London, 384–414.

WHITAKER, W. 1899. Anniversary address of the President. Water-supply and sanitation. *Quarterly Journal of the Geological Society of London*, **55**, 53–83.

WHITAKER, W. 1912. *The water supply of Surrey, from underground sources, with records of sinkings and borings*. Memoirs of the Geological Survey of England and Wales, HMSO, London.

WHITAKER, W. & REID, C. 1899. *The water supply of Sussex from underground sources*. Memoirs of the Geological Survey of England and Wales. HMSO, London.

WHITAKER, W. & THRESH, J. C. 1916. *The water supply of Essex from underground sources*. Memoirs of the Geological Survey of England and Wales, HMSO, London.

WILSON, H. E. 1985. *Down to earth: One Hundred and Fifty years of the British Geological Survey*. Scottish Academic Press, Edinburgh, 189p.

WOODWARD, H. B. 1897. *Soils and sub-soils from a sanitary point of view: with especial reference to London and its neighbourhood*. Geological Survey England and Wales, Memoirs, HMSO, London.

WOODWARD, H. B. 1910. *The Geology of Water-Supply*. Edward Arnold, London.

YEATS, T. 1826. Section of a well sunk at Streatham Common, in the county of Surrey. In a letter addressed to – Brown Esq. secretary to the Westminster Fire-Office; and by him communicated to the Geological Society. *Transactions of the Geological Society of London, series.2*, **2**, 135–136.

YOUNGER, P. L. 2004. 'Making water': the hydrogeological adventures of Britain's early mining engineers. *In:* MATHER, J. D. (ed.) *200 Years of British Hydrogeology*. Geological Society, London, Special Publications, **225**, 121–158.

The water-related work of William Smith (1769–1839)

H. S. TORRENS

Lower Mill Cottage, Furnace Lane, Madeley, Crewe CW3 9EU, UK
(formerly University of Keele, ST5 5BG, UK)
(e-mail: gga10@keele.ac.uk)

Abstract: In 1797 Smith recorded his first known 'Order of Strata'. This was based on his work as a land, colliery and canal surveyor around Bath, Somerset. It already shows a clear awareness of the occurrence of spring lines (especially in the Fuller's Earth, soon to cause such problems in the construction of the Somerset Coal Canal). In his better known June 1799 version, Smith much extended this, with a new third column showing springs, now tabulated for five of his 23 strata. Smith thus had a keen awareness of both the importance of, and the problems raised by, how water was, or was not, retained in rocks and how it was released at stratigraphically controlled spring lines. This paper briefly reviews five of his involvements with 'water-related' geology. The first was as canal engineer. Here one of the two branches of his first canal later had to be abandoned because it could not be made to retain water where it passed over the Dolomitic Conglomerate. The second was as a land drainer. This he first attempted at Camerton in about 1796. This skill brought him most of his early employments after his dismissal from canal work in June 1799. Third, Smith was next a significant exponent of the art of creating water meadows, particularly in Bedfordshire and Norfolk. Smith was active next in a fourth field, erecting sea defences along the east coast of England. Finally he was often consulted on how to find or control new water supplies, as at Swindon or Scarborough. It was this last work which used his stratigraphic skills to their fullest extent.

William Smith has had the triple misfortune to

(a) bear the commonest name in England and thus to have suffered endless confusions with other Smiths, in particular in the world of water, with James Smith (1789–1850) of Deanston.
(b) to have recently attracted the attention of journalists and ex-airline pilots, one of whom has been relied on in a recent *Biographical Dictionary* article which is sadly full of errors (Skempton *et al.* 2002).
(c) to have lived in a now-forgotten period when water was the commonest and most important source of power, as well as providing, as it still does, the basis of human life.

Hydrogeology was a word that Smith never used, although it was one invented quite early in his career. But this was a) in French and b) used in a completely different sense from today, by Jean Baptiste Lamarck (1744–1829) in 1802 (Carozzi 1964, p. 18). He then used it (in separation from meteorology – the theory of the atmosphere) to encompass his theory of the (largely physics of the) earth's interaction with water over the earth's external crust. From this use it did pass briefly into use in English in 1824 (*Oxford English Dictionary*). So I shall carefully avoid using such a confusing word in any Smithian contexts here.

In 1818 Smith, then in London, was facing imprisonment for debt. He and his London friends, particularly John Farey (1766–1826) and Thomas Tredgold (1788–1829) drew up, and distributed, various versions of a document recording 'Mr Smith's Geological Claims'. These were in hopes of seeking some immediate financial reward for Smith's discoveries. Smith's claims, in the field of what we only now call hydrogeology, were noted as the seventh claim, worded as follows:

> 'Having by the same persevering attention to the surface in connection with its Strata beneath it, ascertained the true source of the supply of all Springs of Water, to be the superficial water (of rains, or streams, pools etc) percolating downwards through porous Strata or Alluvia, until intercepted by water-tight strata, or by Faults or patches of clayey alluvia, or by water already stagnated, in such porous masses; – and having deduced and applied in a extensive practice, then commenced, these investigations and conclusions, as to the strata and springs, to the Draining of Land, wherein Mr. Smith has been employed, in most of the improving agricultural districts in the kingdom, since about the beginning of this century' (Farey 1818).

Smith's own version of this 'Claim' reports his:

> 'having discovered and put into practice very extensively a New Art of Draining and improving Land founded on a knowledge of the Strata and of the Springs they produce and also a new mode of supplying Canals with Water, derived from the same Principle; and in thence deducing a correct Theory of Springs or an accurate knowledge of the receptacles and Currents of Water in the Earth; which accords with the practice of Mining

From: MATHER, J. D. (ed.) 2004. *200 Years of British Hydrogeology*. Geological Society, London, Special Publications, **225**. 15–30. 0305-8719/04/$15

and Draining, and is thus rendered extremely useful in obtaining water for Canals, Brew-houses etc, requiring a large supply; which in some cases may be obtained without Machinery' (Sheppard 1920, p. 219–220).

Smith's training

Smith, born in 1769, was orphaned by the death of his father in 1777. In 1787 he became an assistant to a remarkable man, Edward Webb (1751–1828), land surveyor and engineer of Stow-in-the-Wold (Anon. 1828), who then trained Smith. Smith in much later reminiscences claimed only to have 'admired the talent of my master, his placid and ever unruffled temper, and his willingness to let me get on, [as] I required no teaching'. None the less Webb clearly introduced Smith to an extraordinary range of skills.

Apart from land surveying and map making (on which see Richeson 1966), Webb was also busy and skilled as an engineer. Phillips has recorded how Webb, despite being:

> 'self taught . . . [was] possessed of great ingenuity and skill in mechanics, mensuration, logarithms, algebra and fluxions. His practice included many things now conceded to the engineer, such as the determination of the forces of water, and planning machinery' (Phillips 1844, p. 5).

A later pupil, of both Webb and Smith, the Anglo-American geologist and engineer, Richard Cowling Taylor (1789–1851), confirmed this. He added that Webb:

> 'was somewhat eccentric, much given to mount some favourite hobby to his own pecuniary loss; yet ever esteemed a man of strict integrity and of correct sense of every moral obligation . . . That he did not succeed more completely was owing to the want of early scientific instruction, and a knowledge of what had been already accomplished by others. Thus he would labour very hard and assiduously to produce some improved mechanism, which had been matured and in use long previously, yet perfectly unknown to him' (Mitchell 1874, p. 7).

Webb was almost certainly responsible, for example, for the piped water-supply system built for the inhabitants of Stow before 1803 (Rudge 1803, vol. 1, p. 179). In 1807, Thomas Rudge gave an equally enthusiastic summary of the:

> 'most curious and scientific mode of open drainage which has [yet] been effected [in Gloucestershire. This again was by Webb, at Kempsford, near Lechlade, beside the Thames and Severn Canal. It was] new, and interesting, and reflects credit on the ingenuity of Mr Edward Webb of Stow, under whose direction the plan was carried into execution' (Rudge 1807, p. 263–264 and plate).

The work here involved carrying off a large body of water and Webb used the principal drain which collected the greatest part of this water to drive a waterwheel by which more water was also raised, to be drained from a lower level. Webb fitted two wheels onto one shaft; the broad wheel was six feet wide by eight feet high, while the narrow wheel was one foot wide and fifteen feet high. This is clear testimony to Webb's abilities in what we today would regard as engineering. But there is no evidence that Webb was then aware of any directly geological possibilities or connections.

Smith's early work in Somerset with water in canals and mines

Soon after Smith was sent to Somerset by Webb in 1791 he became involved in a number of ways with what we would now group within the 'geology of water'. In his pioneering stratigraphical studies he had soon become keenly aware of the importance of separating water-bearing rock units, which yielded spring water, from those which did not. In his first known 'Order of Strata' of 1797, Smith noted this property of two of the rock units he then separated. These were:

> 'no. 9 – Fullers Earth [which was] found in great abundance in many of the wet sidelaying Lands near the Tops of the Hills about Bath [and] never found far from that tier of springs which Issue from the upper Stratum of Freestone [Great Oolite. Similarly] no. 13 – Sand and Sandburs [Midford Sands] have large open joints between them disposed in the same regular manner as was observed in the Freestones and Chalk and serve as Channels for that water which is the abundant supply of the numerous springs Issuing from the bottom of the sand' (Douglas & Cox 1949).

In his better known 'Order of Strata' prepared in June 1799, immediately after his final dismissal by the Somerset Coal Canal (= SCC) Company, Smith much extended this information. His table now included a new, third, column, in which springs were tabulated, as present in five of his 23 strata. This Order, although widely distributed then, was only first published in 1815 (Smith 1815) and reprinted by Fitton in his partly foiled attempts to do further justice to Smith between 1818 and 1821, after Smith's 'Claims' had been issued (Torrens 2002, ch. 10).

Smith's famous note of January 1796, first quoted by Phillips (1844, p. 17–18), shows that he was also already well informed of the distributions of fossils

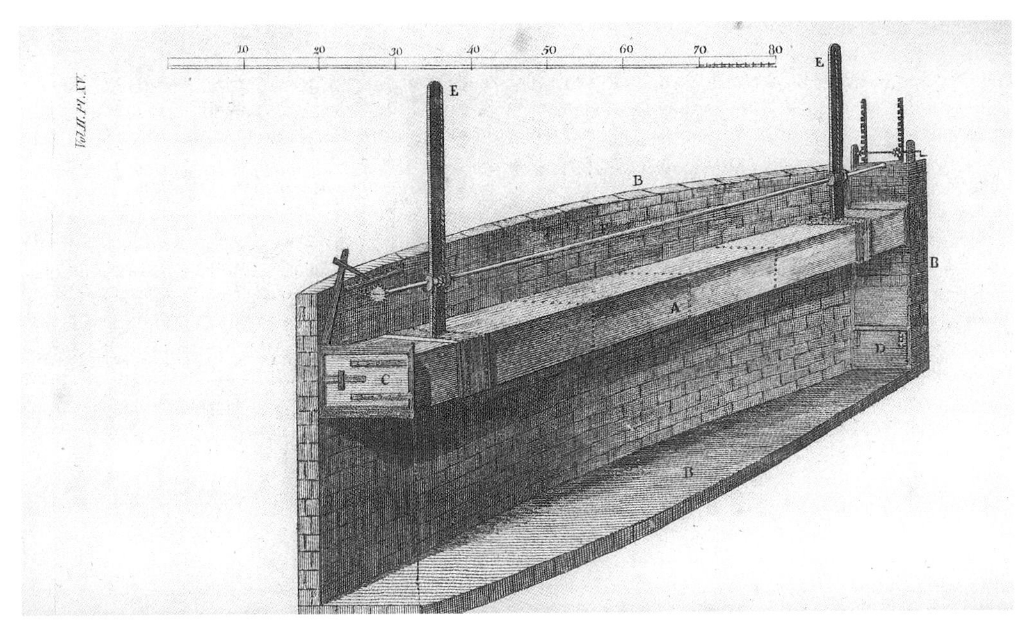

Fig. 1. Robert Weldon's caisson (from Billingsley 1798, p. 317–319).

that were to be found in many, but not all, of these Bath strata and how he had now realized their utility in identifying such ordered strata. These two realizations enabled Smith to start separating repetitious lithologies against this standard stratigraphic order. This he seems to have been the first to achieve. He was also able to start colouring stratigraphic maps, only some of which survive (Cox 1942, p. 25).

All of Smith's work as a canal engineer was stimulated by his appointment in 1793 as the well paid Surveyor and/or Engineer to the SCC Company. Any canal engineer then had to be highly aware of the problems of water supply and control. Unlike roads and railways, canals have to be flat and any change in level must be carefully planned, executed and controlled by locks, inclined planes or, in Smith's case, the use of a novel, but hydrogeologically problematic, and troublesome caisson (Torrens 1975).

This caisson was a brave piece of 'new technology' but it was finally defeated by the choice of stratum in which it was sited (the lower Fuller's Earth), one with the unfortunate ability to change volume when wet, and the then-less-than-perfect ability to make water-retaining structures, using blocks of worked Bath Stone (not brick) to retain these water-bearing Fullers Earth clays, before the invention of concrete. The strata in which this device was excavated, to a depth of over 60 feet, were recorded by Smith in his 1799 'Order' as 'numbers 8, Blue and 9, Yellow Clay and 10, Fuller's Earth and 11, Bastard ditto and sundries [and were characterized] as visible at a distance, by the slips on the declivities of the hills round Bath'. These Bath hills comprise 'one of the most intensely land-slipped areas in Britain' (Kellaway & Taylor 1968, p. 65).

The caisson was excavated through the full thickness of the last unit, the Lower Fuller's Earth Clay. The only SCC caisson completed was in the end sabotaged by the extraordinarily wet nature of 1799, the year in which it finally failed, and in which Smith was dismissed (Torrens 2001, p. 65). The SCC proprietors' resolution to replace the caisson on 5 June 1799 was defeated on the same day that Smith's employment was finally terminated. I agree with Joan Eyles (1969*b*, p. 154) that it was the failure of the caisson which must have caused Smith's dismissal, since he must have sided with the anti-caisson party, knowing full well that it was built in a geologically unstable place. This geological instability, and large volume changes in these Fuller's Earth clays between when wet and dry, has been investigated by a team led by Brian Hawkins (Hawkins *et al.* 1986, 1988). The final resolution to abandon the caisson was not passed until 11 February 1800. There is simply no evidence to support the idea that Smith's dismissal was due to his having purchased property at Tucking Mill, as has been suggested by others (Cox 1942, p. 73).

The operation of this caisson, involving the principles of water and buoyancy enunciated by Archimedes, has been discussed by Billingsley (1798, p. 317–319) & Brown (1990). It worked by allowing a loaded canal boat to enter a sealed box (the wooden trunk or caisson). This had to be completely submerged in a taller water-filled caisson cistern. The box could then be moved up and down by admitting or discharging water from it. Brown has pointed out that the water pressure on the submerged box when at the bottom of the caisson lift shaft would have been 3140 lbs/square foot! He also notes that the cross section of this box, into which the canal boats were floated, would have needed to be kept as small as possible.

The dimensions of this box have been variously quoted. But obviously any wooden box capable of withstanding such pressures would have had quite different internal and external dimensions. The *Bath Herald*'s 1798 figures (Torrens 1975), 80 feet long, 10 and a half wide and 11 and a half high, must refer to the external dimensions. Arthur Young in 1798 gave 80 by 7 by 7 and a half (Young 1798, p. 79) which must record internal ones. An Irish visitor in the summer of 1797 recorded 80 by 7 by 11 (Taylor [1797]) which confirms the crucial width was 7 feet.

Debate continues about the site of the only caisson completed (Paget-Tomlinson 2003). But, as was pointed out nearly 30 years ago (Torrens 1975), its location has long been clear (Torrens 2003). This was confirmed by the discovery of a 1804 SCC map in the Public Record Office, Kew by Roger Halse (Halse & Castens 2000, p. 36). As a result of this confirmation, it is now clear that the 'mystery tunnel', leading south from the original enlarged canal basin close by Caisson House (Clew 1986, p. 168–119), and shown on the map in Torrens (1975), must be directly connected with the Caisson. This tunnel's dimensions are 17 feet long, 5 feet 4 inches wide and 4 feet 3 inches high from ground level to its roof, which is here flush with the canal-basin's retaining wall. It cannot be a drainage tunnel, as Clew claimed.

In view of the water levels involved, it is more likely to have been a supply tunnel. Since it leads so directly into the site of the one caisson built, one might suppose it could have been the actual, later walled-up, boat entrance to the Caisson. The difficult question is what were the precise dimensions of SCC boats used in this? Dr. Adrian Padfield has located a drawing in the British Library, showing the half section of a boat, labelled 'the dimensions [now] agreed by the K&A, Wilts and Berks and SCC companies'. This records that the full beam dimensions of boats were then 6 feet 7 inches, without any rubbing strake. But this drawing must date from after March 1801, when these three companies agreed to issue specifications for future boats and only now use those made to an approved design, having a capacity of 35 tons (Anon. 1801; Clew 1986, p. 49). Evidence from a sale notice of January 1803, which offered for sale 24 unwanted boats previously built for the SCC, gave the same width. Thirteen of these boats were square ended and '23 feet in length, four feet in depth and 6 and a half feet wide' (Anon. 1803; Clew 1986, p. 61). Both these widths are too big to have passed through this surviving tunnel.

But before these dates the beam of boats using the caisson may have been even smaller. Weldon speaks of his caisson boats having only 25 to 30 tons burthen (Billingsley 1798, p. 318). So this tunnel needs further investigation. It at least allows us to see and understand the materials used in the construction of the caisson cistern (carefully dressed blocks of Bath Stone) and the problems faced both in constructing it and trying to make it water-proof. Water would certainly have had to be brought down to it from the large caisson reservoir lying above (the site now beneath the tennis courts at today's Caisson House). But it seems unlikely, because of the water levels involved, that this tunnel can have been used to channel merely water.

We should also record here how any investigation of this vital period in Smith's career has been enormously complicated by the later destruction of all the SCC Company committee's minute books. This seems to have occurred because of the later connection between the SCC and Samuel George Mitchell (1823–1907), Bath builder and Widcombe Bridge Company director. He occupied the SCC's former engineers' headquarters, Caisson House at Combe Hay, after the death of the SCC engineer William Hill (1776–1868) senior, between at least 1884 and 1895 (according to Bath *Directories*). Mitchell then must have acquired these minute books, which had clearly been left behind in that house. But 'a large quantity of [his] company records' were destroyed around 1945, by his niece Mrs F. E. Lambert who then occupied his later home, Calton Villa in Bath), save for the cover of the first 'Coal Canal Committee Book 1793' – which is now in Bath Public Library (according to letters dated 3 July and 3 August 1973, from A. M. Burgess to K. R. Clew – copies in HST archives). In another letter, from A. G. C. King of the Bath firm of solicitors, Stone, King and Wardle, to Desmond Donovan (20 May 1963 – copy in HST archive) he also confirmed that 'the Coal Canal papers were among those which were destroyed some 30 years ago'.

Luckily we have a fine contemporary source, by one of Smith's closest friends, which discusses canal building in some detail. This is the article 'Canal' for Rees' *Cyclopaedia* by John Farey (1766–1826) senior (Farey 1806). Farey's introduction to this noted that to 'Mr William Smith, engineer of Buckingham Street, London, we are indebted for many valuable hints and information given on many points, as we are also to Mr Benjamin Bevan, engineer of Leighton Buzard, Beds' (see below). Much of Smith's later work on other canals concerned more his skills as a surveyor and engineer of canals than the actual maintenance of water in them, as with the Sussex Ouse (Phillips 1844, p. 63–64).

Fig. 2. The surviving caisson tunnel at Combe Hay (photographed by Ray Bibby).

One further point of importance when considering Smith's canal work is his role as a teacher. One of this country's later most active and important canal engineers, the above mentioned Benjamin Bevan (1773–1833), had been a brewer and land surveyor in Leighton Buzzard, Bedfordshire. He was encouraged to become a canal engineer after meeting Smith in 1801, when Smith taught him the rudiments of stratigraphy on tour with Farey (Phillips 1844, p. 39; Fitton 1833, p. 33 & 42–43; Eyles 1985).

The biggest problem for any canal builder was the control of water and here some unexpected but, very real, problems had to be countered. Sometimes, for example, too much water could enter the canal, as with the SCC. Here Smith built a number of stone-lined culverts to remove excess water. These were specifically referred to by Farey (1806). Farey recorded (under Somerset Coal Canal) how 'in several places this canal was cut through strata disposed to slip, but by the small tunnels or soughs which Mr William Smith constructed, for drawing off the springs, the same was prevented'. Some are still visible under the SCC and a fine photograph of one appeared in 1988 (Allsop 1988, p. 23). This is between Tucking Mill and Midford and its caption rightly records it was 'probably a drainage sluice', although it only runs beneath the towpath, not across the entire canal.

The nearby Kennet and Avon Canal (= K&AC) soon suffered similar problems already noted by Smith in 1803 (Cox 1942, p. 88). He was finally called to help in 1811 (Phillips 1844, p. 55, 68 & 86). The historian of this canal has noted how:

> 'the section between Limpley Stoke and Bradford-on-Avon has always been liable to landslips, from the time that construction first began at that point when some 7 acres came down in one slip near Bradford. The reason is that the canal is cut on the side of a hill between these two points and lies on an oolite formation in which many fissures exist. In times of rainy weather water permeates into the rock, forcing air upwards under pressure, which is likely to cause a "blow" in a section of the canal bed and possibly a landslide' (Clew 1973, p. 60).

Clew noted how the canal was leaking by 1804 (Clew 1973, p. 68) and how the canal section from Dundas aqueduct to Bath had had to be repaired before 1810 (Clew 1973, p. 73). These were the reasons that Smith was then called in. Some fine circa 1910 photographs show how long these problems continued on this section of the K&AC. It was:

> 'the most notorious length [of the western canal] in terms of its susceptibility to leaks, landslips and blow-outs [where] it passes over deep-fissured [Inferior Oolite] limestone and it was the underground movement of water associated with this that caused the problems' (Allsop 1987, p. 87).

Smith's Diary for 14 January 1812 refers to his 'writing on K & A Canal etc' (Smith Diary, Oxford University Museum of Natural History = OUM). From this date on, Smith wrote at least three manuscript Reports, all of which survive (Smith 1812*a*, *b*, ?1814). They give a real indication of the problems facing a water engineer on this Canal. The first Report, directed to the Committee is dated from 7 Northumberland Place, Bath, 20 July 1812. Smith had spent the period 29 June to 26 July there working on this and other projects. It discussed the springs between the Dundas Aqueduct and Bath:

> 'A complete knowledge of Springs is certainly one of the most intricate parts of Natural History – as it must involve a complete knowledge of the different Strata which they flow from and also of the various positions of the Strata and the most particular knowledge of the distortions and dislocations of each stratum. This from my very extensive practice in Draining I must acknowledge to have been one of my most intricate pursuits' (Smith 1812*a*).

The most troublesome springs along here came from the surfaces of the strata beneath the upper and under great 'Rocks of Oolite' [i.e. Great and Inferior varieties]. Smith significantly noted how:

> 'The making of such canals previous to the executions of the K&AC and SCC was a new thing in all the southern and western parts of the island and never before attempted in any of the strata where the greatest difficulty has been experienced, except in the tunnel and summit Level of the Thames and Severn Canal . . . excavated through the same strata as that part of the K & A now under consideration.'

Smith then noted the 'need to intercept these springs' [as he had had to at Combe Hay on the SCC] (Smith 1812*a*, p. 10, 21 & 23). He notes also the problems caused by plants breaking through the clay used to puddle the floor of the canal (Smith 1812*a*, p. 13). This was *Equisetum palustris*, the effect of which on drains Sir Joseph Banks had described in 1800 (Torrens 2002, ch. 5, p. 72). They had already 'choaked up Mr Elkinton's Drains in the Prisley Bog', (Bedfordshire, on which see below). Smith next noted that the 'Stratum of loamy Sand [was] far from being watertight' (Smith 1812*a*, p. 15). This is the Midford Sands, whose water-yielding problems Smith had also long recognized. He then cited the property of the bed above this [the Inferior Oolite] which when pressurized by 'air passing through chasms' caused further problems (Smith 1812*a*, p. 18). Smith concluded that only by 'cutting away the rocks which supported the foot' of land slips here and, then by properly draining them, would the problems they had caused be resolved. Smith urged that 'the canal be divided into sections' (Smith 1812*a*, p. 27–29) and promised to send 'plans and sections which will be finished in a few days'. There seems to have been some delay in fulfilling this promise.

His second report 'on the strata which produce the springs . . . between Bath and Bradford[-on-Avon]' (Smith [1812*b*]) is not dated. It concerns the springs which had caused breaches, leaks, slips and other damage to the Canal here and how these may be prevented in future by again collecting and retaining the water from such springs for the proper use of the canal. The springs arose from the stratum at the base of Claverton Hill. Since this Report notes an October 1812 breach in this canal it must date from November 1812 at the earliest, when Smith intended to write on 'K & A springs and leaks'. His diary notes that he had been in Bath 4–24 September, 18–30 October and had on '27 October 1812 – walked to breach in K & A canal near Stoke and along it to Bath. Very wet' (Smith Diary – OUM).

The third Report 'on the course of the K & A around the base of that remarkable hill between the Dundas Aqueduct and Bath' is also undated. It concerned the two principal rocks which were again causing problems, both the Great and Inferior Oolites with spring lines below. Since this report includes an annotated copy of the printed first 'Order of the Strata . . . proved prior to 1799' it is thought to date from 1814 (Smith [?1814]). All these problems show why the water supply at this troublesome western end of the canal had had to be augmented by a new pumping house supplying river water at Claverton from 1813 (Clew 1973, p. 71 & 80).

The occurrence of leaks and water-loss from canals did not always happen as soon as water was let into a canal. Leaks, or queaches, might appear later, depending on the ability of the puddle used to retain water. As Farey noted, a canal company should 'employ some professional man expressly for this purpose who should by a judicious application of his experience and knowledge of the *strata* in every place, apply that particular method of draining which every spot may require' (Farey 1806).

Sometimes however there proved to be too little water for a particular canal. The best example of this to involve Smith occurred on the southern Radstock branch of the SCC from Radstock to Midford, which had had water introduced to it by 1804 (Clew 1986, p. 68). But by September 1814 the SCC Committee of Management had convened a special meeting of the Proprietors to receive and consider John Hodgkinson (1773–1861)'s recent report on the bad state of this line (Anon. 1814*a*). In 1813 a 30 hp steam engine had been installed at Radstock to supply more water but by June 1815 this was for sale, having 'worked only two summers'. This was because in September 1814, it had been decided to abandon this Radstock branch as a canal (Anon. 1814*b*). In July 1815 a new rail road from Radstock to Midford was laid to replace this canal completely (Anon. 1815 *a*, *b*).

The problem here was the nature of the water supplied by the geology over which this line of the canal had been purposefully passed. The Dolomitic Conglomerate, here a very wonderful aquifer, had been chosen to supply the water since:

> 'the cavernous structure of this rock forms vast reservoirs of water, but . . . they are not calculated to afford a constant supply, but when once tapped may soon be exhausted. This was experienced in the branch of the SCC near Radstock which was carried through the conglomerate in order that it might be fed by these natural reservoirs; their whole contents however soon ran off; and they defeated instead of answering the intended purpose, by draining off the water of the canal; which was consequently obliged to be puddled along the whole line' (see Torrens 1976).

Puddling the whole line well proved equally problematic. This crisis came to a head long after Smith had been dismissed by the SCC Company in 1799 but it graphically shows the problems then faced by canal engineers, when dealing with the largely unknown geology of water. It is possible that Smith had been responsible for the idea that the Dolomitic Conglomerate would serve as a good water source. He was certainly aware of the stratigraphic position of this Millstone unit by 1797.

Similar problems with this same rock unit had before hit the adventurers seeking coal, under Smith's direct mining supervision, at Batheaston (Torrens 2001, p. 69–70). The first water influx here came from the Blue and White Lias before March 1808. In 1811 a second and much greater influx of water, at a depth of 128 metres, came from this same Dolomitic Conglomerate aquifer. This was named 'The Cataract' and it caused these Batheaston coal sinkings and borings to be abandoned in 1813 (Kellaways 1991, p. 27). Rev. Joseph Townsend (1739–1816), who had been closely involved both with Smith and work on the K&AC, noted early in 1813:

> 'should the Kennet and Avon Company take a hint from the experience of these unfortunate coal adventurers, and, in Claverton, sink through the blue marl into the lyas, and then into the mill-stone bed [=Dolomitic Conglomerate], there is a probability that they will have the same springs, for a never-failing supply of water in the continuation of their canal between Bath and Bristol' (Townsend 1813, p. 312).

But the problems of supplying such water to a canal were not so simple. As we have seen, a major problem for the southern branch of the SCC was to keep the water puddle intact when it passed directly over such a 'cavernous' rock aquifer as the Dolomitic Conglomerate.

Smith's work with water between 1799 and 1808

In January 1808, while at Bath, Smith prepared a list of over 50 of his chief private employments between his dismissal by the SCC Company in 1799 and 1808. It lists the names of those for whom he had been employed to stop

(a) landslips.

It then carefully separated those commissions which had involved

(b) draining (i.e. the removal of water)

from commissions involving

(c) irrigation (i.e. supplying it).

It ended by listing

(d) 'embanking work by machinery' in East Norfolk and

(e) other water-related work (Smith Archive, Box 40, OUM).

This list is worth putting on record here in numbered order, as given by Smith, although categorized differently.

Since the problem of slides [=landslips] involved the most direct use of the skills Smith had previously acquired while working on the SCC these are listed first. Late in 1799 there had been very heavy rain in the Bath area, flooding the city to a greater depth than for over 50 years. This water gave Smith all his first employments in the Bath area after his dismissal by the SCC Company (Torrens 2001, p. 65), even if it had also caused the most likely reason for that dismissal, through the failure of the caisson. Kellaway and Taylor (1968) give a useful survey of the landslipping problem in the Bath area.

The 1808 list reads

(a) SLIDES [p. 2]

1. Late [William] Davis Esq Combe Grove [Bath]
   This was work done for Elizabeth née Jenkins (died 1810), the widow of William Davis (died 1798) (Anon. 1798), in 1800 at this elegant mansion which lay near the boundary of the Fuller's Earth and overlying Great Oolite, the level at which Smith had noted springs in 1797. Smith obliquely referred to his work here in 1801 (Cox 1942, p. 87–88) and his first report of February 1800 on the work needed here was reproduced by Phillips (1844, p. 33 & 60). Smith here tunnelled into the hill and intercepted the springs, as he had on the SCC (Townsend 1813, p. 130). But problems still persist here and in 1988 I was approached in an attempt to discover if this tunnel survived.
2. Charles Crook Batheaston [nr Bath]
3. Revd Mr [Richard] Warner [Cottage, Bath]
   (see Torrens 2001, p. 64)

Smith noted that all these land-slides [between June 1799 and 1803] he had 'stopped by the same method [as] many on the Coal Canal'. As we have seen, he there built culverts to lead away the water which caused landslips. The same problems also affected the K&AC.

Smith's work as land drainer and irrigator

Smith had been active as a private drainer of land at least from 1796. His first commission in this field was for the SCC Chairman, James Stephens (1748–1816), on his Camerton estates, Somerset (Torrens 2002, ch. 3, p. 230–231 & 241). This commission marked the start of Smith's third career, as a Water Drainer/Irrigator. At Camerton he:

> 'drained a piece of clayey and stony ground, which was levelled by James Stephens Esq. in his park at Camerton, on whose estates I first put into practice my ideas of draining derived from a knowledge of the strata, . . . by moving the soil and laying the sod on again' (Smith 1806, p. 54).

Smith's later 1799–1808 commissions in this field make an impressive list. The 1808 list reads here:

(b) DRAINING [listed on p. 1 and part of 2]

4. Late Marquess of Lansdown Bowood [Wiltshire]
5. Charles Gordon Grey Esq. Tracey Park [Glos]
   'at Tracey Park near Bath, where many hundred pounds had been expended, without the least advantage, by surface draining, Mr. William Smith cut off the springs, which had poisoned the estate, conducted them to a distant meadow, which stood in need of water, and thereby doubled the intrinsic value' Townsend 1813, p. 418–419.
6. Thom[a]s Crook Tytherton[-Kellaways, Wilts]
   (see Torrens 2001, p. 66–67)
7. Earl of Ilchester Melbury [Dorset]
   This work was carried out by Smith early in 1801 (see Dorset Record Office D/FS 1/Box 189). It provides a fascinating example of Smith's ability to already separate quite disparate rocks when faulted together. Here [Jurassic] Oxford clay is faulted against [Cretaceous] Greensand and Chalk and fossils ranging from the Greensand right down to the Cornbrash are duly recorded from here in his geological collection (Natural History Museum, London).
8. Marquis of Bath Longleat [Wiltshire]
9. Duke of Bedford Woburn Park and Prisley Bog [Beds]
   (see Smith 1806)
10. Earl Talbot Ingestry [Staffordshire]
11. Lord Anson Shugborough [Staffordshire]
12. Lord Eliot [Down Ampney] Gloucestershire
13. W[illia]m Child[e] Esq Kinlet, Salop
14. W[illia]m Colhoun Esq Wretham [Norfolk]
15. W[illia]m Wyndham Esq Dinton Park [Wilts]
16. Duke of Manchester [Kimbolton, Hunts]
17. Rob[er]t Doughty Esq Hanworth Hall [Norfolk]
18. Thos W[illiam] Coke Esq M.P. [Holkham Hall, Norfolk]
19. Late Miles S. Branthwayte Esq Taverham [Hall, Norfolk]
20. [Late] John Old Goodford Esq Near Yeovil [Somerset]
21. John Daniels Esq Near Yeovil [Somerset]
22. —— Hay Esq near Tiverton, Devon
23. Mr [Eleazer] Pickwick Bathford [Bath]
24. Barne Barne Dunwich [Suffolk]
25. Revd W[illia]m Phillips [Phelips] Montacute [Somerset]
26. Revd W[illia]m Johnson Ellingham House [Norfolk]

27. W[illia]m Northey W.P. Box [Wiltshire]
28. Rob[er]t Sparrow Esq Warlingham Hall [Suffolk]
29. J[ohn] Motteux Esq [Beechamwell near Swaffham] Norfolk
30. S[amuel] Lloyd Harford Esq [of Bristol]
31. —— Gordon Esq
32. —— Bush Esq Lilliput [near Bath]
33. George Whitmore Esq Pucklechurch [Glos] (see Torrens 2001, p. 68–69)

Some numbers including 6, 13, 1, 5, 4, 7 and 8 have later been renumbered in pencil as numbers 1 to 8, perhaps reflecting the larger draining operations which Smith had undertaken? As can be seen, such work had taken Smith all over the country.

(c) IRRIGATION [p. 3]
34. Duke of Bedford Prisley Bog[Bedfordshire] (see no. 9, Smith 1806; Brown 1999]
35. Duke of Queensbury Amesbury [Wiltshire]
36. Paul Methuen Esq Chittern [Wilts]
37. M. S. Branthwayte Esq (see no. 19) Taverham
38. Thomas W[illia]m Coke Esq (see no. 18) Wrighton and Lexham
39. G[eorge] R[obert] Eyre[s] Esq Lynford Hall [Norfolk]
40. Earl of Peterborough Dauntsey [Wilts]
41. Thom[a]s Crook Esq (see no. 6) [Tytherton, Wiltshire]
42. Mr Sam[ue]l Davis Silton Duke of Somerset's land [Dorset]
43. John Motteux Esq (see no. 29) Norfolk
44. Mr Barker Ashford, Kent
45. Mr Creed Ashford, Kent
46. S[ilvanus] Bevan Esq [Riddlesworth, Norfolk]
47. B[arne] Barne Esq (see no. 24) [Dunwich]
48. W[illia]m Child[e] (see no. 13) Kinlet Hall [Salop]
49. Earl Thanet Ashford, Kent
50. Rich[ar]d Gurney Esq [Keswick Hall, Norfolk]
51. [in pencil] Mr J. Grant A[ll] Cannings [Wilts]

The cause for, and the extent, of Smith's work as a land drainer and/or irrigator needs to be carefully considered. In 1799 Smith had been dismissed as a canal engineer but, although he was clearly well aware of the vital importance of the geological discoveries he had already made, he was then and for some years afterwards quite unable to persuade people of this (Torrens 2001, p. 74–77).

As noted above, the appallingly wet season of 1799 had come at exactly the right time to force Smith to start a new career as a land drainer/irrigator. In May 1800 he was draining for Thomas Crook at Tytherton Kellaways, Wiltshire and here Thomas William Coke (1752–1842), later the Earl of Leicester, an improving landowner, came to see Smith's work there. Coke was immediately impressed and invited Smith to come to far-away Holkham in Norfolk in October 1800. This Smith did. Phillips quotes Smith's notes on his journey there (Phillips 1844, p. 35). The need for Smith to seek new professional openings, after his dismissal from the SCC, was now solved by Thomas Coke's invitation to work for him in Norfolk. Coke was a remarkable man. The monument erected to his memory at Holkham between 1845 to 1850 shows the abundant means whereby Coke sought to increase both agricultural cultivation and production in Norfolk. It records how 'the Arts lament in him a liberal and fostering patron; and Agriculture to which from early manhood to the close of life he dedicated time, energy, science and wealth' (Stirling 1908, vol. 2, p. 492 & plate opp.). John Martin reminds me that its basal frieze shows many of the noteworthy people that Coke had dealt with in improving his large estates and that the third section specifically relates to Smith's Irrigation work (Stirling, 1908, vol. 2, plates opp. p. 488 & 490). The figures shown include William Smith.

Smith was soon persuaded, wrongly as it emerged, that if he were to publish his methods of making water-meadows (Smith 1806), this would prove to be a sure money-spinner. As a result 2000 copies of this book, dedicated to Coke, were printed (Eyles 1969*a*, p. 90). This is a useful source for Smith's work in this field. But, although it gives few clues on any detailed hydrogeological aspects, Smith does note how few existing drainers had 'any general notions of strata' and how his 'numerous and extensive journies for the purposes of business and collecting information on the strata' had revealed 'some of the places which appear to be . . . calculated for improvement by irrigation'. He claimed that his drainage method derived 'from my knowledge of the strata [and] is already reduced to a science, substantiated by a number of practical proofs of its utility in different parts of the kingdom' (Smith 1806, p. xiv, 21 & 80).

Smith's most important commission as a land drainer and irrigator was that at Woburn for Francis Russell (1765–1802), the 5th Duke of Bedford to whom he was introduced by Coke in the autumn of 1801. Smith was busy there for some time and a mass of data in the Russell archives survives in the Bedfordshire Record Office to document this. But the Duke died in March 1802. This death proved a major disaster for Smith's hoped-for advancement, through the sudden ending of that Duke's patronage (Phillips 1844, p. 39–44).

Smith's draining work at Woburn had been further complicated by the earlier involvement there of Joseph Elkington (1739–1806), farmer and another pioneer of land drainage, from Warwickshire. Elkington had there discovered, at Princethorpe, the method of land drainage for which he is remembered. He had then discovered, by accident, how some strata were porous and pervious to water while others were not, and that he could locate the former with the augur (or rod) then used to explore for marl and coal. His methods were introduced in Warwickshire in about 1780. By 1793 Elkington's services were here in such demand that 'his crow bar was compared to the rod of Moses' (Prothero 1961, p. 366). So a Lancashire group, along with Sir Joseph Banks, brought Elkington's methods to the attention of the new Board of Agriculture early in 1794. They felt that Elkington, who was an epileptic, needed encouragement to 'disclose his secret . . . before any accident may deprive the public of so usefull a member'. Banks in particular felt this 'discovery [was] a matter of great national interest but not sufficiently known' (Sutro Library, San Francisco, MSS F 4:83). But he equally believed that Elkington's rights should be those of any patentee, based on priority of discovery.

The Board's president, Sir John Sinclair, was more enthusiastic and late in 1794 the Duke of Bedford allocated a trial site at Woburn, where the Board could study Elkington's methods. King George III confirmed on 25 June 1795 that he would award a sum not exceeding £1,000 as an inducement to Elkington, if he would divulge his methods. This sum Parliament agreed to repay. This was to be the 'first [sum] ever granted by Parliament for any discovery of importance to husbandry' (Anon. 1795). But, because Elkington worked primarily by instinct, and failed to keep records of how, or where, he had worked, the Board had to employ James Johnstone (died 1838), land surveyor of Edinburgh, to study and report on Elkington's system for him, through Sinclair's intercession. Johnstone observed Elkington at work in the spring of 1796. In autumn 1797, Johnstone published the first edition of his *Account* (of at least five in English) (Johnstone 1801).

Despite this, Elkington was still felt not to have explained his techniques sufficiently clearly to obtain the grant and payment was still being withheld in 1800. This was because, from October 1795, counter claims of originality had been submitted by James Anderson (1739–1808), that he was the system's true originator and that he had published it before Elkington (Anderson 1797).

In 1798 Elkington was taken ill which left his trials at Woburn 'unfinished'. He was soon however granted a lease of improvable lands at Madeley, Staffordshire. This was clearly made as a replacement for the sum which had been earlier voted to him, but which was never paid to him, by Parliament. His place in history remains complex. Donaldson noted that Elkington's 'fallacious principle has long since vanished. It is surprising that it was ever entertained at all' (Donaldson 1854, see also Torrens 2004). Smith's friend, John Farey, who had been closely involved with him, regarded Elkington's attempt at Woburn as a complete failure (Farey 1813, p. 362–383). Smith was the man called in to rectify this work.

Smith published a number of other items, apart from his 1806 book, about the Woburn drainage (Eyles 1969*a*, items 2, 3 & 6) and was awarded a silver medal by the Board of Agriculture for his work here. Rev. Thomas Wright (1756–1815), originally of South Cerney in Gloucestershire had written an earlier book on making water-meadows (Wright 1789). His last edition of this includes some not very penetrating 'Remarks on Mr Smith's Book on Water Meadows' (Wright 1808, p. 117–154), which reveal how little he was aware of the geological significance of Smith's work.

Smith remained active as a water-drainer well after 1808, although details are less available. One of his most important commissions in this period was the work he carried out at Weymouth. In July 1812 Smith was called in by the Weymouth Corporation to drain the Backwater there. His work there involved the Lord of the Manor, and recently deposed M.P., Gabriel Tucker Steward (c.1768–1836). Smith duly went to Weymouth late in July and surveyed the Backwater Marsh there which was proposed to be drained and improved. The outcome of this work is still not clear.

Smith's work on embanking

This was Smith's fourth field of involvement at the interface between water and geology. This was work which does not much concern hydrogeology as understood today, since it would now be regarded more as coastal engineering, however much it involved geological materials. It aimed to keep the North Sea (or German Ocean) out of the marshes of East Norfolk, by sand embankments. Smith's work here extended from 1801 to 1809. A near contemporary source, William Howitt, rather slightingly recorded that:

> 'the irruption of the sea, through the breaches in the dunes of sand, in the neighbourhood of Eccles, Horsey, Waxham etc having been accompanied with serious inconvenience and spoliation, caused a body of highly respected and influential gentlemen to be appointed Sea-Breach Commissioners, and in the year 1804, they engaged an eminent Engineer [Smith], since deceased, who among other information, gave it as his opinion, that if the shallows were all filled up, and the beach kept on an inclined plane, the sea would never gain on the Norfolk coast. He did not however point out how such an assertion could be substantiated' (Howitt 1844, p. 50).

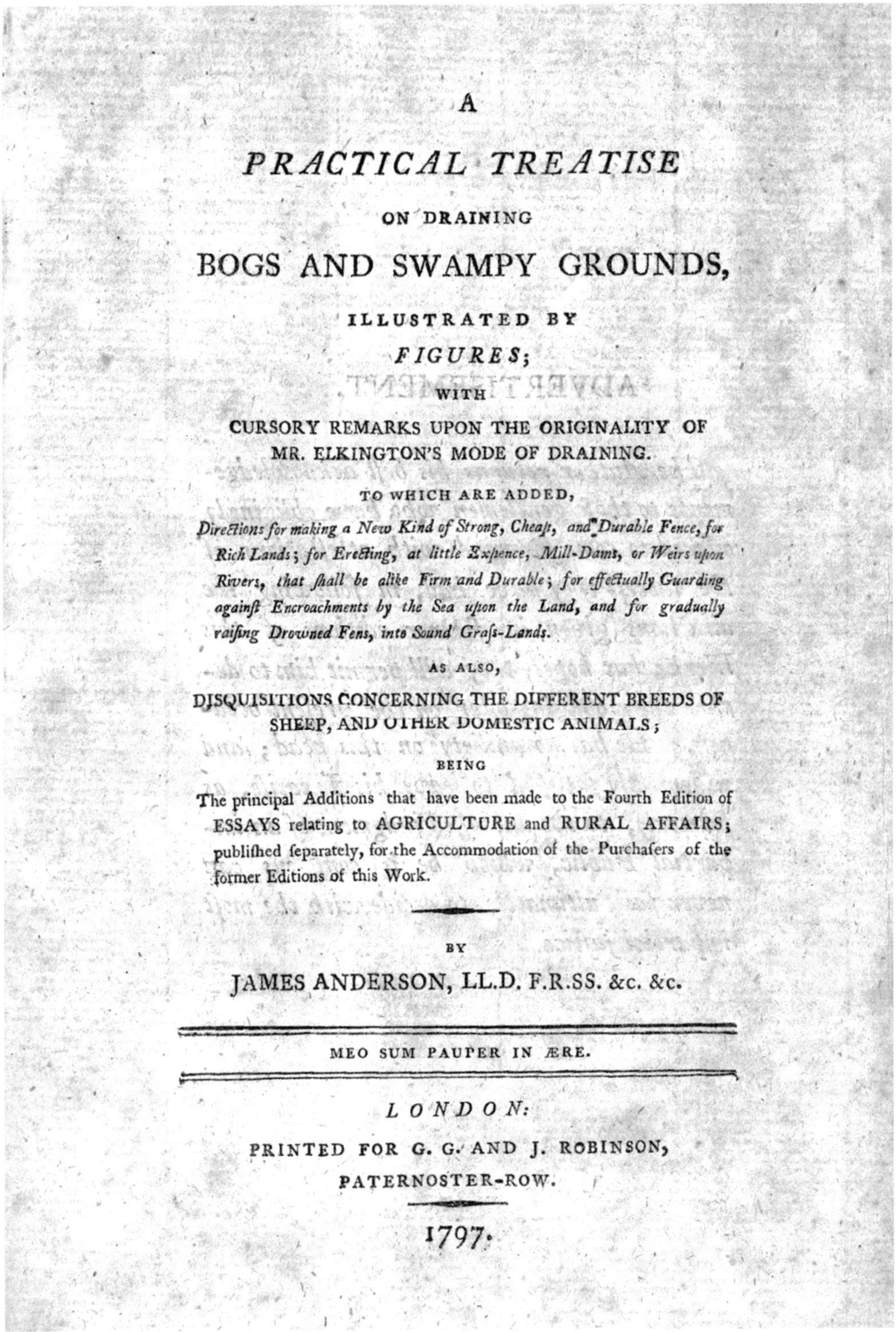

A

PRACTICAL TREATISE

ON DRAINING

BOGS AND SWAMPY GROUNDS,

ILLUSTRATED BY

FIGURES;

WITH

CURSORY REMARKS UPON THE ORIGINALITY OF MR. ELKINGTON'S MODE OF DRAINING.

TO WHICH ARE ADDED,

Directions for making a New Kind of Strong, Cheap, and Durable Fence, for Rich Lands; for Erecting, at little Expence, Mill-Dams, or Weirs upon Rivers, that ſhall be alike Firm and Durable; for effectually Guarding againſt Encroachments by the Sea upon the Land, and for gradually raiſing Drowned Fens, into Sound Graſs-Lands.

AS ALSO,

DISQUISITIONS CONCERNING THE DIFFERENT BREEDS OF SHEEP, AND OTHER DOMESTIC ANIMALS;

BEING

The principal Additions that have been made to the Fourth Edition of ESSAYS relating to AGRICULTURE and RURAL AFFAIRS; publiſhed ſeparately, for the Accommodation of the Purchaſers of the former Editions of this Work.

BY

JAMES ANDERSON, LL.D. F.R.SS. &c. &c.

MEO SUM PAUPER IN ÆRE.

LONDON:

PRINTED FOR G. G. AND J. ROBINSON, PATERNOSTER-ROW.

1797.

Fig. 3. The title-page of James Anderson's book (which claimed he had preceded Elkington in discovering the principles of land-drainage – author's collection).

Smith's 1808 list continues with

(d) EMBANKING by machinery in East Norfolk [p. 2]

52. N. Micklethwayte Esq and others	1,600 [acres]
53. G. Cubit Esq and others	200 [acres]
54. J.B. Huntingdon Esq and others	1,500 [acres]
55. Sir Geo[rge] Brograve Bart	1,400 [acres]

and [p. 4 further notes that he had been busy]

'Stopping the Sea out of 50,000 acres of land on the Eastern Coast of Norfolk with loose sand & shingle – laid together in imitation of the Beach'.

Such sea breaches were a recurring problem here. Smith was to have read a paper on his work here at the British Association meeting at Edinburgh in 1834 but sadly this paper was read only in title (Eyles 1969*a*, item 41). Luckily Phillips gives a good summary of this work which was highly successful (Phillips 1844, p. 50–54), despite Howitt's claim. One of those who knew Smith's work here better, Joshua Trimmer (1795–1857), more accurately recorded that Smith's sea-breach work, 'by means of sand and shingle was one of the triumphs of applied geology in the hands of Smith' (Trimmer 1847, p. 473–474).

Smith's 1808 list adds finally

(e) OTHER EMBANKING [p. 4]

56. W.A. Maddox Esq [Tremadoc] N. Wales

This last entry on Smith's 1808 list has alone been crossed out. It concerns an operation of which Phillips merely wrote 'in September 1802, we find [Smith] examining the ancient slaty rocks of Wales in the vicinity of Dolymelynllwyn near Dolgelle, on occasion of a visit to inspect Mr Maddocks's successful embankment across the Traethbychan at Tanyralt (Tremadoc)' (Phillips 1844, p. 54). Sadly this is a period for which no Smith diary survives. William Alexander Madocks (1773–1828) having purchased the Penmorfa Estate in north Wales in 1798, made it the basis for a remarkable experiment in regional planning. Having obtained an Act of Parliament in 1797, he embanked it against the sea thereby reclaiming 1,000 acres to use for scientific agriculture (Beazley 1967). The date of Smith's visit suggests that he may have been much more than a mere visitor to Madocks' remarkable experiment, and may have been involved in Madocks' first attempts at embanking.

Smith as a seeker of new water supplies

Smith, as we have seen, had already being consulted privately as a water drainer, while still employed by the SCC. Late in 1796, he was also consulted to survey the water-supply to the west end of the town of Heytesbury, Wiltshire and how it might be put to use to power an intended clothing mill. Smith's report shows he was highly aware already of the possibilities of both improving local land and ensuring a source of much needed energy by a proper use of water supplies here (Rogers 1976, p. 245).

A fine example of the water-finding skill which Smith had acquired as a result of his grasp of stratification comes in 1802. While he, John Farey and Benjamin Bevan were engaged together on their late January 1802 'Journey to investigate the strata' in Bedfordshire at the Duke of Bedford's expense (Eyles 1985) they:

> 'accidently met with the Reverend Mr. [Thomas] Le Mesurier [(1756–1802)] rector of Newton-Longville near Fenny-Stratford [Buckinghamshire] who related his having undertook to sink a well, at his parsonage house, within a mile or two of which, no good and plentiful springs of water were known, but finding clay only at the depth of more than 100 feet, was about to abandon the design; Mr Smith on looking into his map of the strata, pointed out to us, that Newton-Longville stood upon some part of the clunch clay strata [=Oxford Clay], and that the Bedford lime stone [=Cornbrash] appeared in the Ouse river below Buckingham, distant about 8 miles in a north west direction, and he assured Mr. L. that if he would but persevere, to which no serious obstacles would present themselves, because all his sinkings would be in dry clay, he would certainly reach this lime stone, and have plenty of good water, rising very near to the surface; Mr. L. accordingly did persevere . . . and at 236 feet beneath the surface . . . [found] the limestone rock which produced a plentiful jet of water' (Thompson 1820, p. 297–298).

After 1808, Smith's dual involvements with water and geology further diversified. An interesting example is the problems with water which Smith encountered with the Batheaston Mining Company and its sinkings and borings for coal. These had started in 1804 and continued, ultimately unsuccessfully, until 1813 (Torrens 2001, p. 69–70). The attempts here hit major problems with water, some of which Smith and his foreman, William Hill (1776–1868) senior, did manage to overcome. When the project was abandoned in 1813 Hill immediately found new employment, as chief engineer to the SCC (Clew 1986, p. 76). This transfer of employments shows the then close proximity between mining and canal engineers, as they were both so much concerned with the control of water. Hill, who was born at Stow-on-the-Wold, was long a close associate of Smith's; being engaged with him in an number of other mining ventures. As one would expect, such an association was marked by Hill making his own fine collection of rocks and fossils, duly arranged in Smithian stratigraphic order. These

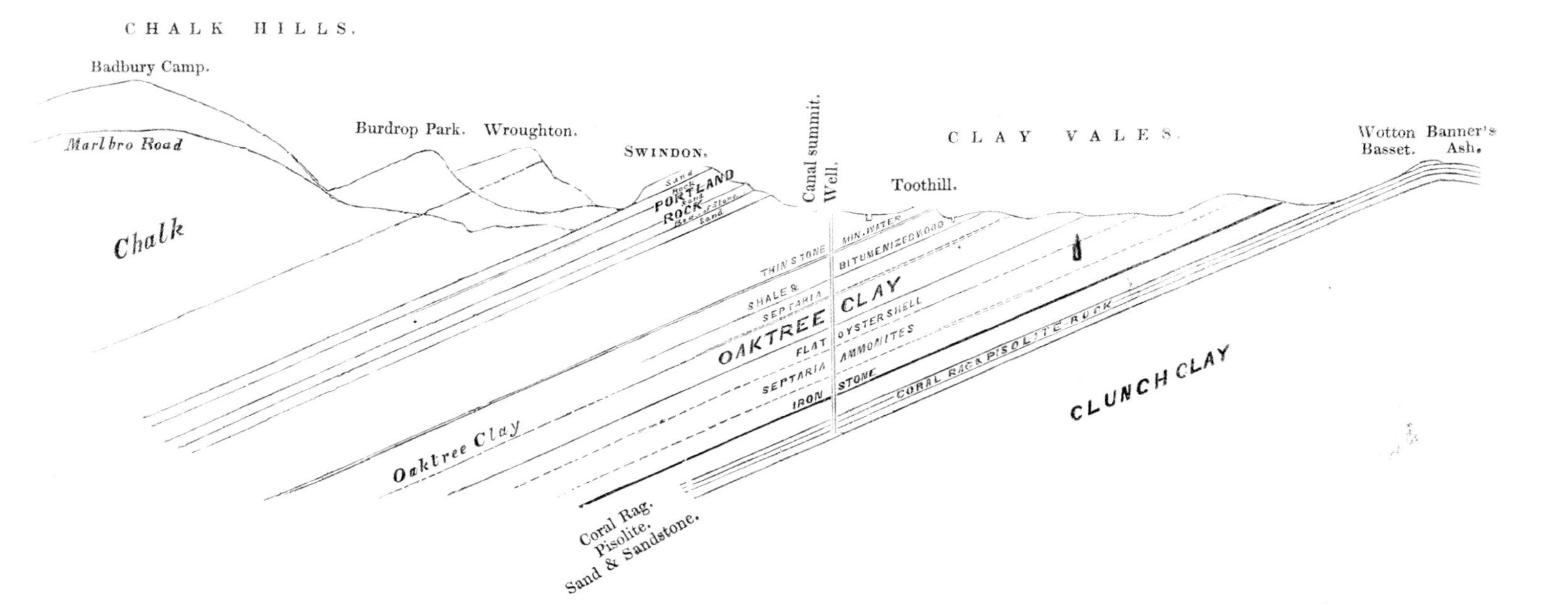

Fig. 4. Smith's section of strata – North Wilts (from Phillips 1844, p. 85).

he donated to the Bath and West of England Agricultural Society in 1819' (Torrens 1977, p. 482).

These Batheaston water problems had been much complicated by their soon proven connection with the partial failure of the Bath Hot Springs (Kellaways 1991), on which the then Bath tourist industry so depended. Smith was here involved, with his brother John Smith, between 1808 and 1811, both with vital remedial work on the Hetling Spring and with improving conditions for users of the water at the King's Bath. Smith was first called in 1808 to work on the Hetling Spring and then turned his attention to the nearby King's Bath. Kellaways gives details of what little is known of Smith's work here. The remarkable point is that Smith correctly believed the hot-springs had a deep source. The success of Smith's work here remains problematic as these hot springs effectively failed in 1812 (Mitchell 1869). It was only when work on the Batheaston coal shafts was abandoned in 1813 that the hot springs supply returned, in final proof that the first had indeed effected the second. Bath's citizens could heave a sigh of relief that coal was not going to be found at Batheaston.

1812 was a particularly busy, and watery, year for Smith. Apart from his work on the Minsmere Marshes drainage in Suffolk with which he was busy all that year (Phillips 1844, p. 55 & 73–74), and his work at Weymouth referred to above, his diary records that in January he was to be involved with 'Lady Rivers and the Supply of Bath with water'. This is Lady Martha Rivers, née Coxe (1749–1835) whose brother, the historian Rev. William Coxe (1747–1828), was a subscriber to Smith's 1815 *Geological Map*. She, through her former husband Sir Peter Rivers, later (from 1767) Gay, Bart (1721–1790), owner of the large manor of Walcot (Collinson 1791, p. 73), became 'possessed of very extensive property in Bath, in houses and ground rents' (Anon. 1835). Today's Rivers and Gay streets in Bath are a legacy of this. Part of this Rivers estate included a private Rivers Waterworks in Crescent Fields. This was supplied from a spring on Beacon hill, Bath (Bartrum 1878) and it was this water supply and distribution network which Smith was now called in to try and improve. He was busy with this throughout 1812, but again details are sparse.

Other examples of Smith's work as a water-seeking geologist include the work he did in seeking a supply of water for Swindon in 1816. He was then asked to find canal water there and, as Phillips records (1844, p. 80), when he noted that to 'direct the attempt for such in new situations requires the aid of *science*, as distinguished from *experience*'. Phillips also reproduces a portion of Smith's Report on this dated 13 April 1816. Smith felt that water might be found, at the same site but which had so far met only failure, if only a suitably water-bearing rock could be found there (Phillips 1844, p. 80–85). This was the same approach Smith had suggested in 1802 in Buckinghamshire. His new water-bearing rock was here the Coral Rag of which Smith had learnt only recently from some other futile coal-hunting searches at Bagley Wood, in Oxfordshire (Torrens 2001, p. 74). Smith's successful work, if only in the short term, at Swindon became the basis for a new lithographed 'section of Strata North Wilts' which Smith had published by 1817 (Eyles 1969*a*, item 17). It was intended to 'show the application of W. Smith's principles of obtaining water for canals' (Cox 1942, p. 60–61).

In February 1818 Smith was in Monmouth where again one of his tasks was to seek 'water for the Town of Monmouth' (Eyles 1969*a*, item 21). After his London imprisonment for debt in 1819 Smith was forced into a Yorkshire exile. Here again he was busy with water-supply problems, in this case for his newly adopted town of Scarborough (Smith 1827; Phillips 1844, p. 112–113; Rowntree 1931, p. 358).

Smith's last involvement with water came in 1837 when he was called as a witness in the Harrogate Sulphur Wells case held at York in March (Richardson 1891, p. 110–111; Sopwith 1994, p. 97, 147 & 192). A former owner of the Crown Hotel had sunk a well close to the Pump Room there and accidentally tapped into the Old Sulphur Well (Cooper & Burgess 1993, p. 77–79). A law-suit was begun but the case was never concluded and so Smith's report became his last to be published, but first issued 25 years after his death (Eyles 1969*a*, item 51).

Conclusion

Smith's friend and supporter Rev. Joseph Townsend provides a fitting summary of the value of Smith's pioneering work on water and geology. In his extraordinary publication of 1813, Townsend devoted a whole chapter to the subject of springs (Townsend 1813, p. 304–320), in which he noted that:

> 'of all the countries with which I am acquainted, no one is interesting to the geologist as the vicinity of Bath, because in no other are so many strata exposed to view. Neither is there any other country enriched with such a number and variety of *springs*, for within five miles we have six distinct sets of them, independently of the hot springs, by which the city is distinguished'.

Townsend, who had travelled very widely throughout the continent of Europe and had considerable mining experience, especially in South Wales, noted how this:

> 'makes Bath an excellent school for civil engineers, and I may add, for lawyers; if the former wish to discover hidden supplies of water, and the latter would qualify themselves to settle the numerous disputes which either

now subsist, or may arise respecting *springs* [He concluded] when, if ever, W. Smith shall be enable to publish his inestimable discoveries, and his observations upon *troughs* [synclines] the science of drainage will attain perfection' (Townsend 1813, p. 304 & 309).

He returned to the subject of drainage in his later chapter on 'The Great Importance of Geology'. Here he rightly concluded that a knowledge of the behaviour of water was of great importance to:

(1) owners of landed property,
(2) civil and canal engineers and
(3) builders.

(Townsend 1813, p. 418–425).

He ended by noting that:

'a treatise on this new art of [deep] drainage [based on a knowledge of stratigraphy] will come with more propriety from its inventor, W. Smith, but the general principles, on which this art is founded, must be obvious to every one, who has paid attention to what I have delivered on strata and on springs. It is the application only that requires sagacity, and in this Mr. Smith excels' (Townsend 1813, p. 418).

This work has been in part, and appropriately, funded by BP Exploration Ltd. J. Fuller and J. Martin have persuaded me, whenever I have doubted, that Smith is a significant, if difficult, figure in the history of geology. I owe grateful thanks to S. Brecknell (OUM) for much help with the Smith archive, and to W. Cawthorne (Geological Society, London), D. Donovan (London), R. Halse (Chippenham), A. Mathieson (Portishead), A. Padfield (Sheffield), F. Pole (Combe Hay), S. Pierce (Wincanton) and P. Wakelin (Telford). The staff of the Sutro Library, San Francisco facilitated my work on the Joseph Banks papers in their care.

References

ALLSOP, N. 1987. *Images of the Kennet and Avon, 100 years in camera, Bristol to Bradford-on-Avon*. Redcliffe, Bristol.

ALLSOP, N. 1988. *The Somersetshire Coal canal rediscovered*. Millstream, Bath.

ANDERSON, J. 1797. *Two Letters to Sir John Sinclair 1796*, reprinted (p. 1–66) in his *A Practical Treatise on draining Bogs and Swampy Grounds . . . with cursory remarks upon the originality of Mr. Elkington's mode of Draining*. G. G. and J. Robinson, London.

ANON. 1795. *Annals of Agriculture*, **24**, 563.

ANON. 1798. *Gentleman's Magazine*, **68** (2), 814.

ANON. 1801. *Salisbury and Winchester Journal*, **23 March 1801**, p. 1, col. 2.

ANON. 1803. *Bath Chronicle*, **10 January 1803**.

ANON. 1814*a*. *Bath Journal*, **15 September 1814**.

ANON. 1814*b*. *Bath Herald*, **24 September 1814**.

ANON. 1815*a*. *Bath Chronicle*, **20 July 1815** & *Bath Herald*, **22 July 1815**.

ANON. 1815*b*. *Salisbury and Winchester Journal*, **24 July 1815**, 4.

ANON. 1828. Obituary of Edward Webb. *Jackson's Oxford Journal*, **13 September 1828**, 3.

ANON. 1835. Obituary of Lady Rivers. *Bath Chronicle*, **27 February 1835**.

BARTRUM, J. S. 1878. *History of the present State of the City Waterworks*. Dawson, Bath.

BEAZLEY, E. 1967. *Madocks and the Wonder of Wales*. Faber, London.

BILLINGSLEY, J. 1798. *General View of the Agriculture of the County of Somerset*, 2nd edition. Cruttwell, Bath.

BROWN, D. 1999. Reassessing the influence of the aristocratic improver: the example of the fifth Duke of Bedford (1765–1802). *Agricultural History Review*, **47**, 172–195.

BROWN, D. K. 1990. The Combe Hay Caisson Lock: An early submarine. *Five Arches*, **9**, 5–6.

CAROZZI, A. V. 1964. [translation of] *Hydrogeology by J.B. Lamarck* [into English]. University of Illinois Press, Urbana.

CLEW, K. R. 1973. *The Kennet and Avon Canal*, 2nd edition. David & Charles, Newton Abbot.

CLEW, K. R. 1986. *The Somersetshire Coal Canal and Railways* (second edition). Bran's Head, Frome.

COLLINSON, J. 1791. *The History and Antiquities of the County of Somerset*, vol. 1. Cruttwell, Bath.

COOPER, A. H. & BURGESS, I. C. 1993. *Geology of the country around Harrogate*. HMSO, London.

COX, L. R. 1942. New light on William Smith and his Work. *Proceedings of the Yorkshire Geological Society*, **25**, 1–99.

DONALDSON, J. 1854. *Agricultural Biography*. For the author, London.

DOUGLAS, J. A. & COX, L. R. 1949. An early list of strata by William Smith. *Geological Magazine*, **86**, 180–188.

EYLES, J. M. 1969*a*. William Smith (1769–1839): a bibliography of his published writings etc. *Journal of the Society for the Bibliography of Natural History*, **5**, 87–109.

EYLES, J. M. 1969*b*. William Smith: Some aspects of his Life and Work. *In:* SCHNEER, C. J. (ed.) *Toward a History of Geology*. MIT Press, Cambridge, Mass, p. 142–158.

EYLES, J. M. 1985. William Smith, Sir Joseph Banks and the French Geologists. *In:* WHEELER, A. & PRICE, J. H. (eds). *From Linnaeus to Darwin: Commentaries on the History of Biology and Geology*. Society for the History of Natural History, London, p. 37–50.

FAREY, J. 1806. Canal. *In:* REES, A. (ed.). *The Cyclopaedia*, volume **6**, (unpaginated). Longman etc, London.

FAREY, J. 1813. *General View of the Agriculture and Minerals of Derbyshire*, vol. 2. McMillan, London.

FAREY, J. 1818. Mr Smith's Geological Claims stated . . . *Philosophical Magazine*, **51**, 173–180.

FITTON, W. H. 1833. *Notes on the Progress of Geology in England*. Richard Taylor, London.

HALSE, R. & CASTENS, S. 2000. *The Somersetshire Coal Canal: a Pictorial Journey*. Millstream Books, Bath.

HAWKINS, A. B., LAWRENCE, M. S. & PRIVETT, K. D. 1986. Clay mineralogy and plasticity of the Fuller's Earth formation. *Clay Minerals*, **21**, 293–310.

HAWKINS, A. B., LAWRENCE, M. S. & PRIVETT, K. D. 1988. Implications of weathering on the engineering properties of the Fuller's Earth formation. *Géotechnique*, **38**, 517–532.

HOWITT, W. 1844. *An essay on the Encroachments of the German Ocean along the Norfolk Coast*. For the author, Norwich.

JOHNSTONE, J. 1801. *An Account of the Mode of Draining Land according to the system practised by Mr Joseph Elkington*, 2nd edn. McMillan, London.

KELLAWAYS, G. A. (ed.) 1991. *Hot Springs of Bath*. City Council, Bath.

KELLAWAY, G. A. & TAYLOR, J. H. 1968. The Influence of Landslipping on the Development of the City of Bath. *Proceedings of the 23rd International Geological Congress*, **12**, 65–76.

MITCHELL, W. S. 1869. The Failure of the Bath Waters in 1812. *Bath Chronicle*, **11 March 1869**.

MITCHELL, W. S. 1874. How the study of Stratigraphical Geology commenced near Bath. *Bath Chronicle*, **5 March 1874**, 7.

PAGET-TOMLINSON, E. 2003. Robert Weldon's hydrostatic lock. *Waterways World*, **32**(2), 71.

PHILLIPS, J. 1844. *Memoirs of William Smith, LL.D.*. Murray, London [reprinted 2003 by the Royal Literary and Scientific Institution, Bath].

PROTHERO, R. E. (Lord Ernle) 1961. *English Farming: Past and Present*, 6th edn. Heinemann, London.

RICHARDSON, B. W. 1891. *Thomas Sopwith*. Longmans, Green and Co., London.

RICHESON, A. W. 1966. *English Land Measuring to 1800*. MIT Press, Cambridge, Mass.

ROGERS, K. H. 1976. *Wiltshire and Somerset Woollen Mills*. Pasold Research Fund, Edington.

ROWNTREE, A. (ed.) 1931. *The History of Scarborough*. Dent, London.

RUDGE, T. 1803. *The History of the county of Gloucestershire*. For the author, Gloucester.

RUDGE, T. 1807. *General View of the Agriculture of the County of Gloucester*. Phillips, London.

SHEPPARD, T. 1920. *William Smith: his maps and Memoirs*. Brown and Sons, Hull.

SKEMPTON, A. W., CHRIMES, M. M., COX, R. C., CROSS-RUDKIN, P. S. M., RENNISON, R. W. & RUDDOCK, E. C. (eds). 2002. *A Biographical Dictionary of Civil Engineers in Great Britain and Ireland: volume 1, 1500–1830*. Thomas Telford, London.

SMITH, W. 1806. *Observations on the Utility, Form and Management of Water Meadows and the draining and irrigating of Peat bogs with an account of Prisley Bog*. Bacon, Norwich.

SMITH, W. 1812*a*. [Manuscript] *Report on the subject of Springs between the Dundas Aqueduct and Bath*. 20 July 1812 (for the K&ACC), Public Record Office, Kew, Rail 842/74 (25) no. 3.

SMITH, W. [1812*b*]. [Manuscript] *Account of the Strata which produce the Springs and occasion Breaches, Leaks, Slips and other damage to the Kennet and Avon Canal between Bath and Bradford*. Undated (for the K&ACC), Public Record Office, Kew, Rail 842/74 (25) no. 1.

SMITH, W. [?1814]. [Manuscript] *Report of the course of the Kennet and Avon Canal around the base of that remarkable hill between the Dundas Aqueduct and Bath*. Undated (for the K&ACC), Public Record Office, Kew, Rail 842/74 (25) no. 2.

SMITH, W. 1815. *A memoir to the Map and Delineation of the Strata*. Cary, London.

SMITH, W. 1827. On retaining water in the Rocks for Summer Use. *Philosophical Magazine*, New series **1**, 415.

SOPWITH, R. 1994. *Thomas Sopwith surveyor*. Pentland Press, Edinburgh.

STIRLING, A. H. W. 1908. *Coke of Norfolk and his friends*. John Lane, London.

TAYLOR, E. [1797]. [Manuscript] 'Journal of a Tour in England and Wales'. National Library of Wales, Ms. 3479B p. 72.

THOMPSON, P. 1820. *Collections for a Topographic and Historical Account of Boston*. Longman & Co, London.

TORRENS, H. S. 1975. The Somersetshire Coal Canal Caisson Lock. *Bristol Industrial Archaeological Society Journal*, **8**, 4–10.

TORRENS, H. S. 1976. The Radstock arm of the Somerset Coal Canal. *Bristol Industrial Archaeological Society Journal*, **8**, 37.

TORRENS, H. S. 1977. The Bath Geological Collections. *Newsletter of the Geological Curators' Group*, **1** (10), 482–483.

TORRENS, H. S. 2001. Timeless Order: William Smith (1769–1839) and the search for raw materials 1800–1820. *In:* LEWIS, C. L. E. & KNELL, S. J. (eds). *The Age of the Earth: from 4004 BC to AD 2002*. Geological Society, London, Special Publications, **190**, 61–83.

TORRENS, H. S. 2002. *The Practice of British Geology 1750–1850*. Ashgate, Aldershot.

TORRENS, H. S. 2003. [Robert Weldon's] Caisson lock located. *Waterways World*, **32**(4), 90.

TORRENS, H. S. 2004. Entry for Joseph Elkington. *In: Oxford Dictionary of National Biography*. University Press, Oxford.

TOWNSEND, J. 1813. *The Character of Moses established for Veracity as a Historian recording Events from the Creation to the Deluge*. Bath, Gye.

TRIMMER, J. 1847. On the Geology of Norfolk. *Journal of the Royal Agricultural Society of England*, **7**, 444–485.

WRIGHT, T. 1789. *An Account of the Advantages and Method of Watering Meadows by Art*. Rudder, Cirencester.

WRIGHT, T. 1808. *The Formation and Management of Floated Meadows; with Corrections of Errors found in the Treatises of Messrs . . . and Smith*. Mackenzie, Northampton.

YOUNG, A. 1798. A Farming Tour in the south and west of England 1796 [and 1797]. *Annals of Agriculture*, **31**, 79–80.

Dr John Snow and an early investigation of groundwater contamination

MICHAEL PRICE[1,2]

[1]*School of Human and Environmental Sciences, The University of Reading, Whiteknights, PO Box 227, Reading, Berkshire, RG6 6AB, UK (e-mail: m.price@reading.ac.uk)*
[2]*Present address: Water Management Consultants Ltd., 23 Swan Hill, Shrewsbury, Shropshire, SY1 1NN, UK (e-mail: mprice@watermc.com)*

Abstract: John Snow was a physician but his studies of the way in which cholera is spread have long attracted the interest of hydrogeologists. From his investigation into the epidemiology of the cholera outbreak around the well in Broad Street, London, in 1854, Snow gained valuable evidence that cholera is spread by contamination of drinking water. Subsequent research by others showed that the well was contaminated by sewage. The study therefore represents one of the first, if not the first, study of an incident of groundwater contamination in Britain. Although he had no formal geological training, it is clear that Snow had a much better understanding of groundwater than many modern medical practitioners. At the time of the outbreak Snow was continuing his practice as a physician and anaesthetist. His casebooks for 1854 do not even mention cholera. Yet, nearly 150 years later, he is as well known for his work on cholera as for his pioneering work on anaesthesia, and his discoveries are still the subject of controversy.

The life of John Snow shows several similarities with that of William Smith (Torrens 2004). Like Smith, Snow came from a humble background and – at least in his early life – had little formal education. Both men pioneered new fields of study and achieved fame and respect but not without some controversy and reluctance on the part of their contemporaries to accept some of their views. Both also followed dual careers: Smith as a surveyor and geologist, Snow in two new fields of medicine – epidemiology and anaesthesiology. In the latter field he was accepted and respected in his own lifetime; in the former, in which he is probably best remembered today, at least by those outside the medical profession, he faced opposition and sometimes ridicule.

Snow's childhood

Much has been written about Snow and his work but biographical details are lacking. There is one contemporary account of Snow's life and work, written by a close friend and colleague, Dr (later Sir) Benjamin Ward Richardson, and published within two months of Snow's death. This was written as a Memoir to preface Snow's work *On chloroform and other anaesthetics* (Snow 1858) which he had just completed at the time of his death in 1858 and whose publication was overseen by Richardson (Ellis 1994). It has to be remembered that Richardson wrote this work at the age of 29, when he was still mourning the death of the man he regarded as a close friend and mentor. Consequently it is affectionate in tone and some of the detail may not be totally reliable. More recent biographies include the summary by Ellis (1994), the book by Shephard (1995) and papers by Roberts (1999) and Stephanie Snow (2000*a,b*). All of these are forced to draw heavily on relatively limited personal details (mostly Richardson's work) but from them a picture of Snow and his life emerges.

John Snow was born in York on 15 March 1813. Ellis (1994) notes that Richardson erroneously records the date as 15 June. He was the oldest of nine children of William and Frances (Fanny) Snow who had married in 1812. His mother had been born in 1789, the illegitimate daughter of John Empson and Mary Askham, who married when Fanny was nearly three (Snow 2000*a*). Fanny's brother, Charles Empson, was to have a major influence on John Snow's early life.

At the time of Snow's birth, the family lived in North Street, alongside the River Ouse in a poor part of York that was badly drained and prone to flooding. His father's occupation was recorded as 'labourer'. Given these inauspicious beginnings, the family's subsequent history and improvement are remarkable (Ellis 1994; Snow 2000*a*). Of the six sons of William and Fanny, the youngest died in infancy and nothing is known of the third, Charles. Of the other four, John became a doctor, William a businessman, Robert a colliery manager and Thomas a teacher and subsequently a vicar. Of John Snow's three sisters, Mary and Hannah both became teachers, remained spinsters and eventually ran their own school before retiring to Harrogate. His youngest sister, Sarah, married a farmer.

Five or six years after John Snow's birth, his

From: MATHER, J. D. (ed.) 2004. *200 Years of British Hydrogeology*. Geological Society, London, Special Publications, **225**. 31–49. 0305-8719/04/$15

father changed his occupation from labourer to driver and sometime between 1821 and 1823 the family moved their home a short distance to Wellington Row (Snow 2000*a*). There appears to have been a major improvement in the family's fortunes by the time of the birth of Hannah, Snow's younger sister, in March 1825, for it was then that William Snow purchased a house in Queen Street. In 1832 William Snow is listed as a farmer in Queen Street, and Richardson describes John Snow as the son of a farmer. In the 1830s William also bought four more houses in Queen Street and rented them out. In 1841 the family moved to a farm at Rawcliffe, then a village to the NE of the city.

This transition from unskilled manual worker to farmer, from labourer to landowner, albeit perhaps on a small scale, is notable. Snow (2000*a*) speculates that William Snow may have made money to buy the farm at Rawcliffe by selling the Queen Street property to a railway company, but where did he get the money to buy these properties in the first place? It is clear that William and Fanny (or at least one of them) had energy and the foresight to ensure that their children were educated and provided with the best possible start in life given their circumstances. Even so it would have been difficult for a labourer or cart driver to amass the money needed to buy property, especially while trying to feed, house and clothe a wife and several children. Snow (2000*a*) suggests the possibility of a generous benefactor.

Richardson records that John Snow was educated at a 'private school at York'. Ellis (1994) is at pains to point out that 'private school' in this context does not refer to anything approaching a modern British public school with its expensive fees, but something more akin to what would later be called a 'dame's school'. This would have been some form of local school, organized by parents from the poorer classes who were trying to give their children the elements of an education. Snow (2000*a*) agrees and comments that education at such a school would probably have cost around 6d (2.5 pence) per week, a substantial outlay from William's probable weekly wage of around 15s (75 pence or £0.75). This is an indication of the determination of Snow's parents to improve the lot of their children.

Ellis suggests that the education would have given him little more than the basic reading, writing and arithmetic. He notes that at some point in his later medical training Snow would have needed Latin and possibly Greek. Snow (2000*a*) speculates that he might have acquired this in York. However, this seems at odds with Richardson's comment that at this school 'he learned all that he could learn there' with the implication that this was not enough for his later life.

Snow enters medicine

In 1827, at the age of 14 and with his elementary education as complete as it could be, Snow left York to seek training in medicine. At that time, there were four classes of medical practitioners – physicians, surgeons, apothecaries and a group referred to as 'those in practice prior to the 1815 Act' (Ellis 1994). The physicians alone were entitled to be called 'Doctor'. They were the medical élite and enjoyed high social status. In England they belonged to the Royal College of Physicians of London, to join which they had to be graduates of Oxford or Cambridge universities. Physicians could prescribe medicines but not dispense them. They regarded themselves as greatly superior to the surgeons, who were regulated by London's Royal College of Surgeons and were Members of that College. Most combined surgery with midwifery and were able to prescribe medicines for non-surgical conditions. A few 'pure' surgeons were eligible to be Fellows of the College.

The surgeons in turn were superior to the apothecaries, mere pill-makers (Ellis 1994). They were licensed to practise by the Worshipful Society of Apothecaries in London, were allowed to prescribe and supply medicines, and by established custom advised on the treatment of non-surgical illnesses. The fourth group, those with no formal qualifications in any branch of medicine but established in medical practice before the Apothecaries Act of 1815, were allowed to continue in practice until their retirement. The 1815 Act was part of the progress of medical reform that took place in the early 19th century and led to the development of the class that would become known as general practitioners.

Clearly, for John Snow the route via Oxford or Cambridge was out of the question. Instead, on 22 June 1827, he became apprenticed to William Hardcastle in Newcastle. Snow's reasons for choosing a medical career are unknown, though Snow (2000*b*) points out that at the time medicine, entered via an apprenticeship, still offered one of the cheaper routes for the lower classes to rise up the social scale. However, the whole episode of his training gives rise to many questions. To begin with, why go to Newcastle, some 80 miles from home, when he could presumably have trained in York? Ellis (1994) suggests that his uncle, Charles Empson (his mother's younger brother) may have been the influence, because he had business connections in Newcastle and later had business premises near the centre of the town. Snow (2000*b*) however points out that Empson did not establish his business in Newcastle until after Snow had settled there with Hardcastle; she points out that Hardcastle had served his own apprenticeship in York and that a connection could have been established then.

A more intriguing question is how the Snow family found the money for this apprenticeship. Snow's apprenticeship fee was 100 guineas (£105), an almost inconceivable sum for a labourer's family to find. In addition there would have been the outlay for books and instruments and the considerable cost of travel between York and Newcastle. The most likely source of the money, and probably the reason for the choice of Newcastle, seems to be Charles Empson.

Snow generally seems to have got on well with his master, who appears to have been well respected in the local medical community. (He was among other things doctor to the family of Robert Stephenson who lived at Killingworth and was a friend of Snow's uncle, Charles Empson (Snow 2000*b*). In addition to the training he would have received under Hardcastle, Snow attended classes at a rudimentary medical school that was established in Newcastle on 1 October 1832. It was at Killingworth Colliery that he saw at first hand the effects of cholera.

In 1833, his term of apprenticeship with Hardcastle completed, Snow took a post as assistant to Mr Watson of Burnop Field, near Newcastle. This may have been to gain more clinical experience before going on to formal medical training, or it may simply have been because he lacked the funds for that education. Study at a formal institution required the student to find the money to pay for accommodation, food and books as well as tuition fees, putting such education generally beyond the reach of the poorer classes.

Snow appears to have been less happy with Watson than he had been with Hardcastle, and left the post after just a year. Richardson (1858) remarks that Snow 'worked too hard for his money'. As Snow's later life reveals that he was not one to avoid hard work, it seems that Watson must have been a demanding master.

Snow left Burnop Field to return to Yorkshire, again to work as a medical assistant. His principal this time was Joseph Warburton, who practised at Pateley Bridge in Nidderdale, which at that time was a small lead-mining and quarrying town with a population of about 1500 (Ellis 1994). Farms were small, the land was steep, the climate cold and wet and the living hard, but Snow seems to have enjoyed himself there; Richardson states that he thought highly of Mr Warburton. In 1836 he returned to York to stay with his family for a few months. In this period Richardson records that he made long walking expeditions into the countryside, 'collecting all kinds of information – geological, social, sanitary, and architectural'. Although he never seems to have had any formal geological training, it is noteworthy that later, in London, he was to show evidence of clear thinking on geological matters.

Later in 1836 Snow moved to London to take up his formal medical education. He made the journey on foot, via Liverpool, Wales and Bath where he visited Charles Empson. He arrived in London in October and enrolled at the Hunterian School of Medicine where he studied for one year. The cost of a complete course of lectures, demonstrations and practicals was 29 guineas (£30.45); by paying £34 a student could be entered as 'perpetual' with the right to attend future courses of lectures without further charge (Ellis 1994). The Hunterian School was in Great Windmill Street, on the western edge of Soho, an area that Snow would come to know well. In the very heart of London, it was within walking distance of several of London's teaching hospitals, including the Westminster Hospital at which Snow continued his studies in 1837–1838 by walking the wards (Ellis 1994; Snow 2000*b*).

During his time at the Hunterian School, Snow needed convenient and cheap lodgings. The once-fashionable area of Soho was moving socially down market, as the genteel classes who had once been its principal residents moved further west to Piccadilly and beyond. Many parts of Soho were occupied by poor families who welcomed lodgers as a source of income and who did not charge the rents that would have been asked nearer to Piccadilly. Snow took up lodgings at 11 Bateman's Buildings, running south from Soho Square. Today, Bateman's Buildings is a narrow, non-descript, dank alleyway, gloomy and smelling of urine. In Snow's day, we can assume that it was less attractive. Similar alleys and 'courts' formed a large part of the residential property of Soho. Snow shared his lodgings with a fellow student at the Hunterian, Joshua Parsons (Snow 2000*b*).

On completing his hospital practice in 1838, Snow applied for the post of apothecary in the Westminster Hospital, which became available in July of that year. He had successfully sat the examination of the Royal College of Surgeons in May 1838, but ran into the immediate problem that he had not yet passed the examination of the Society of Apothecaries, who because of a technicality would not allow him to sit the examination until October of that year. Snow sat and passed the Apothecaries' examination on 4 October, but the chance of a hospital appointment, with its attendant kudos and the basis it provided for establishing a private practice, was lost (Snow 2000*b*). Instead, at the beginning of September 1838, he moved from his student lodgings in Bateman's Buildings around the corner to a house at 54 Frith Street, where he set up in practice.

For the first few years of his life as a medical practitioner, Snow supplemented his income by working in the outpatient department of Charing Cross Hospital. He attended meetings of medical societies and continued to gain medical qualifications (Ellis

1994). In November 1843 he became a Bachelor of Medicine of the University of London, and in December 1844 he attained that university's doctorate, the MD. In June 1850 he passed the examination to become a Licentiate of the Royal College of Physicians, a qualification equivalent to the Member (MRCP) status of today.

Ellis points out that none of these qualifications was necessary for Snow to practise as a family doctor, so his acquisition of them probably indicated a desire to continue to better himself. Perhaps also they may indicate a need to prove that he, despite his humble origins, was as good as the products of the traditional training; perhaps it was a feeling of insecurity, and the desire to ensure that he was firmly established in his profession. With or without these qualifications, it was possible that Snow would have continued in practice as a family doctor in Soho, competent and respected, but unremarkable and unremembered, but for two events; the introduction of anaesthetics and the return of cholera.

The advent of anaesthesia

The first recorded use of anaesthesia is by a Boston dentist, William Morton, who administered sulphuric ether to perform a dental extraction in his own practice on 30 September 1846. Two weeks later, on 16 October, he gave the first public demonstration of the use of ether in surgery at the Massachusetts General Hospital in Boston (Ellis 1994). Initially, Morton attempted to maximize the financial benefit of his discovery by keeping the details secret. This secrecy was largely counter-productive, because it led to a slow acceptance of the use of ether in the USA. Instead the idea spread to Britain, where it was first used on Saturday, 19 December 1846 by another dentist, James Robinson. Ellis records that within a few days, Snow had heard of the use and visited Robinson's home in London to see anaesthesia in practice.

Snow realized the significance of Robinson's demonstration. He also realized that for anaesthesia to work effectively and safely, its use must be on a scientific basis. He designed an inhaler for the administration of ether. He also studied his patients closely and in late 1847 published 'his small but classic textbook' (Ellis 1994) *On the inhalation of the vapour of ether in surgical operations*. This book was published by John Churchill, who was later to publish much of Snow's work on cholera.

Within a few years he had established himself as London's leading expert in the science and practice of anaesthesiology. He administered chloroform to Queen Victoria at the births of Prince Leopold (1853) and Princess Beatrice (1857). By then, he had embarked on another area of study.

The coming of cholera

The word cholera derives from the Greek word for bile, the illness being characterized by severe diarrhoea and bilious vomiting. An illness described by various names such as 'British cholera' had long been recognized in Britain but in the 19th Century Western medical men became increasingly aware of a much more serious, often fatal, illness that originated in Asia and was termed 'Asiatic cholera' to distinguish it from the milder, home-grown version.

Snow wrote in 1849 that cholera had first been recognized in India in 1817. In 1855, he commented that:

> 'The existence of Asiatic Cholera cannot be distinctly traced back further than the year 1769. Previous to that time the greater part of India was unknown to European medical men; and this is probably the reason why the history of cholera does not extend to a more remote period.'

A few years later, Macpherson (1867) recorded that the Portuguese reached the coast of India in 1497 and began settling there around 1502, with Goa founded in 1510. In 1503, during a campaign, they noticed cholera and smallpox proving fatal to Europeans. Then, in 1543, they reported 'an epidemic of cholera of frightful intensity at Goa'.

It seems now accepted that, however long cholera may have been present in India, the major outbreak that really drew it to wider notice occurred in Bengal, in and around the Ganges Delta, in 1817. (It is possible that the bacterium responsible may have undergone a mutation at around this time.) It then began to spread rapidly. Some geologists have linked this spread to the displacement of populations caused by poor harvests following the eruption of Tambora on 10 April, 1815 (Robock 2002). Other workers attribute it more prosaically to the expansion of the British Empire, the increase in trade between India and Europe, and the general increase in movement of people and commodities (see, e.g. Evans 1987). By 1819 it had reached Mauritius and by 1824 it covered the whole of South-East Asia. Westward, it was carried by traders across Afghanistan. It was then halted by a military *cordon sanitaire* around Astrakhan in 1823, but reappeared in Persia and again crossed the Caspian Sea. This time the *cordon sanitaire* failed to work and it spread up the Volga, eventually reaching Moscow in 1832. By July 1831 it had reached Riga on the Baltic (Evans 1987).

A namesake of John Snow, writing on the other side of the Atlantic, recorded its progress as reaching Berlin and Vienna in August 1831. From Berlin it spread to Hamburg and thence to Sunderland, on the coast of NE England, where the first case was recorded on 24 October 1831 (E. M. Snow 1857). Stephanie Snow (2000*c*) reports his name as William Sproat. The arrival from Hamburg is significant. In the subsequent epidemic of 1848–1849, John Snow recorded that the first case in London 'was that of a seaman named John Harnold, who had newly arrived by the *Elbe* steamer from Hamburgh, where the disease was prevailing' (Snow 1849, 1855). In 1857, the influential journal *The Builder*, fearing the return of the disease from Europe, remarked 'Hamburg has ever been our warning'.

Snow (1855) remarked perceptively that cholera 'travels along the great tracks of human intercourse, never going faster than people travel and generally much more slowly. In extending to a fresh island or continent, it always appears first at a sea-port.'

In the present day it is almost impossible to imagine the impact that cholera had on populations in the 19th Century. Fatal diseases, including smallpox, tuberculosis and typhoid, were no strangers, but they generally acted slowly or at least spread by obvious contact with infected persons. Cholera came suddenly and without warning; its victims could be perfectly healthy in the morning and dead by nightfall. Perhaps worse was the manner of their demise. Those with tuberculosis showed few outward symptoms of their ailment – as Evans (1987) remarks 'on the whole they merely became pale and interesting'.

Cholera typically begins with a vague feeling of being unwell, followed rapidly by internal cramps and intense and prolonged vomiting and diarrhoea. The diarrhoea continues, the evacuations usually becoming almost colourless and odourless and having the milky appearance of water in which rice has been boiled, and referred to as 'rice-water evacuations'. The victim goes through a period known as the 'blue, cold stage'. This is brought about by the massive loss of body fluid (up to 25%), which results in the blood becoming thickened and ceasing to circulate properly. The reduction in ability to transport oxygen and heat means that the patient takes on a blue colouration, and feels cold and clammy to the touch. The general loss of fluid volume results in the skin becoming corrugated, and the eyes becoming sunken and dull. The victims are frequently described as having apparently aged many years. In about half of the cases this cold blue stage was fatal, and gave rise to the common name for Asiatic cholera of 'The Blue Death'.

In some cases there was no preliminary feeling of being unwell, and the cramps and evacuations were the first sign of the illness. In rare cases, the *cholera sicca* or dry cholera, there were no evacuations, the intestines being found after death to be distended with the watery excretion. This was apparently the fate of Gustav von Aschenbach in *Death in Venice*; Thomas Mann spared him the indignity that, for the Victorians, was probably one of the most distressing aspects of cholera symptoms – the loss of control, often in public, of bodily functions.

Patients who survived the blue stage went on to the 'warm pink stage' in which they exhibited fever and began to sweat. Normal colour returned and with the taking of fluid recovery was relatively rapid.

If the symptoms of cholera were bad, the treatment was if anything worse. Medical science had no clear idea of the causative agent of the disease. The general belief was that the disease entered the victim's bloodstream and caused the diminution of the blood by expelling fluid into the intestines. If the body was trying to reduce the volume of blood, it was argued that the best course for the physician to follow was to encourage it. Thus bleeding, either by opening a vein or by the use of leeches, was considered one of the safest courses of medical action.

One of the leading cholera 'experts' of the day was John George French. He comments that the 'first object for the relief of the patient is the diminution of the circulating fluid' and goes on to say that workers in India agree 'that bleeding is beneficial and that a very large quantity of blood should be abstracted in order to derive advantage from this operation' (French 1835). Other treatments recommended included purging (it is not quite clear what would be left to purge) and electric treatment. How severely ill patients responded to such treatment does not seem to be recorded.

The communication of cholera

As there was no effective treatment for cholera, it was obviously best to avoid catching it. The problem was that, as with the treatment, there was no real idea on how the disease was transmitted. One of the most frightening things about cholera was the way it would suddenly appear in a community and then proceed to strike simultaneously people who had never met. In recent decades, only AIDS in its early days can have had anything like the same ability to terrify.

Two competing schools of thought developed to explain the way cholera was spread. One, the contagionists, argued that cholera could clearly be transmitted from one individual to another who had been in close contact. The other group, the miasmatists, argued that for the illness to appear suddenly in a district and strike unconnected people simultaneously, there must be a different mechanism. Noting that cholera was often more prevalent in low-lying areas with bad drainage, or where sanitary facilities were lacking, they claimed that the disease was transmitted by a 'miasma' or foul air. Hence those living near overflowing sewers or cesspits were considered especially vulnerable.

When cholera first arrived in Britain in 1831, it

spread quickly from Sunderland to other towns and settlements in the NE of England. William Hardcastle, Snow's master in Newcastle, was one of two doctors appointed to tend the poor who were suffering from cholera. One locality that was badly affected was the area around Killingworth Colliery, the base of George and Robert Stephenson. Snow accompanied Hardcastle, or was sent by him, to Killingworth. He was thus among the first medical men in the country to see the disease at first hand and it is clear from his later writings that the episode stayed in his memory.

Cholera returned to Britain in 1848. There is some dispute over its first appearance, but Snow (1849) reports that 'the first decided case . . . in London' was that of the seaman John Harnold, who brought it from Hamburg as described above. He went to live in Horsleydown, 'was seized with cholera on 22 September, and died in a few hours'.

It is not clear when Snow began to think about the way cholera is communicated. He himself wrote (Snow 1855, p. 125) that he had formed his views in the latter part of 1848. He had seen cholera at first hand in the epidemic of 1831–1832 and took an early interest when it returned to London, studying the circumstances of several cases in detail. Snow set out his evidence (Snow 1849) and his arguments in a pamphlet 'On the mode of communication of cholera', which was published by John Churchill. He formed the theory that rather than being caused by some 'poison' that alters the blood so that its watery and saline parts begin to be extruded, 'it is more consistent with observation to say that exudation begins as a result of irritations of the mucous membranes of the alimentary canal' (Snow 1849).

He went on to say (Snow 1849, p. 8–9) that:

> 'if the disease is communicated by something that acts directly on the alimentary canal, the excretions of the sick at once suggest themselves as containing some material which, being accidentally swallowed, might attach itself to the mucous membrane of the small intestines, and there multiply itself by the appropriation of surrounding matter, in virtue of molecular changes going on within it, or capable of going on, as soon as it is placed in congenial circumstances.'

This certainly makes it possible that the disease can be transmitted through contact, by for example, persons handling soiled clothing or bed linen and then handling food or putting their fingers in their mouths (a particular problem with small children). This could explain how the infection is transmitted between members of the same family, within lodging houses or down coal mines and would satisfy the contagionist argument. But it also opens up for consideration 'a most important way in which cholera may be disseminated' – by sewers emptying into sources of drinking water (Snow 1849, p. 11). Snow showed that in all the cases he had considered, there was clear evidence that the disease was associated either with direct or indirect personal contact or with contamination of drinking water by human excrement. Thus, without resorting to some fanciful 'miasma' it was possible to explain how the illness could strike many people, who had never been in contact, simultaneously in a town or village.

He considered this theory 'less dreary' than the miasmatic theory, because it offered some way of checking the disease – by hygiene and the provision of a proper water supply. The most urgent need was in south and east London for a supply 'from some source quite removed from the sewers' (Snow 1849, p. 30).

Snow's 1849 pamphlet was reviewed by an anonymous reviewer in the *London Medical Gazette* who reported that the *experimentum crucis* would be for the suspected water to be conveyed to a distant locality and there produce the disease in all who drank it (Shephard 1995; Snow 2000*c*). Otherwise, there was no evidence that infection had not been passed from one victim to another or that all victims had not suffered from some common cause such as a 'miasma'. It was to take five years and cost more than six hundred lives, but that experiment was to take place.

Cholera returns

After the 1848–1849 epidemic, cholera returned to London in 1853. By this time neither the nature of the disease nor its cure were generally recognized. Theories of its communication were multiplying, though the idea of the 'miasma' seemed to hold sway. One interesting idea was the 'geological theory' propounded by John Lea of Cincinnati (1850). Lea had noticed that when cholera arrived in the United States, it seemed more prevalent in limestone areas than in those underlain by sandstones – 'it passed around the arenaceous, and spent its fury on the *calcareous* regions'. He thus concluded that calcareous water was 'a *proximate cause* of that disease' – that calcium or magnesium salts were necessary for it to act. Most hydrogeologists would conclude that the effect arose because sandstones are able to filter bacteria from groundwater, whereas limestones, with predominantly fissure permeability and porosity, do not do so; the apertures are large enough to allow bacteria to be transported and the flow speeds great enough that the micro-organisms do not die. John Snow heard of this theory and, though no geologist, was quick to point out that sandstones have the effect of 'oxidizing and thus destroying organic matters; while the limestone might not have that effect'. He acknowledged the likelihood of the truth of Lea's assertion that people who used rain water escaped almost entirely the effects of cholera.

Ironically, by 1854, both the true nature of the disease and an effective treatment were known, but not widely. The German scientist, Robert Koch, is generally credited with identifying *Vibrio cholerae*, the 'comma' bacillus and the causative agent of the disease, in 1885, but an Italian, Filippo Pancini, is now acknowledged to have discovered it in 1854 (Evans 1987). The modern treatment for cholera includes intravenous rehydration with saline solution; this technique was suggested in 1831 by William O'Shaughnessy and first used by Thomas Latta in 1832. Snow was clearly aware of the use of injection of saline solution and described its effectiveness in overcoming the symptoms of cholera, commenting that 'the patient is able to sit up, and for a time seems well'. This last phrase suggests that the treatment was not a permanent cure; it seems that it was not widely used because it was regarded as dangerous, something to be tried only as a last resort on patients who were near death. Usually it seems to have been administered too late, so that in Snow's day it had not often been successful and was not in favour.

By 1853, water closets and sewers were replacing privies and cesspits in many towns and in many parts of London, in response to the work of Edwin Chadwick and his fellow sanitary reformers. Unfortunately, in many cases the cesspits were neither emptied nor removed, but merely covered over (*The Builder* 1854). Perhaps more unfortunate, until Bazalgette installed his great 'interceptor' sewers, these apparent advances served mainly to convey raw sewage quickly into the tidal reaches of the River Thames, from which many individuals and some companies were still drawing their water supplies. In dry weather in particular, when the flow of the Thames was reduced, the action of the tide served to move this sewage up and down the river through the capital, mixing it thoroughly with the river water from which part of London derived its water supplies.

The Metropolis Water Supply Act of 1852 prohibited companies, after 31 August 1855, from taking water from the Thames below its tidal limit at Teddington Lock (Binnie 1981). It also required that within five years every company must lay on a constant supply of water, for it was still common for many properties to receive water intermittently. Occupiers of buildings stored it in tanks and cisterns, where it was open to disease.

One company, the Lambeth Water Company, had anticipated that act by moving its intakes from opposite Hungerford Market up river to Thames Ditton, above the tidal limit of the Thames and above the worst-polluted stretch of the river. In the 1849 epidemic, Snow had noticed that the southern districts of London appeared to suffer disproportionately from the cholera. Many professionals attributed this to factors such as poverty, overcrowding and bad ventilation, but Snow recognized that it struck in particular houses supplied by two water companies – Southwark and Vauxhall Water Works and the Lambeth Water Company. By the time cholera returned in 1853, the Lambeth Company was operating its new intakes, while the Southwark and Vauxhall was still taking water from its intake at Battersea Fields, on the tidal reach of the Thames with its burden of sewage. Snow realized that the comparison of deaths among the populations served by the Lambeth Water Company and the Southwark and Vauxhall Company was a major piece of evidence in his campaign to identify the means by which cholera was communicated (Snow 1855). The two companies served the same areas. In Snow's words:

> 'The pipes of each Company go down all the streets, and into nearly all the courts and alleys. A few houses are supplied by one Company and a few by the other, according to the decision of the owner or occupier at that time . . . The experiment, too, was on the grandest scale. No fewer than three hundred thousand people of both sexes, of every age and occupation, and of every rank and station, from gentlefolks down to the very poor, were divided into two groups without their choice, and, in most cases, without their knowledge; one group being supplied with water containing the sewage of London, and, amongst it, whatever might have come from the cholera patients, the other group having water quite free from such impurity . . . To turn this grand experiment to account, all that was required was to learn the supply of water to each individual house where a fatal attack of cholera might occur.' (Snow 1855)

Snow duly turned the 'grand experiment to account' and found, among other things, by comparing fatal attacks of cholera with source of water, that in every 10000 houses supplied by the Lambeth Water Company there were 5 attacks, and in every 10000 houses supplied by the Southwark and Vauxhall Company there were 71 attacks (Snow 1855, 1857).

Cholera in St James's Parish: the 'Broad Street pump'

In 1852 Snow had moved from Frith Street to more prestigious premises at 18 Sackville Street, off Piccadilly. He seems to have retained many of his contacts and patients in Soho, some of whom, according to Richardson, he treated for little or no payment.

In September 1854, while busy with his investigation into the outbreak in south London and the relative quality of the water supplied by the two water companies, Snow became aware of what he described as the 'most terrible outbreak of cholera which ever occurred in this kingdom . . . the mortality in this limited area probably equals any that was ever caused in this country, even by the plague'.

The outbreak occurred in the Parish of St James, and was particularly severe in Broad Street. There had been a few cases of cholera in the neighbourhood during August 1854, but a dramatic increase occurred on the night of 31 August. The weather had been hot and dry; the weather reports in *The Times* indicate day after day of dry, fine weather, often with cloudless skies and with the daytime temperature in the high sixties to low eighties Fahrenheit (around 20 to 29° Celsius). These are conditions that, as Snow points out, tended to favour the spread of cholera in England. He attributed this partly to the fact that the British, unlike the French and other Europeans, tended not to drink cold water except when the weather was warm, preferring hot drinks at other times.

Snow seems to have become acquainted with the outbreak on 3 September and immediately suspected contamination of the water of a pump in Broad Street, which was much used by the local people. He examined the water on the evening of 3 September but found little visible evidence of any contamination. Over the next two days, he continued to examine the water and found varying amounts of organic impurity, visible to the naked eye, in the form of small white flocculent particles.

Snow obtained a list of deaths from cholera registered during the week ending 2 September in the sub-districts of Golden Square, Berwick Street and St Ann's Soho. In these three sub-districts there had been 89 deaths; six had occurred in the first four days of the week, four on Thursday, 31 August and the other 79 on the Friday and Saturday, 1 and 2 September. Snow therefore regarded the outbreak as having started on the Thursday and made detailed enquiries about the 83 deaths that occurred on the Thursday, Friday and Saturday.

He found that nearly all the deaths had occurred within a short distance of the Broad Street pump. Perhaps more significant, there were only ten deaths in houses situated decidedly nearer to another pump. In five of those, the families of the dead told Snow that they always obtained water from the pump in Broad Street, as they preferred the water to that from the nearer pump. In three other cases the victims were children who went to school near the Broad Street pump. Two had been known to drink the water and the third was believed likely to have done so. The other two deaths, some way from the pump, Snow regarded as merely representing the background mortality from cholera that was occurring before the outbreak took place.

Of the 73 deaths that occurred near the pump, Snow found that 61 of the victims had been known to drink water from the pump in Broad Street, either continually or occasionally. Six of the others had been known not to drink the water, and in the other six cases he could get no information because everyone connected with the dead individuals had either also died or had left the area.

Snow therefore concluded that 'there had been no particular outbreak or increase of cholera, in this part of London, except among the persons who were in the habit of drinking the water of the above-mentioned pump' (Snow 1855). He sought a meeting with the Board of Guardians of St James's Parish, in which the pump was situated, on the evening of Thursday 7 September. What was said is not recorded. Richardson (1858) records Snow telling him that his explanation for the disease was treated initially with incredulity but the Guardians nonetheless had the handle removed from the pump on the following day.

Earlier in the week, *The Times* (1854*a*) had reported that:

> 'The severe outbreak of cholera in part of St James's parish, and in the adjacent parts of the parish of St Anne, Soho, has been promptly met by the sanitary and other preventive measures carried out by the boards of guardians, under the advice of the General Board of Health.'

It went on to say that the President of the Board of Health (Sir Benjamin Hall):

> 'has directed special attention to the supply of water in the several localities, the source of supply, and whether filtered or not before supply; and when two companies supply in any one district, the inspector is to state whether the disease is more prevalent in one district than the other, having due regard to similar classes of dwellings.'

It seems either that Snow's message was getting through, at least in some quarters, or that other people were thinking along similar lines or possibly that the authorities were anxious not to leave themselves open to criticism for not covering all possibilities.

Figure 1 shows the deaths from cholera in St James's Parish during August and September 1854. (When using this figure it is important to remember that it shows the dates of the recorded deaths, which typically occurred two or three days after the date of infection). It is clear from the figure, and Snow readily admits, that the outbreak was in decline by the time that the handle was removed. Generally speak-

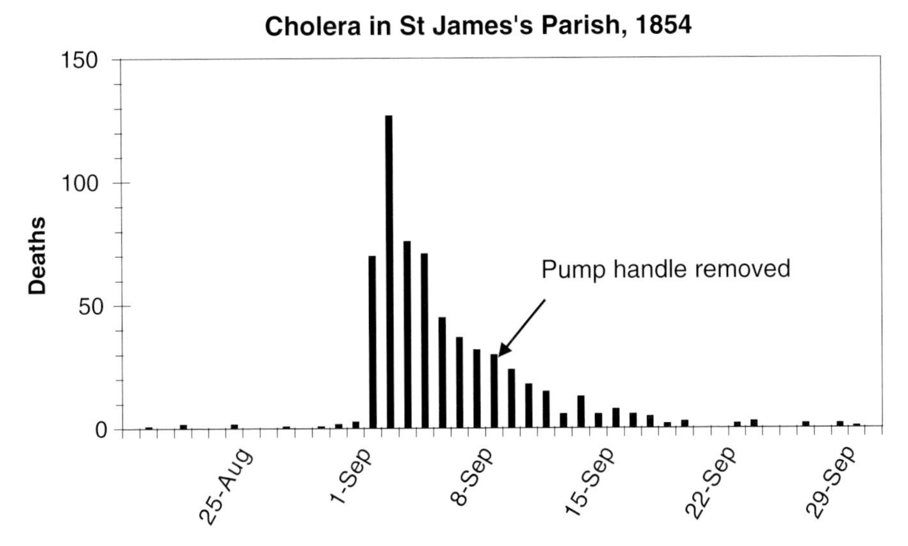

Fig. 1. Deaths from cholera in St James's Parish, 1854. Data from Snow's (1855) monograph.

ing, all those who were able to do so left the area fearing for their lives. The matter received more coverage in the press (*The Times*, 1854*b*). *The Builder* (1854) was not slow to voice its opinions, but with so many possible causes and treatments, it is hard to know what the general public would have been left to think. At the same time, the correspondence columns of *The Times* carried a vigorous debate, with daily contributions, as to the efficacy of castor oil as a treatment for cholera, together with complaints about profiteering by the chemists who sold it.

Snow then says that he was busy with 'an inquiry in the south districts of London' (the one concerning the Lambeth and the Southwark and Vauxhall water companies) and, by the time he was able to return to the Golden Square problem, so many of the people had moved away that it was impossible to establish exactly what had happened. He did find that the 83 deaths he had investigated were an underestimate; some people had been removed to hospital and died there, their deaths being registered elsewhere. Other deaths on the Friday and Saturday had not been recorded until the following week, so that altogether 197 people had died on 1 and 2 September (Fig. 1).

Snow also found several facts that seemed to confirm his theory that the source of the outbreak had been the pump in Broad Street. In Poland Street was a large Workhouse. It was largely surrounded by houses in which many inhabitants died, yet of its own 535 inhabitants only five (all elderly) died from the disease, despite the fact that sick people were taken there and died. Snow found that the Workhouse had a supply of water from the Grand Junction Water Works and its own deep well (probably into the Chalk) and that its inmates never used water from the Broad Street well.

In Broad Street itself, near the pump, was a brewery owned by a Mr Huggins. It employed more than seventy men, yet none had died of cholera. The brewery had its own deep well and a supply from the New River company. Mr Huggins believed that the men never drank water from Broad Street, and probably never drank water at all, being given a free allowance of the brewery's produce! In contrast, on the other side of the pump was a percussion-cap factory owned by Mr Eley. This employed two hundred people; for them to quench their thirst, two large tubs of water were kept there, filled from the pump. Of these two hundred, 18 died of cholera.

As Snow went around, asking questions, checking facts, he kept finding evidence of the link between water from the well and infection with cholera. Perhaps the most striking example was drawn to his attention by a colleague, Dr Fraser. This was a perfect answer to the anonymous reviewer who, following Snow's 1849 pamphlet, had said that the *experimentum crucis* would be if suspect water could be conveyed some distance away from an outbreak and there produce cholera in those who drank it. Fraser drew Snow's attention to a death recorded in Hampstead on 2 September. This was of the widow of a percussion-cap maker (probably, though Snow does not say so, of the former owner of the factory in Broad Street; Chave (1958) states that she was Susannah Eley).

Snow found from the lady's son that she had not been in Broad Street for many months, but that she had a liking for the water from the Broad Street well. A cart went from Broad Street to Hampstead every day and conveyed to her a large bottle of the water. Water was taken to her on Thursday 31 August and she drank some of it on that evening and the next day.

She was attacked by cholera on Friday evening and died on Saturday 2 September. Her niece, who was visiting her, also drank some of the water before returning to her home in Islington; she too died of cholera.

Snow recorded his findings in the second edition (1855) of his work *On the mode of communication of cholera*. This is a substantial enlargement on the 1849 edition, and deserves the name monograph rather than pamphlet. Richardson records that its publication cost Snow £200 and that only 56 copies were sold, for which he received the princely sum of £3.12s (£3.60).

As part of his work, Snow produced a map of the area (Snow 1855) on which he marked the occurrence of each death from cholera (Fig. 2). There has been some speculation about this map. It has been widely stated that Snow drew the map and inferred from it that the deaths from cholera occurred around the Broad Street pump. (I admit to the same error (Price 1985)). Brody *et al.* (2000) point out that, in all likelihood, Snow did not draw the map until he was preparing his monograph of 1855 and his contribution to a report to the Cholera Inquiry Committee of St James's Parish. The map was therefore drawn to illustrate his argument that the disease had spread from the pump rather than as the means of deducing the fact. Brody *et al.* are almost certainly correct in this assertion. However, Snow (1855) says that he suspected the pump as soon as he 'became acquainted with the situation and extent of this irruption of cholera'. This implies geographical knowledge, which of course Snow had having lived in the area for 18 years. In effect then, the map was in his head. He had no need physically to plot it for himself, but he did need it to convince others.

Here, and elsewhere, we have a problem. We do not have Snow's notes of his investigation. Snow was known to be a meticulous worker. Richardson reports that:

> 'He kept a diary in which he recorded the particulars of every case in which he had administered chloroform or other anaesthetic . . . He kept a record of all his experiments and short notes of observations made by his friends.'

We know also that Snow kept case books in which he recorded the details of his operations, administration of anaesthetic and other day-to-day work of a physician. Presumably, therefore, he kept similar detailed notes of his work on cholera. Unfortunately, no such notes survive. When Snow died it seems that his family handed over to Richardson his medical papers, which presumably included three volumes of casebooks. Those three volumes were found among Richardson's papers when he himself died in 1896 (Ellis 1994). They were duly passed on and in 1938 were given by Richardson's daughter to the Royal College of Physicians. They have been edited by Ellis and published. They are virtually all that survives of Snow's papers and they contain no mention of cholera beyond the fact that in 1849 Snow administered chloroform to cholera patients to help relieve the pain of the cramps. In the sections of the books that deal with the summer and autumn of 1854, there is no mention of cholera whatever but it is clear that Snow, in addition to his work on cholera in south London and on the outbreak in St James's, was continuing with his routine medical work. He was clearly a man of considerable energy.

In looking at Snow's maps and others of the time, and relating them to present-day London, there are various points to bear in mind. The street plan remains largely unaltered – Snow would probably still be able to find his way around the area – but some of the street names have changed. Broad Street is now Broadwick Street (though it is still 'broad') and many of the buildings have been renumbered. Cambridge Street, which met Broad Street at right angles, is now Lexington Street, and the public house on the corner, which Snow would have known as the 'Newcastle Arms', is now called the 'John Snow'. The site of the workhouse in Poland Street is now occupied by a multi-storey car park and Huggins's Brewery has disappeared, though it is possible to identify clearly the site that it occupied.

The map that Snow incorporated in his monograph contains an interesting error (Brody *et al.* 2000). The pump is marked as being exactly on the corner of Broad Street and Cambridge Street, whereas in reality it was a little further along, outside what was then number 40 but is now 41. In his report to the parish, Snow incorporated a slightly revised version of the map (Fig. 3). This has the pump in the correct position and it incorporates a dotted line inside which it was quicker to walk to the Broad Street pump than to any of the other street pumps. This was even more conclusive in indicating that the deaths from cholera seemed to have been related to the Broad Street well.

For all his certainty that the Soho outbreak had its origin at the Broad Street pump, Snow could not explain how the well had itself become infected, or for how long the infection remained in it. He commented that the area contained a great mix of social classes, from the well-off of Poland Street and Great Pulteney Street, who lived one family to a house, through intermediate areas such as Broad Street, to streets occupied largely 'by the poor Irish'. Tailors, employed in making clothes for the fashionable shops of Piccadilly and around, made up a large proportion of the population. The disease had struck among them with indifference. The deaths columns of *The Times* for the period record several deaths from the better-off streets such as Great Pulteney

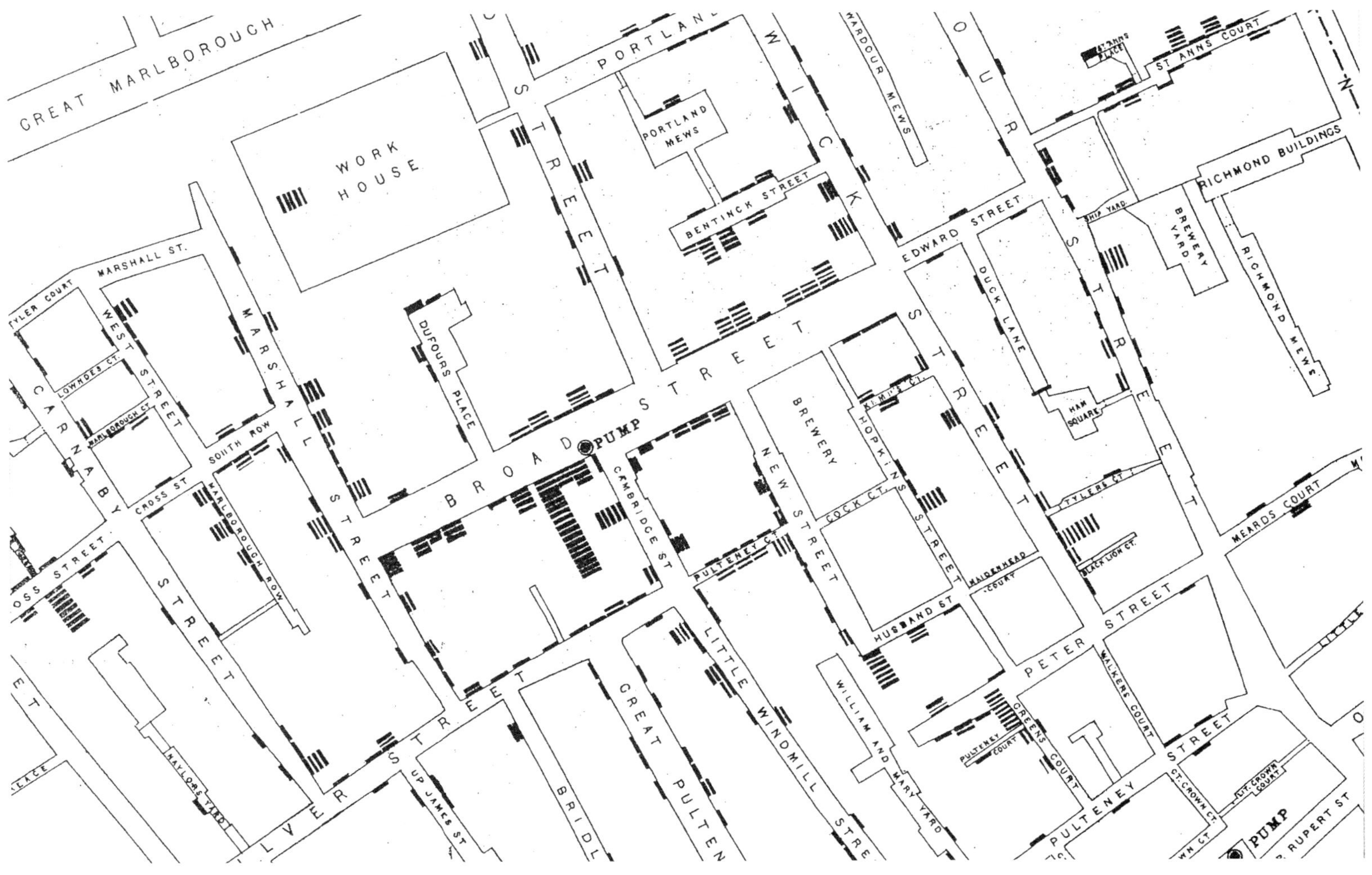

Fig. 2. Part of John Snow's map (Snow 1855) showing deaths from cholera. Each black bar represents one death.

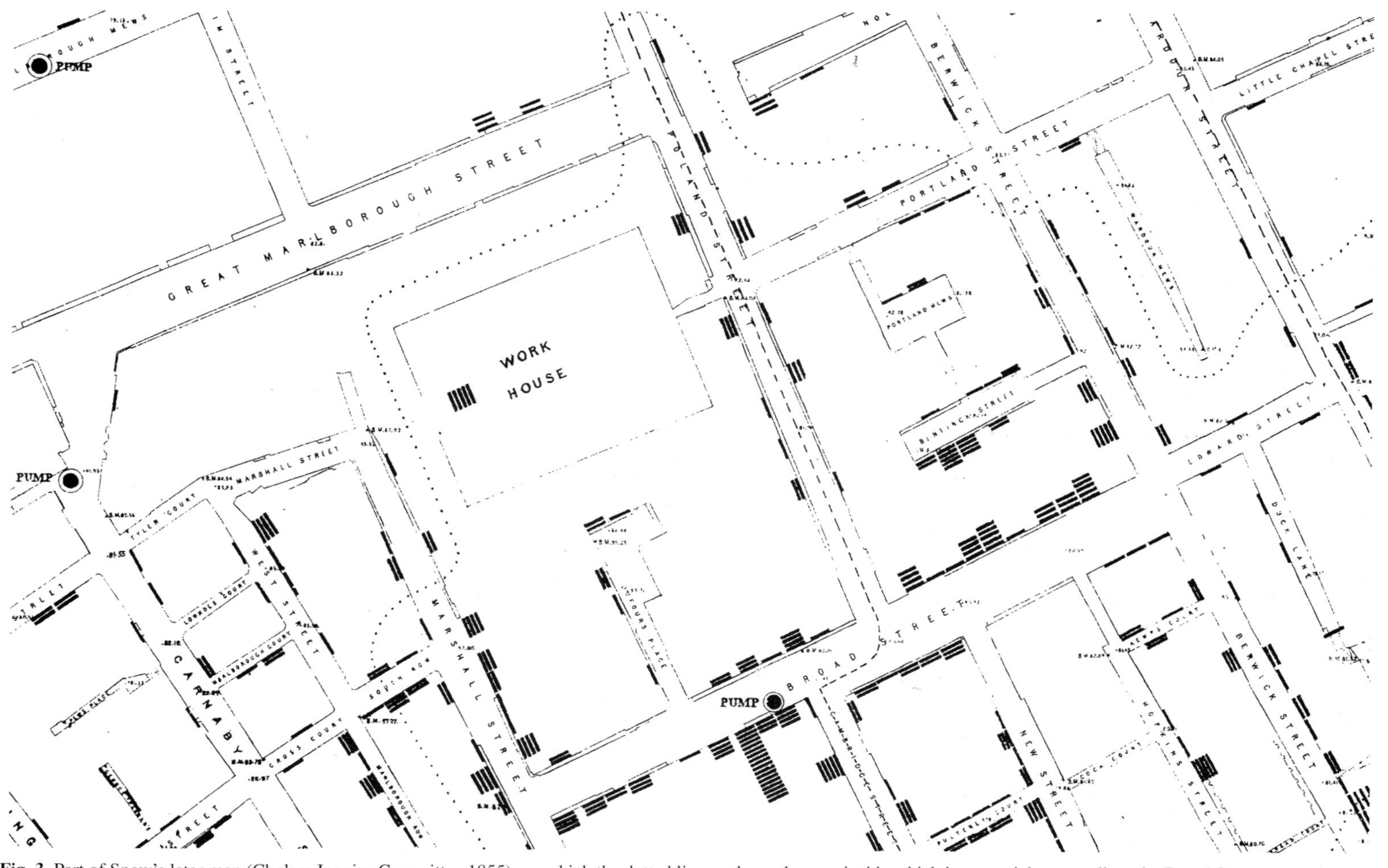

Fig. 3. Part of Snow's later map (Cholera Inquiry Committee 1855), on which the dotted line encloses the area inside which it was quicker to walk to the Broad Street pump than any other pump.

Street, Poland Street and Berwick Street, as well as Broad Street. Cholera was rarely mentioned – the usual form was to refer to death 'after a sudden illness' or 'a few hours' illness'. But however indiscriminate the appearance of the disease, Snow was well aware that once the cholera came among the poor, who lived in overcrowded conditions, it was more likely to spread than among the rich, who were less crowded, had better sanitary conditions and were more able to nurse the sick in rooms separate from those where the rest of the family was living.

Snow described the well in Broad Street as being 'from twenty-eight to thirty feet in depth' (about 9 m) and passing through gravel to the surface of the clay beneath. This is a good description. The latest geological map (British Geological Survey 1993) shows that the Soho area is underlain by Lynch Hill Gravel, a post-diversionary Thames River Terrace Deposit, resting on London Clay. Snow ascertained that a sewer passed at a depth of 22 feet (about 7 m) 'within a few yards of the well.' He described how the well had been opened and inspected, and that the brickwork contained 'no hole or crevice . . . by which any impurity might enter'. This reveals some ignorance on the part of the inspectors. The well was later described as being lined with brick laid without mortar, so that water could enter, and presumably so too could contamination. It seems that in the 1850s, as now, many people felt that groundwater (and contaminants) would flow only through sizeable conduits.

Subsequent investigation records the well as 6 feet (1.8 m) in diameter, and the water level about 21 feet (6.5 m) below ground level. There is no record of this well, or any of the others mentioned as being in the vicinity, in the British Geological Survey National Well Collection, nor are any mentioned by Barrow and Wills (1913).

Snow's monograph was not greeted with the acclaim that he, and we, might have expected. It seems strange that a man who was hailed as at the forefront of the medical profession as an anaesthetist and who had given medical services to the queen, should be disregarded in this way. Perhaps the Victorians reacted to a man stepping outside his acknowledged area in much the same way as people today might react to a footballer pronouncing on the Middle East or the Archbishop of Canterbury on nuclear physics. People should stick to what they know. In any event, he was at best largely ignored, at worst opposed or ridiculed. In an editorial, *The Lancet* (1855) derided him for saying that offensive smells were not in themselves injurious, commenting that 'the fact is that the well whence Dr Snow draws all sanitary truth is the main sewer'.

The committee of the General Board of Health appointed to inquire into the cause of the outbreak dismissed Snow's arguments and the idea of contaminated water as possible causes of the cholera outbreak (Snow 2000*c*). Fortunately, help came in the form of Edwin Lankester, FRS. He established a Cholera Inquiry Committee to look in more detail into the outbreak in St James's. The committee dated its report 25 July 1855 (Cholera Inquiry Committee 1855) but it may have been some time before it was published. *The Builder* (1855) greets its report in its edition of 20 October. The report was reprinted as Appendix 10 of the Rivers Pollution Commission (1874).

The report contained a new map (Fig. 4), apparently based partly on that produced in September 1854 by Edmund Cooper, an engineer for the Metropolitan Commission of Sewers. The Commission, and Cooper, were anxious to show that the outbreak was not linked to the recent excavations for sewers or to the presence of gases emanating from gully-holes and gratings (Brody *et al*. 2000). The map in the report had several advantages over Snow's earlier maps. First, it showed the street numbers of the houses and other buildings. Secondly, it included deaths that Snow had not known about. It showed, for example, the deaths of workmen involved in building 'Model dwelling houses', between Hopkins Street and New Street behind the Huggins Brewery, to replace some 'of the lowest and filthiest description' (Whitehead 1854) that had been demolished. Thirdly, it distinguished between deaths of residents and those of non-residents, particularly useful in cases like that of the percussion-cap factory. Incidentally, the factory is shown on this map as being at 38 Broad Street, but Snow describes it as being No 37.

One of Snow's initial opponents was the Reverend Henry Whitehead, the curate of St Luke's Church in Berwick Street, who was a member of the committee of inquiry. Whitehead (1854) wrote that the cholera was sent by God as a chastisement for sin, the sin apparently being to allow the poor to live in such wretched conditions. (Quite why the poor themselves had to be so heavily punished to meet God's purpose, Whitehead does not make clear). Whitehead knew many of the local families intimately, had visited the sickbeds of many of the deceased and was able, probably better than anyone, to know of their habits. At first dismissive of Snow's arguments (he had himself drunk water from the Broad Street well on the evening of 3 September without ill effect) he then began to realize, like Snow, the connection between the well and the illness (Whitehead 1867; Chave 1958), and he became an ardent supporter of Snow and a tireless worker in the fight against cholera. He later wrote

> 'Few habitual drinkers of the pump water, to my knowledge, escaped with impunity. Few survivors were able to assure me that they so drank it regularly during the week ending September 2nd. But on and after that day several

Fig. 4. Part of the map from the Report of the Cholera Inquiry Committee, 1855. The grey shaded area represents part of an area on which the Earl of Craven had houses constructed during the Great Plague of 1665–1666 for people suffering from the illness and which was also used for burial of plague victims. It became known as the pest-field and some residents claimed that the cholera epidemic was in some way linked to the pest-field or to its disturbance for the construction of new sewers.

persons, who had not been in the habit of drinking it, began doing so, at least occasionally, from a notion (due to the enormous quantity of it taken by some patients who revived from collapse) that it was "good for cholera." More of these drinkers than of habitual drinkers of the pump-water escaped with impunity, I myself being among the number as I drank some of it at 11 p.m. on September 3rd, though not from any idea of its beneficial qualities.' (Whitehead 1867)

Whitehead then discovered what had eluded Snow – the probable original source of the infection in the well. He noticed that among the returns of death was that of an infant on 2 September. The child, a girl of five months, had lived at 40 Broad Street, the house outside which the well was situated. The child's mother had survived the epidemic and told Whitehead that the child had been taken ill with diarrhoea and that, beginning on 28 August, she had washed the child's napkins and emptied the water into the cesspool at the front of the house (Fig. 5). She had continued to do this on 29 and 30 August. Whitehead reported that a man who did not normally drink from the Broad Street well had done so at noon on 31 August and had been seized with cholera at 9 a.m. on 1 September. From this Whitehead concluded that it could have taken no more than three days for the discharged water to reach the well. (The warm weather may again have been a factor here: the intermittent supply of mains water meant that water stored in cisterns would become warm and less palatable, causing people who might otherwise rely on mains water to seek the cooler, fresher water from the pump).

There was some debate about the cause of the child's death, for the illness was recorded only as exhaustion after diarrhoea, not specifically cholera. The committee were generally satisfied, however, that given that cholera symptoms are rarely well marked in young children, and that diarrhoea during a cholera epidemic can generally be regarded as cholera, this was a true case of cholera. The question still remains, however, of how a child so young came to be infected with cholera in a district where it was not widely prevailing at the time.

As part of the investigations into the outbreak, and as a direct result of Whitehead's discovery, a Mr Jehosephat York, evidently a local surveyor who was secretary to the local Cholera Inquiry Committee, opened the well again in April 1855 (Chave 1958). It appears that the well in Broad Street (Fig. 6) was like many others in the area in being sunk to the top of the London Clay. 'The sides are built in brick, laid dry, through which the water readily enters, the arches are turned over with brick laid in mortar or cement, and covered in with a key stone, also secured in mortar or cement' (Cholera Inquiry Committee 1855). The Committee reports that some of the wells in the parish are 'rapidly fed from the water bed in the sand, that they cannot be pumped dry' while most, including the Broad Street well, can be 'laid dry by continuous pumping in four or five hours'. It is significant that the water level in the Broad Street well is shown as 6.5 m below ground level. This is about the maximum depth from which a surface pump can lift water, so if the Broad Street pump was of the lift type – the cylinder being at the surface instead of down the well – the pump would effectively cease to lift with a small depression of the water level. There is no information available on the type of pump that was fitted.

The Committee also noted that if the natural supply into the wells was limited, there was more likelihood of contamination entering. This seems unlikely, in that the contaminants would probably arrive from above, through the unsaturated zone, rather than through the saturated zone.

York's examination showed that the well was very close to the vault (cellar) of 40 Broad Street. The main drain from the house to the sewer beneath the street passed through the vault and had a flat bottom, about 0.3 m wide, with brick sides about 0.3 m high, covered with 'old stone'. The base of the drain was filled with 'soil' about 5 cm deep. On clearing this away, York found that the bottom was 'like a sieve' and that the sewage from the house must have been able to pass through it with ease. In the front area of the house (the sunken area open to the sky) there was a 'convenience' with a cesspool about a metre deep that was intended to serve as a trap to prevent sewage, vermin and gases from entering the house but which had been wrongly constructed (apparently a common fault of the time) and which instead was filled with 'soil' and preventing the easy outflow of sewage from the house. It was into this cesspool that the mother of the sick child had thrown the infant's evacuations.

York found that the brickwork of the cesspool was also badly decayed, so that the bricks could be easily lifted from their bed. He found that the earth between the cesspool and the well was black and saturated. Perhaps the worst aspect was that it transpired that the same cesspool had been found to be contaminating the well 17 years previously.

Analysis of the water from the Broad Street well in November 1854 showed that it contained 96 grains per gallon (1370 mg l^{-1}) of total solids. A more detailed analysis in June 1855 showed the chloride level to be about 11 grains per gallon (about 160 mg l^{-1}). This was taken as confirmation that the well was polluted by 'débris, refuse and excreta' (Cholera Inquiry Committee 1855).

Whitehead also raised an interesting point about the length of the cholera outbreak. It seemed clear that the outbreak was declining by the time the handle was removed from the pump. Snow's advice did not, therefore, halt the epidemic. But Whitehead

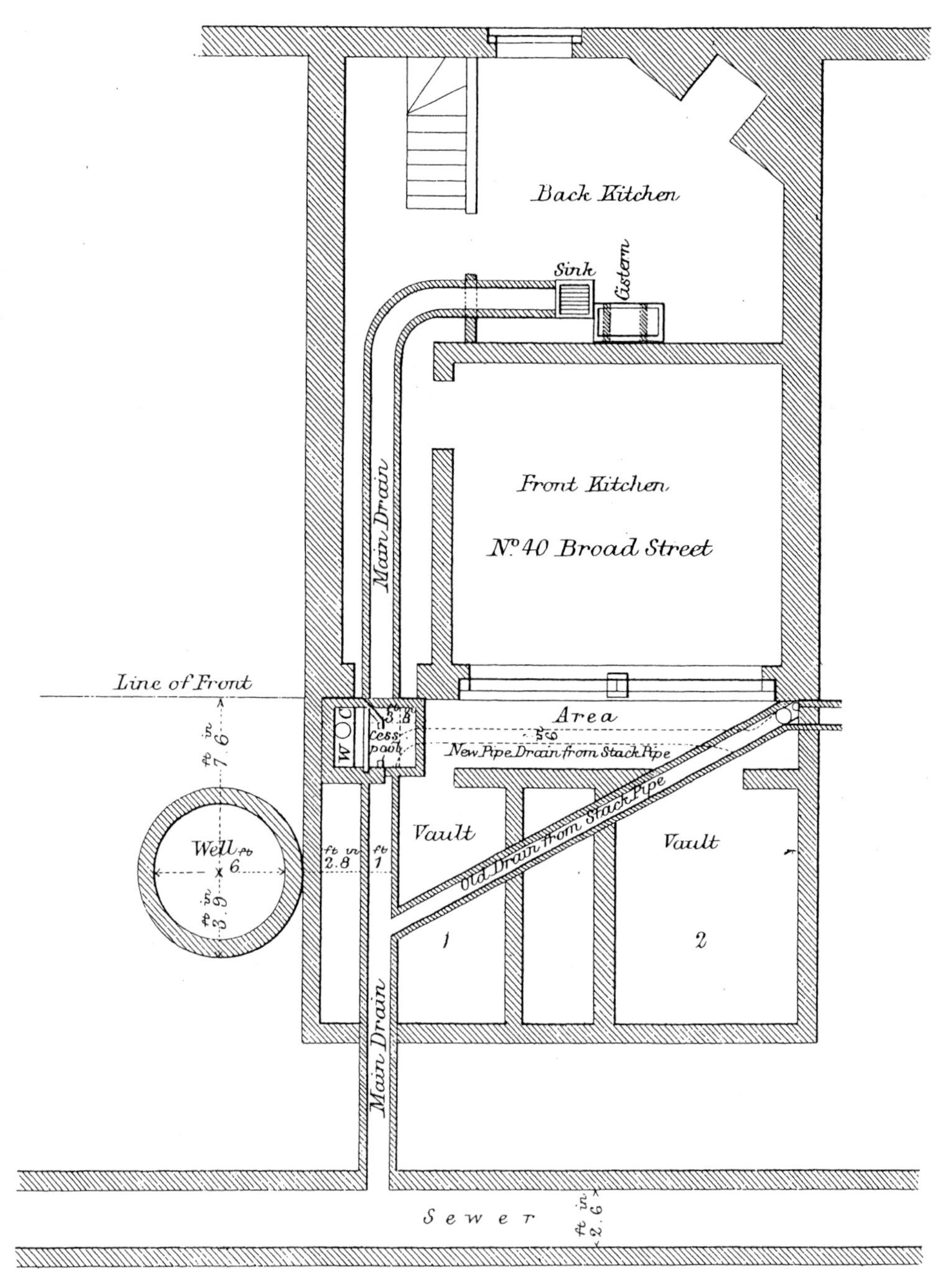

Fig. 5. Plan of 40 Broad Street, showing the Broad Street well (from the Report of the Cholera Inquiry Committee 1855).

discovered that cholera continued to be present in the house at 40 Broad Street, so that the well could constantly have been re-contaminated from the cesspool and the drain. Fortunately, the first three victims of the outbreak lived on the upper floors of the house, and their relatives saved themselves the trouble of carrying their 'discharges' down to the cesspool by the simple expedient of throwing them out of the upstairs back windows into the yard below. But the next victim, the father of the child who was supposed to have been the source of the outbreak and who slept in the same kitchen, was taken ill on 8

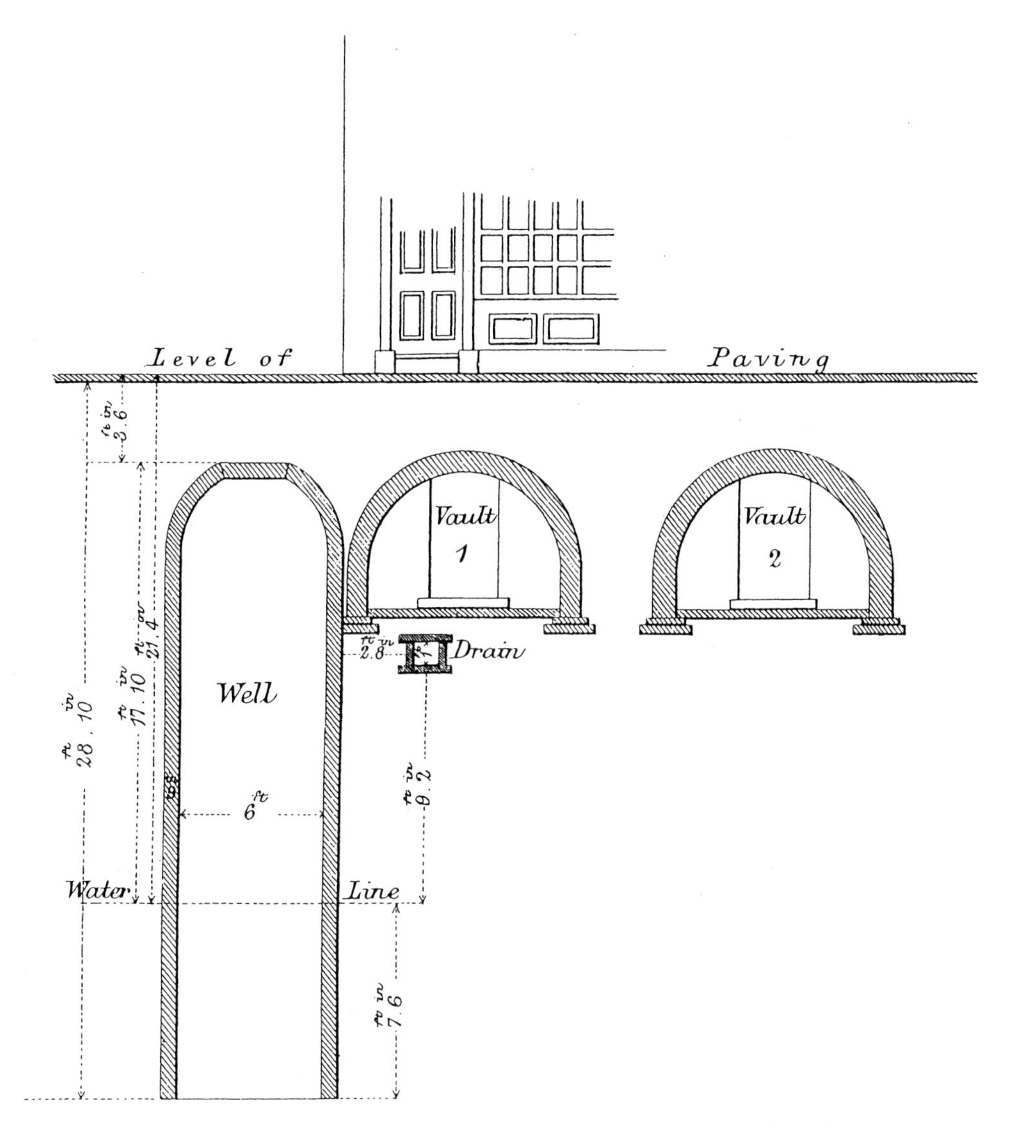

Fig. 6. Section through the Broad Street well (from the Report of the Cholera Inquiry Committee 1855).

September, the day the pump handle was removed. His evacuations undoubtedly went into the cesspool and re-contaminated the well, so that it was only the removal of the handle that prevented the cycle from being continued (Whitehead 1867).

John Snow: the man and his memory

It is always pleasing for a biographer to report that his subject was a charismatic individual, handsome, eloquent and witty. By contemporary description, John Snow lacked some of these qualities. Richardson (1858) records that 'He laid no claim to eloquence, nor had he that gift. A peculiar huskiness of voice, indeed, rendered first hearing from him painful'. He also notes that Snow's long periods as a student and in comparative isolation had made him reserved with strangers. However, Richardson also notes that he had a kindly nature, that with friends he was always open and of 'sweet companionship' and that in later life he had regrets that he had never married, 'the fates had been against him permanently on that score'. As he became famous and more popular, he was less reserved. All this suggests shyness. But there was also the fact that 'he moderated every enjoyment, and let nothing stand in the way of his scientific pursuits'.

At the age of 17 Snow became a vegetarian (Richardson 1858), and firmly committed to the vegetarian diet. For the first eight years, he supplemented his vegetable diet with eggs and dairy products and was proud of the fact that he could equal or better his meat-eating friends at physical pursuits, especially swimming at which he excelled. Then, when he was sharing lodgings, a fellow student (probably Joshua Parsons) questioned him as to the nature of the 'vegetables' (milk) he was taking for breakfast. It was probably meant in fun, but Snow took it seriously and became a Vegan. In his thirties he began to suffer a renal disorder which was attributed to the Vegan diet and he was forced to resume eating animal products.

At around the same time that he became a vegetarian, Snow also became a teetotaller, and firmly committed to the temperance cause. He remained a member of the York Temperance Society until his death but from about 1845 was persuaded to take a little wine as an aid to digestion.

Richardson describes him as 'of middle height, of somewhat slender build, and of sedate expression' though it seems that in his later years he filled out slightly.

Snow disliked reviews (perhaps the memory of that anonymous reviewer in 1849 stayed with him). He believed that 'a good book carried the review of its own merits. If it were bad, it were better left untouched' (Richardson 1858).

Snow's studies seemed absolute confirmation of the fact that cholera was transmitted by water. It would then be gratifying to record that by the time cholera again appeared in Britain, in 1866, the mode of transmission was fully understood and the authorities fully prepared. Sadly, this was not the case. The miasmatists and contagionists stuck to their theories, the one group warning against inhaling the breath of cholera patients and the other of the dangers of bad smells from drains. In 1866, Alexander Hamilton Howe MD wrote that cholera was attributable to 'a peculiar state of the atmosphere' and that the only cosmical body that can influence the atmosphere is the moon. He then went on to link the outbreaks of 1831–1832, 1848–1849 and 1866 to the 18-year lunar cycle (conveniently overlooking the epidemic of 1854!).

But by then, John Snow was dead. On the morning of 9 June 1858, he wrote the last sentence of his book, *On chloroform and other anaesthetics*, laid down his pen, and suffered a severe stroke from which he did not recover. His death certificate records the date of death as 16 June 1858.

A year after Snow's death, Richardson wrote to the *British Medical Journal* inviting contributions to a fund to place a monument over Snow's grave in Brompton Cemetery. The monument was duly put in place, but it has not been without its own history.

Fig. 7. John Snow, MD (1813–1858) (Courtesy of The Wellcome Institute).

Evidently the original did not stand up well to the local conditions, because an inscription on the plinth records that it was restored 'by Sir Benjamin W Richardson FRS and a few surviving friends' in 1895, the year before Richardson's own death. Another inscription was added in 1938 to record that the original inscriptions had been restored again. Finally, the whole memorial was destroyed by bombing in April 1941 and a replica put in its place in 1951. It may be significant that this was placed by anaesthetists rather than epidemiologists.

Apart from this, there are few tangible monuments to Snow. In gloomy Bateman's Buildings there is no memorial. The house where he lived and practised in Frith Street has been replaced, though a blue plaque does commemorate his time there. There was a similar plaque on his house in Sackville Street, but the house was demolished in the 1970s and replaced by the store of Austin Reed. In Broad Street there is a small plaque, and a pink granite kerbstone outside what is now No 41 marks the alleged site of the infamous pump. A replica pump, without a handle, has been set up in the wrong place, on the opposite side of the street. Otherwise, all that is left in his memory is the renamed public house, the former 'Newcastle Arms', now the 'John Snow'. For a man who was a teetotaller and a member of the temperance movement, it seems a strange memorial.

Acknowledgements

I am indebted to staff of the British Library, the Library of CEH Wallingford, and especially of the Wellcome Library in London, for help in discovering material about John Snow. Dr S. Snow kindly provided copies of her papers. Professor J. Mather guided me to the reproduction of the report of the Cholera Inquiry Committee, and Dr E. P. Loehnert (formerly of the University of Munster) introduced me to Robert Evans's book on cholera in Hamburg. My grateful thanks to all of them, and to Dr M. Rivett and Dr S. Snow for reviewing the manuscript.

References

BARROW, G. & WILLS, L. J. 1913. Records of London wells. *Memoirs of the Geological Survey of England and Wales*. HMSO, London.

BRITISH GEOLOGICAL SURVEY. 1993. *Geological Survey, 1:50 000* Sheet 256. British Geological Survey, Keyworth.

BINNIE, G. M. 1981. *Early Victorian water engineers*. Thomas Telford, London.

BRODY, H., RIP, M. R., VINTEN-JOHANSEN, P., PANETH, N. and RACHMAN, S. 2000. Map-making and myth-making in Broad Street: the London cholera epidemic. *The Lancet*, **356**, 64–68.

CHAVE, S. P. W. 1958. Henry Whitehead and cholera in Broad Street. *Medical History*, **2**, 92–109.

CHOLERA INQUIRY COMMITTEE. 1855. *Report on the cholera outbreak in the Parish of St James, Westminster, during the Autumn of 1854*. John Churchill, London [Reprinted as Appendix 10 of Rivers Pollution Commission, 1874].

ELLIS, R. H. 1994. *The case books of Dr. John Snow (*Medical History, Supplement No. 14*)*. Wellcome Institute for the History of Medicine, London.

EVANS, R. J. 1987. *Death in Hamburg: Society and politics in the cholera years 1830–1910*. Clarendon Press, Oxford.

FRENCH, J. G. 1835. *The nature of cholera investigated*. J. G. and F. Rivington, London, 35 p.

HOWE, A. H. 1866. *Reflections on cholera*. Robert Hardwick, 192 Piccadilly.

LEA, J. 1850. *Cholera, with reference to the geological theory*. Wright, Ferris and Company, Gazette Office, Cincinnati.

MACPHERSON, J. 1867. On the early seats of cholera in India and in the East. *Transactions of the Epidemiological Society of London*, **3**, 53–84.

PRICE, M. 1985. *Introducing groundwater*. George Allen and Unwin, London.

RICHARDSON, B. W. 1858. *Memoir of John Snow*. *In:* SNOW, 1858.

RIVERS POLLUTION COMMISSION. 1874. *Sixth Report of The Commissioners appointed in 1868 to inquire into the best means of preventing the pollution of rivers. The domestic water supply of Great Britain*. HMSO, London.

ROBERTS, S. 1999. John Snow (1813–1858) and Benjamin Ward Richardson (1828–1896): a notable friendship. *Journal of Medical Biography*, **7**, 42–49.

ROBOCK, A. 2002. Volcanic eruption, Tambora. *In:* MUNN, T. (ed.) *Encyclopedia of global environmental change*. John Wiley and Sons Limited, Chichester.

SHEPHARD, D. A. E. 1995. *John Snow: Anaesthetist to a Queen and Epidemiologist to a Nation*. York Point Publishing, Cornwall, Canada.

SNOW, E. M. 1857. History of the Asiatic Cholera in Providence. *Providence Journal*, 31 December 1857.

SNOW, J. 1849. *On the mode of communication of cholera*. John Churchill, London.

SNOW, J. 1855. *On the mode of communication of cholera, 2nd edition*. John Churchill, London.

SNOW, J. 1857. Cholera, and the water supply in the south districts of London. *British Medical Journal*, 864–865.

SNOW, J. 1858. *On chloroform and other anaesthetics: with a memoir of the author by Benjamin W Richardson*. John Churchill, London.

SNOW, S. J. 2000*a*. John Snow MD (1813–1858). Part I: A Yorkshire childhood and family life. *Journal of Medical Biography*, **8**, 27–31.

SNOW, S. J. 2000*b*. John Snow MD (1813–1858). Part II: Becoming a doctor – his medical training and early years of practice. *Journal of Medical Biography*, **8**, 71–77.

SNOW, S. J. 2000*c*. Death by water: John Snow and cholera in the 19th century. *Medical Historian*, **11**, 5–19.

THE BUILDER, 1854. **12**, No. 605, 473–474.

THE BUILDER, 1855. **13**, No. 663, 493–494.

THE BUILDER, 1857. **15**, No. 767, 593.

THE LANCET, 1855. 23 June, 1855, 634–635.

THE TIMES, 1854a. September 6, 1854, 5. London.

THE TIMES, 1854b. September 9, 1854, 9. London.

TORRENS, H. 2004. The hydrogeological work of William Smith (1769–1839).

WHITEHEAD, H. 1854. *The cholera in Berwick Street*, 2nd edition. Hope and Company, Great Marlborough Street, London.

WHITEHEAD, H. 1867. Remarks on the outbreak of cholera in Broad Street, Golden Square, London in 1854. *Transactions of the Epidemiological Society of London*, **3**, 99–104.

William Whitaker (1836–1925) – geologist, bibliographer and a pioneer of British hydrogeology

WILLIAM H. GEORGE

11 Sterry Road, Barking, Essex IG11 9SJ, UK

Abstract: William Whitaker was employed by the Geological Survey from 1857 until 1896 and subsequently worked as a consultant until his death in 1925. This paper examines the background to the era in which he worked and why he merits detailed consideration. Whitaker's personal life, career in the Geological Survey, contribution to learned societies and field clubs, work in retirement and his death are detailed. His contradictory personality, contribution to hydrogeology and his claim to the title of 'father of English hydrogeology' are assessed.

William Whitaker was actively employed as a geologist for nearly 70 years, from 1857 to 1925. He worked as a geologist with the Geological Survey of England and Wales from 1857 until his retirement in 1896. He was then self-employed as a geological consultant specialising in water supply until his death in 1925. When Whitaker started his career the two main national geological institutions were the Geological Society of London, founded in 1807 and celebrating its 50th anniversary, and the Geological Survey, formed in 1835 which was in its 22nd year of existence. He played a leading role in both bodies. Whitaker served as President of the Geological Society in 1898–1900 and was awarded the Murchison Medal (1886), Prestwich Medal (1906) and Wollaston Medal (1923). At the Geological Survey he rose from Assistant Geologist (1857) to Geologist (1863) and District Surveyor (1882). Whitaker served as President of the Geologists' Association (founded 1858) in 1900–1902 and 1920–1922 and led 52 excursions.

He followed in the footsteps of William Smith (1769–1839) and George Bellas Greenough (1778–1855) who had both published the first really detailed geological maps of England and Wales in 1815 and 1819 respectively. Whitaker geologically mapped, at one inch to the mile, the greater part of London and SE England, including Essex, Suffolk and parts of west Norfolk.

His greatest legacy to hydrogeology is the vast body of his published work. The majority of the 19 Geological Society Memoirs, which he wrote or to which he contributed, listed in Table 1, contain details of wells. He produced 56 water supply papers and reports, which date from 1867–1924 and are listed in Table 2. His contributions to 15 Water Supply County Memoirs, totalling 3653 pages, which date from 1899–1928, are listed in Table 3. This totals 90 substantial pieces of work.

Whitaker merits detailed consideration because of his prolific work as a field geologist, hydrogeologist and bibliographer. He also made major contributions to learned societies and field clubs.

His family background and education

His parents were William Whitaker (1803–1893), a wholesale perfumerer, brush manufacturer and wine merchant of Crutched Friars, and Margaret Burgess Michie (1806–1894), both Londoners. They were married by licence at St. Andrew's Church Holborn, on 8th February 1833. William Whitaker (1836–1925), their only child, was born on 4th May 1836 at 69, Hatton Garden, London at a time when the City of London was a residential quarter, and baptized on 30th May 1836. An 1840 engraving of the west side of Hatton Garden shows the three and a half storied terraced house of 'Whitaker & Co. Importers of Sponge & General Factors' (Marryat & Broadbent 1930, p. 78).

William Senior was a moderately successful businessman who enjoyed a reasonable standard of living. He died of senile decay and exhaustion, aged 90, on 12th August 1893 at 33, East Park Terrace, Southampton. His estate was valued at £1513.27. His wife, Margaret, was the daughter of John and Sarah Michie. She was baptized at St. Dunstan's Church, Stepney on 6th July 1806. Margaret Whitaker died of senile decay and bronchitis, on 10th January 1894, at 33, East Park Terrace, Southampton.

William Whitaker was educated at a boarding school in St. John's Wood in 1846. Physicians considered he was showing signs of consumption and advised Whitaker's parents that their son should pursue an open-air life (Thompson 1925, p. 95). Accordingly Whitaker spent two years at farms in Kent before he was transferred to St. Alban's Grammar School, Hertfordshire as a boarder at £6 a year. The Headmaster at this time was Henry Hall of Magdalene College, Cambridge. Whitaker studied French, German, drawing and drill, and other

From: MATHER, J. D. (ed.) 2004. *200 Years of British Hydrogeology*. Geological Society, London, Special Publications, **225**. 51–65. 0305-8719/04/$15 © The Geological Society of London.

Fig. 1. William Whitaker's birthplace, 69 Hatton Garden, London. From an engraving of 1840 (Marryat & Broadbent 1930 p. 78).

modern subjects, which Hall introduced (Leach 1908, p. 67). At the time of the 1851 Census the 14 year old Whitaker was boarding with the headmaster Henry Hall, a 'clergyman without a cure' aged 31, and 28 other pupils at Fish Pond Street, St. Albans (HO 107/1713 folio 449 p. 34 schedule 158).

Whitaker entered University College, Gower Street in 1852 at the age of 16, with the intention of taking up civil engineering (Davies 1925, p. xxxix). Here he studied chemistry and then geology under Professor Morris (Anon. 1907, p. 49) and took a BA with honours in chemistry from University College, London in 1855 aged 19 (Anon. 1926).

William Whitaker married Mary (1846–1916), daughter of Thomas Keogh, publisher, at the Consul's Office, Calais, on 25th August 1869. Whitaker was 33 when he married his 23-year-old wife. They separated before 5th April 1891, but never divorced. His wife Mary died, aged 70, on 6th November 1916 of cirrhosis of the liver, at 26, Kent Road, Southsea, Portsmouth. At the time of her death she was living at 53, Darlington Road, Southsea. Her moderate gross estate was valued at £764.21.

William and Mary Whitaker had three children. Their eldest child was Ellen Freda Whitaker (1874–1893), who was born on 25th December 1874 in Ipswich, Suffolk. She was recorded as a scholar on the 1891 census (RG 12/914 folio 25 schedule 269). Ellen sadly died, aged 18, on 16th November 1893, of enteric fever and pneumonia, in Southampton.

Their second child, Mary De Fraine Whitaker (1876–1932), was born on 11th September 1876 in Ipswich, Suffolk. She was recorded as an 'Art Student' living at 3, Campden Road, Croydon on the 1901 census (RG 13/636 folio 102 schedule 171). She married Ernest Willington Skeats (1875–1953) on 23rd December 1904 at Croydon, Surrey. They immediately emigrated to Australia where Skeats was appointed to the Chair of Geology and Mineralogy at the University of Melbourne. She died, aged 56, at 'Vectis' the University Grounds, Carlton, Melbourne, Australia on 7th November 1932, of sarcoma of the pharynx and was buried at Melbourne General Cemetery, Carlton on 8th November 1932.

Their youngest child and only son, Harry Lynn Whitaker (1882–1931), was born on 20th March 1882 in Kings Lynn, Norfolk. He was educated at Dulwich College, leaving in July 1901 (Ormiston 1926, p. 346). Harry married Agnes Catherine Stewart, a spinster aged 22, the daughter of Dr. Robert Wallace Stewart, at St. George's Presbyterian Church, Oakfield Road, Croydon, by licence, on 16th April 1908. He was a private secretary at the time of his marriage

Harry Whitaker did not achieve much in his career. He assisted his father in the writing of some of the water supply memoirs. He died, a boot manufacturer's clerk, aged 49, of double pneumonia at 2, Sydenham Road, Croydon on 13th April 1931. Harry was buried on 15th April at Mitcham Road Cemetery, Croydon in the same grave plot as his father. Administration of Harry's moderate £2281.01 estate was granted on 25th June 1931, to his widow Agnes and his bank manager.

His career in the Geological Survey

Field geologist

Whitaker's interest in geology developed in his early student days. He made the acquaintance of Professor Thomas Rupert Jones (1819–1911), Assistant Secretary to the Geological Society and editor of their *Quarterly Journal* and became Jones' volunteer assistant in 1856–1857. Whitaker's parents presumably supported him financially during this period. Shortly afterwards he was temporarily engaged as an assistant in the Geological Society's museum, based at Somerset House, where he arranged collections of recent and fossil mollusca and worked in the library (Woodward 1907, p. 309). It was at this time that Whitaker developed his lifelong interest in geology and bibliographical research that governed his later career.

William Whitaker joined the English staff of the Geological Survey as an Assistant Geologist on 1st

April 1857 at the age of 20. At this time 24 field geologists were employed by the Survey. His first work with the Survey was with Edward Hull (1829–1917), Thomas Roxburgh Polwhele (1831–1909) and Hilary Bauerman (1834–1909) on the geology of parts of Oxfordshire and Berkshire (Hull & Whitaker 1861) and around Brill, in Buckinghamshire (Flett 1937, p. 63). This was followed by work with Henry William Bristow (1817–1889) on Berkshire and the northern part of Hampshire (Bristow & Whitaker 1862).

Whitaker was promoted to the position of Geologist in 1863. He then took up the original survey of the geology of the London Basin, especially the Chalk and Eocene beds. This work was briefly interrupted in the summer of 1865 when Whitaker accompanied Archibald Geikie (1835–1924) and his younger brother James Geikie (1839–1915) on an official field trip to Arctic Norway to study the evidence of recent glaciation (Geikie 1882, p. 127–166; Geikie 1924, p. 107). Whitaker superintended the Drift Survey of the London area that started in 1869 under the direction of Bristow. He later worked at Walton-on-the-Naze, Harwich and Ipswich. The geological surveying for the original one-inch maps of London and the Thames Basin was largely the work of his hands (Strahan 1925*a*, p. 129). This work took him 35 years.

Whitaker was a pioneer in working out the detail of the Tertiary strata and superficial deposits of the south and SE of England, second only to Sir Joseph Prestwich (1812–1896) who had laid the broad outines (Strahan 1925*a*, p. 129). He named several stratigraphical units, including the Chalk Rock (Whitaker 1861*a*, p. 166), Whitaker's Three-Inch Flint Band (Whitaker 1865, p. 395), Thanet Base-Bed (Whitaker 1866, p. 405), Woolwich Bottom Bed (Hull & Whitaker, 1861), *Corbula regulbiensis* Bed (Whitaker 1872, p. 171), Blackheath Beds (Whitaker 1872, p. 239) and Oldhaven Beds (Whitaker 1866, p. 413), and 'Clay with Flints' (Whitaker 1861*b*, p. 54–55). Shortly after Whitaker's death the term 'Whitakerian' was suggested as a 'break-name' for the gap between the Cretaceous and Eocene periods (Martin 1926, p. 55). This term was never adopted.

It has been recorded that the Director-General Sir Roderick Impey Murchison (1792–1871), who was well aware of Whitaker's industry and excellent fieldwork on the softer sedimentary rocks of southern England, said to Andrew Crombie Ramsay (1814–1891), then Director for England and Wales, 'Don't you think, Ramsay that Whitaker has worked long enough in the Eastern Counties, where he gets nothing but soft squashy materials, and no good hard rocks to hammer? We might send him to the North or West, and give him some decent *solid geology* instead of this interminable Chalk, London Clay, and Drift' to which Ramsay replied, 'But, Sir Roderick, I assure you Whitaker *likes to work where he is*.' 'If that is really the case' said Sir Roderick, 'pray let him stay there, but I am sorry for him all the same.' However, Whitaker was able to study the Palaeozoic rocks from the first deep borings under the Tertiary and Secondary Strata of the south of England (Strahan 1925*a*, p. 129). From these he was able to sketch broadly the form of the Palaeozoic floor. He also predicted in 1889 'that Coal Measures are likely to occur somewhere along the line of the Thames Valley, or in the neighbouring tracts. . . . It is rash to attempt to foretell the future; but it seems to me that the day will come when coal will be worked in the south-east of England' (Whitaker 1889, vol. 1, p. 46).

The information and knowledge that Whitaker gained from this fieldwork made him the ideal candidate to superintend the making of a large model showing the geology of London. The model was made of wooden frames with machinery for raising moveable divisions to show geological cross sections. This was made at the scale of six inches to the mile and was first exhibited in the Museum of Practical Geology in 1873 (Flett 1937, p. 90). Details of the model and its elaborate construction were printed in Whitaker's *Guide to the Geology of London and Neighbourhood*, which first appeared in 1875.

Whitaker started surveying the Mesozoic and Tertiary rocks in East Anglia following completion of his work in London and the southeast of England (Flett 1937, pp 90–91). The whole of Essex and Suffolk and parts of west Norfolk and Cambridgeshire were surveyed by him with assistants working under his direction. Bailey (1952, p. 133) thought that Whitaker's East Anglian Geological Memoirs were some of the best produced during Archibald Geikie's Director Generalship.

In 1882, following the retirement of William Talbot Aveline (1822–1903), Whitaker was promoted to District Surveyor (Bailey 1952, p. 101), despite Bristow's Annual Confidential Report of 1880. This report noted of Whitaker 'Inadequate amount of work, difficult to move, obstructive and needlessly controversial, a kind of mutineer. Does not obey instructions. Makes absurd excuses for staying far too long in one place' (Wilson 1985, p. 126). In the previous three years 1877–1879 Whitaker had surveyed 63 square miles, compared to Ussher's 468 square miles (Wilson 1985, p. 33). Bristow was undoubtedly irritated by Whitaker's meticulous and time consuming attention to detail.

Whitaker then moved to the region south of the River Thames in the mid 1880s to survey the counties of Berkshire, Hampshire, Surrey and Sussex (Flett 1937, p. 113). In the later part of his career with the Survey, he was based at Southampton and personally surveyed some parts of South Hampshire. If Whitaker had a Hampshire Basin geology

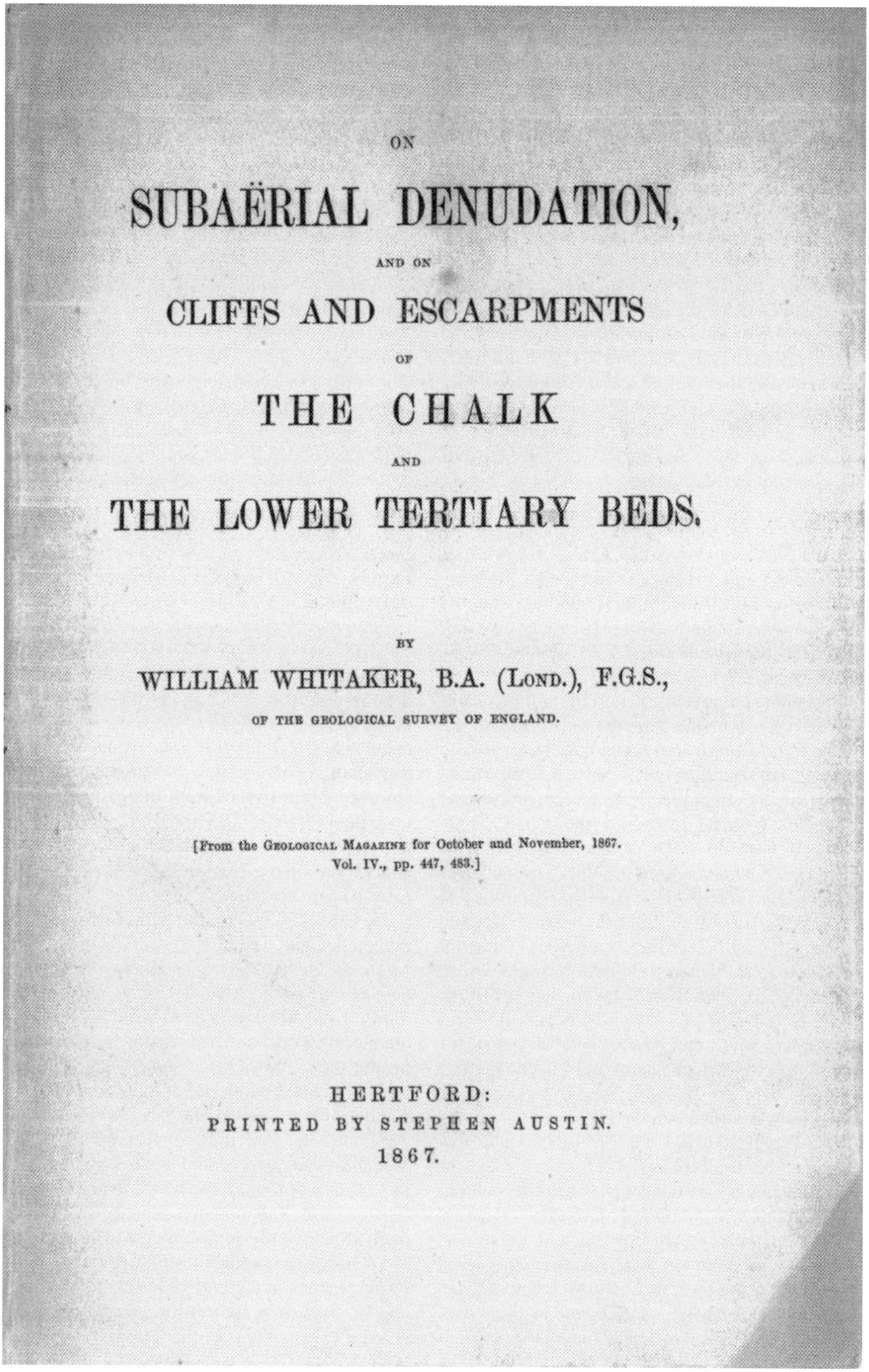

ON

SUBAËRIAL DENUDATION,

AND ON

CLIFFS AND ESCARPMENTS

OF

THE CHALK

AND

THE LOWER TERTIARY BEDS.

BY

WILLIAM WHITAKER, B.A. (LOND.), F.G.S.,

OF THE GEOLOGICAL SURVEY OF ENGLAND.

[From the GEOLOGICAL MAGAZINE for October and November, 1867.
Vol. IV., pp. 447, 483.]

HERTFORD:
PRINTED BY STEPHEN AUSTIN.
1867.

Fig. 2. Title page of Whitaker's paper 'On Subaerial Denudation'. The Geological Society rejected this paper and it was published in the *Geological Magazine*. Whitaker subsequently distributed offprints to Fellows of the Geological Society.

problem he would visit John William Elwes (1850–1918) in Broadmoor Criminal Lunatic Asylum where he had been sent in 1891 after he had shot his mother (Torrens *et al.* 1978, p. 117). At his house in East Park Terrace, Southampton Whitaker had his private 'collection of rocks and minerals chiefly from the London basin' (Dale 1888, p. 24–25). Additionally he was superintending the re-surveys in progress of no less than seven other counties, Bedford, Berkshire, Buckinghamshire, Dorset, Northants, Oxford and Wiltshire.

Whitaker's most important work was undoubtedly centred on London and the Thames Valley. This work led to the publication of two major works in 1872 and 1889. Aubrey Strahan (1852–1928) described the later, *The Geology of London and of part of the Thames Valley*, as probably the most detailed account of any region that had ever been published.

Whitaker's time at the Survey was taken up with applied geology in surveying, mapping and the writing of district memoirs and water supply papers. This work both confirmed and greatly expanded Prestwich's research undertaken in the 1840s and 1850s. Whitaker, who followed in Prestwich's footsteps, acknowledged his huge indebtedness to his pioneering work (Whitaker 1872, p. 395; Prestwich 1899, p. 80). Shortly before his death Whitaker again acknowledged his debt to Prestwich when he said, upon receiving the Wollaston Medal from the Geological Society 'I have had to follow along a line in which Prestwich was perhaps the pioneer, that is the application of geology to questions of water-supply and kindred practical subjects' (Seward 1923, p. xlviii). Prestwich was himself vaguely following the path beaten by Dr. James Mitchell (1785–1844), who had worked on the Tertiary and superficial deposits around London and also on well records in the 1820s and 1830s. It was Whitaker however who published many of Dr. Mitchell's notes, including his well records, in the 1889 London Memoir (Woodward 1907, p. 135; Whitaker 1889, Vol. 2).

Whitaker's field experience made him the ideal candidate to help train the large addition to the field staff secured by Murchison in 1867 and 1868 (Flett 1937, p. 83). Flett described Whitaker as one of Ramsay's geologists 'endowed with natural gifts of the highest order' (Flett 1937, p. 95).

Whitaker occasionally found himself in trouble with his superiors at the Geological Survey. In 1882, for example, Geikie instructed that nothing should be noted on the six-inch and one-inch field maps that was not to appear on the published geological maps. Whitaker, who wished to do his mapping on the larger six-inch scale, but publish on the one-inch scale, thought this was silly and arranged for about 50 petitions to be sent in from contacts in Hampshire, where he was then mapping. Geikie

Table 1. *List of Whitaker's Geological Survey Memoirs*

Date	Geological Memoirs	Pages
1861	Oxfordshire and Berkshire (parts)	57
1862	Berkshire and Hampshire (parts)	51
1864	Middlesex etc.	112
1872	London Basin	619
1875	London & neighbourhood [6th edition 1901]	72
1877	Walton-on-the-Naze and Harwich	32
1878	Essex (NW); Cambridgeshire (parts); Suffolk (parts) and Hertfordshire (parts)	92
1881	Stowmarket	26
1885	Ipswich, Hadleigh and Felixstowe	156
1886	Aldborough, Framlingham, Orford and Woodbridge – additions only	59
1886	Bury St. Edmunds and Newmarket	27
1887	Halesworth and Harleston	41
1887	Southwold and Suffolk Coast: Dunwich to Covehithe	88
1889	London and Thames Valley (part) 2 Vols.	908
1891	Cambridgeshire and Suffolk (parts)	127
1893	Norfolk (SW) and Cambridgeshire (N)	178
1899	The Wash	146
1901	London and Neighbourhood guide 6th ed.	102
1902	Southampton -Contribution	70

threatened Whitaker with dismissal from the service if he repeated this conduct. Geikie also stated in a letter to the Department of Science and Art 'this incorrigible Whitaker gives more trouble than all the rest of the staff' (Wilson 1985, p. 103). Whitaker however publicly expressed his preference for six inch mapping. This was sensibly shortly after he retired from the Geological Survey (Whitaker 1899, p. lxxx–lxxxiii). In the 1890s, Whitaker was practically in general charge of the Survey work, especially in the districts south of the Thames and along the south coast (Flett 1937, p. 123–124). The 19 memoirs he published, alone or jointly, are listed in Table 1.

In his Presidential address to the British Association (Section C – Geology) in 1895, delivered shortly before he retired,Whitaker publicly stated, with humour that perhaps masked some bitterness, '. . . an official life of over thirty-eight years has led me to do what I am told to do, and to suppress my own ideas of what is right.' (Whitaker 1895, p. 666). Whitaker's 26 Geological Field Notebooks, which date from 1857 until his retirement are held in the British Geological Survey Archive. He crossed out his notes when the information had been published. The notebooks contain much unpublished miscellaneous information such as notes on churches (Vol 12 1864, p. 57–65; Vol. 13 1865, p. 69–84); notes of a holiday or excursion in France (Vol. 13 1865, p. 51–68); details of his expenses at Walton-on-the-Naze, Essex (Vol. 16 1868–1872, p. 45). Whitaker retired from the Survey on 22nd October 1896, aged 60, to pursue economic geology (Strahan 1925*a*, p. 129).

Whitaker tended his resignation in a letter addressed to 'Dear Sir Archibald' on 3rd October 1896. He stated 'I am sorry that my last official time on the survey should have been so spoilt by my illness; but I hope to make up for this, as there are many bits of work that I should like to see to later on'. He kept his promise and laboured to within a few weeks of his death some 30 years later! Geikie acknowledged this in his preface, dated 25th January 1901, to the sixth edition of Whitaker's *Guide to the Geology of London*. He wrote 'The work has now again been brought up to date by Mr. Whitaker, who, although he retired from the Geological Survey in 1896, continues to show his active interest in its welfare' (Whitaker 1901, p. vi).

Aubrey Strahan recorded in his obituary notice that although much of Whitaker's work was of a more or less statistical character, there stands to his credit a great record of original research (Strahan 1925*a*, p. 129). Strahan was particularly thinking of his papers on Subaerial Denudation, on the Chesil Beach, and on the Water Supply from Chalk (Strahan 1925*a*, p. 129). He published much of this while employed by the Geological Survey.

Whitaker was early in the field as an ardent student of subaerial denudation and read an excellent paper on 'Cliffs and Escarpments of the Chalk and Tertiary Strata' to the Geological Society on 8th May 1867. However it became a 'Rejected Address' and the Society declined to publish it (Woodward 1907, p. 233). Whitaker however wished the paper to be published in full and sent it to the editor of the *Geological Magazine* who published it and enabled Whitaker to have a sufficient number of offprints produced for distribution amongst Fellows of the Geological Society. His paper, which Bailey referred to as masterly (Bailey 1952, p. 71), was theoretical, original and closely researched. It finally ended the belief that escarpments were formed by marine erosion (Sowan 2002). The paper was highly commended and praised by Charles Darwin (Darwin 1883, p. 234).

Hydrogeologist

Whitaker's main contribution to hydrogeology was the way in which he systematically collected and published a huge databank of wells and boreholes (see Table 2). He was a geologist and water supply engineer with an encyclopaedic knowledge of wells and boreholes in SE England. He did not produce significant original work or show aptitude for any specific aspect of hydrogeology. Flett believed the initial impetus for the Survey pioneering the devel-

Table 2. *Whitaker's Water Supply Papers and Reports*

Date	Subject
1867	London Wells and Borings
1869	Boring at Crossness Pumping Station
1877	London Water Supply
1877	Well Section at Holkham Hall, Norfolk
1881	Well Section at Stonehouse, Plymouth
1881	Report for Lambeth Waterworks
1883	Metropolitan Sewage Discharge
1883	Report for Lambeth Waterworks
1884	Chalk Water Supply
1884	Southampton Water Supply
1884	Deep Well at Norwich
1886	Addington, Croydon Water Supply
1886	Ballard v Tomlinson
1886	Goldstone Bottom, Brighton Water Supply
1887	Purity of Water Supply
1887	Margate Water Supply
1888	Windsor Water Supply
1888	Sutton, Surrey Water Supply
1891	London Water Supply
1892	King's Lynn, Norfolk Water Supply
1893	Metropolitan Water Supply
1893	Chalk as a Source of Water Supply
1893	Swallow Holes in Chalk of London Basin
1893	Water in Northern Kent
1893	Chalk of the London Basin and Water Supply
1893	Ware, Hertfordshire Water Supply
1894	Croydon Water Supply
1895	Hastings Water Supply
1895	Caterham Asylum Water Supply
1897	Weeley Borehole (Coal)
1897	Portsmouth Water Supply
1898	Shedfield, Hampshire Water Supply
1898	Chalk Water in Hertfordshire
1899	Southampton Waterworks
1899	Kent Waterworks
1899	Rushden Water Supply
1899	Geology in Reference to Water Supply
1900	Staffordshire Clays
1901	Devizes Water Supply
1901	Woolwich Workhouse Well
1901	Langport Water Supply
1902	Swindon Water Supply
1902	Howdenshire, Yorkshire Water Supply
1903	Malmsbury Water Supply
1903	On the Shortage of Water Available for Supply
1904	Cambridgeshire Wells
1905	Surrey Wells
1905	Weston-Super-Mare Water Supply
1905	Lincoln Water Supply
1907	Purton Water Supply
1907	Hemel Hempstead Water Supply
1909	Coasts of England and Wales
1917	South Hampshire Wells
1919	Surrey Wells
1923	Vange Mineral Wells, Essex
1924	Geology of the Winchester District

opment of English hydrogeology came from Whitaker but that other geologists on the English Staff were also fully aware of its importance (Flett 1937, p. 129). This view was reiterated by Bailey (1952, p. 85) writing more than 25 years after Whitaker's death who readily acknowledged that the Geological Survey's 'usefulness in regard to all aspects of underground water has steadily grown, but the lion's share of the credit still belongs to Whitaker'. Wilson (1985, p. 124) however believed two of Whitaker's colleagues Joseph Lucas (1846–1926) and Charles Eugene De Rance (1847–1906) 'seem to have been given less recognition than they deserve'

Lucas was certainly the first person to emphasise the need for a systematic survey of all sources of underground water and he first used the term 'hydrogeological' in 1874. He also produced the first ever British hydrogeological maps in 1877 and 1878 following on from the pioneering work of Delesse in France in 1858. However, in reality, several individuals such as the Reverend James Charles Clutterbuck (1801–1885), Prestwich, Lucas and Whitaker developed the concepts and laid the first scientific foundations of the subject of hydrogeology in the 19th century on which the sophisticated models of today are based (Mather 1998, p. 194).

Whitaker's interest in water supply from wells started early in his Geological Survey career and he began collecting details of well records from 'water companies, well-sinkers and local borough surveyors' soon after his appointment (Wilson 1985, p. 126). Strahan, sixty years earlier had recorded that Whitaker was indefatigable in collecting records of temporary sections, wells and boreholes, which abounded in London (Strahan 1925*b*, p. 129).

As early as June 1864 Whitaker published two appendices of well sections, detailing 109 wells, to his Geological Survey Memoir on parts of Middlesex, Hertfordshire, Buckinghamshire and Surrey (Whitaker 1864, p. 102–105). In the same publication he asked for details of new well sections. He also requested details of any old wells that he may have missed and added that he trusted the table of well sections would be carried on in the Memoir to be written to accompany Sheet 1 in the eastern part of London (Whitaker 1864, p. 1 & 103). Whitaker was interested in well sections because of the help they gave him in his geological surveying

Shortly afterwards, in 1866, Whitaker prevailed upon the Medical Officer of the Privy Council to allow him to produce an appendix on the surface geology of London based on wells and bores (Whitaker 1867*a*, p. 346–366; Mather 1998, p. 193). In 1869 Whitaker reported on the boring operations at the Crossness Pumping Station (Anon. 1907, p. 58). In addition Mather has stated 'most of the geological memoirs covering south-eastern England published after 1871 are adorned with records of wells and temporary exposures which William Whitaker had a passion for collecting' (Mather 1998, p. 194). For example the 1872 London Memoir had some 150 pages of records of well and bores (Bailey 1952, p. 85). The revised and extended 1889 edition of this 1872 memoir had a separate volume of 352 pages devoted to lists of well sections, investigation boreholes and temporary exposures (Whitaker 1889; Bailey 1952, p. 126; Mather 1998, p. 193).

Whitaker readily collaborated with other workers in hydrogeology and systematically published the results of his work. In 1874, at the Belfast Meeting of the British Association, a Committee of Enquiry was set up to investigate the circulation of underground waters in the New Red Sandstone and Permian Formations of England. This remit was later extended to cover all permeable formations of the country. This committee sat for 20 years. De Rance was appointed to be the Secretary. The Chairman was Edward Hull, also of the Geological Survey. Whitaker sat on the committee and produced two relevant bibliographies, the first in 1887 followed by a second in 1895. These valuable works of reference contained 695 items (Bailey 1952, p. 85–86). In 1882 De Rance produced *The Water Supply of England and Wales*. This was largely a compilation based on material supplied to him in his position as Secretary of the British Association Committee of Enquiry. He was later forced to resign from the Geological Survey in 1898 for 'inefficiency and addiction to drink' (Wilson 1985, p. 125). De Rance's hydrogeological legacy has undoubtedly been tarnished by his personal problems.

Whitaker however continued to produce a most voluminous body of hydrogeological information almost up until his death in 1925. He managed to juggle his Geological Survey employment with private consultancy work. For example in 1877 Whitaker produced short reports on 'Some Questions relating to the Water Supply of London' and on a well section at Holkham Hall, Norfolk (Anon. 1907, p. 56). Shortly after this, in 1881, Whitaker produced a short report on a well section at Stonehouse, Plymouth (Anon. 1907, p. 56).

In the early 1880s Whitaker acted as a private consultant to the Lambeth Waterworks. He produced reports in 1881 and 1883 about the prospects of the company obtaining water from wells in the Chalk at Tooting, Lewisham, Beckenham and Ditton (Anon. 1907, p. 58). Whitaker also gave evidence to the Royal Commission on Metropolitan Sewage Discharge on 13th December 1882. His evidence related to water supply from the Chalk, wells, and the geology of the London district. In the mid and late 1880s Whitaker (Anon. 1907, p. 58) produced many reports on water supply which are listed in Table 2. He also published a paper detailing the outcome of a court case, concerning the contamination of water from wells, where he had been employed as a consultant (Whitaker 1886). The case related to a brewery well in Brentford, west London, which was found to be polluted from another well nearly 100 yards away which had been used as a cesspit for a printing works. The case Ballard v Tomlinson, was finally settled in the Court of Appeal, which found that no owner had the right to pollute a source of water supply common to his own and other wells (Mather 1998, p. 193). Whitaker continued to undertake private commissions before his retirement from the Geological Survey. In 1891, with Alexander Henry Green (1832–1896), he produced, for the London County Council, a preliminary report on the possibility of obtaining a supply of water for London within the Thames Basin (Anon. 1907, p. 58).

Whitaker gave detailed hydrogeological evidence to the Royal Commission on Metropolitan Water Supply in 1892. His boss and contemporary, Sir Archibald Geikie, sat on the Commission. Whitaker stated that he was the senior officer of the English Staff of the Geological Survey and that he had 35 years of official knowledge of the Thames and Lea Valleys (*Royal Commission on Metropolitan Water Supply* 1893, p. 349). His evidence, which was given verbatim on 20th October and on 16th December 1892, occupies some 18 printed folio pages (p. 349–356 & 438–447). Four papers entitled 'Chalk as a Source of Water Supply'; 'Memorandum on Swallow-holes in the Chalk of the Thames Basin'; 'On the Water (chiefly from the Chalk) in Northern Kent and 'On the Chalk of the London Basin' appeared as Appendices. This last paper was reprinted in the *Geological Magazine* and covered such topics as the covered thickness of the Chalk, divisions in the Chalk, water flow and the relation of the structure and position of the Chalk to Water Supply (Whitaker 1895*b*).

In contrast, Joseph Lucas's evidence, which was given on 1st June 1892, occupied just over 3 pages (p. 91–94). He stated his credentials thus 'I have paid attention to the chalk water system for 20 years' (p. 91). In his evidence he gave woolly details about a well published 'in Mr. Whitaker's Geological Survey Memoir of 1879, I think it is' (p. 92). Charles Eugene De Rance gave his evidence to the Royal Commission on 21st October 1892. This occupied 10 pages (p. 357–367) and he explained that he had paid attention to the subject of underground water since 1866 (p. 357).

In the Royal Commission Report the assistance of Whitaker's statements and evidence were acknowledged (1893, p. 16, 40, 42, 43 & 47–48). The Royal Commission Report however disagreed and dissented from Whitaker's opinion, which he had expressed 'with some caution and hesitancy' about the quantity of underground water obtainable by pumping from the Lea Valley (1893, p. 52). This contradicts Matthews's statement that 'Mr. Whitaker was always listened to with the greatest deference when appearing as a witness before Parliamentary Committees and other tribunals, on account of his transparent honesty of purpose, and the tolerance with which he treated the evidence of those who differed from him' (Matthews 1925, p. 152–153).

Bibliographer

Whitaker meticulously compiled and published bibliographies on the geology of English counties from 1870. He also carefully collected details of wells and borings which then appeared as appendices to official publications. When Whitaker sent Geikie some references about Scottish geology on 21st May 1873 he stated 'When I first began to look out English Geological bibliography I thought Scotland and Ireland might come in my work as well. I soon found that unity was enough and so gave up all hope of the Trinity (though I may be damned for it)'. Whitaker must have changed his mind however because he produced a 39 page bibliography on the geology, mineralogy and palaeontology of Wales in 1880 (Whitaker 1880). His bibliographical publications are shown in Table 4.

This bibliographical work, which may have proved tedious to many, was undertaken in his spare time. For six years (1874–1879) he edited the *Geological Record* but was compelled to give up this task due to lack of support from others. Horace Woodward (1907, p. 258) stated, in the official centennial history of the Geological Society, that no individual Fellow of the Geological Society helped more than Whitaker in the production of numerous geological bibliographies of counties and special subjects. Earlier, both Horner in 1846 & 1861, and Lyell in 1866 had mentioned the problems of keeping abreast of geological literature (Woodward 1907, p. 258). Whitaker's work was crucial in ensuring that the 'great talus heap' of geological literature was not buried and lost sight of. Bailey noted (1952, p. 86–87) that William Topley's Geology of the Weald, which appeared in 1875 'along with many other memoirs of the English branch [of the Survey] owes much of its bibliograpical completeness to Whitaker's collecting instinct'.

Contributions to learned societies and field clubs

Whitaker was a great supporter of national learned societies and local groups. This side of his activity was succinctly summarized by Henry Woodward who noted 'No man has been more assiduous in helping others individually or by taking part in the ordinary meetings, councils, and committees of various learned societies and field clubs' (Anon. 1907, p. 52). He was frequently elected to presidential chairs and presented numerous addresses.

Whitaker was elected a fellow of the Geological Society in early 1859, when aged 23, and first served on its Council in 1873 and intermittently for many years afterwards. Whitaker sat as President of the Geological Society in 1898–1900 and as Vice-President in 1901–1902. For two weeks in 1900, Whitaker was president of the Geological Society and the Geologists' Association, a unique achievement. He received the Geological Society's Murchison Medal (1886). When Professor Bonney presented this he said 'Your papers on the western end of the London Basin and on the Lower London Tertiaries of Kent deserve to be ranked with the classic memoirs of Prestwich as elucidating the geology of what I may call the Home District' (Bonney 1886, p. 32–33).

20 years later Whitaker was a most appropriate recipient of the second Prestwich Medal which was awarded in 1906. The citation by Marr emphasized that the medal was awarded 'as an acknowledgement of the value of your researches among the Tertiary Strata of the London and Hampshire Basins . . ., which were advanced in a high degree by the founder of the medal. You have also followed in the footsteps of Prestwich in matters of economic geology – the question of water supply and the study of underground geology – for which you, the recipient of the medal, like its founder, have done so much' (Marr 1906, p. xlvii–xlviii). From the award Whitaker gave £12.30 to the Prestwich Fund and £6.65 to the Geological Relief fund (Woodward 1907, p. 316). The final and highest honour Whitaker received from the Geological Society was the award of the Wollaston Medal, the blue ribbon of British geology, in 1923. When presenting this Seward beautifully stated 'There is a genius of the heart as well as of the brain; and, if I may say so you are fortunate in possessing both' (Seward 1923, p. xliv–xlv).

He became an honorary member of the Geologists' Association in 1875, served two terms as president (1900–1902 and 1920–1922), and led 52 excursions. His first excursion was in 1872 to Watford and his last in 1921 to the Mole Valley at the age of 84 (Sweeting 1958, p. 41). He always proved to be a popular, ready and most able director of excursions (Anon. 1907). Whitaker pointed out in 1921 that the Geologists' Association was a sort of nursery for the Geological Society and had been the means of making many and good additions to the fellowship (Sweeting 1958, p. 17). Sweeting, who wrote the centennial history of the Geologists' Association recorded that Whitaker was 'One of the best-loved figures in our history' (Sweeting 1958, p. 122). He added that Whitaker's picturesque personality was more than familiar at the excursions and meetings of the Geologists' Association 'it was the life and soul of those gatherings' (Sweeting 1958, p. 41). When he led a Geologists' Association excursion to Oxted, a few years before his death, following the customary vote of thanks the party joined in singing 'For he's a jolly good fellow', 'an unprecedented indiscretion' (Davies 1925, p. xxxix).

Whitaker was elected a Fellow of the Royal Society in 1887 for his contribution to Geological Survey Memoirs, for papers in the Quarterly Journal of the Geological Society – especially on Cretaceous and Tertiary Beds and 'on the subject of water supply'. His editorship of the *Geological Record* and his geological bibliographies were also cited.

Whitaker was at one time delegate to the yearly meetings of the British Association. He presided over Section C of the British Association at Ipswich in 1895 and gave an illuminating address on the underground geology of that part of England. Whitaker was elected an honorary member of the Institution of Water Engineers in 1899. According to Matthews he 'always took the keenest interest in its work and objects, was a very regular attendant at its meetings, frequently took part in, and made valuable contributions to, the discussions upon the papers read' (Matthews 1925, p. 151). It was due to Whitaker's influence that the Institution was for many years able to enjoy the privilege of using the rooms of the Geological Society for its London meetings.

Whitaker was an Associate of the Institution of Civil Engineers. He was a member of the Council of the Royal Sanitary Institute and its treasurer (Matthews 1925, p. 153). In 1890 Whitaker received a silver medal from the Society of Arts for his paper on coal in the southeast of England (Whitaker 1890).

Whitaker was president of several local societies including the South Eastern Union of Scientific Societies in 1899 (Young 1925, p. xlvi–xlvii), the Croydon Natural History Society in 1899–1900 and again, 1911–1912. In his presidential address at Croydon, he summarized the work that had been published on the geology and meteorology of Surrey in the period 1890–1911. He also contributed five papers about Surrey wells. He was President of the Essex Field Club in 1911–1914. Whitaker was a member of the Belgian, Liverpool, Manchester, Norfolk and Yorkshire Geological Societies, the Hampshire Literary and Philosophical Society, Hampshire Field Club, Hertfordshire Natural History Society, Philosophical Society of York and correspondent of the Academy of Natural Science of Philadelphia (Strahan 1925*a*, p. 130)

Fig. 3. William Whitaker aged 81. Taken in 1918 (Thompson 1925, p. 93).

His work in retirement and his death

Before he retired Whitaker's principal employment was with his official duties at the Geological Survey. However his private consultancy work was continuing to expand in line with his growing reputation. As a consulting geologist on sanitation generally, and on questions of water supply especially, he attained a high reputation. When he retired, aged 60, his main employment switched to his consultancy work, but he continued to work voluntarily for the Survey producing an impressive series of water supply memoirs.

On his retirement, in 1896, he moved back to the London area. Here he practised as a consulting geologist and water engineer and 'was undoubtedly the greatest authority on the Water Supply from the Chalk and Tertiary formations' (Matthews 1925, p. 152). His knowledge of nearly every deep well and boring from Hertfordshire to the English Channel ensured there was no better opinion on the site for any building, housing development or sewer. He had a considerable private practice as a water engineer after his retirement. He was one of the leading authorities upon questions of water supply for towns and public buildings, principally in the eastern and southeastern counties of England. He also reported on the Kent coal borings.

In his retirement Whitaker took pleasure in rendering service to the Geological Survey (Strahan 1925*b*, p. 129). Bailey (1952, p. 167) recorded that much of the work of producing the county water supply memoirs fell to Whitaker in his retirement. He collaborated with others including the famous meteorologist, Hugh Robert Mill and John C. Thresh, for many years the County Medical Officer of Health for Essex (Bailey 1952, p. 167 & 173). He found time to write or assist in the production of no less than 15 major official water supply publications. This long series of memoirs on county water supplies from underground sources, testifying to the diligence with which he collected records of wells and springs and to the skill with which he interpreted them, is listed in Table 3. This work he continued almost until he died (Strahan 1925*b*, p. 129).

Table 3. *Water Supply County Memoirs*

Date	Water Supply County Memoirs	pages
1899	Sussex	123
1902	Berkshire	115
1904	Lincolnshire (contributions)	229
1906	Suffolk	177
1908	Kent	399
1910	Hampshire (including Isle of Wight)	252
1911	Sussex-supplement	255
1912	Surrey	352
1916	Essex	510
1921	Buckinghamshire and Hertfordshire	368
1921	Norfolk	185
1922	Cambridgeshire, Huntingdonshire and Rutland	157
1925	Wiltshire	133
1926	Dorset – Wells and springs	119
1928	Somerset – bibliography of Bath Thermal Waters only	279

The first of these county Water Supply Memoirs was co-authored with Clement Reid, in 1899, and covered the county of Sussex (Mather 1998, p. 194). It mostly contains well records with little supporting text. Later memoirs in the series contain much more information. For example the *Water Supply of Kent* (Whitaker 1908), has some 63 pages relating to shafts and galleries, geological succession, geology, rainfall (by Mill), springs, swallow holes and intermittent streams (Bailey 1952; Mather 1998, p. 193).

Whitaker's Water Supply of Essex was printed in 1916 but suppressed until the Great War (1914–1918) had ended (Thompson 1925, p. 93). He also collaborated with George Barrow (1853–1932) to publish some Middlesex Well sections (1907). Whitaker's last two memoirs on Wiltshire (1925) and Dorset (1926) were posthumous and appeared as joint publications with Francis Hereward Edmunds (1893–1960) and Wilfrid Edwards (1897–?) respectively (Bailey 1952, p. 201).

Bailey, who joined the Geological Survey in 1902, some 6 years after Whitaker had retired, repeated the anecdote that Whitaker 'willingly accepted the pres-

Table 4. *Whitaker's Bibliographical Publications*

Date	Title	Pages
1870	Devonshire Geology Mineralogy & Palaeontology	33
1871	Books & papers by J. Bete Jukes	6
1873	Hampshire Geology Mineralogy & Palaeontology	19
1873	Wiltshire Geology Mineralogy & Palaeontology	14
1873	Cambridgeshire Geology	15
1874	Warwickshire Geology Mineralogy & Palaeontology	10
1875	Cornwall Geology Mineralogy & Palaeontology	49
1876	Hertfordshire Geology	5
1876	Cheshire Geology Mineralogy & Palaeontology	20
1880	Wales Geology Mineralogy & Palaeontology	39
1883	Oxfordshire, Berkshire and Buckinghamshire Geology and Palaeontology	40
1883	Cumberland and Westmoreland Geology	26
1886	England & Wales coast changes and shore-deposits	43
1886	Staffordshire, Worcestershire and Warwickshire Geology Mineralogy and Palaeontology	33
1888	England & Wales Underground Water	30
1889	Shropshire Geology Mineralogy & Palaeontology	29
1889	Essex Geology	23
1895	England & Wales coast changes and shore-deposits	4
1895	England & Wales Underground Water	8

idency of any local society which was prepared to reproduce a presidential address consisting largely of the details of well sections' (Bailey 1952, p. 85). Whitaker's own statements do not confirm this. He mischievously said before his presidential address to Section C of the British Association 'Of all the inventions of later years I look upon the presidential address as perhaps the worse' (Whitaker 1895*a*, p. 666). He reiterated this in his Presidential Address to the Geologists' Association in 1921 when he said 'A few years ago it occurred to me that, having evolved sixteen presidential addresses to nine different bodies, I might be spared from inflicting another on my friends' (Whitaker 1921, p. 183).

For 20 years, until his death, Whitaker was co-opted vice-chairman of Croydon libraries committee. Much of the success of the Croydon Libraries was due to the fact that Whitaker was a man of first-rate knowledge and a good man of business. He was also an enthusiast for libraries and the kindliest and most companionable of men (Anon. 1925*d*, p. 28). Whitaker's spare alert figure was often to be seen in Croydon as he briskly walked to and from the Library, at which he was a constant visitor (Topley 1925, p. 29). He made many donations to Croydon Library. These, in the form of books, pamphlets, maps and illustrations numbered over a thousand, and he regularly presented the transactions of eight scientific societies to the Library (Topley 1925, p. 29). The Library still has ten beautifully red leather-bound volumes with gold tooling of 'Papers, Memoirs, etc. by or partly by William Whitaker (1859–1908)' which he presented.

Whitaker died of cancer of the pancreas, aged 88, at his home in Wellesley Court, Croydon, on 15th January 1925 and was buried in Croydon Cemetery (grave 13143/G1 – not marked), Mitcham Road, on 19th January. His will dated 5th December 1923 was proved on 4th March 1925. The gross value of his estate was £9,854 (*Times* 10th March 1925). He bequeathed £50 each to his son, daughter and daughter-in-law. The residue was to be divided between his son (40%), daughter (40%) and daughter-in-law (20%).

When Whitaker died 'All those who knew him will feel that they have sustained a personal loss, by the removal of one so universally esteemed as well for his attainments as a Geologist as for his genial and kindly personality' (Matthews 1925, p. 151). Boswell reported the loss of 'a charming and ever-helpful friend' (Boswell 1926, p. 198).

Aubrey Strahan noted that his many friends felt Whitaker's death as a deep personal bereavement although it had been known for some weeks that he was dying (Strahan 1925*a*, p. 129). Croydon Library still has a small bound volume entitled 'Whitaker in Memoriam' which contains a few obituaries and press cuttings relating to him.

Discussion

Whitaker had a most interesting and sometimes apparently contradictory personality. The majority of Whitaker's contemporaries saw him as a delightful, genial, kindly and humane character and regarded him as unselfish, honest, kind, good-tempered, unaffected, transparent and truly loveable.

Others, such as his superiors Bristow and Geikie at the Geological Survey, saw Whitaker as incorrigible, obstructive and mutinous. Wilson perhaps best summarized Whitaker as 'a strong-minded and determined man who saw the value of his work and refused to have it compromised' (Wilson 1985, p. 126). Whitaker's contradictory personality can probably be best explained as him challenging the actions of his superiors at the Geological Survey. He was financially independent, had a growing reputation as a consulting geologist and he was accordingly able to express his dissent and take the consequences.

Whitaker's relationship with his peers has been summarized by Boswell as being 'Too broadminded and too kindly in disposition to be severely critical' (Boswell 1926, p. 199). Davies observed that Whitaker 'was great enough to be humble, though not diffident. Keenly critical of opinions, there was nothing personal in his criticisms; they were full of common sense and humour, and void of offence' (Davies 1925, p. xxxix). Whitaker's apparently contradictory nature was evident following the death of the younger Searles Valentine Wood (1830–1884) when Whitaker described Wood's work on Pleistocene deposits as 'too complicated' and having 'too many hypotheses'. Whitaker confessed his 'inability to follow' some of Wood's reasoning and thought and added 'Mr Wood . . . has . . . been led astray by over theorising'. Whitaker concluded 'Surely when a simple explanation accords with observed facts, it is needless to go further afield for a complicated and absolutely bewildering one'. This attack was published in a Geological Survey Memoir (Whitaker 1889, vol. 1 p. 363–364).

Whitaker was however generally liked by his contemporaries. According to Strahan 'Whitaker made many a friend, but never an enemy' (Strahan 1925*a*, p. 130). Boswell recalled how Whitaker 'was a real geological father to many of us' (Boswell 1926, p. 199). Speaking from personal experience Boswell noted 'He was most particular in sparing no effort to encourage, by gifts of books, maps, information, and by voluminous correspondence, any young geological worker who came to his notice'(Boswell 1926, p. 199). Strahan who stated 'he encouraged younger geologists by sharing knowledge and imparting enthusiasm' confirmed this sentiment. Aubrey Strahan recorded that 'the attainment of the truth was the dominant motive with him, and it gave him as much pleasure that it should be attained by others as by himself' (Strahan 1925*a*, p. 130). Seward emphasised this point when awarding Whitaker the Wollaston Medal in 1923 (Seward 1923, p. xliv–xlv). His comraderie was confirmed by Matthews who observed 'While amongst engineers and others, associated with him, it was always felt they were acting rather with a friend than a professional advisor' (Matthews 1925, p. 153).

Commentators invariably describe Whitaker's appearance as an old man. According to Strahan, Whitaker was a picturesque figure, alert in body and mind in his eighties whose geniality endeared him to all (Strahan 1925*a*, p. 129). In his later years his long white hair and beard dominated Whitaker's appearance and, despite his age, he never needed glasses for reading (Thompson 1925, p. 94). Whitaker was a Christian, a staunch liberal and a member of the National Liberal Club and Croydon Liberal Association. However he took no active part in local politics (Topley 1925, p. 28). Whitaker's personality can perhaps be summed up as an 'exceptional character, with outstanding abilities and a long life's work' (Sweeting 1958, p. 122)

Whitaker's contribution to geology in general and hydrogeology in particular was characterized by observation and meticulous recording of detail. He was a methodical, systematic, well-organized and exceptionally industrious field geologist who followed in the footsteps of others confirming their findings by filling in the detail of their original work. Whitaker was not incisive, innovative or pioneering. He was not given to theorising or drawn into controversy (Strahan 1925*a*, p. 129). Boswell confirmed this opinion when he noted 'Accurate observation and careful recording of facts were characteristic of his work; he rarely speculated or theorised, and his comments were seldom devoid of humour' (Boswell 1926, p. 199). William Topley Junior astutely observed that Whitaker was 'Not prone to enter into controversy or to theorising, his incisive mind pierced through unimportant details to the core of the matters under discussion' (Topley 1925, p. 28). Matthews added that Whitaker 'was never dogmatic, nor would he allow anyone to tempt him to express opinions upon any matter which, in his opinion, came within the province of any profession other than his own' (Matthews 1925, p. 153). These obituary writers kindly overlooked the disagreement and dissent that the 1893 Royal Commission on Metropolitan Water Supply had from Whitaker's opinion about the underground water obtainable from the Lea Valley.

Whitaker never considered himself the 'father of English hydrogeology'. This epithet was posthumously awarded to him by Wilson in 1985, 60 years after Whitaker's death (Wilson 1985, p. 125). His claim to fame lies with the systematic way in which he amassed, over an astonishing period of 67 years, a vast database of details relating to wells and water supply. Initially he collected this information in his official capacity as a field geologist with the Geological Survey to help him interpret the structure of the areas he was mapping. He soon realized the tremendous value of this information and pressed for it to be published in Geological Survey Memoirs.

The volume of his published work is outstanding. For example the majority of the 19 Geological Society Memoirs listed in Table 1 contain details of wells. His 56 water supply papers and reports, which date from 1867–1924 are listed in Table 2. Finally his contribution to 15 Water Supply County Memoirs, totalling 3653 pages, which date from 1899–1928, are listed in Table 3. This totals 90 substantial pieces of work. His 1887 Royal Society Fellowship was in part awarded for his water supply papers.

This colossal body of work, however useful and worthy, does not necessarily qualify Whitaker to the title of 'father of English hydrogeology'. He was a most prolific worker and undoubtedly published more than other contenders to this title. However, his work was largely descriptive and lacked a rigorous theoretical basis. He was a compiler of information, a task which he carried out with enthusiasm and conviction. The raw data were available for others to analyse and interpret. As his obituarists record he was not given to theorising and, if the hydrogeologists of today require a father for their subject they should probably look elsewhere.

Conclusion

From 1857 to 1924, 67 years,William Whitaker was employed as a geologist. He worked for the Geological Survey from 1857 until he retired as a District Surveyor in 1896. For 35 years he geologically surveyed London and southeastern England on the one-inch scale, putting fine details on the maps of William Smith and George Bellas Greenough. He worked out the detail of the Tertiary and Pleistocene deposits in the London and Hampshire Basins outlined by Joseph Prestwich's pioneering work. Whitaker wrote or contributed to 19 Geological Survey Memoirs including two district memoirs for London which appeared in 1872 and 1889.

He did not consider himself the father of English Hydrogeology. His main contribution to hydrogeology was systematically collecting and publishing a huge database of records of wells and boreholes. As early as June 1864 Whitaker published two appendices of well sections, detailing 109 wells, to his Geological Survey Memoir on parts of Middlesex, Hertfordshire, Buckinghamshire and Surrey (Whitaker 1864, p. 102–105). He used this information to understand the geology of southeastern England. He made reasonable conclusions about groundwater- based water supply but his well boring and water supply knowledge was developed more by precedent and experience than theory or hypothesis.

Whitaker was the first systematic collector of groundwater related records and produced 56 water supply papers and reports which date from 1867 to 1924 (see Table 2). His contributions to 15 Water Supply County Memoirs totalling 3653 pages, which date from 1899 to 1928, are listed in Table 3. He was recognized as a leading authority on questions of water and served on numerous committees and commissions dealing with both private and public supplies. Flett recorded in his official history of the Survey that Whitaker's knowledge of water supply was unequalled and his industry was exceptional (Flett 1937, p. 177).

The Institution of Water Engineers awarded the first Whitaker Medal in 1930 in honour of 'the greatest authority on water supplies from the Chalk and Tertiary formations' (Hobbs 1954, p. 4). In addition to collecting details of wells and boreholes Whitaker also collected and published 18 bibliographies on the geology of English counties (Table 4). Whitakers' contribution to national learned societies and local field clubs was long-lasting and outstanding. He played a leading role in the Geological Society of London and the Geologists' Association, serving uniquely as President of both for a short period in 1900. Local field clubs benefited greatly from his support, where he would serve as president or give a lecture.

Whitaker had a most interesting and sometimes contradictory personality. He was strong willed and increasingly prepared to challenge his superiors at the Geological Survey as his financial independence and reputation developed. Following his retirement from the Geological Survey Whitaker worked as a water supply consultant and produced County Water Supply Memoirs for the Survey. In his mid 80s he could journey to Wales in winter and tramp the mountains in order to advise on questions of water supply (Davies 1925, p. xxxix) He was working until a few months before his death on the water supply of Somerset.

References

ANON. 1907. Eminent Living Geologists: William Whitaker. *Geological Magazine. New Series Decade V*, **4**, 50–58 & plate. [Written by H. Woodward & W. Whitaker]

ANON. 1925*a*. Obituary of William Whitaker. *The Times*, **23 January 1925**.

ANON. 1925*b*. Obituary of William Whitaker. *Croydon Advertiser*, **24 January 1925**.

ANON. 1925*c*. William Whitaker, B.A., F.R.S. *Geological Magazine*, **62**, 240.

ANON. 1925*d*. Brevities – The late Mr.William Whitaker. *Reader's Index: the bi-monthly magazine of The Croydon Public Libraries*, **27**, 28.

ANON. 1926. *University of London. The Historical Record (1836–1926)*, 2nd edition. University of London Press.

BAILEY, E. B. 1952. *Geological Survey of Great Britain*. Thomas Murby & Co, London.

BONNEY, T. G. 1886. Award of the Murchison Medal. *Proceedings of the Geological Society*, **42**, 32–33.

BOSWELL, P. G. H. 1926. William Whitaker, B.A. F.R.S. *Proceedings of the Liverpool Geological Society*, **14**,198–199.

BRISTOW, H. W. & WHITAKER, W. 1862. *Geology of parts of Berkshire and Hampshire*. Memoirs of the Geological Survey.

DALE, W. 1888 List of private collections in Hampshire. *Proceedings of the Hampshire Field Club and Archaeological Society*, **1**, 24–25.

DARWIN, C. 1883. *The Formation of Vegetable Mould through the action of Earthworms*. Murray, London

DAVIES, G. M. 1925. William Whitaker, B.A., F.R.S., F.G.S 4th May 1836–15th January 1925. *Proceedings of the Croydon Natural History Society*, **9**, xxxix & portrait.

DEWEY, H. 1926. William Whitaker. *Proceedings of the Geologists' Association*, **37**, 231–235.

EVANS, J. W. 1925. William Whitaker, *Quarterly Journal of the Geological Society*, **81**, lxi–lxii

FLETT, J. S. 1937. *The First Hundred Years of the Geological Survey of Great Britain*. HMSO, London.

GEIKIE, A. 1882. *Geological Sketches at Home and Abroad*. MacMillan, London.

GEIKIE, A. 1924. *A Long Life's Work: An Autobiography*. MacMillan, London.

HOBBS, A. T. (ed.) 1954. *Manual of British Water Supply Practice*, 2nd edition. The Institution of Water Engineers, Heffer, Cambridge.

HULL, E. & WHITAKER, 1861. *Geology of parts of Oxfordshire and Berkshire*. Memoirs of the Geological Survey, 57p.

LEACH, A. F. 1908. St. Alban's School. *Victoria History of the County of Hertfordshire*, **2**, 47–69.

MARR, J. E. 1906. Award of the Prestwich Medal. *Proceedings of the Geological Society*, **62**, xlvii–xlviii.

MARRYAT, H. & BROADBENT, U. 1930. *The Romance of Hatton Garden*. James Cornish & Sons. London.

MARTIN, E. A. 1926. Break-names in Geological History. *South-Eastern Naturalist*, **31**, 52–56.

MATHER, J. 1998. From William Smith to William Whitaker: the development of British hydrogeology in the nineteenth century. *In:* BLUNDELL, D. J. & SCOTT, A. C. (eds) *Lyell: the Past is the Key to the Present*. Geological Society, London, Special Publications, **143**, 183–196.

MATTHEWS, W. 1925. Obituary of William Whitaker. *Transactions of the Institution of Water Engineers*, **29**, 151–153.

ORMISTON, T. L. 1926. *Dulwich College Register 1619–1926*. J.J. Keliher, London, Record 5311.

PRESTWICH, G. 1899. *Life and Letters of Sir Joseph Prestwich*. Blackwood, Edinburgh.

ROYAL COMMISSION ON METROPOLITAN WATER SUPPLY. 1893. *Report of the Royal Commission appointed to inquire into the Water Supply of the Metropolis* p. 1–72; *Minutes of Evidence taken before the Royal Commission on Metropolitan Water Supply* p. 1–555.

SEWARD, A. C. 1923. Award of the Wollaston Medal. *Proceedings of the Geological Society*, **79**, xliv–xlv.

SOWAN, P. W. 2002. The North Downs Subaerial denudation controversy of the 1860s. *Bulletin of the Croydon Natural History Society*, **115**, 5–6.

STRAHAN, A. 1925*a*. Mr. W. Whitaker FRS. *Nature*, **115**, 129–130.

STRAHAN, A. 1925*b*. William Whitaker 1836–1925. *Proceedings of Royal Society Series B*, **97**, ix–xii & portrait.

SWEETING, G. S. 1958. *The Geologists' Association 1858–1958*. Bernham & Company.

THOMPSON, P. 1925. William Whitaker. *Essex Naturalist*, **21**, 93–96.

TOPLEY, W. W. 1925. William Whitaker. *Reader's Index: the bi-monthly magazine of The Croydon Public Libraries*, **27**, 28–29.

TORRENS, H. S., GETTY, T. A. & CRANE, M. D. 1978. Collections and Collectors of Note 7 John William Elwes. *Newsletter of the Geological Curators Group*, **2**, 117–120.

WHITAKER, W. 1861*a*. On the 'Chalk-Rock', the topmost bed of the Lower Chalk in Berkshire, Oxfordshire, Buckinghamshire, etc. *Quarterly Journal of the Geological Society*, **17**, 166–170.

WHITAKER, W. 1861*b*. *The Geology of parts of Oxfordshire and Berkshire*. Memoirs of the Geological Survey.

WHITAKER, W. 1864. The Geology of parts of Middlesex, Hertfordshire, Buckinghamshire, Berkshire and Surrey. Memoirs of the Geological Survey of Great Britain.

WHITAKER, W. 1865. On the Chalk of the Isle of Thanet. *Quarterly Journal of the Geological Society*, **21**, 395–398.

WHITAKER, W. 1866. On the Lower London Tertiaries of Kent. *Quarterly Journal of the Geological Society*, **22**, 404–435.

WHITAKER, W. 1867. On Subaerial Denudation, and on cliffs and escarpments of the Chalk and Lower Tertiary Beds. *Geological Magazine*, **4**, 447–454 & 483–493.

WHITAKER, W. 1867*a*. Note on the Surface-Geology of London; with Lists of Wells and Borings, showing the Thickness of the Superficial Deposits. *Report of the Medical Officer of the Privy Council for 1866*, Appendix, 346–366.

WHITAKER, W. 1872. *The Geology of the London Basin. Part 1 – The Chalk and Eocene Beds of the Southern and Western tracts*. Memoirs of the Geological Survey, vol. 4, Longmans, London.

WHITAKER, W. 1875. *Guide to the Geology of London and the Neighbourhood. An explanation of the Geological Survey Map of London and its environs and of the Geological model of London, in the Museum of Practical Geology*. Memoirs of the Geological Survey, 2nd edition 1875; 3rd edition 1880; 4th edition 1884; 5th edition 1889; 6th edition 1901.

WHITAKER, W. 1880. List of Works on the geology, mineralogy, and palaeontology of Wales (To the end of 1873). *Report of the British Association for the Advancement of Science*, 397–436.

WHITAKER, W. 1886. On a recent legal decision of importance in connection with water supply from wells. *Geological Magazine*, **3**, 111–114.

WHITAKER, W. 1888. Chronological list of works referring to underground water, England and Wales, Appendix in 13th Report of the British Association Committee appointed for the purpose of investigating the circulation of underground waters in the permeable formations of England and wales and the quantity and character of water supplied to various towns and districts from these formations. *Report of the 57th meeting of the British Association, Manchester, August/September 1887*, John Murray, London, 388–414.

WHITAKER, W. 1889. *The Geology of London and Part of the Thames Valley*. Memoirs of the Geological Survey, vol. 2, appendices.

WHITAKER, W. 1890. Coal in the South East of England. *Journal of the Royal Society of Arts*, **38**, 543–553.

WHITAKER, W. 1895*a*. 1. Second chronological list of the works referring to underground water, England and Wales. p. 394–402. 2. Underground in Suffolk and its Borders. Presidential Address to Section C. Geology. p. 666–675. *Report of the 65th meeting of the British Association for the Advancement of Science held at Ipswich in September 1895*. John Murray, London.

WHITAKER, W. 1895*b*. On the Chalk of the London Basin in regard to Water Supply. *Geological Magazine, Decade 4*, **2**, 360–366.

WHITAKER, W. 1899. Anniversary Address to the Geological Society. *Proceedings of the Geological Society*, **55**, lxix–lxxxiii.

WHITAKER, W. 1901. *Guide to the Geology of London and the Neighbourhood*, 6th edition. Memoirs of the Geological Survey.

WHITAKER, W. 1908. *The Water Supply of Kent with Records of Sinkings and Borings*. Memoirs of the Geological Survey.

WHITAKER, W. 1921. Geologists and the Geologists' Association. *Proceedings of the Geologists' Association*, **32**, 183–188.

WHITAKER, W. 1924. Geology of the Winchester District. *Transactions of the Institution of Water Engineers*, **27**, 17–23.

WHITAKER, W. & REID, C. 1899. The Water Supply of Sussex from Underground Sources. Memoirs of the Geological Survey.

WILSON, H. E. 1985. *Down to Earth: One hundred and fifty years of the British Geological Survey*. Scottish Academic Press, Edinburgh.

WOODWARD, H. B. 1907. *The History of the Geological Society of London*. Geological Society, London.

YOUNG, G. W. 1925. William Whitaker, B.A., F.R.S., F.G.S. *South-Eastern Naturalist*, **30**, xlvi–xlvii.

Joseph Lucas (1846–1926) – Victorian polymath and a key figure in the development of British hydrogeology

J. D. MATHER[1], H. S. TORRENS[2] & K. J. LUCAS[3]

[1]*Department of Geology, Royal Holloway University of London, Egham, Surrey, TW20 0EX, UK.*

[2]*School of Earth Sciences, University of Keele, Staffs, ST5 5GG, UK.*

[3]*Consolidated Information Services, 641 West Queen's Road, North Vancouver, B.C., V7N 2L2, Canada.*

Abstract: Joseph Lucas joined the Geological Survey in 1867 and spent almost 9 years mapping in Yorkshire. Forced to resign in ignominious circumstances, for the rest of his life he earned his living advising on groundwater supplies. In 1874 he was the first to use the term hydrogeology in its modern context and defined this new subject in a series of papers in the 1870s. He drew the first British maps showing groundwater contours and described how to carry out a hydrogeological survey. For many years he lobbied for such a survey to be carried out over the whole country and for it to be used as a basis for water resource planning. He was an accomplished linguist, translating material from a variety of European languages, and wrote books on natural history and genealogy. He and his family lived at Tooting, in south London, where he is buried in the Churchyard of Saint Nicholas.

Little attention has so far been given to the history of the practice of geology in Britain. As Rupke has pointed out, in the early part of the nineteenth century the 'English School of Geology', centred on the Geological Society of London, 'regarded the economic aspect of geology as of little interest . . . since it was not thought to merit academic rank' (Rupke 1983, p. 18 & 200). Tweedale also addressed this real problem. He pointed out (Tweedale 1991), when discussing a contemporary of Lucas, how industrial geology was a neglected area which had been little studied, partly because many 'industrial geologists left relatively few papers and sometimes never published their results'. As a result 'the literature so far lacks detailed case studies of the careers and work of applied geologists'. Tweedale discussed a number of industrial geologists and specifically noted two who were much involved with questions of water supply – William Whitaker (1836–1925) and William Topley (1841–1894). Torrens (2002) has also discussed the problems concerning the two faces of geology; the academic and the practical.

Prior to 1871 the Geological Survey had shown little interest in water supply although some of the early memoirs (e.g. De La Beche 1839) had extensive sections on economic geology. As early as 1850 the Reverend James Charles Clutterbuck (1801–1885), Vicar of Long Wittenham in Berkshire, bemoaned the fact that a geological survey was being carried out in a remote region of North Wales, where no urgent need existed, instead of in the metropolitan areas where there was a real need to discover the structure of the deep water-bearing strata (Clutterbuck 1850). Only Edward Hull (1829–1917) seemed to recognize that underground water was a worthy topic for Survey publication (Hull 1865, 1868), although Survey staff were probably already answering geological enquiries related to the siting of wells. The then Director General, Sir Roderick Murchison (1792–1871), himself a prominent member and former President of the Geological Society, had little interest in the economic applications of geology (Flett 1937) and may have been responsible for the lack of any significant work within the Survey. In the absence of such inputs and interests the foundations of the science of hydrogeology in Britain seem to have been laid by mining engineers (Younger 2004), not least Robert Stephenson (1803–1859) (Preene 2004), and a few gentlemen geologists such as the Reverend James Clutterbuck and Sir Joseph Prestwich (1812–1896) (Mather 1998).

Following the death of Murchison, in 1871, and the appointment of Andrew Ramsay as Director, there was an increase in Survey activity related to water. As well as Edward Hull, three other geologists William Whitaker, Joseph Lucas and Charles Eugene de Rance (1847–1896) began to take an interest in water supply. Some of this work was part of their official duties, but particularly in the case of Lucas, much was carried out unofficially. Early historians of the Survey give all the credit to Whitaker (Flett 1937; Bailey 1952). However, Wilson (1985) recognized the significant part played by both Lucas and de Rance although still considering Whitaker to be the 'father of English hydrogeology'.

Of the three men certainly the most innovative

From: MATHER, J. D. (ed.) 2004. *200 Years of British Hydrogeology*. Geological Society, London, Special Publications, **225**. 67–88. 0305-8719/04/$15

contribution came from Joseph Lucas. According to one of his obituarists his work was 'marked by a distinct originality of thought which at once brought him to public notice' (Anon 1926*a*). He was the first person to use the term 'hydrogeology' in a modern context (Mather 2001) and defined this new subject in a series of papers in the 1870s. The Oxford English Dictionary credits him with the first modern usage of this term in 1877 although he first used it as early as 1874. He drew the first British maps to show water-level contours and described the objectives and methods of conducting a hydrogeological survey (Lucas 1874). He was awarded a Silver Medal by the Society of Arts in 1879 and the Telford Medal by the Institution of Civil Engineers in 1880, but remained largely unrecognized by the geological community.

His interests extended much more widely than geology and he had a considerable knowledge of natural history, languages and literature. It is the objective of the present paper to chronicle the life and achievements of this Victorian polymath. His writings and history suggest that he had a difficult personality and a rather unsettled career but his achievements as a hydrogeologist deserve much greater recognition than they have so far received.

Family background

Joseph Lucas came from a prosperous middle-class family which traced its origins back to a prominent family of millers and maltsters from Hitchin in Hertfordshire. His ancestor William Lucas (d. 1704) was an early follower of the Society of Friends but his great-grandfather Rudd Lucas (1747– 1810) was the last to be buried a Quaker. The next generation seems to have dispersed from Hitchin and his grandfather William Lucas (1777–1832) farmed at Stapleton Hall, near Hornsey in Middlesex. His father, also Joseph Lucas (1811–1903), was born at Stapleton Hall and attended Edinburgh University, qualifying as a solicitor in 1833. Joseph (Snr) married his cousin Sarah Lucas Judkins in 1835 and came to live at Stapleton House, Upper Tooting in south London about 1843, residing there until his death. Joseph (Snr) practised from offices at 1, Trinity Place, Charing Cross and latterly from 21, Surrey Street off the Victoria Embankment.

Joseph (Snr) and Sarah had 16 children, 12 surviving beyond infancy, of whom Joseph was the third surviving son. The two eldest sons Frederic William (1842–after 1931) and Edgar (1845–1932) qualified as solicitors in 1865 and 1872 respectively. The fourth, Samuel Francis (1847–1875) was a student of the Inner Temple before his premature death. The fifth, Bernard John (1853–1910) was an artist who painted landscapes in both oils and watercolour. The sixth, Charles Arthur (1858–1884), was a medical student at the time of his death but little is known of the seventh son Henry Warter (1859–1922). The five surviving daughters, to use the language of the time, all made 'good marriages'. Emily Sarah (1837–1870) married the Rev. Albert Alston; Frances (1839–1938) married an Austrian, Charles Redl, who became rector of the Royal College in Mauritius; Clara (1844–1916) married Sir Charles Bruce who had a distinguished career in the Colonial Office retiring as Governor of Mauritius in 1903; Constance Mary (1855–after 1942) married Sir Joseph Hutchinson who became the Chief Justice of Ceylon in 1906 and Margaret Helen (1857–after 1931) married the publisher George Macmillan. Joseph was thus a member of a large successful family which had both money and considerable status in Victorian England.

Joseph himself was born on Christmas Eve 1846 at Upper Tooting. He attended Westminster School in 1863 and 1864 where he had the distinction of rowing, at number 4, in the last Eton and Westminster boat race (Barker & Stenning 1928; Anon. 1926*b*). What stimulated his interest in geology is a matter for conjecture. He may have been influenced as a boy by his relative William Lucas (1804–1861) of Hitchin, a brewer and banker. William had been apprenticed to a pharmacist in the Haymarket in London and had become aquainted with the leading scientists of the day, including Charles Lyell, Henry de la Beche and Adam Sedgwick. William's diaries (Bryant & Baker 1934) show that he regularly attended meetings of the Geology Section of the British Association. William was at Birmingham in 1839 where he participated in the famous excursion to Dudley and at Oxford in 1860 when discussion centred on Darwin's new book on the origin of species. A reference in William's diaries to visiting 'J Lucas's, Charing Cross' in 1858 (Bryant & Baker 1934, vol. 2, p. 517) and the fact that Joseph prepared reports on the water supply of Hitchin (Lucas 1888) and of the Cotswold Brewery in Cirencester (Lucas 1881*a* & 1886*a*), owned by Messrs Bowly and Son (relatives of William's mother Ann Bowly (1769–1853)), all suggest that the Tooting and Hitchin branches of the family were in contact and that Joseph would have known William's family.

Another real connection with the then small world of geology comes through Joseph's great uncle James Lucas (1780–1839), a Captain in the Merchant Navy, whose fourth daughter, Rosa (1813–1873) married the Rev. George Ferris Whidborne. There is evidence that Joseph's family maintained close contact with this other branch of the family (Whidborne1917). Their son, also the Rev. George Ferris Whidborne (1845–1910), who was almost the same age as Joseph, became a leading expert on Devonian fossils (Anon. 1910; Watts 1911) having from an early age shown a great interest in natural history. Joseph later carried out a geological survey of Whidborne's estate, Hammerwood Park, which lay on the Northamptonshire Ironstone (Lucas 1888).

Whatever the source of his interest, Joseph chose to forsake the legal profession, espoused by most of the male members of his family, and became a geologist.

Employment with the Geological Survey

Joseph Lucas was recommended for appointment to the Geological Survey of Great Britain, as an Assistant Geologist, on November 10th 1866, when only 19 years of age. His Certificate of Qualification was issued by the Civil Service Commission on January 17th 1867 and he must have started work shortly afterwards, some 3 weeks after his 20th birthday. He was proposed as a Fellow of the Geological Society on January 23rd 1867 and formally elected on 20th February. His proposers were his senior officers at the Survey, Andrew Crombie Ramsay (1814–1891) and Henry William Bristow (1817–1889), along with John Morris and James Tennant. Morris (1810–1886) was involved in teaching geology at University College, London from 1853 and Tennant (1808–1881) occupied a similar position at King's College, London from the same year. The likelihood is that the young Lucas attended their lectures after leaving school in 1864 (which may provide another clue as to why he chose to become a geologist).

His first year at the Survey was marred by an unspecified accident on 4th May 1867 and he was on sick leave until November 27th of that year. In a letter dated August 17th 1867 and preserved in the BGS Archives, the Director General, Sir Roderick Murchison, wrote 'Mr Lucas is really a very fine young man of good abilities, good health in general and of active habits.' The tone of the letter suggests that Murchison probably knew Lucas and his family.

Lucas was assigned to the group of geologists engaged on the 6 inch to the mile Geological Survey of Yorkshire. Initially his career seems to have gone well and he was promoted to 'the staff of Geologists' on 22nd March 1873. Between 1867 and 1872 he was engaged mainly in the mapping of Nidderdale and the country around Harrogate. His name appears as a surveyor on many of the published 6-inch maps and he seems to have been particularly involved with the Millstone Grit as on a number of sheets he is credited with its resurvey. Following completion of the Nidderdale work he was surveying in the Bradford/Leeds area in 1872/3.

On 21st March 1872 Lucas married Elizabeth Storie Mackean (1852–1911) in London at Saint Peter's, Eaton Square. His bride was the daughter of John Mackean, a timber merchant originally from Inchinnan in Renfrewshire, where Elizabeth was born on 26th July 1852. Sometime between 1861 and 1871, the family had left Scotland and moved to the north Yorkshire coast, residing at Picton House, Redcar. The Yorkshire home of George Macmillan, later to marry Joseph's sister Margaret, was at Botton Hall, Danby, some 21 km from Redcar. It seems possible that whilst he was mapping in Yorkshire Joseph was a regular visitor to Danby where he might have been introduced to the Mackean family.

Family tradition suggests that the new Mrs Lucas used Storie, her mother's maiden name, as her given name. However, in the 1901 census she is recorded as Ellen. Their first child (Millicent Storie 1873–1956) was born almost a year later at Epsom in Surrey and a second (Beatrice Mary 1874–1945) at Sandsend between Redcar and Whitby on the Yorkshire coast. By this time, June 1874, Lucas had moved from the Bradford area and was mapping Jurassic rocks in North Yorkshire under Henry Hyatt Howell (1834–1915) as his District Geologist.

Lucas also began to publish in the scientific literature. His first contribution was a short letter published in the Geological Magazine commenting on the absence of Boulder Clay south of the Thames Valley (Lucas 1869). This was followed by 3 papers based on his mapping work in Yorkshire. Lucas (1872) comments on the supply of salts to the Permian Sea and particularly on the origin of the iron which imparts a red colour to the sediments. Two further papers (Lucas 1873*a* & 1873*b*) discuss the origins of the clay-ironstones of the Coal Measures. He concluded that every bed of clay-ironstone 'marks a terrestrial horizon as much as every coal bed does, or as every bed of sediment which bears the imprint of raindrops' (Lucas 1873*a*, p. 367). Andrew Ramsay, the new Director General of the Geological Survey was present when Lucas presented the paper on clay-ironstone to the Geological Society on 12th March 1973 and commented that the paper 'exhibited considerable ingenuity', although he did not agree that ironstone was never deposited in marine strata (Lucas 1873*a*, p. 368).

During his mapping work in Yorkshire Lucas also became interested in groundwater. Later in his life Lucas said that his interest was first aroused by Joseph Prestwich's Presidential Address to the Geological Society in 1872 (Lucas 1888). Prestwich had been a member of the Royal Commission on Water Supply, which had reported in 1869 and he chose for part of his address the subject 'Our springs and water supplies' (Prestwich 1872). Lucas realized that little information existed on the water-bearing formations in the south of England and in January 1873 he started to make measurements to the south of London to fill 'the hiatus in our knowledge of the subterranean water systems'(Lucas 1888, p. 19). After 18 months work he published a quarto volume of 86 pages which proposed a scheme for the improvement of London's water supply using what he called 'horizontal wells' (Lucas 1874). These were visualized as galleries driven along the strike at the base of the main water-bearing formations to capture groundwater that was otherwise discharged as springs or seepages. The scheme was far-fetched and no attempt was made to implement it. However, the volume was significant for two reasons; firstly it contained the first British map to show groundwater contours and secondly it used the term hydro-geological for the first time in a modern context (Mather 2001).

Groundwater contours were plotted on the Chalk aquifer over an area to the south and west of Croydon in Surrey (Fig. 1). Two sets of contours are shown – 'contours of upper surface of water in the Chalk early in 1873' and the same 'at the end of 1873' (Lucas 1874, map A). No colour is used and the area contoured is less than 50 square miles, nevertheless the data are all original and their collection must have involved a considerable amount of work at a time when Lucas was supposedly mapping in Yorkshire. The term hydro-geological is used once only as the heading of a section entitled 'Objects and Mode of Constructing a Hydro-geological Survey of the Water-bearing Formations' which is in an appendix to the main report (Lucas 1874, p. 57). In this appendix Lucas discusses the parameters to be measured: rainfall, evaporation, percolation, spring discharges and the 'height of the water line', and considers how these can be used to find the quantity of water passing under the overlying impervious beds.

In 1875 Lucas provided advice to Dr Richard Thorne Thorne of the Medical Department of the Local Government Board in his enquiry into the reasons for the large mortality from 'fever' in the Registration District of Guisbrough in Yorkshire. This District included the Urban District of Redcar, the home of Lucas's wife, and so was of particular interest to Lucas whose own report is embodied in the final report to the Assistant Medical Officer, Edward Seaton (Thorne 1875). Lucas divided the superficial deposits into an older boulder clay overlain by sands and gravels which were in turn overlain by an upper red clay. Many supplies were derived from wells which tapped water in beds of sand and gravel between the beds of clay. Lucas suggested that 'since these intermediate sand and gravel beds nowhere form a gathering ground, they must of necessity derive their water by soakage through the clay'. He further suggested that 'the permeability of the red clay is . . . rendered the more certain by the deep cracks which form in it in dry weather' (Thorne 1875, p. 4). Using this information Thorne concluded that most of the wells were liable to be polluted and that they should be closed. This must be one of the first publications to comment on groundwater recharge through supposedly impermeable clays and is an early example of a groundwater vulnerability assessment.

Unfortunately for Lucas the time which he was spending on non-Survey activities must have begun to affect his mapping and his behaviour certainly started to attract the attention of his senior officers. The first indication of trouble is a letter, preserved in the British Geological Survey Archives, dated 20th November1874, from Henry Bristow, who had been appointed Senior Director (England and Wales) in 1872, asking Lucas to send to the Office fossils 'which you have collected during the time you have been mapping the Yorkshire Oolites – no material has been received since you have been in your present district.' This was followed by a further letter, dated 30th November, responding to Lucas' reply which has not survived, in which Bristow writes that neither Professor Ramsay nor I ever gave orders to do your mapping in your present district 'on purely physical principles'. In a footnote Bristow adds 'It is such a thick fog I can hardly feel to write'.

Over the following year Lucas's position within the Geological Survey continued to deteriorate. On 7th June 1875 Bristow wrote to him asking for an explanation of why he was absent from his field station and why he had extended his leave without permission. Clearly the reply was not satisfactory as he wrote again on 26th June:

> 'Your movements are so erratic and your place of residence so uncertain that it has of late been impossible to tell where you were to be found; in addition to which you have apparently chosen your stations more with regard to your own personal convenience than the requirements of the Survey. Besides this the accounts of your work are so unsatisfactory that I feel myself, at last, unwillingly compelled to take notice of this. I request that, on receipt of this, you will immediately place yourself under Mr Howell's orders and proceed with your field work in accordance with his wishes and instructions. As I told you verbally, when you were in London, you are not permitted to absent yourself from your station without previously applying for permission to do so in the regular way.'

He threatened to place the matter before the Director General unless 'some manifest improvement takes place'.

Maybe there was some temporary improvement but it seems that relations between Howell and Lucas soon reached breaking point. On 7th January 1876 Bristow wrote again to Lucas, this time addressing him as 'Sir' rather than 'My dear Lucas':

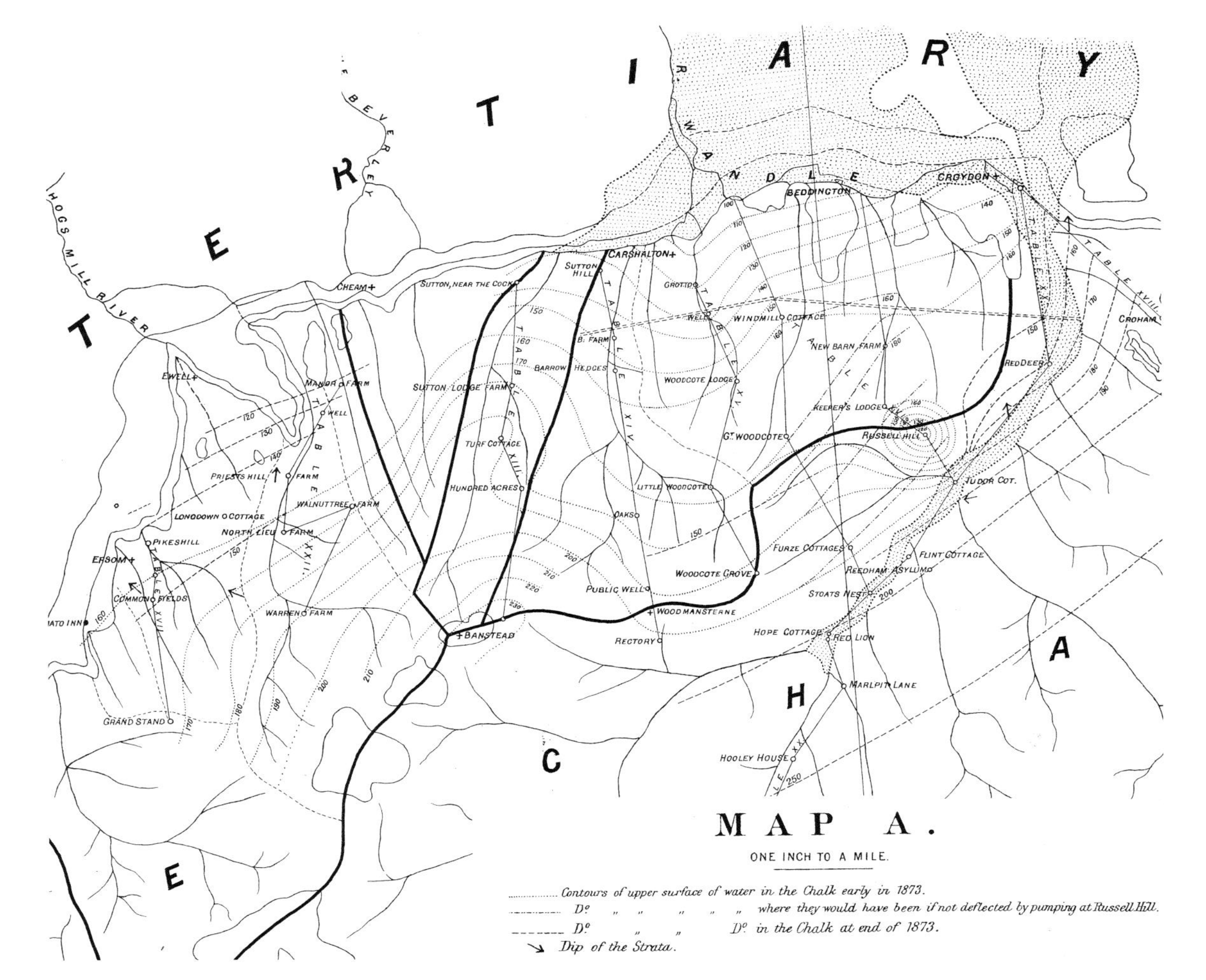

Fig. 1. Map of the Chalk aquifer between Croydon and Epsom in Surrey, the first British map to show groundwater level contours (extract from map A in Lucas 1874).

> 'Mr Howell having reported to me that you assaulted him while officially engaged on Survey business, I hereby suspend you from duty until your conduct has been enquired into. On receipt of this letter I desire you to come to London and to report yourself at this office with the least possible delay. It is essentially necessary that you should bring with you all the maps and the other Survey documents in your posession relating to the district in which you have been engaged.'

Lucas must have come down to London rapidly as on January 17th he sent to Bristow, from Picton House Redcar, the home of his in-laws, a simple resignation letter beginning 'In accordance with the advice which you have been kind enough to offer me. . .'. The brief reply, addressed to Lucas at Upper Tooting, confirmed that his resignation had been accepted by the Director General dating from 7th January 1876, the date of his suspension from duty.

Thus Lucas's Survey career came to an ignominious end after almost nine years as a field surveyor. A review of some of his geological field slips held in the British Geological Survey archives show that he was not the most skillful and diligent of field geologists. His field slips are scruffy and have few observations compared with those of his colleagues. Annotations on them include 'The enclosed field-maps in this sleave may have suspect geological observations' and 'Bits of Lucas of no value, small amount of work in. Entirely replaced by Cameron's map, made under Mr Howell's inspection'. Clearly his colleagues put little value on his observations and his dismissal was probably of ultimate benefit to both the Survey and himself as he was now able to concentrate on the hydrogeological work to which he was more suited.

The productive years 1876–1882

As soon as he was relieved of his Survey post Lucas returned to London and at some time during 1876/1877 moved to Tooting Graveney, a few kms south of his parents home in Upper Tooting. Between 1876 and 1883 his address is recorded in the list of members of the Surveyor's Institution merely as Tooting Graveney SW. However, the 1881 census gives the address as 6, Defoe Road, Lower Tooting. From 1883/1884 the Surveyor's Institution lists him at 11, Defoe Road and from 1885/1886 at 21, Defoe Road returning to 11, Defoe Road in 1904/1905. Whilst it is possible that he moved house it is much more likely that he remained at the same property which was progressively renumbered as his house became gradually surounded by further properties. At this time Tooting Graveney or Lower Tooting was a small village with a manor house and spring-fed fish ponds and cress beds. Defoe Road was opposite The Broadway right in the centre of the village. On the Ordnance Survey 6-inch map published in 1874, but based on a survey of 1869, it is marked as New Road suggesting that it had only recently been built up.

The Victorian villas, in one of which the Lucas family lived, at the Broadway end of Defoe Road (now called Garratt Lane) still stand. The date on the group of four, called Wendover Villas, closest to The Broadway is 1879, which fits in well with the period at which the family moved. Prior to 1850 Defoe Road was a meadow and it is probable that at this time much of this land was owned by the Lucas family. Hurley (1947) records that in the early 1850s the site at the corner of Garratt Lane (now Garratt Terrace) and Tooting High Street was the most desirable corner plot in Tooting and that 'Mr Joseph Lucas . . . who lived at Stapleton House, Upper Tooting, conveyed the plot to the Rector in 1853 to build Infant Schools in the Parish . . . The site cost £250. No doubt . . . a generously low price considering the value of sites in Tooting Broadway in these days' (Hurley 1947).

By the time of the 1938 Ordnance Survey 6-inch map, the spring-fed watercress beds on the River Wandle had disappeared and the whole area had become urban. Today Garratt Lane (Defoe Road) is the route of the A 217. Tooting Broadway is at a major crossroads and the site of an underground station on the Northern Line. The whole area would now be totally unrecognisable to Joseph Lucas.

Lucas's first scientific article since being relieved of his duties with the Geological Survey was published in *The Architect* on 29th July 1876 (Lucas 1876). It was addressed to architects who had to design a house, find water for it and drain it, often within a very limited area. The article emphasised the measures which needed to be taken in order to avoid contamination of well waters.

His small book on horizontal wells (Lucas 1874) proved of interest to engineers involved in water supply work and on the 28th November 1876 he made his first significant presentation at a meeting of the Institution of Civil Engineers in London. The paper entitled 'The Chalk Water System' created enormous interest; the discussion went on for three nights and occupied 55 pages of the Proceedings compared with 13 pages for the text of the paper itself. The paper described groundwater conditions over a large tract of ground to the south of London, stretching from the Darent in the east to the Wey in the west (Lucas 1877*a*). According to Lucas 'The observations of the Author extend over four years, and range over about 200 square miles of country, on which almost every accessible well has been measured. The measurements number many hundreds.' (Lucas 1877*a*, p. 71). Remember that for three of these four years he was supposed to have been mapping in Yorkshire. The paper was illustrated by a

'Hydrogeological Map shewing Water Contours on the Chalk Water System' and by a number of hydrogeological sections across the area. The term hydrogeology was used throughout, now without a hyphen, and is defined in the first sentence as 'Hydrogeology . . . takes up the history of rainwater from the time that it leaves the domain of the meteorologist, and investigates the conditions under which it exists in passing through the various rocks which it percolates after leaving the surface.' (Lucas 1877*a*, p. 70).

At this time most engineers believed that a limited portion of the rainfall was consumed by evaporation, supported vegetation or was discharged as springs. However, they considered that the greater proportion descended deep into the Chalk where it was held up by the underlying Gault Clay and eventually discharged to the River Thames, through the shingle and sand which covers the coast or into the bed of the sea itself. (Stephenson 1840; Barlow 1855; Homersham 1855). Thus they believed that a vast amount of water could be pumped from deep wells in the Chalk and that if such wells were made watertight to a depth of 60 to 70 feet (18 to 21.5 m) they could be pumped with no effect on surface springs and streams. These views had been vigorously opposed by the Reverend James Clutterbuck in papers presented to the Institution of Civil Engineers (Clutterbuck 1842, 1843 & 1850). The debate following Lucas's paper continued the dispute with Homersham, the engineer, and Clutterbuck, the scientist, both speaking. The conclusions of Lucas's work, in which rainfall was related to changes in water level, strongly supported Clutterbuck's view, and was to lead him into many acrimonious discussions with engineers in the future.

On 26th February 1877 Lucas read another paper, this time at the Surveyor's Institution (now the Royal Institution of Chartered Surveyors) in which he further developed his ideas (Lucas 1877*b*). He proposed a more concise definition for his new science of hydrogeology: 'Hydrogeology takes up the history of rain water from the time that it touches the soil, and follows it through the various rocks which it subsequently percolates.' (Lucas 1877*b*, p. 154). He also introduced the term hydrogeologist. The paper included a discussion on ponds, pointing out that most successful ponds were spring-fed, and giving examples particularly from the Upper Greensand which he concluded was 'as a water system, . . . quite distinct from that of the chalk.' (Lucas 1877*b*, p. 167). He returned to this paper some years later in a letter to *The Times*. In response to correspondence suggesting ponds could be dug on fallow land for the rearing of carp, he pointed out that the maintenance of ponds for the breeding of fish depended on the springs which fed them. He asked 'Will you allow me to very respectfully point out that the greater number of the apparently suitable sites for fishponds must inevitably fail of that object from lack of a sufficient supply of spring water?' (Lucas 1884*a*).

On 9th May 1877, at the Society of Arts, Lucas read a third paper in which he described the artesian system in the Chalk aquifer to the south of the Thames (Lucas 1877*c*). It was illustrated by a hydrogeological map on which he plotted both the supposed natural area of overflow and the 'modern' (i.e. 1877) area of overflow. He visualized that 'The water in the chalk, which is syphonised by the curve of the strata beneath the tertiary beds, presses upwards against their undersurface, and, on being liberated by a boring, rises to just such a height as is due to the pressure at that point.' (Lucas 1877*c*, p. 597). He called the upward limit of ascent of the liberated water, the artesian plane, which could be represented by contours. By comparing these with contours on the surface of the ground the limits of the area of overflow could be defined. The most prolific overflow area was that in the Wandle Valley around Tooting, close to where Lucas had been born. He was able to show that the number of boreholes had risen from 15 or 20 in 1850 to at least 110 in 1877, with a concomitant reduction in yield, to one tenth of the original volume in one case, and a reduction in the size of the overflow area. However, Homersham who took part in the subsequent discussion still 'did not agree with Mr Lucas that there was a gradual depression in the chalk water in this district' believing that 'if they were to cease pumping the water would rise to its normal level' and that there was no permanent effect (Homersham in discussion of Lucas 1877*c*, p. 605).

Lucas's next task was to put all the information which he had collected together in the form of a map and Sheet 1 of his Hydrogeological Survey was duly published on the 8th December 1877 (Lucas 1877*d*), together with an explanatory booklet (Lucas 1877*e*), by Edward Stanford of Charing Cross London. The base map used was the Ordnance Survey New Series 1-inch sheet 270 (mainly the area described in the Society of Arts paper) which had been enlarged by photography to a scale of 1.5 inches to a mile. Hydrogeological information was then superimposed using coloured ornament. Water levels in the two principal aquifers, the Chalk and the overlying Thanet Sands, were contoured separately at 10 feet intervals. In the unconfined areas the contours are blue, dark blue for minimum levels and light blue for maximum levels, and in the 'artesian districts' dark and light reds are used. Impervious clays are coloured grey with various symbols being used to represent different types of well. The explanation begins with an outline of the objects and origins of the hydrogeological survey and emphasises that the work has been carried out 'for upwards of two years . . . at my sole expense' (Lucas 1877*e*, p. 4).

The fourth and last of Lucas's ground-breaking papers was given at the Surveyor's Institution on the 29th April 1878. It examined the hydrogeology of Middlesex and part of Hertfordshire to the north of the area examined previously (Lucas 1878*a*). The paper summarizes some two years of observations which he had made since leaving the Geological Survey. In it he emphasizes the relationship between faulting and folding in the Chalk and the occurrence of groundwater, with better yields in synclinal environments, although deeper boreholes were needed. The measurements made became the basis for Sheet 2 of his Hydrogeological Survey which was published on the 18th May 1878 and covered an area to the north of Sheet 1 (Lucas 1878*b*). Just prior to that, on 1st May 1878, a 2nd edition of Sheet 1 had appeared (Lucas 1878*c*) and at the same time a further explanatory leaflet to accompany these two sheets was published (Lucas 1878*d*). The differences between the two editions of Sheet 1 relate solely to the colours used. In the second edition the artesian contours relating to the Thanet Sands, as well as the symbols used to show the sites of wells penetrating them, were changed from red to yellow to distinguish them from the Chalk. Similar colouring was used on Sheet 2 with yellow lines being used to denote the 'sand springs' of the Thanet Sands and the Woolwich and Reading Series (now known as the Lambeth Group).

In a period of some 18 months Lucas completely changed the direction of groundwater studies in England and Wales. His four papers provided a wealth of information, carefully collated in a series of tables, and summarized in two coloured maps. However, despite the claims of some writers (e.g. Wilson 1985), these were not the first ever hydrogeological maps. In France a map of the City of Paris, which showed water-level contours on three aquifers had been published as early as 1858 at a scale of 15mm to 100m (Delesse 1858). Interestingly the contours were coloured red the same colour used by Lucas. A second map was published in 1862 covering the Department of the Seine in which water-level contours were plotted for five separate 'nappes souterraines' (Delesse 1862). This is a magnificent map, multicoloured, in two sheets and at a scale of 1:25000. The author, Achille Ernest Oscar Joseph Delesse (1817–1881), held the Chair of Geology at the Sorbonne and was also Inspector of the Quarries of Paris during the period when these maps were produced. References in the discussion of Lucas (1877*a*) and in the Explanation of Sheet 1 of his Hydrogeological Survey (Lucas 1877*e*) show that Lucas knew about these French maps, and the red used for the artesian contours suggests that they may have had some influence on him.

On the 21st and 22nd May 1878, shortly after Lucas's maps appeared, a National Water Supply Congress was held at the Society of Arts at the instigation of its President, HRH The Prince of Wales. The idea was to have an open discussion on 'how far the great natural resources of the kingdom might, by some large and comprehensive scheme of a national character, . . . be turned to account, for the benefit not merely of a few large centres of population, but for the advantage of the general body of the nation at large.' Lucas was one of many engineers and scientists asked to give his views. In a short presentation he exhibited his maps and concluded that

> 'I am of the opinion . . . that a national survey of the water-bearing strata is the measure that would best serve the present requirements of the country in general. It is an essential forerunner to the execution of any scheme of a constructive or legislative kind, and even in my opinion, to the sucessful framing of such. Government would best minister to this popular want by investigating and providing the necessary data, alike to the landowner, the ratepayer, and the engineer, which no interested party can be expected to undertake on a comprehensive scale.' (Lucas 1878*e*, p. 776).

The Times agreed and in a leading article on 25th May 1878 reporting on the Conference, commented that:

> 'But among the most solid contributions to a solution of the question is Mr Lucas's suggestion that, before anything is decided on, the water-bearing strata of the country should be systematically surveyed. We must take stock first of our water wealth before we can know what subvention it can afford to distant localities without starving itself.'

The Congress adopted a resolution urging the Government to appoint a small permanent Commission to investigate and collect facts connected with water supply 'in order to facilitate the utilisation of the national sources of water supply, for the benefit of the country as a whole'.

During the rest of 1878 Lucas pursued his suggestion of a 'National Hydrogeological Survey'. He read a paper at the Congress of the National Association for the Promotion of Social Science, at Cheltenham, in which he urged the Association to impress on the Government '. . . the importance of instituting a preliminary survey as a forerunner to serious legislation.' (Lucas 1879*a*) In 1878 Lucas was elected to membership of the British Association and in August he read a paper at the Dublin meeting (Lucas 1879*b*) In September a further paper appeared in *Nature* (Lucas 1878*f*). In each case the value of a survey of groundwater resources was emphasized, the tangible product of which would be a map showing the 'capabilities of each river basin' (Lucas 1878*f*, p. 495).

The Dublin paper was published in full along with other contributions given at a special session devoted to 'Rivers Conservation' (Lucas 1878*g*). In opening the session Edward Easton, the President of the Mechanical Engineering Section suggested that:

> 'A new department should be created – one not merely endowed with powers analogous to those of the Local Government Board, but charged with the duty of collecting and digesting for use all the facts and knowledge necessary for a due comprehension and satisfactory dealing with every river-basin or water-shed area in the United Kingdom – a department which should be presided over, if not by a Cabinet Minister, at all events by a member of the Government who can be appealed to in Parliament' (Easton 1878, p. 21).

Lucas did not agree with the President over this considering that:

> '. . . history and experience show that the sudden creation of large institutions has in the past proved to be a failure. Everything that suceeds develops itself gradually . . . it must begin in some comparatively small measure, which shall contain within itself the essential faculties for self-development.' (Lucas 1878*g*, p. 113)

He envisaged starting from a simple survey from which would gradually evolve a system of management and administration which would ultimately take the form of a Water Board.

In May 1879 a further Congress to discuss National Water Supply was held at the Society of Arts. At this meeting the Society offered a gold medal and three silver medals 'for the best suggestions, founded upon evidence already published, for dividing England and Wales into watershed districts, for the supply of pure water to the towns and villages of each district'. Lucas submitted an essay (Lucas 1879*c*), for which he was awarded a silver medal: one other siver medal was awarded but no gold medal. Six of the essays were published, five of which suggested that boundaries should be drawn along watersheds to form areas containing one or more river basins. Lucas proposed six districts making the further suggestion that they should be grouped into three to form northern, midland and southern regions across the country (Fig. 2). This meant that each region had both easterly and westerly flowing rivers, mountains and lowland, as well as a share of both the wettest and driest areas. It was not to be until 1965, some 85 years later, that large units were formed for water supply purposes, and there is a valid argument that Lucas' scheme had advantages over that eventually adopted. At the 1879 Congress Lucas also presented a paper comparing the surface and subsurface catchment boundaries on the Chalk from the Medway through Surrey and Hampshire to the south coast (Lucas 1879*d*). He demonstrated that differences in position between 'surface and subterranean water-shed ridges' meant that there were considerable differences between the real area of chalk contributing to many catchments and the apparent area calculated from the surface topography.

In 1879 Lucas also read a paper to the British Association, this time at its meeting in Sheffield (Lucas 1879*e*). Entitled 'On the Quantitative Elements of Hydrogeology' it discussed percolation and made recommendations on procedures to be adopted when making future observations. The following year at the Congress of the Sanitary Institute he gave another paper with an almost identical title in which he discussed other quantitative aspects of his subject, in particular the drawing of contours and sections (Lucas 1880*a*).

The last of Lucas's major regional papers, describing the hydrogeology of the Lower Greensands of Surrey and Hampshire, was published by the Institution of Civil Engineers in 1880. It summarized 3 years of observations and described the contrasting properties of the Hythe, Sandgate and Folkstone Beds (Lucas 1880*b*). The paper discussed the folding and faulting which affect groundwater flow and the various ponds some of which date back to Roman times. The paper included a superb hydrogeological map of the Lower Greensand from Dorking in the east to Farnham in the west at a scale of one inch to a mile and resulted in the award of a Telford Medal and Telford Premium to Lucas by the Institution.

By now Lucas must have become increasingly frustrated at the lack of action on his and many others' call for a national survey to collect data which could be used to improve water supplies. The situation was particularly bad in London, where Lucas lived, and in November 1880 he directed a polemic at the local ratepayers. He considered that it was his duty 'to make plain to the uninitiated public what is already well known to scientific men and engineers, that no real progress whatever can be made in this Water question until a proper Survey has been taken of the sources of supply.' (Lucas 1880*c*, p. 4). He emphasized that the survey should be made by, or on account of, the Government so that it would be impartial and also that if London wanted a good water supply it would have to pay for it. He calculated that '. . . the Londoner is paying one third for his water of what he would have to pay if he lived in a well-favoured village in a gravitational district, among the great Devonian hills of Exmoor; but in the latter case the quality of the water is perfect, whereas in London it is not so.' (Lucas 1880*c*, p. 13). He concluded that:

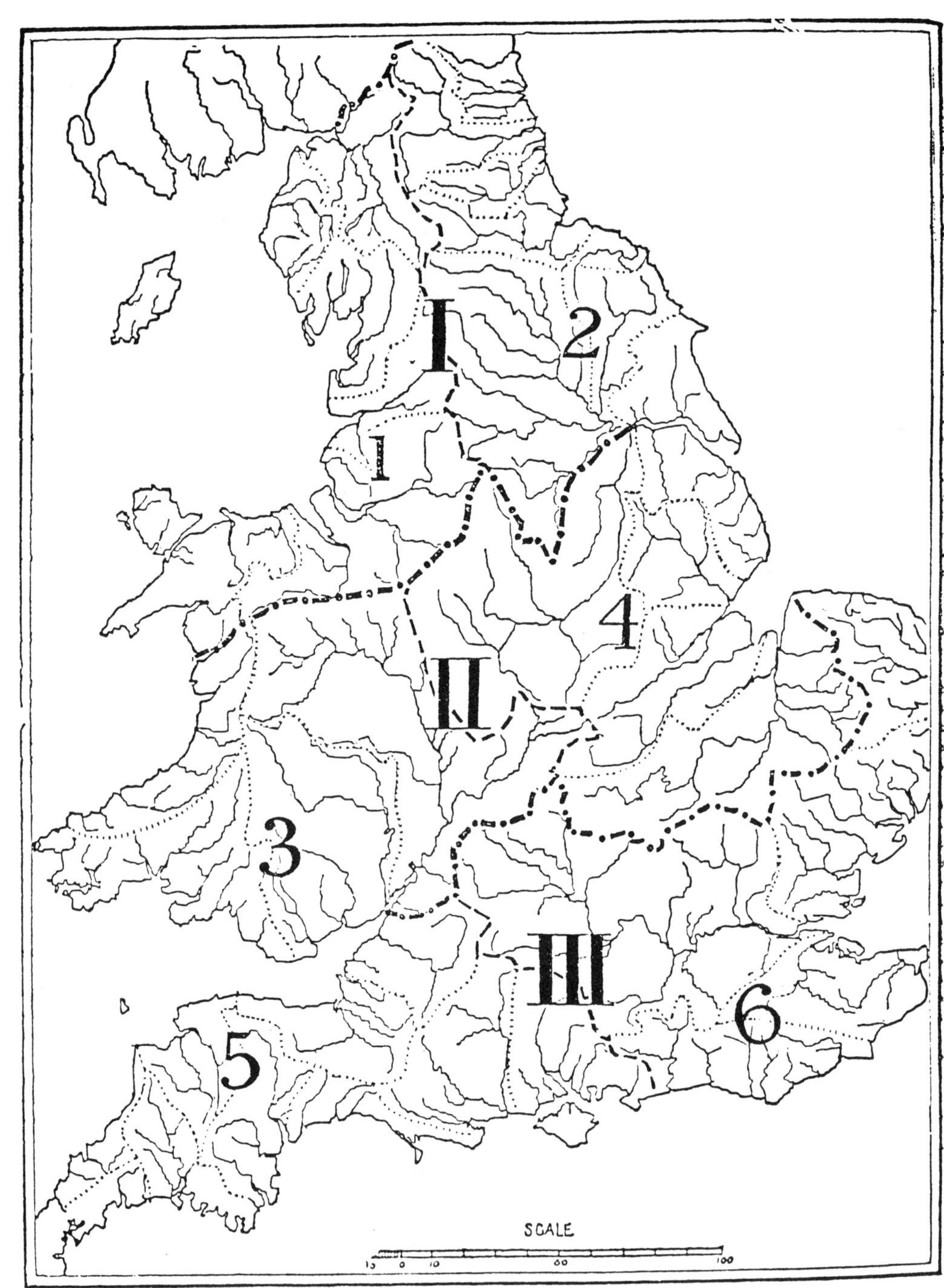

MAP OF DISTRICTS PROPOSED BY JOSEPH LUCAS.

Boundaries of Commissioners' Districts (I, II, III), thus :— ▪ • ▪ • ▪
,, of Sub-Commissioners' Districts' (1, 2, 3, 4, 5, 6), thus : — — — —
,, of Groups of Basins, thus :— ......

Fig. 2. Map of watershed districts proposed by Lucas in his presentation to the National Water Supply Congress in 1879 (Lucas 1879*c*), for which he was awarded a silver medal.

'1. the villager pays for quality
2. London has not yet spent enough money on water to obtain a good article
3. London can well afford to go further afield in search of pure water
4. London does not know where to go for it
5. Government should make a proper survey of the sources of supply, not only for London, but in the interest of the nation at large
6. when proper sources have been found, London still has a large margin left before spending so much on water, in proportion to rateable value, as has actually been disbursed by many of our poorest villages.'

In January 1881 Lucas read another paper at the Surveyor's Institution which took up the issue of rural water supply (Lucas 1881*a*). The Public Health Act of 1878 put a requirement on every Rural Sanitary Authority to see that every occupied dwelling house within their district had an available source of wholesome water. Lucas pointed out that as rural supplies were mostly derived from springs and wells the Authorities required a knowledge of the hydrogeology of their districts. Then came yet another plea for a Government funded survey of the whole country to provide the necessary information and the setting up of Watershed Boards to administer water supplies.

Later in the year, as his call for a Government survey continued to fall on deaf ears, Lucas decided that he would try and collate information on water levels from observers in different parts of the country with a view to publication. Accordingly on 21st April he wrote to *The Times* (Lucas 1881*b*):

'Sir, – As there are now a good many observers of the variations of water-levels in wells throughout the country, and as such registers are of great value in questions of water supply, especially when taken in conjunction with rainfall observations, it would be desirable that all these observations should be brought together and others added to them for the purpose of publication at such annual, monthly, or other regular intervals as may be found to be most economical. May I ask whether you will be so kind as to give publicity to this important subject by means of your widely-circulated journal, and to allow me to say that I shall be happy to receive any such registers, and to supply owners of wells with proper forms and a most simple and inexpensive gauge for the purpose.'

He received seven replies to his letter and initially observations from six wells in Dorset, Berkshire, Surrey, Cambridgeshire and Hertfordshire. These were published in September 1881 as number 1 in a series entitled 'Wells, Springs and Rivers of Great Britain', together with an extended preface by the editor (Lucas 1881*c*). In this preface Lucas makes reference for the first time to the cost of the surveys which he had carried out for some seven years, commenting that 'the rate of progress of the field work soon outstrips the legitimate appropriations of one individual towards the expenses of publication.' and 'Material support has never been asked for, but it is evident that if this interesting work is ever to expand to a scale of existence which shall be at all commensurate with its dignity and importance, material support will have to be forthcoming.' (Lucas 1881*c*, p. 3). He publicized this work in a paper given to the British Association in York on September 7th 1881, emphasizing that he would publish returns monthly until some better arrangement could be made (Lucas 1882*a*).

Number 2 was issued in October 1881 and included articles on 'East Brent Waterworks' by George Anthony Denison and 'Artesian Wells' by James Clutterbuck as well as a table of measurements (Lucas 1881*d*). The 3rd and last number was issued in December 1881 and the editorial commenced with the words 'A valuable series of river gaugings will commence in No. 4. They are all taken at mills, weirs, or gates, the dimensions of the waterways having been all carefully measured by the Editor.' (Lucas 1881*e*, p. 25). Clearly it was the intention to publish further issues but no more appeared, perhaps because there were too few observers and/or because at a price of 6d per issue there were insufficient sales to make production economic.

During his mapping work in Nidderdale and other parts of Yorkshire Lucas had made what he describes as 'voluminous notes' which he had began to write up culminating in the publication of a book *Studies in Nidderdale* in 1882 (Lucas 1882*b*). His geological field slips have notes in pencil on the birds seen during his field mapping and occasionally comments such as 'Hooded crow Oct.28th/70' are inked in as if they have some geological significance. The first paper appeared in September 1879 in *The Zoologist* in two parts; the first gives a general introduction to Nidderdale, the origin of some of the names and descriptions of the flora. The second describes the ornithology of the district. His interest in philology – the science of language – emerges for the first time and there is considerable discussion on the origin of many of the place names (Lucas 1879*f*).The subject of the Nidderdale flora was taken up in a further paper presented to the British Association meeting at York in 1881 (Lucas 1882*c*), and published by the Yorkshire Geological and Polytechnic Society (Lucas 1881*f*). In the paper he considers evidence that the last vestiges of ancient forests can be identified in Nidderdale, concluding that 'a lowering of some hundreds of feet has taken place in the forest line within the last thousand years. (Lucas 1881*f*, p. 373).

Studies in Nidderdale (Lucas 1882*b*) consolidates information from previous papers, whilst adding a wealth of new observations, to produce a rather disjointed collection of antiquarian, zoological, botanical and philological information. The following paragraph is typical of the style used (Lucas 1882*b*, p. 28):

> 'The *Bukker, Bink* or *Binch*, is a large flagstone "which is leant against the side of a wall," and is used to "bray" sand upon for floors. The name *Bukker* (*pron.* Booker) is here misapplied, as it properly belongs to the instrument with which the sand is brayed. Swed. *Bokare*, a *breaker, Boka*, to bray sand – whence Fr. *Bocarder*. It is probably from *Bok*, beech, the original Bukker being a beech stump, from its hardness'. Etc.

The book includes a glossary of more than 1000 words, chiefly collected between 1867 and 1872, of the dialects of Nidderdale. Many references to this list appear in the Oxford English Dictionary and Lucas is credited under words such as sumpy, tiffany, ruckle, hurlot and clint as well as hydrogeology. The book also includes an epic poem, written by Lucas and based on the story of the Shepherd Lord, Henry de Clifford, who lived as a shepherd for some 30 years after forfeiting his estates after the battle of Towton in 1460. However, lines such as;

> 'The battle had been lost and won!
> Into Barden Tower there ride
> Lord De Clifford and his bride.
>
> Sayes, "now the race is over
> Let the good stead eat his clover."'

(Lucas 1882*b*, p. 204) do not suggest that Lucas was a naturally gifted poet.

At about the same time Lucas became interested in the gypsies and following a historical study of the history of the petty Romany kingdoms, dukedoms and clans scattered around Europe outside Romania (Lucas 1880*d*) he published a small book developing his ideas on their history (Lucas 1882*d*). This is again primarily concerned with language and the origin of words. Lucas records that he visited the gypsy settlement of Yetholm (to the north of the Cheviot Hills) and Queen Esther faa-Blyth, to whom the book is dedicated, in March 1880. The text traces the origin of the various gypsy tribes back to India and shows how many of their words have their origins in languages such as Sanscrit and Hindustani.

By the end of 1882 Lucas had a wealth of published material to his credit in the form of papers, books and maps. He now had four more children; Joseph (1877–1956), Claud Vernal Warter (1879–?), Janet Dorothy (1880–1937) and Miran Cathlin (1882–1960). The 1881 census records that there was also a 24-year-old servant, Sarah Clark, living in the house. His profession is recorded as 'Scientific Hydrogeologist' and he is almost certainly the first person to use hydrogeologist as his job description.

Although residing at Tooting Graveney the address which he often used was 21, Surrey Street, Victoria Embankment, which was the business address of his father. It was presumably from here that he operated the consultancy business by which he had gained an income since leaving the Geological Survey. A list of projects with which he was involved is given in Lucas (1888) but few of the reports which he must have produced have survived or were printed and all but one date from after 1882. The exception is a report to the Directors of the Southwark and Vauxhall Water Company commenting on possible sites for wells within their district which dates from 1878 or maybe 1879. This report was subsequently bound with other geological reports and submitted to the Royal Commission on Metropolitan Water Supply some 14 years later to support their evidence (Southwark and Vauxhall Water Company 1892). Lucas was later cross-examined by the Commissioners on his estimate of the long term yield of a deep borehole in Streatham sited on the basis of his report (Royal Commission on Metropolitan Water Supply 1893, p. 91–94). Some of the other projects with which he was involved are used as examples in Lucas (1881*a*) and include work for Breweries and Water Companies. His Hydrogeological Maps were sold through Stanfords at a price of 21s (a guinea) per sheet and must have sold reasonably well as they were recorded as being out of print for many years in 1893 (Royal Commission on Metropolitan Water Supply 1893).

Another source of income came from his appointment as Examiner in Geology for the Professional Examination of the Surveyor's Institution. He had been elected a Professional Associate of the Institution on 14th May 1877 some 10 weeks after reading a paper at one of their meetings (Lucas 1877*b*). His appointment as an examiner dates from 1881 and the syllabus for the course is given in the Transactions as follows;

> Classification of rocks as aqueous and igneous
> Aqueous formations – Succession of the strata of
> – Characteristic fossils of
> Explanation of geol terms such as dip, strike, synclinal, anticlinal, escarpment etc
> Agents of denudation, disintegration and chemical decomposition of rocks
> Various conditions of deposition in fresh water and the sea
> Glacial drift
> Water-bearing strata and flow of u/g water
> Artesian wells

The recommended text books were: Lyell – Students elements, Geikie – Primer on Geology and Text Book of Geology, Jordan – Table of Strata, and Ramsay – Geological Map.

The water engineer 1883–1911

By the close of 1882 Joseph Lucas had just turned 36 years of age and had made a considerable impact as a hydrogeologist – a subject which he himself had more or less invented. He should have had many more productive years in front of him. However, as an obituarist recorded 'although he continued to practice as a water engineer for some years, he ceased after this to make any further publications of general scientific interest' (Anon 1926*b*). Was this because he ran out of ideas, became frustrated with the lack of any response to his lobbying, or more simply because he ran out of money?

There are a number of lines of evidence which suggest that the latter may at least have been a contributory factor. Firstly, the comments about the need for material support in his first editorial in *Wells, Springs and Rivers of Great Britain* (Lucas 1881*c*) and the fact that this publication died after only three issues. Secondly, the failure to issue the third sheet of his Hydrogeological Survey, which he apparently had prepared for publication in 1877 (Lucas 1877*e*, p. 4). Thirdly, he was removed from Fellowship of the Geological Society to which he had been elected in 1867 soon after joining the Geological Survey. He is listed in the 1881 list of members, but his name appears in the report for 1882 among those 'removed'. The usual reason such action was taken was for non-payment of fellowship dues. There is also no doubt that his publishing activities would have been expensive and after 1882 these were curtailed significantly. There were no more presentations at the annual meetings of the British Association, although he did not give up his membership of the Association until 1906. He also ceased to be a Fellow of the Meteorological Society which he had joined in the 1870s.

Certainly if there was a financial crisis it was not one that affected his family as a whole as they owned a lot of valuable building land in the Tooting area. He records in his report to the Southwark and Vauxhall Water Company in 1878/1879, when discussing the availability of land in Tooting for the site of a well, that 'I believe a site of half-an-acre may be had for £1000, but my father is better informed, being the owner' (Southwark and Vauxhall Water Company 1892). He himself owned property in Hornsey, which he may have inherited from his grandparents, and which was still in his possession at the time he made out his will in February 1912. Presumably he would have sold this in 1882 had he been under severe financial pressure.

Whether or not he suffered some sort of financial crisis is unclear. However, 1882 certainly marked a watershed in his career, so much so that the 1910 Library Catalogue of the British Museum (Natural History) suggested that he died around 1881. This was partly because, in most of his early publications, he referred to himself as Joseph Lucas, FGS. When he dropped the FGS after 1882, he lost this identity and although the Museum Catalogue of 1933 corrected his date of death it now listed two separate people, Joseph Lucas, FGS (1846–1926) and Joseph Lucas of Tooting Graveney.

Lucas continued to live in Tooting Graveney where the size of his family grew with the birth of Ronald (1883, died within a few days), Ronald Fairfax (1884–1973), Maud Elizabeth (1886–1971) and Patience (1888–?). His consultancy work took him around the country and the reports which he produced (listed in Lucas 1888) show him working near Whitby, Leeds and Alnwick in the north, Bath and Cirencester in the west as well as many areas of southern England. Although most of his work involved the evaluation of water resources he also advised on iron ore workings as far afield as Shetland and North Wales as well as coal prospects in Durham and Oxfordshire. Unfortunately, most of his reports were made in manuscript and will need to be hunted down in archive offices. However, a few were printed, presumably as a small number of copies by the customer, and demonstrate the type of work in which he was engaged.

In a document dated December 23rd 1885 Lucas reports on a survey of the quantity of water which might be obtained from gravels on the North Frith Estate near Tonbridge in Kent (Lucas 1885*a*). He mapped out the extent of the gravels, suggesting that they were not river gravels but rather subaerial deposits composed of the debris of Weald Clay Beds from which the finer material had been washed out. He notes the existence of a number of springs and 'lines of ooze' at the base of the gravels and calculates the yield which might be obtained.

In another report he reviews the prospects of finding coal under the estate of Lieut-Colonel William Gregory Dawkins (1825–1914) at Over Norton in Oxfordshire (Lucas 1891*a*, *b*). There are two editions of this report, the first dated January 23rd 1891 with amendments on the second dated February 26th 1891. Lucas outlines the evidence for the existence of Coal Measures beneath Over Norton plotting two sections, one east/west and one north/south-west (Fig. 3). These led him to conclude that there was an 'absolute certainty' that the Coal Measures were present at a depth of about 1000 feet beneath the surface. The thickness of the Middle Coal Measures was estimated to be 500 feet which were likely to contain four or five beds of coal with a combined thickness of perhaps 15 feet. The first

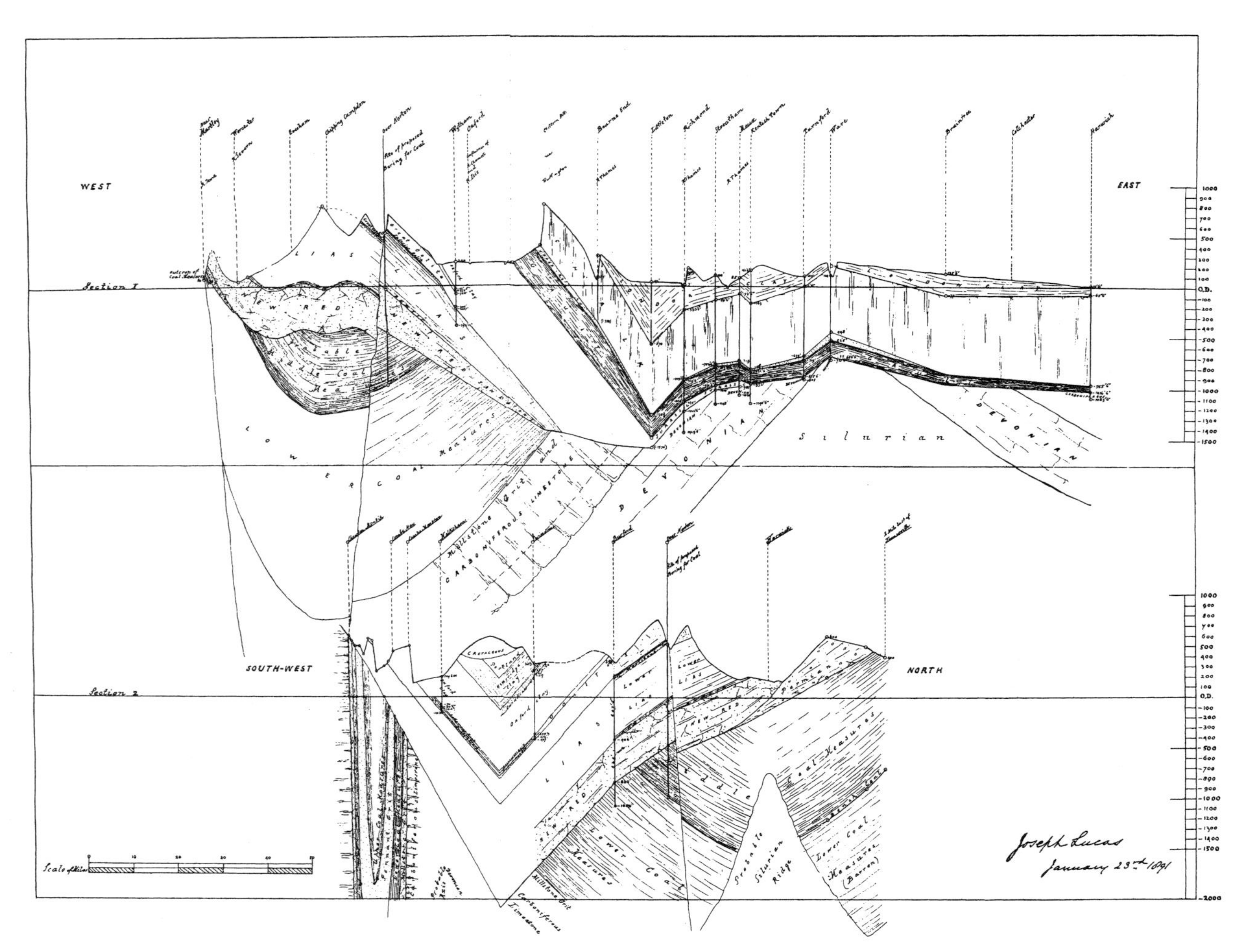

Fig. 3. Geological sections from the report to Lieut.-Colonel W. G. Dawkins on the prospects of finding coal at Over Norton, Oxfordshire (Lucas 1891*a*, *b*).

edition of the report proposed a site for a borehole; this was moved in the 2nd edition to bring it closer to a brook from which a continuous supply of water could be obtained for drilling. The 2nd edition also contains an interesting specification for the work in which an inclusive estimate for drilling costs, to include mobilization, provision of temporary casing, etc., is given as £1 per linear foot for the first 500 feet rising to £4 per linear foot between 1900 and 2000 feet below the surface (Lucas 1891*b*).

The reports on both the North Frith and Over Norton Estates are well written and are competent pieces of work. A further printed report is listed by Lucas (1888) relating to work carried out for the Alnwick and Canongate Local Board of Health, in Northumberland, on the prospects of boring for water at Rugley Wood and Thorntree Well. Although this report has been referenced (Lucas 1885*b*) a copy has not been located, despite considerable research by the authors.

In July 1884 the Society of Arts held another conference on Water Supply, this time in conjunction with the International Health Exhibition. Lucas presented a paper on water from the Chalk in which he returned to his call for a national survey of the sources of water supply and gave examples to show why such a survey was necessary and important (Lucas 1884*b*). He records in this paper that he had been laid up from ill health for a few months but no details of his illness are given. Two years later in a short note published by the Surveyor's Institution he gives details of some recent borings for water with which he had been involved including a scheme for a water supply for a new country house designed by the London architect Benjamin Edmund Ferrey, who was also his first cousin (Lucas 1886*a*). The only other original article which he published during the 1880s was on the lead mines of Grassington in Upper Wharfedale in Yorkshire (Lucas 1885*c*). This is a short historical account and a description of a visit made to the deserted workings with an old miner. It would appear that Lucas was also continuing to make observations on the Chalk as in 1888 he records that a description of the Chilterns, together with a map of the hydrogeology of the Chalk country between the Thames and the Cam, 70 miles in length and 1100 square miles in area, based on 8 years observations, awaited publication (Lucas 1888). However, this never appeared.

During the 1880s Lucas had further developed his interest in languages by undertaking a series of translations, some of hydrogeological works but many of interest to a wider audience. Works were translated into English from Norwegian, Swedish, Danish, Dutch, German, French and Russian. A list of the most substantial of these translations dating from before 1888 is given in Lucas (1888). His brother Edgar had married Alice Erichsen whose parents were Danish and resided at Upper Tooting. Under the name of Mrs E. Lucas, Alice did one of the first and most popular translations of Hans Christian Anderson's fairy tales from Danish into English and must have been of some help to Joseph in his work on the Scandanavian languages. He also records that he had travelled in Norway, the Hartz Mountains, Switzerland and the Italian Lakes but his knowledge of some of the languages which he translated must have been derived from books and perhaps visitors from overseas. Some translations from the Norwegian were made in 1872, about the time he travelled in Norway according to references in *Studies in Nidderdale*, but most date from 1885 and subsequent years.

Printed copies of two of these translations survive. In 1886, on behalf of the engineering consultants Joseph Quick and Sons, he translated a series of papers relating to the water supply of Moscow. One extract on the *Geology and Hydrogeology of Moscow* by Aleksyei Nikolaevich Petunnikoff was printed (Lucas 1886*b*). Marked 'Private and Confidential. Unpublished' only 50 copies were produced, presumably for internal distribution within the company. However, his most quoted translation and the work for which he is best known outside hydrogeology is the account of the visit to England by the Swedish naturalist, Pehr Kalm (Lucas 1892). According to Lucas (1888) he had translated this into 5 manuscript volumes as early as 1886 but it was not published until 1892 by his relatives the Macmillans. Pehr Kalm, a pupil of Linnaeus, had been commissioned by the Swedish Government in 1747 to visit North America in order to report on any useful plants found there which might be expected to thrive in northern Europe. On his way to and from America Kalm spent some weeks in England staying at Gravesend, London, Woodford and Little Gaddesden in Hertfordshire, studying the social life and natural history of the districts. The English portion of his journeys had never been translated until completed by Lucas who did not follow Kalm's day-to-day arrangement but grouped the information according to subject matter. A reviewer in *The Times* thought that the translation was of great value as a contribution to the history of rural economy and that it was skilfully executed. However, he noted that 'his habit of interlarding his text with words and sentences in the original Swedish is rather tiresome.' In his translators preface Lucas records his thanks to his eldest brother, Frederic William, who had contributed a number of historical notes to the text. Frederic himself was an historian who published two books on the discovery and exploration of North America (Lucas, 1891, 1898) and may also have assisted him with his translations.

During this period of his life Joseph Lucas addressed a number of letters to *The Times*. His letter on fishponds has already been referred to (Lucas

1884*a*), but he also wrote about the northern limit of the birch and beech in Europe (Lucas 1887), London's water supply (Lucas 1891*c*) and, under the pseudonym 'Hydrogeologist, Tooting Graveney', two letters on artesian wells in south-west London (Lucas 1884*c*). The latter referred to a deep experimental well in Streatham, pumping of which had cut off the supply of water to artesian wells in Tooting. The tone of the 1887 letter is particularly overbearing, full of quotations from obscure publications in a variety of European languages.

When the position of Professor of Geology at Oxford became vacant following the retirement of Joseph Prestwich, who himself had made major contributions to the subject of water supply, Lucas stood for the position. His printed application, dated January 24th 1888, consisted of 16 testimonials, a memorandum outlining his career, together with an appendix listing his geological and hydrogeological books, maps and papers and certain literary papers and translations (Lucas 1888). The testimonials come from a variety of sources including civil servants, polititians, lawyers, civil engineers and surveyors for whom he had worked, Whitaker, his old colleague in the Geological Survey, Prestwich himself, and two of his cousins Benjamin Edmund Ferrey and George Ferris Whidbourne, later famous for his work on Devonian fossils. Lucas states in his memorandum that he had been advised to stand for the chair but this advice was perhaps ill-conceived and probably assumed the post to go again to someone with an interest in water supply. The Electors must surely have been made aware of his past history with the Geological Survey and his lack of Fellowship of the Royal Society might also have stood against him. In the event the post went to his old mapping colleague in Yorkshire, Alexander Henry Green (1832–1896), who had left the Survey in 1874 on his appointment as Professor of Geology in the Yorkshire College at Leeds and who had been elected FRS in 1886.

In 1894 Lucas again demonstated his wide interests with the publication of a slim volume of some 41 pages, containing translations into Latin of a number of poems. (Lucas 1894). These included Gray's Elegy (translated as *Elegia in coemeterio rustico scripta*), Serenade by Shakespeare (*Verbo Mane Salutandi*) and Alexander Selkirk by Cowper (*Alexandri Selkirk in insula Juan Fernandes Soliloquium*). The first line of the latter 'I am monarch of all I survey' was translated *as 'Omnia quae video regione monarcha guberno'*. He was clearly interested in such translations as he notes in the preface that he had met with eight published Latin versions of Gray's Elegy. He records that four of his translations dated from 1890, whilst all the others were made after March 1893.

Two years later he published what was to be his most monumental work, on the genealogy of the family of Bayne of Nidderdale (Lucas 1896*a*). According to the prospectus for this work, which is bound into the back of the copy of the book held in the British Library, it was originally intended to have formed part of his *Studies in Nidderdale* (Lucas 1882*b*). However, the difficulty of obtaining the materials for such a study meant that it was another ten years before, with the support of John Baynes of Ripon, he was able to return to the work. The book was published in an edition of 200 numbered copies but at 635 pages was only 'a portion of the work as matured'. The rest was to be issued in a second volume which in December 1886 was already in manuscript. Like many of his previous publishing ventures this second volume never appeared. The book contains lists of pedigrees, details of deeds and comprehensive surveys of estates formerly owned by various branches of the Bayne family in Nidderdale. The tradition was that the family was descended from Donald Bane, King of Scotland from 1093–1097 through the migration to Nidderdale from Scotland in 1182 of one Walter, 5th in descent from Donald Bane. Through a detailed study of coats of arms, Lucas was able to demonstate that the Nidderdale family was in fact descended from the family of the Norman Baron Hugo de Bayeux who lived in the reigns of Henry II and Richard I. The documentary evidence which he assembled to prove his point was impressive, although why he became interested in the Bayne family which seems to have had no link with his own is unclear. The book is not for the casual browser and a real interest in genealogy is required of the reader. As a review in the *Leeds Mercury* of 20th January 1897 states 'Mr Lucas . . . is a stickler for facts, not the facts of his memory, but of registers, records, seals, arms, etc., and that we presume, ought to be considered as the chief virtue of a book of this kind.' 20 copies were also printed of a slimmer volume which reproduced the principal genealogical trees of the family of Bayne of Nidderdale (Lucas 1896*b*).

Throughout this period of his career Lucas maintained his membership of the Surveyor's Institution and even the title page of his book on 'Bayne of Nidderdale' styled him as Joseph Lucas, P.A.S.I. (Professional Associate of the Surveyor's Institution). He was a regular attender at their meetings and took part in the discussion of papers on geology and water supply. In the 1885–1886 session he seconded the vote of thanks following a paper on the 'Geology of the Surface' (Fream 1886, p. 392–398). Fream had used 'alluvia' as a plural of 'alluvium' and with his philological background Lucas took exception to this, examining the derivation of the word from its Latin root. He also contributed to discussions following papers by Grantham (1888, p. 202–205); Shires (1900, p. 296–300); Middleton (1901, p. 131–135) and Grahame & Bidder (1907, p. 344).

Fig. 4. Sketch of Shakespeare's Cliff, made in 1863 when Joseph Lucas was 16 years old and a pupil at Westminster School (from Lucas 1908*a*).

He amplified his comments on the paper by Grahame & Bidder in the Professional Notes of the Institution (Lucas 1908*a*). He considered that the expression 'known and defined channel' should never be applied to the movement of underground water and argued strongly that the presence of a recurrent bourne did not establish the presence of an underground stream. He thought that the expression 'vast storage reservoir' applied to an artesian system was absurd, giving as a example the declining yield of the Wandle Basin around Tooting where he lived. He also noted that the height of bench marks on 6 inch maps published in 1895 were about one and a quarter inches below those given on older maps, suggesting that this could be the result of 'the liberation of so much water of support which formally upheld the surface of the ground . . .' (Lucas 1908*a*, p. 374), an early recognition of possible land subsidence resulting from groundwater abstraction.

The Professional Notes of the Institution also contained a section on professional queries and replies to which Lucas was a regular contributor. For example in 1901 Lucas reduced a question about the reservation of underground rights to a discussion on 'is water matter?'. Some years he would comment on 2 or 3 queries, but in other years he was either too busy or none of the questions were sufficiently interesting to provoke him into a reply.

In 1908 his final paper was published in the Transactions of the Surveyor's Institution (Lucas 1908*b*). In this Lucas discusses the geology and hydrogeology of the Upper, Middle and Lower Chalk (Fig. 4) together with the Gault of the Dover Basin and makes some comments on the East Kent Coalfield. At that time shafts were being sunk at Guildford, Tilmanstone and Snowdown Collieries and Lucas interpreted the evidence available as confirming his view, expressed in his reports to Lieut-Col. Dawkins (Lucas 1891*a*, *b*), that the Belgian coal basin made a great bend and turned northwest towards the Yorkshire Coalfield and not across to the Bristol Coalfield. However, as pointed out by the resident engineer to the Kent Coal Concessions in the discussion, '. . . the facts were equally available to prove a totally opposite conclusion'.

Fig. 5. Photograph of Joseph Lucas, probably in his 60s (photograph in the possession of the Lucas family, Vancouver, Canada).

The Final Years 1912–1926

The paper on the Dover Basin appears to have been Lucas's last publication of either a scientific or literary nature. He continued his membership of the Surveyor's Institution until his death but although he may have attended meetings there is no record of him contributing to the discussions, even to papers in which he might have been expected to have an imput. His final comment on a query in the Professional Notes was in 1911 when he replied to a question on riparian rights.

His wife died on the 31st January 1911 and soon after this, at the age of 65 (Fig. 5), he moved to live at St Leonard's-on-Sea in Sussex. His will, dated the 7th February 1912, just after his wife's death, describes him as 'of 11 Defoe Road Tooting Graveney, Surrey . . . and now residing at 22 East Ascent St

Leonards-on-Sea, Sussex'. The list of members of the Surveyor's Institution lists his address as 11, Defoe Road until 1914/1915 and as 22, East Ascent from then until his death. The house in St Leonard's-on-Sea appears to have been a lodging house and perhaps he did not move there permanently until around 1916 when he would have been 70 years old. East Ascent is close to the sea near the Marina and Grand Parade and number 22 is now a listed building within the Conservation Area. It would not have been an unattractive location for his final years.

His connection with St Leonard's-on-Sea is unclear. His wife and two of his daughters (Beatrice and Patience) were staying at a lodging house, run by a Miss Lucy Lock, at 17, East Ascent at the time of the 1901 census and it may be that the connections with St Leonard's were through her family. His son Claud went to live in St Leonard's but the timing of his move is uncertain. Claud was still living in Tooting Graveney in 1901 where he is listed on the census as articled to an accountant, so would not have been the reason for the visit of the three women at this time. However it is possible that Joseph later moved to St Leonard's to be close to Claud.

Of his nine surviving children, the eldest Millicent became a nurse and worked for some time at Great Ormond Street Childrens Hospital in London. However, in the 1901 census both she and her brother, Joseph are listed as actors. According to one family record Joseph eventually became a civil servant and his father notes that in 1908, when he would have been just over 30 years old, he was maintaining a record of measurements of wells in the Wandle Basin (Lucas 1908*a*) so was presumably still resident in Tooting. After this Joseph's whereabouts become obscure and it seems possible that his lifestyle may have been too unconventional for the rest of his family. Beatrice became a civil servant, working in the Scottish office in Whitehall. Claud qualified as a chartered accountant and the youngest son Ronald emigrated to Canada in about 1910.

The four youngest daughters all became musicians, studying in Prague and Vienna. On their return to England they formed a string quartet, The Lucas Sisters, giving their first concert at the Bechstein Hall in 1909. A review in *The Times* of 25th May 1909 was complimentary reporting that 'They have been so thoroughly trained in the special art of quartet playing that their ensemble is remarkably well finished in every detail, their tone is pure and beautifully balanced, and while they have plenty of vigour at command it is in the cantabile movements that they specially excel'. As the elder sisters either died or retired the two youngest, Maud and Patience, joined the BBC Symphony Orchestra where they ended their careers.

None of the daughters married and of the sons only Ronald, who married Norah Brighty in Edmonton, Alberta in 1915, had children. Ronald's son, also Ronald Fairfax Lucas and now aged 88, and daughter Katherine, now Mrs Power and aged 85, live in British Columbia. One of Ronald's two daughters, Katherine Jane, a co-author here, has never married and will be the last of the descendants of Joseph Lucas to bear the Lucas name.

Fig. 6. Photograph of the grave of Joseph Lucas, his wife Elizabeth and daughter Janet in the Churchyard of St Nicholas, Tooting Graveney (photograph by J. D. Mather).

Joseph continued to live in St Leonard's-on-Sea until his death on 20th April 1926. He is buried along with his wife and daughter Janet in the churchyard of Saint Nicholas Tooting amongst a group of graves of similar age, immediately to the south of the Church (Fig. 6). The gravestone, of quartzitic sandstone, was chosen well and has escaped damage from the acid deposition which has affected adjacent limestone monuments. Obituaries were published in *Nature* (Anon 1926*a*) and in the *Journal of the Surveyors Institution* (Anon 1926*b*). A brief acknowledgment also appeared in the *Proceedings of the Geological Society* (Blather 1927) and in *The Times* of 21st April.

Discussion and conclusions

In Hitchin Museum, in a boxfile of notes made by Reginald Hine when compiling his book *Hitchin*

Worthies (Hine 1932), is a hand-written note describing the character of some of the male members of the Lucas family, '. . . some of us are rash and reckless talkers who let their tongues run away with them. Many Lucases were very shy and some even taciturn who took all but gave nothing in return and not so much dropped a subject but smashed it.' If an uncommunicative nature was a typical Lucas trait, Joseph was atypical. He put his views across forcibly and was not afraid to disagree with even the most eminent speaker. When he joined the Geological Survey as a bright young geologist in 1867 there must have been every expectation that he would prosper. However, he was not cut out to be an establishment man. His concentration on unofficial work to the detriment of his official duties soon ensured that he attracted the wrong sort of attention. When reprimanded he seems to have made no changes to his behaviour and by ending up assaulting his senior officer, took insubordination on to a higher plane.

His anti-establishment stance is also evident from some of his publications. For example in commenting on two bills recently introduced by Parliament in 1878 he stated:

> 'Is it likely that the landowners will consent to be cajoled by the first, or compelled by the second, to go to the expense of making reservoirs or wells,when they see no provision for the aquisition of that technical knowledge of the natural sources that has led to so many disastrous failures in the past? As is usual in these cases, the collection of this class of knowledge begins with the individual, whence it is first reflected upon the interested professions and afterwards upon that all-powerful minority which has been called the "landed aristocracy". This important body includes in its numbers the Capitalists and the Legislators of the country, so that when it once becomes stamped with an impression it is not long in transmitting it to the Government.' (Lucas 1879*a*, p. 519).

Comments such as this and his polemic to the ratepayers of London (Lucas 1880c) suggest that he was leaning towards socialism in his political views.

Lucas also appears as very much a lone worker. Once he left the Survey there is no evidence that he worked with anybody else. He is the sole author of all his published papers and printed reports. Although he belonged to Societies and Professional Bodies he never became involved in their governance and, as far as is known, was never invited to take part in the deliberations of Government Committees or Commissions. In 1874 the British Association formed a Committee to investigate the circulation of underground waters which continued for some 20 years. The Committee was dominated by his old Survey colleagues. Hull was the Chairman, De Rance the Secretary and the members included Whitaker, Green his old mapping colleague and Howell with whom he had the fracas in Yorkshire. Lucas never became involved despite his considerable expertise, which was far superior to that of most members of the Committee.

Reading his papers one is left with the impression of a difficult and perhaps rather arrogant man. His application for the chair at Oxford (Lucas 1888) shows that he had a high opinion of his abilities and of the value of his work. However, he seems to have been an individualist with little respect for the views of others. This is clear from his contributions to discussions, particularly in his latter years, when he used expressions such as 'fallacious' and 'absurd' to describe concepts with which he did not agree (e.g. Lucas 1908*a*).

He does seem to have had a mischievious sense of humour, christening his second son who was born on the 25th March close to the spring equinox, Vernal. Also, one of the translations listed in Lucas (1888) is by Platon Lucashevitch. Is this a reference to himself, identifying him with the great Greek philosopher?

The major puzzle with his career is the hiatus which occurred around 1882. Prior to this he produced a range of innovative maps and scientific papers, contributed to meetings all over the country and generally established himself as the leading figure in the new science of hydrogeology. After 1882 the flow of innovative papers stopped, he rarely contributed to meetings except those held at the Surveyor's Institution and concentrated much more on his translations and philological and genealogical interests. It was suggested earlier in this paper that financial problems may have contributed to this change in direction, with a need to earn his living becoming of higher priority than continuing his pioneering scientific work. However, the evidence for this view is not overwhelming. It is also possible that the death of his infant son Ronald shortly after birth on 25th January 1883 put a strain on his family and that he chose to curtail his scientific work and spend more time at his home in Tooting.

Certainly, Joseph Lucas made a significant contribution to the application of geology to the study of groundwater and water supply. His colleague Whitaker emphasised this in the testimonial he wrote when Lucas applied for the Oxford chair, 'I think that your career is a most marked instance of the successful application of geological knowledge to practical purposes . . .' (in Lucas 1888). Lucas was the first person to use the term hydrogeology in a modern context and almost certainly the first to describe himself as a hydrogeologist. His hydrogeological maps were not bettered for another 90 years. His suggestions for a hydrogeological survey of the whole country were ignored by Government but if carried forward would have defined groundwater resouces and had a major impact on the provision of water supplies. As well as geology and hydrogeology his interests included natural history, philology, translation and genealogy and he made significant published contributions in each of these fields. He was a true Victorian polymath : well read, with wide interests and seemingly abundant energy.

Much of the family history in this paper is derived from a typewritten document, in the possession of the Lucas family in Vancouver, British Columbia, Canada, which chronicles the descendents of Richard Lucas buried at Hitchin, Hertfordshire on the 25th April 1572. The history, referred to by the family as the Lucas Book, was probably drafted in the 1920s and regularly updated into the 1930s. Sincere thanks are due to the staff of Hitchin Museum, Northumberland Record Office, The British Library, the Library of the Royal Institution of Chartered Surveyors, G. McKenna and R. Bowie at the British Geological Survey and in particular to W. Cawthorne at the Geological Society for their considerable help in locating reference material and obscure tracts.

References

All known publications by Joseph Lucas have been referenced in the text and appear in the following list. This includes all printed reports and tracts. Following references to the more obscure publications, an indication of where they may be consulted is given in [——]. Abbreviations used are as follows [BL] British Library, [BGS] British Geological Survey Library, [GSL] Geological Society Library.

ANONYMOUS.1910. Obituary – Reverend George Ferris Whidborne (1846–1910). *Geological Magazine*, **7**, 141.

ANONYMOUS. 1926*a*. Mr Joseph Lucas (Obituary). *Nature*, **117**, 730.

ANONYMOUS. 1926*b*. Obituary Joseph Lucas. *Journal of the Surveyor's Institution*, **5**, 670–671.

BAILEY, E. B. 1952. *Geological Survey of Great Britain.* Thomas Murby and Co., London, 278p.

BARKER, G. F. R. & STENNING, A. H. 1928. *The record of Old Westminsters.* Chiswick Press, London.

BARLOW, P. W. 1855. On some peculiar features of the water-bearing strata of the London Basin. *Proceedings of the Institution of Civil Engineers*, **14**, 42–65.

BLATHER, F. A. 1927. Anniversary Address of the President. Joseph Lucas. *Quarterly Journal of the Geological Society*, **83**, lx.

BRYANT, G. E. & BAKER, G. P. (eds) 1934. *A Quaker Journal Being the Diary and Reminiscences of William Lucas of Hitchin (1804–1861) A Member of the Society of Friends.* 2 vols. Hutchinson and Co, London.

CLUTTERBUCK, J. C. 1842. Observations on the periodical drainage and replenishment of the subterraneous reservoir in the Chalk Basin of London. *Proceedings of the Institution of Civil Engineers*, **2**, 155–160.

CLUTTERBUCK, J. C. 1843. Observations on the periodical drainage and replenishment of the subterraneous reservoir in the Chalk basin of London – continuation of the paper read at the Institution, May 31st 1842. *Proceedings of the Institution of Civil Engineers*, **3**, 156–165.

CLUTTERBUCK, J. C. 1850. On the periodical alternations, and progressive permanent depression, of the Chalk water level under London. *Proceedings of the Institution of Civil Engineers*, **9**, 151–180.

DE LA BECHE, H. T. 1839. *Report on the Geology of Cornwall, Devon and West Somerset.* Longman, Orme, Brown, Green and Longmans, London.

DELESSE, A. E. O. J. 1858. *Carte hydrologique de la Ville de Paris.* Publiée d'apres les ordres de M. le Baron G. E. Haussmann senateur, prefet de la Seine, Conformément a la délibération du Conseil Municipal du 8 Novembre 1857 et exécutée par M. Delesse ingénieur des mines inspecteur des carrieres du départment de la Seine.

DELESSE, A. E. O. J. 1862. *Carte hydrologique du Département de la Seine.* Publiée d'apres les ordres de M. le. Baron C. E. Haussman, Senateur, Prefet de la Seine.Conformément a la délibération de la commission Départementale et exécutée sur la carte topographique, Gravée sons la Direction de M. L'Ingénieur en Chef des Ponts et Chaussées, par M. Delesse Ingénieur de Mines du Départment de la Seine.

EASTON, E. 1978. Address by Edward Easton President of Section G. *In: Rivers Conservation. Address and Papers read before the British Association at Dublin, 1878.* P. S. King, London, 1–24.

FLETT, J. S. 1937. *The first hundred years of the Geological Survey of Great Britain.* HMSO, London, 280p.

FREAM, W. 1886. The geology of the surface in its practical aspects. *Transactions of the Surveyor's Institution*, **18**, 361–402.

GRAHAME, W. V. & BIDDER, H. F. 1907. Underground water a discussion of certain recent enactments affecting water rights. *Transactions of Surveyor's Institution*, **39**, 307–346.

GRANTHAM, R. F. 1888. Notes on water supply with special reference to villages and country mansions. *Transactions of the Surveyor's Institution*, **20**, 129–162 & 193–218.

HINE, R. L. 1932. *Hitchin Worthies.* George Allen & Unwin, London.

HOMERSHAM, S. C. 1855. The chalk strata considered as a source for the supply of water to the metropolis. *Journal of the Society of Arts*, **3**, 168–182.

HULL, E. 1865. On the New Red Sandstone as a source of water supply for the central towns of England. *Quarterly Journal of Science*, **2**, 418–429.

HULL, E. 1868. Chapter 12 The Bunter Sandstone as a source of water supply. *In: The Triassic and Permian rocks of the Midland Counties of England.* Memoirs of the Geological Survey of England and Wales, 115–120.

HURLEY, A. J. 1947. *Days that are gone, milestones I have passed in south-west London. Memories of fifty years of local newspaper work, public and social life.* A. J. Hurley, Tooting, 281p.

LUCAS, F. W. 1891. *Appendiculae historicae, or, Shreds of history hung on a horn*. H. Stevens, London.

LUCAS, F. W. 1898. *The annals of the voyages of the brothers Nicolo and Antonio Zeno: in the north Atlantic about the end of the fourteenth century, and the claim founded thereon to a Venetian discovery of America; a criticism and an indictment*. H. Stevens Sons and Stiles, London.

LUCAS, J. 1869. The Boulder-Clay and the Thames Valley. *Geological Magazine*, **6**, 188.

LUCAS, J. 1872. The Permian beds of Yorkshire. *Geological Magazine*, **9**, 338–343.

LUCAS, J. 1873*a*. On the origin of clay-ironstone. *Quarterly Journal of the Geological Society*, **29**, 363–369.

LUCAS, J. 1873*b*. On clay-ironstone. *Iron*, **April 26th**, 453.

LUCAS, J. 1874. *Horizontal Wells. A new application of geological principles to effect the solution of the problem of supplying London with pure water*. Edward Stanford, London. [BL, BGS, GSL]

LUCAS, J. 1876. The construction of wells. *The Architect*, **July 29th**. Reprinted in *Van Nostrand's Engineering Magazine*, 1876, **15**, 469–471.

LUCAS, J. 1877*a*. The Chalk water system. *Proceedings of the Institution of Civil Engineers* [for 1876], **47**, 70–167.

LUCAS, J. 1877*b*. Hydrogeology: one of the developments of modern practical geology. *Transactions of the Institution of Surveyors*, **9**, 153–184.

LUCAS, J. 1877*c*. The artesian system of the Thames Basin. *Journal of the Society of Arts*, **25**, 597–619.

LUCAS, J. 1877*d*. *Hydrogeological Survey, Sheet 1*. Edward Stanford, London. [BL]

LUCAS, J. 1877*e*. *Hydrogeological Survey, Explanation accompanying Sheet 1*. Edward Stanford, London. [BL]

LUCAS, J. 1878*a* The hydrogeology of Middlesex and part of Hertfordshire, showing the original position of the artesian plane, and its present position over the metropolitan area of depression, as lowered by pumping. *Transactions of the Institution of Surveyors*, **10**, 279–316.

LUCAS, J. 1878*b*. *Hydrogeological Survey, Sheet 2*. Edward Stanford, London. [BL, GSL]

LUCAS, J. 1878*c*. *Hydrogeological survey, Sheet 1, 2nd edition*. Edward Stanford, London. [BL, GSL]

LUCAS, J. 1878*d*. *Hydrogeological Survey, Explanation accompanying Sheet I, 2nd edition and Sheet II*. Edward Stanford, London. [BL, GSL]

LUCAS, J. 1878*e*. What the difficulty is; what has been done to meet it; what remains to be done. *Journal of the Society of Arts*, **26**, 275–277.

LUCAS, J. 1878*f*. Hydrogeological Survey of England. *Nature*, **18**, 494–495.

LUCAS, J. 1878*g*. Hydrogeological Survey of England. *In: Rivers Conservation. Address and Papers read before the British Association at Dublin, 1878*. P.S. King, London, 68–71. [BL]

LUCAS, J. 1879*a*. Hydrogeology in its relation to Water Supply. *Transactions of the National Association for the Promotion of Social Science, Cheltenham Meeting 1878*. Longmans, Green and Co., London, 518–520.

LUCAS, J. 1879*b*. On the Hydrogeological Survey of England. *In: Report of the 48th Meeting of the British Association, Dublin August 1878*. John Murray, London, 692.

LUCAS, J. 1879*c*. Suggestions for dividing England and Wales into Watershed Districts. *Journal of the Society of Arts*, **27**, 715–727.

LUCAS, J. 1879*d*. Watershed lines. Subterranean water – ridges. *Journal of the Society of Arts*, **27**, 829–831.

LUCAS, J. 1879*e*. On the Quantitative Elements of Hydrogeology. *In: Report of the 49th Meeting of the British Association, Sheffield August 1879*. John Murray, London, 499–501. Reprinted in *Symon's Meteorological Magazine*, 1879, **14**, 146–147.

LUCAS, J. 1879*f*. The Naturalist in Nidderdale. *The Zoologist, 3rd series*, **3**, 353–370 & 403–417.

LUCAS, J. 1880*a*. On the Quantitative Elements in Hydrogeology. *In: Report of the 3rd Congress of the Sanitary Institute of Great Britain*, 195–202.

LUCAS, J. 1880*b*. The hydrogeology of the Lower Greensands of Surrey and Hampshire. *Proceedings of the Institution of Civil Engineers*, **61**, 200–227.

LUCAS, J. 1880*c*. *What the Ratepayers really want the Government to do for London Water Supply*. Vacher and Sons, London, 14p. [BL]

LUCAS, J. 1880*d*. Petty Romany. *Nineteenth Century*, **8**, 578–592.

LUCAS, J. 1881*a*. Rural water supply; with especial reference to the objects of the Public Health Water Act, 1878. *Transactions of the Surveyor's Institution*, **13**, 143–178.

LUCAS, J. 1881*b*. The rise and fall of water in wells. Letter in *The Times*, **22nd April**.

LUCAS, J. (ed.) 1881*c*. *Wells, Springs and Rivers of Great Britain. Returns of gaugings and other observations on surface and subterranean water economy, by various observers*. No 1, September 1881, Vacher and Sons, London. [BL]

LUCAS, J. (ed.) 1881*d*. *Wells, Springs and Rivers of Great Britain. Returns of gaugings and other observations on surface and subterranean water economy, by various observers*. No 2, October 1881, Vacher and Sons, London. [BL]

LUCAS, J. (ed.) 1881*e*. *Wells, Springs and Rivers of Great Britain. Returns of gaugings and other observations on surface and subterranean water economy, by various observers*. No 3, December 1881, Vacher and Sons, London. [BL]

LUCAS, J. 1881*f*. Vestiges of the ancient forest on part of the Pennine Chain. *Proceedings of the Yorkshire Geological and Polytechnic Society, New Series*, **7**, 368–372.

LUCAS, J. 1882*a*. On an organisation for the systematic gauging of the wells, springs, and rivers of Great Britain. *In: Report of the 51st Meeting of the British Association for the Advancement of Science, York, August/September 1881*. John Murray, London, 781.

LUCAS, J. 1882*b*. *Studies in Nidderdale: upon notes and observations other than geological, made during the progress of the Government Geological Survey of the District, 1867–1872*. Elliot Stock, London; Thomas Thorpe, Pateley Bridge, 292p. [BL]

LUCAS, J. 1882*c*. On some vestiges of the ancient forest of part of the Pennine Chain. *In: Report of the 51st Meeting of the British Association for the Advancement of Science, York, August/September 1881*. John Murray, London, 680–681.

LUCAS, J. 1882*d*. *The Yetholm History of the Gypsies*. J. and J. H. Rutherfurd, Kelso, 152p. [BL]

LUCAS, J. 1884*a*. Waste Lands and Fish Ponds. Letter in *The Times*, **13th November**.

LUCAS, J. 1884*b*. Water from the Chalk. *Journal of the Society of Arts*, **32**, 859–861.

LUCAS, J. [Under Pseudonym 'Hydrogeologist, Tooting Graveney'] 1884c. Artesian Wells in South-West London. Letters in *The Times*, **25th September** & **7th October**.

LUCAS, J. 1885*a*. *Report to Edward Hales, Esq. On the quantity of water to be obtained from the bed of gravel, so called, on the higher part of the North Frith Estate.* Printed by Mr Hales, 8p. [BL, GSL]

LUCAS, J. 1885*b*. *Report on the prospects of boring for water in Rugley Wood and at Thorntree Well.* Alnwick and Canongate Local Board of Health, Alnwick. [Not located]

LUCAS, J. 1885*c*. Grassington Lead Mines. A decayed industry. *Old Yorkshire, 2nd Series*, **1**, 49–53.

LUCAS, J. 1886*a*. On some recent borings for water. *Professional Notes of the Surveyor's Institution*, **1**, 14–18.

LUCAS, J. 1886*b*. *Petunnikov on the Geology and Hydrogeology of Moscow*, Translated from the Russian by Joseph Lucas, Privately Printed, London, 59p. [BL]

LUCAS, J. 1887. The home of the birch and the beech. Letter in *The Times*, **28th September**.

LUCAS, J. 1888. *Testimonials with Memorandum and Appendix [submitted by J. Lucas as a candidate for the Professorship of Geology in the University of Oxford].* Privately printed, 31p. [BL, BGS]

LUCAS, J. 1891*a*. *Report to Lieut.-Colonel W. G. Dawkins, on the prospect of finding coal under the estate of Over Norton, Oxfordshire (in the Parish of Chipping Norton).* Privately printed, 16p. [BL, GSL]

LUCAS, J. 1891*b*. *Report to Lieut.-Colonel W. G. Dawkins, on the prospect of finding coal under the estate of Over Norton, Oxfordshire (in the Parish of Chipping Norton)*, 2nd edition. Privately printed, 20p. [Copy owned by H. S. Torrens]

LUCAS, J. 1891*c*. Water supply for London. Letter to *The Times*, **17th October**.

LUCAS, J. 1892. *Kalm's account of his visit to England on his way to America in 1748, [Extracted from 'En Resa til Norra America].* Translated by Joseph Lucas, Macmillan and Co, London, 480p. [BL]

LUCAS, J. 1894. *Carmina Barbarica. Latine Reddita.* Printed for the author by William Fraser, Tooting Graveney, 41p. [BL]

LUCAS, J. 1896*a*. *Historical Genealogy of the family of Bayne of Nidderdale, showing also how Bayeux became Baynes.* William Harrison, Ripon, 635p. [BL]

LUCAS, J. 1896*b*. *Principal genealogical trees of the family of Bayne of Nidderdale.* William Harrison, Ripon. 21p. [US Library of Congress]

LUCAS, J. 1908*a*. On Messrs. Vaux Graham and Bidder's Paper on 'Underground Water'. *Professional Notes of the Surveyor's Institution*, **14**, 368–376.

LUCAS, J. 1908*b*. The Hydrogeology of the Dover Basin, Dover Harbour, and the Channel Tunnel: the position of the East Kent Coalfield. *Transactions of the Surveyor's Institution*, **40**, 455–482.

MATHER, J. D. 1998. *From William Smith to William Whitaker: the development of British hydrogeology in the nineteenth century. In:* BLUNDELL, D. J. & SCOTT, A. C. (eds) Lyell: the Past is the Key to the Present. Geological Society, London, Special Publications, **143**, 183–196.

MATHER, J. D. 2001. Joseph Lucas and the term 'hydrogeology'. *Hydrogeology Journal*, **9**, 413–415.

MIDDLETON, R. E. 1901. The future of the London Water Supply. *Transactions of the Surveyor's Institution*, **33**, 23–194.

PREENE, M. 2004. Robert Stephenson (1803–1859) – the first groundwater engineer. *In:* Mather, J. D. (ed.) *200 Years of British Hydrogeology*. Geological Society, London, Special Publications, **225**, 000–000.

PRESTWICH, J. 1872. The Anniversary Address of the President. *Quarterly Journal of the Geological Society*, **28**, 29–90.

ROYAL COMMISSION ON METROPOLITAN WATER SUPPLY. 1893. *Minutes of Evidence taken before the Royal Commission on Metropolitan water Supply*. HMSO, London.

RUPKE, N. 1983. *The great chain of history: William Buckland and the English School of Geology 1814–1849.* Clarendon Press, Oxford.

SHIRES, J. 1900. Underground water. *Transactions of the Surveyor's Institution*, **32**, 253–308.

SOUTHWARK AND VAUXHALL WATER COMPANY. 1892. *Royal Commission, May 1892, to consider the question of the Metropolitan Water Supply. Geological Reports by Professor Prestwich, Professor Ansted, Professor Lucas, Professor Ramsay to the Southwark and Vauxhall Water Company*, Waterlow and Sons, London, 12p.

STEPHENSON, R. 1840. *Report to the Provisional Committee of the London and Westminster Water-Works, Etc., Etc.* Reproduced in the *Morning Advertiser*, **December 29th**.

THORNE, R. T. 1875. *Report to the Local Government Board on the Sanitary Condition of certain portions of the Guisbrough Registration District, with special reference to the prevalence of infectious diseases in them.* HMSO, London.

TORRENS, H. S. 2002. *The practice of British geology 1750–1850.* Ashgate, Aldershot.

TWEEDALE, G. 1991. Geology and industrial consultancy: Sir William Boyd Dawkins (1837–1929) and the Kent Coalfield. *British Journal of the History of Science*, **24**, 435–451.

WATTS, W. W. 1911. Obituary of Rev. George Ferris Whidborne (1846–1910). *Quarterly Journal of the Geological Society*, **67**, lvii.

WHIDBORNE, M. 1917. *George Ferris Whidborne [III] (1890–1915).* Privately Published, Glasgow.

WILSON, H. E. 1985. *Down to earth: One Hundred and Fifty years of the British Geological Survey.* Scottish Academic Press, Edinburgh, 189p.

YOUNGER, P. J. 2004. 'Making water': the hydrogeological adventures of Britain's early mining engineers. *In:* Mather, J. D. (ed.) *200 Years of British Hydrogeology*. Geological Society, London, Special Publications, **225**, 000–000.

19th century studies of the hydrogeology of the Permo-Triassic Sandstones of the northern Cheshire Basin, England

JOHN H. TELLAM

Hydrogeology Research Group, Department of Earth Sciences, School of Geography, Earth and Environmental Sciences, University of Birmingham, B15 2TT, UK (e-mail: J.H.Tellam@bham.ac.uk)

Abstract: About a dozen workers were active in researching the hydrogeology of the Permo-Triassic sandstones of the northern Cheshire Basin, UK, in the 19th century. They were mostly amateur geologists, members of the geological societies of Liverpool and Manchester. Spurred by the water resource requirement of the two cities and by the formation of the societies, research burgeoned from the mid 1800s. Over the latter part of the century, a conceptual model of flow in the sandstones was developed which has most of the essential features of a present-day conceptual model, including intergranular and fracture flow, fault influence on flow, recharge reduction by drift and urban land cover, overspill recharge, influent river recharge (including estuarine intrusion), and cross-boundary flows. Water balances were undertaken to assess aquifer yield and attempt to understand well yields. Pollution from sewers, river water, estuary water, and graveyards was considered, as was water/rock interaction. Experimental work demonstrated: the proportionality between flow rate and pressure difference (1869); the importance of fractures to well yields (1850); the principle of specific yield (1869); the effects of lamination on unsaturated flow (1877); and the shape of breakthrough curves is sigmoidal (1878). Most of these findings were independent of previous work elsewhere. Whether this is because of the local, essentially amateur environment in which the researchers were operating, or whether this lack of communication was essentially a feature of research at this time, is uncertain. However, it does have implications for general theories of science.

This paper concerns historical hydrogeological research undertaken on the Permo-Triassic sandstones at the northern end of the Cheshire Basin, in the area between Manchester and Liverpool (Fig. 1). The work described is almost all from the 19th century, though some reference is made to work in the late 18th century and just into the 20th century. During this interval, geological research was calling into question many established assumptions about the universe and the place of humans within it, and was at the centre of the intellectual life of Britain and many other countries (e.g. Hallam 1989). For example, it was in this century that the great age of the Earth, with all its attendant implications, was starting to be recognized. In a rather less profound way, groundwater had long been seen as being of importance, but the recognition of hydrogeology as a separate discipline only slowly emerged over the century. For example, the term 'hydrogeology', in the sense that it is now used, was only introduced in the mid 1870s (Lucas 1874, 1877; see Mather 2001, 2004), and is not mentioned in any of the historical papers cited in this paper, yet much hydrogeological work had been undertaken before that date.

Table 1 lists all the groundwater publications identified which were undertaken in the region during the 19th century. There are about a dozen main workers. There is a sudden increase in the numbers of publications from the middle of the century. There are two main reasons for this (though in addition the British Association for the Advancement of Science also played a role). The first is the rise in major public interest in geology resulting from the revolutionary ideas emerging in this period (e.g. the realization that geological (Lyell 1830–1875) and biological (Darwin 1859) evidence suggested that the Earth was of great age). This public interest lead to the formation of the Manchester Geological and Mining Society in 1838 and the Liverpool Geological Society in 1859 (Hewitt 1910). The main preoccupation of the Manchester Geological Society was the Coal Measures and sandstone hydrogeology papers occur more frequently in the Liverpool Geological Society's publications. The Liverpool Geological Society consisted of 65 members in 1862, and this membership remained almost constant until at least the last decade of the 19th century. For most of the members, geology was an interest rather than a profession.

The second main reason for increase in hydrogeological work in the middle years of the century was the need to plan for the future water supplies for Liverpool and Manchester. In the case of Liverpool, the Town Council started planning just before the middle of the century, at least 30 years before the building of the Vrnwy dam in North Wales and the abandonment of groundwater as the main source of public water supply.

The authors of the hydrogeological papers in the

From: MATHER, J. D. (ed.) 2004. *200 Years of British Hydrogeology*. Geological Society, London, Special Publications, **225**. 89–105. 0305-8719/04/$15

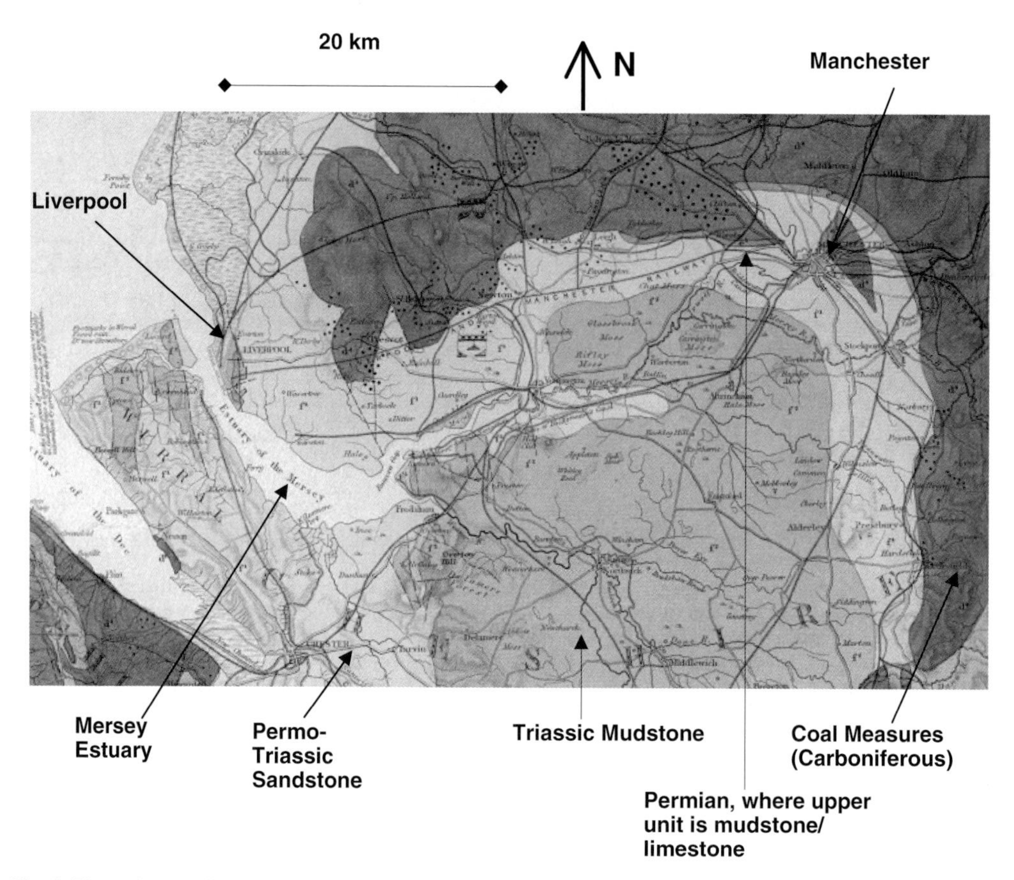

Fig. 1. The study area (from the map by G.B. Greenough dated 1865).

publications of the two geological societies were often local inhabitants who were interested in a wide range of geological topics. For example, George H. Morton (Fig. 2a) was a house painter and decorator, as well as being a founder of the Liverpool Geological Society, and an evening class lecturer in Geology for the Queen's College. He published papers not only on hydrogeology, but also on the geology of the New Red Sandstone, the Carboniferous Limestone, the Coal Measures, and the Quaternary geology deposits of the region. A few of the authors had some connection with geologically related professions. Isaac Roberts was involved with quarrying and borehole drilling and T. Mellard Reade (Fig. 2b) was an architect/civil engineer who 'in the course of his professional work . . . was first lead to take an interest in geological questions' (Hewitt 1910) (and later made a significant contribution to the debate about the age of the Earth with a calculation based on denudation rates (Reade 1878*b*)). However, these 'amateurs' were not completely isolated from direct contact with professional geological researchers. How much contact is difficult to assess, but certainly Charles de Rance and Edward Hull from the Geological Survey were both influential [de Rance appears to have known at least some of the local members of the Society well, as he refers to 'my friend, Isaac Roberts' in one of his papers (de Rance 1886)]. A very influential figure in the awakening of interest in the hydrogeology of the area was Robert Stephenson, the engineer who was commissioned in 1850 by the Liverpool Town Council to investigate various possibilities for the long term water supply for Liverpool (see also Preene 2004). Hence the intellectual environment in which the hydrogeological publications listed in Table 1 were produced can be viewed as partially closed – most of the work was completed in ignorance of work being carried out elsewhere, especially in other countries, but contact with professional workers with national/international experience occasionally occurred (though it is uncertain whether even these latter workers were aware of many of the advances being made in other countries). As a result, the research includes rediscoveries and near misses, as well as new discoveries.

Table 1. *Hydrogeological research in the Liverpool Manchester area in the 19th century*

Date	Conceptual models	Recharge/ water balances	Hydraulics	Chemistry/pollution	Number of Papers
<1800		Townley, Dalton		Houlston	2
1800–1809					0
1810–1819					0
1820–1829					0
1830–1839					0
1840–1849	Cunningham	Cunningham		Smith, Cunningham, Ormerod	3
1850–1859	Stephenson		Stephenson	Braithwaite, Stephenson	3
1860–1869	Hull, Morton, Roberts, Bostock	Roberts, Bostock	Hull, Roberts, Bostock	Hull, Bostock, Binney	5
1870–1879	Binney, Dawkins, Reade, Boult, Morton	Morton, Boult, Binney, Dawkins, Reade	Binney, Dawkins, Boult	Smith, Boult, Reade, Binney, Dawkins, de Rance, Brown, Morton, Roberts	11
1880–1889	Hull, Reade, de Rance, Fox	de Rance, Hull	de Rance, Reade, Roberts	de Rance, Hull, Reade, Roberts	14
1890–1900			Moore	Hewitt, Holland	2
1900–1909				Moore, Rhodes	3
1910–1919				Greenwood & Travis	2

The emphasis of this paper is on the development of the ideas rather than a purely chronostratigraphical description. Initially the conceptual model brought together by Stephenson in 1850 is presented. This leads to consideration of the development of ideas on the sources of the groundwater, the hydraulic properties and behaviour of the sandstones, and the groundwater chemistry.

Conceptual model at ~1850

Robert Stephenson (who, with his father George, built the Liverpool/Manchester railway line in the 1820s, including the first railway station and railway junction in the world), was commissioned by Liverpool Town Council to report on the long-term viability of supplying Liverpool with water from the sandstone aquifers below the city, and/or from a proposed reservoir at Rivington, 22 miles (35 km) to the northeast of Liverpool centre. This he did in 1850 (four years before the first Geological Survey map of the area (Morton 1866)). The report shows that he took evidence from a wide range of local people, several with their own theories on flow in the sandstones and then brought his considerable national experience to bear on the problem. At this time there were few descriptions of groundwater flow systems, and even concepts such as a regional water table and the source of groundwater were still debated (see below). It should be noted that there had been some published hydrogeological work relating to the area prior to Stephenson's (1850) report, including Houlston (1773) (see below) and Cunningham (1847). Cunningham (1847) (who, incidentally, had given talks on what he thought were a fossil medusa, tortoise, and vegetable matter from the sandstones (Morton 1870)) recognized that the main groundwater unit was the sandstone sequence rather than the Coal Measures, that north-south and east-west faults broke the sandstone up into blocks, that intrusion from the Mersey Estuary was occurring and that a water balance could be used to determine the likely yield of the sandstones.

However, the dominant publication of the mid 1800s was Stephenson's (1850) report. His conceptual model for the sandstone groundwater system can be summarized as follows:

- 'Large sheets of water may be conceived as spread out one over the other, being retained in their positions by intermediate beds more or less porous'.
- An 'infinite series of fractures' destroys insulation between the sheets. Some of his witnesses considered the fractures to be filled with clay, which Stephenson thought 'may to some extent be true'. Stephenson undertook an experiment on the Bootle 'Well', a public supply well in Liverpool, in order to determine fracture interconnections. The 'well' consisted of collector reservoirs – dug wells – with large numbers of boreholes in the base: by switching on the well pumps with all boreholes except one plugged, and unplugging the boreholes one by one, he determined that almost all the boreholes were

Fig. 2. (**a**) George H. Morton, painter, decorator, and geologist (Hewitt 1910) (**b**) T. Mellard Reade, architect and geologist (Hewitt 1910) (**c**) Dr John Dalton of Manchester, scientist (Smith 1856).

connected by fractures (for more details, see Preene 2004).

- Rainfall was the source of all groundwater, except for some local Mersey intrusion.
- The water in the sandstone 'forms an inclined plane towards the easiest outfall, the angle varying slightly with the horizon according to the variations of the season'.
- Pumping causes small indentations in the water table, with the influence being felt up to at least 2 miles (3.2 km) distant from the pumping well.
- When the sandstone was full, overflow occurred to the Mersey. Normally the Mersey Estuary waters are kept back by pressure of freshwater, but locally intrusion occurs, sometimes through faults.

It is clear that Stephenson's model has many features in common with a present day conceptual model, and that issues which were of concern to him, in particular faults and fractures, are still of concern today. However, his views on the source of the groundwaters in wells was not universally accepted at the time, and aspects of the hydraulic behaviour and the chemistry of the groundwaters were also strongly debated. The development of the ideas in these three areas is described in the following sections.

The source of the groundwater

The hydrogeological papers from around the time of Stephenson's report in 1850 (1847–1862) all assume implicitly assume that rainfall is the dominant source of groundwater. Subsequently, when water balances were attempted, some authors decided that other sources, including sea water and even subterranean condensation must be more important. That precipitation was the dominant source of groundwater had been established in the 1600s by Pierre Perrault (1674) and Edmé Mariotte, both working in the Seine catchment (e.g. Davis & de Wiest 1966; Pinneker 1983). This result had been confirmed locally in 1799 by John Dalton, the famous atomist and discoverer of the law that now bears his name, who lived in Manchester (Fig. 2c). Dalton (1799) used a water balance approach, as had Perrault and Mariotte (Smith 1856). None of this research, including Dalton's, appears to have been known to the other authors publishing on the hydrogeology of the area in the latter half of the 19th century.

Cunningham (1847), assuming that infiltration was about one ninth of rainfall, calculates that over the 80 square miles (200 km^2) around Liverpool the sandstone should be able to supply 12.7 Mgpd (58 Ml/d) (enough for Liverpool's needs). However, he noted inflow from the Mersey Estuary and suggested the possibility that the salt was being removed by filtration, thus affording a potentially infinite supply of water (Morton 1870).

Bostock, 'an excellent practical geologist from Birkenhead . . .' (Reade 1878*a*), a baker by trade, published a water balance calculation for a typical well in his paper of 1869. He used figures for infiltration derived from observations carried out using apparatus designed by John Dalton in the last decade of the 18th century.

Dalton's apparatus consisted of a 10 inch (0.25 m) diameter cylinder, 3 feet deep (0.9 m), open at the top and closed at the base: it was filled with local soil and sunk into the ground so that the top of the cylinder was level with ground surface. Dalton filled the apparatus with local soil which, according to de Rance (1889), was probably 'vegetable matter mixed with the stiff Lancashire Boulder Clay'. Again according to de Rance (1889), Dalton found 25% of rainfall was absorbed by the soil. Bostock (1869) summarizes this and other work with the Dalton apparatus by stating that 11–27% of rainfall infiltrates to 3 ft (0.9 m) depth in the last two and the first two months of the year 'according to the nature of the soil used and the state of the atmosphere'.

Bostock (1869) then assumed that, for a typical well of depth 70 yards (65 m), the catchment radius was at most 350 yards (320 m). This estimate seems to have been based on intuition rather than hard evidence.

Bostock (1869) next needed rainfall data. According to de Rance (1889), the first systematic collection of precipitation data appears to have been by a Mr Townley, a Lancashire squire from Townley, near Burnley, in the mid 1600s (presumably the Townley of Townley's hypothesis which later became Boyle's Law). By the mid 1800s, under the instigation of G.J.Symons, there were hundreds of rainfall stations across the country. Bostock (1869) chose a value of 34 inches per year (865 mm/y). He then assumed that one sixth of this infiltrated, allowing him to calculate that for a well catchment radius of 350 yards the yield would be ~34 700 gallons per day (0.160 Ml/d). A survey of wells in the region indicated that well yields were 10 to 90 times greater than this figure, and, rather than re-examining his other assumptions, he concluded that there must be some other source of the groundwater. This he suggested to be sea water. To get around the problem of the salinity of sea water, Bostock (1869) suggested that groundwater derived from sea water would have to be stripped of its salt, but he was uncertain how this happened. 'Whether a chemical change takes place when the particles are divided . . . whilst slowly filtering through the sandstone . . . is not at present known; but it is a fact, that when sea . . . water . . . has undergone such filtration, it is deprived of its salt'. [the word 'particles' arises as Bostock was writing well before many had accepted that the idea – strongly expounded by Dalton – of atoms, and even less ions, was anything more than a convenient imaginary construct]. This idea of sea water filtering goes back to classical Greece (e.g. Davis & de Wiest 1966).

Roberts (1869), in a paper read to the Liverpool Geological Society in the session following Bostock's (1869) paper, agreed with Bostock (1869) that the abstraction rates were greater than could be accounted for by rainfall alone and that sea water must be entering the aquifer. Morton (1870), a painter and decorator from Liverpool, agreed, but tentatively suggested that the water-balance problem could be avoided by simply increasing the assumed well catchment area. He also points out that some water might come across the faulted boundary with the Coal Measures.

Boult (1875) used a similar approach and data and concluded that an area of 60 square miles (150 km^2) would be needed to supply Liverpool's estimated 19 Mgpd (86 Ml/d) requirement. This assumed that the rainfall could enter the sandstones freely, but because much of the area was covered with boulder clay or with urban 'impermeable' surfaces, some water would be lost from runoff. In addition, some water would be lost to springs, and some taken up by plants. He concluded, though does not provide detailed reasoning, that this would increase the required catchment area to 150 'miles' (square miles) (385 km^2). An area of 150 square miles would be included in a 10 mile (16 km) radius semi-circle centred on Liverpool (Boult tacitly assumes that the Mersey provides a hydraulic boundary). However, this area would include outcrop of Coal Measures and some heavily abstracting existing wells. Hence the actual area required to feed the well would have to extend further, and in doing so intercept more wells and thus be extended even further. With this hare-and-tortoise argument, he concludes that there is not enough water simply from rainfall. It is not clear from his paper where he thought that the water abstracted did come from, though he does suggest distillation in 'nature's great subterranean laboratory', an idea which again goes back at least to the ancient Greeks. In the published discussion of the paper, which was the second in a series of two, there is the following rather rude comment from the local driller, C. H.Beloe. 'What Mr Boult's own opinion was was a mystery as far as the first paper was concerned, and it still remained a mystery in the second. (Laughter.)'. However, despite some quirkily iconoclastic arguments, Boult (1875) does raise important and very relevant issues to do with recharge and well catchment areas which seem to have escaped Beloe.

de Rance (1876), writing as the secretary to the British Association for the Advancement of Science Committee for the Investigation of Underground Waters, agreed that another recharge source in addition to rainfall was needed. He argued that all the water abstracted from wells in Manchester – around 6 Mgpd (27 Ml/d) – cannot come from infiltration of rainfall in an urban area. He concludes that infiltration is occurring from the rivers Irk, Medlock and Irwell (an argument originally suggested by Hull (1863) and used without cross-reference by de Rance (1888)).

Dawkins (1876) decided that rainfall should be able to supply enough water for the wells, and set out to demonstrate this using essentially the same infiltration data as the previous researchers. He noted that runoff from till would eventually infiltrate and, as de Rance (1876) had done, that indirect rainfall recharge via rivers might also occur. He also preempts later work on induced recharge in response to falling water levels:

> 'If large quantities of water were abstracted . . . , the vacuum would rapidly be filled up by increased percolation.'

By these processes, Dawkins (1876) suggested that the catchment area could be increased in effect to 800 square miles (2000 km^2), which would yield a possible 314 Mgpd (1430 Ml/d)!! Reade (1878*a*) considered that the main source of groundwater was precipitation. In his opinion, none of the other hypotheses 'hold water', and present difficulties 'ten times' those of assuming rainfall is dominant. He also states that 'many Cassandra-like water-prophets cry out that because the water-level is reduced in Liverpool, therefore we are drawing on water capital'. His calculations suggested a well would need a catchment area of radius 1.5 miles (2.4 km) to get a pumping rate of 3 Mgpd (14 Ml/d), and that therefore fears were 'groundless'.

Roberts (1879), picking up from his assertion a decade earlier that sea water must be entering the aquifer (Roberts 1869), also proposed that the salt was removed from the sea water as it passed through the sandstone. To investigate this further he carried out a column experiment (Fig. 3). He took two 13 inch (0.33 m) high and 12 inch (0.30 m) square blocks of sandstone and cut a 9 inch (0.23 m) square by 1 inch (0.025 m) deep recess in the upper face of each block. He 'thoroughly dried [the blocks] in air' and coated their vertical faces with black varnish and covered them with oil-cloth. Finally he mounted one block immediately above the other (Fig. 3a). He then poured settled half-ebb water from the Mersey Estuary at Rockferry, Liverpool, into the recess in the upper block, sampled the water which dripped out of the bottom block and analysed it for chloride. The resulting breakthrough curve is shown in Figure 3b. It took 93.5 fluid oz (2.66 l) for the Mersey water to breakthrough to effectively 100% of the initial concentration. He then air dried one block for one month and flushed it with spring-water. All the chloride was flushed out by 92 fluid oz (2.61 l). He suggested that the salinity was removed by some physical process of finite capacity and tentatively proposed that the capillary attraction for the Na, Ca and Mg chlorides was greater than for pure water. He appears to have recognized that all the salt which was taken up was released during the spring-water flushing phase. He concludes that his experiments explain the fact that some wells close to the Mersey abstract water of quality intermediate between estuary water and fresh groundwater. It appears that it was not until 1884 that Roberts' interpretation was comprehensively reassessed. de Rance (1886) cites W.M.R.Nichols of the Boston [USA] Society of Engineers (*On the filtration of certain saline solutions through sand, Boston*) who, by repeating Roberts (1879) experiments using sand and also 'Ohio sandstone', showed that the apparent retention of the salt was due to the presence of moisture within his sandstone blocks at the start of the flushing – i.e. they were not completely dry to start with. Nevertheless, this experiment must have been one of the very first solute transport experiments ever reported.

de Rance (1880*d*) agreed that if infiltration was ~10 inches (0.25 m), Liverpool public wells must obtain some of their water from outside the area of Liverpool. Hull (1882), calculating infiltration (I) using

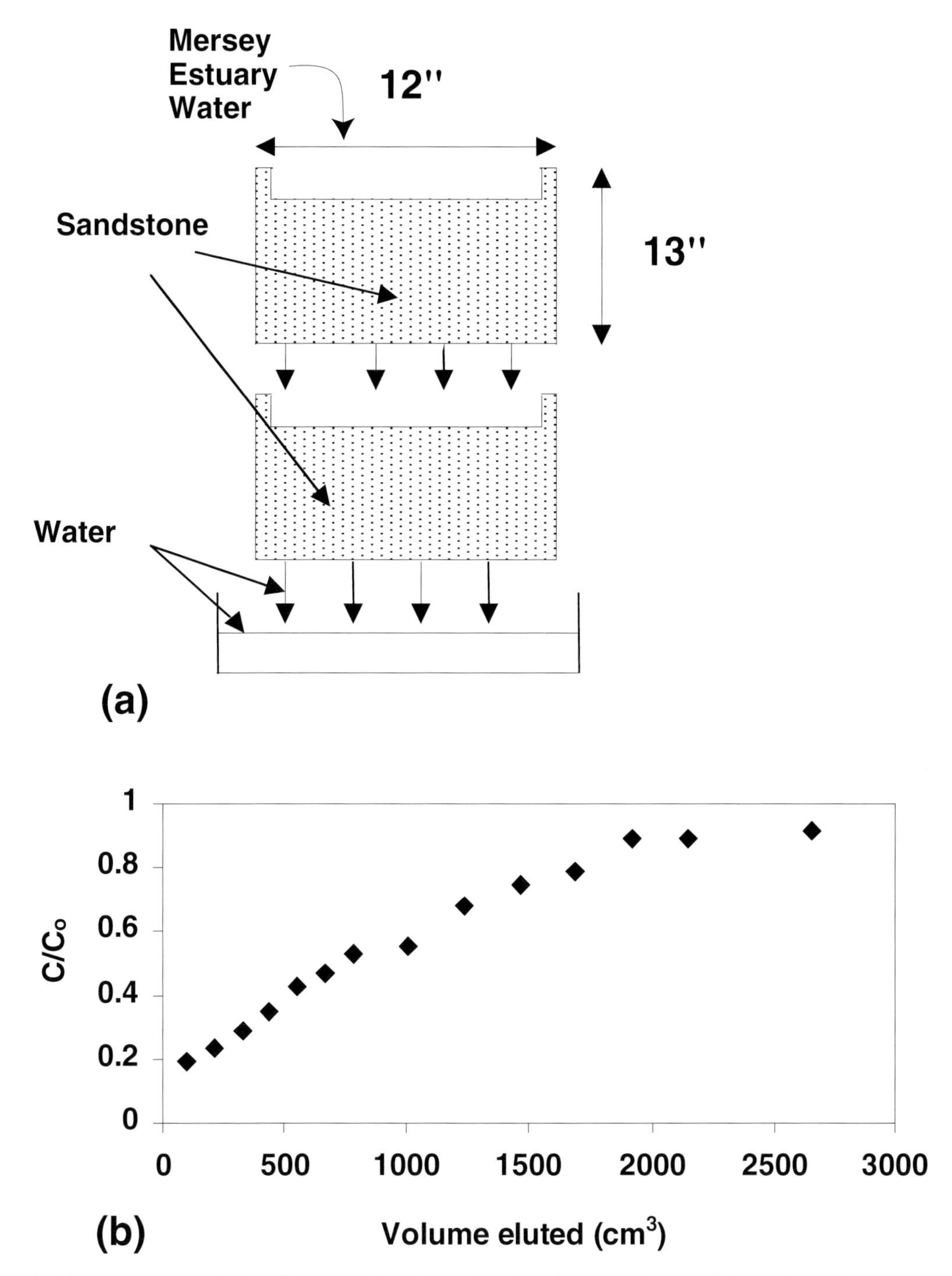

Fig. 3. (**a**) The apparatus Roberts (1879) used for his breakthrough column experiments (**b**) The breakthrough curve obtained using the apparatus shown in (**a**): note that the sandstone was partially saturated at the start of the experiment. Saturated pore volume is ~3.33 l.

Table 2. *Recharge studies, 1847 to 1889, and 1981. P = precipitation*

Author	Infiltration Rate Assumed	Where Infiltration Rate Applied
Cunningham (1847)	P/9	Everywhere, except till, urban
Bostock (1869)	P/6	Everywhere, except till, urban
Boult (1875)	P/4	Everywhere, except till, urban
Dawkins (1876)	P/3	Everywhere including till, urban areas, rivers
Reade (1878a)	P/3	Everywhere, except Mersey
de Rance (1880d)	P/3	Everywhere, except till, urban
Hull (1882)	P/8	Everywhere
de Rance (1889)	P/3	Only where permeable
Bham Univ/ NWWA (1983)	*P/3*	*Overall average, including mains leakage*

$$I = \text{Area pervious} \times \text{infiltration rate for pervious areas} + \text{Area impervious (sic)} \times \text{infiltration rate for impervious areas},$$

concluded that infiltration was less than abstraction, and that this resulted in the observed falling water levels and possibly the increasing hardness of the groundwater. He suggested that if pumping was stopped, recovery might occur within 2 to 3 years.

Even in 1887, when presenting one of his papers, de Rance was prompted into an exasperated reply to a question from a member of the audience to say that in recent years in 'paper after paper' he had stated that the main source of groundwater was rainfall (de Rance 1887).

Overall, it is clear that the uncertainty associated with the methods for estimating recharge were such that the conclusion as to the source of groundwater could not be proved. However, the consensus was gradually reached that the dominant source of groundwater was indeed rainfall, but that the abstraction rates were such that water levels were falling and water was being drawn into the sandstones from the Mersey Estuary and other surface water bodies. There was an awareness that the water quality was changing, or likely to change in response to the intruding water, but the extent of water/rock interaction was uncertain. Thus the argument had progressed from implied assumption of the dominance of rainfall as the source of the groundwater, through realization that the water balances might be in deficit, to induced sea water intrusion with filtration of the 'saline particles', to, possibly, a rather more sophisticated understanding of falling water levels and intruding surface waters. The convergence of ideas was driven more by intuition than by hard evidence.

Table 3. *Laboratory hydraulic property measurements*

	'Porosity'	Method (n)	'Sp Yield'
Anon (<1869) cited in Bostock (1869)	>50%	? (?)	
Bostock (1869)	16%	Sorption (?)	
Roberts (1869)	11%	Sorption (?)	1%
Reade (1884)	12–14%	Sorption (4)	
Moore (1898)	35–36%	Saturation (2)	
	7 & 10	Saturation (2)	
Campbell (1982), Allen et al. (1997)	*25%*	*Saturation (many)*	*lab ~1%*

Does the final consensus accord with the conclusions of modern investigations? Table 2 lists the infiltration estimates from the 19th century for the Liverpool area together with values from a major study in the 1980s (University of Birmingham/ NWWA 1983). It is clear that the infiltration rates are in many cases similar to those in the most recent regional studies. Even the inference of sea water intrusion is correct. Howard (1987) shows for the region between Liverpool and Manchester that estuary water infiltration volumes are a significant proportion of the total volume abstracted over the period 1850–1980 (436 Mm^3 compared with 2856 Mm^3, i.e. ~15%). In Liverpool, it is highly likely that Mersey intrusion also provided a significant component of the water balance (e.g. Rushton *et al.* 1988), and certainly the historical chemical evidence seems to suggest this (Tellam 1996).

Hydraulic behaviour and properties

Porosity and specific yield

Bostock (1869) reports porosity values on rocks from the region that are rather larger than is likely (> 50%). The method used for these measurements is unknown and it is uncertain who had carried them out. The measurements made in the 1860s-1880s produced rather low porosity values, presumably because they were undertaken by 'sorption' (uptake of water by a dried sample) (Table 3). Moore (1902) maps the variation in porosity across a fault, finding lowest values in the fault plane itself.

Meinzer (1928) is usually credited with the introduction of the concept of elastic storage, but it is clear that others were thinking along the same lines much earlier (e.g. Grabham 1910). Specific yield

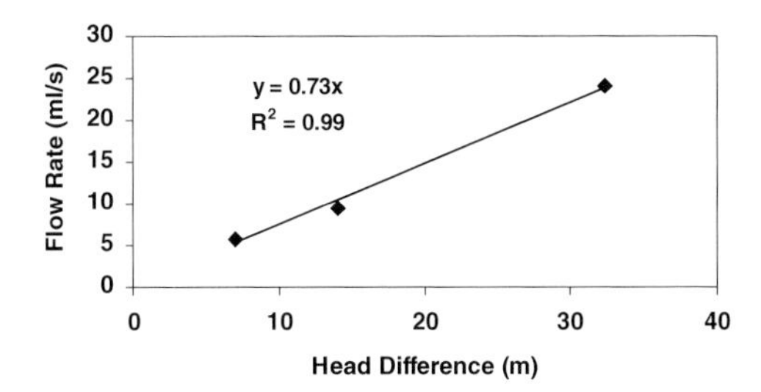

Fig. 4. The data Roberts (1869) obtained from flow experiments on a block of Triassic sandstone. The permeability is ~0.2 m/d.

was a more obvious concept, and even in the 19th century, the idea that only some of the water in a rock drained out under gravity was generally accepted. Bostock (1869) says that there is 'no means of knowing' how much of total water the sandstone contains would be parted with by drainage (de Rance (1889) puts this nicely: 'constant struggle between gravity and capillarity'). Roberts (1869) undertook a drainage experiment on a block of sandstone and found that 1% by volume drained out. This is in keeping with laboratory-determined values obtained more recently (see summary in Allen *et al.* (1997)), though almost an order of magnitude too low for field values where the thickness of the unsaturated zone is much greater (e.g. Rushton & Howard 1982).

Permeability

Darcy's (1856) flow experiments were unknown to the researchers in the study area, up until at least the end of the century [Bostock (1869), as in the case of specific yield, is pessimistic. 'No means of knowing' how much water will pass through in a given time]. Roberts (1869), as always the experimentalist, undertook Darcy-type experiments on blocks of the sandstone:

> 'I found by experiment that one square foot of compact sandstone 10.5 inches in thickness and of average coarseness, passed the following quantities of water through per hour:-
> At a pressure of 10 lbs. to the square inch, 4.5 gals.
> At a pressure of 20 lbs. to the square inch, 7.5 gals.
> At a pressure of 46 lbs. to the square inch, 19 gals.
> The increase being nearly directly as the pressure.'

Figure 4 indicates that his experimental technique was reasonable (and that the permeability of the block of sandstone was 0.2 m/d, a lowish value possibly affected by incomplete saturation). Roberts (1880) also undertook a basic pumping test at a well he drilled at his new home in Maghull to the north of

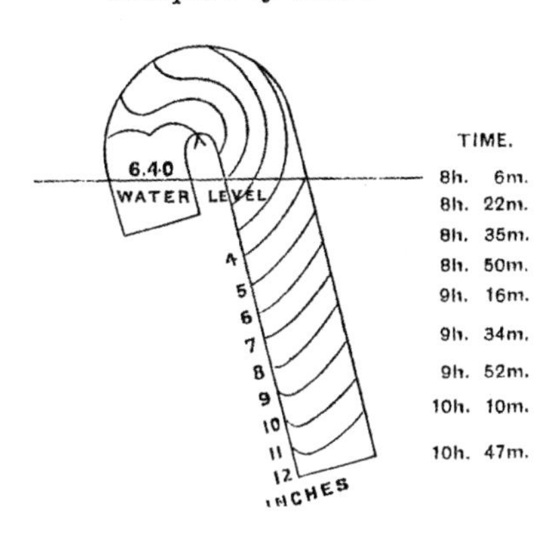

Fig. 5. Reade's (1878*a*) 'stone siphon', a walking-stick shaped block of sandstone, square in cross-section, which he used in experiments on the movement of groundwater.

Liverpool, with two men pumping by hand for two days to give a drawdown of 10 feet (3 m). However, he did not attempt an analysis, though had developed a Dupuit (1863) steady-state radial flow equation which had been developed only seven years after Darcy had published his experiments (Darcy 1856).

Reade (1878*a*) uses Roberts' (1869) findings without comment, quite happy with the direct relationship between pressure difference and flow. Likewise, de Rance (1889) shows that he is also quite at ease with the Darcian concept. 'Such water flows with "head", due to the difference of vertical level of the "area of outcrop" to that of the "area of discharge", less the frictional resistance of the fragments of the rock through which it passes'. Elsewhere, he talks of permeability, albeit in a qualitative sense. [Incidentally, de Rance's (1889) paper has a title which might be from the present day – *Notes on groundwater-supply and river floods.*]

Behaviour of flow in unsaturated rock

Reade (1884) also experimented on rectangular blocks of sandstone. He examined the water uptake by dry samples, including the effect of layering and the size of the capillary fringe. He even produced a 'stone siphon', a sandstone block of square cross-section shaped like a walking stick (Fig. 5). Having dried it, he inserting one end of the siphon in water, and mapped the zone of wetting as it rose up the block (Fig. 5). He then measured the flow rate from the end of the siphon. He also studied water uptake by rectangular blocks of sandstone, including the effects that lamination had on the process. He then argued that the siphon was an analogue for flow from a river into the sandstone, and that the experiments also indicated that the sandstone could suck water from the overlying boulder-clay. (He had clearly met with some scepticism of his somewhat bizarre, though interesting, experiments, as his parting comments are that these conclusions 'may to some extent answer those troublesome people who are always asking Cui bono?' Virtually no more work was undertaken along these lines until the late 1990s (Digges la Touche 1998; Frey 1999)).

Conceptual models of regional scale groundwater flow

Dawkins (1876) recognizes water levels are dependent 'in a great degree on the surface contours'. He also finds that there is no indication that well yields from depth are less than those from near surface. Despite Robert's (1869) 'dried peas' model of flow in the sandstones and his flow experiments, Boult (1875), in his usual individualistic manner, says that although the sandstone is 'cellular' (i.e. porous), the cells are almost all filled with 'impermeable matter'. The pore sizes are so small as to be invisible even when using a high powered glass. Hence flow in the sandstones is almost all through fractures, with 'subterranean reservoirs, beds, feeders, and currents'. 'Whether those streams and reservoirs are to be found in every description of rock is not yet apparent, but until the contrary is rendered probable, it may be assumed that they are.' [Mr Beloe (see above), in the discussion of the paper, comments with the caustic humour he appears to have displayed on the day the paper was presented: '[I] did not understand Mr. Boult's cellular theory . . . Probably, however, it was correct.'!] Boult was, however, a maverick, and most workers accepted that both intergranular and fracture flow occur.

A number of authors mention fill material in fractures, a topic again in vogue (Wealthall *et al.* 2001; Pearce *et al.* 2001): de Rance (1880b) discusses fill material and fault permeability in the context of the type of wall rock, Holland (1896) gives a chemical analysis of a joint infill, and Hewitt (1898) describes fissure infills as seen in excavations in Liverpool. In the latter case, the fissures are sub-vertical, north-south in orientation, and have a width from a fraction of an inch to 4 or 5 feet (1.2–1.5 m), and filled with sand, or sandstone rubble, or sand and clay with mixed material from remote sources. Fox, F.(1886) reports that during the driving of the Mersey Railway Tunnel, only one fissure was encountered. This was 10 inches (25 cm) across and filled with disintegrated sandstone and clay. According to Fox, D. (1886), before the construction of the tunnel, 'everybody was quite certain that the sandstone was full of fissures, and that, through those fissures, the River Mersey would immediately come in upon the tunnel'. There was at least one dissenter – Sir John Fowler, a past president of the Institution of Civil Engineers – who claimed that the fissures would be 'dry' as indeed was the case. Although fissures were uncommon, fractures appear to have been much more common, especially on the Liverpool side of the Mersey (which had softer sandstone) (Fox, F. 1886).

Hull (1882) suggests that mudstones are important for giving rise to springs.

Although Roberts' (1869) Darcy experiments were accepted almost as a matter of course (see above) (e.g. Reade 1878*a*), the concept of head, however, was not understood, pressure head being substituted for total head:

> 'And it must be remembered that if [the sandstone is] rendered more compact at great depth by the pressure of the superincumbent rocks, that pressure would also cause the water to traverse them more easily than it would do at the surface.' (Dawkins 1876).

In fact the meanings of head were not to be properly put to rest until Hubbert (1940).

In this sense, it is unclear what most of the authors thought about the relationships between water level measured in a well and the pressures required in a laboratory experiment to force water to flow through the rock. Indeed the whole idea of regional water levels was questioned by Boult (1875):

> 'The really important question is on the existence of a district water level, by sinking to which, water of some amount will certainly be obtained. I am unable to see any evidence that there is such a level; for water seems obtainable at any practicable depth, and in almost any stratum, but distributed capriciously; that is, under some law at present unknown.'

However, most accept that regional water levels exist. Hull (1882) presents a hydrogeological map and cross-section of the area just to the east of Liverpool, the former showing the zone where the piezometric surface lies above ground level, and the latter showing a regional water level which generally reflects the topography. He then spoils his interpretation by suggesting that the water level is held up by capillary pressure. By the late 1880s, de Rance (1887) is aware that the location of the water table is dictated by rainfall.

Roberts (1869) claims that many members of the Liverpool Geological Society believed that groundwater came from Denbighshire, Yorkshire, or Derbyshire. Though Morton (1870) disputes this, the concept of a distant source for the groundwater was certainly seriously discussed (but dismissed) by Stephenson (1850) (and the Derbyshire version was still an hypothesis which was advanced by local well-owners when I was engaged in groundwater sampling in ~1980 in the region).

Dawkins (1880) talks of flows across the Cheshire Basin to the wells in Liverpool, implying that he thought that flow at distances of 100 km would be affected by the pumping. Dawkins (1876) talks of 800 mile^2 (2000 km^2) well catchments and Hull (1863) suggests flow down dip into the centre of the Cheshire Basin (to discharge where?). It is clear that ideas concerning flow at a large scale were rather speculative.

Well hydraulics

Given the limited understanding of the hydraulic behaviour of the sandstones, it is not surprising that the understanding of well hydraulics was also rather limited in the 1800s. Stephenson (1850) suggested that the water level around a well was in the form of an inverted cone. With increasing time, the cone height and the apex angle became larger until there was a balance between the friction experienced by the water flowing down the water level gradient, and gravity.

Boult (1875) begs to differ (again!). He argues that the flow is more likely to be vertical than horizontal, effectively because in the horizontal direction there is only capillarity acting, but in the vertical direction there is also gravity. Again he is wrong, but again he has thought one step further than many, and has touched on what might eventually lead to a more sophisticated understanding.

Reade (1878*a*) considered flow to a well in some detail. He shows that flow to wells is greater than predicted from calculations based on Roberts' (1869) experimental results, and concludes that the sandstones are

> 'in fact, a large rock-filter, with veins and ramifications extending in various directions, which enable us to tap and draw off the supply; and it is this freer circulation than what would take place through homogeneous rock that enables us to draw in some cases those immense supplies, such as is obtained at Green Lane, of 3,243,549 gallons per diem [14.8 Ml/d] as a maximum . . .'

Roberts (1869) notes that the area of the base of the inverted ('irregular') 'water cone' expands until recharge equal to the pumping rate is intercepted. He thus had an explanation for the fall in water level in a well when pumping begins, but does not discuss any change in the angle of the apex of his 'water cone'. By the late 1880s, de Rance (1887) was using the term 'cone of exhaustion' in place of Roberts' (1869) 'water cone', but there is no evidence as to whether the understanding has significantly improved.

Fox, F. (1886) includes some interesting details of the drilling of some large diameter (40 inch/1m) boreholes for part of the lift systems for stations associated with the Mersey Tunnel Railway stations. Designs for tubbing of various shafts sunk for the Tunnel relied upon expertise from the mining engineers of Durham (see Younger 2004).

Groundwater quality

Fresh groundwater

The first major study of fresh groundwater chemistry from the region was that of Houlston (1773) (or possibly by Worthington, mentioned by Houlston as having published on experiments with the Liverpool Spa water during the printing of Houlston's own paper). After a discourse on different types of spring waters – saline, sulphurous, and metallic – Houlston (1773) describes a sequence of 29 experiments designed mainly to understand the iron-rich chemistry of a spring issuing from a quarry in Liverpool. He rightly concludes that CO_2 plays an important role in the dissolution of Fe in this section, a chemical process that had only recently been proposed (Lane 1769). Little more of CO_2 is heard of in the local literature, but by 1880 its importance for groundwater chemistry had been accepted (de Rance 1880*b*), and by 1908 it was being implicated in climate change by researchers publishing in the Liverpool Geological Society proceedings (Dwerryhouse 1909). Houlston (1773), again emphasizing the long pedigree of many present day issues, finishes his paper with a discussion of lead pollution citing major sources as pipes, paint and cosmetics (and undertakes experiments which lead to the suggestion that meat might be preserved by excluding oxygen). Stephenson (1850) tentatively suggests (on rather inadequate data) an increase in hardness with time of pumping for wells in Liverpool. Reade (1878*a*, 1884) confirms this, but it is unclear what either author thinks is the mechanism. One possibility is influx of through-drift recharge containing significant amounts of Ca and SO_4 (Tellam 1996). This agrees with evidence cited by Boult (1875) which indicates harder water overlying softer water. Hull (1863) talks of hardness and relates it to the carbonate content of the rocks. He also discusses Fe. He points out that the quality of the groundwaters is much more constant than that of surface waters and uses data collected by R. A. Smith to infer the source of some of the groundwaters in Manchester. By 1882, Hull is suggesting

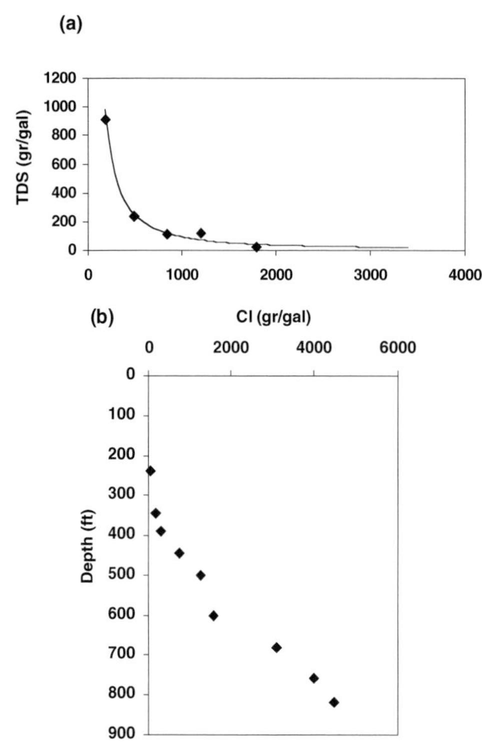

Fig. 6. (**a**) The drop in total dissolved solids as a function of distance from the Mersey according to the data collated by Roberts (1869) (1200 grains per gallon is ~17000 mg/l; 4000 yards is ~3.7 km) (**b**) The increase in chloride concentration as a function of depth at Dallam Lane Forge well (de Rance 1879) (5000 grains per gallon is ~71300 mg/l; 900 feet is ~275 m).

that the decline in water levels and the increased hardness are both indications of abstraction being in excess of recharge, though he does not explain the reason for the chemical change.

Saline intrusion

Saline groundwater was reported in Liverpool as early as 1668 (Moore (1668) cited in Touzeau (1910) as reported by Allen (1969)). Roberts (1873) indicates that some of this saline groundwater may have originated from a pool, marked on Leland's 1539 map, that was connected via a sluice to the estuary. Stephenson (1850) discusses the line of a creek which could also have introduced saline water into the aquifer. Stephenson (1850) collated chemical analyses, and undertook new sampling. He attempted to use the data to determine the saline groundwater zones and their movement in time from 1846 to 1850 and concluded that the chemistry of the water was consistent with an Estuary origin. However, he noticed that the hardness did not precisely agree with the chloride concentrations, and in a discussion of Braithwaite's (1855) paper, which compared the groundwater chemistry of London with that of Liverpool, Stephenson (1855) suggests that the Ca-Cl composition of the well waters close to the Mersey may have been changed by reaction with 'earths', as discussed by Gustav Bischoff in his 'valuable "Treatise on Chemical Geology"' (Cavendish Society). This was possibly the first time ion exchange had been suggested as a process during intrusion of the Mersey water, a century before Hibbert (1956) was to return to the idea. Roberts (1869, 1871) collated data on salinity variations with distance from the Mersey (Fig. 6a).

Deep saline groundwaters

Although saline groundwater underlies the fresh groundwater throughout much of the area (Tellam *et al.* 1986), few records seem to exist in the 19th century. Binney (1862) and Dawkins (1876) note saline water at 460 feet (140 m) in the Permian Sandstone in a well in Ordsall, Manchester, and assumed its origin to be the Permian mudrock sequence. Ormerod (1848) notes a saline water spring at Woolden Hall, to the west of Manchester, and links it to the nearby (Mercia Mudstone Group) saltfield. de Rance (1879, 1880*a*, *b*, c, 1886) presents salinity data collected during the sinking of the Dallam Lane Forge well in Warrington (Fig. 6b), and subsequent work by the erstwhile National Rivers Authority on borehole core porewaters and geophysical log profiles has confirmed that the profile obtained at Dallam Forge is typical (Brassington 1992; Brassington *et al.* 1992; Tellam *et al.* 1986; Tellam 1995*a*). From the location and concentration, it was clear that the saline water was not of recent marine origin. de Rance (1880*a*) at first suggested that the saline water had migrated from the salt sequences in the Mercia Mudstone Group via a north-south fault, but later changed his mind in preference of a source in the Carboniferous (de Rance 1880*b*, *c*, 1886). Part of his reasoning for this change of mind was that he could not imagine the saline water travelling along a fault without being 'absorbed' by the surrounding sandstone – an early expression of the dual porosity model as it relates to solute migration (de Rance 1880*b*). This new theory for the origin of the saline waters provoked a good deal of discussion in his audience, much of it in favour of the Triassic evaporites as a source, and later work on the chemical composition of the sandstone brines strongly suggests that they come predominantly from the Triassic evaporites rather than from the Coal Measures (Tellam 1995*a*).

Pollution

R. Angus Smith presented a summary of a great deal of research on pollution of air, rain water, and cistern waters in the Liverpool/Manchester area in his ground-breaking 1872 book on 'chemical climatology' (Smith 1872). He also undertook chemical analyses of well waters (Smith 1849; Hull 1863). Several authors are worried by pollution from sewage. Boult (1875) reports that when Green Lane Well in Liverpool started abstracting, 63 private wells were drained, 'most of which probably became cesspools afterwards', and that one was a cattle market well. Reade (1878*a*) noted that the water quality of the Windsor Well, again in Liverpool, improved after a sewerage system had been installed across the area. He also presents analyses of public well water samples taken in 1868 with nitrate concentrations of 12, 4 and 64 mg/l (Bevington Bush, Copperas Hill and Soho wells respectively). He interprets the pollution to be from sewage and in one case talks of the 'urine and other sewage matter with which the ground around the well is saturated'. Boult (1875) cites an 1875 report by the public analyst for Liverpool, Dr Campbell Brown, that states that the deepest water in the Liverpool Town Council Bootle Well (at ~1334 feet (400 m) one of the deepest boreholes in the UK (de Rance 1879)) was 'almost as free from nitrates as river water', demonstrating, if demonstration were needed, that high nitrate was an urban hydrogeology problem well before it became a rural groundwater problem. Dawkins (1876) even suggests a diaphragm that could be inserted in a well to spread the catchment and reduce local sewage influxes. Brown, the public analyst for Liverpool, found that the deeper groundwater contained less organic matter than that nearer the surface (Boult 1875; de Rance 1889), an observation rarely investigated since (an exception is Stagg *et al.* 1997). Graveyards were another concern (de Rance 1887). Bostock (1869) suggests that it would be unlikely that a well would be sited in a graveyard, but that the siting of a graveyard near a well 'would appear to be quite another matter'. Hull (1863) claimed that the sandstones filter 'all noxious impurities' from the water. The Royal Commission into the Pollution of Rivers suggests that the New Red Sandstone is one of the most effective filters known, and a destroyer of organic matter:

> 'for being a ferruginous rock, it exerts a powerful oxidizing influence upon the dissolved organic matter which percolates through it.'

(de Rance 1889). This is an issue which is only now being revisited. [In an interesting aside, de Rance (1886) notes that zero valent Fe was used in the Antwerp Waterworks for getting rid of organic matter, over a century before its re-introduction in permeable reactive barriers.]

Geochemistry

Apart from a few chemical analyses of the sandstones (Brown 1878), the first geochemistry papers in the region emerged around the turn of the century. Rhodes (1900) describes the chemistry of an industrial waste deposit in Widnes, just to the east of Liverpool, and this must be one of the first contaminated land geochemistry papers ever published. The deposit, 10 million tons of it, covering an area of ~2 km^2 to a depth of 12 feet (3.7 m), consisted mainly of calcium, iron and sodium sulphides, calcium hydrate, and calcium carbonate. Partial oxidation was indicated by the presence of large selenite crystals (which may explain the high sulphate concentrations noted by Carlyle *et al.* (2004) for one well close to the Mersey). Holland (1896) and Hewitt (1898) present chemical analyses of rock samples, including samples of fill material from fractures. Moore (1898, 1902) presents some interesting data on the chemistry of red and buff coloured sandstones and on concretions. Travis and Greenwood (1911) and Greenwood and Travis (1914) published detailed descriptions of the mineralogy and geochemistry of the core of a deep borehole at Heswall, Wirral. One of their many findings was that carbonate had been dissolved out from the upper parts of the sandstone profile, a feature of major significance to groundwater chemistry and pollutant mobility in the sandstones across many parts of the UK (e.g. Tellam 1994, 1995*b*). They also note halite in the lower parts of the core.

Conclusions

After 50 years of consideration, the conceptual model at the turn of the century included:

- rainfall is the source of all groundwater except intrusion from surface water bodies;
- recharge is reduced by till and urban areas, but runoff may recharge at the edge of low permeability cover;
- in some places the recharge will be head-dependent;
- both intergranular and fracture flow occurs, and faults can sometimes be barriers to flow;
- well interference occurs, increasing total drawdown, and well yields are strongly dependent on fractures;
- seawater is modified during intrusion;
- urban pollution is occurring, with high NO_3 in

groundwater resulting from sewage infiltration;
- brines occur at depth in the sandstones, and their source is the underlying Carboniferous deposits.

With the exception of the last part of the last conclusion, all the other points are in agreement with current interpretations for the region, and the basic conceptual model has changed little (cf. Schwartz & Ibaraki 2001). Indeed, some of the hot topics from the 19th century are still unresolved today, the role of faults and fractures and the quantification of recharge being two main examples. Where the understanding of the 19th century workers differed most from that of today was in the quantification of flow systems and the details of the chemistry.

In addition to this conceptual model, the researchers determined, from experimental work, that flow rate is proportional to pressure difference, only a proportion of the water in the rock can be drained by gravity, lamination has a significant effect on unsaturated flow and that contaminant breakthrough curves are sigmoidal in shape. Most of these conclusions had already been made by researchers elsewhere in the world, but this was not known by the workers in the Liverpool/Manchester area, and their achievement should not therefore be considered diminished. It is unclear whether the amateur and partially closed research environment in which they worked was ultimately responsible for the repetition of effort, or whether the generally less efficient communications of the time were to blame. Either way, the work of the amateur hydrogeologists of the Liverpool and Manchester areas demonstrates a parallelism or repetition in the development of the subject, a parallelism which must have been occurring at many locations around the world. The role of time-dependent communication on the development of science appears to be a topic which is under-represented in most theories on the way science works.

References

Note that for the Liverpool and Manchester Geological societies there is often a delay between the reading of a paper and its publication. The date a paper was read is often known quite precisely, but sometimes the date of publishing is less precisely known. In general, where possible the dates given below are publication dates.

ALLEN, A. D. 1969. *The hydrogeology of the Merseyside area.* PhD thesis, University College, London University.

ALLEN, D. J., BREWERTON, L. J., COLEBY, L. M., GIBBS, B. R., LEWIS, M. A., MACDONALD, A. M., WAGSTAFF, S. J. & WILLIAMS, A. T. 1997. *The physical properties of major aquifers in England and Wales.* Environment Agency R&D Publication 8 / British Geological Survey Technical Report **WD/97/34**.

BINNEY, E. W. 1862. Geology of Manchester and its neighbourhood. *Transactions of the Manchester Geological Society*, **3**, 350–365.

BOSTOCK, R. 1869. The New Red Sandstone as a source of water supply. *Proceedings of the Liverpool Geological Society*, **1**, 58–71.

BOULT, J. 1875. An inquiry into the source of water in the New Red Sandstone (Part 2). *Journal of the Liverpool Polytechnic Society*, 18–47.

BRAITHWAITE, F. 1855. On the filtration of salt-water into the springs and wells under London and Liverpool. *Minutes of the Proceedings of the Institution of Civil Engineers*, **14**, 507–523.

BRASSINGTON, F. C. 1992. Measurements of head variations within observation boreholes and their implication for groundwater monitoring. *Journal of the Institution of Water and Environmental Management*, **6**, 91–100.

BRASSINGTON, F. C., LUCEY, P. A. & PEACOCK, A. J. 1992. The use of downhole focussed electrical logs to investigate saline groundwater, *Quarterly Journal of Engineering Geology*, **25**, 343–350.

BROWN, J. C. 1878. Analyses of rocks from the 1300 feet deep bore-hole, at Bootle. *Proceedings of the Liverpool Geological Society*, **session twentieth**, 63.

CAMPBELL, J. E. 1982. *Permeability characteristics of the Permo-Triassic sandstones of the Lower Mersey Basin.* MSc thesis, University of Birmingham.

CARLYLE, H. F., TELLAM, J. H. & PARKER, K. E. 2004. The use of laboratory-determined ion exchange parameters in the prediction of field-scale major cation migration over a 40 year period. *Journal of Contaminant Hydrology*, **68**, 55–81.

CUNNINGHAM, J. 1847. On the geological conformation of the neighbourhood of Liverpool, as respects the supply of water. *Proceedings of the Literary and Philosophical Society of Liverpool*, **3**, 58–.

DALTON, J. 1799. Experiments and observations to determine whether the quantity of rain and dew is equal to the quantity of water carried off by the rivers and raised by evaporation. *Memiors of the Literary and Philosophical Society of Manchester*, **V**, 346.

DARCY, H. P.-G. 1856. *Les fontaines publiques de la ville de Dijon.* Victor Dalmont, Paris.

DARWIN, C. R. 1859. *The origin of species by natural selection.* John Murray, London.

DAVIS, S. N. & DE WIEST, R. J. M. 1966. *Hydrogeology.* John Wiley and Sons, New York, 463p.

DAWKINS, W. B. 1876. On the water supply in the red rocks of Lancashire and Cheshire. *Transactions of the Manchester Geological Society*, **14**, 133–374.

DAWKINS, W. B. 1880. On the Palaeozoic and secondary rocks of England, as a source of water-supply, for towns and districts, discussion, *Transactions of the Manchester Geological Society*, **14**, 449–450.

DE RANCE, C. E. (Reporter) 1876. First Report of the Committee for investigating the circulation of the Underground Waters in the New Red Sandstone and Permian Formations of England, and the quantity and character of the water supplied to various towns and districts from these formations. *In: Report of the Forty-Fifth Meeting of the British Association for the Advancement of Science held at Bristol in August 1875*, John Murray, London, p. 114–141.

DE RANCE, C. E. (Reporter) 1879. Fourth Report of the

Committee for Investigating the Circulation of the Underground Waters in the Jurassic, New Red Sandstone and Permian Formations of England, and the Quantity and Character of the Waters supplied to various Towns and Districts from these Formations, with Appendix by Mr Roberts, on the Filtration of Water through Triassic Sandstone. *In: Report of the Forty-Eighth Meeting of the British Association for the Advancement of Science held at Durham in August 1878*. John Murray, London, p. 382–419.

DE RANCE, C. E. 1880*a*. Notes on some Triassic borings. *Transactions of the Manchester Geological Society*, **15**, 90–112.

DE RANCE, C. E. 1880*b*. Further notes of Triassic borings near Warrington. *Transactions of the Manchester Geological Society*, **15**, 388–398.

DE RANCE, C. E. (Reporter) 1880*c*. Sixth Report of the Committee consisting of . . . appointed for investigating the Circulation of the Underground Waters in the Permian, New Red Sandstone and Jurassic Formations of England, and the Quantity and Character of the Waters supplied to Towns and Districts from those Formations. *In: Report of the Fiftieth Meeting of the British Association for the Advancement of Science held at Swansea in August and September 1880*. John Murray, London, p. 87–106.

DE RANCE, C. E. 1880*d*. On the Palaeozoic and secondary rocks of England, as a source of water-supply, for towns and districts. *Transactions of the Manchester Geological Society*, **14**, 403–447.

DE RANCE, C. E. 1886. On the occurrence of brine in the coal measures, with some remarks on filtration. *Transactions of the Manchester Geological Society*, **18**, 61–81.

DE RANCE, C. E. 1887. Notes on Lancashire water supply. *Transactions of the Sanitary Institute of Great Britain*, **9**, 6p.

DE RANCE, C. E. (Reporter) 1888. Thirteenth Report of the Committee appointed for the purpose of investigating the Circulation of Underground Waters in the Permeable Formations of England and Wales, and the Quantity and Character of the Waters supplied to Towns and Districts from those Formations. *In: Report of the Fifty-Seventh Meeting of the British Association for the Advancement of Science held at Manchester in August and September 1887*. John Murray, London, p. 358–414.

DE RANCE, C. E. 1889. Notes on underground water-supply and river floods. *Proceedings of the Yorkshire Geological and Polytechnic Society*, **11**, 200–216.

DIGGES LA TOUCHE, S. V. 1998. *The unsaturated zone of a Triassic sandstone aquifer*. PhD thesis, University of Birmingham.

DUPUIT, J. 1863. Etudes théoriques et pratiques sur le mouvement des eaux dans les caneaux découverts et a Frauers les terrains. Dunod, Paris, 304pp.

DWERRYHOUSE, A. R. 1909. Carbonic acid. *Proceedings of the Liverpool Geological Society*, **10**, 289–308.

FOX, D. 1886. The Mersey Railway, discussion. *Minutes of the Proceedings of the Institution of Civil Engineers*, **86**, 105.

FOX, F. 1886. The Mersey Railway. *Minutes of the Proceedings of the Institution of Civil Engineers*, **86**, 40–59.

FREY, A. 1999. *Groundwater recharge and pesticide leaching in a Triassic sandstone aquifer in south-west England*. PhD thesis, University of Exeter, 330p.

GRABHAM, G. W. 1910. Notes on some recent contributions to the study of desert water-supplies. *The Cairo Scientific Journal*, **No. 46**, Vol. IV, 9p.

GREENWOOD, H. W. & TRAVIS, C. B. 1914. The mineralogical and chemical constitution of the Triassic rocks of Wirral, PtII. *Proceedings of the Liverpool Geological Society*, **12**, 161–188.

HALLAM, A. 1989. *Great geological controversies*, 2nd edition. Oxford University Press, 244p.

HEWITT, W. 1898. Notes on some sections exposed by excavations on the site of the new technical schools, Byrom Street, Liverpool. *Proceedings of the Liverpool Geological Society, session the thirty-ninth*, **8**, 268–273.

HEWITT, W. 1910. *The Liverpool Geological Society A retrospect of fifty years' existence and work*. C. Tinling and Co., Liverpool.

HIBBERT, E. S. 1956. The hydrogeology of the Wirral Peninsula. *Journal of the Institution Water Engineers*, **10**, 441–469.

HOLLAND, P. 1896. Notes on analyses of Permian and Triassic rocks from the neighbourhood of Liverpool. *Proceedings of the Liverpool Geological Society, session the thirty-fifth*, **7**, 443–452.

HOULSTON, T. 1773. *Essay on the Liverpool Spa water*. Williamson, Liverpool, 73p.

HOWARD, K. W. F. 1987. Beneficial aspects of sea-water intrusion. *Ground Water*, **25**, 398–406.

HUBBERT, M. K. 1940. The theory of groundwater motion. *Journal of Geology*, **48**, 785–944.

HULL, E. 1863. *On the New Red Sandstone and Permian formations as sources of water-supply for towns*. Memoirs of the Manchester Literary and Philosophical Society, II(series 3), 256–276.

HULL, E. 1882. *The Geology of the country around Prescot, Lancashire*, 3rd edition. Memoir of the Geological Survey, HMSO, London, 65p.

LANE, 1769. *Philosophical Transactions of the Royal Society*, **222**, cited by HOULSTON (1773).

LUCAS, J. 1874. *Horizontal wells, a new application of geological principles to effect the solution of the problem of supplying London with pure water*. Stanford, London.

LUCAS, J. 1877. Hydrogeology: one of the developments of modern practical geology. *Transactions of the Institution of Surveyors*, **IX (VII)**, 153–184.

LYELL, C. 1830–1875. Principles of Geology, many editions. John Murray, London.

MATHER, J. D. 2001. Joseph Lucas and the term 'hydrogeology'. *Hydrogeology Journal*, **9**, 413–415.

MATHER, J. D. *et al.* 2004. Joseph Lucas (1846–1926) – Victorian polymath and key figure in the development of British hydrogeology. *In:* MATHER, J. D. (ed.) *200 Years of British Hydrogeology*. Geological Society, London, Special Publications, **225**, 67–88.

MEINZER, O. E. 1928. Compressibility and elasticity of artesian aquifers. *Economic Geology*, **23**, 263–291.

MOORE, C. C. 1898. The chemical examination of sandstones from Prenton Hill and Bidston Hill. *Proceedings of the Liverpool Geological Society, session the thirty-ninth*, **8**, 241–267.

MOORE, C. C. 1902. The study of the volume composition of rocks, and its importance to the geologist. *Proceedings of the Liverpool Geological Society*, **9**, 129–162.

MORTON, G. H. 1866. On the position of the wells for the supply of water in the neighbourhood. *Proceedings of the Liverpool Geological Society, session seventh*, 27–30.

MORTON, G. H. 1870. Anniversary address by the President. *Proceedings of the Liverpool Geological Society, session twelfth*, 1–29.

ORMEROD, G. W. 1848. Outline of the principal geological features of the saltfield of Cheshire and adjoining districts. *Quarterly Journal of the Geological Society of London*, **4**, 262–288.

PEARCE, J. M., HOUGH, E., WILLIAMS, G. M., WEALTHALL, G. P., TELLAM, J. H. & HERBERT, A. W. 2001. Sediment-filled fractures in Triassic Sandstones – pathways or barriers to contaminant migration? *In:* KUEPER, B. H., NOVAKOWSKI, K. S. and REYNOLDS, D. A. (eds) *Proceedings of Fractured Rock 2001, Toronto, Canada*. CD publication (http://www.fracturedrock.org/ proceedings.html).

PERRAULT, P. 1674. *De l'origine des fontaines*. Translation by La Roque, A. 1967. Hafner Publishing Co., New York, USA.

PINNEKER, E. V. 1983. *General hydrogeology*. Cambridge University Press, Cambridge, 141p.

PREENE, M. Robert Stephenson (1803–1859) – the first groundwater engineer. *In:* MATHER, J. D. (ed) *200 Years of British Hydrogeology*. Geological Society, London, Special Publications, **225**, 000–000.

READE, T. M. 1878*a*. On the South-Lancashire Wells. p. 66–72. *In:* DE RANCE, C. E. (compiler). Third Report of the Committee for investigating the circulation of the Underground Waters in the New Red Sandstone and Permian Formations of England, and the quantity and character of the Water supplied to various towns and districts from these formations. *Report of the Forty-Seventh Meeting of the British Association for the Advancement of Science held at Plymouth in August 1877*. John Murray, London, p. 66–72.

READE, T. M. 1878*b*. President's address [chemical denudation in relation to geological time]. *Proceedings of the Liverpool Geological Society, session eighteenth*, **3**, 211–235.

READE, T. M. 1884. Experiments on the circulation of water in sandstone. *Proceedings of the Liverpool Geological Society, session twenty-fifth*, 434–447.

RHODES, E. 1900. Notes on crystals found in alkali waste, Widnes. *Proceedings of the Liverpool Geological Society*, **8**, 479–482.

ROBERTS, I. 1869. On the wells and water of Liverpool. *Proceedings of the Liverpool Geological Society*, **1**, 84–97.

ROBERTS, I. 1871. Effect produced by red sandstone upon salt water. *Proceedings of the Liverpool Geological Society*, **2**, 66–68.

ROBERTS, I. 1873. Section of strata above the boulder-clay at Whitechapel, Liverpool. *Proceedings of the Liverpool Geological Society, session fourteenth*, 32–34.

ROBERTS, I. 1879. Experiments on the filtration of sea water through Triassic Sandstone. *In:* DE RANCE, C. E. (reporter). Fourth Report of the Committee for Investigating the Circulation of the Underground Waters in the Jurassic, New Red Sandstone and Permian Formations of England, and the Quantity and Character of the Waters supplied to various Towns and Districts from these Formations, with Appendix by Mr Roberts, on the Filtration of Water through Triassic Sandstone. Report of the Forty-Eighth Meeting of the British Association for the Advancement of Science held at Durham in August 1878. John Murray, London, p. 397–401.

ROBERTS, I. 1880. Notes on the strata and water-level at Maghull. *Proceedings of the Liverpool Geological Society, session twenty-second*, 233–236.

RUSHTON, K. R. & HOWARD, K. W. F. 1982. The unreliability of open observation boreholes in unconfined aquifer pumping tests. *Ground Water*, **20**, 546–550.

RUSHTON, K. R., KAWECKI, M. W. & BRASSINGTON, F. C. 1988. Groundwater model of conditions in Liverpool sandstone aquifer. *Journal of the Institution of Water and Environmental Management*, **2**, 67–84.

SMITH, R. A. 1849. On the Air and Water of Towns. *In: Report of the Eighteenth Meeting of the British Association for the Advancement of Science held at Swansea in August 1878*. John Murray, London, p. 16–31.

SMITH, R. A. 1856. Memoir of John Dalton and history of the atomic theory up to this time. *Memoirs of the Literary and Philosophical Society of Manchester*, H. Bailliere, London, 316p.

SMITH, R. A. 1872. *The beginnings of a chemical climatology*. Longmans, Green, and Co., London, 600p.

STAGG, K. A., TELLAM, J. H., BARRETT, M. H. & LERNER, D. N. 1998. Hydrochemical variations with depth in a major UK aquifer: the fractured, high permeability Triassic Sandstone. *In:* BRAHANA *et al.* (eds) *Gambling with groundwater*. Proceedings of the International Association of Hydrogeologists Conference, Las Vegas, 53–58.

STEPHENSON, R. 1855. On the filtration of salt-water into the springs and wells under London and Liverpool, discussion. *Minutes of the Proceedings of the Institution of Civil Engineers*, **14**, 521.

STEPHENSON, R. 1850. *On the supply of water to the town of Liverpool*. Bradbury and Evans (printers), London.

SCHWARTZ, F. W. & IBARAKI, M. 2001. Hydrogeological Research: Beginning of the End or End of the Beginning? *Ground Water*, **39**, 492–498.

TELLAM, J. H. 1994. The groundwater chemistry of the Lower Mersey Basin Permo-Triassic Sandstone aquifer system, UK: 1980 and pre-industrialisation/ urbanisation. *Journal of Hydrology*, **161**, 287–325.

TELLAM, J. H. 1995*a*. Hydrochemistry of the saline groundwaters of the lower Mersey Basin Permo-Triassic sandstone aquifer, UK. *Journal of Hydrology*, 165, 45–84.

TELLAM, J. H. 1995*b*. Urban groundwater pollution in the Birmingham Triassic Sandstone aquifer. *Proceedings of 4th Annual IBC Conference on Groundwater Pollution, London, 15–16th March, 1995*, 10p.

TELLAM, J. H. 1996. Interpreting the borehole water chemistry of the Permo-Triassic sandstone aquifer of the Liverpool area, UK. *Geological Journal*, **31**, 61–87.

TELLAM, J. H., LLOYD, J. W. & WALTERS, M. 1986. The morphology of a saline groundwater body; its investigation, description and possible explanation. *Journal of Hydrology*, **83**, 1–21.

TOUZEAU, J. 1910. *The rise and progress of Liverpool from 1551 to 1835*. Liverpool Booksellers Co., Liverpool.

TRAVIS, C. B. & GREENWOOD, W. H. 1911. The mineralogical and chemical constitution of the Triassic rocks of Wirral. *Proceedings of the Liverpool Geological Society*, **11**, 116–139.

UNIVERSITY OF BIRMINGHAM/NWWA, 1983. *Lower Mersey Basin saline groundwater study*. Final Summary Report, 81p.

WEALTHALL, G. P., STEELE, A., BLOOMFIELD, J. P., MOSS, R. H. & LERNER, D. N. 2001. Sediment filled fractures in the Triassic Sandstones in the Cheshire Basin: observations and implications for pollutant transport. *Journal of Contaminant Hydrology*, **50**, 41–51.

YOUNGER, P. L. 2004. 'Making water': the hydrogeological adventures of Britain's early mining engineers. *In:* MATHER, J. D. (ed) *200 Years of British Hydrogeology*. Geological Society, London, Special Publications, **225**, 121–158.

Robert Stephenson (1803–1859) – the first groundwater engineer

M. PREENE

Arup, Rose Wharf, 78 East Street, Leeds, LS9 8EE, UK

Abstract: From a humble background in the mining communities of Tyne and Wear, with little academic education, Robert Stephenson followed in the footsteps of his father, George, and became one of the foremost civil and mechanical engineers of the early 19th Century. While he is primarily associated with railways, Robert Stephenson had considerable dealings with groundwater during his professional life, applying a rational, empirical approach that would be familiar to modern practitioners. Stephenson's approach to groundwater issues was probably shaped largely by the years spent battling water-bearing quicksands during construction of the Kilsby Tunnel near Rugby on the London to Birmingham Railway. Careful observations allowed him to conclude that local drainage by use of arrays of wells was possible, without the need to drain the whole aquifer body. Later in his career he advised on public water supplies from the Chalk for London and the Sherwood Sandstone for Liverpool. His careful observations and reasoned interpretation, allowed him to advance the concept of a 'cone of influence' around a pumped well and to develop tests and monitoring programmes to assess the impact of new abstractions on existing water features. Today, his work may seem basic, even obvious, but, in the days before the work of Darcy and Dupuit, there were many who disputed his findings. Stephenson preferred to let the facts to speak for themselves, but where this was not possible he vigorously publicised the benefit of applying a scientific approach to the management and control of groundwater.

By any reasonable measure, Robert Stephenson was one of the foremost civil and mechanical engineers of his time. Furthermore, he was one the most prominent figures in professional society. There can be few contemporaries from science and engineering who were so fêted with awards and honours. Stephenson was elected a Fellow of the Royal Society, was President of the Institution of Civil Engineers in 1855–1857, was awarded the *Legion de Honneur* by the French Emperor, received an honorary Doctor of Civil Law at Oxford University (in company with Isambard Kingdom Brunel and the explorer Dr Livingstone), and was twice elected as the Member of Parliament for Whitby, North Yorkshire, serving there until his death in 1859. Even in death he was honoured, being laid to rest in Westminster Abbey. A contemporary obituary (Anon. 1860) describes the funeral as being 'conducted with fitting solemnity . . . in the presence of several thousand persons, amongst whom were not only his professional and private friends, but also the most distinguished representatives of the literature, science and art of Great Britain in the nineteenth century'.

The esteem with which Stephenson was held, by business, science and the public was based on his work on the development of the railway network and locomotive technology in Britain, building on the pioneering work of his father George. This is fully documented in contemporary and modern biographies (Jeaffreson 1864; Rolt 1960). However, what is much less well known is that Robert Stephenson developed considerable practical expertise in the management of groundwater, both as a source of supply and when it posed an impediment to construction works. That Stephenson developed groundwater sources and groundwater lowering systems which would be recognizable to modern practitioners is a tribute to his skills, especially when it is considered that most of his work pre-dates even the publication of Darcy's law (Darcy 1856).

Stephenson's groundwater work had three main elements: fighting against groundwater during the construction of the Kilsby Tunnel on the London to Birmingham Railway (1835–1838), work on developing water supplies for the metropolis of London (1841 onwards), and study of the water supply for the City of Liverpool (1850). That this work is not widely known among the hydrogeological community is perhaps because Stephenson did not publish papers in the scientific or learned journals – one of his few formal contributions being the section on 'iron bridges' in the *Encyclopaedia Britannica* of the time. However, Stephenson produced numerous professional reports and contributed to technical discussions minuted by the Institution of Civil Engineers and others. The remainder of this paper will use those sources to document the range of Stephenson's work with groundwater. This will include his work developing sources in major aquifers such as the Chalk and Sherwood Sandstone, his epic groundwater lowering exploits at Kilsby and his conceptual understanding of the 'cone of depression' around a well source and the interaction between wells.

To better understand how Robert Stephenson was

From: MATHER, J. D. (ed.) 2004. *200 Years of British Hydrogeology*. Geological Society, London, Special Publications, **225**. 107–119. 0305-8719/04/$15 © The Geological Society of London.

perhaps uniquely equipped to manage and manipulate groundwater to his will, with little contemporary theoretical support, it is worthwhile considering his background and early work.

The making of Robert Stephenson the engineer

It is well documented that Robert Stephenson did not have an especially academic upbringing (Jeaffreson 1864; Rolt 1960). Born in 1803 at Willington, Tyne and Wear, Robert was the only son of George Stephenson, who at the time was the brakesman of a local colliery engine. Robert was involved with his father's work at the pit from an early age, and in this way received much education of a practical and mechanical nature. On leaving school he was apprenticed to his father's friend Nicholas Wood, the mining engineer at Killingworth Colliery, but was released from this on grounds of ill health. In 1822–1823 Robert spent one term (probably of less than six months duration) at the University of Edinburgh studying natural philosophy, chemistry and natural history. These studies included geological excursions with Professor Jamieson, which Stephenson recalled fondly in later life. An award by Professor Leslie in recognition of his mathematical skills was perhaps a sign of a latent analytical skill that would prove useful later.

In the early years of his professional career, Robert (Fig. 1) was naturally in the shadow of his father, who by the 1830s had become the driving force behind the development and implementation of railway systems and steam locomotive technology. During this time Robert travelled to Venezuela and Colombia to act as Engineer to the Colombian Mining Association. This was not the most fruitful relationship and after three years he returned to Britain to continue his work as a railway engineer. His chance to establish himself in his own right came from 1830 during the construction of the London to Birmingham Railway, where he was Engineer-in-Chief. At the time he could not have predicted how groundwater would play such an important role in the construction of the project.

Fig. 1. Robert Stephenson as a young man (from Jeaffreson 1864).

The London to Birmingham railway

The idea of a railway linking London and Birmingham had first been seriously considered in the 1820s. In 1830 the London and Birmingham Railway Company was formed, and subsequently appointed the company of George Stephenson and Son to make plans and surveys and to carry the line through the parliamentary process. The surveys and plans were made by Robert himself. In 1833 the Railway Act was passed by the Commons and the Lords and Royal Assent was granted. In the same year Robert Stephenson was appointed Engineer-in-Chief at an annual salary of £1500 plus expenses of £200 per year. The directors of the company stated that 'they are persuaded that to no one could this charge be more safely or more properly confided' (Jeaffreson 1864). This view was to be ultimately confirmed by the works at Kilsby.

The London to Birmingham Railway was the first of the railways in the modern pattern, allowing high speed travel between major cities. The railway covered 180 km between the cities, with its metropolitan terminus on a vacant piece of land at Euston Square, now Euston Station. The quality of the chosen alignment, minimising steep gradients and sharp curves, is illustrated by the fact that the railway now forms part of the West Coast Main Line. Until recently, very few major engineering works have been necessary to allow trains to travel in excess of 160 km per hour along a line where the original rail traffic was to average 30 km per hour.

It was the desire to have a line with minimal gradients that resulted in the alignment with a number of substantial cuttings, tunnels, embankments and viaducts to carry the railway through or over geographical obstacles. With hindsight the most notable and challenging of these was the tunnel at Kilsby.

The Kilsby Tunnel

The construction of the Kilsby Tunnel has been thoroughly documented by Lewis (1984). It was an epic undertaking. Required to carry the railway through a ridge of high ground, the tunnel itself, located approximately 8 km southeast, of Rugby is 2217 m long, 7.6 m wide and 8.5 m high, with two huge ventilation shafts (Fig. 2). The tunnel was up to 63 m below the summit of the ridge. Constructed between 1835 and 1838 at a final cost of £291 030, 1250 labourers were employed at the height of the works. 26 men lost their lives building the tunnel.

As Engineer-in-Chief Stephenson was determined that the project would be planned, programmed and costed in a rational manner, an approach that would be familiar to the modern construction industry. At the time this was a revolutionary approach. The entire line was divided into 30 separate contracts, each with their own drawings, estimates and specifications. Most of the works were supervised by an Assistant Engineer (responsible for a section of line), reporting to Stephenson as Engineer-in-Chief. The contract for the construction of the Kilsby Tunnel was let to the contractor Joseph Nowell and Sons of Dewsbury, Yorkshire in May 1835.

As will become apparent, groundwater posed a great obstacle to the construction of the tunnel, but a key question is – had Stephenson considered this in advance? It seems likely that he had, at least to some degree; his alignment had been pushed slightly to the west to avoid 'quicksand' problems encountered nearby during the earlier construction of the Union Canal. Furthermore, Stephenson indicated in correspondence that 'symptoms of quicksand made their appearance [in trial borings]' (Lewis 1984). Very early on in the site works four trial shafts were sunk to allow the contractor to judge the nature of the ground. Very little water was found in these shafts, and the tunnelling work was expected to be easy. The reality was to prove rather different.

Fig. 2. Great ventilating shaft, Kilsby Tunnel (Bourne 1839: courtesy of the Institution of Civil Engineers Library).

Early construction problems

A number of working shafts were to be constructed, from which the tunnel would be driven. However, early in the works the contractor hit water in the working shafts and was flooded out. Conditions were worst between two of the trial shafts, leading Stephenson to believe that a basin of sand 400 m long had been missed by the investigation.

Pumping engines were hurriedly procured and two shafts were sunk to the level of the water, with a 180 m long drift to aid draining of the sand. There were a number of inflows of sand and water into the drift during construction, including one event where the drift was filled with sand over a length of 80 m.

Matters were complicated by the death of the contractor, Joseph Nowell, in January 1836. By late February it was clear that Nowell's sons could not carry on the contract in the necessary manner. In March 1836 Joseph Nowell and Sons relinquished their contract and construction works were taken over by London and Birmingham Railway Company. Stephenson was now fully in charge.

Geological setting

Accounts based on contemporary records (such as Boyd-Dawkins 1898) describe the strata through which the tunnel was driven as Inferior Oolite, based presumably on the original records and geological mapping of the time.

However, modern mapping (British Geological Survey 1980) indicates a different geology. This indicates the tunnel was driven through Middle Lias Silts and Clays comprising silts, mudstones and thin silty limestones. These strata form the ridge of high ground which was the obstacle requiring the railway to pass through a tunnel. The ridge is shown to be capped by sand and gravel and glacial till, and to be underlain by the Lower Lias, comprising mainly

Fig. 3. Working shaft, Kilsby Tunnel (Bourne 1839: courtesy of the Institution of Civil Engineers Library).

mudstones with a few very thin limestone bands. The modern mapping is apparently consistent with William Smith's early mapping (Smith 1815), which pre-dates the construction of the tunnel and identifies the strata at Kilsby as 'Blue Marl of the Lias'.

A crude conceptual model of the ground through which Stephenson drove his tunnel might be as follows. The tunnel was driven through the Middle Lias Silts and Clays, which contained horizons or lenses of water-bearing silts and sands. These strata were underlain by the Lower Lias, which is generally of low permeability, so the water in the Middle Lias Silts and Clays would have been perched above the Lower Lias. Stephenson's tunnel would have intercepted this water and the flow through the silts and sand horizons would have given the quicksand conditions so feared by the tunnellers. That the flows did not significantly diminish with time is consistent with the presence of a mantle of sand and gravel over part of the ridge. This may have acted as a reservoir providing recharge to the silt and sand layers, with vertical flow perhaps facilitated by cambering features.

Stephenson's observations

By the end of July 1836, a total of 11 shafts had been constructed, some for tunnelling (where water was not a problem, Fig. 3), but mostly in wet ground where pumping was necessary. Stephenson's reports, quoted in Lewis (1984) state:

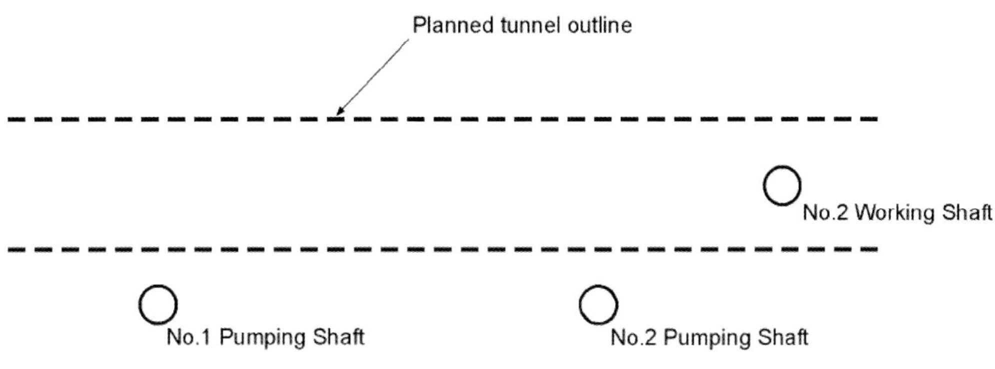

Fig. 4. Kilsby Tunnel, location of Working and Pumping Shafts.

'The distance between the shafts to be sunk in the quicksand must be governed by the advances made in the draining. It is probable that distances will not exceed 100 yards [91 m] from shaft to shaft . . . The diagram [Fig. 4] shows the line of the tunnel in the quicksand and the relative positions of the three shafts by which drainage has been attempted, and its progress observed. When one pump was put to work in No.1 Pumping Shaft, the water immediately fell in No.2 Working Shaft, but No.2 Pumping Shaft being nearer the pump than No.2 Working Shaft, the water level fell lower than in the [other] shaft. This is precisely what might have been anticipated, and demonstrates in the most conclusive manner that the supply of water at this level to the quicksand is very moderate – and that the sand throughout it is so open as to admit freely the passage of water.'

Stephenson then outlines his plans to complete the works:

'From the commencement of pumping on even a small scale, we have been able to lower the general level of the water in the quicksand gradually, notwithstanding any falls of rain which have from time to time taken place. When we first attempted the drainage by drawing water in buckets, a rapid thaw took place, with a considerable quantity of snow on the ground – in this instance the water rose in the shaft. Nothing of the kind has occurred since the pumps were worked; on the contrary, the general level of the water has been lowered at the rate of 10 inches [250 mm] per week. Previously when only one pump was worked, the fall was at the rate of 6 inches [150 mm] per week . . . That local drainage by shafts will be effectual seems placed beyond any reasonable doubt by the complete sympathy existing between the levels of the water in the shafts at a considerable distance from each other, and by the simultaneous effect of one pump in different shafts. As an example in sinking the two pumping shafts, the sinkers were compelled to work in them alternately, for the moment one shaft was sunk a few inches deeper than the other the water flowed into it and left the other in a fit state for sinking. It was in this manner that they were sunk to their present depths. Precisely the same thing has taken place within the last few days in No.2 Working Shaft. The men were some time since driven out of the shaft by the rapid influx of water. The general level of the water having gone on lowering, it is now quite dry, and fit for sinking again to commence.'

Stephenson's method of 'local drainage by shafts', where it is recognized that several wells acting in concert achieve more drawdown than from a single well, and where the spacing between wells is an important parameter, controlled by ground conditions, is the ancestor of modern 'construction dewatering' methods. Shafts may have been replaced by boreholes and pumps are now driven by electricity, rather than by steam (Fig. 5), but the principle remains exactly the same.

Resolution

Stephenson's use of shafts and pumping engines allowed the tunnel to be completed in June 1838, at a cost of £291 030 (nearly three times the original estimate).

It must have been a very heavy workload for Stephenson to be Engineer-in-Chief for the entire railway, while personally managing the works at Kilsby following the loss of the contractor. Many men would have lost themselves in action and would have had little time for reflection on what could be learned from the observations at Kilsby. However, in Stephenson's case, the evidence suggests he gave considerable thought to groundwater flow through the sand. The following statement of his Kilsby experience is from one of his reports on the water supply to London (Stephenson 1841).

'I was soon much surprised to find how slightly the depression of the water level in the one shaft influenced that of the other, notwithstanding a free communication existed between them through the medium of the sand,

Fig. 5. Pumps for draining the Kilsby Tunnel (Bourne 1839: courtesy of the Institution of Civil Engineers Library).

> which was very coarse and open. It then occurred to me that the resistance the water encountered, in its passage through the sand to the pumps, would be accurately measured by the angle or inclination which the surface of the water assumed toward the pumps, and that it would be unnecessary to draw the whole of the water off from the quicksand, but to persevere in pumping only in the precise level of the tunnel, allowing the surface of the water flowing through the sand to assume that inclination which was due to its resistance.
>
> If this view were correct, it was evident that no extent of pumping whatever would have effected the complete drainage of the bed of sand. To test it, therefore, boreholes were put down at about 200 yards [183 m] from the line of the tunnel, when it was clear that, notwithstanding that pumping had been going on incessantly for twelve months, and for the latter six months of this period, at the rate of 1800 gallons per minute [137 $l s^{-1}$], the level of the water in the sand, at a distance not exceeding 200 yards [183 m] had scarcely been reduced.
>
> The simple result, therefore, of all the pumping, was merely to establish and maintain a channel of comparatively dry sand in the immediate line of the intended tunnel, leaving the water heaped up on each side by the resistance which the sand offered to its descent to that line on which the pumps and shafts were situated.'

Put in modern language, Stephenson had identified that water flowing through granular soil experienced a resistance (which we now call hydraulic conductivity). When water is abstracted at a point, the water flowing to that point must overcome the resistance. The groundwater head distant from the abstraction point must be greater than at the abstraction itself, with the slope of the groundwater head (which we call the hydraulic gradient) being controlled by the hydraulic conductivity of the material. This might be interpreted as an alternative formulation of Darcy's Law, 15 years before it was formally published (Darcy 1856). Stephenson had also identified that the lowering of groundwater levels extended only a finite distance from the point of abstraction and reduced with distance. This may seem self-evident today, but at the time there were many who disputed it.

The water supply for the metropolis

London's water in the early nineteenth century

In the first part of the 19th century the water supply to the population of London was a mess. Much of the city's drinking water was drawn from the Thames, which also received almost all the sewage from the metropolis. The river became a foul, stinking pool, the tides conveying putrescent material up and down the tidal stretch of the river. The city's inhabitants were literally drinking dilute sewage (Halliday 1999).

But society was changing and London was growing rapidly; there was need for more wholesome water and lots of it. Barlow (1854) paints a vivid picture:

> 'This marvellous extension [the growth of London's population from 865000 in 1801 to 2362000 in 1841] would have demanded improved means of supply, even if there had not been any change of social habits; but when civilisation and luxurious habits advance with equal strides, baths become necessaries of life for all classes, and sanitary measures enforce the copious use of water in all ways'.

The opinions of the time are articulated in Stephenson's own words 'the present mode of supplying London with water, has, for a length of time, been anything but satisfactory to the public' (Stephenson 1840). Following his first hand experience of groundwater at Kilsby, from around 1840 Stephenson became involved in schemes to supply water for the Metropolis, by means of wells. At this time there were many schemes to improve the water supply and drainage of the capital. Many companies were formed in the speculative hope of making money from this opportunity. Stephenson's contribution is recorded in his two reports to the London Westminster and Metropolitan Water Company (Stephenson 1840, 1841) and various responses and rebuttals by others.

Stephenson's first report

Stephenson's first report (Stephenson 1840) was published in the *Morning Advertiser* of 29 December, 1840. The report proposed a groundwater supply to the city on the basis that:

> 'Nature has supplied us with the means of substituting a pure and unceasing flow of spring water for the outpourings of filthy drains, and that this can be done without encountering difficulties of any but an ordinary nature'.

Following the Royal Commission of 1828 the distinguished engineer Thomas Telford had been appointed to report on new sources of water for London. Other engineers and companies had also proposed water supply schemes. Stephenson begins his report by commenting on these schemes and says 'it is indeed surprising that . . . every scheme, including Mr Telford's, should have contemplated using the water of streams which are all subject to be affected by the surface drainage of a more or less extensive tract of country, and, consequently, only a very few degrees better than that already in use'.

Stephenson differentiates these surface water schemes from proposals for groundwater use, which 'obtain the water by perforating the London clay' – in other words schemes to sink wells into the confined aquifer beneath central London. However, Stephenson does not look favourably on these proposals, and he explains why this is so using what we would nowadays call a 'conceptual model' of the hydrogeology presented in the report:

> 'The group of strata, designated as the lower tertiary, or eocene and consisting of two divisions, the upper called the London clay, and the lower composed of various coloured sands, and argillaceous deposits, distinguished as the plastic clay [in modern terms the Lambeth Group and Thanet Sand Formation], lying immediately upon the chalk formation, may in general terms be described as a huge mass of clay resting upon a still more extensive bed of chalk . . . the surface of the country occupied by the clay is surrounded on all sides by a belt of chalk, excepting to the east, where the German Ocean [the North Sea] for some distance interrupts the continuity, and you will perceive that this cretaceous circle is, generally speaking, higher in level that the deposit of clay which fills the centre of the basin.
>
> It is almost needless that I should inform you, that of the water which descends as dew or rain upon the surface of the London clay, little, if any, can be considered as absorbed into the earth, and that whilst a part either again reascends into the atmosphere as vapour, or enters into the composition of animal and vegetable bodies, by far the greater portion flows off into the main drain of the district of the river Thames.
>
> In this respect there is a most material difference from that portion of the surface where the chalk comes to light, divested of any covering which could intercept the passage of the moisture; being not only extremely porous, but also full of fissures in every direction, a very rapid absorption takes place, and we accordingly find that there are but few streams carrying off the surplus surface water, and that these are insignificant, and, indeed, many of them dry during the greatest part of the year . . . The lower part of the cretaceous group, and the gault which immediately succeeds it, again presents an impermeable stratum of clay, causing the water to accumulate through the lower regions of the more porous chalk. An enormous natural reservoir of water has thus been formed, and the level to which it might be considered as quite full of water is the lowest point where it can find a vent and overflow; therefore, as the chalk communicates under the coasts of Norfolk, Suffolk and Essex with the ocean, this level, in the present case, may be considered as the same as the mean height of the sea.
>
> That there is, however, an extensive accumulation of water above this level will be obvious, when it is considered that the friction, which from the nature of the small fissures and pores must exist, will necessarily prevent the water from exerting rapidly its hydrostatic pressure . . . The greater or lesser facility, which from lines of

fissures, soft strata, and pores, the water may encounter if flowing towards the centre of the basin, will also govern its surface, and cause it to assume an inclination, the angle of which will represent the friction; and in this manner we may readily account for the different levels, which often appear anomalous, at which water will be found to stand in wells.'

Stephenson does not claim to have done the geological investigation and study that allowed this conceptual model to be developed (he refers to work by Dr Buckland) – his skill was interpreting what this meant in practical terms. His comment on the inclination of the groundwater surface (and by implication the hydraulic gradient) being related to the friction or resistance to flow of the aquifer pores and fissures (a measure of hydraulic conductivity) results from his careful observations at Kilsby and, again, could be interpreted as a formulation similar to Darcy's law.

Stephenson then uses this conceptual model to present his view against the proposal to obtain water supplies by perforating the London Clay and installing wells into the Chalk. He states that because the Chalk beneath the clay cannot receive recharge easily, large and excessive drawdowns would result:

'whenever a large quantity is extracted, the wells in the vicinity, which derive their water from the same strata, are sensibly affected . . . the level for some distance around this focus will be temporarily reduced'.

From our modern perspective, Stephenson seems to have been remarkable prescient. The excessive lowering of piezometric levels beneath London during the 20th century is indisputable (Marsh & Davies 1983).

Being unhappy with the idea of abstraction from the confined Chalk, Stephenson became involved with a scheme to abstract water from the unconfined Chalk north of Watford, proposed by a Mr R Paten. Stephenson states 'The abundance of springs which overflow into the Colne valley, above Watford, and the apparent purity of the water, had long attracted [Mr Patens's] attention'. Various investigations had been done, and Stephenson was 'requested to examine whether the experiments were well grounded'. The remainder of the report seeks to 'explain the proposed method of procuring the water and conveying it to London; and lastly, to submit such remarks as will enable you, in my opinion, to present the project before Parliament'.

In essence, the proposal was for large-scale abstraction from shallow wells in the unconfined Chalk in the Colne Valley. A modern geological map (British Geological Survey 1999) shows this area is near the feather edge of the Tertiary deposits overlying the Chalk, an environment where we would expect enhanced transmissivities and the potential for high yielding wells. Stephenson's report makes it clear that he understood the geology of the area. However, his estimate of the effective recharge to the chalk is less satisfactory. Stephenson uses data from others to assess precipitation, and conjectures that two thirds of the precipitation would be available as recharge. Today we would expect only around one third of the precipitation to act as recharge. From his recharge calculations, plus his belief that the bed of the river was formed of low permeability alluvium, Stephenson believed that the proposed abstraction would not 'produce any visible effect upon the springs which feed the Colne'. While parts of his analysis may be questionable, Stephenson's next step – a practical experiment – was not.

A well of 10.4 m depth, and 3.81 m diameter at its base, was sunk in Bushey Hall meadows, near the Colne and equipped with four pumps, powered by two steam engines. The test, whereby 'The water of the well was now repeatedly pumped out, as low as the power of the engines admitted, and the height of the Colne at those times carefully noted' is clearly a pumping test of the modern pattern. Stephenson concluded that pumping of the well had no discernable effect on the Colne, and that the well could yield approximately 5 Ml/day. Further borings showed that, below 24 m depth, the water supply became 'prodigiously plentiful'. Stephenson also noted that the water was very clear and

'indeed, there was abundant ocular demonstration ([the water] was so beautifully transparent as to admit of the bottom of the well being seen when the water was upwards of 30 feet [9.1 m] deep'.

This is a key point as the alternatives to groundwater sources were from surface waters that were often turbid or worse.

Stephenson concluded that the scheme to abstract water from the Chalk near the Colne had many advantages in forming 'the supply to London with facility and economy' and 'in making use of the enormous reservoir which nature has supplied us with in the chalk, and effecting this at a spot where no existing interests can be injured'.

The response to Stephenson's first report

There were objections to the proposals. Barlow (1854) later summarized the arguments as:

'This project was opposed by the millowners in the neighbourhood, who contended, that this quantity of water [46 Ml/day] must be obtained from the same source whence the springs which feed the mill-streams were derived . . . On the other side of the question it was argued, that the springs were derived only from the

> upper stratum of chalk, and that if the well was made watertight to the depth of 60, or 70 feet [18–21 m], it would have no effect on the surface streams; the supply being obtained from the lower water-channels, which had no connection with the upper supply'.

As was the convention of the day, the objectors produced pamphlets (such as Anon. 1841) stating why Stephenson's opinions of his first report were wrong. Parts of the pamphlets can seem vitriolic by modern standards, attacking Stephenson in quite a personal way. It is useful to view the major objections in light of Stephenson's response.

Stephenson's second report

Stephenson answered his, apparently numerous, detractors in a second report (Stephenson 1841). One of the objectors, a Mr Webster, had a rather major objection, in that Stephenson says Mr Webster 'denies, emphatically, that the chalk underlying the stratum of London clay is the 'great water-bearing stratum''. Stephenson rebuts this from a geological perspective, and then from a practical one, by outlining how the cities of Winchester, Arundel, Brighton, Dover, Deal and Walmer, Canterbury, Gravesend and St Albans derive their water supply from the Chalk.

Some of the objectors had added a slightly demeaning air; Stephenson's practical upbringing did not tolerate this well, although he preferred the facts to speak for themselves. When Mr Webster (he who did not consider the Chalk to be the major aquifer beneath London) said of Stephenson 'many are old in years, yet young in geology', Stephenson responded thus of Mr Webster:

> 'Some are old in years, yet young in mining. When Mr Webster has spent as much of his time underground as I have done, and not till then, will he understand the exact truth of this remark . . . However, . . . I am convinced any observations of a personal nature never, in a question of this sort, can be substituted with proprietary, for sound arguments or fair deduction'.

One of the principal objectors was the Reverend J. C. Clutterbuck (Clutterbuck 1841), who was concerned for local interests that may be injured by the proposed abstraction. Stephenson says that Clutterbuck and his relatives owned property in the vicinity of the abstraction. In response to Clutterbuck's objections Stephenson (whose previous report had said that there would be no noticeable effect on river and spring levels) invited Clutterbuck 'to suggest any experiments which he conceived were calculated to decide whether the views I entertained were correct or incorrect'. According to Stephenson, Clutterbuck proposed continuous pumping from the well for three days, having first accurately determined initial levels in surrounding water features, with levels monitored during the test. These tests were carried out in March 1841 and Stephenson reports that there were no perceptible effects on water levels in any of the wells that were monitored.

Stephenson expressed the hope that Clutterbuck would accept that his and other local interests would not be affected, but this was not to be the case. Apparently, in addition to water level observations taken jointly by Clutterbuck and Stephenson and his assistants, Clutterbuck had some observations taken by him alone. Stephenson believed, not withstanding the unwitnessed nature of some of the observations, that Clutterbuck was using the data selectively, ignoring nearby wells that were not affected, and attributing movements in water levels in distant wells directly to the influence of the test pumping. Stephenson is dismissive of Clutterbuck's hypothesis:

> 'He adopted a theory at variance with all hydrostatic laws. To discuss it, would really be a waste of time, and I think I shall be justified in treating it thus summarily, by stating that it follows, from his theory, that pumping at any point in the chalk is not so likely to affect the neighbouring as the distant wells . . . he gives a diagram where the level of the water in the chalk is shewn to be influenced to a far greater extent at fifteen miles [24.1 km] distant than at one mile [1.6 km] distant'.

Stephenson, by reference to his observations at the Kilsby Tunnel, reiterates his view that the effect of pumping must reduce with distance from the point of abstraction. He then makes the following statement:

> 'The result of pumping at a deep shaft, in chalk or other similar porous material, would be the drainage of a portion of the district, represented by an inverted cone, the point being at the bottom of the shaft; the upper surface, or what is generally designated the base of the cone, occupying an area depending on the inclination which the fluid assumed in passing through such porous mass.'

This statement is remarkable in that it almost perfectly describes the concept of a circular 'cone of influence' around a source, which is implicit in all radial flow models beginning with Dupiut (1863) and the workers that followed. As with many things that Stephenson did, a modern perspective makes Stephenson look remarkably precescient.

Later developments

The variance of views between Stevenson and Clutterbuck did not end there, with Clutterbuck responding in another pamphlet (Clutterbuck 1842) and also in a number of papers and discussions in the *Minutes of the Proceedings of the Institution of Civil Engineers*. Nevertheless, later workers have supported the conclusion of Stephenson that large-scale abstraction from the unconfined chalk was viable and that the impacts of abstraction will reduce with distance. It is not clear whether the scheme proposed by Stephenson was ever implemented, but Boyd-Dawkins (1898) states that Stephenson's views 'bore fruit at the time, in the sinking of numerous wells in the chalk, and notably in those of the New River and Kent Companies'.

Work in Liverpool

Stephenson was appointed in 1850, by Liverpool Town Council, to report on and recommend the best plan to secure an adequate supply of water for the town. His report (Stephenson 1850), describes the sources of water at the time, and discusses options to procure additional water, not just from groundwater, but also from surface water schemes, such as expansion of the Rivington Works.

As with his previous studies, Stephenson did not do any original geological work himself, but invited representations on such matters by others and discussed them in his report. What he did commission specifically was a series of pumping tests on wells and groups of wells and the gathering of much more detailed groundwater level data than was previously undertaken.

The wells supplying Liverpool and the hydrogeology of the sandstone

Stephenson's report contains considerable detail of water levels at the seven 'pumping stations' where wells supplied the town. These were: Green Lane; Windsor; Park (also known as Water Street); Hotham Street; Soho; Bush (also known as Bevington Bush); and Bootle. Records are also presented from 63 private wells, where Stephenson was interested in the impact of pumping from the public wells. Stephenson's data show the maximum yield of all the public wells as 23.5 Ml/d, although he interprets the data to give a more typical output as 18.2 Ml/d.

In his review of the geological submissions put before him, Stephenson states that his aim was:

> 'to ascertain correctly the quantity of water yielded by the existing wells, the influence which they exert upon each other, and the mode by which the water contained in the mass of sandstone is transmitted from one place to another'.

The details of all the submissions are not recorded, but Stephenson says that some of the evidence was:

> 'very conflicting, some of the witnesses maintaining, that however large a quantity might be pumped from one well, little or no effect was found to be produced upon those in the vicinity; and of this several authenticated instances were certainly adduced, but a careful consideration of the whole mass of facts leads me to believe that these cases form rather the exception than the rule; and that they are occasioned by local geological faults, partially or wholly water tight, which are known to be interspersed throughout the new red sandstone formation [now known as the Sherwood Sandstone Group] in the neighbourhood of Liverpool'.

In the face of conflicting views on the practical importance of fissures, Stephenson believed, as we do today, that fissures can significantly affect the hydraulic properties of the sandstone, and states:

> 'Different degrees of porosity unquestionably exist, satisfactorily accounting in my mind for the different degrees of influence which wells are found to exert on each other. The facility with which the water will pass from one part of the sandstone to the other, depends principally on the size of the fissures, their character and direction; and hence it is quite consistent with the existence of a very large number of fissures, that two wells at a great distance may affect each other while two that are near may show little or no connection'.

Stephenson concludes that the sandstone forms a very productive aquifer. Furthermore, based on various pumping tests and consideration of monitoring data, he estimates that the maximum long-term yield of a well in the sandstone is unlikely to be greater than 4.5 to 5.4 Ml/d.

Interference between wells

When planning to develop new groundwater resources, Stephenson realized that the spacing between wells, and their influence on each other, was an important factor. His report re-states the concept of an inverted cone of influence centred on a well, with the diameter of the cone being controlled by the frictional resistance to flow (i.e. the hydraulic conductivity of the aquifer). He expands on this concept and introduces the idea of non-circular areas of influence:

> 'looking upon the area drained by a well as represented on the surface by a circle is not strictly correct, because its form will be of course modified by the relative sizes, characters and directions of the fissures through which

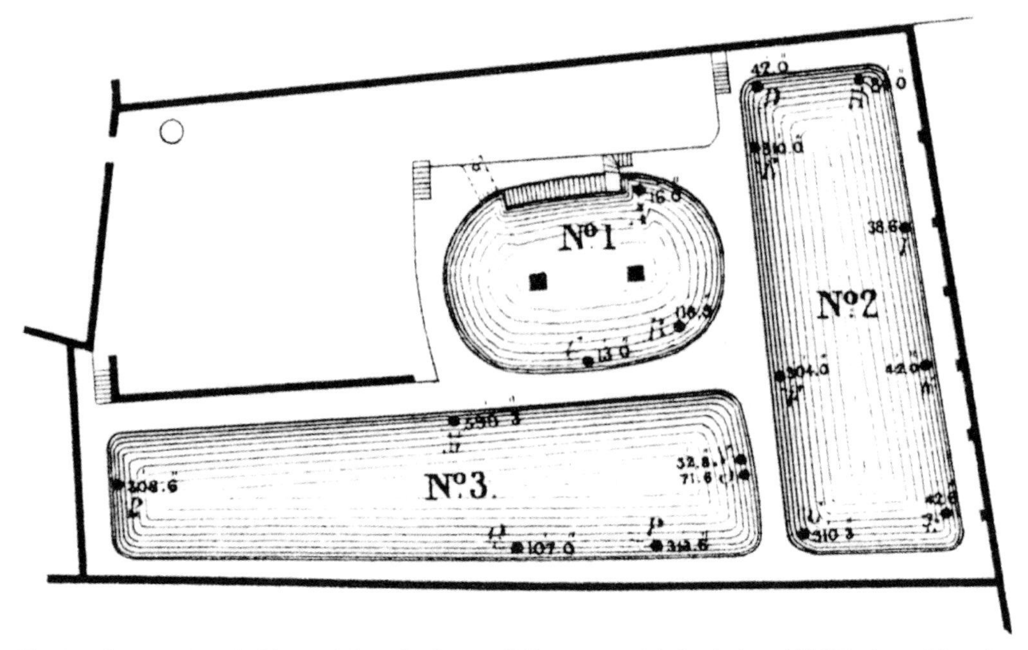

Fig. 6. Well group at Bootle, Liverpool (from Stephenson 1850: courtesy of the Institution of Civil Engineers Library).

> the water finds its way to the well. The area . . . will, therefore, most probably be very irregular in outline'.

Again this statement would not seem out of place in a modern textbook on modelling of zones of depletion around wells.

The report considers the increase in total yield that could be obtained by drilling additional wells relatively close to existing ones. Stephenson did a test on the 16 boreholes at the Bootle works, which flowed into the base of the reservoir (Fig. 6). The test involved temporarily capping all the boreholes. The first was unplugged, yielding 4.2 Ml/d. Then, in the form of a step test, the other boreholes were opened in turn, with the yield measured at each stage. Stephenson determined that the final yield, from all 16 bores was 4.7 Ml/d, little more than 10% greater than the yield from one well. He considers that, unless wells or bores are widely spaced, drilling of additional wells is unlikely to result in a dramatic increase in total yield, as the new wells will merely tap into the same network of fissures as the existing ones and that their areas of influence will overlap and interfere. He then draws an interesting analogy:

> 'This group of boreholes at Bootle present a complete epitome of what is actually going on upon a large scale throughout the town of Liverpool. The difference is only one of degree, consisting in the intervention of a large mass of rock between the wells, which offers more difficulty to the free passage of water from one to the other'.

Stephenson also arranged an extensive series of pumping tests on the existing wells, devoting much effort into obtaining accurate flow measurements by the timed volumetric method. He surmises that the yield of an individual source in the sandstone is unlikely to be greater than 4.5 to 5.4 Ml/d, and that wells should be located considerable distances apart to ensure such yields can be attained. These studies led Stephenson to recommend that the additional water needs of the town be met by 'a system of independent wells, placed throughout the district, and lying generally to the east of Liverpool', rather than by drilling additional wells beneath the town itself, an area already well populated with abstractions.

Saline intrusion

The report also dealt with the question of brackish or salty water appearing in the wells of Liverpool. A review was made of the presence and concentration of chloride of sodium in the wells across the town to, according to Stephenson, determine 'the disputed point regarding the existence and extent of the connection between the river and wells'. Stephenson concludes that the salinity of the wells is due to connection with the river, albeit with the process of saline intrusion being very gradual, and that the particularly high salinity of a group of wells was the result of them being in or near a buried river channel. Comparisons are made between data from 1846 and

Fig. 7. Robert Stephenson (from Jeaffreson 1864).

1850 and Stephenson interprets that levels of salinity were rising, and attributed the deterioration of water quality to the continued and extensive pumping from beneath Liverpool. Stephenson recommended against deepening existing wells beneath the town, which would create greater drawdown and draw in saline water. His preferred option was to locate new wells to the east, away from the River Mersey.

Conclusion

Robert Stephenson's achievements as a civil and mechanical engineer are manifold. However, his more arcane work to control and exploit groundwater also reflects well upon him.

The works to complete the Kilsby Tunnel through the treacherous quicksands were an epic engineering task of its day. Stephenson (Fig. 7) was able to observe and interpret characteristics of groundwater flow, that allowed him to develop a groundwater lowering system, based on principles that are still valid today.

His later work on water supply for the cities of London and Liverpool showed his understanding of the concept of the 'cone of influence' around a well and of the importance of the interference of drawdown between wells. He devised and carried out pumping tests and monitoring to assess the impacts of abstractions, using approaches that would be familiar to modern practitioners.

That his work was done before the publication of the work of Darcy and Dupuit is remarkable. The methods and conclusions of Stephenson may seem obvious today, but the objections he faced at the time remind us that Stephenson's rational approach was many years ahead of its time.

The staff of the library and archives of the Institution of Civil Engineers were of great assistance in the preparation of this paper. The author is grateful to M. Osborne of Arup for his comments on the geological setting of the Kilsby Tunnel.

References

ANON. 1841. *Observations on a report made by Robert Stephenson, Esq., civil engineer, to the proposed London and Westminster Water Company*. John Green, London.

ANON. 1860. Obituary, Robert Stephenson. (1860). *Minutes of the Proceedings of the Institution of Civil Engineers*, **19**, Session 1859–1860, 176–182.

BARLOW, P. W. 1854. On some peculiar features of the water-bearing strata of the London Basin. *Minutes of the Proceedings of the Institution of Civil Engineers*, **14**, Session 1854–1855, 42–65.

BOURNE, J. C. 1839. *Drawings of the London and Birmingham Railway*. Collection of the library of the Institution of Civil Engineers, London.

BOYD-DAWKINS, W. 1898. On the relation of geology to engineering, James Forrest lecture. *Minutes of the Proceedings of the Institution of Civil Engineers*, **134**, Part 4, 2–26.

BRITISH GEOLOGICAL SURVEY. 1980. *Sheet 185 Rugby 1 : 50000 Series*, Solid and drift edition. Ordnance Survey, Southampton.

BRITISH GEOLOGICAL SURVEY. 1999. *Sheet TQ19NW Watford 1 : 10,000 Series*, Solid and drift edition. Ordnance Survey, Southampton.

CLUTTERBUCK, J. C. 1841. *A letter to Sir John Sebright, Bart. on the injurious consequences likely to accrue to a portion of the County of Hertford, if the London and Westminster Water Company should carry into effect their project of supplying the Metropolis with water from the Valley of the River Colne*. Simpkin and Marshall, London.

CLUTTERBUCK. J. C. 1842. *Supply of water to the Metropolis from the valley of the Colne. A few words in answer to Mr Stephenson's second report to the directors of a proposed water company*. J. Peacock, Watford.

DARCY, H. 1856. *Les fontaines publique de la Ville de Dijon*. Dalmont, Paris.

DUPUIT, J. 1863. *Etudes théoretiques et practiques sur les mouvement des eaux dans les canaux decouverts et a travers les terrains permeable*. Dunot, Paris.

HALLIDAY, S. 1999. *The Great stink of London – Sir Joseph*

Bazalgette and the cleansing of the victorian capital. Sutton Publishing Limited, Stroud.

JEAFFRESON, J. C. 1864. *The life of Robert Stephenson, with descriptive chapters on some of his most important professional works by W. Pole.* Longman, London.

LEWIS, G. F. 1984. *The constructional history of the Kilsby Tunnel.* MSc dissertation, City University, London.

MARSH, T. J. & DAVIES, P. A. 1983. The decline and partial recovery of groundwater levels beneath London. *Proceedings of the Institution of Civil Engineers*, **74**, 263–276.

ROLT, L. T. C. 1960. *George and Robert Stephenson: the railway revolution.* Longman, London.

SMITH, W. 1815. *A delineation of the strata of England and Wales with part of Scotland; exhibiting the Collieries and Mines, the Marshes and Fen Lands originally overflowed by the Sea, and the varieties of Soil according to the Variations in the Substrata, illustrated by the most descriptive names by W. Smith.* Cary, London.

STEPHENSON, R. 1840. *Report to the Provisional Committee of the London and Westminster Water-Works, etc, etc. Morning Advertiser* **December 29** [included as an appendix in Anon. (1841)].

STEPHENSON, R. 1841. *London Westminster and Metropolitan Water Company. Mr Stephenson's second report to the Directors.* Privately printed, London.

STEPHENSON, R. 1850. *Report of Robert Stephenson, civil engineer: on the supply of water to the town of Liverpool.* Bradbury and Evans, London.

'Making water': the hydrogeological adventures of Britain's early mining engineers

PAUL L. YOUNGER

School of Civil Engineering and Geosciences
University of Newcastle Upon Tyne, UK

Abstract: The earliest detailed technical descriptions of British mining practices still in existence (which date from the late 17th and early 18th centuries) dedicate many paragraphs to the problems posed by the unwanted ingress of ground water into underground workings. Excessive water in working areas seriously hinders production. More importantly, sudden inrushes of ground water to underground workings are a significant mortal hazard. In view of the problems experienced with water ingress to workings, the main preoccupations of the early mining engineers were utterly practical, focusing on the efficient removal of water which could not be prevented from entering the workings (by simple bailing, by adit drainage or by pumping), and on efforts to minimize water ingress in the first place (by the use of tubbing in shafts and the use of rock barriers and dams in working areas). Occasionally, the mining engineers took time to reflect upon the origins of the water they encountered in their work. In their writings we find some of the earliest accurate conceptualizations of issues of ground water origin, driving heads, hydraulic gradients (including vertical upward gradients) and natural heterogeneities in water quality. So successful were these early mining engineers in their endeavours that they bequeathed most of the technological basis for the development of large-scale public-supply ground water abstractions, and much of the basis for the geotechnical control of ground water during construction projects, from about 1820 onwards. By the late 19th Century, mining engineers concerned with ground water management became gradually isolated once more within their own specialist domain, where they went on to develop a vernacular hydrogeology of their own, replete with its own key concepts and vocabulary. Nevertheless, occasional interchanges of experience between mining and the water industry have continued to enrich both sectors down to the present day.

Mines do not only produce minerals, they also 'make water'. In mining terminology, the total water yield of a mine or a specific district of a mine is known as its 'water make', and the mine or district is said to be 'making water'. Most of the water being 'made' by the mine is, of course, ground water. Excessive water makes have long been the enemy of miners. In responding to this enemy, mining engineers accidentally became pioneers in the field of ground water engineering. As a by-product of their efforts in relation to water management, the early mining engineers inadvertently laid the foundations for much of the modern water industry, so that in a very real sense the practice of mining hydrogeology can be said to have been 'the making of water' as a technologically advanced industry. The miner's struggle with water also paved the conceptual and technological way for much of modern geotechnical practice in the control of ground water in the construction industry. While a full account of the many ways in which mining engineering fathered both modern water engineering and modern ground water control practices would require a book in itself, this paper has the more modest aim of drawing together some of the key experiences of the early mining engineers with ground water and what they made of them in both practical and scientific terms. It should be noted that the coverage of this paper is largely restricted to events and scientific developments which occurred prior to 1913 (the year which marked the peak of production in most major British coalfields), with only a few minor references to more recent events which happen to shed light on earlier historical developments. The paper also ignores those pre-1900 mining developments which were undertaken to deliberately stimulate ground water movement, with the purpose of facilitating brine production in the Cheshire salt basins; a recent review of the hydrogeological setting and specific mining practices of that area has been presented by Cooper (2002).

Ground water problems in mining

'. . . How silently in former ages all this water had
found its way,
perhaps drop by drop, into the stony reservoirs!
How silently it had lain there, under solid strata,
no one suspecting its existence!
But now at length, man must trouble the peaceful
waters –
must rudely unseal their rocky caskets, and, lo!
He shall have no more peace for them –
no more quiet shall there be in that vicinity.

From: MATHER, J. D. (ed.) 2004. *200 Years of British Hydrogeology*. Geological Society, London, Special Publications, **225**. 121–157. 0305-8719/04/$15 © The Geological Society of London.

The fountains of the deep in the hollow places of the earth
have been broken up by rude hands, and they shall now pour forth unceasingly
thousands of gallons after thousands of gallons,
and these, too, minute by minute, – torrent after torrent – rush, rush, rush!
Ah, foolish man! what hast thou done,
evoking a Spirit of the floods thou canst not lay? . . .'
(Leifchild 1853)

While water has long been of benefit to certain aspects of mining operations, by providing an energy source for winding and pumping, or acting to suppress dust release, it is more often remembered in mining circles as at least a nuisance, if not a mortal hazard.

Nuisance water

In its more mundane manifestations, as gradual drippers or sustained feeders entering working areas, excessive water is often a nuisance, for it can:

(1) make the work of the miners more laborious and uncomfortable
(2) degrade floor and roof strata, thus hindering haulage and maintenance activities
(3) accelerate the corrosion of mining equipment
(4) necessitate the use of water-resistant machinery and expensive, waterproof explosives, and
(5) add substantially to the weight of the run-of-mine product, making it more difficult and expensive to handle.

All of these problems can be overcome to some degree by means of well-planned and well-managed dewatering and related drainage activities. For the individual miner, the inconvenience caused by wet conditions was eventually assuaged to some degree when bonus payments for 'wet working' were introduced. In terms of the overall financial performance of a mine, the costs of managing large water makes (principally by means of incessant pumping) can make all the difference between a profitable and bankrupt mine. By the mid-19th Century, there were already many collieries which were raising far more water than coal to daylight (Taylor 1858), the record at that time being held by the James Pit at Wylam (Northumberland), which raised 30 tonnes of water for every tonne of coal. This same mine in its flooded state continues to deliver a very large, perennial, ferruginous discharge to the River Tyne (at grid reference NZ 123647), at a point very close to the cottage in which the great mining engineer and railway pioneer George Stephenson was born. Stephenson's famous achievements in steam engineering were based on his early experiences in the James Pit, helping to maintain the substantial steam-driven dewatering pumps (Leifchild 1853). Here is another happy by-product of unwanted ground water ingress to mine workings: the fostering of engineering skills which were to prove key to the furtherance of the industrial revolution. This is a point to which we shall return on several occasions later in this narrative.

Dangerous inrushes

'. . . the watt'ry 'wyest', mair dreedful still, alive oft barries huz belaw:
Oh dear! it myek's yen's blood run chill! May we sic mis'ry niver knaw!
Te be cut off frae kith and kin, the leet o' day te see ne mair
and left frae help and hope shut in, te pine and parish in despair . . .'
(from the Tyneside dialect poem 'The Pitman's Pay'; Wilson 1843)

In the form of violent inrushes to underground workings, water has claimed thousands of lives worldwide, both by drowning and by entrapment of miners in isolated mine voids, where death comes more slowly by asphyxiation or starvation. Few miners survive such inrushes, a notable recent exception being the major Quecreek Mine inrush in Pennsylvania (USA) on 24th July 2002, from which all nine trapped miners were rescued after 77 hours underground. Less happy outcomes have almost always been the norm. Nevertheless, the fatality rate from water inrushes is relatively modest in comparison to other causes of death in underground mines, at around 1.5% of all fatalities (Hyslop *et al.* 1927). This is nevertheless of the same order of magnitude as the fatality rate from explosions of methane and/or coal dust (3%), which are generally accorded far more attention in popular conceptions of the hazards of mining. In fact, the predominant causes of death in mining are far less sensational than either inrushes or explosions and they take their toll day-by-day: falls of ground (50% of all fatalities), mishaps with haulage equipment (25%) and errors in blasting procedures (10%) (Richards 1951).

Table 1 summarizes all major inrushes in Britain from 1648 to 2002. As the table reveals, there are two principal sources of inrushes: flooded old workings and natural bodies of water (aquifers or surface water bodies). Of the two, inrushes from old workings are by far the more common. The risk from old workings was particularly acute prior to 1872, the year in which a statutory requirement to deposit mine plans for future reference was finally introduced. According to Hyslop *et al.* (1927) a further 15 years elapsed before surveying procedure had

Table 1. *Notable mine water inrushes in Britain, 1648–2002*

Date	Mine[a]	Cause of inrush	No. of fatalities	Further information[b]
1 Aug 1648	Two unnamed collieries (Durham)	Workings were rapidly flooded when the River Wear burst its banks	None known	See Archer (1992)
May 1658	Galla Flatt	'Breaking in of water from an old waste'; the victims' bodies were found and buried in 1695	2	Dunn (1848); Galloway (1898); Doyle (1997)
17 Nov 1771	Wylam (Northumberland)	Water flooded down shafts when the River Tyne burst its banks in its greatest-ever recorded flood	None known	See Archer (1992)
17 Nov 1771	North Biddick, Chatershaugh and Low Lambton Collieries (Durham)	Water flooded down shafts when the River Wear burst its banks in its greatest-ever recorded flood	None known	34 pit ponies recorded as drowned. See Galloway (1898, p. 273), and Archer (1992)
08 Sep 1796	Slaty Ford (Northumberland)	No detailed documentation of cause located	6	DMM archive
27 Mar 1807	Discovery Pit, Felling (County Durham)	An advancing bord holed into unsuspected flooded old workings.	3	Hair (1988, p. 72–73)
30 Jun 1809	East Ardsley (Yorkshire)	No detailed documentation of cause located	10	DMM archive
03 May 1815	Heaton Main (Northumberland)	Inrush from old flooded workings of Jesmond Colliery, despite precautionary borings having been made without reaching water	75	Dunn (1848); recently recounted in detail by Doyle (1997)
13 July 1828	Towneley Main Colliery (Durham)	River Derwent (a tributary of the Tyne) burst its banks and flooded the Star Flat Pit	1	14 pit ponies also drowned. See Archer (1992)
3 Dec 1829	Willington (Northumberland)	Gas explosion holed into old workings	4	Hair (1844, p. 16)
7 March 1832	Beamish Colliery (Durham)	'Pit inundated'	2	Galloway (1898, p. 504)
15 Oct 1832	Houghton Colliery (Durham)	Failure of cast-iron tubbing in the main shaft gave rise to a major inrush from the Magnesian Limestone aquifer. Although all the miners escaped, all the ponies in the pit were drowned.	None	Galloway (1898, p. 505)
1833	Workington Lady and Isabella Pits (Cumbria)	Inundated from old workings	4	Galloway (1898, p. 504)
1835	St Helens (Lancashire)	River water burst into mine	17	Galloway (1898, p. 504)
28 Jul 1837	Workington (Cumbria)	Pillar working beneath the sea bed led to cracking of roof strata to sea floor, and a whirlpool in the bay which drowned the mine in seconds	27	Galloway (1904)
4 July 1838	Silkstone (Yorkshire)	Surface water entering an inclined drift ('futterail') during a thunder-storm trapped children against a downward-opening ventilation door.	26	Galloway (1904)
02 May 1839	Kingswood (Somerset)	'Irruption of water'	11	Galloway (1904)
23 Oct 1840	Farnacres, Bensham (Durham)	'Drowned from old workings'	5	DMM archive
'1840–1'	Stronne (Lanarkshire)	Flooded after holing into old workings	4	Galloway (1904)
9 Dec 1842	Fenwick, Belford (Northumberland)	Flooded after holing into old workings	2	Galloway (1904)

Table 1 *(cont.)*

Date	Mine[a]	Cause of inrush	No. of fatalities	Further information[b]
09 Oct 1843	Pasture Hill (Northumberland)	Flooded after holing into old workings	7	DMM archive
14 Feb 1844	Landshipping Colliery, Haverford-west (Glamorgan)	Void migration from shallow workings intercepted bed of River Dunleddy, causing major inrush of river water to mine	40	Dunn (1844); Galloway (1904)
Feb 1845	Haye's Wood (Somerset)	'Irruption of water'	10–11	Galloway (1904)
Dec 1847	Ince Hall Colliery (Lancashire)	River Douglas burst its banks near Wigan and flowed through old workings to the working colliery	Several; number uncertain	Galloway (1904)
10 May 1852	Gwendraeth (Glamorgan)	Quicksand burst into workings from overlying strata	26	DMM archive
11 Jun 1861	Clay Cross No 2 Pit (Derbyshire)	Flooded old workings of No 1 Pit, thought to be 50 yards from working face, were actually only two feet away when the coal gave way	23	See Judge (1994)
10 Feb 1865	Coneygree (Staffordshire)	Advancing mine workings holed into old shaft which was not suspected to pierce the strata at this point	3	See Appendix XVIII in Hyslop *et al.* (1927), in which it is wrongly spelled 'Conygre'
1 Apr 1867	North Levant Tin Mine (Cornwall)	Water entered stope from adjoining old workings of Wheal Maitland	3	See Vivian (1990)
19 Jan 1871	Wheatley Hill (Durham)	Water entered colliery from adjoining old workings of Thornley Colliery	5	DMM archive
24 Jan 1871	Seaham (Durham)	Inaccurate plans (in any case inadequately consulted) in faulted strata resulted in flooded workings being much closer than estimated	2	See Appendix XVIII in Hyslop *et al.* (1927)
30 Mar 1871	Highbridge (Staffordshire)	Sudden large water make from post-Carboniferous strata above unconformity	3	See Appendix XVIII in Hyslop *et al.* (1927)
4 Dec 1875	Penygraig (Glamorgan)	A large fault seen in nearby workings was assumed to separate old and new workings, but the throw of the fault in fact petered out before it reached the Penygraig workings	2	See Appendix XVIII in Hyslop *et al.* (1927)
11 Apr 1877	Tynewydd, Rhondda (Glamorgan)	Mistaken interpretation of old plans in area of confusing faulting	5	See Llewellyn (1992)
10 Mar 1881	Page Bank (Durham)	Flooded when River Wear burst its banks and spilled into shaft	Not known	See Archer (1992)
25 May 1883 29 Oct 1884	Hodbarrow Iron Ore Mine (Cumbria)	Successive inrushes of ground water carrying loose sand from overlying drift deposits. Although these thankfully claimed no lives, they led to major surface subsidence and damage to infrastructure	None	Harris (1970)
03 Jun 1885	Newbottle Margaret Pit (Durham)	Failure of a supposed 200 yard barrier of intact coal, suggesting that the relevant plans were highly inaccurate, even though others of the same period in this area were known to be accurate	14	See Appendix XVIII in Hyslop *et al.* (1927)
20 Dec 1885	Hodbarrow Iron Ore Mine (Cumbria)	Further inrush of ground water carrying loose sand from overlying drift deposits, associated with development of major surface depression inland	None	Harris (1970)

Table 1 *(cont.)*

Date	Mine[a]	Cause of inrush	No. of fatalities	Further information[b]
3 May 1892	Ashton Vale (Somerset)	Advancing mine workings holed into old shaft which was not suspected to pierce the strata at this point	2	Circumstances similar to those at Lofthouse 81 years later. See Appendix XVIII in Hyslop *et al.* (1927)
04 Aug 1892	Ravenslodge (Yorkshire)	Failure of tubbing in upcast shaft allowed sudden inrush of strata water	6	See Command report 6902
14 Jan 1895	Audley (Staffordshire)	Inaccurate plans led to failure of what had been intended as a 80-yard wide barrier separating the mine from old workings of Diglake Colliery	77	See Appendix XVIII in Hyslop *et al.* (1927)
09 Dec 1896	River Level (Glamorgan)	Illegally worked area of coal not shown on plans by miscreants	6	See Command report 8465
26 Mar 1897	Devon Colliery, Furnacebank No. 1 Pit (Clackmannan-shire)	Failure of an access door in an old dam which held water back in abandoned workings	6	See Command report 8637
06 May 1897	East Hetton (Durham)	Inrush from unrecorded old workings	10	DMM archive
18 May 1898	Hodbarrow Iron Ore Mine (Cumbria)	Major inrush of ground water carrying loose sand from overlying drift deposits and sea bed, forming a surface crater in the inter-tidal zone	None	Harris (1970)
4 Sept 1902	Navigation (Gloucester)	Error in current plans led to holing into old workings	4	See Appendix XVIII in Hyslop *et al.* (1927)
16 Nov 1903	Sacriston (Durham)	Sudden failure of a barrier holding back known flooded old workings	2	1 miner survived trapped above water line for 88 hours; see Purdon (1979)
26 Aug 1901	Donibristle (Fife)	Inrush of liquefied peat from the Moss Morran peat bog	8	See Command report 851
27 July 1903	Dudley Wood (Worcestershire)	Advancing mine workings holed into old shaft which was not suspected to pierce the strata at this point	4	See Appendix XVIII in Hyslop *et al.* (1927)
1903	Easington (Durham)	During the sinking of the South Shaft through the Permian Yellow Sands Aquifers by means of ground freezing, an inrush estimated to total about 60 m^3 entered the shaft from an unfrozen pocket	1	See Temple (1998, p. 60–61) and Emery (1992, p. 84)
28 Jan 1908	Roachburn (Cumbria)	Inrush of liquefied peat from overlying bog	3	See Robertson (1997)
15 Feb 1908	Brereton (Staffordshire)	Sudden large water make from post-Carboniferous strata above unconformity	3	See Appendix XVIII in Hyslop *et al.* (1927)
16 Dec 1909	Podmore Hall Minnie Pit (Staffordshire)	Sudden large water make from post-Carboniferous strata above unconformity	1	See Appendix XVIII in Hyslop *et al.* (1927)
24 May 1911	Podmore Hall Minnie Pit (Staffordshire)	Sudden large water make from post-Carboniferous strata above unconformity	1	See Appendix XVIII in Hyslop *et al.* (1927)
23 Dec 1911	Bamfurlong (Lancashire)	Advancing mine workings holed into old shaft which was not suspected to pierce the strata at this point	1	See Appendix XVIII in Hyslop *et al.* (1927)
16 Jun 1913	Car House (Yorkshire)	Unfortunate failure of precautionary boreholes, drilled in accordance with regulations, to detect flooded void	8	See Appendix XVIII in Hyslop *et al.* (1927)
09 Jul 1918	Stanrigg And Arbuckle (Lanarkshire)	Inrush of liquefied peat from surface peat bog	19	See Command report 146

Table 1 *(cont.)*

Date	Mine[a]	Cause of inrush	No. of fatalities	Further information[b]
21 Apr 1923	Shut End (Staffordshire)	Sudden release of water from old goaf through which new roadways had been driven two weeks previously	4	See Appendix XVIII in Hyslop *et al.* (1927)
25 Sep 1923	Redding (Stirlingshire)	Inaccurate plans (in any case inadequately consulted) in faulted strata resulted in flooded workings being much closer than estimated	40	See Command report 2136
10 Mar 1924	Harriseahead (Staffordshire)	Illegally worked area of coal not shown on plans by miscreants	None	See Appendix XVIII in Hyslop *et al.* (1927)
27 Nov 1924	Killan (Glamorgan)	Inaccurate plan falsely implied a wider barrier to flooded old workings than existed	5	See Appendix XII in Hyslop *et al.* (1927)
30 Mar 1925	Scotswood Montague Pit (Northumberland)	Inadequate availability of mine plans led to flooding from documented old workings nearby	38	See Command report 2607
27 Dec 1925	South Hetton	Fault led to unanticipated connection with flooded old workings	None	See Appendix XVIII in Hyslop *et al.* (1927)
07 Sep 1950	Knockshinnock Castle (Ayrshire)	Liquefied peat entered mine from surface bog	13	See Command report no 8180
21 Mar 1973	Lofthouse (Yorkshire)	Longwall panel holed into flooded shaft	7	Last fatal inrush in a British colliery. See Command report 5419 (Calder 1973)
1987 (problems developed over several months)	Sherburn-in-Elmet Gypsum Mine (Yorkshire)	The mine struck a fissure which was in open communication with the Magnesian Limestone Aquifer (Brotherton Formation) at depth below the workings. Despite two attempts at pressure grouting, the battle to save the mine was finally lost	None	Cooper, A. H. British Geological Survey, pers. comm. 2003.
23 Mar 2002	Longannet (Fife)	Sudden failure of a dam which had long held back water in very old workings released some 75 Ml of water in ten minutes	None	Direct cause of closure of this last deep mine in Scotland. Pers. comm. from mine staff.

[a] all are coal mines unless otherwise noted: while every effort has been made to include all inrush records which have come to the author's attention, it must be stressed that records were far more systematically kept for coal mines than for metalliferous/industrial mineral mines and that these categories of mine are almost certainly under-represented in this list. The author will be grateful for any additional records submitted by readers for inclusion in subsequent editions. [b] These are not necessarily the primary references for these incidents. 'Command' reports are official government reports by HM Inspector of Mines, which were usually ordered by the government in the cases of major accidents in mines. 'DMM archive' refers to the archive of mining disasters maintained, on-line and with free access, by the Durham Mining Museum (*www.dmm.org.uk*). This includes entries gathered from old press reports and local history accounts as well as official records

become sufficiently standardized for the deposited plans to become generally reliable. Hence plans dating from before 1887 should be viewed only as 'evidence that old workings existed in the neighbourhood', rather than as indicating the actual position of workings (Hyslop *et al.* 1927).

Significant insights into mine hydrogeology can be gained by reflecting on the circumstances of some of the incidents listed in Table 1. For instance, the inrush at Tynewydd Colliery (Rhondda Valley, South Wales) in 1877 serves to demonstrate just how lowly-permeable Coal Measures shales can be. Although four miners were drowned immediately after an advancing roadway unexpectedly encountered flooded old workings, ten others survived below the water table for a week, in air pockets which were trapped by the rising water table against shale roofs in isolated up-dip workings. Even under substantial excess head from the surrounding water mass, the trapped air did not dissipate into the shales. One of the ten trapped miners was sadly killed at the moment of rescue, when the release of compressed air forced him into the hole dug from above by the rescue brigade, fracturing his skull. The remaining nine men survived, although all suffered from the bends (believed to be the first medically-documented cases of this affliction) following the rapid decompression (Llewellyn 1992). As in the case of Quecreek (Pennsylvania) in 2002, the Tynewydd inrush was an international media sensation of its age, and was quite probably the inspiration for the dramatic inrush entrapment passages which form Part 7 of the classic novel '*Germinal*' by Émile Zola (1885).

Miners gradually devised ever-more effective precautions against devastating inrushes. In fact, the British mining industry has now become so good at mining safely under bodies of water that it has established a world-leading expertise in this field, with an unequalled safety record in workings extending in excess of 10 km offshore (Orchard 1975; Aston and Whittaker 1985). Nevertheless, accidents have continued to occur sporadically, even in onshore coalfields, down to the present day. The most serious inrush in living memory was that at Lofthouse Colliery (West Yorkshire) in 1973, when an advancing longwall face holed into a flooded shaft, taking seven lives (Calder 1973). Most recently of all, the ultimate trigger for closure of the last underground coal mine in Scotland (Longannet) in March 2002 was a major inrush due to failure of an old underground dam (This fortunately failed to claim any lives, as it occurred on a Saturday when few miners were in the pit).

Engineering responses to ground water ingress prior to 1900

Both nuisance water and dangerous inrushes have necessarily received a lot of attention from mining engineers and mining geologists over the centuries. In the following paragraphs, a brief history of some of the methodologies and technologies developed in response to these problems is presented.

The very earliest mines, being little more than scrabblings from conspicuous surface outcrops on hillsides, were largely free from water problems. It was only as mines were developed to greater depths in pursuit of dipping seams and plunging veins that they passed below the local base level of drainage and began to incur the penalty of a heavy water make. The evolution of responses to high water makes followed a four-step progression which can be summarized as follows:

(1) simple bailing of bell pits and other shallow workings using buckets
(2) under-drainage by long adits linking the sole of the workings to the deepest local valley
(3) the pumping of mine water from great depths up to surface watercourses
(4) the use of physical barriers to minimize water ingress to shafts and active workings.

Bailing and its limitations

The most obvious response to the unwanted flooding of any hole with water is to bail the water out using a bucket or similar vessel of larger dimensions. This practice, which is generally termed 'winding water' in mining circles, may be all that is required to enable profitable mining to continue, particularly above the water table in low permeability strata. The very oldest surviving mine workings in Britain (i.e. Neolithic flint mines (Russell 2000), Bronze Age copper mines in North Wales (Jenkins & Lewis 1991; Timberlake & Jenkins 2001) and the Roman gold mine at Dolaucothi in South Wales (Burnham 1997)) are all above the local water table and would never have required anything more than bailing to secure access to working faces. The same is true of the vast majority of pre-1600 coal mine workings, which were typically in the form of modest bell-pits well above the local base-level of drainage (Levine & Wrightson 1991). Even after the progression of coal mines below the water table, shaft haulage of water to surface using kibbles (i.e. large buckets used in shaft sinking and maintenance) continued to play a significant role in mine dewatering in many places. Eye witness accounts record this practice at certain Tyneside pits in 1724, 1740 and 1848 (Clerk 1740; Dunn 1848; Atkinson 1966). Throughout the mid-19th Century, winding water was common practice in several Midlands coalfields, with the simple kibble (or 'bowk' as they were locally called) often being replaced by purpose-designed side-discharging tanks running on cage-guides (Knipe per. comm. 2003). Even in modern times, pumping is still occasionally augmented under exceptional circumstances by winding water (Sinclair 1958) using shaft hoisting equipment. For instance, in February 1921 East Pool tin mine (Camborne, Cornwall) closed, and by a terrible coincidence its main pumping shaft collapsed five months later. After about two years, the East Pool workings had flooded up to the level of the lowest decant route into the adjoining workings of South Crofty mine. In April 1923, virtually the entire former East Pool water-make began to flow unhindered into South Crofty, and in the struggle to cope with the sudden doubling of the latter mine's water make, the winding engines were used for water hoisting for seven months, until sufficient extra pumping capacity could be installed (Buckley 1997, p. 116–118).

Clearly there are severe practical limitations on the use of bailing as the principal means of dewatering a mine which is accessible only by shafts. In *The Compleat Collier* of 1708, manual bailing is mentioned as the first option for coal mine drainage, with horse gins being used for bailing from shafts deeper than about 50 m ('J.C.' 1708, p. 20). In a study of the development of the Wet Earth Colliery (Lancashire) during the 18th and 19th centuries, Banks and Schofield (1968) used *prima facie* reasoning to quantify the limiting water make beyond which bailing via a shaft becomes impractical if a colliery is to continue producing coal without frequent interruptions. The figures they obtained are based on shaft and kibble dimensions which were typical for underground mines throughout 19th Century Britain, and therefore are likely of generic applicability. By consideration of the motive power available from horse-powered whim gins and a kibble capacity on the order of 600 gallons, for the 48 m shaft at Wet Earth Colliery they obtain a maximum manageable inflow rate of about 4.5 ls^{-1}. For deeper collieries, other studies cited by Banks and Schofield (1968) suggest that the limiting inflow rate beyond which bailing will be insufficient is around 2 to 3 ls^{-1}. For inflows greater than this, mines will either have been pumped or else abandoned.

In some parts of Britain the progression in ground water control technology was directly from bailing to pumping. This was particularly so in areas such as the Grassington Moor lead orefield (North Yorkshire), where the topography was not suited to the development of adit drainage (Morrison 1998). However, in the majority of mining districts, the development of efficient pumping technology lagged some years behind the desire to delve deeper (e.g. 'J.C.' 1708), and adit drainage succeeded simple bailing as the preferred means of mine dewatering.

Drainage adits

Adits are simply long tunnels which are driven from some hillside to intersect wet underground workings for the purpose of under-draining them by free gravity flow to the portal of the tunnel. Most adits give the appearance of being horizontal in disposition, but careful surveying generally reveals them to slope gently towards their portals (1-in-500 is a common grade), as this favours free drainage and the avoidance of inconvenient ponded areas.

Adit technology has been around for a very long time. For instance, Buckley (2000) has noted the existence of an apparent reference to the excavation of drainage adits in the Book of Job. In this Old Testament biblical text, a succinct summary of metalliferous mining includes the following lines:

> '. . . Men dig the shafts of mines . . . Men dig the hardest rocks,
> dig mountains away at their base . . . As they tunnel through the rocks
> they discover precious stones. They dig to the sources of rivers
> and bring to light what is hidden . . .'
> (Book of Job, Chapter 28, verses 4, 9 and 11, as in the United Bible Societies' Translation of 1976)

Although the dating of specific texts in the Book of Job is a significant challenge for Old Testament scholars, it is safe to say that the above quotation refers to mining practices which were taking place more than 2500 years before present (Anderson pers. comm. 2002). Archaeological evidence has revealed that adit drainage technology was well-known in the Roman Empire. Davies (1935) documents Roman drainage adits in gold mines of Spain, Greece and Slovakia, some of which reached lengths as great as 2 km.

Metal mine drainage adits prior to 1600

At none of the known sites of prehistoric metalliferous mining in Britain have workings been found to have extended deep enough for adit technology to be necessary. Only at one suspected Roman lead mining site at Greenhow Hill (North Yorkshire) is there a possibility that a surviving gallery (the Jackass Level) at one time served in part as a drainage adit, though if this was indeed the case it has long-since been dewatered as the water table was lowered by adjoining drainage adits which were driven in the 19th century. Despite their known familiarity with adit technology, therefore, it remains a moot point whether, during their 400-year occupation of the island, the Romans dewatered their metalliferous mines in Britain by this means. According to documentary evidence recently reviewed by Buckley (2000) it is not until the late 13th century that clear evidence emerges for the installation and maintenance of drainage adits in metalliferous mines. A document dating from 1308 provides what is possibly the earliest recorded instance of roof-fall clearance works in a drainage adit. These works were undertaken to facilitate the re-opening in 1292 of silver mines on the eastern bank of the River Tamar in Devon. The work specified included 'clearing of adits and workings of dead work (rubble) . . . [and] penetrating a blockage of rock, in length 22½ fathoms, which had been stopping the water in the highest part of the mine of Furshill, where it touched the minerals being worked' (Buckley 2000, p. 26). Another very early example comes from the rather modest lead mining district of Rossendale (Lancashire) in relation to which financial accounts covering the year 1304–1305 include an entry for payment 'to the miners for making a certain trench underground to draw the water off from other trenches' (see Raistrick & Jennings 1965, p. 71). In the more substantial lead mining district of Weardale (County Durham), the driving of a 'watergate' (a north country term for a drainage adit) at a site called

Balkden is recorded in a financial account dated 1426 (Raistrick & Jennnings 1965, p. 71). Sporadic further references to the use of drainage adits in the metal-mining districts of Devon and Cornwall continue throughout the 14th and 15th Centuries, mirroring the development of metal-mining technology elsewhere in Europe up to the mid-16th Century state-of-the-art as described in Agricola's *De Re Metallica* of 1556 (Hoover & Hoover 1950).

Coal mine drainage adits prior to 1600

Coal was used to a limited extent in ancient times, but archaeological evidence has long indicated that its early use was largely restricted to metallurgical applications (Taylor 1858; Galloway 1898). For instance, while it has been found in association with metal-working remains in the Roman Camps along Hadrian's Wall (Northumberland), it has long been noted that conspicuous exposures of coal in close proximity to several of the camps show no evidence of having been worked on a scale consistent with domestic use during Roman times (Taylor 1858). Given the apparent indifference of the Romans in Britain to the possibilities of coal as a major fuel source, it is not surprising that no firm evidence exists for dewatering adits of Roman origin in Britain. Ancient mine roadways with a disposition which suggests they may once have served as drainage adits are occasionally unearthed during opencast mining. Some of these features have dimensions similar to those of known Roman adits in Mediterranean countries, which has prompted speculation that they may indeed be of Roman origin. However, no firm dating evidence (from timbers etc) has ever been found to substantiate such conjectures. On balance it seems unlikely that the Romans experienced such severe shortages of firewood that they needed to mine coal far underground. Such ancient mine roadways are thus more likely to be of mediaeval origin.

Little changed in the scale of exploitation of coal, and hence in associated dewatering needs, in the first millennium following the departure of the Romans. For instance, in the third decade of the 8th Century, in writing the first history of the English-speaking people, the monk Bede (746) noted the presence of 'rich veins of metal – copper, iron, lead and silver' in Britain, but had only this to say about the coal which is so abundant in the vicinity of the Tyneside monastery where he spent his entire life: 'This is a glossy black stone which burns when placed in the fire; when kindled it drives away snakes' (The fact that coal was noted as burning 'when placed in the fire' indicates that it was not the normal domestic fuel of the day, which must still have been wood in that period). Indeed, it is not until January 26th 1357–1358 that coal mining had advanced to the extent that problems of coal mine drainage could make their first appearance in the documented history of Britain, in the form of a complaint from the Prior of Tynemouth to the King of England to the effect that 'the men of Newcastle were digging in his moor of Elswick, and endeavouring to demolish the drain (*seweram*) from his mine in Elswick Moor, which was the chief form of sustenance of himself and his house' (Galloway 1898, p. 42). Although Taylor (1858) suggests that pipe rolls dated 19th February 1367 can be interpreted as implying the driving of drainage adits at coal mines in the vicinity of Birtley and Winlaton (northern County Durham), the earliest unequivocal records of the deliberate construction of coal mine drainage adits (as opposed to their destruction, as described in the 1357–1358 Elswick case) are found in the 15th Century financial records of Finchale Priory, a Benedictine monastery in the Wear Valley some five kilometres north of Durham City. In 1428, the monks of Finchale drove a 'water-gate' at nearby Coxhoe, a venture which proved reasonably successful in comparison with further drainage adit projects in the Durham area which they undertook subsequently at Softley (1433–1434) and Baxtandfordwood (1442–1443). In Scotland, records dating from 1531 document monks of Newbattle Abbey planning the construction of coal mine drainage adits to carry water down to the sea in the vicinity of Prestongrange (Midlothian), a few kilometres east of Edinburgh (McKechnie & MacGregor 1958) (Interestingly, during 2002 the Prestongrange sea-adits became the focus of renewed attention, during the investigation of mine water impacts on transport infrastructure in the area).

Mine drainage adits after the advent of gunpowder

It is essential to note that all drainage adits driven in Britain before the mid-17th Century, in both metal- and coal-mines, were excavated without the aid of explosives (Ford & Rieuwerts 2000, p. 24). This is the principal reason why, as Raistrick and Jennings (1965, p. 71) have noted, 'adits for mine drainage did not come in to general use until the seventeenth and eighteenth centuries'. It was only after the widespread adoption of gunpowder for blasting in the decades following 1630 that drainage adits commonly began to extend over distances of more than one or two kilometres. Table 2 summarizes some of the more significant drainage adits constructed in Britain between 1600 and 1900.

It is immediately evident from Table 2 that Derbyshire was Britain's pioneering region in the development of explosives-driven drainage adits (Ford & Rieuwerts 2000), with many of the major drainage adits in that County dating from the 17th Century. The principal reason why gunpowder-driven adits were developed in Derbyshire much earlier than elsewhere is quite simple: the majority of lead ore deposits in the Derbyshire orefield are

Table 2. *A summary of some of the more importanta drainage adits driven in Britain between 1600 and 1900*

Year driveage commenced/ first noted	Name of adit	Location	Type of mine drained by adit	Comments	Source of information
1617	Black Burn Watergate	Durham	Coal	Driven to underdrain the world's first industrial-scale collieries on Whickham Fell, Gateshead. Still flowing in 2002 despite considerable under-drainage by pumping at Kibblesworth Colliery nearby	Levine & Wrightson (1991); Clavering (1994)
1627	Weet Sough	Derbyshire	Lead	Located at Winster; the earliest known extant drainage adit in Derbyshire	Ford & Rieuwerts (2000)
1632	Dovegang Sough	Derbyshire	Lead	Driven from Cromford Hill by Sir Cornelius Vermuyden; widely (if erroneously) quoted as the first drainage adit in England, but was almost certainly the earliest drainage adit to extend for more than a few hundred metres	Ford & Rieuwerts (2000)
1653	Aspull (Haigh) Sough	Lancashire	Coal	Originally driven by Sir Roger Bradshaigh (1025 m) from 1653 to 1670; subsequently extended a further 1575m by about 1856. Continues to be a source of pollution to the present day	Unpublished Coal Authority records
1654	Tideslow Sough	Derbyshire	Lead	Had reached a length of 800 m by the winter of 1685	Ford & Rieuwerts (2000)
1657	Longe Sough	Derbyshire	Lead	Also known as Cromford Sough; also extends several hundred metres	Ford & Rieuwerts (2000)
1657	Bates Sough	Derbyshire	Lead	The third major 17th Century sough in the Cromford area	Ford & Rieuwerts (2000)
1687	Winster Sough	Derbyshire	Lead	Drains the Portaway Pipe deposit	Ford & Rieuwerts (2000)
@ 1670	Delaval Drift	Northumberland	Coal	Draining the pits of West Kenton, Newbiggin and Whorlton down to the Tyne at Benwell. The portal of this drift (no longer flowing due to subsequent under-mining and sustained pumping at Kibblesworth, Gateshead) was found and preserved in the 1980s during redevelopment of the west end of Newcastle (see Fig. 1a)	Galloway (1898, p. 161)
1693	Hannage Sough	Derbyshire	Lead	Drains the mines north of Wirksworth to its portal at Willowbath Mill	Ford & Rieuwerts (2000)
1700	Stanton Sough	Derbyshire	Lead	Noted as being 'a mile in length by 1700'	Ford & Rieuwerts (2000)
1711	Pool Adit	Cornwall	Copper	Thought to be the first major adit in Cornwall to be driven entirely using gunpowder blasting technology. Later largely under-drained by the Dolcoath Deep Adit (see below)	Buckley (2000)
1723	Apes Tor Sough	Staffordshire	Copper	The first of five major levels driven to drain the major Cu-Pb mine complex of Ecton Hill, North Staffordshire. Total length approximately 340 m	Robey & Porter (1972)
1729	Tanfield Watergate	Durham	Coal	An adit running for several kilometres through the Beckley and Tanfield collieries, for which a wayleave of £2000 was paid, and which contributed to under-draining the Stanley area of Co. Durham to depths in excess of 120 m	Galloway (1898, p. 250)
1737	Clayton Level	Staffordshire	Copper and lead	The second of five major drainage adits to be driven into Ecton Hill, in this case to under-drain the Clayton Pipe Vein. Total driveage approximately 0.6 km	Robey & Porter (1972)
1742	Great County Adit	Cornwall	Copper and Tin	The pre-eminent drainage adit amongst all the metalliferous mining fields of the UK, eventually totalling more than 55 km of driveage, and remaining a continued source of polluted drainage to this day, with an average flow of around 0.4 m^3s^{-1}; see text for further details	Buckley (2000)
1750	Fordell Day Level	Fife	Coal	Originally driven from portal near Fordell Castle (by Inverkeithing) 3.2 km to Drumcooper Pit (reached about 1800); by 1850, further extended by 2.3 km to William Pit. Major source of mine water pollution at the present day	Unpublished Coal Authority records

@ 1751	Tailrace Level	Durham	Lead	This level was apparently initiated at the commencement of mining in the Scarsike Veins of Rookhope, Weardale. It had certainly attained a length of 1.5 km by the early 19th Century, when a 1 km extension of the level to Grove Rake Mine was initiated. The level is currently a major source of polluted drainage to the Rookhope Burn	Fairbairn (1996); Johnson & Younger (2000)
1759	Chaddock Level	Lancashire	Coal	Underground extension of the Bridgewater Canal; a continued source of polluted mine drainage. See text for further details	Mullineux (1988)
@ 1760	Dolcoath Deep Adit	Cornwall	Copper/tin	Driven from Roscroggan on the Red River, through Roskear to Dolcoath; this adit was the last drainage adit in Cornwall to form part of a working mine (South Crofty, closed 1998) and remains a source of river pollution	See Younger (1998); Buckley (2000); Adams & Younger (2002);
1765	Elginhaugh Daylevel	Midlothian	Coal	Portal on the River North Esk, drains ancient workings (some Mediaeval, associated with Newbattle Abbey) in the Whitehill Rough Seam. Estimated 1.6 km length into area of major workings	Unpublished Coal Authority records
1766	Hill Carr Sough	Derbyshire	Lead	Took more than 20 years to under-drain main target vein; so much water was released that the level was later used for barge haulage of ore	Ford & Rieuwerts (2000)
1770	Kitty's Drift	Northumberland	Coal	More than 3 km in length, this adit provided drainage and haulage of coals to the river for a number of large collieries to the west of Newcastle city	Galloway (1898, p. 268)
1772	Meerbrook Sough	Derbyshire	Lead	Driven to under-drain mines in the Cromford and Wirksworth areas, it was 3 km long by 1805, and was later driven further between 1842 and 1882. Continues to yield a major flow of good quality ground water	Ford & Rieuwerts (2000, p. 53)
1774	Ecton Deep Level	Staffordshire	Copper	Driven approximately 400 m from the banks of the River Manifold, this adit under-drained the earlier Apes Tor Sough to open up more workable ground in the Ecton Pipe Vein	Robey & Porter (1972)
1776	Nent Force Level	Cumberland	Lead and Zinc	Underdrains the mines of Nenthead and the Nent Valley down to a portal adjacent to Alston. By 1842, some 8 km had been driven, achieving some 220 m of drawdown at its greatest extension. The Nent Force Level remains today a major element in the hydrology of the area (mean flow: 0.02 m^3s^{-1}), and a significant source of Zn pollution to the Rivers Nent and South Tyne	See Nuttall & Younger (1999); also Wilkinson (2001)
1782	Gillfield Level	North Yorkshire	Lead	The first of three major drainage adits which contribute to the drainage of the Greenhow Hill lead mining field. Along its 1.9 km driveage it intersects three veins (it is driven on lode in the Coldstone Sun Vein) and continues to yield a significant flow of good quality water to this day	Dunham and Wilson (1985); Everett (1997); Gill (1998)
@ 1800	Golynos Watercourse	Torfaen (South Wales)	Coal and ironstone	Originally an ironstone drift mine in the Bottom Vein Minestone of the Abersychan district, it was extended over the years to its current 1km length to provide drainage to extensive coal workings in the Garw Seam	Walker, P., Big Pit National Mining Museum of Wales, pers. comm. 2003.
1818	Halkyn Deep Level	Clwyd (N Wales)	Lead	Also known as the 'Old Drainage', this adit was driven from the Nant-y-Fflint at some 55 m above sea-level, finally totalling more than 8 km in length prior to losing its functionality when it was under-drained by the even deeper Milwr Tunnel in the early 20th Century (see final entry in this Table)	Williams (1997)
1825	Perseverance Level	North Yorkshire	Lead	The second of three major drainage adits to under-drain Greenhow Hill. Totalling more than 800 m driveage, this adit yields a substantial flow of ferruginous water to this day	Dunham and Wilson (1985); Everett (1997); Gill (1998)

Table 2 *(cont.)*

Year driveage commenced/ first noted	Name of adit	Location	Type of mine drained by adit	Comments	Source of information
1825	Eagle Level	North Yorkshire	Lead	The third of the major adits under-draining Greenhow Hill. Totalling 2.4 km of driveage by completion in 1850, Eagle Level failed to open up significant new ore reserves, and also failed to significantly draw the water table down near its forehead, despite achieving a water make of some 0.06 m^3s^{-1} (which was until recently used for public supply by Yorkshire Water)	Dunham & Wilson (1985); Everett (1997); Gill (1998); Younger (1998).
1853	Lucy Tongue Level	Cumberland	Lead	Driven to under-drain (and later to receive water pumped from deeper workings in) the great Greenside Mine, Glenridding. 1.7 km in total length. Remained in constant use until the mine closed in 1962	Shaw (1970)
1855	Blackett Level	Northumberland	Lead	Driving continued until 1912, reaching a total length of some 7 km. Substantially under-drains the major lead mining field of East Allendale and Allenheads, sustaining a permanent drawdown on the order of 180m, with a mean flow of around 0.1 m^3s^{-1}. See Figure 1b.	Raistrick & Jennings (1965, p. 222–223); Dunham (1990, p. 161–165).
1864	Sir Francis Level	North Yorkshire	Lead	One of the few long adits in the Swaledale mining district, driving continued until 1877, totalling 1.4 km. The adit allowed gravitational and pumped under-drainage of the Gunnerside Gill mines	Raistrick (1975); Gill (2001).
1870	Bullhouse Water Drift	South Yorkshire	Coal	600 m long 1:300 grade drainage adit with portal on the banks of the River Don, draining a large volume of workings in the Halifax Hard Seam. Now the site of a major Coal Authority mine water treatment system	Unpublished Coal Authority records
pre-1880	Old Meadows Waterloose	West Yorkshire	Coal	1:150 grade purpose-driven drainage adit extending 343 m from the banks of the River Irwell into the Lower Mountain seam. Now the site of a major Coal Authority mine water treatment system	Unpublished Coal Authority records
1880	River Arch Level	Torfaen (South Wales)	Coal and Ironstone	Approximately 1 km in length, incorporating a culverted section of a local river (the Afon Lwyd) in the vicinity of Blaenafon, and connected into several pre-existing mine entries, including the Woods Level and the Forge Level. Drains coal and ironstone workings to the N, W and E of the Big Pit (Pwll Mawr), the National Mining Museum of Wales (Blaenafon), for which it also provides secondary egress	Walker, P., Big Pit National Mining Museum of Wales, pers. comm. 2003.
1897	Milwr Tunnel	Flintshire (N Wales)	Lead and Zinc	A major adit with portal invert just above sea-level, providing under-drainage for the limestone-hosted Pb-Zn vein deposits of Halkyn Mountain. It under-drained and wholly superseded the earlier Halkyn Deep Level (driveage commenced 1818; see entry above). The mines drained by the adit were worked until 1987; to this day, the Milwr Tunnel sustains an average flow of some 1.2 m^3s^{-1} (with peak flows up to 4.4 m^3s^{-1}), making it the most prolifically yielding mine drainage adit in Britain (and one of the most prolific in the world), reflecting the karstified nature of the limestone country rock of the Halkyn Mountain Orefield.	Ebbs (1993, 2000); Lynch, R., MCG Consultancy Services Ltd, pers. comm. 2003.

Whether an adit is deemed sufficiently 'important' to include here is judged on the grounds of its antiquity, scale, length, or continued importance as a drainage route at the present day.

hosted within permeable, locally karstified Dinantian limestone aquifers (e.g. Ford & Rieuwerts 2000), whereas such prolific aquifers are only sporadically in contact with similar ore bodies in the North Pennines (Dunham & Wilson 1985; Dunham 1990; Johnson & Younger 2002) and are utterly absent from the orefields of Cornwall and Wales. Fortunately, the deeply-incised landscape of the Derbyshire lead orefield lent itself perfectly to the installation of relatively short, deep drainage adits. By contrast, 'in Yorkshire the shape of the ground called for much longer soughs, and only a few were ever driven at great capital cost, and these mainly towards the end of the eighteenth and in the early nineteenth century' (Raistrick & Jennings 1965 p. 132). The sheer volumes of water encountered in the Derbyshire mines also led to the development of many adits dedicated solely to drainage (the term 'sough' denoting a drainage adit only, and never referring to a multiple-purpose adit). Again in contrast, the major adits in the North Pennine orefields were typically multiple-purpose, serving as exploration levels and horse haulage levels as well as drainage galleries (Dunham & Wilson 1985; Dunham 1990; Raistrick & Jennings 1965). Being the first district to embrace gunpowder for adit driveage, the Derbyshire mining engineers acquired skills and experiences which equipped them to profit from transferring their adit-driving techniques to the coalfields (e.g. Aspull Sough in Lancashire and the Delaval Drift near Newcastle; Table 2). By the middle of the 18th Century, such techniques were apparently commonplace throughout the British coalfields, as Clerk (1740, p. 19) noted in relation to the development of drainage adits in the collieries of central Scotland: '. . . now all who know the true method of running mines have fallen into the German manner of blowing up the hard strata in their way with gunpowder . . .'. Notwithstanding the widespread adoption of powder blasting techniques, Clerk (1740) was of the opinion that drainage adits much longer than about 900m were unlikely to be economic to drive or maintain. As Table 2 reveals, this analysis was not borne out by later practice, with coal mine drainage adits in Clerk's native Scotland exceeding 5 km in length within the century following his declaration.

Gunpowder-based adit driving was introduced to Cornwall from the Mendip lead mines in 1689 (Buckley 2000). By 1711 the stage was set for the perfection of the technique in the hard-rock mining terrain in the vicinity of Camborne (e.g. Pryce 1778; Buckley 2000) with the inauguration of the Pool Adit (Table 2). By the start of the 19th Century the mining engineers of Cornwall and Devon had so honed their skills that they came to be in demand even in Derbyshire, the erstwhile heartland of British adit-driving technology. Thus we find the famous Cornish engineer Richard Trevithick engaged in improving the dewatering operations associated with Hill Carr Sough (Table 2) in 1801 (Ford & Rieuwerts 2000, p. 54), and the great engineer of the Devon copper mines John Taylor (Burt 1977) deeply involved in mine drainage and development works at Alport, Derbyshire in the 1830s and 40s (Ford & Rieuwerts 2000, p. 54–55).

Many of the adits listed in Table 2 continue to flow to the present day (Fig. 1), representing permanent changes in the ground water regimes in their host catchments (e.g. Banks *et al.* 1996; Younger 1998). While some of these adits represent sustained, point sources of water pollution in their host catchments (e.g. Younger 1998), others are of such good quality that they have been harnessed for public water supply use (Banks *et al.* 1996).

Even where the flow of such adits diminished or ceased altogether after they were subsequently under-drained by later, deeper workings, they can take out a new lease on life after the deeper dewatering eventually ceases. Two recent examples of this phenomenon are as follows:

(1) the Tunnel Pit, near Standish in southern Lancashire, was served by a 1 km coal mine drainage and barge-haulage adit, which was originally in direct communication with the Wigan-Liverpool Canal. Before the end of the 19th Century the Tunnel Pit had been under-drained by pumping in nearby shafts, and it was therefore decommissioned and its portal sealed and buried. Mining in the vicinity finally ceased in the 1960s and 30 years later water began to flow once more from the then-forgotten, buried tail reaches of the Tunnel Pit. To prevent excessive head build-up (with the risk of a sudden outbreak) and to drain the water to a nearby reed-bed which affords some treatment of the water, a number of relief wells have been recently drilled behind the old Tunnel Pit portal plug (Whitworth 2002).

(2) the Tailrace Level (Table 2) is a major drainage adit in the heart of the North Pennine lead orefield. Throughout living memory, the Tailrace Level had discharged an average of 0.86 Mld^{-1} of fairly good quality water. Following the 1999 closure of the last mine in the area, which lies 2.5 km upstream of the Tailrace Level portal, the flow rate of the Level increased by more than 1 Mld^{-1}, and the quality of the water deteriorated such that it carried tens of milligrammes per litre of each of the contaminant metals Fe, Mn and Zn (Johnson & Younger 2002).

To fully appreciate the scale of disruption of natural hydrogeological conditions wrought by the more extensive adit systems, it is worthwhile considering in

(a)

(b)

Fig. 1. Two examples of long-established drainage adits remaining long after the closures of their respective mines. (**a**) A 17th Century coal mine drainage adit: the portal of the Delaval Drift on Scotswood Road, Newcastle Upon Tyne, which was driven from around 1670 onwards to under-drain extensive workings to the west and north of the city of Newcastle (see Table 2). (Photo: P L Younger). (**b**) A 19th Century lead mine drainage adit: the Blackett Level, Allendale Town, Northumberland in 1997 (see Table 2). (Photo: A Doyle).

a little more detail two of largest drainage adits ever constructed in Britain: the Great County Adit of Cornwall (which drained the tin and copper mines of western Cornwall) and the Chaddock Level, which drained the coal mines of the Worsley area (Lancashire) into the Bridgewater Canal.

The driving of the Great County Adit was initiated in the Carnon Valley of western Cornwall in 1742 (Buckley 2000), with the initial aim of achieving a total driveage of about 4.2 km (2.5 miles) to drain the mines of Poldice all the way to their westernmost extremity. By 1760, the adit had already achieved this aim and was soon extended by means of various branch-levels to underdrain adjoining mines. Extensions to the County Adit system continued to be added until around 1870. By that year the scale of the entire County Adit system was truly staggering (Buckley 2000). Although no part of the Great County Adit was ever to lie at any greater linear distance than 9.2 km from the portal, the ramifications of the adit system meant that far greater distances than this could be travelled by ground water as it made its way from the outermost drained workings to daylight in the Carnon Valley. Overall, the full dendritic network of adit passageways totalled more than 55 km, and under-drained an area in excess of 33 km^2. Although the mines under-drained by the Great County Adit had mostly been abandoned by 1900 (Buckley 2000) it remains a major element of the ground water – surface water interactions of west Cornwall to this day, for it continues to intercept between 40% and 60% of the total effective precipitation falling on the overlying ground surface (Younger 1998).

The name 'Bridgewater Canal' is truly applicable only to an artificial surface waterway which links Manchester with Runcorn in Cheshire (Mullineux 1988). However, this canal came to be so intimately and directly linked with a large system of underground coal workings in the vicinity of Worsley in Lancashire that the name 'Bridgewater Canal' is now also loosely used to refer also to Britain's greatest-ever combined drainage adit and subsurface coal barge haulage system (Whitworth 2002). Extension of the Bridgewater Canal into the subsurface commenced in 1759, in part to supercede a number of more modest pre-existing drainage adits in the vicinity. The subsurface section of the Bridgewater Canal is called the Chaddock Level (Mullineux 1988), and this eventually extended more than 6 km from its portal at Worsley, linking the extensive underground workings of three collieries (Chaddock, Queen Anne and Henfold), all of which were thus able to ship their coals directly from the subsurface into canal barges, which were as long as 14 m by 1.37 m wide (Whitworth 2002). At the height of its extension around the beginning of the 19th Century, the Chaddock Level comprised more than 86 km of underground canal-ways capable of accommodating barges developed on four distinct levels with major systems of underground locks and self-acting haulage (Galloway 1898, p. 330) by means of which the weight of a full barge was used to pull an empty barge up to a higher level. The Chaddock Level remained in dual usage until 1887, when rail haulage of coal at surface was introduced to the district (Mullineux 1988). However, Chaddock Level did retain a useful drainage function until the closure of the nearby Mosley Common Colliery in 1968. To this day, the perennial discharge of ferruginous water from the Chaddock Level remains an important source of discoloration in the Bridgewater Canal.

Mine water pumping

The full story of the evolution of water-lifting technology has been documented elsewhere (e.g. Hill 1984; Walker 1995) and a reiteration of that documentation is beyond the scope of this paper. Only a partial summary, focused on the exigencies of the mining sector, will be given here, with an emphasis on technologies employed before 1881, when underground electrification began to be introduced in British collieries (Hill 1991).

In one sense, the pumping of water from mine workings is a mere sub-plot of the wider history of water-lifting technology. The earliest pumps in the world were almost certainly developed for purposes of irrigation (Hill 1984), and many of the devices invented for that purpose formed the basis for the technology used to this day in mines worldwide. For instance, a clear reflection of the Archimedean Screw of the ancient Mediterranean cultures (Hill 1984) is to be found in the robust Mono™ pumps which have been the technology of choice for pumping the most turbid of coal mine waters since the 1950s (Sinclair 1958, p. 98). However, certain types of pump which are abundant in other contexts, (such as the simple 'swape' or *shaduf* (Hill 1984), in which a bag of water is raised on a long pole passing over a fulcrum beyond which the short end of the pole is weighted), are never mentioned as being applied in the context of mining, no doubt due to the shortage of space for swinging the loaded pole in a cramped mine gallery. Up to about 1700, it is probably fair to say that mine water pumping technology was a mere sub-set of that used elsewhere. However, between 1700 and 1900, the development of automated pumping in the mining industry far outpaced that in any other industry, so that it is also fair to say that mining provided the technical wherewithal for the development of the world's first large-scale public water supply systems based on the pumping of ground water.

Classical references to mine water pumping principally relate to Roman mines in Mediterranean lands. It is clear from both contemporaneous accounts and from archaeological finds (especially in Spain; Davies 1935; Flores Caballero 1981) that the Romans employed a wide range of technologies, including simple force pumps, a rotating bucket device known as the *Tympanum*, Archimedean screws, and large-diameter, manually- or treadmill-operated bucket-wheels known as *noria* (Hill 1984). Just as the evidence for Roman mining in Britain is scant, so we are as yet bereft of direct evidence that these technologies were used in mines here.

At a later stage of historical development, the well-known 16th Century German mining engineering treatise *De Re Metallica* by Georgius Agricola (Hoover & Hoover 1950) includes the world's earliest descriptions of various adaptations of the *noria* and newer pumping technologies such as the rag-and-chain pump which we know for certain were in use in Britain around the same time. Probably the earliest reference to an automated water-lifting device in use in a British mine comes from the Abbey of Finchale (County Durham) whence expense records for the year 1486–1487 (i.e. 70 years before the first appearance of Agricola's *De Re Metallica*) register the expenditure of £9 25s. 6d. '*de le pompe*', relating to the construction of a new pump, apparently driven by a horse gin, for the dewatering of a colliery at a site named Moorhouseclose. Annual expenditures recorded in the following years attest to the ongoing use and maintenance of this pump (Galloway 1898, p. 71). That this was not an isolated development is suggested by another surviving expense record dating from 1492, which registers a payment for 'two great chains' to be employed in drawing water and coals from a mine in the manor of Whickham, some 12 km north of Finchale (Levine & Wrightson 1991, p. 13). While this reference to the dual use of the chains for drawing coals suggests a large-scale bailing operation, we will never know for sure what form the pumps at Moorhouseclose and Whickham took. Bailing operations and tantalisingly-undescribed pumps are mentioned a century later in Cornwall (in Richard Carew's Survey of Cornwall in the 1580s, as summarized by Buckley 2000, p. 28). However, by the 1600s it is at last evident that British mines in general, and the collieries of NE England in particular, were being pumped by rag-and-chain pumps (Clavering 1994). 'Chain engines', as rag-and-chain pumps were then colloquially known, 'drew water up a standing wooden pipe by means of discs mounted on a continuous chain' (Clavering 1994). *Prima facie* reasoning based on the weight of the chains and the water, the strength of chains and the frictional resistance in the wooden riser pipe, suggests that each individual chain pump was incapable of lifting water more than about 30 m (Clavering 1994). Multiple-stage lifts were therefore necessary if water needed to be lifted more than 30 m.

Water powered mine pumps

Although horse gins were used to drive these chain engines in some places and wind power was also considered as an option (Sinclair 1672; 'J.C.' 1708; Clerk 1740), waterwheels were the preferred power source wherever this was feasible. Indeed waterwheels driving mine pumps became so widespread in the coalfields that they came to be denominated by a specific term: 'coalmills' (Clavering 1994). In the 17th Century, colliery owners found coalmills so much more cost-effective than horse gins that they went to considerable lengths to divert water around hillsides for many kilometres to ensure a sufficient supply for their purposes. One such water diversion system was The Trench, a still-visible canal which led water for some 3.5 km from a small dam on a tributary of the Black Burn (NZ 223598) to a three-stage coalmill at Ravensworth Close (Co Durham) (NZ 238594). An unusually detailed contemporaneous account of this system is given in an Appendix to an early hydraulics book entitled 'The Hydrostaticks' (1672) by George Sinclair of Edinburgh. The rarity of such accounts (see Levine & Wrightson (1991) and Clavering (1994) for relevant reviews) means that Sinclair's description merits quoting at length. The Ravensworth Close coalmill comprised three chain engines, each driven by its own waterwheel. Sinclair (1672, p. 299–300) described the three-wheel system thus:

> '. . . For procuring a fall of water, which may serve the wheels
> of all the three sinks [i.e. shafts], [Sir Thomas Liddel has] erected
> the first work [i.e. water wheel] upon pillars like a wind-mill,
> pretty high above ground, from which the water falling
> makes the second go close above ground. And to make the water fall
> to the third, the whole wheel is made to go within the surface of the
> ground, which terminates at a river under the works [i.e. a tailrace adit],
> which mine is of a considerable length . . .'

Elsewhere in his text, Sinclair (1672, p. 298) gives some details of subsurface arrangements of a coalmill, which though not specifically stated to relate to the Ravensworth Close system can be confidently assumed to do so:

> '. . . there is first one [shaft] 40 fathom deep from the grass.
> Another in a right [i.e. straight] line from that, of 24.

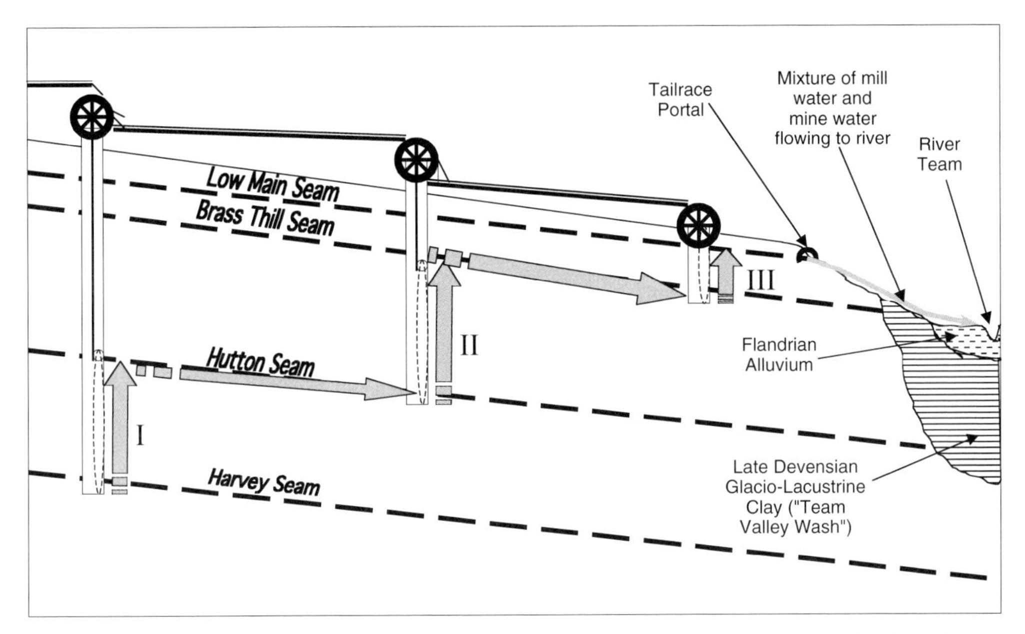

Fig. 2. Ravensworth Close coalmill of 1672: a geological cross-section of the site of this three-wheel chain engine dewatering system. The section has been developed by adding modern stratigraphic knowledge to the verbal description of the system given by Sinclair (1672). Notes: I, II and III are the three vertical lifts by the rag-and-chain pumps, which are represented by the dashed lines in each of the three shafts. The grey block arrows show the progress of the pumped water from the deepest workings to daylight at the portal overlooking the River Team.

Another of 12; upon all which there are water-works [i.e. water wheels].
In the first sink the water is drawn from the bottom 12 fathom,
and thence conveyed into a level or mine, which carries it away to
the second sink. By the second work, the water is drawn out of the
second sink 14 fathom, from the bottom, and set in by a level
to the third sink, which being only 12 fathom deep
the water-work sets it above ground . . .'

Figure 2 reproduces this description in the context of our modern geological understanding of the area (as summarized on British Geological Survey 1:10560 Sheet NZ 25 NW). From the stratigraphic configuration now deduced for this site, it is clear that the Ravensworth Close coalmill was well located to dewater several square kilometres of workings to the west of the hydrogeological barrier represented by the Team Valley Wash, a buried valley immediately to the east of the site which is plugged with glacio-lacustrine clays (Fig. 2).

Wheels were not the only device used to harness water power for pumping from mines. Bucket-operated balance beam engines were apparently widely used in the mid- to late-18th Century, when they were documented in operation in Cornwall, Staffordshire and Scotland, operating in both coal and metalliferous mines (e.g. Robey and Porter 1972, p. 27–30). Vernacular terms applied to these devices include 'flop-jack' (in Cornwall), 'balance bob', 'bucket engine', 'tub engine', and (in Scotland) 'Bobbin' John'. A sole surviving example of the genre remains to this day, at the former Straitsteps lead mines, Wanlockhead (Scotland) (Although records of bucket-operated beam engines at this site date back to 1745, the surviving example is probably of mid-19th Century construction; Anon. 2002). The Straitsteps engine consists of a stone column fulcrum on which is balanced a long timber beam. At one end of the beam is a large bucket, which was positioned to receive surface water diverted via a launder from the nearby Wanlock Burn. At the other end of the beam, a pivot provided a connection to a pump plunger which worked a reciprocating pump deep underground in a mine shaft. As the water bucket at the free end of the beam filled to capacity, it outweighed the pump plunger at the far end of the beam and sank downwards, shedding its load. By this means, the pump plunger was raised within the rising main, drawing water up to the point where it decanted from the main into a shallow adit, where it mixed with the water spilled from the bucket and flowed away to the Wanlock Burn.

Reciprocating pumps, such as that operated by the Straitsteps beam engine, had largely supplanted the 'chain engine' design by the mid-1700s. However, although the subsurface components changed, the use of water power as the motive force for pumping did not cease everywhere as soon as steam-driven pumps became available. For instance the reciprocating pump at Wanlockhead continued in use until about 1931. Even in less remote areas, water power remained in use well after the advent of the Newcomen Engine (in 1712). For instance, a waterwheel-driven reciprocating pump was recorded in action in 1844 at Beamish Colliery, County Durham (Hair 1844; Glendinning 2000) (albeit it was later replaced by a steam engine prior to 1857; Horne 1993). The water which was used to drive this waterwheel was itself mine water, led from the outfall of a nearby drainage adit known as the Telford Drift. At the Laxey Lead Mine on the Isle of Man, what was to be world's largest-ever mine drainage waterwheel (a 22 m diameter overshot wheel named the Lady Isabella) was commissioned as late as September 1854 and remained in service until 1929 (Kniveton 2000). The pumps driven by the Lady Isabella routinely raised 19 ls^{-1} of water from a depth of 365 m, and it has been estimated that this required only 20% of the wheel's maximum power output. The magnificent Lady Isabella and associated beam engines remain in full working order and today serve as a major tourist attraction.

Mine dewatering in the age of steam

In recounting the surprisingly late persistence of waterwheel-driven pumps, it is not intended to understate the importance and overwhelming popularity of steam-driven pumps for the dewatering of mines all over Britain. The first tentative steps towards the economically-viable raising of water by steam power were taken in the 17th Century, with the granting of a patent on 17th January 1631 to one David Ramsay for an invention 'to raise water from low pits by fire' (Galloway 1898, p. 196). However, nothing seems to have come of Ramsay's invention. Experiments by physicists working in the Low Countries and France, most notably Huyghens and Papin (Galloway 1898, p. 236–238), paved the way for the next attempt at full-scale harnessing of the power of heat, which was made public on 25th July 1698, when Thomas Savery obtained a patent for his 'Miner's Friend'. This steam engine met with a certain degree of success in large-scale experiments, but it proved both inefficient and dangerous in the real working environment of coal mines. In particular it proved incapable of raising water more than about 10m without running the risk of a boiler explosion, and since this was far less than the 73m limit of water-driven coalmills then in existence, the 'Miner's Friend' never enjoyed a reciprocal amity from the miners themselves, with only one engine ever being ordered by a colliery (Galloway 1898, p. 197–198). We know that the author of the *Compleat Collier*, writing at the start of the 18th Century, was aware of the disappointing performance of Savery's engine. In reflecting on this, that author lamented ('J.C.' 1708, p. 18–19):

> '. . . were it not for water, a collery [sic] in these parts,
> might be termed a Golden Mine to purpose, for dry colleries
> would save several thousand pounds per Ann. which is
> expended in drawing water hereabouts . . .
> If it would be made apparent, that as we have it noised abroad,
> there is this and that invention found out to draw out all great
> old waists, or drown'd collieries, of what depth soever; I dare assure
> such artists, may have such encouragement as would
> keep them their coach and six, for we cannot do it by our engines,
> and there are several good collieries which lye unwrought and drowned
> for want of such noble engines that are talk'd of or pretended to . . .'

As it turned out, the author of the *Compleat Collier* did not have long to wait, for it was in 1712 that Thomas Newcomen, of Dartmouth in Devon, constructed the world's first, full-scale steam-based engine to successfully dewater a colliery (the 'Dudley Castle' engine at Coneygree Colliery, Tipton, South Staffordshire; see discussion in Rolt & Allen 1997). Within the next two years, further Newcomen engines were installed at Bilston (near Wolverhampton), Hawarden (Flintshire, North Wales) and at Griff Colliery (Warwickshire). The engine at the latter site cut the annual costs of dewatering (which had previously been achieved using horse gins) by more than 80% (Galloway 1898). Such major cost savings soon became trumpeted throughout the land, and within the following eight years, Newcomen Engines were constructed in most of the mining districts of Britain. The 'drowned collieries' of NE England were in the vanguard of the rush of investment in the new technology, fulfilling the desires expressed not ten years previously in *The Compleat Collier* ('J.C.' 1708). The earliest known pictorial representation of a Newcomen engine is dated 1715 and hails from Tanfield Lea Colliery in County Durham (Atkinson 1966, p. 26). Newcomen Engines were erected at several other NE collieries before 1720 (at Washington, Ravensworth, Byker and Elswick) and by 1724, a northern agency for the construction of steam engines had been established at Chester-le-Street (Galloway 1898, p. 243). By 1740, Newcomen engines in Scotland were documented as raising water over vertical distances as high as 146 m (Clerk 1740, p. 24), albeit such a high lift required the use of four stages of 36.5 m each.

Given the overwhelming association in popular imagery of steam engine houses with Cornwall, it is ironic that the uptake of Newcomen Engines was much more widespread in the coalfields of northern England and Scotland than in the copper and tin mines of Cornwall before 1741. This disparity was due to the fact that the sea-borne coal needed for steam-raising was heavily taxed by the English Government at the time (whereas land-scale coal was not and was in ready supply in the coalfields). When the sea coal taxes were suspended in 1741, Cornwall experienced a boom in the construction of Newcomen Engines (Buckley 2000, p. 35). By about the same time, Newcomen Engines were being erected in France (Galloway 1898, p. 244), and soon thereafter in other countries of the European mainland. From small beginnings, nothing short of a revolution had hit the field of mine dewatering throughout the industrialized world, and the Newcomen Engine was to reign supreme for some 60 years.

The next major step in the development of steam-based pumping came in the 1760s in Scotland, when James Watt reflected on the relatively wasteful use of steam in the Newcomen process. The design improvements which Watt developed resulted in the double-acting rotative steam engine, which at a stroke improved the power and economy of the steam engine. Following on from the first full-scale demonstration of the new engine for dewatering of the Burn Pit, Kinneil (Scotland) in 1784, the history of steam technology took on a new trajectory (McKechnie & MacGregor 1958). With steam power no longer restricted just to mine dewatering duties, the age of steam haulage had dawned, which by 1804 had led to the development of locomotives for surface mineral haulage and by 1822 was laying the foundations for the modern passenger rail network (Leifchild 1853).

The new opportunities afforded by the availability of powerful, reliable steam-driven pumps allowed the exploitation of coal and ore reserves on a scale not hitherto possible. One hydrogeological consequence of this boom in mining was the creation of regionally-interconnected systems of mined voids, which in some instances eventually came to be connected over straight-line distances in excess of 50 km (Younger *et al.* 2002). With mined systems interconnected on this scale, it soon became logical to push for regional coordination of mine dewatering operations. For instance, this was advocated as early as 1857 by T. John Taylor, a prominent member of the North of England Institute of Mining Engineers, who devised a specific plan for the integrated dewatering of the entire Tyne Coal Basin between Newcastle and Tynemouth (Taylor 1857). The substantial water management problems of this district in the first half of the 19th Century, to which the proposals of Taylor (1857) were a response, have been documented by Galloway (1904, p. 7–8). Taylor's original proposals were not implemented in the Tyneside coalfields at the time, largely due to the chronic fragmentation of interests amongst the many mine owners operating in the area.

Far more successful was the integrated drainage of the South Staffordshire Coalfield (Redmayne *et al.* 1920). By the mid-19th Century, some 320 km^2 of inter-connected workings in this area were experiencing a total water make on the order of 2.6 m^3s^{-1} (Waller 1867). Prior to the 1870s, individual contributions to this total water make were handled individually by each of the collieries in the area. However, disastrous inrushes at Coneygree Colliery (near Tipton) in 1865 and at Highbridge Colliery in 1871 (see Table 1) provided overwhelming evidence in favour of a regional-scale dewatering operation. On 21st July 1873, the first of what would eventually be eight specific Acts of Parliament to facilitate the establishment of an integrated dewatering system for the South Staffordshire Coalfield was enacted. Four distinct drainage areas were identified in the 1873 Act, within each of which a number of individual 'pounds' (i.e. underground catchments) were mapped, providing the geometric basis for planning the deployment of pumping engines. Nearly 50 years after the establishment of this impressive dewatering system, a thorough review of the operational success of the scheme resulted in the recommendation of only a few minor changes to the drainage arrangements (Redmayne *et al.* 1920); that the original design had proven to be so robust is a ringing endorsement of the sound understanding of mine water hydrology possessed by South Staffordshire mining engineers in the second half of the 19th Century.

District pumps within mines

Thus far in considering pumping we have focused on the final act of removing water from a mine by pumping from a shaft sump. However, much of the pumping effort in a mine is expended in gathering water from remote districts of the workings and delivering it to the principal dewatering sumps. As manual bailing soon becomes impractical within cramped underground workings, small pumps have long been used for gathering water from working areas (Sinclair 1958). We know the Romans used Archimedean screws and *noria* for such purposes in mainland Europe (Flores Caballero 1981). Simple hand-operated pumps on the rag-and-chain principle were certainly used in British mines until at least the mid-19th Century (Fig. 3a). Where the quantity of water to be handled was substantial and deep adits were available for gravity drainage of the lifted water, permanent underground waterwheel installations were used. A splendid reconstruction of one of

(a)

(b)

Fig. 3. (**a**) Portable rag-and-chain pump in use to locally drain a working area in the early 19th Century (from Taylor 1858). (**b**) A portable 'Pulsometer' steam pump in use at the tail of the water in coal workings in the late 19th Century (from Pamely 1904).

these underground waterwheels may be viewed in operation within the Park Level Mine at Killhope, the North of England Lead Mining Museum (Weardale, County Durham).

Nowhere was the use of underground waterwheels more intensively developed than in the lead mines of Allenheads (Northumberland), where no fewer than six installations were in use by the end of the 19th Century (1923). Figure 4 summarizes the Allenheads dewatering system. Water was

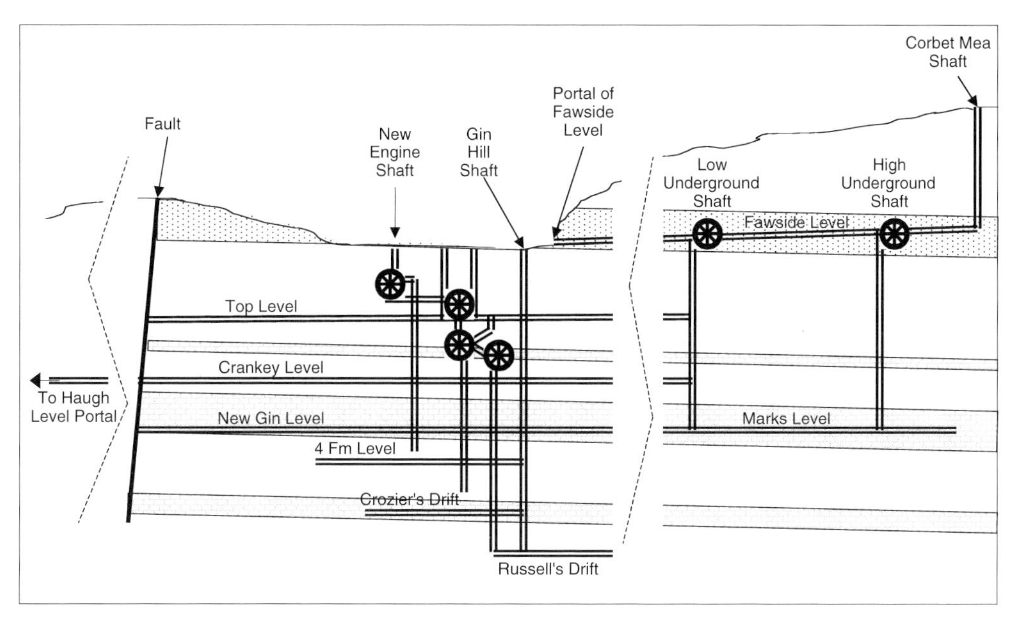

Fig. 4. Simplified synoptic cross-section showing the use of underground waterwheels in the Allenheads lead mine, Northumberland, in the late 19th and early 20th Centuries, developed from more detailed information on specific portions of the whole system presented by Smith (1923) (details in the vicinity of the Gin Hill Shaft) and Dunham (1990) (details between Corbet Mea Shaft and the Fawside Level portal). For an explanation of the sequence of operations see the text. The geological ornament shown relates to specific aquifers. The stippled beds are sandstones of the Firestone Sill. The brick ornament denotes three limestone aquifers, which (from highest to lowest) are the Little Limestone, the Great Limestone and the Four Fathom Limestone.

deliberately introduced to the mine from a surface reservoir via pipes in the Corbet Mea Shaft, and carried in-bye to drive two waterwheels (at the High and Low Underground Shafts). Beyond the Low Underground Shaft, the driveage water and mine water which had been raised by the pumps from Marks Level in the Great Limestone aquifer) was allowed to flow by gravity along the Fawside Level, whence it reached daylight at the adit portal, immediately east of the main Allenheads mine yard. After being used for mineral washing on the nearby dressing floor, the same water was despatched underground once more via the New Engine Shaft (Fig. 4), whence it fell over four separate waterwheels in series, each of which drove a separate reciprocating pump which together lifted water more than 120 m from the deepest workings up to the Crankey Level. Here the pumped water mixed with the wheel-driving water and flowed by gravity for a further 2 km underground, to daylight at the portal of the Haugh Level (an old drainage adit dating to 1684).

As we have seen, steam power was adopted for pumping shaft sumps as early as 1712. However, the difficulties of building, operating and maintaining bulky steam pumps in cramped mine workings appears to have hindered the adoption of steam power for localized pumping within mines until the late 19th Century. In surface-based pumping arrangements (using either waterwheels or steam engines to supply the motive force) it was relatively common practice by the 1850s to operate pumps at some distance from the power source using overland rods which changed the sense of reciprocating movement by means of t-shaped bobs. (An excellent example of this approach survives in working order at Laxey, Isle of Man; Kniveton 2000). Taking a cue from this practice, Moore (1872) described the installation and successful use of a 700 m system of underground rods in Kinneil Colliery, Scotland (incidentally the same site where James Watt first demonstrated his new steam engine design), which transmitted reciprocating action from the vertical rods in the shaft across a number of high- and low-points in a series of mine roadways, to work pumps installed in certain down-dip workings which were prone to flooding. Although successful, the use of extensive underground systems of rods occupied space in cramped mine roadways which could better be used to facilitate coal haulage etc. To overcome this limitation, a truly automated steam-driven pumping technology for localized pumping within mine workings was finally introduced in 1878, in the shape of the 'Pulsometer Pump'. In the low-lift applications for which it was designed (Fig. 3b), the Pulsometer Pump amounted, at last, to a successful implementation of the core principle of Savery's 'Miner's Friend' of 1698 (Galloway 1898, p. 198; Pamely 1904; Hill 1991). Although the Pulsometer remained in use at many collieries into the early 20th century (e.g. Pamely 1904), by then the onward march of electrification was already beginning to make its mark on the field of mine water pumping, albeit the required electricity was generated on-site using steam-powered generators (e.g. Darling 1901).

Physical barriers to minimize water ingress

The well-known axiom that 'prevention is better than cure' was not lost on the early mining engineers, and they devoted considerable thought and physical effort to minimizing the ingress of unwanted water to underground workings. In describing the measures which they devised, it is necessary to distinguish between those implemented during the sinking and equipping of shafts, and those implemented during active working of coal or other minerals.

Minimizing water ingress during shaft sinking

As evidenced above, by the late 17th Century mine water pumping techniques had become sufficiently well advanced that it was possible (in all but the most permeable ground) for mines to proceed to depths of several hundreds of metres below the pre-mining water table. However, as the most easily-worked strata were gradually exhausted (Levine & Wrightson 1991), mine exploration inevitably migrated to locations where it was necessary to sink shafts through considerable thicknesses of saturated, permeable strata before mining could commence. One of the earliest settings in which this took place was in the axis of the Tyne Valley, where unconsolidated sand and gravel deposits of Quaternary age overlie the Coal Measures. As these permeable deposits are not exposed above the high water mark (being extensively blanketed with late Devensian glacio-lacustrine clays) it appears that their presence was at first unsuspected by engineers sinking shafts along the banks of the Tyne. The earliest attempted sinkings through these heavily-watered sands and gravels met with failure and records dating from 1637–1638 report that three pits sunk on the Tyne floodplain near the first colliery boomtown of Whickham had succumbed to closure on account of insuperable water makes in the preceding two years (Levine & Wrightson 1991, p. 40–41).

Methods for coping with these problems soon emerged. Where possible, an auxiliary pumping shaft would be sunk into the problematic saturated sands and gravels (which were often encountered as running sands or 'quicksand' during shaft sinkings) alongside the main sinking. The water make arising from the sands and gravels would be diverted into the auxiliary shaft and pumped to surface ('J.C.' 1708, p. 13), thus avoiding two problems:

- the difficulties and discomforts associated with allowing this water to cascade into the working shaft;
- the greater expense and inconvenience associated with pumping the same quantity from much greater depth at the foot of the working shaft.

The use of auxiliary shafts in this manner is undoubtedly the first documented instance of an approach to water management in mining which is nowadays termed 'advance dewatering' or 'external dewatering' (Younger *et al.* 2002, p. 206–211).

As each sinking proceeded below the foot of its auxiliary pumping shaft, further measures were introduced to minimize water ingress during the sinking of shafts. Thus the author of the *Compleat Collier* stated ('J.C.' 1708, p. 18–19):

> '. . . For framing back our Shaft Feeders, we make use of Wood, but chiefly Firr [sic], because . . . we think it swells with the Water lying against it . . . we make use of Sheep-Skins with the Wool on . . . which, being well Wedged in between the frame and such rough Mettle, &c . . . we find to perfect our Design of stopping the Water . . .'

Already, somewhat more sophisticated approaches were under development. It would appear that stone coffering was in use in some of the coal mines of Shropshire by 1698 (Galloway 1898, p. 223). By 1708, iron frames were introduced to add strength to the timber packing at Harraton Colliery (near Birtley, Co. Durham) ('J.C.' 1708, p. 13). However, for most of the 18th Century, iron frames were not widely used for shaft lining (Galloway 1898, p. 258). Instead, advances in the engineering of wooden linings proceeded to the point that well-coopered timber linings were proving capable of holding back water with driving heads of 70 to 100 m (Galloway 1898, p. 311). These timber linings comprised:

(1) horizontally-oriented 'cribs' or curbs of 0.2 m-square timber, which were fastened to the rock walls of the shaft at vertical intervals of 0.6 to 0.9 m, forming closed rings around the circumference of the shaft, to which were nailed;
(2) tightly-jointed, vertically-oriented planks of wood (up to 75 mm thick and 3 m long).

Because of its resemblance to well-made wooden tubs, this form of shaft lining came to be known as 'tubbing', a name which persisted even after the preferred lining material ceased to be wood.

At shallow depths, or where the strata being excavated yielded little water, the use of brick and mortar shaft linings became commonplace during the early 19th century. For lining-out unconsolidated deposits and water to depths of ten to twenty metres, brick walling founded on timber cribs was used in combination with a sealing fill of puddled clay, which was pressed into the annulus between the brick wall and the rock face of the shaft. This was the system employed, for instance, in the sinking of the Plas Power Collieries near Wrexham (North Wales) in the 1870s (Griffith 1876). By the end of the 19th Century, brick-lined shafts were being constructed to considerable depths (Pamely 1904, p. 101–103), but rarely with much success and never cheaply where the sinking had to contend with substantial aquifers. For such heavy-duty applications, cast-iron tubbing became the favoured option for most 19th century mining engineers.

The pioneer of cast-iron tubbing was John Buddle, one of the great early 'Viewers' (i.e. colliery engineers and managers) of the 18th and early 19th Century coal trade in NE England. During the sinking of the Wallsend A Colliery (Northumberland) in 1792, Buddle overcame the now-familiar problems with saturated Tyne Valley sands by designing and installing cast-iron tubs, which in this first application were cylinders manufactured to closely fit the excavated circumference of the shaft, (Galloway 1898, p. 311). Three years later, the nearby King Pit at Walker (Northumberland) was lined in the same manner. However, as the use of cast-iron tubbing spread, it soon became evident that single-piece, cylindrical iron tubs were extremely difficult to manipulate except in the shallowest of applications. Buddle responded to this by designing segmented iron tubbing which could be easily lowered into place then joined together *in situ* until the entire circumference of the shaft had been lined. This improved technique was first implemented during the sinking of the adjoining Percy Main Colliery (Northumberland) in 1796–1799 (Galloway 1898, p. 312).

Notwithstanding the success of cast iron tubbing in Buddle's sinkings, it had its drawbacks. Foremost amongst these was cost, which was considerably more than for conventional wooden tubbing. Secondly, although it was introduced on the grounds that it was far stronger than wooden tubbing, it was not infallible, as evidenced by the Houghton inrush of 1832 (see Table 1). Furthermore, when employed in upcast ventilation shafts, many of which in the early 19th Century were still driven by shaft-foot furnaces and therefore carried damp air as hot as 170°C, cast-iron tubbing rapidly rusted. For these reasons, wooden tubbing continued to be used in shaft sinking well into the 19th Century (Galloway 1898, p. 295).

Cement was also proposed as an alternative to cast-iron tubbing, on the grounds that it was cheaper and far less prone to degradation under the conditions likely to be encountered in upcast shafts (Watson 1861). This proposal caused considerable controversy (Anon. 1861; Atkinson 1861), principally because the attainable strength of cement rings was far less than that required for the safe tubbing of very high water pressures then being encountered in deep sinkings in the concealed coalfield of Durham (Atkinson 1861; Atkinson & Coulson 1861). Using experimental data provided by Watson (1861) on the strength of cement rings, Atkinson (1861) was able to demonstrate that under 180 m head of water cement rings would crumble, whereas cast-iron tubbing had by then been known to withstand water heads far in excess of this figure for more than 150 years. In the event, it was to be well into the 20th century before that offspring of simple cement, mass concrete, was successfully developed as the shaft-lining of choice (e.g. Forster Brown 1924).

Throughout the 19th century, the mining engineers of the Durham Coalfield were in the vanguard of endeavours to develop major coal reserves inferred to be present beneath major, bedrock aquifers. In the concealed, eastern half of the Durham Coalfield, the strata which unconformably overlie the Coal Measures are of Permian age, and consist of up to 200 m of dolomitic limestones (the Magnesian Limestone) overlying a highly variable thickness (from zero to tens of metres, over distances as short as one or two kilometres) of well-sorted, poorly-cemented sands of aeolian origin (the Yellow Sands). Both the Magnesian Limestone and the Yellow Sands are prolifically permeable, providing between them a substantial component of the public water supply in east Durham to this day (Younger 1995). The experiences gained in battling the Magnesian Limestone and the Yellow Sands in the Durham Coalfield not only paved the way for later sinkings through the Bunter Sandstones of Yorkshire, Derbyshire, Nottinghamshire, Staffordshire and Lancashire, but also provided many of the crucial insights upon which 20th century undersea mining was so successfully founded.

A full account of the many mishaps suffered and lessons learned by the bold pioneers who joined battle with the aquifers of the Durham Permian would require a separate paper and thus only a brief summary can be given here. The earliest attempted sinking through the Durham Permian was at Hetton in 1818. So severe were the water problems encountered when the shaft entered the Yellow Sands that the sinking had to be abandoned. After a re-think, a new start was made in 1821. Despite thorough tubbing, the water make which had to be dealt with was as high as 150 ls^{-1}. By the end of the 19th

Century, a total of 20 successful sinkings had been made through these aquifers (Bell 1899; Atkinson 1902), virtually all of which had been achieved only after major struggles at the limits of contemporaneous pumping and tubbing technologies (e.g. Potter 1856; Anon. 1857; Wood 1857).

During the first half of the 19th Century, the use of tubbing to exclude water from deep shafts proceeded according to 'close topped' designs, in which each tier of tubbing would be completely sealed top and bottom against the water-bearing strata. This was achieved by building each tier of tubbing up from a foundation ring (termed a 'wedging crib') which was firmly set into the shaft wall and sealed against the base of the tubbing structure by means of a gasket. Installation of such close topped tubbing is well-described from a number of sinkings in the eastern half of the Durham Coalfield (e.g. Atkinson & Coulson 1861; Pamely 1904), and generally proceeded as follows:

(1) at a point where a wedging crib was required, the diameter of the shaft would be increased (by manual pickwork, rather than blasting) by 30 or 40 cm to provide a groove into which the crib could be slotted;
(2) segments of iron wedging crib (typically 15 cm hollow square section girder) would be inserted into this groove and wedged together (using oak wedges) until it formed a complete ring which tightly fitted the circumference of the shaft and protruded by about 50 mm or so to provide a ledge upon which tubbing could be founded;
(3) the first course of cast-iron tubbing segments would then be installed on the ledge. 'Sheeting' made of bark would be first laid on the crib, which would deform under the weight of the tubbing segments and form a tight seal. The tubbing segments were typically 0.6 to 0.9 m in vertical dimension and sized and shaped so that ten to twelve of them would complete the circumference of the shaft. Neighbouring segments were designed with inter-locking flanges so that they would fit tightly together without the need for bolting;
(4) five or six succeeding courses of tubbing would be built up, forming a tier between 3 and 6 m in total height, and the uppermost course of tubbing segments in the tier would be tightly sealed (using more bark sheeting) against the sole of the overlying wedging crib (which formed the base of the next tier);
(5) pressure-relief holes were usually present in each segment, which allowed water and gas to escape until the entire tier of tubbing was complete, after which wooden wedges and plugs would be installed to make the tub watertight.

Such tubbing was installed selectively in any given shaft, to case-out any loose and/or saturated strata which would otherwise collapse and/or feed water into the shaft. It was common practice to raise several tiers of close topped tubbing through selected intervals of problematic strata. These tubbed intervals would generally be interspersed with unlined sections of shaft, or sections only lightly dressed with a skin of brickwork or timber lining.

As this technique was implemented against ever-greater hydraulic heads during sinkings through the Magnesian Limestone and Yellow Sands in east Durham, severe problems began to be encountered, amounting to 'the most critical and dangerous source of failure in the tubbing employed to shut off large feeders of water often encountered in the sinking of mines; the blowing out of the sheeting, and displacement of the tubbing, by the pressures of the water and gas confined behind it, in cases where [the tubbing] is close topped' [i.e. tightly sealed top and bottom] (Atkinson & Coulson 1861). At the very least, this sort of displacement made the shaft walls uneven and therefore hazardous for cage haulage operations. In some cases, the pressure of water and/or gas was so great that segments of the tubbing were entirely dislodged and fell down the shaft, breaking shaft fittings and threatening the lives of miners below. Observations of such cases by Atkinson & Coulson (1861) revealed that once the pent-up pressure had obtained some vent, further disturbance of the tubbing was not experienced. A number of strategies were therefore developed to relieve excessive pressure before it could damage the tubbing. These are of hydrogeological interest because they demonstrate a working knowledge of the non-hydrostatic nature of ground water head in heterogeneous aquifers. The three strategies were:

(1) to open up small bleed holes through the tubbing and allow a certain amount of water to enter the shaft, and fall to the sump where it would be picked up by the pumps. It was found that only a relatively small proportion of the feeder which had originally been tubbed-out needed to be allowed to enter the shaft in order for substantial local relief of pressure behind the tubbing to be realized;
(2) to exploit natural differences in head between different tiers of tubbing to provide pressure relief without drainage into the shaft. This was achieved by the use of 'pass pipes' connecting the top of one tier of tubbing to the base of the next, and so on to the top of the tubbed interval, where the uppermost tier was always open-topped (there being no overlying wedging crib into which the uppermost course could have been sealed). This effectively destroyed 'the

isolation of the water behind the different lifts or tiers, [making] them common, one to the other; and thus, in effect, [rendering] the whole of the tubbing open topped, through the medium of the uppermost lift, which is so, in fact' (Atkinson & Coulson 1861, p. 10);

(3) in cases where build-up of gas was suspected to be the principal threat to the tubbing, a pipe would be inserted through the tubbing and raised up the shaft to such a point at which the water level settled. This pipe then allowed gas to bubble up and escape through the open end of the water pipe without the inconveniences associated with draining the water into the shaft.

Each of these strategies was implemented on various occasions by Atkinson & Coulson (1861). In relation to the first strategy, during the sinking of Castle Eden Colliery in southeastern Co. Durham (which took place in 1836; Wade 1998) a 128 m interval of tubbing had been installed to deal with feeders originating in the Magnesian Limestone. No new feeders were encountered in the first nine metres of sinking below the basal crib of this tubbing interval. Then a new feeder of some 11.3 ls^{-1} was encountered, which was quickly tubbed back up to the overlying basal crib and sealed tightly. However, within 36 hours, the build-up of pressure behind the new tubbed interval dislodged some of the tubbing segments. Repeated repairs were made, but the dislodgement always recurred. Eventually, a tap was fitted through the tubbing which allowed a flow of 0.01 ls^{-1} to enter the shaft (i.e. less than 0.1% of the rate of the feeder which had been tubbed-out). No further dislodgement of tubbing ever occurred during a 25 year period of observation. A similar experience at Harton Colliery (South Shields, Co. Durham), where the main source of excessive pressure was methane being released from the saturated strata, met with success when a number of taps through the tubbing were allowed to drain methane and some 0.2 ls^{-1} of water into the shaft (Atkinson & Coulson 1861).

A second attempted sinking in the vicinity of Castle Eden (SE Durham) provides an instance of the second strategy, i.e. inter-connecting separate tiers of tubbing so that they all act as if they were a single open-topped tub. Brick walling to a depth of about 26 m had been used to exclude drift deposits from the shaft. About 3.6 m into the underlying Magnesian Limestone, a water strike occurred, which was eventually tubbed-out by an interval of tubbing founded at 88 m depth and carried up for 62 m to the base of the brick walling. Some 7 m below the basal crib of this tubbing, a horizontal tunnel was driven through the Magnesian Limestone to intersect the base of a shallow, secondary shaft being sunk nearby. This tunnel encountered no water whatsoever. However, on deepening the main shaft a further 5 m below the invert of the tunnel inset, water was encountered once more, this time yielding a substantial feeder of some 38 ls^{-1}. The close topped tubbing which was installed to exclude this water from the shaft was inter-connected with the uppermost interval of tubbing by means of two 230 mm diameter 'pass pipes'. This precaution having been taken, no displacement of the tubbing occurred.

By 1850 'Sinker Coulson' (as William Coulson was called locally) had successfully sunk 11 shafts through the Magnesian Limestone and Yellow Sands of Co. Durham (in addition to a further 19 shafts on the exposed coalfield in Northumberland and Durham) and his reputation was beginning to spread far beyond his native County (Wade 1998). In 1855, Coulson's expertise was called upon by mining entrepreneurs in the Ruhr Coalfield of Germany, where sinkings were now migrating to the exposed coalfield of the Emescher Valley, where the Cretaceous Chalk aquifer overlies the Coal Measures. On March 17th 1855, by way of a St Patrick's Day Celebration for his dedicated team of Anglo-Irish miners, Coulson began the sinking of the Hibernia Colliery in Gelsenkirchen, near Bochum. Extensive tubbing was employed to get the shaft through the main body of the Chalk with its prolific feeders of water, the shallowest of which entered the sinking at a depth of only 11 m below ground level. It was on entering the Greensand aquifer at 100 m below ground level that Coulson and his team found their opportunity to really make history. Encountering a feeder of 42 ls^{-1}, well-sealed close topped tubbing was installed throughout the Greensand interval, a pipe range was connected to a tap through the tubbing and led back up the shaft. Coulson was going to see just how much head was driving this feeder. To his surprise and delight, the water continued to flow from the end of the pipe range even after it had been raised 3.6 m above ground surface (and some 15 m above the shallowest feeder encountered during the sinking of the shaft). Not only had Coulson demonstrated the presence of supra-hydrostatic head in the Greensand aquifer at this location, he had also provided the new colliery with a free source of good quality water, which was captured in a reservoir and used to supply the needs of the site for many years (Atkinson & Coulson 1861; Wade 1998).

Having demonstrated that water could flow upwards from depth where the head conditions were appropriate, Sinker Coulson proceeded to demonstrate his working knowledge of the hydraulics of perched aquifers during the sinking of the South Wingate Colliery in the southeasternmost extremities of the Durham Coalfield. The Quaternary sequence at this site comprises 11 m of clay overlying 20 m of saturated sand (which yielded 5.3 ls^{-1} to the sinking shaft). This sand is in turn underlain by a further 15 m of clay, below which some 49 m of unsaturated Magnesian Limestone were sunk through before the water table was finally reached 95 m below ground level. Realising the opportunity which this situation afforded, Coulson arranged to capture all of the water from the Quaternary sands in a pipe which was led down the shaft and connected into the open top of the tubbing which was installed through the saturated zone of the limestone. The Magnesian Limestone was sufficiently permeable that the addition of more than 5 ls^{-1} of water from the Quaternary aquifer made no difference to water levels behind the tubbing. Water which would otherwise have been pumped at considerable cost was therefore discharged into the Magnesian Limestone aquifer free of charge (Atkinson & Coulson 1861, p. 16–17).

Following the death of Sinker Coulson in 1865, the use of the close topped tubbing system which he had perfected continued to enjoy enthusiastic uptake. The technique reached its apogee in 1869, with the sinking of the Shireoaks Colliery in Nottinghamshire (near the point where that county meets Derbyshire and South Yorkshire). The excavation of the 553 m Shireoaks shafts through two prolific aquifers (the Triassic Sherwood Sandstone and the Permian Magnesian Limestone) was achieved only after the installation of 155 m of cast-iron tubbing (Pamely 1904). The technique subsequently continued to be applied throughout Britain well into the 20th Century.

Having introduced vented, close-topped tubbing to the Ruhr Coalfield in 1855, mining engineers in NE England were the first in Britain to benefit from the return of a favour when, at the very end of the 19th Century, newly-developed German techniques for shaft sinking through aquifers by means of ground freezing were introduced for the sinking of two shafts at the Washington Glebe Colliery, County Durham. The hydrogeological challenges at this site related to the presence of some 24 m of unconsolidated, highly permeable sands of Quaternary age overlying the Coal Measures (Anon. 1902). The contract for ground-freezing at the site was awarded to the specialist firm Gebhardt und König, of Nordhausen, Germany. The contractors started work on the site in the autumn of 1901 and ground freezing commenced on 23rd March 1902 by means of a variant of the Poetsch method (Pamely 1904, p. 83–85). A steam driven compressor was used to cool liquid ammonia, which was then used in heat exchangers to chill magnesium chloride brine to between −8 and −11°C. The brine was then circulated through circular arrays of boreholes drilled beyond the full depth of the saturated sand around the perimeters of the two proposed shafts. The sand was frozen to a diameter of about 1.5 m around each of these boreholes, forming a dam of solid ice. Within the centre of each shaft, the sand remained unfrozen and was therefore easily excavated, with the 24 m of sinking to the underlying clay horizon being accomplished in only two weeks.

The second successful application of ground-freezing to shaft sinking in the UK was in the far more challenging setting of a new coastal colliery at Dawdon in East Durham (Wood 1907). The Dawdon shafts were sunk on a cliff-top plateau overlooking the North Sea. At this site, 109 m of Magnesian Limestone and 28 m of Yellow Sands overlie the Coal Measures (the two being separated here by a 1m thickness of mudstone known as the Marl Slate). Ground water in the Magnesian Limestone at this point was found (from its salinity and tidal piezometric behaviour) to be in hydraulic continuity with sea water. Sinking through the Magnesian Limestone proceeded by the time-honoured techniques of cast-iron tubbing (68 m of which were required) and pumping at rates of up to 533 ls^{-1} (Wood 1907). The two Dawdon shafts (named the Castlereagh and Theresa Shafts) had reached a point just above the Marl Slate by December 1902, at which point the sinking through the Yellow Sands was placed in the hands of Gebhardt und König, fresh from their successful sinking at the Washington Glebe Pit. The freezing process proceeded in precisely the same manner, with the exception that the greater target depth (and therefore higher geothermal temperature) necessitated cooling the brine to even lower temperatures than had been used at Washington, specifically between −13.5°C to −17°C. Seven months passed before the ice wall around the shafts was deemed to be complete, and sinking through the Yellow Sands re-commenced in the Castlereagh Shaft on November 7th 1904. In contrast to the case at Washington, the sands in the centre of the shaft were found to be so heavily frozen that explosives had to be used to proceed with the sinking. The sinking to the Coal Measures (whence conventional sinking could resume) was completed after 10 months work in each of the two shafts. Temporary wooden linings were used in both shafts to minimize the risk of rock falls during working. Apart from a short-lived scare when a pocket of highly-pressurized unfrozen water was struck by a drilled sump hole, resulting in a temporary 6 m-high fountain of cold water in the base of the Castlereagh Shaft, both sinkings proceeded without hindrance. Basal wedging cribs were installed in Coal Measures shale, 2 m below the base of the Yellow Sands. Cast-iron tubbing was carried in tiers back up from this crib to the base of the lowermost tubbed interval in the Magnesian Limestone. The annulus between the backs of the tubbing segments and the shaft wall was filled with concrete to an average thickness of 11 cm. Thawing of the ice walls around both was undertaken carefully, with active warming of the air in the shaft by slowly raising and lowering of a brazier, and by circulation of warm brine in the former freezing boreholes. After allowing for venting of any trapped gases from behind the tubbing, the shaft lining was sealed and found to be secure throughout (Wood 1907).

Since these two pioneering efforts, ground freezing has been used for sinking many more shafts in the UK. While the efforts sometimes ended in tragedy due to inrushes from unfrozen aquifers (see the case of Easington in 1903, in Table 1) the technique has generally been implemented with sustained success, most recently in the development of the Selby Coalfield (North Yorkshire) in the 1980s.

Following the consideration and rapid rejection of cement as a component for pre-cast tubbing in the early 1860s (Watson 1861; Anon. 1861; Atkinson 1861), as discussed above, it seems that no concerted efforts were made to implement cement-based solutions for the exclusion of water from shafts until 1911, when the use of cementation was introduced to British shaft-sinking circles by Mr Albert François, who had perfected the technique when sinking to coals beneath the Chalk aquifer in Belgium and France. The essence of the technique is to sink an array of sub-vertical boreholes, splaying radially from the projected shaft centre, into which grout is injected under pressure, sealing up the native fissures in the rock such that an effectively impermeable annulus is formed within which the shaft can be sunk with minimal difficulty (Forster Brown 1924). Hatfield Main Colliery near Doncaster was one of the first shafts to be sunk in this manner, successfully sealing back otherwise formidable feeders in the Permian dolomites and sandstones of that district. Over the years, variations on this approach were gradually developed, culminating in the mass-concrete lined shafts typical of the most modern collieries today, in which concrete liners are cast *in situ*, with back-grouting sealing any remaining gaps between the liners and the adjoining rock mass.

Minimizing water ingress to active workings

The evolution of methods to minimize water ingress during working was experientially driven, largely by reflections on some of the more devastating of the inrushes listed in Table 1. Common sense suggested that 'supported' methods of mining, in which pillars are left in place to support the roof, would be far less likely to result in significant induction of water inflows from above than the alternative 'caving' methods of mining, in which pillars were extracted (or never left in the first place) and the roof was allowed to fall. Events such as the 1837 inrush at Workington, in which winnowing of pillars below the sea-bed led to a devastating inrush, only served to underline this intuition. As experience grew, regulations were developed governing the 'safe' dimensions of supported workings and/or the minimum cover needed for caving workings. The first time such regulations were codified in a law was in 1877, following the inrush at Tynewydd (Table 1), in an amendment to the Regulation and Inspection of Mines Act of 1860. This new regulation stipulated that 'where a place is likely to contain a dangerous accumulation of water, the working approaching that place shall not at any point within forty yards of that place exceed eight feet in width' (Coulshed 1951). It was not until the 1920s that regulations governing safe working distances beneath the sea bed were introduced.

From the earliest days of industrial scale mining, it had been widespread (though not universal) practice for miners driving roadways in areas where old workings might reasonably be expected to drill ahead in an attempt to prove any flooded workings before the roadway got so close that the barrier of intervening ground might fail catastrophically. For instance, in writing about the copper and tin mines of Cornwall in the three decades of rapid development of deep workings following the dramatic expansion of steam-driven pumping in the County in the 1740s, Pryce (1778) commented that:

> '. . . In some places, especially where a new adit is brought home to
> an old mine, which has not been wrought in the memory of man,
> they have unexpectedly holed to the house of water, before they
> thought themselves near to it, and instantly perished . . . I think
> where they are tolerably acquainted with their situation, much danger
> may be avoided, by keeping three or five borier [sic] holes before them,
> radiated or displayed above and below, to the right and to the left, from
> the center [sic] of the adit . . .'

However, even where such precautions were taken, complete safety could not be assured. For instance, the devastating inrush at Heaton Colliery (Newcastle Upon Tyne) in 1815 (Table 1) took place despite precautionary drilling having been implemented (Dunn 1848). Another problem associated with drilling ahead of advancing workings in the early 19th Century was the problem of the boreholes themselves providing a route for a devastating inrush. This problem was later overcome by the invention of non-return valves on drilling strings (the fore-runners of the blow-out preventers now routinely used also in the petroleum industry) and by 1860 drilling ahead of advancing workings became a legal requirement in all situations where old workings were even remotely likely to be present

(Coulshed 1951). Eventually, mining engineers became sufficiently skilled in such drilling techniques that they were developed beyond a mere precautionary tool to a means for tapping into, draining and putting back into production areas of old workings which had previously been flooded and abandoned. Wilson (1901) gives a detailed account of the draining down of workings at Wheatley Hill Colliery (SE Co. Durham) which had previously been abandoned following the inrush of water which occurred on 19th January 1871 (Table 1). This was achieved using the Burnside Boring Machine, a local invention which included one of the world's earliest blowout preventers (patented by a Mr G Burnside in 1891, a few months after a similar device was patented by a Mr J Cowey; see Hughes 1904, p. 34–35).

Beyond the establishment and proving (by drilling) of natural rock barriers between active workings and bodies of water (natural or impounded in old workings), the technology of shaft tubbing gradually came to be expanded into geotechnical endeavours to dam unwanted waters back in old workings. At first, these endeavours were limited to closing shaft insets to worked-out seams and installing tightly-wedged tubbing at the interface with the shaft (e.g. Taylor 1858). However, given that the permeability of the flooded workings would offer no sort of 'throttle' on the release of water into the workings in the event of a tubbing failure, this practice was eventually discontinued, to be replaced by the construction of more substantial dams of simple design. One such 19th century example was described from the re-opening of George Stephenson's old workplace at Wylam Colliery, Northumberland, in the early 20th Century (Heslop 1994). Early dam designs tended to comprise two brick walls with clay packed between them. By the 20th Century, more formal designs involving reinforced concrete and rock grouting had largely supplanted these *ad hoc* approaches. Details of such techniques are beyond the scope of this paper, but technical details of such dams as constructed at the close of the 19th century are given by Pamely (1904).

Mine water quality – some early descriptions

At the start of the 21st century one of the principal drivers for sustained interest in the hydrogeology of mined ground relates to issues of water pollution (Younger & Robins 2002), arising from the fact that many mine waters carry dissolved loads of contaminants (especially metals such as Fe, Al, Zn etc.) at concentrations which are much higher than are generally permitted in industrial effluents entering receiving watercourses. These concerns, though very much in keeping with the contemporary environmental agenda of the developed world, are of course nothing new. During the very first phase of coal mining on an industrial scale, at Whickham (Gateshead, Co. Durham) in the early 17th century, numerous grievances concerning the effects of rapidly-expanding coal mines on the natural drainage of the area were raised in a legal deposition dated April 1620 (Levine & Wrightson 1991, p. 110–116). While many of these concerns related to the drying up of wells and springs under-drained by the new adits, water quality issues were also of major concern to the plaintiffs:

> '. . . two hundred acres and above [of good *meadow land had been*] quyte spoiled and cankered with the water that issueth out of the colewaists . . .'

Similar complaints mentioned the 'unwholesome, cankered and infectious' water that flowed from the drainage channels of the mines, which were deemed to have polluted the receiving land surface so badly that grass would no longer grow there, resulting in streams that were so polluted that even the beasts of the field would refuse to drink their waters or to eat grass grown by such streams. Nor were these water quality problems restricted to the Great Northern Coalfield. Further north still, in the Midlothian Coalfield (Scotland), Sinclair (1672) noted that the miners were in the habit of referring to mine water as 'the blood of the coal', a clear allusion to the red tinge associated with high concentrations of iron.

Whereas the coal miners of the north were quick to recognize the 'unwholesome' nature of their local mine drainage, a century later (1760) we find the even more polluted mine waters of Parys Mountain (Anglesey, North Wales) being lauded for their supposed medicinal virtues (Rutty 1760; Rowlands 1966):

> '. . . [the mine waters are considered] as a powerful detergent, repelling, bracing, styptic,
> cicatrizing, anti-scorbutic and deobstruent medicine, as hath appeared
> by the notable cures they have effected, not only by external use in
> inveterate ulcers, the itch, mange, scab, tetterous eruptions, dysenteries,
> internal haemorrages [sic], in gleets, the fluor albus, and diorhea [sic],
> in the worms, agues, dropsies and jaundice . . .'

Not a bad set of claims for Britain's most acidic mine waters, which to this day have a pH of 2.3 and carry extremely elevated concentrations of many toxic metals!

A few years later, in the copper and tin mining districts of Cornwall, Pryce (1778) was also disposed to celebrate the salubrious quality of several local mine waters, lionising in particular those of then-abandoned copper mines named Pednandrea and Huel Sparnon. The effluent from several abandoned adits was captured and used for domestic water supplies, apparently to no ill effect (Pryce 1778, p. 11):

> '. . . in twenty-four years of acquaintance with the practice of medicine,
> I have not met with any one patient, whose disorder I could attribute
> to the most trifling unwholesomeness in our mine waters . . .'

Pryce (1778) then goes on to admit that not all Cornish mine waters are as agreeable, noting that mines excavated into veins below the 'gossany bed' (i.e. zone of oxidation) 'do produce water fit for no use but driving mill or engine wheels. Such water is quite noxious, and palpably vitriolick [i.e. acidic] to the taste, particularly at the mines of North Downs, Chacewater and Huel Virgin'. Interestingly, these are amongst the major mines which to this day contribute to the acidic drainage emanating from the Great County Adit in the Carnon Valley (Buckley 2000; Table 2).

One particular experience recounted by Pryce (1778) is worthy of mention here for the echoes it provokes in those familiar with Cornwall's last working mine (South Crofty, which closed in 1998; see Adams & Younger 2002). Pryce (1778) notes: 'In Huel-Musick and Huel-Rofe, the writer has stood with one foot in the warm and the other in the cold water, and has divided and diverted them different ways'. The present author had precisely the same experience at more than 800 m below ground level in South Crofty Mine in 1997, where at one point in a roadway traversing the Great Crosscourse, one hand could be held in a warm dripper while the other received cold water from an immediately adjoining dripper of entirely different water quality.

Perhaps the earliest reference to a direct economic use for the dissolved constituents of mine waters, (long pre-dating the various instances listed by Banks *et al.* 1996) dates from the early 1770s and relates once more to the Parys Mountain copper mines (Rowlands 1966). It was around that time that it was realized that native copper could be precipitated readily from the highly acidic mine waters simply by inserting scrap iron into the water. In only a few seconds, native copper was reduced from the Cu^{2+} to the Cu^0 state by the zero-valent iron (which was itself oxidized to the Fe^{2+} and Fe^{3+} forms in the process). This early example of reductive precipitation, which nowadays underpins the use of zero-valent iron in permeable reactive barriers and similar remediation facilities, was subsequently harnessed for economic copper production at this site and at the Avoca Mines in Ireland.

One of the first thorough descriptions of hetereogeneous water quality in mines was made by Armstrong (1856) in relation to the mine waters of Wingate Grange Colliery, in the southeastern Durham Coalfield. He distinguished between three types of mine water:

- saline waters, of calcium chloride facies, which he considered may be genetically connected to sea water;
- ferruginous ('chalybeate') waters, which he unhesitatingly ascribed to pyrite oxidation processes, and;
- clean fresh waters not noticeably different from ordinary ground waters in the area.

The elevated iron content of the second type of water developed in a particularly unequivocal manner. Clean water had been induced to enter the pit from the overlying Yellow Sands following pillar removal in underlying seams. This water was not in itself ferruginous, but it flowed into the mine at variable rates which at times were so prolific (up to 140 ls^{-1}) that the instantaneous pumping capacity of the mine was exceeded and the water had to be stored in an area of standage which was set aside in old workings in the Low Main seam. When the water was released from this standage, it was found to have become acidic and highly ferruginous. The concerns which this caused were not related to the environmental consequences of its disposal (as would be the case with such waters today) but solely to the problems of clogging and corrosion of pumping and other equipment to which it gave rise (Armstrong 1856). The clogging of pipes by iron oxide precipitates was then (e.g. Swallow 1891) as now (e.g. Parker 2002) a considerable problem in mine water management and as such it prompted considerable efforts to minimize the effort and expenditure needed to counteract it

Ochre deposits were not the only precipitates observed to clog pipes. In a number of collieries in Northumberland and Durham, rapid precipitation of barytes took place within pipe ranges due to the mixing of two types of mine water (Edmunds 1975):

(1) deep, saline mine waters which were highly enriched in dissolved barium, but devoid of significant sulphate, and;
(2) shallower-sourced mine waters which had acquired high concentrations of sulphate through contact with oxidizing pyrite.

As barium rapidly precipitates to form barytes in the presence of sulphate, with barytes being in equilibrium with only about 1 mgl^{-1} of dissolved barium, circumferential precipitates of barytes developed within wooden 'water boxes' (Dunn 1877) and circular, cast-iron pipes (Clowes 1889). Inspection of these barytes precipitates revealed them to be zoned in seven-layer cycles, with six black-stained layers followed by one off-white layer. These layers correspond to the six working days in the mine, when the air and water are heavy with coal dust, and the well-observed Victorian Sabbath (Sunday), when pumping continued in the absence of coal production. In recognition of this pattern, the miners named these deposits 'Sunday Stone'.

In the course of describing the draining-down of old flooded workings at Wheatley Hill Colliery (Co. Durham) in 1900, Wilson (1901) left to posterity one of the first detailed descriptions of the processes of the oxidation of sulphide and iron following aeration of mine water which had long been trapped in old workings (where bacterial sulphate reduction had evidently occurred). On discharging the water into ventilated workings, an odour of hydrogen sulphide was noticed (and its identity proved using lead acetate test papers). The water also proved slightly acidic to litmus paper. To touch, the water was 'soapy feeling' and it had a 'sickly taste'. Initially it was milky white in appearance, due to the presence of tiny bubbles formed as gases came out of solution, but later ran clear. As it flowed along the mine roadway, it first precipitated a 'white slimy deposit' (which later analysis showed was elemental sulphur) and then, downstream, red iron hydroxides. These sequential precipitation reactions, described by Wilson (1901) for what was almost certainly the first time in relation to mine waters, now form part of the armoury of geochemical reactions which present-day engineers are attempting to harness for purposes of passive treatment of polluted mine waters (e.g. Younger *et al.* 2002).

Technology transfer: sinking, pumping and tubbing enter civil engineering practice

The strong links between early mining engineering and the development of civil engineering could form the subject of an entire book. Rieuwerts (1962) provided the first formal reflection on this topic. 40 years later, much detail remains to be added to this account, particularly in relation to interactions between the mining and water engineering sectors. Here, only a brief introduction to this topic can be given, highlighting a few activities in which the mining engineers can be demonstrated to have transferred their technology and skills to the nascent water and geotechnical engineering sectors.

Shaft sinking and adit driving techniques developed for mining purposes were already being widely adopted for purposes of developing public-supply ground water sources by the middle of the 19th century. To many mining engineers, the joys of developing shafts in which they did not have to achieve the exclusion of nearly all of the surrounding ground water had an irresistible attraction. During the second half of the 19th century purpose-made water-supply shafts were being sunk by mining engineers (to typical mining dimensions) across the outcrop of the Magnesian Limestone Aquifer in Co. Durham. With the exception of the Ryhope pumping station, at which the ground water became saline due to sea water ingress, all of these pumping shafts remain in use today as part of Northumbrian Water's resource base. At a number of mid-19th Century colliery sites in the English Midlands, mine shafts shafts sunk through the Sherwood Sandstones were deliberately designed to intercept the fresh ground water and divert it to public supply use. Many of these shafts remain in use to this day by Severn-Trent Water, and interesting issues are now beginning to arise in relation to their long-term future once their host collieries are abandoned.

Mining technology was also adopted in the driving of horizontal adits from the sumps of large-diameter wells, particularly in the Chalk. However, it remains to be determined whether the adits in the Chalk were developed by engineers with personal experience of mining, or whether the form of these structures is an instance of 'convergent evolution'.

One key element of technology which undoubtedly originated in the mining sector and was subsequently transferred to the water supply sector was steam-driven pumping. The evolution of steam pumping technology has already been discussed above. Following the innovations of Watt and co-workers in the final quarter of the 18th Century, the cost of steam-pumping technology had fallen by about 1800 into a price range which made it economically attractive for water supply purposes. It then enjoyed rapid uptake throughout the UK water industry, interestingly aided by the professional inputs of Robert Stephenson (Waller 1872), the famous son of the great George Stephenson, who had first liberated the steam engine from the bounds of collieries to serve the wider needs of society. Throughout the 19th Century, there was no professional dividing line between mine water pumping technology and water supply pumping technology (Watkins 1979). Thus it is no surprise to find that one of the earliest forums in which the performance of water supply pumping stations was presented for technical discussion by a *coterie* of interested engineers was the North of England Institute of Mining and Mechanical Engineers (Waller 1867,1872).

The nascent geotechnical engineering sector of the 19th century enjoyed vigorous interchanges of both methodologies and personnel with mining engineering. Two examples typify opposite ends of the spectrum in terms of the successful integration of mining-derived techniques in subsurface construction projects: the Thames Tunnel and the Severn Tunnel.

The first Thames Tunnel, which connects Rotherhithe and Wapping, is famous for its associations with Marc Brunel (1769–1849) and his world-famous son Isambard Kingdom Brunel (1806–1859) (Rolt 1957). In one of the earliest attempts to get the tunnel underway (1807), the renowned mining engineer Richard Trevithick (1771–1833) engaged the latest Cornish high pressure steam pumping technology and shaft sinking techniques. This attempt proved to be one of the few failures of Trevithick's otherwise dazzling career. The problem was that ground conditions in the Thames floodplain, where highly permeable Devensian gravels underlie a thin veneer of silty alluvium of Flandrian age, were quite unlike the conditions in the wettest of Cornish mines, where old workings in hard rock are invariably the source of hefty water makes. Thus in Cornwall, as long as you can deal with the quantity of water, the strata will require little assistance to remain upstanding, whereas tunnelling in the Thames floodplain means grappling with unconsolidated deposits which are prone to collapse and enter any pumped void along with the water they release. A new approach was needed, and Marc Brunel devised this: a large, cast iron tunnelling shield which could be jacked forward as the forehead of the excavation advanced. Even with the assistance of this shield, the tunnel still took 18 years to complete – very slow progress in comparison with conventional mine drainage adits which were under construction around the same time (Table 2). The finished product has stood the test of time, and remains in use to this day as a train tunnel.

50 years later, at the other extremity of the Great Western Railway for which Isambard Kingdom Brunel had since become so famous, the construction of the Severn Tunnel commenced in 1873. This project resulted in a far longer sub-water table tunnel (7.5 km) than Marc Brunel's Thames Tunnel, and it dealt with sustained water feeders of several hundreds of litres per second. However, this 13-year project was eventually accomplished by the patient application of the cast iron tubbing and conventional steam-powered pumping techniques which had served the master sinkers of the coalfields so well over the preceding seventy years (Walker 1891).

Concepts of ground water occurrence amongst early mining engineers

Having surveyed the practical responses of the early mining engineers to ground water problems which they encountered, it is fitting to ask what, if anything, these engineers made of the phenomenon of ground water occurrence. As these men were not generally given to philosophical musings, it is no surprise to find that many of them preferred to adopt their views of natural phenomena 'off-the-shelf', from existing strands of thought current in their times. To understand how the mining engineers' conceptualizations of ground water changed over time, therefore, it is necessary to briefly consider how hydrological processes were conceived in the wider world. To 17th and 18th Century engineers in Britain, there were basically three philosophical traditions from which they might choose their preferred explanation of ground water occurrence (Biswas 1970):

- the concepts of the ancient Greek philosophers, particularly those of:
  (1) Plato who considered all springs to be derived from sea water by some mysterious, unobservable process of subsurface distillation within the bowels of the earth, with waters being drawn inland from the seafloor and purified as they pass upwards (contrary to gravity) to reach the Earth's surface as fresh water springs, and;
  (2) Aristotle, who considered that subterranean water was derived by the condensation of mysterious vapours entering in caves and other voids in the subsurface.
- the Judaeo-Christian creation myths, as expounded in the Book of Genesis.
- emerging scientific concepts of the hydrological cycle, in which evaporation of sea water was (correctly) considered to be sufficient to explain the origins of virtually all observed terrestrial ground- and surface waters.

The world's earliest treatise on mining engineering, Agricola's *De Re Metallica* of 1556, admits that some proportion of ground water arises from infiltration of surface waters, but largely adheres to the concepts of subsurface condensation as proposed by Aristotle (Hoover & Hoover 1950, p. 46–48). Thus Agricola claims that steam is generated deep underground and migrates upwards towards the Earth's surface, where it eventually condenses to form water: 'In this way water is being continually created underground'.

In 1708, the author of *Compleat Collier* gave the following explanation of mine water occurrence:

> '... all which [mine] Water we suppose to come from the sea, and so being
> fed by that inexhaustible fountain, we call it by the name of a Feeder,
> and that it may rise to the top of any mountain we are subject to believe
> no great Matter of Wonder, because we are so often, by the Curious and Learned,

told, That the Sea, this Fountain Head, is higher than
the Earth . . .'

('J.C.' 1708, p. 17–18)

One presumes that the author of this passage had Genesis 1 (verses 6–7) in mind when he considered mine water provenance, since that verse of scripture ascribes the blue sky to the presence of a vault holding back primordial waters from the earth below.

70 years later in Cornwall, Pryce (1778) briefly reviewed the scientific concepts of seawater evaporation and condensation to form rainfall as advanced by Sir Edmund Halley (the astronomer of comet fame), only to summarily dismiss these (without further explanation) as being 'overturned . . . by Mr Derham's perennial spring in the parish of Upminster, and various others in different parts'. He then goes on to exhume the ancient opinions of Plato and dress them up in what was then modern-sounding apparel, arguing that (Pryce 1778, p. 13):

'. . . the only true origin of perpetual springs [is] the Ocean . . .
our hypothesis is that in the formation of perpetual springs
they not only derive their waters from the sea,
by ducts and cavities running from thence through the bowels
of the earth . . . but that the sea itself acts like a huge
forcing engine, or hydraulick machine to force and protrude
its waters from immense and unfathomable depths, through those
cavities, to a considerable distance inland . . .'

Fortunately for his posthumous reputation, Pryce (1778) had the common sense to temper this claim with explicit recognition later in his text that a substantial component of the water make of mines in west Cornwall could be related directly to seasonal rainfall:

'The waters with which our mines abound, are derived
from both temporary and perennial fountains . . .
and are . . . distinguished by the names of Top and Bottom Water.
Shallow mines have very little water, more than comes from
the surface . . . Our very deep mines are subject to water from
both the sources before mentioned . . . in the depth of winter, when
all the earth is drenched as it were with moisture, we are visibly
affected by the concurring streams both of Top and Bottom Water
. . . The deepest of our mines are not much affected by the influx
of Top Water, before the depth of winter; as it takes
till that time, to fill the interstices of the earth or strata,
and protrude its redundant stream to the deep bottoms . . .'

(Pryce 1778, p. 16)

By the 19th Century, most accounts of ground water occurrence in and around mines showed signs of a nascent scientific understanding. Leifchild (1853) in his poetic description of the origins of subsurface waters (quoted above, at the start of the section entitled 'Ground water problems in mining') clearly understood that the waters encountered in deep mine workings were ultimately of atmospheric origin. The engineering specialists of his day, such as Nicholas Wood, William Coulson and John Atkinson, clearly understood the contribution which aquifer heterogeneity can make to vertical variations in hydraulic head within what otherwise would be regarded as a single aquifer. This understanding is evident in the following explanation of why head in the Magnesian Limestone was sometimes found during tubbing to be far greater than would be anticipated simply from the depth below water table (Atkinson & Coulson 1861, p. 10–11):

'. . . [usually] the water behind each lift [i.e. tier] of tubbing
is naturally connected with that behind all the other lifts
in the same shaft, by means of the cavities and gullets in
the strata; so that the pressure of the water on the tubbing,
is that due to the depth from the level of the source of
the highest feeder met with in sinking, to the part of the tubbing
where the pressure is exerted. In many instances, however, this
is practically not the case, owing, probably, to the very long,
tortuous, and contracted channel through which the connection of
the different feeders of water, the one with the other, is established . . .'

This is an eloquent expostulation of the causes and consequences of heterogeneous flow distributions more than a century before these phenomena were formally described and quantified in the mainstream ground water literature (e.g. Freeze & Witherspoon 1967). Nor were Atkinson & Coulson (1861) alone in conceptualizing non-hydrostatic head profiles in this manner. A lively debate which took place at the North of England Institute of Mining Engineers on May 7th 1857, on the occasion of a discussion of the paper by Potter (1856) on the epic sinking at Murton through the Magnesian

Limestone and Yellow Sands aquifers (Anon. 1857), reveals that most of the great mining engineers of the period (Nicholas Wood, T.Y. Hall, Matthias Dunn and George Greenwell) shared the views of Sinker Coulson and his associates.

As we have already seen, Atkinson & Coulson (1861) were sufficiently confident in their conceptualization of heterogeneous flow patterns that they were able to exploit vertical variations in head to install pass pipes and other measures intended to avoid local head build-up behind tubbed sections at depth. They also clearly appreciated the concept of regional drawdown, as revealed by the following comment which they offered on the temporal evolution of water pressure behind close topped tubbing which they installed in the shaft of Harton Colliery (South Shields, Co. Durham): 'Since this shaft was sunk, the establishment of water works at Cleadon has lowered the water behind the tubbing, to the extent of several fathoms' (Atkinson & Coulson 1861, p. 15) (The former Cleadon pumping station lies some 2 km from the site of the Harton shafts).

By the end of the 19th Century, therefore, the mining engineers of Britain had developed and applied a sophisticated understanding of the occurrence and movement of ground water in some of our major aquifers. Unfortunately, their expertise was rarely made accessible to the wider hydrogeological community, and mine water specialists remained largely within their own discrete professional circle. Like many isolated social groups, they developed and maintained their own shared sets of concepts and values, which were often quite distinctive (at least terminologically and sometimes even in terms of physical understanding) from that shared in mainstream hydrogeological circles. This is clearly evident in the language and concepts which were in common use for describing mining-related aspects of ground water flow systems in the first half of the 20th Century. Particularly influential in this period was Saul (1936, 1948, 1949, 1959, 1970), whose classic descriptions of processes of ground water ingress to active workings in the Durham and Yorkshire coalfields influenced the thought and practice of several generations of mining engineers. Saul's papers describe the major factors governing 'normal inflows' (as opposed to catastrophic inrushes) to deep coal mines in the UK. Some of the major conclusions of Saul's work include the following:

(1) mined ground is not so much a porous medium as a network of interconnected 'breaks', principally vertical water-bearing fractures (mostly corresponding to dip-parallel faults and joints), and 'laterals' (such as beds of sandstone or worked seams) (Saul 1948, 1949).
(2) Water makes its final entry into mine workings in a highly localized manner, appearing either as 'feeders' (like underground springs) or as 'drippers' (resembling 'rainfall' from a small area of the roof). The discrete nature of most inflows to mines reflects the pattern of 'breaks' through which water moves through mined ground before it meets a shaft or roadway.
(3) In the absence of adjoining shallow workings or a steeply-dipping permeable sandstone in the sequence, new mine voids deeper than about 140 m (or more than 140 m below the sea bed, or the base of overlying aquifers as appropriate) are unlikely to encounter major feeders (Saul 1948), save where faults provide short-circuiting connections to higher horizons (Saul 1970).

Because of this very particular conception of mining hydrogeology, porous media-based mathematical models were only sporadically adopted and used in mine water management circles. Even Darcy's Law did not remain inviolate, as the meaning of key terms in the formula were bent to fit the peculiar circumstances of deep mining (see Younger & Adams 1999, for further discussion).

The peculiar hydrogeological vocabulary of the mining sector was mentioned in the introductory paragraphs of this paper: 'water makes', 'feeders', 'drippers', 'swallys' / 'swilleys' and 'holings' are only a few of the florid terms which the mine water specialist uses every day, but which are largely foreign to the mainstream hydrogeologist (Younger *et al.* 2002). However, most of these terms have a lineage that greatly pre-dates the more mundane synonyms with which they might be replaced. For instance, the term 'feeder' was used freely by the author of *The Compleat Collier* in 1708, was commented upon for its expressiveness by Leifchild (1853) and remains in common parlance to this day in English-speaking mine water circles. For how many other 'mainstream' hydrogeological terms can such ancient pedigree be claimed?

The author gratefully acknowledges access to the magnificent Nicholas Wood Memorial Library, located in the headquarters of the North of England Institute of Mining and Mechanical Engineers, which has the distinction of being the earliest professional association of mining engineers to be established anywhere in the world. It is an honour and a pleasure to be able to draw on the resources of such an eminent organisation. Dr A. Doyle not only directed me to useful resources in the Nicholas Wood Memorial Library, but also sparked my curiosity with his helpful comments on many aspects of this work. P. Walker of *Pwll Mawr* (Big Pit), the excellent National Mining Museum of Wales near Blaenavon, and R. Lynch of MCG Consultancy Services Ltd (Monmouth), both provided very useful information on Welsh drainage adits for Table 2. I am also indebted to three Keiths, for three different things: to K. Parker of the Coal

Authority for sending me a copy of the House of Commons Water Dangers Committee Report of 1927 (Hyslop *et al.* 1927); to K. Whitworth for bringing Clerk (1740) to my attention, and to K. Anderson for putting me straight on the minimum age of the Book of Job. Finally, the excellent referee comments of D. Banks and C. Knipe are gratefully acknowledged.

References

ADAMS, R. & YOUNGER, P. L. 2002. A physically based model of rebound in South Crofty tin mine, Cornwall. *In:* YOUNGER, P. L. & ROBINS, N. S. (eds) *Mine Water Hydrogeology and Geochemistry*. Geological Society, London, Special Publications, **198**, 89–97.

ANON. 1902. Excursion meeting of the North of England Institute of Mining and Mechanical Engineers, held at Washington Colliery, May 13th 1902. *Transactions of the Institution of Mining Engineers*, **23**, 258–261.

ANON. 1857. Monthly meeting of the North of England Institute of Mining Engineers, May 7 1857. *Transactions of the North of England Institute of Mining Engineers*, **5**, 1147–159.

ANON. 1861. General meeting of the North of England Institute of Mining Engineers, Thursday April 18th 1861. *Transactions of the North of England Institute of Mining Engineers*, **9**, 141–171.

ANON. 2002. *Scottish Lead Mining Museum, Wanlockhead*. On-line Educational Resources, url: *www.leadminingmuseum.co.uk*

ARCHER, D. 1992. *Land of singing waters. Rivers and great floods of Northumbria*. The Spredden Press, Stocksfield, 217p.

ARMSTRONG, W. 1856. On the constitution and action of the chalybeate mine waters of Northumberland and Durham. *Transactions of the North of England Institute of Mining Engineers*, **4**, 271–281.

ASTON, T. R. C. & WHITTAKER, B. N. 1985. Undersea longwall mining subsidence with special reference to geological and water occurrence criteria in the north-east of England. *Mining Science and Technology*, **2**, 105–130.

ATKINSON, A. A. 1902. Working coal under the River Hunter, the Pacific Ocean and its tidal waters, near Newcastle, in the State of New South Wales. *Transactions of the Institution of Mining Engineers*, **23**, 622–667.

ATKINSON, F. 1966. *The Great Northern Coalfield 1700–1900. Illustrated notes on the Durham and Northumberland Coalfield*, 1968 edition. University Tutorial Press Ltd, London, 76p.

ATKINSON, J. J. 1861. On the strength of tubbing in shafts, and the pressures or forces it has to resist. *Transactions of the North of England Institute of Mining Engineers*, **9**, 175–184.

ATKINSON, J. J. & COULSON, W. 1861. On the precautions to be adopted in order to secure the stability and prevent the displacement or failure of close-topped tubbing in the shafts of mines. *Transactions of the North of England Institute of Mining Engineers*, **11**, 9–17.

BANKS, A. G. & SCHOFIELD, R. B. 1968. *Brindley at Wet Earth Colliery: an engineering study.* David & Charles, Newton Abbot, 156p.

BANKS, D., YOUNGER, P. L. & DUMPLETON, S. 1996. The historical use of mine-drainage and pyrite-oxidation waters in central and eastern England, United Kingdom. *Hydrogeology Journal*, **4**, (4), 55–68.

BEDE, 746. *An Ecclesiastical History of the English-Speaking People.* (Translation as quoted is that of John Gregory, as published in 'The Illustrated Bede' by John Marsden, published 1989 by Macmillan, London).

BELL, T. 1899. Notes on the working of coal-mines under the sea, and also under the Permian feeder of water in the County of Durham. *Transactions of the Manchester Geological Society*, **26**, 366–399.

BISWAS, A. K. 1970. *History of hydrology.* North-Holland Publishing Co., Amsterdam. 336 p.

BUCKLEY, J. A. 1997. *South Crofty Mine – a history*, 2nd edition. Dyllansow Truran, Truro, 208p.

BUCKLEY, J. A. 2000. *The Great County Adit.* Penhellick Publications, Camborne, 144p.

BURNHAM, B. C. 1997. Roman Mining at Dolaucothi: the implications of the 1991–3 excavations near Carreg Pumsaint. *Britannia*, **28**, 323–336.

BURT, R. 1977. *John Taylor mining entrepreneur and engineer.* Moorland Publishing Company, Buxton, 91p.

CALDER, J. W. 1973. *Inrush at Lofthouse Colliery, Yorkshire. Report on the cause of, and circumstances attending, the inrush which occurred at Lofthouse Colliery, Yorkshire, on 21 March 1973.* Department of Trade and Industry (Command 5419), HMSO, London, 26p. plus plans.

CLAVERING, E. 1994., Coalmills in Tyne and Wear Collieries: the use of the waterwheel for mine drainage 1600–1750. *Bulletin of the Peak District Mines Historical Society*, **12** (3), 124–132.

CLERK, J. 1740. *A dissertation on coal.* Privately published, Edinburgh 37p. [Preserved in manuscript by the National Archives of Scotland, Reference no. GD18/1069].

CLOWES, F. 1889. Deposits of barium sulphate from mine water. *Proceedings of the Royal Society*, **46**, 169–187.

COOPER, A. H. 2002. Halite karst geohazards (natural and man-made) in the United Kingdom. *Environmental Geology*, **42**, 505–512.

COULSHED, A. J. G. 1951. Water dangers and precautions. *In:* MASON, E. (ed) *Practical coal mining for miners*, 2nd edition. Volume I, Virtue and Company, London. 310–318.

DAVIES, O. 1935. *Roman Mines in Europe*. Clarendon, Oxford, 291p. (Reprinted 1979 by Arno Press, New York).

DARLING, F. 1901. Electric pumping plant at South Durham collieries. *Transactions of the Institution of Mining Engineers*, **23**, 267–269.

DOYLE, A. 1997. *Sacrifice, achievement, gratitude. Images of the Great Northern Coalfield in decline*. County Durham Books, Durham, 115p.

DUNHAM, K. C. 1990. *Geology of the Northern Pennine Orefield, Volume 1 Tyne to Stainmore (2nd Edition).* Economic Memoir of the British Geological Survey, sheets 19, 25 and parts of 13, 24, 26, 31 & 32 (England and Wales), HMSO, London, 299p.

DUNHAM, K. C. & WILSON, A. M. 1985. *Geology of the Northern Pennine Orefield, Volume 2 Stainmore to Craven*. Economic Memoir of the British Geological

Survey, sheets 40, 41, 50 and parts of 31, 32, 51, 60 & 61, New Series (England and Wales), HMSO, London, 247p.
DUNN, M. 1844. *An historical, geological and descriptive view of the coal trade of the North of England; comprehending its rise, progress, current state, and future prospects. To which are appended a concise notice of the peculiarities of certain coal fields in Great Britain and Ireland; and also a general description of the coal mines of Belgium.* W Garrett, Newcastle Upon Tyne, 248p.
DUNN, M. 1848. *A treatise on the winning and working of collieries.* Privately published, Newcastle upon Tyne, 372p.
DUNN, J. T. 1877. On a water-box deposit. *Chemical News*, **35**, p. 140.
EBBS, C. 1993. *The Milwr Tunnel: Bagillt to Loggerheads 1897–1987.* Privately published, Chester, 68p.
EBBS, C. 2000. *Underground Clwyd: The Armchair Explorers' Guide.* Gordon Emery, Chester, 72p.
EDMUNDS, W. M. 1975. Geochemistry of brines in the Coal Measures of north east England. *Transactions of the Institution of Mining and Metallurgy, (Section B: Applied Earth Sciences)*, **84**, 39–52.
EMERY, N. 1992. *The coalminers of Durham.* Alan Sutton Publishing Ltd, Stroud, 210p.
EVERETT, S. A. 1997. *The hydrogeology of the Greenhow mining area and its relationship with the Eagle Level water supply.* M.Phil. Thesis, Department of Civil Engineering, University of Newcastle, UK, 188p.
FAIRBAIRN, R. A. 1996. *Weardale Mines.* British Mining, **No. 56**. Northern Mine Research Society, Keighley, 151p.
FLORES CABALLERO, M. 1981. *Las antiguas explotaciones de las minas de Río Tinto.* (The ancient mine workings of Río Tinto). Excellentisima Diputación Provincial, Huelva (Spain), 93p. [in Spanish].
FORD, T. D. & RIEUWERTS, J. H. 2000. *Lead mining in the Peak District*, 4th edition. Landmark Publishing, Ashbourne, 208p.
FORSTER BROWN, E. O. 1924. Chapter III. The history of boring and sinking. *In:* Anon. (ed.) *Historical Review of Coal Mining.* Mining Association of Great Britain, London.
FREEZE, R. A. & WITHERSPOON, P. A. 1967. Theoretical analysis of regional ground water flow: 2. Effect of water table configuration and subsurface permeability variation. *Water Resources Research*, **3**, 623–634.
GALLOWAY, R. L. 1898. *Annals of Coal Mining and the Coal Trade. Volume 1 (First Series, up to 1835).* Colliery Guardian Company Ltd, London. 534p. [Reprinted 1971 by David & Charles, Newton Abbot].
GALLOWAY, R. L. 1904. *Annals of Coal Mining and the Coal Trade. Volume 2 (Second Series, 1836–1850).* Colliery Guardian Company Ltd, London. 409p. [Reprinted 1971 by David & Charles, Newton Abbot].
GILL, M. C. 1998. *The Greenhow Mines.* British Mining, **No. 60** Northern Mine Research Society, Keighley, 161p.
GILL, M. C. 2001. *Swaledale, its mines and smelt mills.* Landmark Publishing Ltd, Ashbourne (Derbyshire) 174p.
GLENDINNING, D. 2000. *The Art of Mining. Thomas Hair's Watercolours of the Great Northern Coalfield.* Tyne Bridge Publishing/Friends of the Hatton Gallery (University of Newcastle Upon Tyne), 48p.
GRIFFITH, N. R. 1876. On the 'coffering' of shafts to keep back waters. *Transactions of the North of England Institute of Mining and Mechanical Engineers*, **27**, p. 3–11.
HAIR, P. E. H. (ed.) 1988. *'Coals on Rails or The Reason of My Wrighting'. The Autobiography of Anthony Errington from 1778 to around 1825.* Liverpool University Press, Liverpool, 281p.
HAIR, T. H. 1844. *A series of Views of the Collieries in the Counties of Northumberland and Durham, with descriptive sketches and a preliminary essay on coal and the coal trade by M Ross.* Privately published, Newcastle Upon Tyne, 52p. [Republished by Davis Books, 1987].
HARRIS, A. 1970. *Cumberland Iron: the story of Hodbarrow Mine 1855–1968.* D Bradford Barton Ltd, Truro, 122p.
HESLOP, H. 1994. *Out of the old earth.* Bloodaxe Books, Newcastle Upon Tyne, 270p.
HILL, A. 1991. *Coal mining: a technological chronology 1700–1950.* A British Mining Supplement, Northern Mine Research Society, Keighley, 60p.
HILL, D. 1984. *A history of engineering in classical and medieval times.* Routledge, London, 263p. [1996 paperback edition].
HOOVER, H. C. & Hoover, L. H. 1950. *De Re Metallica* by Georgius Agricola. Translated from the 1st Latin edition of 1556. Dover Publications Inc., New York, 638p.
HORNE, M. 1993. Pithill, Beamish. *In:* DURHAM COUNTY ENVIRONMENTAL EDUCATION CURRICULUM STUDY GROUP (eds) *Coal mining in County Durham.* Northern Echo Publishing, Darlington, 28–31.
HUGHES, H. W. 1904. *A text-book of coal-mining.* Charles Griffin & Co Ltd, London, 563p.
HYSLOP, G. P., DOONAN, J., MITCHESON, G. A., MOTTRAM, T. H., STRAKER, H. & WALKER, H. 1927. *Report of the British Parliamentary Committee Appointed to Inquire into the Methods to Prevent Dangers in Mines from Accumulations of Water or other Liquid Matter (1924–1927).* HMSO, London, 13p plus 19 appendices.
'J.C.', 1708. *The Compleat Collier: or the whole art of sinking, getting, and working, coal mines, &c. As is now used in the Northern Parts, especially about Sunderland and New-Castle.* G Conyers, London, 47p. [Reprinted 1990 by Picks Publishing, Wigan. ISBN 0 95164843 1 0].
JENKINS, D. A. & LEWIS, C. A. 1991. Prehistoric mining for copper in the Great Orme. *In:* BUDD, P., CHAPMAN, B., JACKSON, C., JANAWAY, R. C. & OTTAWAY, B. S. (eds) *Archaeological Sciences 1989*, 161–161.
JOHNSON, K. L. & YOUNGER, P. L. 2002. Hydrogeological and geochemical consequences of the abandonment of Frazer's Grove carbonate hosted Pb/Zn fluorspar mine, North Pennines, UK. *In:* YOUNGER, P. L. & ROBINS, N. S. (eds) *Mine Water Hydrogeology and Geochemistry.* Geological Society, London, Special Publications, **198**, 347–363.
JUDGE, T. 1994. *The Clay Cross Calamities.* Scarthin Books, Cromford, 112p.
KNIVETON, G. N. 2000. *Lady Isabella and the Great Laxey Mine. Official guide.* Manx Museum and National Trust, Douglas, 34p.

LEIFCHILD, J. R. 1853. *Our coal and our coal-pits, the people in them and the scenes around them, by a Traveller Underground*, Parts 1 and 2. Longman, Green and Co., London, 243p. [Reprinted 1870 in a single volume].

LEVINE, D. & WRIGHTSON, K. 1991. *The making of an industrial society. Whickham 1560–1765.* Clarendon Press, Oxford, 456p.

LLEWELLYN, K. 1992. *Disaster at Tynewydd. An account of a Rhondda mine disaster in 1877*, 2nd edition. Church in Wales Publications, Penarth, 92p.

MCKECHNIE, J. & MACGREGOR, M. 1958. *A short history of the Scottish coal-mining industry.* National Coal Board, Scottish Division, Edinburgh, 116p.

MOORE, R. 1872. Arrangements of machinery adopted for pumping water in dip workings at the Kinneil Iron Works, at a distance from the shaft. *Transactions of the North of England Institute of Mining and Mechanical Engineers*, **21**, 159–160.

MORRISON, J. 1998. *Lead mining in the Yorkshire Dales.* Dalesman Publishing Company, Skipton, 142p.

MULLINEUX, F. 1988. *The Duke of Bridgewater's Canal.* Eccles and District History Society, Salford.

NUTTALL, C. A. & YOUNGER, P. L. 1999. Reconnaissance hydrogeochemical evaluation of an abandoned Pb-Zn orefield, Nent Valley, Cumbria, UK. *Proceedings of the Yorkshire Geological Society*, **52**, 395–405.

ORCHARD, R. J. 1975. Working under bodies of water. *The Mining Engineer*, **170**, 261–270.

PAMELY, C. 1904. *The Colliery Manager's Handbook. A comprehensive treatise on the laying-out and working of collieries, designed as a book of reference for colliery managers and for the use of coal-mining students preparing for first-class certificates.* Crosby Lockwood & Son, London, 1178p.

PARKER, K. 2002. Mine water management on a national scale – experiences from the Coal Authority. *In:* Nuttall, C.A. (ed) *Mine water treatment: a decade of progress.* Proceedings of a Conference held in Newcastle Upon Tyne, Nov 11–13th 2002, ISBN 0-954-3827-0-6, 102–113.

POTTER, E. 1856. On Murton Winning in the County of Durham. *Transactions of the North of England Institute of Mining Engineers*, **5**, 45–61

PRYCE, W. 1778. *Mineralogia Cornubiensis; a treatise on minerals, mines and mining: containing the theory and natural history of strata, fissures, and lodes, with the methods of discovering and working of tin, copper and lead mines, and of cleansing and metalizing their products; shewing each particular process for dressing, assaying, and smelting of ores. To which is added, an explanation of the terms and idioms of miners.* Privately Published, London. 331p. [Facsimile reprint published in 1972 by D Bradford Barton Ltd, Truro].

PURDON, G. J. 1979. *The Sacriston mine disaster.* Privately published, Beamish, County Durham, 57p.

RAISTRICK, A. 1975. *The lead industry of Wensleydale and Swaledale. Vol. 1. The mines.* Moorland Publishing Company, Ashbourne (Derbyshire), 120p.

RAISTRICK, A. & JENNINGS, B. 1965. *A history of lead mining in the Pennines*, reprinted in 1989. Davis Books Ltd, Newcastle Upon Tyne, 347p. (First published by Longmans, Green and Co, Ltd, London).

REDMAYNE, R. A. S., WATTS-MORGAN, D., DEACON, M., ASHTON, A. J. & KILPATRICK, J. A. 1920. *'South Staffordshire Mines Drainage: Report of the Committee appointed to inquire into the drainage of the mines in the South Staffordshire Coalfield.* Parliamentary Command Report no 969, 57p. plus appendices.

RICHARDS, H. P. 1951. Accidents. *In:* MASON, E. (ed.) *Practical coal mining for miners*, 2nd edition. Virtue and Company, London, Volume I, 337–341.

RIEUWERTS, J. H. 1962. Connections between the pioneers of civil engineering and mining practice. *Peak District Mines Historical Society*, **1** (4), 10–12.

ROBEY, J. A. & PORTER, L. 1972. *The copper & lead mines of Ecton Hill, Staffordshire.* Peak District Mines Historical Society/Moorland Publishing, Leek (Staffs), 92p.

ROLT, L. T. C. 1957. *Isambard Kingdom Brunel : a biography.* Longmans Green, London, 345 p.

ROLT, L. T. C. & Allen, J. S. 1997. *The steam engine of Thomas Newcomen.* Landmark Publishing, Ashbourne (Derbyshire).

ROWLANDS, J. 1966. *Copper Mountain.* Anglesey Antiquarian Society, Llangefni, 200p.

RUSSELL, M. 2000. *Flint mines in Neolithic Britain.* Tempus Publishing Ltd, Stroud, 160p.

RUTTY, J. 1760. Of the vitriolic waters of Amlwch. *Philosophical Transactions of the Royal Society*, **51, Pt II**, 470.

SAUL, H. 1936. Outcrop water in the South Yorkshire Coalfield. *Transactions of the Institution of Mining Engineers*, **93**, 64–94

SAUL, H. 1948. Mine water. *Transactions of the Institution of Mining Engineers*, **107**, 294–310.

SAUL, H. 1949. Mines drainage. *Transactions of the Institution of Mining Engineers*, **108**, 359–370.

SAUL, H. 1959. Water problems in the coalfields of Great Britain. *Colliery Guardian*, **199**, 191–199, 229–234.

SAUL, H. 1970. Current mine drainage problems. *Transactions of the Institution of Mining and Metallurgy (Section A)*, 79, A63–A80.

SHAW, W. T. 1970. *Mining in the Lake Counties.* Dalesman Publishing, Company Ltd, Clapham (North Yorkshire), 128p.

SINCLAIR, G. 1672. *The Hydrostaticks OR The Weight, Force and Pressure of Fluid Bodies, Made Evident by Physical and Sensible Experiments. Together with some Miscellany Observations, the last whereof is a short History of Coal, and of all the common and Proper Accidents thereof, a Subject never treated of before.* Printed by George Swintoun, James Glen and Thomas Brown, Edinburgh, 319p.

SINCLAIR, J. 1958. *Water in mines and mine pumps.* Sir Isaac Pitman & Sons Ltd, London, 130p.

SMITH, S. 1923. Lead and zinc ores of Northumberland and Alston Moor. *Memoirs of the Geological Survey, Special Reports on the Mineral Resources of Great Britain*, **Volume XXV**, HMSO, London, 110p.

SWALLOW, R. T. 1981. Description of a method of removing deposits from the inside of rising main pipes in shafts. *Transactions of the Federated Institution of Mining Engineers*, **3**, 113–118.

TAYLOR, T. J. 1857. Suggestions towards a less local system of draining coal mines. *Transactions of the North of England Institute of Mining Engineers*, **5**, 135–141.

TAYLOR, T. J. 1858. *The Archaeology of the Coal Trade.* Memoirs chiefly illustrative of the antiquities of Northumberland communicated at the annual meetings of the Archaeological Institute of Great Britain and Ireland held at Newcastle-on-Tyne in August 1852. 76p. [Reprinted in facsimile by Frank Graham Publishers, Newcastle Upon Tyne, 1971].

TEMPLE, D. 1998. *The Collieries of Durham, Volume 2.* TUPS Books, Newcastle Upon Tyne, 126p.

TIMBERLAKE, S. & JENKINS, D. A. 2001. Prehistoric mining: geochemical evidence from sediment cores at Mynydd Parys, Anglesey. *In:* MILLARD, A. (ed.) *Proceedings of Archaeological Sciences '97 Conference.* Held at Durham, UK, 2–4 September 1997, 193–199.

VIVIAN, J. 1990. *Tales of the Cornish miners.* Tor Mark Press, Penryn, 32p.

WADE, E. 1998. *William Coulson and the influence of the northern English mining industry on the Ruhrgebiet.* Open University, Newcastle Upon Tyne, 18p.

WALKER, S. C. 1995. *The pump.* Magnus Publications, Ashby-de-la-Zouch, 287p.

WALKER, T. A. 1891. *The Severn Tunnel. Its construction and difficulties 1872–1887.* Privately published. 240p. [Facsimile Edition published 1969 by Kingsmead Reprints, Bath].

WALLER, W. 1867. On pumping water. *Transactions of the North of England Institute of Mining Engineers*, **16**, 135–140.

WALLER, W. 1872. On pumping water – paper no. 2. *Transactions of the North of England Institute of Mining and Mechanical Engineers*, **21**, 123–157.

WATKINS, G. 1979. *The steam engine in industry.* Moorland Publishing Co Ltd, Ashbourne. 128p. [Paperback edition 1994].

WATSON, W. 1861. On the use of cement for walling as a substitute for metal tubbing in shafts. *Transactions of the North of England Institute of Mining Engineers*, **9**, 72–74.

WHITWORTH, K., 2002, Day eyes, level rooms and the importance of sough things for mine water drainage. *In:* NUTTALL, C. A. (ed.) *Mine water treatment: a decade of progress.* Proceedings of a Conference held in Newcastle Upon Tyne, Nov 11–13th 2002, ISBN 0-954-3827-0-6, 224–231.

WILKINSON, P. 2001. *The Nent Force Level and Brewery Shaft.* North Pennines Heritage Trust, Nenthead, 105p.

WILLIAMS, C. J. 1997. *Metal mines of North Wales. A collection of pictures.* Bridge Books, Wrexham, 106p.

WILSON, T. 1843. *The Pitman's Pay and other poems.* William Douglas, Gateshead, 168 p.

WILSON, W. B. 1901. Tapping drowned workings at Wheatley Hill Colliery. *Transactions of the Institution of Mining Engineers*, **23**, 72–84.

WOOD, E. S. 1907. Sinking through the Magnesian Limestone and Yellow Sand by the freezing-process at Dawdon Colliery, near Seaham Harbour, County Durham. *Transactions of the Institution of Mining Engineers*, **32**, 551–578.

WOOD, N. 1857. On the sinking through the Magnesian Limestone at the Seaham and Seaton Winning, near Seaham. *Transactions of the North of England Institute of Mining Engineers*, **5**, 117–129.

YOUNGER, P. L. 1995. Hydrogeology. Chapter 11. *In:* JOHNSON, G. A. L. (ed.). *Robson's Geology of North East England.* (The Geology of North East England, 2nd edition). Transactions of the Natural History Society of Northumbria, **56** (5), 353–359.

YOUNGER, P. L. 1998. Adit hydrology in the long-term: observations from the Pb-Zn mines of northern England. *In: Proceedings of the International Mine Water Association Symposium on 'Mine Water and Environmental Impacts'.* Johannesburg, South Africa, 7th–13th September 1998, Volume II, 347–356.

YOUNGER, P. L. & Adams, R. 1999. *Predicting mine water rebound*, Environment Agency R&D Technical Report, W179, Bristol, UK, 108p.

YOUNGER, P. L. & Robins, N. S. (eds), 2002. Mine Water Hydrogeology and Geochemistry. *Geological Society, London, Special Publications*, **198**, 396p.

YOUNGER, P. L., BANWART, S. A. & HEDIN, R. S. 2002. *Mine Water: Hydrology, Pollution, Remediation.* Kluwer Academic Publishers, Dordrecht, 464p.

ZOLA, E. 1885. *Germinal.* Penguin Books Ltd, London, 499p. [English translation by L. Tancock first published in 1954].

The contribution of geologists to the development of emergency groundwater supplies by the British army

EDWARD P. F. ROSE

Department of Geology, Royal Holloway, University of London, Egham, Surrey TW20 0EX, UK

Abstract: During the 19th Century, the British military pioneered geological mapping and teaching, and the operational use of Norton tube wells. In the First World War, the British army appointed its first military hydrogeologist to serve as such, to develop water-supply maps for Belgium and northern France and guide deployment of Royal Engineer units drilling boreholes into the Cretaceous Chalk of the Somme region and Tertiary sands beneath the Flanders plain. Similar well-boring units were also deployed with geological guidance in the northeastern Mediterranean region. All military geologists were demobilized after hostilities ceased, but wartime experience was quickly drawn together in the first Royal Engineer textbook on water supply. During the Second World War, several British military well-drilling units were raised and deployed, notably to East Africa and North Africa as well as northern France, normally with military geological and sometimes (in Africa) with military geophysical technical direction. A reduced well-drilling capability has since been retained by the British army, through the Cold War to the present day, supported by a small group of reserve army geologists to contribute basic hydrogeological expertise to the armed forces for peace-time projects and war-related operations.

Throughout history, water has been recognized as a vital military resource. Thus the well is a key feature of most fortresses. In medieval castles, it was the military imperative that sometimes caused wells to be dug to depths unusual for the time (Neaverson 1947; Ruckley 1990; Halsall 2000). Even in 1538, the Saxon well within Bamburgh Castle in Northumberland, cut to a depth of 42.7 m through the quartz dolerite of the Whin Sill into Carboniferous sandstones below, was noted as 'of marvellus grett dypnes' (Richard Bellasis, quoted by Dodds (2000)). In general, castles which lacked a well or where the wells were rendered inadequate were vulnerable to siege (e.g. Grosnez on the Channel Island of Jersey in 1373 (Balleine 1950; Rose *et al.* 2002*a*), Edinburgh in Scotland in 1573 (Ruckley 1991; Halsall 2000)). Moreover, troops concentrated as garrisons or for campaigns at times risked death or debilitation as much from disease, water- or arthropod-borne, as from enemy action. As geology developed into a predictive science within the 19th and 20th centuries, so the British military became increasingly aware of its value in the development of groundwater for military purposes.

The first British geologist to be militarily employed as such was John MacCulloch, a lecturer in chemistry at the Royal Military Academy, Woolwich (Flinn 1981; Rose 1996). From 1809 to 1814 he was tasked by the Master General and Board of Ordnance to search Britain for a limestone suitable to make into millstones for grinding gunpowder and its constituents. The military also initiated government-funded geological mapping in Scotland, Ireland and England, largely through the influence of an Ordnance Survey officer, T. F. Colby, R[oyal] E[ngineers], as he rose from the rank of Captain to Major General. MacCulloch (1836) generated the first geological map of Scotland following fieldwork in support of the Ordnance's Trigonometrical Survey. In 1826, Colby appointed Captain J. W. Pringle RE as founding Superintendent of the Geological Survey of Ireland, and Irish geological mapping was conducted intermittently until 1845 by serving Royal Engineer officers (Herries Davies 1974, 1983, 1995; Rose 1997, 1999; Wyse Jackson 1997). In England, the Ordnance Geological Survey was established between 1835 and 1845 under military auspices, and under a director, H. T. De la Beche, who had received a military rather than a university education (Bailey 1952; Rose 1996).

Early in 19th century Britain, geology was thus perceived as a science with potential military applications. Consequently, geological training was provided for cadets or young army officers at the East India Company's military college, Addiscombe, from 1819 to 1835, and again from 1845 to 1861; the Royal Military Academy, Woolwich, from at least 1848 to 1868; the Royal Military College, Sandhurst, from 1858 to 1870; the Staff College, Camberley, from 1862 to 1882 (and on much reduced scale to 1898); and the School of Military Engineering, Chatham, prior to 1896 (Rose 1997). It seems likely from the course books in known use that teaching on the relationship of geology to water supply was introduced to the military curriculum at Addiscombe, following the appointment in 1845 of D. T. Ansted (Fig. 1): his own books (Ansted 1856, 1878) indicate a growing perception of this subject's importance. Later, T. Rupert Jones certainly taught the relationship of geology to topography, sanitation and water supply at Sandhurst and Camberley (Woodward 1907; Rose 1996, 1997).

From: MATHER, J. D. (ed.) 2004. *200 Years of British Hydrogeology*. Geological Society, London, Special Publications, **225**. 159–182. 0305-8719/04/$15 © The Geological Society of London.

Fig. 1. David Thomas Ansted, MA, FRS (1814–1880), an early professor of geology at King's College London, consulting geologist and mining engineer, and lecturer on geology at the East India Company's military college, Addiscombe, 1845–1861. Drawing dated 1850. From Rose (1997), courtesy of the Geological Society, London and the Institution of Royal Engineers.

Geology and military water supply in the nineteenth century

By the mid 19th century it was clear to the military that geology was of value in the siting of boreholes for water supply. Richard Baird Smith, an engineer officer in the army of the East India Company who was later to distinguish himself as commander of the engineers at the Siege of Delhi during the 'Indian Mutiny' of 1857, took borehole site selection as his first example of the practical value of geology in an essay on its military applications (Smith 1849) – at a time when boring (by auger, rotary or percussion means) was replacing the sinking of shafts as a means of reaching groundwater. Smith had trained at Addiscombe from February 1835 to December 1836 (Vibart 1894, 1897), largely after geological teaching had temporarily ceased, in August 1835, and before the appointment of Ansted. His essay constitutes a powerful plea for geology to be re-introduced into the curriculum.

The importance of geology to a military profession is recognized by lengthy articles (Portlock 1850, 1860) on the subject within the massive 3-volume *Aide-mémoire to the military sciences* (Lewis *et al.* 1846–1852, 1853–1862) prepared for officers of the 'technical' corps of the British and East India Company's armies in the mid 19th century. 'Practical applications of geological science' merit a section less than half a page in length, but 'theory of springs' has text and six illustrations that together fill nearly six pages, supplemented by plates to show tools for boring by auger or percussion. These illustrations derive from the work of a Major Baddely RE in Ceylon, and relate to ideas on groundwater flow, the significance of rainwater catchment areas, and geological basin configuration. Brief articles on 'water supply' (Burnell 1853*a*) and 'wells' (Burnell 1853*b*) were contributed by a civilian water engineer, but without geological illustration.

The importance of geology was endorsed by F. W. Hutton (1862, p. 345), who again described well site selection as one of the principal values of geology to military men: 'With respect to the water supply of camps, the geologist is often able to decide whether it is possible to obtain water by sinking wells'. Some of his data on water supply are credited to Ansted's (1856) textbook; others possibly derive from the influence of Rupert Jones. Hutton, a veteran of the Indian campaign of 1857–1858, entered the Staff College, Camberley, in 1860, prior to promotion as a captain in the 23rd (Royal Welsh Fusiliers) Regiment of Foot on 2 March 1862 (Rose 1996).

The Abyssinian war of 1867–1868 arguably provided an opportunity to put theory into practice. Water supply for the troops and their horses or mules was always a serious problem. Near the coast, drinking water was provided from steam ships, by condensation of evaporated seawater. Inland, drinking water had to be obtained from wells, but water from the small local wells was frequently contaminated. The Royal Engineers innovatively developed new supplies of potable water by use of Norton tube wells (Figs 2 & 3) (Porter 1889) – setting a precedent in the use of 'Abyssinian wells' consequently so named that would be followed in both the British and German armies for nearly a century (Rose *et al.* 2002*b*).

Water supply was also at times a problem in the South African wars between 1847 and 1885. For the Bechuanaland Expedition of 1885 it was recorded that 'water supply was a great difficulty' (Porter 1889, p. 42). 'Where water was found it was generally in pools in a river bed, which was fenced round and protected from being fouled by cattle; generally, however, wells had to be sunk and in all cases means for raising water had to be provided.'

In India, military groundwater supply at this time was seemingly from shallow rather than deep wells. Thus at Lucknow, when an extensive scheme of surface drainage for the British military cantonment was carried out between 1877 and 1881, it was reported that water supply from the 'surface' wells was significantly reduced, and water level in all the wells lowered (Anon. 1922*b*, p. 48).

Fig. 2. Training on Norton tube wells in 1868. Photograph courtesy of the Royal Engineers Corps Library, Chatham: box E9, album 3/23, p. 121.

The Corps of Royal Engineers by this time apparently contained at least a few officers experienced in water-supply engineering if not in hydrogeology. One of them, Major Hector Tulloch, was seconded for service as the Engineer to the Local Government Board, and in that appointment generated a substantial report on water supply for the town and fortress of Gibraltar (Tulloch 1890; Rose *et al.* 2004).

By 1889 the School of Military Engineering at Chatham provided training in field fortification, survey, electricity, photography and chemistry, lithography, submarine mining, estimating and building construction, and ballooning. Field fortification included training in field defence, siege works, mining, bridging, railway work and demolitions.

Under this title, cavalry pioneers were given a course of 15 days in which they learnt 'the use of tools; making of fascines; hutting and thatching; water supply, including making of filters, the use of Norton's tube wells and the best arrangements for watering horses and cattle; the defence of a post, the demolition of arches, bridges, railways and telegraph lines; and the construction of military bridges' (Porter 1889, p. 184). Infantry pioneers passed through a longer course, of 55 days, which in addition to the cavalry syllabus included 'the making of gabions, the use of entrenching tools, the making of shelter and siege trenches, the construction of obstacles and redoubts, railway work, the formation of encampments and escalading'. Water supply was thus a component, but not a very large component, of basic engineer training.

The School made some effort to keep abreast of the best civilian practice in the United Kingdom by arranging visiting lectures from engineers then prominent in the water supply industry. Lectures by A. R. Binnie on water supply, rainfall, reservoirs, conduits and distribution generated a book (Binnie 1877) later reprinted. However, neither edition was specifically geological, despite some references to groundwater flow and to prevailing temperature as a factor influencing the amount of water entering the ground. Swindell & Burnell's (1851) rudimentary treatise on wells and well sinking was the textbook still in common use at the time of J. Mansergh's lectures in 1880 on water supply, prospecting for water, well-sinking and boring. The published lectures (Mansergh 1882) include eight pages of text and diagrams based on geology. However, a water supply manual 'Compiled by a committee at the War Office' (Scott-Moncrieff & Tyndale 1909, p. 2) to provide 'information on water-supply problems likely to be met with in military life' (primarily in the UK) and distributed in November 1909 from the War Office by the Director of Fortifications and Works, contained no reference to geology as such – only very

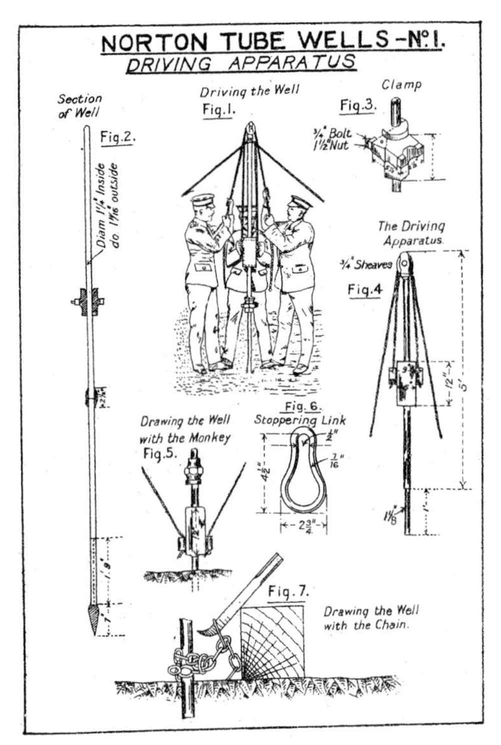

Fig. 3. Norton tube wells: driving apparatus. Techniques of emplacement have changed little from at least 1867 to the present day. From Anon. (1922*b*), courtesy of the Ministry of Defence.

Fig. 4. W. B. R. King (1889–1963), in Royal Welsh Fusiliers uniform, at Versailles in France, in September 1915. The only British military hydrogeologist to serve as such on the Western Front, from June 1915 King held a 'special appointment' as a staff officer to the Chief Engineer (later re-titled Engineer-in-Chief) of the British Expeditionary Force. He was promoted Lieutenant on 1 June 1916, after holding temporary rank from 4 May 1915, and Captain on 27 April 1918, after holding temporary rank from 24 July 1917. From Rose & Rosenbaum (1993*a*), courtesy of Professor C A M King and the Geologists' Association.

short sections on springs, wells, examining wells, deep wells, yield of wells and catchment areas, illustrated by a single simple geological cross section (Scott-Moncrieff & Tyndale 1909, plate 1A, fig. 3).

Geology and water supply in the First World War

The UK declared war on Germany on 4 August 1914. The principal theatre of military operations soon developed as the Western Front, in Belgium and northern France, although British troops also fought elsewhere, notably in the Balkans, Turkey (Gallipoli), Egypt, Mesopotamia (present-day Iraq) and Africa. The best documented uses of geology to promote military groundwater supplies come from the Western Front, Gallipoli and the Mediterranean Expeditionary Force.

The Western Front

The British Expeditionary Force (BEF) was deployed in Belgium and northern France in 1914 effectively as a single army, but reinforced to form five armies before 1918. Serious water supply problems did not emerge until 1915, as troop concentrations increased.

Notes on sources of temporary water supply in the south of England and neighbouring parts of the continent (Strahan 1914) were quickly published by the Geological Survey of Great Britain, for military use. Moreover, in April 1915, at War Office request, the Director of the Survey began to prepare a report on the geological conditions affecting the available sources of water in Belgium and the north of France (Strahan 1915), completed in July, and to train one of the Survey's pre-war geologists (Second Lieutenant W. B. R. King) (Fig. 4) for active service as a military hydrogeologist (Rose & Hughes 1993*a*; Rose & Rosenbaum 1993*a*). In May, a Water Supply Committee was appointed at BEF General Headquarters. In June, the Committee recommended

Fig. 5. Near-contemporary geological sketch map of the British-occupied region of the Western Front. From Anon. (1921), courtesy of the Institution of Royal Engineers.

scales of water supply for both men and animals (Anon. 1921); about this time boring plant was ordered on a small scale, for use in Chalk areas; and Lieutenant (later Captain) 'Bill' King completed training and arrived in theatre (King 1919; Anon. 1922*a*). He was to remain in post throughout the war, benefiting from May 1916 by association with a second geologist (Major, later Lieutenant-Colonel, T. W. Edgeworth David), who arrived with the Australian Mining Battalion and was deployed primarily to support military mining and dug-out excavation (Rose & Rosenbaum 1993*a*, 1998).

The part of the Western Front held by the BEF stretched north approximately from Amiens in France to Nieuport on the Belgian coast (Fig. 5), its position east-west fluctuating with the ebb and flow of battle. Geologically, this region is divisible into three main components (Figs 5 & 6), from south to north:

(1) The Chalk plateau of Artois and Picardy. There is little surface water in the southern part of this area, and groundwater must be extracted from deep boreholes. Artesian wells take their name from its northern part, Artois, a region where hydraulic head causes well water to rise readily to or near ground surface.

(2) The Tertiary sands and clays of the Flanders plain. The only water-bearing strata in the bedrock were a 16-m-thick unit of sands within the Lower Tertiary (Landenian, broadly equivalent to the Thanet Sands of north Kent), commonly lying about 100 m below the surface, beneath thick and impermeable Ypresian clay (broadly equivalent to the London Clay). Pre-war boreholes had been unable to extract groundwater from these sands because of their very fine grain size, and until late in the war, the military relied on piped surface supplies, notably from the River Yser.

(3) Recent dune sands and reclaimed land of the coastal plain. Isolated pumping stations, wells with hand pumps, and Norton tube wells provided sufficient supply during defensive trench warfare conditions, but were inadequate for troops concentrated for attack (in 1917).

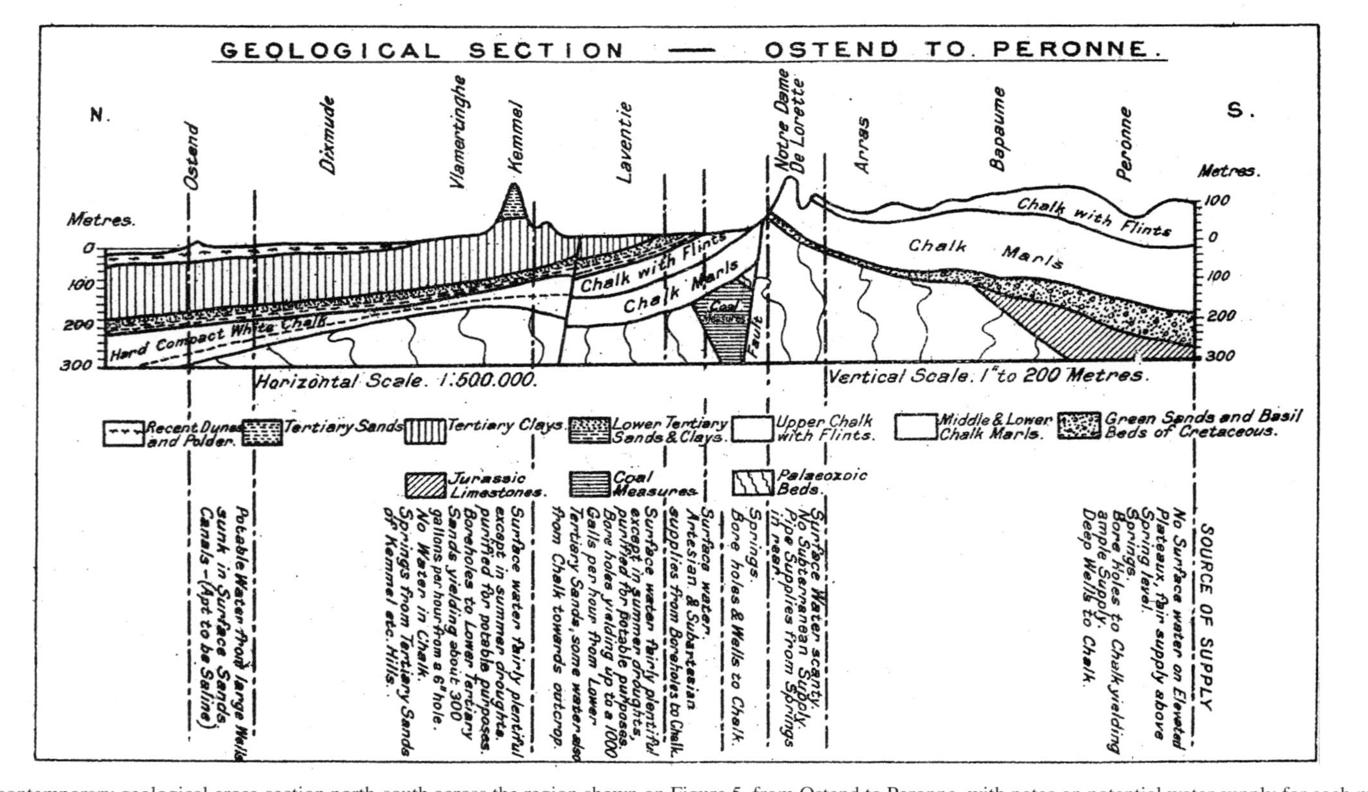

Fig. 6. Near-contemporary geological cross-section north-south across the region shown on Figure 5, from Ostend to Peronne, with notes on potential water supply for each major region sidelined to the right. From Anon. (1922*b*), after Anon. (1921), courtesy of the Institution of Royal Engineers and the Ministry of Defence.

Between 1915 and the armistice of November 1918, military water-supply maps were developed, at three scales:

(1) 1:250000. In 1915, to aid planning for advance, a map was prepared of the enemy-occupied ground in front of the British lines as far as Brussels, indicating potential water quality and quantity, both in terms of three categories. Quality was indicated by colour type (blue = good, purple = medium, red = poor), quantity by colour intensity (dark = abundant, medium = moderate, light = scarce), so mapping indicated regions in terms of nine categories in total. The map was generated from 1:100000 topographical and geological maps, and the records of local water supplies which remained in Allied possession (Anon. 1922*a*, p. 10–11). In 1918, for a similar purpose, a map similar in geographical area distinguished 15 zones on the basis of three criteria: nature of water available for horses; drinking water for troops; and type of water supply plant required (Table 1). The zones were shown by overprinting a standard topographical base map with zone boundaries and numbers in black ink (Anon. 1921, map D).

(2) 1:100000. From 1915, water-supply maps were prepared for the whole of Belgium and the enemy-occupied territory of northern France by plotting information collected about the sites of springs, and their daily yield; location of pipe-lines and adits for collecting water; pumping stations, with daily supply rate; reservoirs, with details as to capacity; sites of boreholes, with depth and yield; number of wells per commune, with range of depth (where possible); and whether the supplies were adequate for the civilian population in peace-time (Anon. 1922*a*, p. 10 & fig. 18).

(3) 1:40000. During the 1916 Battle of the Somme, and afterwards, it was necessary to predict the depth of the water table. Contour lines were drawn at 5 m intervals to show the predicted shape of the water table on tracing overlays (cf. King 1921*b*). Used with contoured topographical maps, these allowed some estimate to be made of potential borehole depths. The overlays were, however, very generalized, and made no allowance for seasonal fluctuation in the height of the water table.

During the first 18 months of the war, the British First and Second Armies operated in a country where adequate supplies of water were obtainable at or near the surface. As the war progressed, it became necessary to exploit groundwater within all three geologically-distinct regions.

In the Chalk plateau, equipment for drilling deep boreholes and pumps for raising water from depth were introduced. The first boring plant was ordered in the spring of 1915, on a small scale, and an officer with considerable boring experience in civilian practice placed in charge (Anon. 1922*a*, p. 23). Beeby Thompson later (1924, p. 110) noted that

> 'Until the European war a small hand-worked or power-driven surging plant [= percussion rig] with a few tools was usually considered adequate for all purposes, and no R[oyal] E[ngineer]s had been trained in the working of mechanical drilling plant of modern design . . .'

When the War Office decided to adopt more modern methods of drilling, the portable drills popular in America were selected (Fig. 7), notably 'Star' and 'Keystone' machines. Boreholes were developed to aid concentration of troops for the Somme offensive in the summer of 1916, and impetus was given to the boring by the first introduction of air-lift pumps, powered by compressors, at this time. Borings were put down to some 100 m depth, through the water-bearing Upper Chalk to the marls of the Middle and Lower Chalk in order to give sufficient submersion for the pumps. General Headquarters BEF approved formation of three Water Boring Sections Royal Engineers in March 1917, later increased to five (one for each army), and supported by a boring section within the Australian Electrical and Mechanical Company. Military boreholes were invariably put down by percussion rather than rotary drilling, the holes being either 6 inches (150 mm) or 8 inches (200 mm) in diameter. (See King (1921*a*) for catalogue and tabulated details of strata thicknesses.) Many new boreholes were put down between April and June 1918, consequent on German attack and the British counter-offensive. In 1918, four Keystone rigs in the Somme region bored a total of 12 000 feet (*c*. 4000 m) in three months, a total of 40 boreholes with an average yield of 6000 gallons (27 m^3) an hour per hole (Anon. 1922*b*, p. 71). Between 1 March and the armistice of 11 November, 1041 pumping plants were issued, the War Office placing large orders for water-supply plant, piping and stores in America since British industry could not meet the demand in the time available. In total, over 470 boreholes for water had been sunk in support of British forces on the Western Front by the close of hostilities.

In Belgium generally, beneath the Flanders plain, the very fine grain size of the water-bearing Landenian sands made it difficult to extract water by means of boreholes. Several methods were attempted: air-lift without strainer; tube-well pump with gauze strainer or horse-hair strainer; and Ashford strainer. The Ashford strainer (Fig. 8), introduced only in mid 1918 after use in India, was the most successful,

Table 1. *Areas distinguished on 1:250000 water-supply map of Belgium and northern France, September 1918*

Area 1 Bruges area
Horses – Plenty of surface water. Quality of Canal de Derivation de Lys bad during flax soaking operations. Shallow wells a few feet deep, will give horse water anywhere on low ground.
Drinking water – Scarce to very scarce. Will have to depend on sterilizing lorries or alum sedimentation and chlorination.
Boring is no good as Thanet sand water is saline.
Plant wanted – Surface punps only. Sterilizing lorries. Alum installations.

Area 2 Antwerp area
Horses – Abundance of surface water, but brackish near Scheldt.
Drinking water – Surface wells on the whole bad. Boring no good except at southern edge where supplies up to about 1000 g.p.h. may be expected from bores up to about 100 feet deep into a 10 feet bed of sand below the Rupelian Clay.
Plant wanted – Mainly sufrace pumps. A few tube well pumps. Sterilizing lorries. Alum plants.

Area 3 Roulers-Tourcoing-Malines area
Horses – Surface water, abundant in winter, but liable to be scarce in summer. Area is similar to the Ypres-Kemmel area. Water sufficient for open warfare but pipe line distribution probably required for stationary warfare. Hill country round Renaix has a few fairly good springs.
Drinking water – Sufficient from springs and wells in hilly country. Borings to Landenian Sands over whole area from about 100 feet deep in southern part to 300 feet deep in northern part of area. Yield about 1000 g.p.h. in west and probably more in eastern part of area. Borings to chalk no good.
Plant wanted – Surface pumps for horses. Ashford strainers and small air lifts or tube well pumps in sands. Sterilizing lorries. Alum plants.

Area 4 Lessines-Hal
Clay overlying Thanet sands on old rocks. Springs in valleys at junction with old rocks. Borings would get some water from Thanet sands.
A fair number of running streams for horse watering.
Plant wanted – Surface pumps. Tube well pumps, not air lifts. Ashford strainers.

Area 5 Brussels area
Area mainly Bruxellian sand with old rocks in the valleys – no chalk. Good springs in the valleys. Town supplies usually from adits in the hills. Borings to old rocks in Brussels yield good water. Borings on high ground 200 feet-300 feet into sand give good supplies.
Plant wanted – Surface pumps for springs. Tube well pumps for bores. Not a good air lift country. Ashford strainers.

Area 6 Vilvorde-Louvain area
Area generally similar to 5. Possibly rather more surface water. Borings to chalk sometimes yield good supply of artesian water from depth of 600 to 700 feet.
Plant wanted – similar to 3.

Area 7 (A) North Lille (B) St. Amand-Mons basin
Thanet sands resting on chalk. Bores to chalk yield good water. Artesian in Scarpe valley and sub-artesian in rest of the area. Surface water abundant.
Plant wanted – Surface pumps. Air lifts or tube well in sub-artesian area.

Area 8 Orchies basin
Blue clay overlying Thanet sands on chalk. Similar to area near Hazebrouck.

Area 9 Orcq area
Area irregular. Consists of marly chalk and sands over clays. Some boreholes will yield good water, but area needs detailed study. Surface water – Escaut river and a few streams, but on the whole rather scarce. Some good bores exist in the area.
Plant wanted – Surface pumps. Tube well pumps. Possibly air lifts.

Area 10 Soignies area
Carboniferous limestones overlaid by Thanet sands and clay. A good number of good springs from the limestone, particularly in old quarries. Probably pipe lines served from limestone springs.
Boring work too slow in limestone.
Plant wanted – Surface pumps.

Area 11 (A) St. Quentin-Cambrai-Valenciennes area (B) Mons basin (C) Lille-Seclin area
Similar to present chalk area, but more running water in 11. (B) Good air lift country.
Plant wanted – Mobile and stationary air lifts, and tube well pumps. Surface pumps.

Area 12 Le Cateau-Rethel
Marly chalk. Numerous springs and rivers flowing at all seasons. Borings only a secondary method of supply, but, where necessary, water can probably be obtained from the green sands (borings up to about 300 feet deep).
Plant wanted – Chiefly surface pumps. A few tube well pumps with filters (brass gauze probably good enough).

Area 13 Charleroi
Coal field and many large towns, with pipe distribution. Water mainly brought by gravity from limestone springs and galleries in sand on north and south sides of area.
Pipe lines required if existing ones destroyed.

Area 14 Maubeuge-Namur-Rocroi
Area of old limestones, slates and sandstones. Good springs from limestones. Streams flow all seasons. Rapid boring no good.
Area between Avesnes and Givet liable to be short of water in summer.

Area 15 Hirson-Mezieres
Oolite limestones overlying clays with some sands. Some springs in valleys from limestones, but on the whole area dry. Borings could get water from sands.
Plant wanted – Surface pumps. Tube well pumps. Air lifts.

Fig. 7. An American-built percussion drilling rig in operation; part of the equipment of a Water Boring Section Royal Engineers during the First World War. © Crown copyright. Courtesy of the Imperial War Museum: photo Q31679.

allowing yields of up to 1000 gallons (4.5 m^3) per hour. Small supplies for dug-outs were also produced from borings made by hand with an earth auger (Anon. 1922*b*, p. 88).

In the coastal plain, fair quantities of water were derived from the dune sand formation by shallow hand-dug wells. A small well of 3 to 5 ft (1 to 1.5 m) diameter or square would yield from 500 to 1000 gallons (2.3–4.5 m^3) per day. The French and Belgians sank a few very large wells: three of 30 ft (10 m) diameter yielded 20 000 to 30 000 gallons (90–135 m^3) per day. Norton tube wells were not a success. The standard 2-inch (50 mm) diameter tube delivered very small quantities of water, while a locally-constructed 4-inch (100 mm) tube was difficult to drive. Hand-dug wells were found to be simpler and better. Additionally, water was piped from Dunkerque to Nieuport, via the longest pipeline system constructed by the BEF during the war.

Gallipoli

The Gallipoli campaign was initiated on 26 April 1915 by an amphibious assault by British, Australian, New Zealand and French troops. This was intended to wrest control of the Dardanelles, the straits controlling passage between the Mediterranean and Black Seas, from the Turks and their German allies. British and Commonwealth troops landed on the Gallipoli peninsula, which formed the western border of the straits and were subject to water-supply problems throughout the campaign. Arguably the success of the Turkish defence was due in large part to the greater availability of water behind their lines. British and Commonwealth forces sought to provide adequate supplies of potable water for their men and animals from three sources: surface water, groundwater and imported supplies (Doyle & Bennett 1999, 2001).

Surface water supplies varied across the region and were mostly insufficient. There were few flowing rivers on the peninsula, and most were seasonal, drying up in the summer months.

Attempts to develop groundwater involved a study commissioned directly from the Geological Survey of Great Britain (Strahan 1919; Bailey 1952; Rose & Rosenbaum 1993*a*). Three former Survey geologists serving as army officers (C. H. Cunnington, R. W. Pocock and T. H. Whitehead) were detailed for geological work. They generated an unpublished report for the War Office, one of the earliest reports on the geology of the peninsula, (although this is currently untraceable, *fide* Doyle & Bennett 1999). Limited quantities of potable water

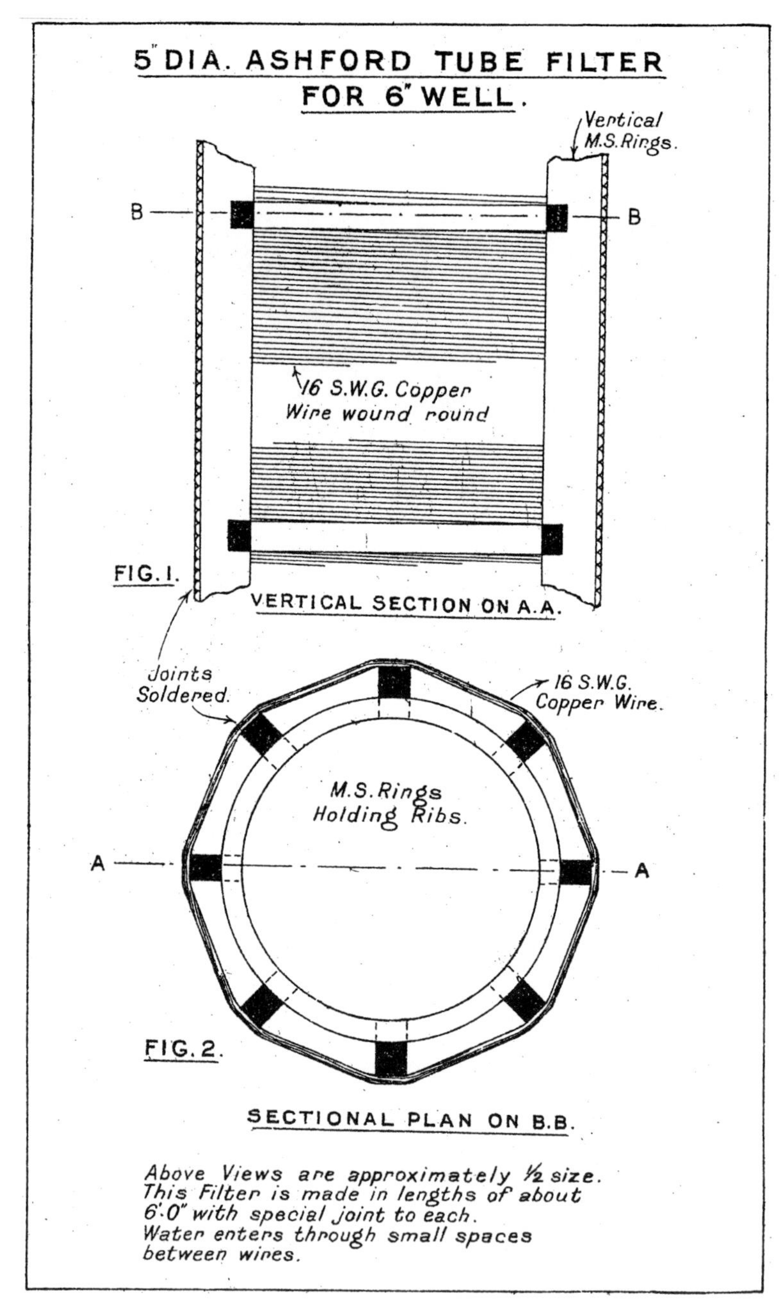

Fig. 8. Diagram of Ashford tube filter. From Anon. (1922*b*), courtesy of the Ministry of Defence.

were obtained from the beach sands at Anzac Cove, and from the alluvium in dry valleys cutting through the cliffs at Cape Helles (Beeby Thompson 1924, p. 92), but wells in the primary landing areas (Doyle & Bennett 1999, fig. 4) were generally sparse, shallow, seasonal and inadequate. The daily water ration from local sources in the 1st Australian Division was rarely more than one third of a gallon (1.5 litres) per man. Attempts to drill boreholes with percussion rigs proved more successful at Cape Helles (where the vegetation was sufficient to provide camouflage) than at Anzac Cove (where lack of vegetation exposed the rig and attracted heavy shelling).

Supplies in many areas were almost entirely from imported water. Water-lighters were towed from the port of Alexandria (Egypt) and from Malta, to be moored alongside piers at Anzac Cove. Initially, the water from them was pumped by hand into storage tanks on the beach, from which it was carried by mules to tanks in the hills and then onward by hand to the front-line troops. Later, mechanical pumps and additional tanks improved distribution. Lessons learnt at Anzac Cove led to the development of more elaborate plans for water supply from lighters for the troops effecting a new landing, at Suvla Bay, in August. However, following the failure of the summer offensives, Allied forces withdrew completely between December 1915 and January 1916.

Beeby Thompson (1924, p. 2) cites Gallipoli as an example of water supply problems which may occur when 'little or nothing is known of the geology or hydrography . . .': 'three army corps were compelled for months to confine their operations to three small and isolated occupied areas enfiladed by enemy artillery and relying mainly upon the perilous expedient of importation of water from Egypt, under the erroneous impression that useful supplies of potable water were non-existent.'

The Mediterranean Expeditionary Force

Between 1915 and 1918, numerous water projects were undertaken in some of the Greek islands, and in Salonika, Macedonia, Greece and Serbia, as well as in Palestine, Sinai, Arabia (including Aden), Somaliland and the Sudan (Beeby Thompson 1924, 1925). These were characterized by an urgent need to develop emergency water supplies and by widespread use of [Norton] tube wells. The water engineer Arthur Beeby Thompson MIMechE, MIMM, FGS was attached to the headquarters staff of the Mediterranean Expeditionary Force from 1915 to 1919, to work closely with the Force's Royal Engineers, but no geologists were appointed to serve as such.

Beeby Thompson (1924, p. 35–42) has documented the relationship between geology and military water supply in Macedonia in the greatest detail. The British army derived water supplies from valley and delta deposits as well as river courses by use of tube wells (driven into near-horizontal sandy clays and loose, unconsolidated sands and gravels), from within a partially-underlying Tertiary bedrock sequence (of inclined sands alternating with clays) by boreholes and from older, nearly impermeable metamorphic basement rocks (near-vertical slates, quartzites, and crystalline limestones) by tapping of surface springs. Tube wells proved to be a particularly significant means of obtaining water, since troops and their base installations were primarily deployed in lowland areas, where alluvial deposits were widespread.

Egypt and the Levant

In Egypt, some Royal Engineer field companies were organized and equipped so as to specialize in water supply (Anon. 1922*b*, p. 15), but did so without military geologist support. Each such field company provided twelve well units, each unit detachment carrying the tools necessary to dig shallow wells. The companies and water distribution system were based on camel transport. In total, a British infantry division in this area required 2200 camels, all carrying two water tanks of 12 gallons (55 litres) capacity each. Photographs held by the Royal Engineers Library at Chatham also prove that near the sea drinking water could be obtained by digging galleries in the coastal sands (Fig. 9), that inland, rather than raise water from shallow wells by the local means of a 'shadouf' (a bucket on a pole), locally-improvised tube wells known as 'spear-points' were sometimes driven into the ground by 56 pound (25 kg) weights, connected to a pump by flexible hose and the water pumped to troughs, and that before the end of the war, Columbia boring rigs had been deployed to Sinai and Palestine.

Between the wars

Hostilities ended in November 1918 and the British army was soon reduced to a peace-time establishment. The geologists were amongst troops demobilized in 1919. British military hydrogeological expertise thus came to an end.

However, the work undertaken by Royal Engineers in the NW European theatre of war was carefully recorded – notably with respect to geology (Anon. 1922*a*) and water supply (Anon. 1921). Wartime experience shaped the water-supply textbook (Anon. 1922*b*) subsequently written to guide the Royal Engineers of the future. It was accepted that 'The Royal Engineers are responsible for the supply and distribution . . . of water for all purposes to all arms and departments of an army in the field' (Anon. 1922*b*, p. 11); that 'The complete conception and execution of a water supply scheme in peace-time calls for the expert direction of a constructional engineer, a mechanical engineer, a water chemist, a bacteriologist, and a geologist' (Anon. 1922*b*, p. 12); that 'A decision as to whether boring will be a profitable means of getting water will depend on primary investigation by a geologist' (Anon. 1922*b*, p. 70); and that 'A boring expert will have a sound training in geology . . . [to] enable him to determine sites for boring and probable results with as much and possibly greater

Fig. 9. Anzac (Australia/New Zealand Army Corps) troops 'digging sumps for water supply for a rest camp on the beach near Jaffa – the water was within a few yards of the sea and although brackish was quite fit to drink, not bad for coffee, but ruined tea. Horses and camels got fond of it.' Photograph (dated to 1916–1919: box F50, album 5/8B, p. 28) courtesy of the Royal Engineers Corps Library, Chatham; wording from original caption.

accuracy than any so-called water diviner' (Anon. 1922*b*, p. 78).

With regard to 'The time taken to sink a bore-hole . . . the experiences of the Great War have upset all preconceived notions so far as English practice was concerned. . . . a 300-foot [100 m] bore-hole can be completed in three or four days, and this brings the system within the scope of the water engineers of an army in the field' (Anon. 1922*b*, p. 70). 'The most suitable plant for work with a field army has been found to be . . . the portable American rig' [e.g. Keystone, Columbia or Star] 'self-contained portable drilling plants capable of drilling up to 1000 feet [330 m] in depth' (Anon. 1922*b*, p. 71). At least one Boring Section Royal Engineers was retained by the army, comprising a headquarters unit plus four drill crews each of a non-commissioned officer plus nine other men. However, the skills of constructing dug wells by hand, and of emplacing Norton tube wells, were also a major feature of the textbook.

When the textbook was later revised (Anon. 1936), in the light of more recent experience (e.g. Martel 1931; Sayer 1932), 'well-borer' was still a recognized service trade within the Royal Engineers. A chapter on 'well sinking and boring' still covered the sinking of wells by hand and the emplacement of Norton tube wells, but the rigs featured for drilling were those of The English Drilling Equipment Company rather than the American 'Keystone' rig of the earlier edition – for such rigs were by then being manufactured in the UK. One of the book's 16 chapters contained a four-page section on 'Geological principles' (Anon. 1936, p. 24–27), with the opening statement that 'In any major campaign geologists will be included in the engineer staff, and part of their duties will be the collection of hydrogeological information'. A statement on the value of investigation by a geologist for potential borehole siting is continued from the 1922 edition. But as regards sites for shallow wells and tube wells, a new principle is introduced: 'In searching for suitable sites for shallow wells and tube wells, the employment of *dowsers* or *water diviners* may save time and possibly fruitless labour in well sinking' (Anon. 1936, p. 42). 'It is not necessary that [the officer commanding a boring section] should be a water diviner, although some people associate this faculty with well-boring' (Anon. 1936, p. 69). With such guiding principles but no geologists, on 4 September 1939 the British army entered the Second World War.

The Second World War

British troops campaigned in many theatres of war between 1939 and 1945. However, the best documented case histories of the military application of hydrogeology come from Home Defence of the United Kingdom, the East and North African Campaigns and Operation Overlord: the Invasion of Normandy.

Home Defence

Following the outbreak of war, W. B. R. King (the British military hydrogeologist of the First World War: appointed to the teaching staff of the University of Cambridge in 1920 but since 1931 Professor of Geology at University College London) was called up during September 1939 from the Army Officers Emergency Reserve and appointed to a Regular Army Emergency Commission in the Royal Engineers (Rose & Hughes 1993*a*; Rose & Rosenbaum 1993*b*). Sent in the rank of 'local major' to France with the British Expeditionary Force, during the winter and spring of 1939/1940 he worked on a variety of problems such as the siting of airfields, the provision of stone and gravel as construction materials and water supply – before the Force's evacuation from Dunkirk [Dunkerque] in June 1940. On return to England, he was attached to Northern Command for a year and then from 1941 to 1943 to GHQ Home Forces.

Hydrogeological studies within the UK during wartime were the responsibility of the Geological Survey rather than the armed forces. The Survey quickly switched manpower from other duties to groundwater work to augment its meager peace-time hydrogeological staff and between 1940 and the end of the war in Europe reports were prepared 'on 305 sites for the Air Ministry, 252 sites for the War Department, 163 sites for the Ministry of Works and 15 for the Admiralty, besides many others for various public and private undertakings' (Bailey 1952, p. 246). Included in this work was the 'provision of maps showing prospects of underground water throughout the whole region of the Army's Southern Command'. A team of 20 geologists with seven assistants generated results that were published in 48 parts as 'Wartime Pamphlets', relating to southern England and much of East Anglia. Bailey (1952, p. 247) records that 'Norfolk is not included with the rest of East Anglia as it was made the subject of a . . . report supplied at the urgent request of the War Office in 1940'.

Royal Engineers well-drilling units were raised in the UK during the war, and some may have served as well as trained in Britain. Thus G. B. Alexander, who was to serve as a military geologist on Gibraltar after the war (Rose & Cooper 1997), was granted an Emergency Commission as a Second Lieutenant, Royal Engineers, in October 1943, and allocated to serve in the United Kingdom, on temporary attachment to 9 Boring Platoon RE from 16 October, before assignment to DCRE East Gloucester from 16 November.

East Africa

The value of geology, especially with regard to siting of boreholes for potable water, was demonstrated in the Ethiopian and Eritrean campaigns by geologists serving with the South African Engineer Corps (Pickard 1946). W. T. Pickard, a graduate of the Royal School of Mines in London (Rose & Rosenbaum 1993*b*), served in the ranks of a British military well-boring unit before being granted a Regular Army Emergency Commission in the Royal Engineers in April 1943. Posted to East Africa, where he finally achieved the rank of major, he put his geological expertise to effective military use, primarily for water supply.

North Africa

In the Spring of 1941, F. W. Shotton embarked for the Mediterranean, to be used as a military hydrogeologist in the British forces in the Middle East, under the Royal Engineer Director of Works, based at Cairo in Egypt (Rose & Rosenbaum 1993*b*). 'Fred' Shotton had joined the Army Officers Emergency Reserve in 1938, whilst lecturing on geology at the University of Cambridge. Called up for active service only in September 1940, he assisted W. B. R. King until posted to the Mediterranean – in the role of a Captain (Electrical & Mechanical) since there was no established post for a geologist as such.

Shotton took responsibility for all geological activities in North Africa and the Middle East, mainly dealing with provision of groundwater supplies and technical direction of the Royal Engineer well-drilling teams. Water supply was one of the main factors influencing movement of the armies in this largely arid or semi-arid zone – a problem solved partly by overland conveyance of water, partly by development of groundwater. Prior to Shotton's appointment, sites for new boreholes had been determined by advice from dowsers, raising concern at both the British and United States Geological Surveys. In January 1942 it was necessary to reassure the House of Commons that with respect to the use of dowsers 'their performance showed a very small percentage of successes, and orders [had been] issued that scientific methods only were to be used' (Bailey 1952, p. 248; Moseley 2000).

Shotton was to oversee extraction of groundwater from the main (sea-level) water table within the near-horizontal Cenozoic limestones and subsidiary marls or clays in the coastal desert of Egypt and northeast Libya (Shotton 1946*a*), from perched water tables within this sequence (Shotton 1944, 1946*b*), by construction of collecting galleries in coastal dune sands and the alluvial gravels that floored major dry river courses (wadis) (Addison & Shotton 1946), by occasional use of cisterns constructed largely in Roman times to store surface runoff and by boreholes into Pliocene sands and gravels of the Nile delta (Shotton 1946*c*). Nearly 50 boreholes and productive wells were put down by the army along the Cairo to Alexandria road to supply airfields and army camps in this region. More were put down through the Nile terraces to the east and south of Cairo, an area also with many military establishments (Shotton 1945*b*, 1946*d*). The mixed sequences of sand, gravel and clay within the terraces nearly always produced satisfactory yields of water, although with variable salinities. Salinities predictably increased on the highest terraces furthest from the River Nile, but there were also seemingly random variations from borehole to borehole. Trial wells demonstrated that salinities tended to increase dramatically downwards, but that changes were abrupt, coinciding with thin layers of impermeable calcrete (Moseley 2000). Waste water could safely be pumped into some sand and gravel horizons, whilst drinking water was pumped out of others.

The 'supreme example of geo-hydrological structure being fully elucidated by the Army' (Shotton 1946*b*) is provided by the Fuka Basin (Fig. 10). Exploratory boreholes by the British prior to the region's capture by the advancing Germans had discovered a limestone some 16 m thick, underlain and overlain by clay, and all folded into an elongate basin whose rim was about 12 m above mean sea level. The perched aquifer within the basin contained groundwater far less saline than that in the main, unconfined aquifer outside. Fuka was therefore the first of the water points to be developed when the Allied counter-attack westwards from El Alamein began in October 1942. On the basis of geological advice it was possible to guarantee: (1) to pinpoint new borehole sites for water exactly along the axis of the syncline, (2) that these boreholes would be 85 ft (27 m) deep and would require 40–45 ft (*c*. 14 m) of casing for the top part, which would be in clay, (3) to promise a yield of 5000 gallons (23 m^3) per hour from a 10-inch (0.25 m) diameter hole and (4), that the water levels were suitable for installation of air-lift pumps, which could therefore be assembled in advance (Shotton 1944, 1946*b*).

Shotton was supported in his role by field surveys undertaken by a team of geophysicists from the South African Engineer Corps led by Major G. L. Paver (Shotton 1945*b*), largely surveys of ground resistivity. Experiences of this unit, the 42nd Geological Section, based on techniques developed in South Africa and on nearly three years of war-time prospecting in the countries covered by the East African and Middle East campaigns, generated a handbook (Anon. 1945) on the location of underground water by geological and geophysical means. This explained the principles of the electrical resistivity method of geophysical prospecting, the interpretation of resistivity curves, organization of electrical resistivity surveys and principles of the magnetometer and interpretation of magnetometric curves.

Well-drilling was carried out during the campaigns by Australian and South African units and those of the Royal Engineers (Fig. 11). Both percussion and mud-flush rotary rigs were employed in the area of Middle East Command, the cumulative length of boreholes drilled finally totalling some 40 km (Shotton 1945*b*). One RE unit was commanded by W. A. Macfadyen (Rose & Rosenbaum 1993*b*), a petroleum geologist and veteran of the First World War who was granted a Regular Army Emergency Commission in May 1941. He served first in England and then in North Africa with well-drilling units, which themselves generated more, junior military geologists (e.g. W. T. Pickard, cited above). Other drillers were recruited from appropriate civilian employment, notably the well drillers commanded by Major G. R. S. Stow, who carried out much of the boring instigated by Shotton, and whose civilian well-drilling company flourished postwar on an international scale. A Royal Engineer boring section was raised directly from the civilian well drilling company of Le Grande, Duke and Ockendon (now DANDO). Volunteer employees were put in uniform and their rig painted a military green to create an almost-instant new unit.

Normandy

The Allied invasion of Normandy began on 6 June 1944. Shotton (1947) has recorded that in the year-long planning stage which preceded it, geological activities included the preparation of water intelligence maps, and that water supply intelligence work continued and developed after D-Day. King (1951, p. 115) has recorded that maps on the scale of '1/50,000 or thereabouts' were prepared for all the bridge-head areas of Normandy before D-Day, showing with respect to the main aquifers: (1) where small springs might be expected but where boring was unlikely to produce large supplies; (2) the main

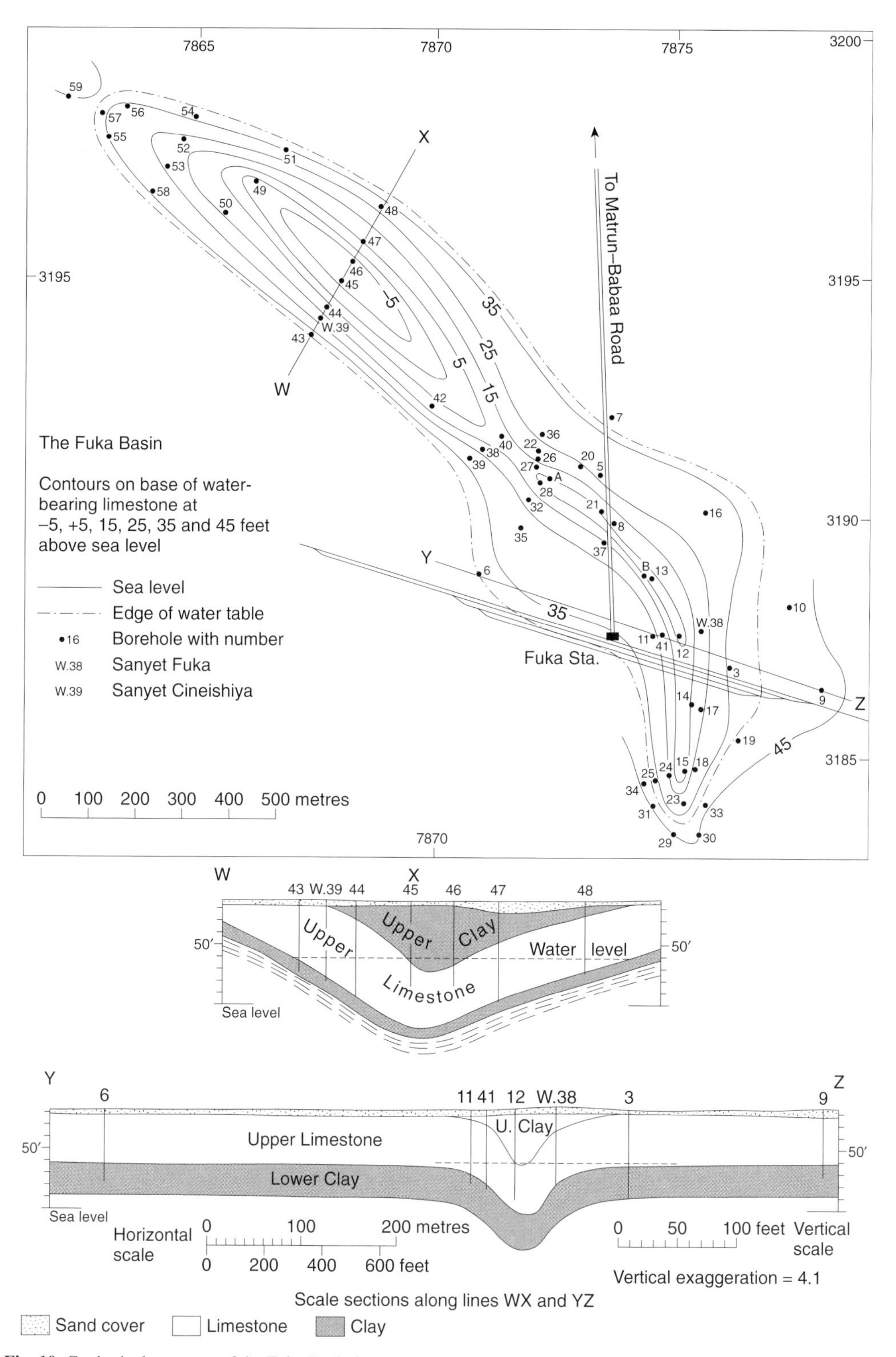

Fig. 10. Geological structure of the Fuka Basin in western Egypt. Redrawn from Shotton (1944, 1946*b*), courtesy of the Institution of Royal Engineers.

Fig. 11. Boring for water in North Africa, 23 January 1943, in support of British troops advancing across Libya. Imperial War Museum photo E21483, published with permission.

outcrop area with, where possible, water table contours; and (3) subsurface contours on the top or base of the aquifer to indicate depth of boring. A distinction was made between groundwater expected to be of good quality and that expected to be saline. Explanatory notes were provided, including geological terms. As acknowledged by King, these maps were a development of the Ground Water Inventory Maps of the Ground Water Provinces of the United States. (See Tolman (1937) for a near-contemporary description of such provinces with key references). Bailey (1952, p. 248) indicates that, by 1942 at least, water supply matters were sometimes of common concern to the British and US Geological Surveys.

By the day following the Allied landings, the existing water supply network in the British-held area had virtually ceased to operate, due to failure of the electricity supply to the pumps from the power station in Caen, still in German hands (Rose & Pareyn 1995, 2003). However, in the region occupied west of the River Orne (Fig. 12), water was available from small rivers and existing deep wells. A network of water purification and storage points was quickly established. These comprised mobile pumping sets, mobile filtration and chlorination plants, and sectional steel storage tanks. Ultimately

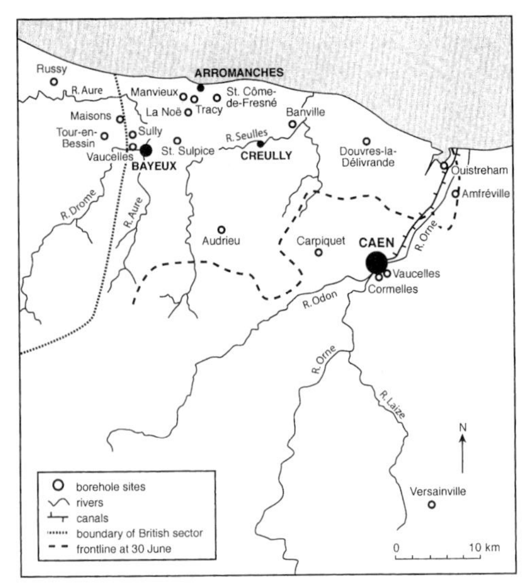

Fig. 12. Map showing positions of main rivers and military borehole sites through Jurassic strata in Normandy, 1944. After Rose & Pareyn (1995), courtesy of Blackwell Publications.

about 50 water points (Fig. 13) were established in the area covered by 1 Corps, from Creully to Caen, 12 of which were operated at any one time. During the first week of battle, each of the three divisions in the British sector required approximately 50 000 gallons (230 m^3) per day from Royal Engineer sources. Existing wells were soon supplemented by new wells, developed with geologist expertise.

33 boreholes were put down in Normandy, according to a summary recorded by Shotton (1945*a*), published in translation by Bigot (1947) after the war. According to Shotton, only one borehole (at Le Haut d'Audrieu) was sunk without geologist advice, by 2nd Canadian Drilling Company. Of the others, six were shallow holes put down in alluvium near the River Drome, in ground unfavourable for deep boreholes. The gravel of the alluvium effected a natural filtration, reducing the number of filtration pumps and plants required. Two of the other boreholes were deep wells through Cretaceous Chalk, in the east of the region soon occupied. One hole was abandoned as crooked, the other due to operational moves, before a satisfactory yield was obtained. Most boreholes (Fig. 12 & Table 2) penetrated Middle Jurassic oolitic and/or bioclastic limestones and subordinate clays, commonly extending into the underlying Liassic marly limestones. Only one reached the Pre-Cambrian basement. Most produced acceptable yields (Table 2).

The wells were drilled by No. 8 Boring Section RE, deployed within Line of Communication Troops during the build-up phase following the initial assault

Fig. 13. The 'well' at Hermanville-sur-Mer: a present-day memorial to a British military water point. Although referenced as a well in some British records, the water source was probably the local pond (*mare*) since the hydrogeological setting and calculated abstraction rate make a borehole source unlikely (see Rose & Pareyn 2003).

(i.e. from 11 June onwards). The section was commanded by a geologist, Lieutenant A. K. Pringle RE, with pre-war civilian experience as an exploration geologist in the Middle East, Australia and New Guinea. Pringle was postwar to become Professor of Applied Geology at Strathclyde University in his native Scotland.

The Cold War 1949 to 1989

The Second World War ended like the first, in that the British military geologists who had served as such were quickly demobilized and returned to civilian life, mostly as university staff. However, W. B. R. King, who had been released from military service

Table 2. *Royal Engineer boreholes for groundwater through Jurassic strata in Normandy, 1944*

Locality	Depth (m)	Yield (m^3h^{-1})
1. Sully	34.75	18.00
2. Amfréville	100.60	3.00
3. Tracy-sur-Mer (La Noë)	97.50	0.45
4. St-Côme-de-Fresné (Buhot No. 1)	105.15	11.80
5. Maisons	26.00	13.50
6. Tracy-sur-Mer (La Rosière)	38.00	18.00
7. St-Côme-de-Fresné (Buhot No. 2)	24.40	13.60
8. Banville	45.70	14.50
9. Douvres-la-Délivrande	39.60	22.70
10. Tour-en-Bessin (Grivilly)	20.70	nil
11. Manvieux	45.70	nil*
12. Ouistreham	33.00	31.90
13. Tour-en-Bessin (Le Coudray)	25.00	9.10
14. Audrieu (Le Haut d'Audrieu)	58.00	3.40
15. Caen (Vaucelles)	73.25	28.70
16. Vaucelles	7.00	13.60
17. Cormelles	62.50	14.00
18. Audrieu (Le Bas d'Audrieu)	30.50	9.00
19. Versainville (Chateau)	70.10	4.50
20. Russy (Russy No. 1)	28.35	3.20
21. Russy (Russy No. 2)	18.30	2.70
22. Russy (Russy No. 3)	18.30	2.70
23. Carpiquet	81.40	4.10
24. St. Sulpice	15.25	1.34
25. Audrieu	45.70	18.20

*Abandoned before completion due to operational necessities. Data from Shotton (1945*a*), Bigot (1947). According to Shotton, most yields were calculated from pumping tests carried out over 24 hours. See Figure 12 for geographical position. After Rose & Pareyn (2003).

in 1943 to become Woodwardian Professor of Geology at the University of Cambridge, was influential in generating a geological textbook for the British army (Anon. 1949) and in founding a unit of military geologists within the reserve army (Shotton 1963)

British reserve army geologists

A 'Pool of Geologists' was established within the Territorial Army during the summer of 1949 (Rose & Hughes 1993*b*). This was to comprise up to eight officers, mostly experienced professional geologists who had already served as army or naval officers (albeit not as military geologists) during the war. Austin Woodland, a senior member of the Geological Survey, was exceptional in this group in that he lacked any previous service with the armed forces – but this deficiency was offset by his experience in hydrogeology.

In 1953 the 'Pool of Geologists' was transferred from the Territorial Army to the newly-formed Army Emergency Reserve. It remained there despite major re-organization of the AER in 1961. In 1967, however, when the reserve army as a whole was re-organized once more, the Geologists' Pool was merged with the AER Works Pool of Officers to form the Engineer Specialist Pool of the newly-created Territorial and Army Volunteer Reserve (Rose & Hughes 1993*c*). The TAVR was later re-organized and again re-named as the Territorial Army, and in 1988 the Engineer Specialist Pool became the Royal Engineers Specialist Advisory Team (Volunteers) within it. No more than six geologists were associated with the Pool or Team in any year.

In peace-time, each reserve army geologist was committed to providing a minimum of 15 days service per year in support of military projects, exercises, or operations world-wide. In war-time, they all had call-out liability for full-time military service – and defence tasks included guiding the preparation of specialist geotechnical maps (such as those showing the trafficability or diggability of terrain) for areas of actual or potential military operations. Additionally, many peace-time tasks related to groundwater development, appraisals of slope stability, and quarrying for aggregates (Rose & Hughes 1993*c*) – the geographical focus of activity changing over the years with changing military priorities.

In the Far East, military geologists were deployed during the 1970s to Thailand to support a Specialist Team Royal Engineers tasked with well-drilling and minor works in aid of the civilian population some 100 km to the west of Bangkok. Neither borehole records nor detailed geological maps were available for the region, so photogeological interpretation and reconnaissance mapping were implemented to guide siting of boreholes in sedimentary basin areas rather than igneous and metamorphic terrain – where village reservoir storage of seasonal surface waters was deemed more appropriate. In Nepal, an alternative to long-distance piped supplies was required for a recruiting centre for British Gurkhas. On geological advice, an expensive (and potentially unsuccessful) deep-drilling project through a thick sequence of coarse clastic sediments was replaced by a programme of hand-dug shallow wells at a fraction of the cost. In Hong Kong, a terrain largely of volcanic rocks intruded by granites, locally deeply weathered, new wells were sited for two villages, to avoid serious pollution problems and to supplement a supply that proved inadequate in the dry season. A large diameter shallow well rather than a narrow borehole was later recommended for a military outpost on one of Hong Kong's islands.

In the Mediterranean region, in the 1960s military geologists appraised potential groundwater supplies for British troops exercising in Libya (Moseley

1963; Moseley & Cruse 1969). On the basis of geology and an analysis of existing wells, they defined the most favourable area in northeastern Libya for groundwater abstraction from the main aquifer and criteria for the siting of new boreholes – in a region underlain by a thick sequence of Cenozoic limestones with subordinate marls. In the 1970s and 1980s military geologists generated a series of reports on water supply for military sites within the British Sovereign Base Areas on Cyprus – an island with a largely igneous (ophiolite) core, surrounded by a sedimentary sequence, mostly of limestones. Saline intrusion was a particular problem in coastal areas, where population density and agricultural water use caused greatest groundwater abstraction. In the 1980s especially, military geologists liaised with hydrogeologists of the then Institute of Geological Sciences (now the British Geological Survey) in groundwater investigations on Gibraltar (Rose *et al.* 2004).

In the Middle East, in the 1960s military operations in Aden necessitated deployment of a military geologist to aid development of secure groundwater supplies for military outposts (Moseley 1966, 1971, 1973, 2000; Shapland *et al.* 1967). Although the basaltic lavas of this region were relatively impermeable, there were ashes and agglomerates of variable thickness within the sequence which could be tapped by boreholes to yield reasonable supplies of groundwater. Water supply problems in the late 1960s generated reconnaissance geological studies of Masirah Island, Oman, then being considered as a possible alternative base to Aden for British troops. A true desert island with less than 24 mm of rain per annum and largely bare rock fringed by gravel fans, the data discouraged development of Masirah as a major base, but were used to guide development of water resources for minor and transient use. In the early 1970s operations on the Oman mainland in support of the Sultan's armed forces included military geological appraisals of potential groundwater sources and a well-drilling programme in the Dhofar region. 11 wells were sunk, the first five producing good yields of potable water, the remaining six proving saline to various degrees but providing water still acceptable locally for irrigation purposes. Total yield of these Dhofar wells was in excess of half a million gallons (2300 m^3) per day.

In more recent years, British military geologists have been deployed to Belize in Central America, as well as to Norway, Germany and the UK, to aid development of secure water supplies for military sites. In European areas geological maps and national expertise were available to help guide deployment of drilling rigs, but in Belize both of these were lacking. Drilling was sited on folded and fractured Cretaceous limestones (largely overlain by Cenozoic mudrocks) only after geological field survey. Military geologists were on call to provide advice on water supply during the Falklands conflict of 1982. More recently, they have been deployed to site boreholes for United Nations troops in the Balkans (Nathanail 1998) and elsewhere.

Textbooks

A joint initiative between the Ministry of Defence and the Institution of Civil Engineers generated a textbook on applied geology (Anon. 1976*a*) for both military and civilian use. Written partly by reserve army geologists, partly by nominees of the Institution, water supply is covered in a single, relatively short, chapter.

When the pre-war military manual on water supply (Anon. 1936) was first revised postwar (Anon. 1956), it was as a textbook to cover the general principles governing both water supply and also petroleum installations and their application in the field. 'The reasons for combining the teaching of water supply and petroleum installations in the one volume are firstly that, broadly speaking, the two R[oyal] E[ngineers] problems are similar, and secondly that much of the standard plant, stores and equipment are common to both supply systems' (Anon. 1956, p. 15). The chapter on well sinking and well boring is much revised: data on sinking of dug wells is much reduced; details of 'Nortons Patent Abyssinian Tube Well' reduced (to a single paragraph by an amendment issued in March 1959) and data on rotary and percussion drilling substantially re-written to emphasize that 'drilling offers great advantages over other means of getting water' (Anon. 1956, p. 162). It is recognized that 'Geological advice should always be sought when siting wells. The scientific basis, if any, in dowsing is not at present understood and it is unwise to rely on the advice of water diviners for the siting of wells. Very few water diviners would claim to be able to predict the salinity of an underground water supply' (Anon. 1956, p. 152). 'In the Service, well drilling is carried out by well boring troops RE, on geological advice' (Anon. 1956, p. 172). Such words of wisdom were carried forward into the later, most recent revision (Anon. 1976*b*).

Well boring by the Royal Engineers

In 1956, a Royal Engineers well boring unit still comprised a headquarters and four boring rig crews (Anon. 1956, p. 172). Each crew then comprised an NCO, four well borers, one fitter, one blacksmith, one cook and one spare hand – thus capable of working as two shifts, each of eight hours duration. The unit was equipped with the

Ruston Bucyrus 22 RW percussion cable drilling machine mounted on a four-wheeled pneumatic tyre trailer, and the Boyles surface drill model BBS2 for rotary drilling – deemed less suitable for military water well use.

'For some time after the Second World War, the subject of well drilling received little attention. Then, in 1960 a few men were trained at 1 ESD and Workshops, Long Marston ([later re-designated] Central Engineer Park) and a 'shadow team' of drillers was formed within the establishment of that unit. This team subsequently [in 1964] became 521 Specialist Team Royal Engineers (Well Drilling)' (Anon. 1971). Initially, it was only fully manned for specific projects. Between projects, a cadre of permanent staff maintained the equipment and investigated the feasibility of potential tasks. Since the Royal Engineers still trained some of its 'sappers' for the trade of well borer before employing them in more general roles, there was thus a pool of trained but inexperienced manpower from which the cadre could be supplemented to form an operational team whenever the need to do so arose. In the 1960s, 521 STRE was operational in this way, notably in Aden. By the end of the decade the army had three types of drilling rig in use: (1) Ruston Bucyrus 22 RW (being phased out after many years of service, the rig was deemed well tried and reliable, but was percussion only – and mounted on a very primitive trailer); (2) Speedstar 55 (a percussion rig with rotary attachment, this had been found heavy and cumbersome, and problems of obtaining spares from its American manufacturer had speeded its replacement); (3) Dando 800 (a British percussion rig with rotary attachment, mounted on its own trailer).

In the 1970s, the STRE was retained as a formed unit, to drill wells to ensure secure water supplies at British bases, notably in Germany, Northern Ireland and Cyprus. It was also deployed on tasks providing 'Military Aid to the Civilian Community', notably for the Overseas Development Agency, for example in Kenya and on the Caribbean island of Anguilla. A separate unit, STRE (Thailand) was briefly operational in 1969 and 1970 to drill wells in the mid-western provinces of Thailand.

From the late 1970s through the 1980s, the primary effort was redirected to providing secure water supplies for Royal Air Force bases, first overseas and then in the UK, although with some additional tasks overseas, notably British military bases in Belize. Not required during the Falklands War of 1982, the unit was operational in Bosnia during 1993–1994, to provide wells for British and, later, other United Nations forces' bases (Wye 1994). (During the Gulf War of 1991, potable water for British troops was obtained by reverse osmosis of saline water drawn from existing boreholes and irrigation pipelines (Walton-Knight 1994).) Equipment is currently the DANDO 250 (top drive rotary rigs purchased in the mid-1980s and adapted to military use), being replaced by air-portable EDECO HE175 and Traveller 60 rigs ('heavy' and 'light' respectively). Recently re-structured to operate in a water development rather than merely well-drilling role, the team has been active world wide, notably in support of troop deployments to Kosovo and Afghanistan.

For many years 521 STRE was complemented by a Territorial Army unit: 520 STRE (V). This began as 502 STRE (Well Drilling), a unit under command of Headquarters Army Emergency Reserve (Resources) and based at Long Marston. In 1964, its command was transferred to HQ AER (Field and Works), based at Cove in Hampshire. This became Central Volunteer HQ when the AER was merged into the newly-formed Territorial and Army Volunteer Reserve in 1967 and 502 STRE was absorbed into 198 (Army) Engineer Park Squadron and later into 111 Engineer Regiment, before re-emerging as a separate unit, with the number 520. However, during re-structuring of the reserve army in the year 2000, 520 STRE was reduced to a sponsored reservist unit – with peace-time training capability, but need for enhancement if required for wartime operational use.

Conclusions

- During the 19th century, British military initiatives and funding promoted early teaching of geology, and geological mapping in Scotland, Ireland and England. The value of geology in siting successful wells was recognized, primarily by officers who had served on the Indian subcontinent, but water supply was given relatively minor coverage in military technical training. Groundwater was abstracted by the British army under operational conditions by sinking hand-dug rather than drilled wells or, from 1867 at least, by emplacement of driven [Norton] tube wells.
- During the First World War, innovations by the British army on the Western Front included deployment of its first military hydrogeologist to serve as such: Lieutenant, later Captain, W. B. R. King; generation of specialist water-supply maps, for parts of Belgium and northern France, at scales of 1:250000, 1:100000 and 1:40000; formation of the first Water Boring Sections within the Royal Engineers, equipped with mobile percussion drilling rigs of American origin, capable of drilling to at least 100 m depth in the Cretaceous chalk of the Artois/Picardy plateau; introduction of air-lift pumps for military use, to maximize yield from

these boreholes; and in 1918, introduction of the Ashford strainer to exploit groundwater from boreholes into the fine-grained Tertiary sands beneath the Flanders plain – an exploitation deemed impracticable pre-war. Royal Engineer boring units were also deployed in Mediterranean campaigns, in Gallipoli and the Balkans with some geotechnical guidance, but Norton tube wells and hand-dug wells remained relatively important features of temporary groundwater supplies for military purposes outside northwest Europe.

- After the First World War, military geologists were quickly demobilized, but the first British military textbook specifically on water supply recognized that borehole drilling could be sufficiently rapid to be of future operational use and that geological advice would be needed to help site wells in any future campaign. However, some credence was also given to water divining as a means of well site selection. Well-drilling equipment of British rather than American manufacture was introduced for military use by 1936.
- At the start of the Second World War in 1939, the hydrogeologist Major, later Lt-Col, W. B. R. King was mobilized from the Army Officers Emergency Reserve for active duty in France and the UK. Captain (later Major) F. W. Shotton was similarly mobilized in 1940, later being deployed for distinguished operational service as a hydrogeologist in the Middle East and North Africa. 42nd Geological Section of the South African Engineer Corps aided prospecting for groundwater in East and North Africa by geophysical means, primarily resistivity but also magnetometer surveys. Specialist water-supply maps, a development of the Ground Water Inventory Maps of the Ground Water Provinces of the United States, were prepared to assist planning of the Allied invasion of Normandy in June 1944. Several Royal Engineer well-boring units were raised, for service in East and North Africa, northwest Europe and the UK, with officers who were commonly either well-borers or geologists by pre-war civilian profession.
- Through the Cold War of 1949–1989 and to the present day, the British armed forces retained limited hydrogeological expertise within a small group of reserve army geologists, mostly although not exclusively recruited from university staff, for short-term deployment world-wide in support of military or humanitarian tasks during time of peace and potential full-time military service in time of war. A well-drilling capability has been maintained with similar role.

Part of the information on which this paper is based is derived from 21 years of service (1969–1990) as a reserve army geologist and discussions within that time with military geologists and well drillers too numerous to conveniently list here. Additionally, I am grateful to librarians at the Royal Engineers Library, Chatham (especially M. L. Roxburgh), the Imperial War Museum, London and the British Geological Survey, Keyworth, for help to access archive data; to Major J. Evans RE and members of 521 STRE (Water Development) for recent helpful discussion; and to two referees for constructive criticism.

References

ADDISON, H. & SHOTTON, F. W. 1946. Water supply in the Middle East campaigns. 3. Collecting galleries along the Mediterranean coast of Egypt and Cyrenaica. *Water and Water Engineering*, **49**, 427–436.

ANON. 1921. *The work of the Royal Engineers in the European War, 1914–19: Water supply*. Institution of Royal Engineers, Chatham.

ANON. 1922*a*. *The work of the Royal Engineers in the European war, 1914–19: Geological work on the Western Front*. Institution of Royal Engineers, Chatham.

ANON. 1922*b*. *Military Engineering Vol. VI: Water supply*. HMSO, London.

ANON. 1936. *Military Engineering Vol. VI: Water supply. War Office Code 7489*. HMSO, London.

ANON. 1945. *Military Engineering Vol. VI – Water supply. Supplement No. 1. The location of underground water by geological and geophysical methods. War Office Code 7490*. HMSO, London.

ANON. 1949. *Military Engineering Vol. XV: Application of Geology. War Office Code 8287*. HMSO, London.

ANON. 1956. *Military Engineering Vol. VI: Water supply and petroleum installations. War Office Code 9017*. HMSO, London.

ANON. 1971. *Military Engineering Vol. VI: Water supply part 1B (provisional). Well drilling. War Office Code 70715*. HMSO, London.

ANON. 1976*a*. *Military Engineering Vol. XV: Applied Geology for Engineers. Army Code No. 71044*. HMSO, London.

ANON. 1976*b*. *Military Engineering Vol. VI: Water supply. War Office Code 71045*. HMSO, London.

ANSTED, D. T. 1856. *An elementary course of geology, mineralogy, and physical geography*. 2nd edition. Voorst, London. [1st edition 1850].

ANSTED, D. T. 1878. *Water and water supply: chiefly in reference to the British Islands; surface waters*. William H. Allen, London.

BAILEY, E. B. 1952. *Geological Survey of Great Britain*. Murby, London.

BALLEINE, G. R. 1950. *A history of the island of Jersey*. Staples, London.

BEEBY THOMPSON, A. 1924. *Emergency water supplies for military, agricultural and colonial purposes*. Crosby, Lockwood & Son, London.

BEEBY THOMPSON, A. 1925. Geology as applied to military requirements. *Royal Engineers Journal*, **40**, 53–63.

BIGOT, A. 1947. Forages pour recherches d'eau dans le Calvados. VII – Forages de l'Armée Anglaise en 1944. *Bulletin de la Société Linnéenne de Normandie*, sér.9, **5**, 130–133.

BINNIE, A. R. 1877. *Lectures. Water supply, rainfall, reservoirs, conduits and distribution, delivered at the School of Military Engineering, Chatham.* Royal Engineers Institute, Chatham [2nd edition 1887].

BURNELL, G. R. 1853*a*. Water supply. *In*: LEWIS, G. G., JONES, H. D., LARCOM, T. A., WILLIAMS, J. & BINNEY, C. R. (eds) 1853–1862. *Aide-mémoire to the military sciences*, 2nd edition. Weale, London. Vol. 3, 721–746.

BURNELL, G. R. 1853*b*. Wells. *In:* LEWIS, G. G., JONES, H. D., LARCOM, T. A., WILLIAMS, J. & BINNEY, C. R. (eds) 1853–1862. *Aide-mémoire to the military sciences*, 2nd edition. Weale, London, Vol. 3, 775–782.

DODDS, G. L. 2000. *Historic sites of Northumberland and Newcastle upon Tyne*. Albion Press, Sunderland.

DOYLE, P. & BENNETT, M. R. 1999. Military geography: the influence of terrain in the outcome of the Gallipoli Campaign, 1915. *Geographical Journal*, **165**, 12–36.

DOYLE, P. & BENNETT, M. R. 2001. Terrain and the Gallipoli Campaign, 1915. *In*: DOYLE, P. & BENNETT, M. R. (eds) *Fields of battle: terrain in military history*. Kluwer Academic Publishers, Dordrecht, 149–169.

FLINN, D. 1981. John MacCulloch, M.D., F.R.S. and his geological map of Scotland: his years in the Ordnance, 1795–1826. *Notes and Records of the Royal Society of London*, **36**, 83–101.

HALSALL, T. J. 2000. Geological constraints on the siting of fortifications: examples from medieval Britain. *In*: ROSE, E. P. F. & NATHANAIL, C. P. (eds) *Geology and warfare: examples of the influence of terrain and geologists on military operations*. Geological Society, London, 3–31.

[HERRIES] DAVIES, G. L. 1974. First official geological survey in the British isles. *Nature*, **249**, 407.

HERRIES DAVIES, G. L. 1983. *Sheets of many colours: the mapping of Ireland's rocks 1750–1890*. Royal Dublin Society, Dublin.

HERRIES DAVIES, G. L. 1995. *North from the Hook: 150 years of the Geological Survey of Ireland*. Geological Survey of Ireland, Dublin.

HUTTON, F. W. 1862. Importance of a knowledge of geology to military men. *Journal of the Royal United Service Institution*, **6**, 342–360.

KING, W. B. R. 1919. Geological work on the Western Front. *Geographical Journal*, **54**, 201–215 & discussion 215–221.

KING, W. B. R. 1921*a*. Résultats des sondages exécutés par les armées britanniques dans la Nord de la France. *Annales de la Société géologique du Nord, Lille*, **45**, 9–26.

KING, W. B. R. 1921*b*. The surface of the marls of the Middle Chalk in the Somme valley and the neighbouring districts of northern France, and the effect on the hydrology. *Quarterly Journal of the Geological Society, London*, **77**, 135–143.

KING, W. B. R. 1951. The recording of hydrogeological data. *Proceedings of the Yorkshire Geological Society*, **28**, 112–116.

LEWIS, G. G., JONES, H. D., NELSON, R. J., LARCOM, T. A., DE MOLEYNS, E. C. & WILLIAMS, J. (eds) 1846–1852. *Aide-mémoire to the military sciences*. Weale, London, 3 vols.

LEWIS, G. G., JONES, H. D., LARCOM, T. A., WILLIAMS, J. & BINNEY, C. R. (eds) 1853–1862. *Aide-mémoire to the military sciences*, 2nd edition. Weale, London, 3 vols.

MACCULLOCH, J. 1836. *Geological map of Scotland*. Arrowsmith, by order of the Lords of the Treasury, London, scale 4 inches : 1 mile, 4 sheets, hand coloured.

MANSERGH, J. 1882. *Lectures. Water supply, prospecting for water, well-sinking & boring. Delivered at the School of Military Engineering, Chatham*. School of Military Engineering, Chatham.

MARTEL, G. LEQ. 1931. Water supply in the field. *Royal Engineers Journal*, **45**, 93–99.

MOSELEY, F. 1963. Fresh water in the North Libyan desert. *Royal Engineers Journal*, **77**, 130–140.

MOSELEY, F. 1966. Exploration for water in the Aden Protectorate. *Royal Engineers Journal*, **80**, 124–142.

MOSELEY, F. 1971. Problems of water supply, development and use in parts of Audhali, Dathina and Fadhli, South Yemen. *Overseas Geology and Mineral Resources*, **10**, 309–327.

MOSELEY, F. 1973. Desert waters of the Middle East and the role of the Royal Engineers. *Royal Engineers Journal*, **87**, 12–23.

MOSELEY, F. 2000. From dowsing to hydrogeology in the Royal Engineers 1939–70. *In*: ROSE, E. P. F. & NATHANAIL, C. P. (eds) *Geology and warfare: examples of the influence of terrain and geologists on military operations*. Geological Society, London, 315–338.

MOSELEY, F. & CRUSE, P. K. 1969. Exploitation of the main water table of north-eastern Libya. *Royal Engineers Journal*, **83**, 12–23.

NATHANAIL, C. P. 1998. Hydrogeological assessments of United Nations bases in Bosnia Hercegovina. *In*: UNDERWOOD, J. R., Jr. & GUTH, P. L. (eds) *Military geology in war and peace*. Reviews in Engineering Geology **13**, Geological Society of America, Boulder, Colorado, 55–66.

NEAVERSON, E. 1947. *Medieval castles in north Wales: a study of sites, water supply and building stones*. Hodder & Stoughton, London.

PICKARD, W. T. 1946. Geological work in the East Africa Command. *East African Engineer*, **July 1946**, 17–19.

PORTER, W. 1889. *History of the Corps of Royal Engineers Vol. 1*. Institution of Royal Engineers, Chatham.

PORTLOCK, J. E. 1850. Geognosy and geology. *In*: LEWIS, G. G., JONES, H. D., NELSON, R. J., LARCOM, T. A., DE MOLEYNS, E. C. & WILLIAMS, J. (eds) 1846–1852. *Aide-mémoire to the military sciences*. Weale, London, Vol. 2, 77–182.

PORTLOCK, J. E. 1860. Geology and geognosy. *In*: LEWIS, G. G., JONES, H. D., LARCOM, T. A., WILLIAMS, J. & BINNEY, C. R. (eds) 1853–1862. *Aide-mémoire to the military sciences*, 2nd edition. Weale, London, Vol. 2, 91–190.

ROSE, E. P. F. 1996. Geology and the army in nineteenth century Britain: a scientific and educational symbiosis? *Proceedings of the Geologists' Association*, **107**, 129–141.

ROSE, E. P. F. 1997. Geological training for British army officers: a long-lost cause? *Royal Engineers Journal*, **111**, 23–29

ROSE, E. P. F. 1999. The military background of John W. Pringle, in 1826 founding superintendent of the Geological Survey of Ireland. *Irish Journal of Earth Sciences*, **17**, 61–70.

ROSE, E. P. F. & COOPER, J. A. 1997. G. B. Alexander's studies on the Jurassic of Gibraltar and the Carboniferous of England: the end of a mystery? *Geological Curator*, **6**, 247–254.

ROSE, E. P. F. & HUGHES, N. F. 1993*a*. Sapper geology. Part 1. Lessons learnt from world war. *Royal Engineers Journal*, **107**, 27–33.

ROSE, E. P. F. & HUGHES, N. F. 1993*b*. Sapper geology. Part 2. Geologist pools in the reserve army. *Royal Engineers Journal*, **107**, 173–181.

ROSE, E. P. F. & HUGHES, N. F. 1993*c*. Sapper geology. Part 3. Engineer Specialist Pool geologists. *Royal Engineers Journal*, **107**, 306–316.

ROSE, E. P. F. & PAREYN, C. 1995. Geology and the liberation of Normandy, France, 1944. *Geology Today*, **11**, 58–63.

ROSE, E. P. F. & PAREYN, C. 2003. *Geology of the D-Day landings in Normandy, 1944*. Geologists' Association Guide No. **64**, Geologists' Association, London.

ROSE, E. P. F. & ROSENBAUM, M. S. 1993*a*. British military geologists: the formative years to the end of the First World War. *Proceedings of the Geologists' Association*, **104**, 41–49.

ROSE, E. P. F. & ROSENBAUM, M. S. 1993*b*. British military geologists: through the Second World War to the end of the Cold War. *Proceedings of the Geologists' Association*, **104**, 95–108.

ROSE, E. P. F. & ROSENBAUM, M. S. 1998. British military geologists through war and peace in the 19th and 20th centuries. *In*: UNDERWOOD, J. R., Jr. & GUTH, P. L. (eds) *Military geology in war and peace*. Reviews in Engineering Geology **13**, Geological Society of America, Boulder, Colorado, 29–40.

ROSE, E. P. F., GINNS, W. M. & RENOUF, J. T. 2002*a*. Fortification of island terrain: Second World War German military engineering on the Channel island of Jersey, a classic area of British geology. *In:* DOYLE, P. & BENNETT, M. R. (eds) *Fields of battle: terrain in military history*. Kluwer, Dordrecht, 265–309.

ROSE, E. P. F., MATHER, J. D. & WILLIG, D. 2002*b* German hydrogeological maps prepared for Operation 'Sealion': the proposed invasion of England in 1940. *Proceedings of the Geologists' Association*, **113**, 363–379.

ROSE, E. P. F., MATHER, J. D. & PEREZ, M. 2004. British attempts to develop groundwater and water supply on Gibraltar 1800–1985. *In*: MATHER, J. D. (ed.) *200 years of British Hydrogeology*. Geological Society, London, Special Publications, **225**, 239–262.

RUCKLEY, N. 1990. Water supply of medieval castles in the United Kingdom. *Fortress*, **7**, 14–26.

RUCKLEY, N. 1991. Geological and geomorphological factors influencing the form and development of Edinburgh Castle. *The Edinburgh Geologist*, **26**, 18–26.

SAYER, A. P. 1932. Water supply in the field. *Royal Engineers Journal*, **46**, 61–70.

SCOTT-MONCRIEFF [G. K.] & TYNDALE, W. C. 1909. *Water supply manual*. HMSO, London.

SHAPLAND, P. C., LITTLE, J. F. & MOSELEY, F. 1967. Well drilling in the Federation of South Arabia. *Royal Engineers Journal*, **81**, 240–245.

SHOTTON, F. W. 1944. The Fuka Basin. *Royal Engineers Journal*, **58**, 107–109.

SHOTTON, F. W. 1945*a*. *Summary of Normandy Boreholes for water*. British Geological Survey archive document dated 21 May 1945, ref. 21 AGp 8779 WK, 5 p. plus covering letter, unpublished.

SHOTTON, F. W. 1945*b*. *Water in the desert*. University of Sheffield, inaugural lecture.

SHOTTON, F. W. 1946*a*. Water supply in the Middle East campaigns. 1. The main water table in the Miocene limestone in the coastal desert of Egypt. *Water and Water Engineering*, **49**, 218–226.

SHOTTON, F. W. 1946*b*. Water supply in the Middle East campaigns. 2. Perched water supplies above the main water table of the Western Desert. *Water and Water Engineering*, **49**, 257–263.

SHOTTON, F. W. 1946*c*. Water supply in the Middle East campaigns. 4. Water supplies adjacent to the Cairo-Alexandria desert road. *Water and Water Engineering*, **49**, 477–486.

SHOTTON, F. W. 1946*d*. Water supply in the Middle East campaigns. 5. The desert between the Nile delta and the Suez Canal. *Water and Water Engineering*, **49**, 529–540.

SHOTTON, F. W. 1947. [abstract] Geological work in the invasion of North-West Europe. *Quarterly Journal of the Geological Society, London*, **102**, v.

SHOTTON, F. W. 1963. William Bernard Robinson King. *Biographical Memoirs of Fellows of the Royal Society*, **9**, 171–182.

SMITH, R. B. 1849. Essay on geology, as a branch of study especially meriting the attention of the Corps of Engineers. *Corps Papers, and Memoirs on Military Subjects; of the Royal Engineers and the East India Company's Engineers*, **1**, 27–34.

STRAHAN, A. 1914. *Notes on sources of temporary water supply in the south of England and neighbouring parts of the continent*. Geological Survey & Museum, London.

STRAHAN, A. 1915. *Report on the geological conditions affecting the available sources of water in Belgium and the north of France*. Unpublished report, Geological Survey of Great Britain.

STRAHAN, A. 1919. Introduction. Work in connection with the war. *Memoirs of the Geological Survey. Summary of Progress of the Geological Survey of Great Britain and the Museum of Practical Geology for 1918*. HMSO, London, 1–4.

SWINDELL, J. G. & BURNELL, G. R. 1851. *Rudimentary treatise on wells and well sinking*. Lockwood & Son, London. [New edition 1882].

TOLMAN, C. F. 1937. *Ground water*. McGraw-Hill, New York & London.

TULLOCH, H. 1890. *Report on the water supply and sewerage of Gibraltar*. Waterloo & Sons, London.

VIBART, H. M. 1894. *Addiscombe: its heroes and men of note*. Constable & Co., London.

VIBART, H. M. 1897. *Richard Baird Smith: the leader of the Delhi heroes in 1857*. Constable & Co., London.

WALTON-KNIGHT, M. P. 1994. Supplying water to the British army during the Gulf War. *Royal Engineers Journal*, **108**, 154–159.

WOODWARD, H. B. 1907. *The history of the Geological Society of London*. Geological Society, London.

WYE, T. W. 1994. Well drilling in Bosnia. *Royal Engineers Journal*, **108**, 149–153.

WYSE JACKSON, P. N. 1997. John W. Pringle (c.1793–1861) and Ordnance Survey geological mapping in Ireland. *Proceedings of the Geologists' Association*, **108**, 153–156.

Groundwater versus surface water in Scotland and Ireland – the formative years

N. S. ROBINS[1], J. R. P. BENNETT[2] & K. T. CULLEN[3]

[1]*British Geological Survey, Maclean Building, Wallingford, Oxfordshire, OX10 8BB, UK (e-mail: nsro@bgs.ac.uk)*

[2]*White Young Green Environmental, 1 Locksley Business Park, Montgomery Road, Belfast, BT6 9UP, UK*

[3]*White Young Green Ireland, Bracken Business Park, Bracken Road, Sandyford Industrial Estate, Dublin 18, Ireland*

Abstract: Celtic interest in groundwater has continued to the modern era in much of Scotland and Ireland, despite abundant good quality surface waters. Groundwater investigation in the 19th and 20th centuries was prompted by the need to remove water from mine workings in Scotland and to provide water for industry in the Midland Valley of Scotland and the Lagan Valley in the north of Ireland. Little development took place in the south of Ireland until relatively recently. Champions of groundwater investigation include the venerable Scottish geologists Ben Peach and John Horne, as well as lesser known advocates of hydrogeology such as John Jerome Hartley in Ireland. These workers were supported by numerous people directly and indirectly involved with developing the understanding of the groundwater resources of Scotland and Ireland.

The groundwater resource potential available in Scotland and Ireland was largely ignored throughout much of the 20th Century. Reviewing the availability of water resources in 1973, the former Scottish Development Department used the report title *A Measure of Plenty* to describe Scotland's water resources with little acknowledgement of any contribution from groundwater (SDD 1973). The same attitude was also prevalent in Northern Ireland, then as now, heavily dependent on Lough Neagh and Silent Valley for public water supplies to its urban population in the Belfast and Newtownards area. In the south of Ireland the cities of Dublin, Limerick and Cork were fortuitously able to make use of large impounded reservoirs originally built by the Electricity Supply Board for hydro-power, so obviating the incentive to look for groundwater supplies. However, groundwater has always been important, not only because of the long-standing social and economic importance of groundwater to many communities but also because surface water low flows all derive from groundwater baseflow.

But what was the overall understanding of groundwater and who took it upon themselves to investigate the availability of groundwater and groundwater quality in these formative years? Understanding was patchy, although surprisingly good in some places. There are a number of workers, some familiar, such as Scotland's famous pair of geologists Ben Peach and John Horne, and some less familiar, such as the engineer/geologist J. J. Hartley who worked in the Belfast area, who deserve mention. This paper describes the historical role of groundwater and the early hydrogeological studies carried out by these workers and others who were to lay down the foundation for the current understanding of groundwater occurrence in Scotland and Ireland.

Historical background

The historical dependence of people in Scotland and Ireland on groundwater may well originate in the Celtic vision of groundwater as a gift from the Otherworld (Robins & Misstear 2000). The Clootie wells (Devil or Witch wells) of Invernesshire, the Black Isle and elsewhere illustrate the reverence in which the rural people held their groundwater sources (Fletcher *et al.* 1996). The alleged curative powers of some springs and wells are illustrated by treatment, for example, of Robert the Bruce who drank the water from Scotlandwell in Fife as a cure for leprosy (Robins 1990).

The Celtic respect for water, both surface and groundwater, highlights the longstanding and important role of water in Scotland and Ireland. Ross (1967) discussed sanctuaries linked to water with the introduction:

> 'Springs, wells and rivers are of first and enduring importance as a focal point of Celtic cult practice and ritual. Rivers are important in themselves, being associated in Celtic tradition with fertility and with deities such as the divine mothers and the sacred bulls, concerned with this fundamental aspect of life.'

From: MATHER, J. D. (ed.) 2004. *200 Years of British Hydrogeology*. Geological Society, London, Special Publications, **225**. 183–191. 0305-8719/04/$15

The respect for, and reliance on, wholesome springs for water supply by the Celts are reflected in many Gaelic place names and named springs. Tober/tubber and poula are gaelic words for a well or spring which are widely preserved in names throughout Ireland and Scotland. The antiquity of some is evident from their meaning, such as Tobernaveen in County Antrim which probably translates to 'Well of the Fianna', the latter being Finn's soldiers of Irish mythology (the same Finn who reputedly created the Isle of Man from Lough Neagh). There is also evidence in Ireland of cooking hearths and other activities of prehistoric man, in proximity to prominent springs, probably frequented by hunting parties or non-settled people.

The importance of groundwater continued through Medieval times with a gradual conversion from pagan cult to Christianity and the modern practice of naming a well after a saint. Brennerman & Brennerman (1996) described this transition:

> 'In addition to votive offerings, other practices that are ongoing at the springs include drinking the water for cures and grace; placing objects into the well; personifying wells that evolve from a cult god to the name of a saint; and presenting for a cure the cloth fragments or clooties that signify disease.'

On the domestic front, most of the older farmsteads in the prosperous communities of, for example, Strathmore in eastern Scotland or the rich Glens of Antrim in Ireland, had a well beneath the flagstone floor of the kitchen. Many of these wells still exist, and some are now a hazard in upgrading and modernizing properties in which the well has long been forgotten.

One of the more celebrated medieval wells is the Castle Well in Edinburgh Castle, situated high up at the Half Moon Battery. Reputedly capable of sustaining the garrison in siege, the well was surveyed in 1913 by HM Office of Works and found to be 34 m deep with a static water level of 15 m below ground level. However, the recharge zone for the well can only have been local, unless there was some artesian connection to deeper strata. Tunnels and shafts believed to connect Holyrood Palace with the Castle along the line of the Royal Mile are now well beneath the water table, reflecting intense local use of groundwater, albeit in a weakly transmissive aquifer, in the city during the 17th and 18th centuries.

In the early 17th century much of the surface water then supplying the many growing urban areas was polluted. The history of water development in Belfast is typical of many early urban water supplies (Barty-King 1992). Contaminated river sources led to the development of springs in Belfast from 1678. The Corporation of Belfast initially exploited springs at Tuck Mill Dam near Divis Street to bring water via wooden pipes to the Boyne Bridge area; the capital cost was some £250. By 1710 springs and wells in the Sandy Row and Fountainville areas were brought into the system to serve the town's population of about 1000. By the end of the 18th century the wooden pipes had become rotten and new piping was installed, new spring sources were commissioned and water charges of a halfpenny for four gallons were introduced. The Belfast Water Commissioners were created under the Belfast Water Act (1840), but from that time onwards the town turned its attention to developing surface water supplies (Plester and Binnie 1995).

Few written reports on groundwater have survived in Scotland and Ireland from the 18th and 19th centuries. Most are catalogues of holy wells and spa wells (e.g. Anon. 1875) and work such as the celebrated essay of Rutty (1757) who took it upon himself to describe the therapeutic characteristics of a number of Irish wells. Many descriptions of this era are picturesque as in Professor Wallace's description of the waters of Pitkeathly Spa at Bridge of Earn near Perth (Anon. 1915):

> 'After a fair trial, I think it is a most agreeable beverage for table use, either by itself or in a combination with wine or spirits, particularly the latter. Unlike all other saline and aerated waters I have experience of, it is not depressing, but tonic in its effects.'

Another early report is that of a new artesian borehole drilled in Lemon Street, Aberdeen. Described in *The Daily Free Press* of Saturday 7 December 1895, the borehole had been drilled to 32 m and penetrated a sequence of Quaternary sands, silts, peats and clays, and exceptionally had an artesian head of 7m:

> 'The sinking of artesian wells is an operation more or less associated with American ideas, but the drilling system is now resorted to nearly everywhere. The enterprise has brought to light a mineral water of great natural potency and its medicinal properties are expected to prove of no inconsiderable account. Analysis showed the water to be absolutely pure and colourless, rich in minerals, having carbonate of iron in solution.'

Edinburgh was wholly dependent on spring sources from the volcanic rocks of the Pentland hills until upland impounded reservoirs were brought into commission. Groundwater users in Glasgow, however, could not wait to abandon the poor quality supplies drawn from the Coal Measures beneath Glasgow once Loch Katrine water had been piped to the city in 1859. Shallow coal workings had encouraged ferruginous material into solution and many boreholes discharged ochre contaminated water and had to be flared of methane before use (Hall *et al.* 1998).

Table 1. *Brewery usage of groundwater in Edinburgh in 1938*

Brewery	Borehole	Yield (gallons per hour)
John Aitchison & Co Ltd Holyrood Road	Nos. 1 and 2 combined	5000
T & J Bernard Ltd	No. 1	3000
	No. 2	3000
	No. 4	4800
	No. 5	2400
	No. 6	1800
Gordon Blair (1923) Ltd Craigwell Brewery		1000
A Campbell, Hope & King Ltd Argyle Brewery		3600
Robert Deuchar Ltd Duddingston		1200
Drybrough & Co Ltd Craigmillar	No. 1	11600
	No. 2	9750
John Jeffrey & Co Ltd	Grassmarket	3960
Heriot Brewery	Ale Brewery	3300
	Lager Brewery	3300
Wm McEwan & Co Ltd	Park Well	4290
Fountain Brewery	Court Well	252000 gallons per week
	Villa Well	252000 gallons per week
Maclachlans Ltd		13000 gallons per week
Castle Brewery, Craigmillar		
J & J Morrison Canongate		2160
W. Murray & Co Ltd	No. 1	3500
Craigmillar	No. 2	4000
St Mary's Brewery Holyrood Road		4000
Steel Coulson & Co Ltd Abbeyhill	Nos. 1 and 2 combined	5000

The invention of the steam engine and the mechanical drilling machine allowed new groundwater sources to be introduced for industry, and sub-water table mining to commence on a large scale. Both the Permo-Triassic sandstone aquifer of the Lagan Valley to the south-west of Belfast and of the Carboniferous and Devonian sediments of the Midland Valley of Scotland were heavily exploited from the 1880s onwards until the decline of water intensive industry in post-war years. Coal and oil-shale mining underpinned the Scottish economy but not that in Northern Ireland where only small pits at Coalisland, Dungannon and Ballycastle were ever operational. There is little evidence of any borehole drilling in what is now the Republic of Ireland until the 1920s and 1930s, reflecting both the lack of industrial development in the Republic and the hardness of the rock strata making progress with early cable and tool machines slow and expensive.

With the exception of Glasgow, the Scottish brewing and distilling industries were largely dependent on groundwater throughout much of the 19th and into the 20th centuries, the distillieries often drawing from spring fed surface waters. In Edinburgh alone, 65 groundwater sources were identified to the Edinburgh Fire Brigade as having potential for fire fighting in 1938 by the brewing industry. Normal abstraction from these sources included the interesting list of breweries and their boreholes shown in Table 1. The borehole yields reflect the modest aquifer properties of the Carboniferous and Devonian strata beneath the city.

One of the more technical discussions of this era was that by Tait (1914, 1926), who reported that the water quality available from the Wardie Shales, although of use as a hydropathic water at six sites in Edinburgh (Spence & Robins 2004), was not suited to the brewers. Better quality water was then being developed in the Upper Old Red Sandstone to satisfy their needs, for example, a new borehole at Craiglockhart apparently had an artesian flow of 65 gallons per minute – a unique occurrence for the Edinburgh area.

The equivalent Irish industries were based almost exclusively on surface water sources, with the most famous stout being almost synonymous with Liffey Water. Only a few minor distilleries appear to have utilized groundwater. On the other hand, within the growing Victorian city of Belfast the world's largest AErated (*sic*) Water Trade developed following lowly beginnings around 1825, based on the discovery that 'the water from the springs in the Cromac district of Belfast was exceptionally pure, free from alkaline matter, and capable of absorbing carbon dioxide to a high degree' (Owen 1921). A firm of chemists introduced a new process, developed in Europe, of artificially charging sweetened and flavoured water with more carbon dioxide gas than it would dissolve under atmospheric pressure.

Following pollution of the Cromac Springs, apparently by migration of 'noxious elements' from the nearby town gasworks, which had been opened about 1868, numerous purveyors of the trade 'moved shop' and drilled deep water boreholes into the Triassic sandstone beneath the city from the 1880s onwards. Thus almost 30 manufacturers were active by 1913 when more than 14 million bottles were exported through the port of Belfast alone. These would have comprised perhaps 23 different beverages, some of them medicinal. The trade ceased after the outbreak of the Great War in 1914 when it was designated 'non-essential' to the war effort and only recovered to a small extent after 1918.

The need to study groundwater at this time had not yet arisen and it was only the coming of industry, not least the mining industry, and of the steam drilling rig which were to focus this need. Paradoxically the driver in Scotland was the need to get rid of groundwater during mining, whereas early studies in Ireland focussed on water supply. Much of the early water supply development took place in the 1870s and 1880s when it was common to use local mining knowledge to dig shafts as far as the water table would permit and to drill on beyond as demand required. The Scottish miners were also employed on such work in Ireland from time to time, displaced temporarily from their main industry for cash inducement.

The geologists and engineers

Towards the end of the Victorian era it was acknowledged that good water supplies were to be had from the sandstones beneath the Lagan Valley, to the SW of Belfast, and beneath Belfast itself and from parts of the Midland Valley of Scotland. In both areas groundwater development began from about 1875 onwards and development only waned in the 1930s when demand for industrial boiler feed water was reduced by the advent of mains electricity. It was also realized that the coal mines being driven into the deeper part of the Glasgow coalfield to the east of the city were increasingly wet, as were the new pits being developed elsewhere in the Midland Valley. It was at this time that the geologists and engineers cooperated to investigate prevailing groundwater conditions.

As with so much of the early geological investigation in Scotland, Messrs Ben Peach and John Horne at the Geological Survey Office in Edinburgh were also active in groundwater investigations. One of the first documented activities were reports prepared by Mr Horne and Mr Hugh Miller at the behest of the Duke of Sutherland in 1887 and by Messrs Peach and Horne in 1900 on the provenance and sustainability of the Spa waters at Strathpeffer (Peach & Horne 1900). The study concluded that the source of the sulphur spring waters was the Spa Beds, fetid limestones within the Devonian bedrock, whereas the chalybeate (iron-rich) spring waters derive from the ancient crystalline schists. The authors concluded that both quality and quantity were justification for further expansion of the Spa facilities, but that the 'Strathpeffer sulphur water is particularly sensitive to exposure to the air, changes in temperature and agitation' all of which 'cause deterioration by loss of sulphur'.

Ben Peach retained an interest in groundwater throughout his long career in Scotland reporting, for example, on the Lochaber tunnel (Peach 1929): 'the dykes have had a considerable effect on the excavation work in the tunnel . . . usually associated with strong springs at their margins'. This role was later taken on by geologists such as T. R. Robertson and A. G. (Archie) MacGregor (Fig. 1), as well as Doug Flett, and in more recent years champions included J. D. Peacock – see the description of the groundwater potential of the Elgin area in the geological sheet memoir (Peacock *et al.* 1968) – and G. H. Mitchell.

The longstanding interest in groundwater by the geologists of the Edinburgh Office of the Geological Survey of Great Britain led, in 1935, to the Secretary of State for Scotland saying (Wilson 1985):

> 'The work of examining and securing the amplification of the information on this subject (underground water) could best be done by the Geological Survey, who have on their staff persons with the necessary knowledge and experience.'

Shortly afterwards, Cumming (1936), who was then the Carnegie Teaching Fellow at St Andrews University, published an eloquent discussion on the geological constraints to groundwater flow and storage in the Midland Valley. This was the first time in Scotland that it was recognized that useful borehole yields were not universally available and that some geological interpretation was necessary to increase the chances of drilling better yielding boreholes. Cumming reveals a detailed knowledge of the water bearing properties of the Palaeozoic sediments and volcanic rocks of the Midland Valley and a comprehensive understanding of the role of till and the important water bearing properties of granular surficial deposits. His discussion highlights the role of pore throat sizes in the relationship between porosity and 'perviousness', the important role of fracture porosity, citing the Chalk of England as an example, and describes the water table as erratic and inconsistent due to the compartmentalized nature of many of the smaller groundwater units typical of the Edinburgh area.

During the Second World War the Department of Health for Scotland became responsible for emergency water supplies. The Survey office assisted

Fig. 1. Archie MacGregor.

with this task by producing a series of wartime pamphlets summarizing the understanding of and presenting the well and borehole inventories for, a series of selected one-inch geological sheets. These pamphlets were later updated during the 1960s as part of the Well Catalogue Series of the Water Supply Papers of the Geological Survey of Great Britain. As such they provided a foundation for the development of modern day hydrogeological investigation. By then, however, groundwater had been utilized for public supply from spring sources, for example at Edinburgh and in the Borders, although there were some small borehole supplies in Fife and elsewhere. Systematic exploitation of the Devonian aquifer commenced during the 1970s in Fife and shortly afterwards also in the Permian aquifer at Dumfries. Industry, however, had always relied heavily on borehole supplies since Victorian times.

The situation in Ireland had been quite different as in the 19th Century little attention was given either by the state-employed geologists or the academics to groundwater. Mather (1998) has pointed out that Edward Hull was the first geologist of the Geological Survey of Great Britain to show an interest in and publish on groundwater in 1865, and later to initiate inclusion of well sections in a survey memoir. It is pertinent to note that Hull transferred to the Geological Survey of Ireland in 1869, became Director later that year, and served with distinction until retirement in 1890. Yet, despite continuing to publish on English groundwater, there is no evidence of Hull having engaged in water supply studies in Ireland. Even his text book (Hull 1891), which included a mineral resource chapter, did not refer to groundwater, suggesting that in those days surface water and spring sources were more than adequate throughout most of Ireland.

It was only when special drift surveys of Dublin, Belfast, Cork, Limerick, Killarney and Londonderry were carried out between 1901 and 1908 that well records began to be collated in some of the accompanying memoirs to the published one-inch maps. For example, the Belfast memoir of 1904 lists some 29 borings to as deep as 199.5 m, while the one for Londonderry of 1908 mentions several. Notably there were no public water supply sources among them. The Dublin memoir of 1903 mentions the town supply being partly from the Grand Canal and partly from numerous wells in gravel, before water was piped in from the Vartry River (Roundwood Reservoir) in Co. Wicklow. Limerick was fed mainly from the Shannon, with a limited supply from deep borings of which no details were available. The information contained in the Cork memoir of 1905 is more interesting. Some 11 borings and a 40 foot long adit are mentioned but most of the results were not satisfactory. The borings included one for the Cold Storage and Pure Ice Company, a name which crops up in Belfast in the same context. However, the most surprising record is that Cork Corporation drew water from an artificial tunnel in gravels in the Lee valley, more than a mile upstream of the city. This had been constructed in 1879, but not used until 1899–1900 when it was adopted 'as an alternative to an expensive scheme of filtering the water drawn directly from the river' (Lamplugh *et al.* 1905). The following account was given:

> 'The tunnel runs parallel to the Lee for about 1,100 feet, 25 feet from its bank on the north side, and at 15 to 20 feet in depth below the surface of the alluvium. At this level the top of the tunnel is just lower than the surface of the water in the river, and somewhat higher than its present bed. The original intention appears to have been to make the tunnel a reservoir for water from the river, partially filtered by passing through the intervening sand and gravel. The present scheme involved the cutting off of the river-water by means of puddled clay where especially necessary, so that the supply might come solely from the gravels at a lower level than the river.'

The tunnel yielded 4 million gallons per day of water of better bacteriological quality and turbidity than that of the river.

After the state geological survey ceased to be effective north of the border following the partitioning of Ireland in 1921, it fell to Professor J. K. Charlesworth and J. J. Hartley, at the Queen's University of Belfast Geological Department, to foster hydrogeological enquiry. This they did most effectively until interest was taken up by the newly created Geological Survey of Northern Ireland in 1947 (Bennett 1979).

John Jerome Hartley (Fig. 2), a Yorkshireman, was a remarkable hydrogeologist who had firstly gained a Bachelor of Engineering degree at Liverpool University in 1906. He then worked for both the Canadian Pacific Railway and the South Eastern & Chatham Railway. He served as a sapper during the Great War, and was employed at Finsbury Technical College from 1920–1926, during which time he also studied geology at Birkbeck and Imperial Colleges obtaining B.Sc. and Masters degrees in 1920 and 1924. He began an eventful and productive career at Queen's University of Belfast in 1927, initially as a Demonstrator in Civil Engineering but moved backwards and forwards between Engineering and Geology until 1939 when he settled in the Geology Department up to his retirement in 1951.

As both a qualified engineer and a geologist, Hartley adopted a rigorous scientific approach to studying groundwater distribution and occurrence and by systematic measurement of piezometry, was able to describe groundwater transport in the sandstone aquifers around Belfast and Newtownards. He also produced an overall assessment of the groundwater resources of Northern Ireland (Hartley 1935).

The excellent official history of the Geological Survey of Ireland, 'North from the Hook' (Herries Davies 1995), does not mention advice on water supply problems until after 1924. They were obviously not popular among the survey staff, whose numbers were very small in the early decades of the new state, as shown by a 1937 letter between colleagues quoted by Herries Davies '. . . and of course Water Reports, like the poor, are always with us. I have half a dozen of these lying in front of me at the

Fig. 2. J. J. Hartley with a student at the fossil locality at Bala Beds, near Pomeroy in Co. Tyrone.

moment.' In 1941, a newly appointed Director of the Survey, D. W. Bishopp, requested approval to appoint officers with various modern geological specialisms, including hydrogeology, but was unsuccessful. Appointed in 1960, C. R. (Bob) Aldwell was the first survey officer to begin to specialize in groundwater supplies.

It is interesting to note that Williams (1971) reported 'The underground water resources of (the Republic of) Ireland are practically unknown,' and that there was then no research in hydrogeology nor any attempt at groundwater management, the main occupation being how to get rid of excess water. The universities in the Republic had not helped and were reputed to maintain a distaste for applied geology, a stance now very much reversed, with applied hydrogeology high on their agendas. In 1967 the Geological Survey of Ireland, hitherto a moribund agency, had been reorganized, and Cyril Williams appointed as its Director. Williams had earlier developed an appreciation of the value of groundwater during his time with the colonial geological surveys, especially in Uganda and he was able to establish a modest groundwater section in the survey during the 1970s which has since gone from strength to strength.

The first public supply borehole was drilled in Northern Ireland in 1923 at Ballycullen, Newtownards (Thompson 1938). Hartley was responsible for the first public-supply borehole drilled at Lisburn in 1934 (Clarke 1935). This borehole still yields 1.5 Ml d^{-1} from the Triassic Sandstone and it was the first of a number of boreholes drilled in the area over the next decade.

However, Thompson also recorded that when Newtownards Council decided, in 1922, to have the first Ballycullen well drilled, the town's main water supply already came from two tunnels, whose yield at certain seasons was insufficient to meet demand. More recent enquiries have established that both tunnels, which converge on Ballycullen service reservoir, are adits beneath Scrabo Hill driven through Sherwood Sandstone and a Tertiary dolerite sill. Little is known about their history, but both were constructed for the purpose of collecting groundwater. For much of the information about the larger tunnel we are indebted to Dr J. Preston (pers.comm.) and his recollections from the 1950s.

The smaller, both in diameter (600 mm) and length (630 m), was the water supply for the local workhouse, later Ards Hospital, and may date back to the Irish Famine in the middle of the 19th century. The Glebe tunnel, so called because it starts in Glebe Quarry on the western side of Scrabo Hill, is 1250 m long and tall enough for a man of average height to walk through, and is unlined except for the floor. The combined discharge from the two tunnels was only around 0.14 Ml d^{-1} in dry weather, more in wet weather, and has flowed to waste since about 1963 when the supply was found to be contaminated by gasworks wastes which had been landfilled on the west side of Scrabo Hill.

In the 1940s the next public-supply initiative came from J. S. Edmondson, Surveyor for Newtownards Borough Council, who began development of the Permian, Triassic and sand/gravel aquifers in the Newtownards and Comber area. By 1964 the total groundwater contribution to public supply in Northern Ireland was 9 Ml d^{-1} (Manning 1972).

Around this time the cause for developing boreholes for public supply was ably championed by survey geologists, notably P. I. Manning and H. E. Wilson, often against acrimonious opposition from local engineers who favoured upland reservoir solutions to meeting increasing water demands. Thus by 1977 the contribution of groundwater to public supply had risen to 51 Ml d^{-1}, while private industry was abstracting at least 25 Ml d^{-1} from boreholes for on-site use (Bennett & Harrison 1980).

The geological survey offices in both Belfast and Edinburgh appointed dedicated hydrogeologists in 1974 and 1977 with the arrival of Peter Bennett and Ian Harrison, respectively. The first hydrogeologist to be appointed to one of the regulators, Tricia Henton (later Chief Executive of the Scottish Environment Protection Agency), joined the Clyde River Purification Board in the early 1970s. In the Republic of Ireland a small groundwater section was established in the Geological Survey of Ireland with the appointment of Eugene Daly in 1971 and Geoff Wright and David Ede in 1975, the latter replaced in 1978 by Donal Daly. Work was also stimulated in the Republic by the return of David Burdon in 1974 to his native Co. Cork after his retirement from a distinguished career overseas, largely serving in UN agencies.

Landmark studies

Given the abundance of surface water, it is unsurprising that little innovative work took place specifically in Scotland or Ireland on which the science of hydrogeology could later draw. There are, however, a number of studies which have a peculiarly northern flavour, and that reported in Symon's Monthly Meteorological Magazine (Anon. 1866) regarding well water temperature is an early example. In addition to providing data for a number of wells in Scotland, the article identifies that the depth to water is critical to the annual range of well water temperature. The reported annual ranges in temperature vary from only 0.3 °F in a well at Cargen near Dumfries (presumably drawing on the Permian Dumfries Basin aquifer) to 7.2 °F in a shallow well at Barry, near Carnoustie.

Reports on groundwater investigation in Scotland are sparse thereafter, but a landmark paper was published as part of a national resources review by Earp & Eden (1961). This report described, for the first time, the likely resource potential of the various rock types available in Scotland and drew on the repository of drilling returns that had been received by the Edinburgh Geological Survey Office since submission had become a statutory obligation in 1946. Earp and Eden recognized the importance of secondary permeability through the occurrence of fissuring in otherwise modestly permeable sandstones, and the strategic importance of the many superficial granular deposits that occur in many valley bottoms.

A notably innovative project was the installation of the first commercial Ranney Well in Britain near Bristol in 1957, and Scotland turned to the Ranney Collector Well system in 1961 for public water-supply to the towns of Banff and Macduff (Anon. 1961). The design yield was 750 000 gallons per day from gravels in the lower Deveron near Fochabers for the Lower Deveron Water Board. Four 8-inch diameter laterals were driven to a total length of 300 feet into the gravel to achieve the design yield. Total cost of the project was £22 000. The main well is 13 feet in diameter and 25 feet deep. Experience from the operation of this well led, in later years, to the development of a well field in the Lower Spey to supply water to the new coastal ring main.

Landmark publications for Northern Ireland, somewhat akin to that of Earp & Eden (1961) for Scotland, are by Manning (1971, 1972). These for the first time attempted a comprehensive review of the groundwater resource potential available in the diverse rock types within Northern Ireland.

References

ANON. 1866. Wells, their temperature and depth. *Symon's Monthly Meteorological Magazine*, 20–21.

ANON. 1875. *Black's Picturesque Guide to Scotland.* Adam & Charles Black, Edinburgh.

ANON. 1915. The ancient wells of Pitkeathly: a modern Scottish enterprise. *The Scots Pictorial*, **11 September**.

ANON. 1961. Ranney collector for the Lower Deveron *Water Board. Water and Water Engineering*, **January**, 12–15.

BARTY-KING, H. 1992. *Water the book, an illustrated history of water supply and waste water in the United Kingdom.* Quiller Press, London.

BENNETT, J. R. P. 1979. Hydrogeological conditions in Northern Ireland. *Papers & Proceedings of Hydrological Meeting, May 1979, Irish National Committee of the International Hydrological Programme*, 71–92.

BENNETT, J. R. P. & HARRISON, I. B. 1980. *Explanatory notes for the International Hydrogeological Map of Europe : Sheet B3 Edinburgh.* UNESCO, Paris.

BRENNERMAN, W. L. & BRENNERMAN, M. G. 1996. *Cross the Circle at the Holy Wells of Ireland.* University Press of Virginia, Charlottesville and London.

CLARKE, R. E. L. 1935. *Recent improvements to Lisburn water supply*. Institute of Civil Engineers, Belfast and District Association, pamphlet 15p.

CUMMING, G. A. 1936. Underground water circulation. Geological factors and complications affecting it in the Midland Valley of Scotland. *Water and Water Engineering*, **June**, 319–322.

EARP, J. R. & EDEN, R. A. 1961. Amounts and distribution of underground water in Scotland. *Natural Resources of Scotland, Transactions of Symposium at the Royal Society of Edinburgh*, 127–134.

FLETCHER, T. P., AUTON, C. A., HIGHTON, A. J., MERRITT, J. W., ROBERSTON, S. & ROLLIN, K. 1996. *Geology of Fortrose and eastern Inverness district*. Memoir for 1: 50 000 Geological Sheet **84W**, HMSO, London.

HALL, I. H. S., BROWNE, M. A. E. & FORSYTH, I. H. 1998. *Geology of the Glasgow district*. Memoir for 1: 50000 Geological Sheet **30E**, The Stationery Office, London.

HARTLEY, J. J. 1935 *The underground-water resources of Northern Ireland*. Institution of Civil Engineers, Belfast and District Association, pamphlet, 31p.

HERRIES DAVIES, G. L. 1995. *North From the Hook*. Geological Survey of Ireland, Dublin.

HULL, E. 1891. *The physical geology and geography of Ireland*, with two coloured maps. 2nd edition. Dublin.

LAMPLUGH, G. W., KILROE, J. R., MCHENRY, A., SEYMOUR, H. J., WRIGHT, W. B. & MUFF, H. B. 1905. *The Geology of the Country around Cork and Cork Harbour*. Memoirs of the Geological Survey of Ireland, Dublin.

MANNING, P. I. 1971. The development of the water resources of Northern Ireland progress towards integration. *Quarterly Journal of Engineering Geology*, **4** (4), 335–352.

MANNING, P. I. 1972. *The development of the groundwater resources of Northern Ireland*. The Institute of Civil Engineers, Northern Ireland Association, pamphlet 31p.

MATHER, J. 1998. From William Smith to William Whitaker: the development of British hydrogeology in the nineteenth century. *In:* BLUNDELL, D. J. & SCOTT A. C. (eds) *Lyell: the Past is the Key to the Present*. Geological Society, London, Special Publications, **143**, 183–196.

OWEN, D. J. 1921. *History of Belfast*. W & J Baird, Belfast.

PEACH, B. N. 1929. The Lochaber Water Power Scheme and its geological aspect. *Transactions of the Institution of Mining Engineers*, **78** (4), 212–225

PEACH, B. N. & HORNE, J. 1900. The Strathpeffer Spa Waters: an appraisal at the end of the 19th century. Report Geological Survey Office, Edinburgh. Report edited 1989 (Robins, N. S. & McMillan, A. A. (eds)), and reprinted. *The Edinburgh Geologist*, **22**, 2–14.

PEACOCK, J. D., BERRIDGE, N. G., HARRIS, A. L. & MAY, F. 1968. *The Geology of the Elgin District, Explanation of Geological Sheet 95*. Memoirs of the Geological Survey Scotland, HMSO, Edinburgh.

PLESTER, H. R. F. & BINNIE, C. J. A. 1995. The evolution of water resource development in Northern Ireland. *Journal Chartered Institute of Water and Environment Management*, **9**, 272–280.

ROBINS, N. S. 1990. *Hydrogeology of Scotland*. HMSO, London.

ROBINS, N. S. & MISSTEAR, B. D. R. 2000. Groundwater in the Celtic regions. *In:* ROBINS, N. S. & MISSTEAR, B. D. R. (eds) *Groundwater in the Celtic regions: studies in hard rock and Quaternary hydrogeology*. Geological Society, London, Special Publications, **182**, 5–17.

ROSS, A. 1967. *Pagan Celtic Britain, Studies in Iconography and Tradition*. Routledge and Kegan Paul, London.

RUTTY, J. 1757. *An essay towards a natural, experimental and mechanical history of the mineral waters of Ireland*.

SDD. 1973. *A Measure of Plenty*. Report Scottish Development Department, Edinburgh.

SPENCE, I. & ROBINS, N. S. 2004. The Scottish hydropathic establishments and their use of groundwater. *In:* MATHER, J. D. (ed.) *200 Years of British Hydrogeology*. Geological Society, London, Special Publications, **225**, 000–000.

TAIT, D. 1914. On bores for water and medicinal wells in the Wardie Shales near Edinbugh. *Transactions of the Edinbugh Geological Society*, **10** (3), 316–325.

TAIT, D. 1926. On a bore for brewing water near Craiglocklart. *Proceedings of the Edinburgh Geological Society*, **11** (3), 271–274.

THOMPSON, J. H. 1938. *Deep water supplies (Newtownards and Lisburn)*. Institute of Civil Engineers, Northern Ireland Association, pamphlet, 8p.

WILLIAMS, P. W. 1971. The management of groundwater resources in the Republic of Ireland. *Quarterly Journal of Engineering Geology*, **4** (4), 334–335.

WILSON, H. E. 1985. *Down to Earth, One Hundred and Fifty Years of the British Geological Survey*. Scottish Academic Press, Edinburgh and London.

Bath thermal waters: 400 years in the history of geochemistry and hydrogeology

W. MIKE EDMUNDS

Oxford Centre for Water Research, School of Geography and Environment, Oxford University, Mansfield Road, Oxford OX1 3TB, UK
(e-mail: wme@btopenworld.com)

Abstract: The importance of the Bath thermal springs in the development of science over some 400 years is explored. Several references to the springs from Saxon times and the Middle Ages give qualitative information of interest for the present day. The springs have drawn some of the most famous philosophers and scientists to test new theories and develop hypotheses on the nature of matter and to develop early ideas in the chemical and geological sciences. Theories on the hydrological cycle and on hydrogeology were tested and the springs have a long history as a site for discoveries in chemistry and natural radioactivity. Interest in the springs continues to the present day. Whilst our knowledge of the origin of the water, the heat, the detailed chemistry and other properties has been resolved, some questions still remain for the attention of future generations and for the application of advancing scientific methods.

The origins of the Bath thermal springs have been a focal point of curiosity for philosophers, writers, physicians and laymen for over two millennia. In the past 400 years they have also been a focal point for advances in medicine and science. In particular the springs have played a major part in the development of the geological sciences and as a field site for testing developments in analytical chemistry. Many of the famous names and pioneers in scientific thinking, including early geologists have published their ideas on the springs.

In this paper the roles played by the springs in the emergence of scientific thinking are described, more especially in the past 400 years but also, where records exist, including the perceptions of the nature of the springs back to Roman times. The emphasis of this paper is on the earlier discoveries, since the properties of the springs have been well documented in the past two decades (Kellaway 1991).

Early stirrings

The springs were known to early settlers and archaeological evidence dates from around 7000 yr BP (Kellaway 1991). The legend of their discovery in Celtic times is attributed to Bladud who, seeing his swine cured of leprosy in the mud surrounding the springs, assigned healing properties to the waters. Bath, like many thermal springs, was celebrated from early times as a religious site dedicated to a female deity, in this case the Celtic goddess Sul. The Romans venerated the site Aquae Sulis and built a temple adjacent to the springs, dedicated to the goddess Sulis Minerva, as noted by the writer Solinus in the 3rd century AD:

> 'In Britain are hot springs adorned with sumptuous splendour for the use of mortals. Minerva is the patron goddess of these . . .'

Roman engineers were the first to harness the springs for their heat, enclosing the spring chamber and creating hot baths – the first example in Britain of the exploitation of geothermal energy.

After the Romano-British era and Saxon invasion in the 6th century the site was abandoned and the Roman bath collapsed (Guidott 1676). A fragment of an early Saxon poem contained in the library of Exeter cathedral reputedly records the situation of the springs at that time, as translated in Earle (1864):

> 'There stood arcades of stone
> The stream hotly issued
> With eddies widening
> Up to the wall encircling all
> The bright bosomed pool
> There the baths were
> Hot with inward heat.
> Nature's bounty that!'

At this time the city was called Akmanchester, or the 'city of sickly people'. A nunnery was founded there in 676AD and shortly afterwards a church. Records of the springs are found from Norman times when John de Villula, a physician from Tours, bought the town of Bath from King John and converted the Abbey Church into a cathedral. The Abbey of Bath like several European cathedrals and abbeys (cf Chartres), was therefore founded on a site of a pre-existing famous spring with healing properties dedicated to a female deity.

From: MATHER, J. D. (ed.) 2004. *200 Years of British Hydrogeology*. Geological Society, London, Special Publications, **225**. 193–199. 0305-8719/04/$15

During the monastic period the springs continued to be used for the care and treatment of the sick. Alexander Necham in the 13th century wrote some verse about the springs:

> 'Igne suo succensa quibus data Balnea fervent
> Aenea subter aquas vasa latere putant
> Errorem figmenta solent inducere passim
> Sed quid? Sulphureum novimus esse locum.

> People think that bronze vessels lie hidden underneath the waters
> and so heated from below by its own fire the baths boil.
> People are wont to drag in errors and imaginative solutions
> all over the place, but why? We know it's a sulphurous spot.'

This fragment of verse suggests that in the early Middle Ages ferric oxides were predominant around the springs, yet in addition, H_2S was also in evidence as well as gas bubbles. This implies the same characteristics as the present day, where the oxidizing and reducing characteristics of the water are one of the main features, suggesting mixing from time to time of the anaerobic thermal water with traces of water from near to surface.

Shortly before the dissolution of the monasteries, John Leland records that at Bath there were three springs of which the Hot Bath was the hottest. Around 1450 a poem by John Boccae (quoted by Guidott 1676) on the Baths gives some insight on the nature of the springs at the end of the Middle Ages:

> 'There runs from underground . . .
> Springs sweet, salt, cold and hot even now as then (the time of Bladud)
> From rock, salt peter, alom, gravel, fen
> From sulphur, iron, lead, gold, silver, brass and tin
> Each fountain takes the force of vein it coucheth in'

This last line is reminiscent of the observations some 14 centuries earlier by Pliny the Elder (Pliny circa 50 AD):

> 'Waters take on the character of the rocks through which they pass'

This may be regarded as the definition of hydrogeochemistry and provides an insight of that period into the water-rock interaction taking place.

Until the 17th century, therefore, Bath Springs were considered of religious significance with healing properties. An air of mystery surrounded their origins, although the experience of miners meeting water in underground workings appears to have influenced early thinking (see also Agricola 1556). Hot springs at this time before the scientific age, however, where accompanied by sulphureous vapours, recalled hell-fire rather than hydrogeology.

The age of enlightenment

The first book or record which specifically considers the origins of Bath thermal waters was published in 1631 by Edward Jorden. Jorden studied medicine and chemistry at Padua at the time of Galileo, before returning to England and taking up residence in Bath. He was interested in the origins of the hot springs and the application of the principles of hydrostatics. He was the first to attribute the temperature of the thermal waters at Bath to the passage of groundwater through hot rocks in the interior of the earth and to explain their reappearance as being due to hydrostatic pressure (Kellaway 1991).

Jorden was the first person to propose the modern hydrological cycle. Until the early 17th century the conventional wisdom was that of the reversed hydrological cycle (Tuan 1968). These ideas mentioned in Ecclesiastes 1(7) and persisting through Greek and Roman periods proposed that seawater was the source of groundwater, losing its salts by underground purification.

> 'All the rivers run into the sea, yet the sea is not full; unto the place from whence the rivers come, thither they return again'

Jorden (1631), like Galileo had done before him, questions the conventional doctrine although 'holding sacred the canon of the scriptures, that all rivers are from the sea':

> 'I persuade myself that there is a natural reason for the elevating of these waters unto the heads of these fountains and rivers, though it hath not yet been discovered . . . My conceit therefore is this, that . . . water being put in at one end will rise up in the other pipe as high as the level of the water – so may it be in the bowels of the earth . . . although this water enters the earth very deepe, yet the level must answer to the superficies of the sea . . .'

Jorden then goes on to suggest that:

> 'the actual heat of Bath and the mineral qualities which they have are derived unto them by fermenting heat. . .'

The distinguished scientist Joseph Glanvill also made interesting early observations on the origin of the springs. Glanvill, who worked with Sir Isaac Newton, was one of the first Fellows of the Royal Society and was also Rector of the Abbey Church in Bath. This was a characteristic of many scientists of this age who combined rigorous scientific study with holy orders. He was one of the first to promote the experimental approach in scientific investigation (Smith 1972). He writes (Glanvill 1669):

> 'The baths are of great antiquity . . . which if so, would give occasion to enquire how consistent with it that hypothesis (concerning the heat of these waters) would be which makes it to be the fermentation of minerals; and whether it be likely that the minerals through which these waters should pass, should be in that state of imperfection so many hundred years, and that the whole disposed matter in those places should not be perfectly concreted in so great a tract of time.'

And:

> 'You doubtless also know the other conjecture, which supposeth the cause of this heat to be . . . that two streams having run through and imbibed certain sorts of different minerals, meet at last, after they have been deeply impregnated, and mingle their liquors, from which commixture arises a great fermentation that causes heat [like as we see it is in vitriol and tartar, which though separately they are not hot, yet when mingled beget an intense heat and ebullition between them]. This seems to me a probable cause of the lastingness of the heat of these waters.'

Thus, Glanvill proposed a hypothesis for the heat of the waters, being from an exothermic chemical reaction caused by the mixing of fluids. He had probably observed the effects of the oxidation of sulphide minerals and seen the process of travertine cementation around springs – this is also therefore one of the earliest accounts of water-rock interaction. A contemporary scientist/philosopher T. Guidott (1638–1733) also discounted a volcanic (subterranean fire) source of heat for the springs (although at a later date others would reconsider this) and favoured a local source of the heat. Guidott (1676) was also one of the first to consider the composition of the waters. He distinguished saline and non-saline components:

> 'the former I shall endeavour to evince to be nitre, common salt and vitriol; the latter to be partly unctuous, as bitumen and sulphur, partly gritty as freestone, partly earthy as marl and ochre . . .'

Although questions on the composition of the thermal springs began to be asked during the 17th and early 18th centuries, the science of modern chemistry had not yet taken off. Robert Boyle, who also considered the composition of the springs, concluded in resignation (Boyle 1684):

> 'To discover the nature of mineral waters is a far more difficult task than those who have not tried it would imagine!'

During the first half of the 18th century very little new scientific thinking on the origins or composition of the springs appeared. This was the era when the mineral waters at Bath and elsewhere became fashionable as spas and most of the writings between 1700–1750 refer to the medicinal properties of the waters.

The late 18th and 19th centuries

The first modern work on the chemistry of the springs appeared around 1750 with the work of Charleton (1754). This was at the time of discovery of the structure of matter and the recognition of the chemical elements; this was also the time of the birth of modern chemistry with the work of Antoine Lavoisier in France (Smith 1972). The work of Charleton just predates this revolution but his work probably contains the best example of the experimental approach of the time. In his opinion:

> 'the ingredients which impregnate these springs are iron, earth, common sea salt, a neutral salt, elementary fire and a sulphureous matter.'

He and others of the time disagree on whether or not sulphur (meaning H_2S) is present. Dr Charles Lucas (quoted in Charleton 1754) asserts that:

> 'there is a subtle acid which flies off in vapour and sometimes sensibly strikes the nose (although elsewhere in his account this is contradicted): water newly drawn up.. has no sensible smell, no more has its vapour.'

This again previews modern work where H_2S is sometimes but not always detectable. Charleton (1754) believed the origin of the thermal water to be:

> 'common spring water running through a bed of pyrites, with its remarkable heat ascribed to elementary fire.. by passage through the earth the particles of these minerals set at liberty this imprisoned matter.'

The experimental approach of the day was to make an artifical Bath water from the proposed constituents, although Charleton points out:

> 'It would be absurd to imagine that chemistry can afford us the power of making so perfect a composition as that which nature produces. It is the province of this art to imitate her operations and explain them.'

A major contribution to the composition and origin of the Baths was made at the turn of the 19th century by George Gibbes. He published several papers mainly on the medicinal properties and chemistry of the waters. He concluded (Gibbes 1800) that

> 'uniformity as to temperature, quality and quantity (of the springs) . . . shows that they are caused by continued and regular operation of extensive agents in the bowels of the earth . . . the cause of the heat is uniform and very deeply situated.'

Gibbes was the first to discover silica in the springs (a grain and a half in one pound of water) and saw this as a clue to their origin. He can therefore be regarded as the unwitting initiator of silica geothermometry:

> 'A still more convincing argument to prove that their heat is intense at a certain depth is their containing so large a proportion of siliceous earth. I believe that siliceous earth is capable of being diffused in water when that water has been subjected to an intense heat . . . and pressure.'

Then in the first half of the 19th century a rush of chemical discovery began with several new analyses appearing. The Bath springs became the focus for testing new analytical techniques and the first reporting of elemental occurrences in nature:

Gibbes (1800)	Si
Scudamore (1820)	Mg
Murray (1829); Cuff (1830)	I
Daubeney (1830)	Br, P
Merck and Galloway (1848)	Mn
Roscoe (1864)	Li, Sr

The first full and definitive analyses of the Bath spring waters, however, were published by Herepath in 1837 and again in 1844 (Herepath 1837; 1844), also by Noad (1844).

The foundation of modern hydrogeology was undoubtedly built by William Smith during the years 1799–1813 during his residence at Bath. Bath and its environs provided the field setting for his work (Kellaway 1991). Visitors to Bath, which had become a fashionable Georgian resort, demanded guidebooks and this resulted in one of the earliest geological maps, covering a five-mile radius of the city (Winchester 2001). Smith's hydrogeological insight came from his work as a mine surveyor and canal engineer. The construction of the 150 m shaft for coal at Batheaston (only 3 km from Bath) was a turning point. This not only provided the first detailed stratigraphy of the Bath region. It penetrated the unconformity at the base of the Triassic and proved 78 m of steeply dipping Carboniferous strata. More importantly it also encountered strong artesian flows of warm water (≅ 17 °C) in the Lower Lias and more strongly at the unconformity. A decrease in the flow of the Cross Bath was recorded at this time.

William Smith reflected accurately on the origins of the springs and especially their strong artesian pressure:

> 'the Bath Hot Springs have a deep source and rise as through a natural borehole into the Blue marl to the surface'

This concept of a spring pipe for the springs conduit was only confirmed during the scientific studies and archaeological work starting in 1978 (Kellaway 1991).

Sir Charles Lyell (Lyell 1864, 1865), in his presidential address to the British Association, held at Bath in 1864 expressed his belief that the Bath Springs:

> 'as in the case of many mineral waters . . . marked the site of some great volcanic convulsion and fracture of the earth's crust, at some not very remote period geologically speaking.'

These views, it was noted (Freeman 1888), were at odds with the majority of contemporary writers, although such an origin was still being considered in the early 20th century (Rastall 1926).

Joseph Priestley lived for a time near Bath and took an interest in the springs. As part of his experiments and observations on different kinds of air (Priestley 1775), he took a pint of water from the springs and expelled its gases by boiling for 4 hours. These experiments showed no abnormal properties, except that the air was:

> 'muted not to the water but to some calcareous matter in the water'

The first comprehensive and quantitative studies on the dissolved gases may be attributed to Daubeny (1834). He measured the mean gas discharge at 223 cu ft in 24 hours. He also recognized the importance of nitrogen and found oxygen in the gases evolved from the springs.

The temperature of the springs

Although the first recorded use of the thermometer was in 1611, the earliest mention of the temperature of the Bath springs by Glanvill (1669) was qualitative:

> 'the hottest spring will not harden an egge'

Accurate records of temperature of the Bath springs date from the late 18th century and these show a consistent range of values up to the present day, with a range of 44.4 to 48.8 °C. The modern average for the King's spring is 44.8 °C and the evidence suggests either that the water temperature may have been higher in the past or that some overestimation was a feature of some earlier measurements.

Discovery of radioactivity in the early 20th century

The discovery of radium in 1893 by the Curies in Paris led to immediate research in Britain into natural radioactivity. In 1898 Lord Rayleigh discovered helium in the springs. In 1903, Strutt, the son of Rayleigh, working with Sir James Dewar, discovered radium first in the deposits from the springs and then in the waters themselves – a first for Britain (Strutt 1904). Sir William Ramsay confirmed the radium content of the springs (Ramsay 1912*a*) and also 'radium emanation', the last undiscovered noble gas, radon, in the springs, as well as neon (Ramsay 1912*b*). Radon was quantified as 2 millionths of a milligram per litre by Munro (1928) with a lower estimate of 2.10^{-12} gml^{-1} by Jacobi (1948). This gas with a half life of only 3.8 days was proof of uranium minerals in the rocks from which the springs emerged.

Thus Bath became once more the focus for new scientific discoveries and testing. This new excitement led to a temporary new lease of life for the spas (halted by the first world war) which had started to enter a period of decline at the turn of the century (McNulty 1991).

The modern era

At the time of the two world wars Bath Spa was in decline and few scientific advances were made in the first half of the twentieth century. Several papers were published referring to the gas composition (Munro 1928) and especially the high radioactivity was again confirmed (Jacobi 1949). A detailed analysis was also published by Judd Lewis (Bath City Council 1938), which contained in particular the most comprehensive listing of trace elements to that date. This was matched by a definitive chemical analysis by Riley (1961), who used the spring as a testing ground for new methods of colorimetric analysis – much in the tradition of the early 19th century chemists. This period coincided however with a general period of decline of British Spas which lasted into the late 20th century; only eight active spas remained in the United Kingdom by the 1960s (Edmunds *et al.* 1969).

The exploration for geothermal resources in the United Kingdom, which commenced in 1977 (Downing & Gray 1986), provided the next real stimulus for new scientific work based on modern hydrogeological concepts, access to deep drilling in the area at Ashton Park (Kellaway 1967) and especially to advances in hydrochemistry. The Bath – Bristol area proved to be an area of below average heat flow with a mean value of 50 $mW\ m^{-2}$ in comparison with other regions such as the Pennines where the values reached 80 or in Cornwall 130 $mW\ m^{-2}$. Thus the emergence of the largest and warmest thermal spring in such a location as Bath seemed anomalous.

An intensive decade of study ensued during which the piezometric records from the Bath-Bristol region were reviewed, detailed chemical and isotopic analysis of the waters and gases took place. This work was interrupted in 1978 by the unfortunate death of a young girl from an amoebic infection (*Naegleria fowleri*) while swimming in the thermal waters (Kilvington *et al.* 1991). If anything, this event redoubled the scientific effort (Kellaway 1991). The spring chamber was drained and in addition an inclined borehole was drilled to intercept the thermal water for supply, some 60 m below the spring chamber, thus avoiding any surface contamination. The results of this decade of hydrogeological work were published in Burgess *et al.* (1980), Andrews *et al.* (1982), Edmunds & Miles (1991), Andrews (1991) and the new knowledge gained at this time may be summarized:

Temperature of the springs: a significant decline in the temperature, from 45.3 measured on the spring to 44.4 °C measured on the inclined borehole was recorded.

Temperature at depth: the silica geothermometer (recording the likely maximum temperature reached at depth) indicated a range between 64–96 °C.

Depth of circulation of the water: using a value for the local geothermal gradient of 20 $°C\ km^{-1}$ a maximum circulation depth of 2.7–4.3 km was derived.

Origin(s) of the water: the thermal water clearly originated as meteoric water as indicated by its stable isotope $\delta^{18}O$ and $\delta^{2}H$ compositions. Noble gas ratios also showed that the temperature at which the water was recharged must have been similar to that of the present day.

Host aquifers: the $\delta^{13}C$ of the dissolved bicarbonate and the overall chemistry of the springs indicated strongly that the water had evolved mainly in the Carboniferous Limestone aquifer. The strontium and sulphur isotope values also indicated that the Limestone or secondary vein minerals in the Limestone were the source of the sulphate.

Age of the water: no conclusive 'age' for the water could be reached from the various indicators (e.g. ^{14}C) since they had undergone extensive modification with the rock. However the $\delta^{18}O$ and noble gas evidence placed an upper limit on the age of post glacial (Holocene) and therefore less than approx 12000 yr BP.

Local situation (mixing and discharge): oscillations in the chemistry (especially the redox conditions) and the occasional detection of tritium in the springs suggested a small component of shallow water (probably from Mesozoic aquifers). The thermal water also created a 'discharge mound' of thermal water in the area around the spring.

Overall composition: the total mineralization of the thermal water (Kings Spring borehole) amounted to 2278 mg l^{-1}. The 1977–1986 studies however produced a comprehensive analysis of the major, minor and trace elements in the springs (following in the 200 year-old tradition).

Most recently during the millennium year (2000) advances have been made in defining further the age of the thermal water using noble gas isotopes, the results of which are shortly to be reported. It is now likely that a minimum age of 1000–1200 years can be assigned to the water, although the upper age remains difficult to resolve. Yet more detailed chemical analyses were possible in this recent study where, using the latest analytical tools (ICP-MS and other isotopic analysis) some 70 chemical elements, isotopic and gaseous components were positively identified in the springs.

Conclusions

The importance of the Bath thermal springs in the development of science over some 400 years has been explored in this paper. The springs have drawn several of the earlier philosophers and their scientific descendents to test new theories and develop hypotheses on the nature of matter, to develop early geological ideas and especially to play a part in the development of chemical sciences and the applications of natural radioactivity. As well as documenting the scientific era several references to the springs from Saxon times and the Middle Ages also give qualitative information of interest for the present day.

Many of the famous names in the history of the natural sciences, including Jorden, Boyle and Priestley visited the spas and recognized their potential to test theories and conduct experiments. William Smith founded concepts of the geological sciences as well as hydrology in the Bath area. The interest in the springs continues to the present day and they still form a testing ground for concepts in hydrogeology and geochemistry. Whilst our knowledge of the origin of the water, the heat, the detailed chemistry and the host rocks through which the springs circulate have been resolved, some questions still remain, notably the exact age of the water, for the attention of future scientific enquiry and for the application of advancing scientific methods

References

AGRICOLA, G. 1556. *De Re Metallica.* (Translation by HOOVER, H. C. & HOOVER, L. H. 1912). The Mining Magazine, London.

ANDREWS, J. N. 1991. Radioactivity and dissolved gases in the thermal waters of Bath. *In:* KELLAWAY, G. A. (ed.) 1991. *The Hot Springs of Bath.* Bath City Council, 157–170.

ANDREWS, J. N., BURGESS, W. G., EDMUNDS, W. M., KAY, R. L. F. & LEE, D. J. 1982. The thermal springs of Bath. *Nature*, **298**, 339–343.

BATH CITY COUNCIL. 1938. *Bath Official Handbook.*

BATTEN, E. C. 1877. On the causes of heat in the Bath Waters. *Proceedings of the. Somerset Archaeological and Natural History Society*, **22**, 52–60.

BOYLE, R. 1684. *Short memoirs for the natural experimental history of mineral waters; with directions as to the usual methods of trying them, and other useful remarks and curious experiments.* London.

BURGESS,W. G., EDMUNDS, W. M., ANDREWS, J. N., KAY, R. L. F. & LEE, D. J. 1980. *The hydrogeology and hydrochemistry of the thermal water in the Bath-Bristol basin. Investigations of the geothermal potential of the UK.* Unpublished Report, Institute of Geological Sciences, 57 p.

CANTON, J. 1762. Observations on the heat of the Bath and Bristol waters. Philosophical Transactions, **12**, 420.

CHARLETON, R. 1754. *A treatise on Bath Waters, wherein are discovered several principles of which they are composed, the causes of their heat, and the manner of their production*, Bath 8vo.

CUFF, C. 1830. On the presence of iodine, potash and magnesia in the Bath waters. *Philosophical Magazine*, Series 2, **7**(2), 9–10.

DAUBENY, C. 1830. Memoir on the occurrence of iodine and bromine in certain mineral waters of southern Britain. *Philosophical Transactions*, **121**, 223–238.

DAUBENY, C. 1834. On the quantity and quality of the gases disengaged from the thermal spring which supplies the Kings Bath in the city of Bath. *Philosophical Transactions*, **124**, 1–14.

DOWNING, R. A. & GRAY, D. A. 1986. *Geothermal energy – the potential in the United Kingdom.* Her Majesty's Stationary Office, London, 187p.

EARLE, J. 1864. *A guide to the knowledge of Bath, ancient and modern.* London.

EDMUNDS, W. M., TAYLOR, B. J. & DOWNING, R. A. 1969. Mineral and thermal water of the United Kingdom. *In: Mineral and thermal waters of the world. A-Europe.* International Geological Congress Prague, Vol.18, 139–158.

EDMUNDS, W. M. & MILES, D. L. 1991. The geochemistry of the Bath thermal waters. *In*: KELLWAY, G A. (ed.) *The Hot Springs of Bath.* Bath City Council, 143–156.

FREEMAN, H. W. 1888. *The thermal baths of Bath: their history, literature, medical and surgical uses and effects, together with the Aix massage and other vapour treatment.* Hamilton Adams, London.

GIBBES, G. S. 1800. A chemical examination of the Bath waters. *Journal of Natural Philosophy, Chemistry and the Arts*, **3**, 359–363.

GLANVILL, J. 1669. Observations concerning the Bath springs. *Philosophical Transactions*, **4**, 977–982.

GUIDOTT, T. 1676. *Discourse of Bath and the hot waters there, and on the St Vincent Rock, near Bristol*. London.

HEREPATH, W. 1837. *Analysis of the King's Bath*, Bath. Report of the British Association for 1836, 70–73.

HEREPATH, W. 1844. Analyses of the Bath waters and of the Bristol Hotwells waters. *Philosophical Magazine*, **24**, 371.

JACOBI, R. B. 1949. The determination of radon and radium in water. *Journal of the Chemical Society*, S314–S318.

JORDEN, E. 1631. *Discourse of natural Baths and mineral waters . . . etc*. London.

KELLAWAY, G. A. 1967. The Geological Survey Ashton Park Borehole and its bearing on the geology of the Bristol district. *Bulletin of the Geological Survey of Great Britain*, **27**, 49–153.

KELLAWAY, G. A. (ed.) 1991. *The Hot Springs of Bath*. Bath City Council. 288p.

KILVINGTON, S., MANN, P. G. & WARHURST, D. C. 1991. Pathogenic Naegleria amoebae in the waters of bath: a fatality and its consequences. *In:* KELLAWAY, G. A. (ed.) *The Hot Springs of Bath*. Bath City Council, 89–96.

LYELL, C. 1864. *Presidential address to the British Association*. Report of the British Association 1864, 60–65.

LYELL, C. 1865. On the mineral waters of Bath and other hot springs and their geological effects. *American Journal of Science*, **39** (Ser.2), 13–24.

MCNULTY, M. 1991. The radium waters of Bath. *In:* KELLAWAY, G. A. (ed.) *The Hot Springs of Bath*. Bath City Council, 65–70.

MERCK, G. & GALLOWAY, R. 1848. Analysis of the water of the thermal spring of Bath (Kings Bath). *Memoirs of the Chemical Society*, **3**, 262–272.

MUNRO, J. M. H. 1928. *Report to the Hot Mineral Baths Committee on 'Radioactivity in the Bath thermal waters'*. Fyson, Bath.

MURRAY, J. 1829. On the discovery of iodine and bromine in the mineral waters of England. *Philosophical Magazine series 2*, **6**, 283–284.

NOAD, M. 1844. Analysis of the Bath Water. *Pharmaceutical Journal*, **3**, 526–532.

PRIESTLEY, J. 1775. *Experiments and observations on different kinds of air*. Vol. 2, London.

RAMSEY, W. 1912*a*. The mineral waters of Bath. *Chemical News*, **105**, 133–135.

RAMSEY, W. 1912*b*. The formation of neon as a product of radioactive change. *Journal of the Chemical Society*, **101**, 1367–1370.

RASTALL, R. H. 1926. Note on the geology of the Bath springs. *Geological Magazine*, **63**, 98–104.

RILEY, J. P. 1961. Composition of mineral water from the hot spring at Bath. *Journal of Applied Chemistry*, **11**, 190–192.

ROSCOE, H. E. 1864. Note on the existence of lithium, strontium and copper in the Bath waters. *Chemical News*, **10**, 158.

SCUDAMORE, C. 1820. *A chemical and medical report of the properties of the mineral waters of Buxton, Matlock, Tunbridge Wells, Harrogate, Bath, Cheltenham, Leamington, Malvern, and the Isle of Wight, London*.

SMITH, A. G. R. 1972. *Science and society in the sixteenth and seventeenth centuries*. Thames and Hudson, London, 216p.

STRUTT, R. J. 1904. Radioactivity of certain minerals and mineral waters. *Proceedings of the Royal Society of London*, **73**, 191–197.

TUAN, Y. F. 1968. *The hydrologic cycle and the wisdom of God*. University of Toronto, Department of Geography.

WINCHESTER, S. 2001. *The map that changed the world*. Viking, London.

Chalybeate springs at Tunbridge Wells: site of a 17th–century new town

J. G. C. M. FULLER

History of Geology Group, The Geological Society, Burlington House, Piccadilly, London W1V 0JU, UK

Abstract: Among Wealden towns Tunbridge Wells is comparatively new. Before the Civil Wars of the 1640s there was no village here, nor any name on a map. Chance finding of chalybeate springs a few miles south of Tunbridge (now *Tonbridge*) attracted attention at Court, and even gynaecological interest. Curiously, this provides explanations both for the supposed virtues of the waters and the founding of a summer resort. By repute, the springs were discovered in 1606, though this story was already 160 years old before it first appeared in print. Verifiable facts indicate that Thomas Neale, FRS, (1641–1699) was the main agent in organizing the nascent resort's amenities, beginning in 1676 with plans to construct a chapel or assembly room. The springs themselves issue from Lower Cretaceous Wealden beds, a few feet above the Wadhurst Clay, in a shallow valley formed by the headwaters of the River Grom. Siderite (iron carbonate or chalybite) abounds in these formations.

'No Spa ever had a more slender claim than this insignificant chalybeate to a high-sounding fame' (Granville 1841)

Tunbridge Wells came into existence in 1676 as a building-site on a nameless patch of land belonging to the manor of South Frith, about 35 miles SE of London and on the county boundary between Kent and Sussex. It acquired its attraction from the presence of chalybeate springs that were alleged to possess special powers. Such springs of iron-rich water abounded everywhere among Wealden rocks in this district, so one asks what were the special powers to be found here? Answers lie partly in a reappraisal of Lord Dudley North's role as the supposed 'discoverer' of these springs and partly in ancient folklore associated with the *Sauvenière* and other springs at Spa, in Belgium.

Careful reading of surviving literature suggests that the date 1606 usually given for the discovery of chalybeate water at Tunbridge Wells is likely to be nine or ten years too early, and that the man credited with its discovery, Lord Dudley North (1581–1666), was neither the flaccid consumptive nor self-indulgent profligate as he has been variously portrayed, but a talented essayist, poet and scholar. Certainly he possessed a somewhat melancholic nature, which he admitted, though his physical ailments undoubtedly arose from excessive use of poisonous defensatives against plague.

A singular taste

According to Thomas Benge Burr's *History of Tunbridge Wells* (1766), the springs from which this town acquired its name were found in 1606 by Dudley, third Baron North (1581–1666). He was said by this earliest-published and most-favoured historical account to have been on his way to London after staying with Lord Bergavenny at Eridge, about two miles from the springs. Young Lord North, said Benge Burr, had gone to Eridge seeking recuperation from the exertions of Court life and to recover his 'capacity for enjoying those pleasures which hitherto he had too freely indulged.'

This date of discovery, 1606, first came to public notice in Benge Burr's book, which was written about 160 years after the event was supposed to have taken place, and for which the only authorities quoted were 'reports of the most aged people'. Lord North's own works, which were substantial, say nothing of this event, though they do contain clues from which Benge Burr could have constructed his story. These clues reside in North's first volume of poetry and prose, which he entitled *A Forest of Varieties* (1645). He wrote much of this book in a state of depression and mental anxiety, which permeated his style and subject matter, and now conveys to us the picture of a man damaged by clinical melancholia, and suffering much from the effects of poisonous medicines. All this he conceded himself (1645, p. 118 & 173):

> 'Being something by nature . . . disposed to a strong melancholy, it was impossible to extricate myself . . . about my age of 25.'

> 'The foundation of my disease laid in the excesse of Treacle [theriac], infinite have been the effects, and my sufferings, which have flowne from that and other concurring circumstances.'

North's age of 25 years at the time he was suffering a setback in health caused by excessive dosage of

From: MATHER, J. D. (ed.) 2004. *200 Years of British Hydrogeology*. Geological Society, London, Special Publications, **225**. 201–212. 0305-8719/04/$15

theriac or Venice Treacle, when added to his year of birth, 1581, gives the year 1606, which is the year when Benge Burr supposed him to have been at Eridge seeking recuperation and 'discovering' the springs. But inconveniently for that story, Lord North's only remarks in *A Forest of Varieties* or his later works, concern medicinal uses for waters that were already identified by the names *Epsom* and *Tunbridge*, not to to any 'discovery' of springs near Eridge 30 or 40 years earlier.

Much of North's writing in *A Forest of Varieties* was composed as a series of essays, afterwards collected and printed in imitation of a journal for the year 1637. The two passages reproduced below were dated November 6th of that year (North 1645, p. 134):

> 'Ordinary physick is but a palliation, nay often an aggravation of the disease. The powder called Kellowayes powder [a defensative against plague], with Gods blessing, is to bee prized; for it goes to the root, it workes at length . . . if such powder breed inconvenience, Epsom* waters, though but a draught in a day at morning, wonderfully allay and rectifie.'

North's asterisk at Epsom in the above quotation directs the reader to a note in the margin saying:

> 'The use of Tunbridge and Epsom waters, for health and cure, I first made known to London, and the Kings people; the Spaw [near Liege, Belgium] is a chargeable & inconvenient journey to sick bodies, besides the money it caries out of the Kingdome, & inconvenience to Religion.'

Under the date of December 11, 1637, North added this to his journal:

> 'I also lately acquainted a Doctor, who received it for good, what a happy alteration I once found by use of a large draught of Epsom waters, mixed with milk, taken fasting, and strongly walking upon it . . . Tunbridge waters are of known good effect, used with good advice.'

Those three passages comprise Lord North's entire discussion of Tunbridge waters. There is no hint that the springs found two miles from Eridge and those subsequently named Tunbridge which is eight miles from Eridge, were in fact one and the same thing.

Another entry (March 15, 1637) indicates that North was again seriously ill with depression after the death in 1612 of the Prince of Wales, at whose court he had been serving:

> 'The loss of the brave Prince Henry [1594–1612], on whom I had laid my grounds, with much sickness soon after, casting me down.'

Prince Henry's death, Lord North's loss of preferment, retirement from Court, and descent into deep melancholic gloom fit well with his presence afterwards at Eridge with Lord Bergevenny. Aside from the story given by Benge Burr's *History*, there are other contemporary links connecting the Bergavennys and the Norths at Eridge. For example, William Byrd's famous collection of keyboard music, *My Ladye Nevells Booke* (Andrews 1926), the original of which was found at Eridge in 1668, contained a note pasted in it, saying:

> 'This Booke was presented to Queene Elizabeth by my Lord Edward Abergevenny called the Deafe, the queene ordered one Sr or Mr North one of her servants to keepe it, who left it to his son who gave it to Mr. Haughton Attorny of Cliffords Inn & he last somer 1668 gave it to me; this mr. North as I remember Mr. Haughton saide, was uncle to the last Ld North.
>
> H. Bergevenny.'

A further fragment, offering a more likely date when Lord North first noticed the chalybeate nature of the spring water beside the road from Eridge, has been preserved in the *Camden Miscellanies* (Almack 1855). It was a handwritten piece, quite possibly by Robert Weller, found among papers formerly belonging to Thomas Weller of Tonbridge (1632–1722). It said:

> 'One day when he was at Eridge Place, my Lord Abergavenny being there, they expected the Lord North (I suppose from London) to come thither, but not coming so soon as was expected, Lord North made this excuse (when my grandfather [Richard Weller, the Steward] was by and heard him), that he had stopt at a place about two miles distant out of curiosity, he having observed a spring by the road, the water of which had made such a tincture in the channel that he suspected it to be a mineral water, and had stayed to taste it, and found by the taste that it was very much like the Spaa water [Spaw or Spa, near Liege], and therefore advised him to send his mason or bricklayer with a few bricks to open the spring, which he believed might be taken notice of; and this very likely is the first notice that was taken of Tonbridge Wells. My uncle [Henry Weller] told me that his father [Thomas Weller of Frant] said he was then about 13 or 14 years old, which would make the time to be in 1615 or 1616.'

The point here is not so much the likely true date of 'discovery', about 1615 or 1616, but Dudley North's appreciation that this mineral-water tasted like the water at Spa. It was a taste that he recognized from his own experience and because Spa waters were said to have particular medicinal powers, such as the *Sauvenière's* power of encouraging fertility, this newly-found spring near Eridge might possess similar medicinal qualities. As he also perceived (1645, p. 134), an additional and potentially huge advantage was that this Kentish spring's benefits might be enjoyed without the fatigue and inconvenience of going to a foreign country.

Lord North's mention of the distinctive taste found in Spa water indicates that he was speaking from personal knowledge, for in 1602 he had left England for the Low Countries, hoping to join one of the military campaigns against the Spanish armies of occupation. While in Europe his travelling companion was infected by plague, and died, which clearly explains North's hypochondriac overdosing on plague antidotes, both theriac and Kellawaye's powder, and his aggravated ill-health consequent on it.

More than that, North's first-hand reports to those around the King in London, informing them that a Spa-quality chalybeate was to be found in Kent, within a day's journey of the capital, was so appreciated by the royal doctors that in 1629 they advised Queen Henrietta Maria, after the sad miscarriage of her first pregnancy, to recuperate her health at this newly-found hygeian marvel in Kent (Oman 1936). This she did, camping in great style for some six weeks on Bishops Down, close by the springs.

The Queene's Wells

Only in 1632, after Queen Henrietta's visit, were the medicinal benefits of this native chalybeate formally published in print. Like the Spa waters in Belgium, Tunbridge waters were credited with special efficacy in matters gynaecological, which is exactly why the Queen's doctors could direct her for treatment to an obscure and uninhabited place where her only accomodation would be a tent.

Chalybeate waters, like the *Sauvenière* spring near Spa, had been recognized since medieval times as possessing powers of heightening fertility in marriage.

> 'Le renom des eaux de Spa, dans les cas de stèrilité, s'est incarné dans la lègende de saint Remacle, attachée à la fontaine de la Sauvenière' (Scheuer 1881).

The legend was not obviously gender-specific or exclusive, which could be understood as allowing that it applied to members of both the sexes. This interpretation was eagerly adopted for the new chalybeate near Tunbridge, according to a small book published in 1632 by a 'Dr of Physicke' practising in Kent. His name was Lodowick Rowzee, and he wrote on both the geological and medicinal features of mineral springs. He assessed directly and objectively the powers inherent in Tunbridge water, arguing that its chief virtue lay in clearing obstructions in the passages of the body, for he had himself visited Spa, and considered that the water there 'hath a great affinitie with that of Tunbridge (Rowzee 1632). He was clearly cognisant of the *Sauvenière* water's reputation – more appropriate to Venus than Hygeia – and continued that line of thought in his description of the Tunbridge chalybeate (p. 46 & 48):

> 'The water helpeth also the running of the reines [kidneys], whether it be *Gonorrhea simplex* or *Veneria*, and the distemper of the *Parastate* arising from thence . . . The greater part of those that drinke of it, are purged by stoole, and some by vomit, as well as by urine . . . In the behalf of women . . . there is nothing better against barrennesse, and to make them fruitfull.'

Given publicity of that kind, it is not surprising that court doctors, with an eye to the safe delivery of royal heirs, dispatched in succession three Stuart Queens, each of whom had experienced setbacks in childbearing – Henrietta Maria, Catherine of Braganza and Queen Anne – to test the restorative and particular gynaecological efficacies of Tunbridge water. Only the first of them could be said to have had any success.

Rowzee's name for the springs, the Queenes Welles, which he introduced to honour Henrietta Maria's visit in 1629, made little impression in this area of Parliamentarian persuasion, and failed to survive the Civil Wars, though during the military dictatorship following, some activity at the springs did continue: A letter supposedly written at Tunbridge by Lady Henrietta Boyle in 1659 speaks of her sisters 'drinkeing the Waters,' while she took to letter-writing (Hickson 1891).

Dr Rowzee's hydrogeology, 1632

Though diminutive, Lodowick Rowzee's book of 79 pages embraced ideas on the nature of mineral waters not seen again until 1674, when Pierre Perrault's *De l'Origine des Fontaines* appeared. Rowzee dismissed Aristotle's classic teaching 'that the generation of water proceedeth from ayre condensated into the same in the bowels of the earth,' commenting simply 'that a man might justly thinke, that the whole element of ayre, which in its owne nature is but very thinne, should scarcely suffice.' In short, there is not enough air for this hypothesis.

'All fountaines and rivers come from the Sea,' he said, 'and are translocated through the veines and porosities of the earth, where in their passage they leave their saltnesse.' The error here, which Rowzee was obliged to accept as having the authority of Scripture, lay not in the teachings of Aristotle, nor in his own mind, but in a too literal interpretation of a sentence found in the first chapter of Ecclesiastes (Eccl.1, v.7), which he quoted:

Fig. 1. An irrestible combination of rural dalliance and water-cure, 1678. Frontispiece to Thomas Rawlin's play *Tunbridge Wells; or a Day's Courtship.*

'All the Rivers runne into the Sea, yet the Sea is not full, unto the place from whence the Rivers come, thither they returne againe.'

But on the special hydrogeology of 'medicinable springs' Rowzee felt free to exercise his commonsense and said what to him was obvious:

'Minerall waters, by their manifold runnings and windings under the ground, are as it were impregnated with diverse vertues and faculties of the severall mineralls, through which they run, and draw with them, either the faculties, or substance of the same, and sometimes both.'

A summertime resort

By 1660, with restored constitutional government, things changed. Visiting the Wells became a royal and popular exercise. The direct route from London lay through Kent by way of Sevenoaks to the Medway river-crossing at Tunbridge, then diverting six miles southward to reach the springs. On arrival, the visitor making the effort of travelling from London would find invigorating and unconventional rewards awaiting him. 'These waters,' said the doctors, 'Are endowed with an admirable and powerful faculty, in rendering those who drink them fruitful and prolifick.' Virtues of this calibre, touted among the rich and gullible, promised even the halt and the lame a glimpse of cupid's gymnasium:

This water encourages: 'a sweet balsamick, spirituous, and sanguineous temperament, which naturally incites and inspires men and women to amorous emotions and titillations' (Madan 1687).

Before the end of the century the springs in summer had become a venue of regular resort, a fashionable place in the country, where nymphs and shepherds might offer members of the company a new and irresistible combination of rural dalliance and water-cure (Fig. 1). Going to the Wells was a novel adventure – the company offered unaccustomed personal freedom and social acceptance and besides, there were diverting entertainments of every sort. One could claim that it was all in pursuit of wholesome salubrity, for the Wells could cure whatever you pleased:

'You ladies who in loose body'd gown,
Forsakes the sneaking City,

And in whole Shoals come trundling down,
Foul, foolish, fair, or wity.
Some for the Scurvy, some for the Gout,
And some for Loves disease,
Know that these Wells drive all ill out,
And cure what e're you please' (Rawlins 1678).

For the half-century before 1676 nearly everything at the Wells was temporary and removable, nothing was permanent, other than a cottage or two belonging to local graziers. During the season, from the end of May until October, huts and removable cabins on Bishops Down were offered at rackrent rates by local people as lodgings for visiting gentry taking the waters. John Lewkenor who was certainly at the Wells before 1684 mentioned with dismay that a three-month's summer rental for a flimsy hut on Bishops Down was charged the same as a year's rent for a lodging in London (Lewkenor 1693). Out of season, from October to the beginning of June, the place was merely an out-of-use camp-site.

The first permanent building at the Wells

Growing awareness of entrepreneurial possibilities on the green walks by the springs, where people of substance tended to congregate in the season, attracted the notice of Thomas Neale, F.R.S., a speculator and overseas trader. He was probably at the Wells in 1674 or 1675, and seems to have appreciated that there were future business opportunities here. He acted at once to obtain control of the land where the springs issued, which was then in the hands of a financially distressed Lady Purbeck, from whom he purchased the manor, persuading her (or helping to persuade her) also to give a plot of land in the adjacent manor of South Frith, close to the springs, on which to construct the first permanent building. This was to be both chapel and place of assembly, functioning also as nearby shelter from inclement weather.

For some time before 1676, when the first recorded solicitations for raising money began, visitors to the springs had realized that a system would have to be devised to organize the place and regulate the increasingly large numbers of pleasure-seekers, water-bibbers, quackery-men, scholars, mountebanks, clerics, politicians and impostors who gathered at the wells; and because there were no permanent buildings in this virtually uninhabited place, the system devised would have to consider shelter, food-supplies, a market, care of amenities, amusements, even a school for children of the poorer sort serving the company.

The scheme began by levying a fee graduated to the subscriber's social standing. Most gave a guinea (a newly-introduced gold piece), which acted as a unit share in the 'company that resorted to the wells,' for all members had an equal right to take part in electing an organising committee or Vestry to run the place. By an unusual turn of good fortune, the names of the original subscribers are known, as well as the amounts they gave (Fuller 2000). The subscribed money was collected on the Walks, properly accounted and spent on various projects, of which the chapel or assembly-room was the most costly. Other expenditures were made to furnish a site-office or 'Vestry-house' and were later directed toward paving the Yard and Market Place by the wells, planting shade-trees along the Walks, erecting a Dial, attending to the convenience places draining into the brook near the Walks and paying a schoolmaster. Figures 2 & 3 illustrate a plan of the wells and market area and a view across the manorial boundary to the chapel.

No architect was named for the chapel and the principal works were made under local contract, using traditional brickwork. The builder seems to have modelled the first structure more or less on a double-cube, measuring inside about 21 feet high, 26 feet wide and 43 feet long (Chesterton 2002). These dimensions, together with numbers calculated from the catalogues of Subscribers, suggest that its capacity would have been about 150 people. Progress was far enough advanced by the season of 1678 for the incomplete structure to be opened for use, the people seating themselves on wooden planks. The building was finished in 1684, though very soon afterward enlargement became necessary, culminating in a doubling or twinning of the original structure.

Who were the people willing to finance a permanent building at the Wells and to furnish it, knowing that it would be used only for a few months of the year? The first catalogue of Subscribers, opened in 1676, lists some 1686 names. A second catalogue opened in 1688, lists a further 937 names. Of the total, just over one third were titled people, the rest were merchants, financiers, traders by sea, lawyers, scholars, virtuosi, politicians and office holders, though all were clearly persons of substance and standing (Fuller 2000). For example, listed among the first subscribers were at least 239 Members the House of Commons, of whom a random count of 100 indicated that their average age was thirty-five, and that only one in four was over 50. The wells seem to have possessed something more attractive than mere salubrity; perhaps, according to John Lewkenor (1693, p. 27), the presence of Tunbridge Fairies.

Also subscribing to this building and all identified with certainty, were 67 Fellows of the Royal Society, a number greater than one fifth of the Society's strength in Britain at the time; and in the same period, between 1676 and 1696, six of the nine

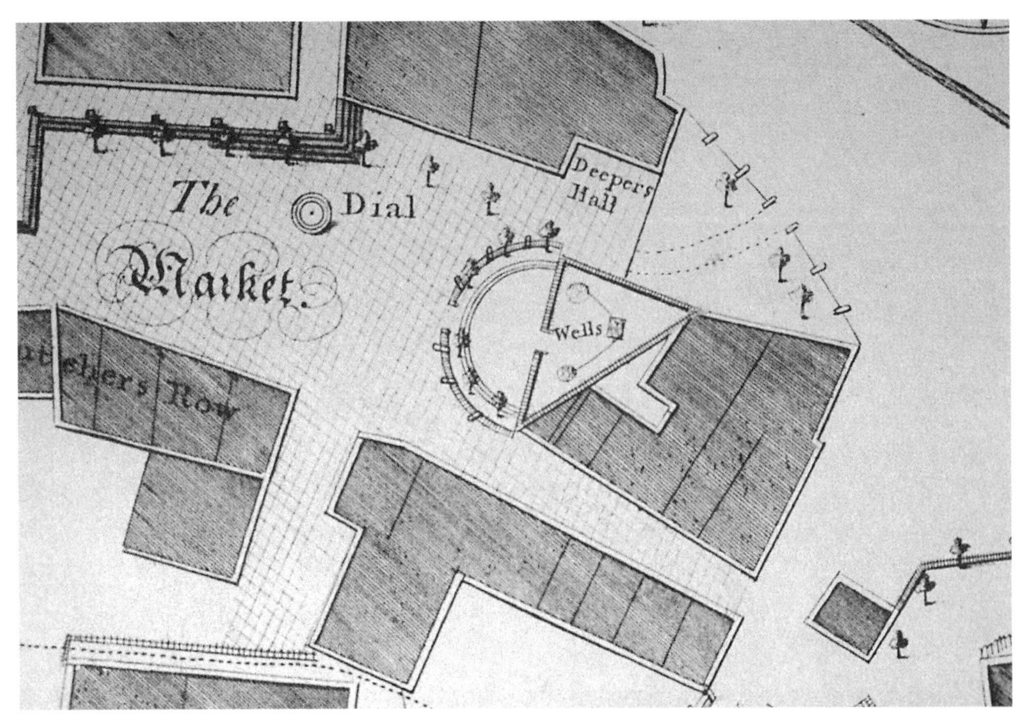

Fig. 2. Plan of the Market Place and triangular enclosure of the wells. Detail from John Bowra's *Plan of Tunbridge Walks*, 1738. For a measure of scale: the distance from the well-enclosure opening to the centre of the dial is 50 feet.

Fig. 3. The two principal springs are within the triangular enclosure. Bishop's Down, in the manor of Rusthall, is to the left. The chapel, to the right, occupies a southern corner of the manor of South Frith Detail of an engraving by Jan Kip from John Harris's *The History of Kent*, 1719.

Fig. 4. A view of the Upper Walk in 1748, looking northeast. Versions of this engraving, based on a drawing by Thomas Loggan, identify persons of note. At the extreme right (No.21) is William Whiston (1667–1752), a man of encyclopaedic contention, chaplain to the Bishop of Norwich, Lucasian Professor after Newton and author of *A New Theory of the Earth* (1696).

presidents then holding office visited the place. There was clearly some geological interest, for in 1684, after a winter of unremitting frost, it was remarked at a meeting of the Society, that 'the Tunbridge wells did not spring in their usual place since the late great frost; but that it was hoped they would be found by digging' (Birch 1756–1757). Actually they had not failed: only balked by the deep severity of frost, starving the aquifer.

Real estate, 1684–1700

The first permanent building, the chapel, was completed in 1684. A few yards away on the Walks, flimsy structures that served as apothecary's shops, curiosity booths, trinket stalls, taverns and other places of entertainment, had begun to appear two years earlier, arising from a leasing agreement made between Thomas Neale, F.R.S., then Lord of the Manor, and local tenant graziers. It was afterwards extended to a sublessee. Yet none of these early buildings could be called a lodging place, for the lease negotiated between Neale and the tenants did not allow for any building of that kind on the Walks, or on Bishops Down, where tenants plied their rack-rent schemes of temporary accommodations for visitors. Even on lands outside Neale's leasing agreement, investment of capital in lodging-houses could not be expected to produce a return on investment out of season, which was three-quarters of the year, for as yet no-one was a settled resident.

After a calamitous fire during 1687 among the haphazard constructions on the Walks, new buildings were aligned at the sides, leaving pedestrian areas free from obstruction, and forming a precinct (Figs 4 & 5), which remains much the same today.

Even before the chapel was finished, huge efforts were made to increase its capacity, first by demolishing one end wall and adding some 15 feet to its length, then further by adding unplanned galleries and eventually by doubling the whole structure. Such alterations offer proof of steeply rising numbers in the annual company and hence of an improvement in the economics of lodging-house construction. These new circumstances seem to have induced Lady Purbeck and her steward, Thomas Weller of Tunbridge (1632–1722), to begin dividing the nearby area of Mount Sion (within the manor of South Frith) into building plots, thus launching a construction boom between 1684 and 1700. Writing at the height of this activity, John Lewkenor (1693, p. 46) was impressed:

> 'Each way you come, some new built Houses stand . . .
> They look like Suburbs of some pleasant Town.

Fig. 5. A view of the Upper Walk in 1786, looking SW. Engraving published in J.Sprange's *The Tunbridge Wells Guide* (1786).

Through these Preliminaries then you go
To th' Upper Walk, divided with a row
Of shady Trees, from that which is below.'

During the next eight years, this first flush of lodging-house building matured. An anonymous doggerel-monger (1701) remarked with amazement on the newly expanded venue of resort and entertainment produced by Lady Purbeck's real-estate venture on Mount Sion:

'Commodious Buildings . . .
And pleasant Lodgings well design'd,
For all Degrees of Human Kind;
For ev'ry Fancy some Device,
Good Music, Dancing, and good Wine:
Fine Beauties to delight your Eyes,
Some vertuous, and some otherwise.'

A proper town

Another century passed before this centre of seasonal indiscretion could begin to be considered as a place fit for permanent residence. Through the 18th century it continued as a stylish summer resort, especially during the middle years between 1735 and 1761, when Beau Nash was Master of Ceremonies and good order was maintained. Records tell of sturdy beggars and noisy drunks confined to a cage and the Walks lit by oil-lamps after dark. Every year saw a new and transient population crowding here in pursuit of pleasure and then departing as winter approached, leaving the site closed and uninhabited. Of course, famous names came and went, grandees long forgotten and worthies still remembered. For example, among those known to geologists, was William Whiston (Fig. 4), author of *A New Theory of the Earth*, Chaplain to the Bishop of Norwich and Lucasian Professor at Cambridge. He preached to any one who would listen, on the Walks and in the chapel, that the end was upon them, the final consummation of all things, including his Bishop and the gaming tables.

Gradually, toward the end of the century, new fashions led by King George III at Weymouth, and the Prince Regent at Brighton, drew pleasure-seekers to other resorts. Enterprising doctors praised the virtues of sea-water and forgot the salubrities of chalybeate.

Writing in 1810, at a time when Tunbridge Wells was clearly becoming more like a place of year-round residence and less like a faded resort, Paul Amsinck felt able to present it as 'a place of considerable wealth, consequence, and respectability,' though not quite a town (Amsinck 1810).

'Little more than half a century ago the season was limited to the short period between Midsummer and Michaelmas. After that time the trades-people themselves migrated, the taverns were closed, the chapel service was discontinued; and the place remained a desert till the following spring. As late as twenty years back, it was very unusual for a family to continue beyond the month of November. Now the case is very

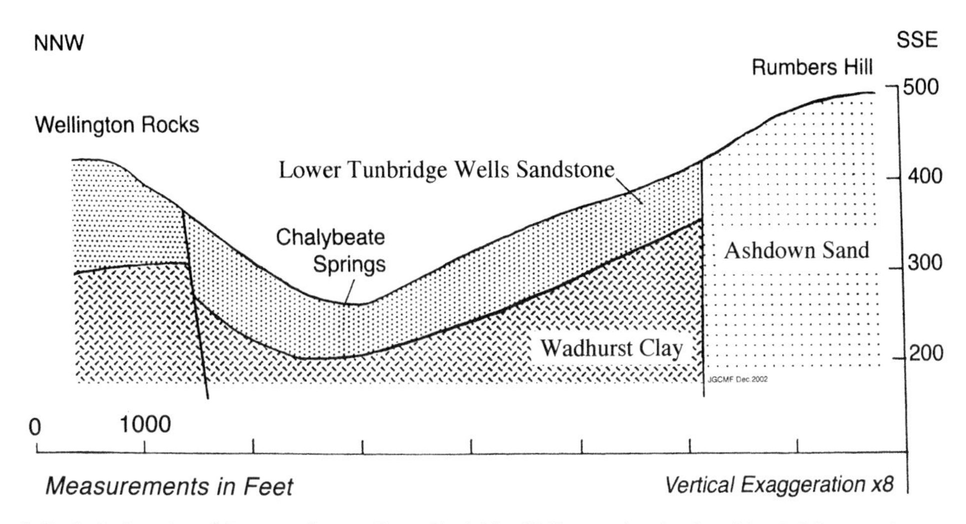

Fig. 6. Geological section of the upper Grom valley at Tunbridge Wells, crossing the site of the chalybeate springs. Rainwater percolating into the Lower Tunbridge Wells sandstone picks up soluble minerals, particularly iron carbonate, during its underground passage downdip to the valley floor, where its downward escape is blocked by impermeable Wadhurst Clay.

different. Many houses, formerly let as lodgings, are now permanently tenanted.'

The resident population of Tunbridge Wells when Amsinck was writing was about 1600, though curiously the census of 1801 was unable record this settlement as a separate place with its own population, chiefly because the census was enumerated by parishes, and Tunbridge Wells was not a parish. That status was achieved only in 1833. 19 years later the railway arrived, hugely accelerating growth.

Geological matters

Amsinck's description of Tunbridge Wells in 1810 attained a fresh standard of narrative and illustration appropriate to the new century. He intended to offer chemical analyses of the waters, but these did not appear for several years, by which time there was a firmer concept of the geology controlling the springs, for instance by Charles Scudamore (1816):

> 'Tunbridge Wells, so celebrated for its Chalybeate Spring, is situated in a valley surrounded on all sides by gently sloping hills. It stands upon a clayey sandstone dipping both from the north and south into the valley, and alternating with beds of clay iron ore. In consequence of this situation most of the wells in its neighbourhood are more or less impregnated with iron, and have a chalybeate taste.'

Figure 6 is a present-day geological section across the upper Grom valley at Tunbridge Wells, passing through the site of the main chalybeate springs and oriented NNW to SSE in line with local dip, which in the Lower Tunbridge Wells Sandstone varies from two to five degrees. Rumbers Hill at the southern end of the section stands at an elevation of 489 feet (149.0 m) above sea-level and Wellington Rocks at the northern end rise to 410 feet (142.7 m). The main chalybeate spring (TQ 5812 3880) emerges at an elevation of 262 feet (79.9 m) in the floor of the downfaulted valley carrying the headwaters of the River Grom. Three rock formations are represented on the section, namely Ashdown Sand (the oldest), overlain in stratigraphical sequence by Wadhurst Clay and Lower Tunbridge Wells Sandstone, from which the chalybeate waters flow. A complete section of this formation had a measured thickness of 131 feet (39.9 m) in a borehole at Culverden Brewery, about a mile north of the spring (Bristow & Bazley 1972), and a partly eroded thickness of 91 feet (27.7 m) at the West railway station (578 384), 600 yards (548.6 m) to the southwest of the spring (Harvey & Matthews 1964). Lithologically, the formation consists of extremely fine-grained sand interbedded with yellow mottled clay and seams of clay-ironstone (siderite or chalybite). It weathers to a mean and heavy soil, brick-hard when dry, deeply impoverished of lime, and known for its 'invariable presence of ferrous compounds' (Hall & Russell 1911).

Rainwater falling on this acidulous terrain gravitates into local declivities, seeps downdip and gathers in a synclinal hollow floored by Wadhurst Clay. The chalybeate springs rise towards the bottom of the hollow, where the water table in the Sandstone comes to the surface, flowing normally at a rate

Table 1. *Rainfall recorded by the Environment Agency at Tunbridge Wells (TQ 5850 3950) millimeters per month during the years 1987 to 1998*

	Jan.	Feb.	Mar.	Apr.	May.	Jun.	Jul.	Aug.	Sep.	Oct.	Nov.	Dec.
1987									39	201	78	28
1988	210	59	84	42	47	12	96	32	49	77	33	17
1989	32	62	80	102	4	47	24	32	31	73	48	133
1990	144	155	6	55	7	58	26	27	36	106	56	65
1991	111	45	40	64	21	121	92	9	48	37	73	25
1992	15	30	64	83	34	13	65	96	60	76	129	74
1993	94	8	24	93	62	48	63	35	109	134	55	147
1994	138	49	57	83	106	57	44	90	100	115	54	114
1995	173	111	68	14	21	16	34	2	137	19	31	102
1996	62	59	46	15	47	16	48	79	33	59	119	14
1997	20	97	16	9	50	154	29	67	10	94	*	103
1998	123	6	63									

* November 1997 record lost

varying from 1pint per minute (34 litres/hour) to 3 gallons per minute (818 litres/hour). The rate responds directly to rainfall variation in the catchment area, and a severe summer drought, as seen in 1995, can cause a complete cessation of flow. Recorded rainfall at Tunbridge Wells during that year illustrates this relationship (Table 1).

In 1995, from the end of March to the beginning of September, precipitation was less than one third of the ten-year average over the years 1988 to 1997. According to figures issued by the Meteorological Office, June to August 1995 was the driest summer for 230 years. At Tunbridge Wells, during those three months of June, July and August only 52 mm of rain was recorded, and in the next month the chalybeate spring ceased to flow. Yet there had been abundant rainfall during the previous winter, about 50% more than the ten-year average amount, which may suggest that the spring's flow rate was responding to a shallow-depth, short-cycle recharge system in which heavy winter rainfall did not ensure continuous summer flow. Further evidence of the spring's near-surface aquifer can be adduced from a second failure, in 1996, when engineering works about 150 feet (45.7m) east of the spring, along the now-buried channel of the Grom brook, disrupted the spring's flow and temporarily polluted the water.

As might be expected of a spring so responsive to rainfall fluctuation, its degree of mineral saturation also varies. Scudamore (1816) noted for example that the spring water in August, 1815, flowing at a rate of 1.1 pints per minute (8.2 litres/hour) contained 2.29 grains per gallon 'oxide of iron', i.e. 25 mg l^{-1} Fe; whereas in March, 1816, when the spring was flowing at a rate of 15 pints per minute (112 litres/hour), the 'oxide of iron' had dropped to 1.63 grains per gallon, i.e. 18 mg l^{-1} Fe.

Analyses in 1857 by John Thomson, made when the spring was flowing at a rate of 27 pints a minute (202 litres/hour) revealed an iron content of 27 mg l^{-1}. An analysis published by Bristow & Bazley (1972) showed an iron content of 37 mg l^{-1}, and more recently (Spencer 1998) another analysis gave 'Iron, as Fe' at 6.7 mg l^{-1}. Neither of these more recent analyses quoted the spring's rate of flow, though it is obvious that there is a very considerable range of analytical result, even for the one singularly critical element in chalybeate water. Nevertheless, as an historical fact one remembers that it was precisely the taste of chalybeate that caused young Lord North to make the original connexion between spring waters found beside the road near Eridge and the medicinal waters at Spa, and to appreciate also the possibility of importing to this place the vital power residing in the legend of the *Sauvenière* (see Appendix).

One may justly say in conclusion that two full centuries passed between the first notice of the chalybeate springs and the emergence of a conventionally settled town. Every year of that time, during an out-of-season period lasting seven or eight months, the place relapsed into a collection of empty buildings, shut up and deserted until the next brief summer commenced for assembling the company, observing the ceremonies, rolling dice, taking the waters, shuffling cards, viewing the tables and exchanging glances of a less than decorous kind:

> 'Don't mention Marriage at Tunbridge, 'tis as much laugh'd at as Honesty in the City: This is a Place of general Address, all Pleasure and Liberty . . . wholly dedicated to Freedom, no Distinction, either of Quality or Estate. This Tunbridge is the Joy of my Life; such Treating, Dancing, Serenading, Raffling and Scandal, I cou'd die here' (Baker 1703).

Appendix

A comparison of the principal constituents found in chalybeate waters at Spa, Belgium, and at Tunbridge Wells, is shown below. At Spa, the oldest named spring is the *Sauvenière*, on the road to Malmédy. It has a higher than average content of iron. Spa waters rise from indurated slaty rocks of Cambro-Silurian age in the Ardennes massif. The Tunbridge Wells water analysis was calculated from figures given by Thomson (1858) These waters issue from Cretaceous Wealden beds, and flow generally at rates varying from one to three gallons per minute, varying somewhat according to seasonal rainfall. They maintain a normal ambient temperature of 50°F (10 °C).

During the tourist season, modern-day 'dippers' continue an old tradition of offering a glass of the chalybeate spring water to passing visitors (Fig. 7).

Comparative Partial Chalybeate Water Analyses, Tunbridge Wells and Spa, mg l^{-1}

	Spa, Marie-Henriette (Comité culturel, no date)	Spa, Sauvenière (Scheuer 1881)	Tunbridge Wells (Thomson 1858)
Iron	21	24	27
Calcium	12	31	11.4
Sodium	9.6	21	18
Magnesium	8.0	11	4.6
Solids	c. 200	c. 215	c. 170
Carbon dioxide	c. 2,300	c. 2,400	c. 166

Fig. 7. At the springs in 1881. The ornate portico over the wells, put up in 1847, has remained virtually unaltered to the present-day. Engraving from *Pelton's Illustrated Guide to Tunbridge Wells.*

References

ALMACK, R. 1855. Papers relating to proceedings in the County of Kent, A.D.1642–A.D.1646. *The Camden Miscellany*, **v.3**, vii, 68 p.

AMSINCK, P. 1810. *Tunbridge Wells and its neighbourhood, illustrated by a series of etchings, and historical descriptions.* William Miller & Edward Lloyd, London, 183p.

ANDREWS, H. [ed.] 1926. *My ladye Nevells booke of virginal music.* London, J.Curwen, 245p. [Reprinted 1969, Dover Publications, New York].

ANON. 1701. *A rod for Tunbridge beaus, a burlesque poem.* London, 30p.

BAKER, T. 1703. *Tunbridge Walks; or the Yeoman of Kent; a Comedy.* Bernard Lintott, London, 70p.

BIRCH, T. 1756–1757. *The history of the Royal Society of London.* [Reprinted 1968, 4 vols, Johnson, USA. (**v.4**, 283)].

BRISTOW, C. R. & Bazley, R. A. 1972. *Geology of the country around Royal Tunbridge Wells.* HMSO London, 161p.

BURR, T. B. 1766. *The History of Tunbridge Wells.* London, 324p. [Thomas Benge Burr was a bookseller and native of the Tunbridge Wells area. This, his only published work, was sold at Tunbridge Wells and London.]

CHESTERTON, R. K. 2002. pers. comm. 2002. By Robert Chesterton, RIBA, consulting architect to the Parish Church of King Charles the Martyr, Tunbridge Wells.

FULLER, J. G. C. M. 2000. *The church of King Charles the Martyr, Tunbridge Wells, a new history.* Friends of the Parish Church, Tunbridge Wells, 104p.

GRANVILLE, A. B. 1841. *Spas of England.* Colburn, London. [A. B.Granville, M.D., F.R.S., was an observant author who wrote from first-hand experience.]

HALL, A. D. & RUSSELL, E. J. 1911. *A report on the agriculture and soils of Kent, Surrey, and Sussex.* HMSO, London, 206p.

HARVEY, B. I. & MATTHEWS, A. M. *et al.*, 1964. *Records of wells in the area of New Series One-Inch (geological) Tunbridge Wells (303) and Tenterden (304) sheets.* HMSO London [Revised version of Geological Survey Wartime Pamphlet No. 10, Pt 7, 1940].

HICKSON, M. 1891. Tunbridge Wells in 1659. *English Historical Review*, **v.6**, 161–162. [This is mainly a disappointing extract from Lord Macaulay's *History.*]

LEWKENOR, J. 1693. *Metellus, his dialogues, the first part containing a relation of a journey to Tunbridge- Wells; also a description of the wells and place.* Tho. Warren, London, 129p.

MADAN, P. 1687. *A Philosophical and Medicinal Essay of the Waters of Tunbridge.* For the Author, London, 26p.

NORTH, Lord Dudley. 1645. *A Forest of Varieties.* Richard Cotes, London, 243 p.

OMAN, C. 1936. *Henrietta Maria.* Hodder & Stoughton, London, 366p. [Documentation of the date 1629 for the Queen's convalescence at the springs is amply referenced, p. 67].

RAWLINS, T. 1678. *Tunbridge-Wells, or a day's courtship.* Henry Rogers, London, 42 p.

ROWZEE, L. 1632. *The Queenes welles, that is, a treatise of the nature and vertues of Tunbridge water, etc.* J. Dawson, London, 79p.

SCHEUER, V. 1881. *Traité des eaux de Spa.* H. Manceux, Bruxelles, 328p.

SCUDAMORE, C. 1816. *An analysis of the mineral waters of Tunbridge Wells with some account of its medicinal properties.* Longman, London, 58p.

SPENCER, J. 1998. pers. comm. 1998. Tourist Information Centre, Royal Tunbridge Wells to Prof. J.D. Mather, *Chemical and Bacteriological Report*, dated 10 April, 1986.

THOMSON, J. 1858. Analysis of the Tunbridge Wells water. *Journal of the Chemical Society*, **10**, 223–229.

The Scottish hydropathic establishments and their use of groundwater

IAIN SPENCE[1] & NICK ROBINS[2]

[1]*17 Kinpurnie Gardens, NEWTYLE, Blairgowrie, PH12 8UY, UK*
(e-mail: Iain.Spence@btinternet.com)
[2]*British Geological Survey, Maclean Building, Crowmarsh, Gifford, Wallingford, UK*
(e-mail: N.Robins@bgs.ac.uk)

Abstract: Scotland today has a plentiful supply of drinking water derived from upland gathering grounds, but groundwater supplied all of its major towns and cities in the past. Pollution of many of the old groundwater sources, as well as the atmosphere, by the massive industrial boom of the mid nineteenth century in central Scptland led to the development of hydropathic establishments to dispense the 'water cure'. Most drew on fresh and pure groundwater sources, and the establishments continued to be a popular source of medical care until the early part of the twentieth century. The groundwater sources were characterized by weak to moderate mineralization unlike the strongly mineralised waters typical of the more traditional spa resorts. A small part of the hydropathic legacy remains in use until this day

Throughout their early growth, most of the major Scottish conurbations depended for their potable water supply on the local groundwater resource. As major industrial complexes developed within and around the cities and the population increased at a rapid rate, the groundwater became polluted and the occurrence rate of ill health increased. Water-borne diseases such as cholera, were endemic in the population which serviced the burgeoning demands of industry. The development of the coal industry, particularly in the Glasgow area, lead to a marked decrease in the quality of the groundwater. Away from the coal fields, for example in Perth and Dundee, which were entirely dependent on wells and springs for their water supply, the levels of contamination grew markedly due mainly to the uncontrolled disposal of sewage and other wastes directly into the ground. Even worse, the abattoir in Dundee was constructed adjacent to the principal water supply well with the well separated from a shallow offal and waste pit only by a low permeable wall.

So bad was the situation that by the middle of the 19th century the Scottish city councils were forced to invest large sums of money to tackle what had become a major health issue. Networks of sewers were constructed to collect the waste water and remove it from the housing areas, dumping it either into rivers or directly into the sea. Groundwater, however, doggedly remained polluted and unsafe to drink and local authorities were again forced to invest, this time in upland reservoirs whose catchments lay in unpolluted areas of high precipitation.

The concept of hydropathic resorts was developed in the middle of the 19th century. These depended on the plentiful supply of fresh air and good quality, weakly to moderately mineralized water in contrast to the Victorian spa resorts such as Strathpeffer which offered sulphur and iron-rich waters. These new resorts were marketed towards the richer class of city dwellers who needed a few weeks away from the grime and unhealthy conditions in the cities to 'take the waters' as a cure for their general state of ill health. In the main the cure involved promenading, bathing and the regular quaffing of clean water. The use of water in medicine was not novel but had been employed by the Greeks and Romans in the days when medical practice used many emetic and astringent drugs combined with the liberal use of leeches. The appeal of hydropathic medicine to the spiritual and physical properties of water combined with healthy exercise and clean air was very attractive and well patronized (Bradley *et al.* 1997).

By contrast, the spas emulated the fashionable European spa resorts and provided mineralised water for drinking and bathing (Edmunds *et al.* 1969). The spas provided Victorian pleasures such as the needle bath, jets of high pressure saline water focussed on the portly spa visitor, a delight which was followed by two, or even three, pint tumbler fulls of heated spa water. They were essentially glamorous holiday destinations and were places to be seen in rather than centres of medical treatment. The Scottish spas were:

- Strathpeffer, north of Inverness but accessible by train from London,
- Pitkeathly Spa, north of Perth,
- Airthrey, Bridge of Allan, north of Stirling, drawing mineralized water draining from a day adit in a former copper mine, but unique in that

From: MATHER, J. D. (ed.) 2004. *200 Years of British Hydrogeology*. Geological Society, London, Special Publications, **225**. 213–217. 0305-8719/04/$15

this was a combined site with a hydropathic unit and Turkish bath,
- Innerlethen (St Ronan's Mineral Wells) in the Southern Uplands,
- Hartfell spa near Moffat in the Southern Uplands, and,
- Melrose in the Borders.

A contemporary report of Airthrey (Anon. 1875) reads:

> 'The primary attraction of Bridge of Allan is the Airthrey mineral water, proceeding from four springs with as many divergencies of medical character. These are generally speaking of saline nature, but with a bitter taste, which may perhaps be accounted for by their being collected in cisterns formed in an old copper mine. The water is generally drunk hot and the usual quantity is two or three large tumblers.
>
> The pump room, or well house, is a handsome building, erected by Lord Abercromby, on the table land above the village. In addition to the mineral baths there are reading and billiard rooms attached. In the same locality is an excellent hydropathic establishment, accompanied by a suite of Turkish baths.'

The heavy metal concentration of the spring water is not recorded!

The Scottish hydropathic establishments

The hydropathic resorts were developed mainly in areas adjacent to the industrial conurbations in central Scotland. Their development was encouraged by the expansion of the rail network with the limiting factor on their growth being ease of access from major cities. Unlike the spa resorts, where it was the high degree of mineralization of the water which was their appeal, the Hydropathic establishments could be developed wherever clean potable water was to be found, with many depending on groundwater as their source. 20 hydropathic resorts were opened between 1843 and 1882, and a further four were planned to be opened in the last decade of the century but were either not completed or were built but not used as hydropathic centres. (Table 1 & Fig. 1). The early resorts were small, caring for at most a dozen patients but as the industry progressed, newer and larger resorts were developed to cater for both medical patients and recreational guests. The use of the hydropathic resorts as centres for medical cures had ended by the middle of the 20th century and the resorts were either closed or converted into tourist hotels. Only five hotels retain their association with the industry by incorporating the title 'hydro' in their name. Of these only the Crieff Hydro stills relies partly on groundwater for its potable water supply.

Table 1. *Location and geology of the Scottish Hydropathic Establishments*

	Hydropathic	Location	Geology
1	Athol	Pitlochry	Dalradian
2	Deeside	Cults	Dalradian
3	Kirn Pier	Dunoon	Dalradian
4	Kyles of Bute	Bute	Dalradian
5	Shandon	Gareloch	Dalradian
6	Peebles	Peebles	Lower Palaeozoic
7	Waverley	Melrose	Lower Palaeozoic
8	Angusfield	Aberdeen	Devonian
9	Aithrey*	Bridge of Allan	Devonian
10	Callander	Callander	Devonian
11	Cluny Hill	Forres	Devonian
12	Crieff	Crieff	Devonian
13	Dunblane	Dunblane	Devonian
14	Glenburn	Rothesay	Devonian
15	Seamill	Seamill	Devonian
16	Wemyss Bay	Skelmorlie	Devonian
17	Gilmourhill	Glasgow	Carboniferous
18	Craiglockhart	Edinburgh	Carboniferous
19	Kilmacolm	Kilmacolm	Carboniferous
20	Moffat	Moffat	Permian

* This was a combined site shared with Aithrey Spa

The geology of the Scottish hydropathic Resorts

The geology of Scotland is divided between three broad regions, the ancient rocks of the Highlands north of the Highland Boundary Fault, the Lower Palaeozoic rocks south of the Southern Uplands Fault and, between the two faults, the Upper Palaeozoic rocks of the Midland Valley (Table 1 & Craig 1991).

The Highlands are an area dominated by metamorphic rocks of the Caledonian Orogeny. The rocks adjacent to the Highland Boundary Fault are low grade slates and phyllites with the grade of metamorphism increasing to the north where schist and psammite are common. They are transected by a series of NE-trending wrench faults. Superimposed upon these basement rocks are small inliers of Devonian, Permian and Jurassic sedimentary strata.

The Southern Uplands is an area of Lower Palaeozoic greywacke, shale and mudstone which have undergone extensive but very low grade metamorphism. These marine sediments are also associated with spilitic lavas and cherts. The lower Palaeozoic rocks are overlain by inliers of Permo – Triassic sediments and a small area of Devonian and Devono – Carboniferous strata.

The Midland Valley graben lies between the Highland Boundary Fault and the Southern Upland Fault. On the northern flank of the graben the rocks are of Lower and Upper Devonian age (the Old Red

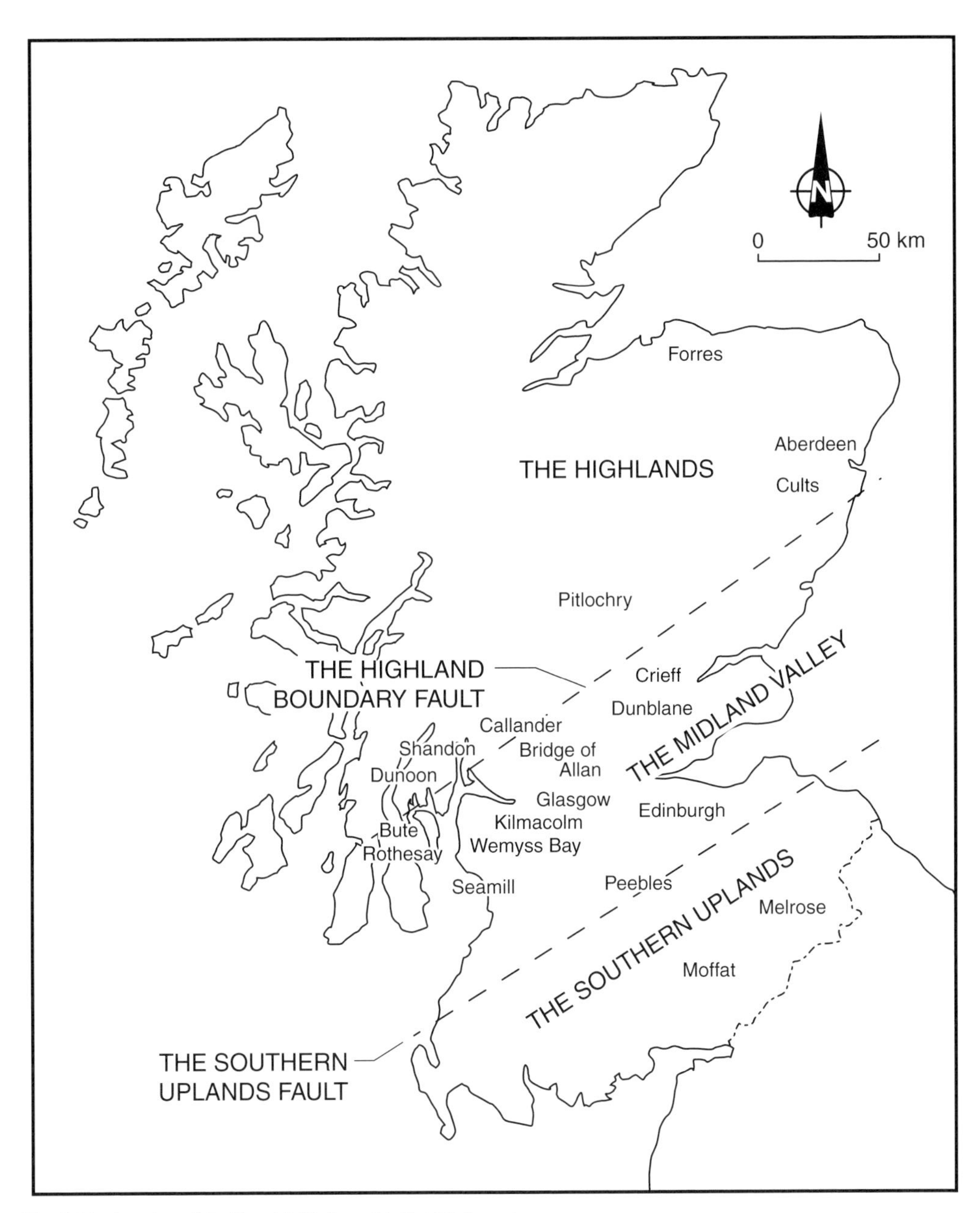

Fig. 1. The location of the Scottish Hydropathic Establishments.

Sandstone sequence). This is a thick sequence of coarse grained sediments associated with the erosion of the mountains to the north. These sediments range from coarse conglomerates through sandstone to fine mudstone. Associated with the deposition of the Lower Old Red Sandstone rocks was a period of extensive volcanism which has not only extruded significant flows of basalt and andesite but also has indurated the sediments. Rocks of Middle Devonian age are absent from the Midland Valley. Overlying the Devonian rocks, in the southern half of the graben, is an extensive suite of Carboniferous rocks, including three major coal basins. It was the exploitation of these coal and associated iron reserves which lead to the major industrial development of Central Scotland.

Quaternary aged sediments derived from a variety of late glacial processes provide a diverse range of

Table 2. *The geochemistry of some of the hydropathic waters (all values mg l^{-1}).* (after Robins 1986)

	Hydropathic	Ca	Mg	Na	K	HCO3	SO4	Cl	NO3
6	Peebles	34.5	106	615	3.9	167	73.2	1786	0.5
9	Bridge of Allan	38.5	9.4	6.0	0.6	124	9.1	7.0	0.1
11	Cluny Hill	63	1.7	20.0	0.9	197	26.0	28.0	0.02
15	Seamill	42.2	21.0	19.0	1.0	217	13.3	25.0	3.2
17	Gilmourhill	107	54.5	31.0	12.0	634	185.5	42.0	0.04
18	Craiglockhart	112	22.4	38.0	5.6	302	73.2	78.0	9.0
19	Kilmacolm	55.8	7.7	12.0	2.0	158	17.2	22.0	5.6
20	Moffat	27.6	6.9	8.0	3.8	70.9	11.1	23.0	5.2

lithologies distributed across the whole of Scotland. These include an extensive till cover of clay and sandy clay, often including lenses of sand and gravel, and coarse grained outwash fans and eskers. Overlain on this are alluvial, lacustrine and marine deposits.

The stratigraphic eras associated with the locus of each of the hydropathic resorts are shown in Table 1. All but seven of the resorts are situated in the Midland Valley adjacent to the major centres of population and close to the Highland Boundary Fault. The first hydropathic establishments, however, were situated in the already popular seaside resorts of Dunoon and Rothesay on the Lower Clyde estuary and were readily accessible to the citizens of the greater Glasgow area by steamer. These were followed by a series of developments which lie close to the northern edge of the Midland Valley along either side of the Highland Boundary Fault. Those immediately to the north of the fault lie in an area of late Pre-Cambrian to early Palaeozoic rocks, all of which have been metamorphosed, extensive folded and fractured. The hydropathic establishments situated to the south of the fault are sited on rocks of Lower Devonian age. These collectively form the Strathmore Aquifer which is a sequence of indurated sediments and lava flows which rely on secondary permeability to transport groundwater. The remaining establishments were:

- at Forres in the north east of Scotland, drawing its water from Middle Devonian rocks; to a magnificently fronted sandstone building now a hotel without reference to its former existence,
- establishments in Glasgow, Edinburgh and Kilmacolm which drew groundwater from the local sandstone of Carboniferous age,
- at Moffat where water was derived from a spring source in the Permian aquifer locally overlain by a thin cover of Quaternary gravel, and,
- other sites in the Southern Uplands which drew water from the Lower Palaeozoic rocks.

Many of the groundwater sources were springs. At Peebles, the Shieldgreen group of springs, yielding 2 l s^{-1}, discharged 'soft' water into a tank from glacial gravel over Silurian shale which is piped to this day to what is now called Peebles Hydro. Other springs in the grounds of the Peebles hydropathic establishment have now ceased to flow. Some of the sites also bottled the water, for example, there was the celebrated Crieff Aerated Water Company which bottled weakly mineralized water from a borehole drilled in 1904 into the Lower Devonian sandstone and conglomerate which yielded 0.3 l s^{-1}. At Pitkeathley Spa the least mineralized water from the Spout Well was bottled as a medicinal and table water by Schweppes until, in 1908, it was decreed too strong for public tastes. It had a recorded total dissolved solids concentration of 4740 mg l^{-1} and would be described as brackish.

Groundwater chemistry

Groundwater in Scotland is mainly weakly to moderately mineralized and dominated by the Ca and HCO_3 ions (Robins 2002). The hydropathic sources are all typical Scottish groundwaters and none offers any remarkable chemical features. The known groundwater chemistry of some of the hydropathic sources is summarized in Table 2 and concentrations outwith the EC Guide Levels are highlighted. Although four sources have SO_4 concentrations which exceed the Guide Lines, none exceed the EC maximum admissible concentration. The same is the case for Na except for the source at Peebles. The supply for the hydropathic establishment at Crieff was treated with chlorine before use in recent years whereas all the other groundwater sources were used without treatment.

The end of an era

The development of the resorts in area of scenic beauty which were known to have readily available sources of good quality groundwater reflects the nature of hydropathic treatment. In their heyday, all the establishments had resident medical staff and visitors were treated as patients. The patients were drawn mainly from the professional, industrial and business middle classes, each seeking respite from the pollution of the industrial cities where both atmosphere and water supply suffered from anthropogenic contamination. The remedy, a four-week stay in the often bracing Scottish countryside, breathing the clean air and drinking the pure water was enough to put most patients back on the road to good health.

Movement of medical practice away from the 'heroic' remedies and the development of the modern urban infrastructure of the sewer network and clean reticulated water supplies drawn from upland gathering grounds finally put the hydropathic resorts into decline. As the general level of health within the urban population improved and medical capability developed only the larger hydropathic establishments which were capable of offering high quality recreational facilities and more traditional water cures, survived. Many of the buildings were converted into holiday hotels or cottage hospitals but the dependency on clean and pure groundwater declined.

By the middle of the 20th century only five of the original hydropathic establishments remained in use. These were the Crieff Hydro, Dunblane Hydro, Peebles Hydro, Seamill Hydro and Pitlochry Hydro. Of these only Crieff and Peebles were groundwater dependent, the others taking their supplies from surface water sources.

References

ANON. 1875. *Black's Picturesque Guide to Scotland*. Adam and Charles Black, Edinburgh.

ANON. 1980. *The quality of water intended for human consumption*. EC Directive, **80/778/EEC**.

BRADLEY, J., DUPREE, M. & DURIE, A. 1997. Taking the water-cure: the hydropathic movement in Scotland, 1840–1940. *Business and Economic History*, **26** (2).

CRAIG, C. Y. 1991. The Geology of Scotland, 3rd edition. Geological Society, London, 612 pages.

EDMUNDS, W. M., TAYLOR, B. J. & DOWNING, R. A. 1969. Mineral and thermal waters of the United Kingdom. *Transactions of the 23rd International Geological Congress*, **18**, 139–158

ROBINS, N. S. 2002. Groundwater quality in Scotland: major ion chemistry of the key groundwater bodies. *The Science of the Total Environment*, **294**, 41–56.

British hydrogeologists in North Africa and the Middle East: an historical perspective

J. W. LLOYD

School of Geography, Earth and Environmental Sciences, University of Birmingham, Edgbaston, Birmingham B55 2TT, UK

Abstract: The main British hydrogeological contributions to a very extensive area are referenced and it is concluded that notable contributions have been made to the understanding of regional sedimentary basin hydrodynamics. The British hydrogeological involvement in much of the area, however, has unfortunately been less than that of the French, principally because of past colonial influences.

Groundwater is clearly a vital commodity in the semiarid and arid countries of the Middle East and is likely to become more important in the future as social and economic development improves and the growing populations become more sophisticated. In the discussion below the British influence on groundwater studies and development is outlined and it is pointed out that despite the extensive political attention that the region experiences this influence has been relatively small. The region discussed is shown in Figure 1.

Most of the hydrogeological work has been carried out by consulting firms and is not recorded in scientific journals, which has posed some problems in identifying projects. Inevitably, some projects will have been missed for which apologies are offered.

The prelude to modern hydrogeology in the region

During the period of major European colonisation in the 19th century, British interest in trade and slavery focused principally upon sub-Saharan Africa and, coupled with the Victorian exploration emphasis on finding the source of the Nile, overshadowed any fledging interest in the arid and semiarid regions of the Middle East. Politically however, the completion of the Suez Canal in 1869 as a trade route to India and the occupation of Egypt by the British to form a 'protectorate' in 1882 until 1921 heightened interest in the region and prompted British scientific involvement.

In the early part of the 20th century, some research in Egypt away from the pharonic Nile began to develop with geographical and geomorphological descriptions of the main oases such as Kharga, Dakhla, Farafra and Bahariya, reported by British members of the Egyptian Survey Department (see Ball 1900; Beadnell 1901*a*,*b*; Ball 1903). Specific hydrogeological accounts of artesian conditions and flowing wells followed (Beadnell 1909, 1911; Ball & Beadnell 1927). During this period the theory that the oases were formed as a result of 'burst' phenomena through high artesian pressures was developed by the Egyptian Survey.

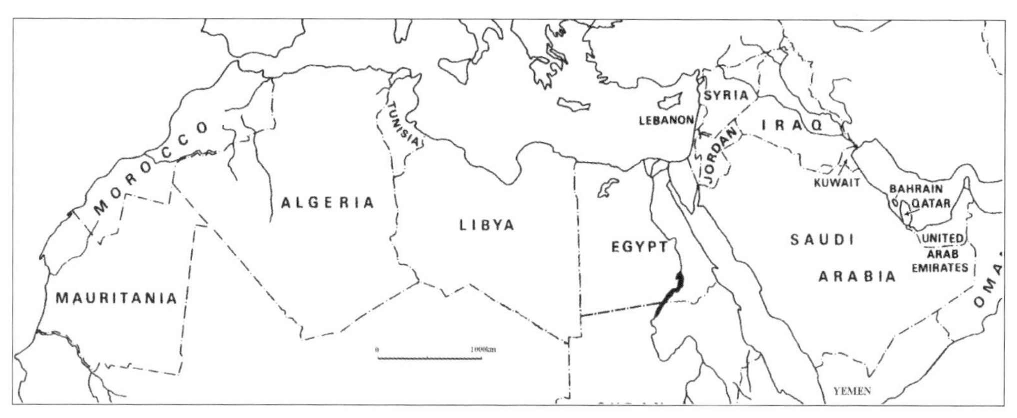

Fig. 1. Country location map.

From: MATHER, J. D. (ed.) 2004. *200 Years of British Hydrogeology*. Geological Society, London, Special Publications, **225**. 219–227. 0305-8719/04/$15 © The Geological Society of London.

Elsewhere in North Africa in the 19th century and early 20th century the French colonized Algeria, Morocco and Tunisia. The Arabian Peninsula and the eastern Mediterranean areas lay under Ottoman rule, as did Libya until taken over by Italy in 1911. The political map changed after the end of the First World War in 1918.

Following the defeat of the Turks and despite promises of independent nationhood to the indigenous Arab peoples quasi-colonization continued with the French annexing Lebanon and Syria in 1920 and the British being given charge of Palestine under the Treaty of Versailles. The Yemen became an Imamate and the British took over areas bordering the Arabian Gulf and Arabian Sea as protectorates. Ibn Saud proved victorious over the Hashamites in the Arabian Peninsula and the Kingdom of Saudi Arabia was fully established in 1926. The Hashamite prince Abdallah had moved to Amman, where significant springs existed, and a new country Trans-Jordan had been created under British mandate in 1920.

During the inter-World War period (1918–1939) geological work was carried out principally by the French in North Africa in the Atlas Mountains and in the eastern Mediterranean. In Palestine, British personnel in the Department of Land Settlement and Water Commissioner undertook geological mapping and surveys of springs and dug wells, which also extended to Trans-Jordan (Blake 1930, 1937; Ionides & Blake 1939). In Egypt British geological and geomorphological work continued with limited studies of the oases areas (Gardner 1932; Caton-Thompson & Gardner 1934) and exploration further a field into the Egyptian-Libyan Sahara (Sandford 1934).

At this time, rivers and springs provided water supplies for major communities. Groundwater was accessed as it had been for centuries, by dug wells for some small settlements and along traditional desert trading routes. Some dug wells were surprisingly deep (~50 m) and had been constructed in stages, probably as a consequence of declining water levels (i.e. the famous Ain Zam Zam in Mecca). In some of the Arabian Gulf areas falages (hand-dug tunnels used for groundwater abstraction) existed, indicating the Persian ganat influence. Only a few machine drilled wells existed in the 1930s. The situation however, began to change with the advent of the Second World War in 1939.

British groundwater investigations during the Second World War

Serious interest in groundwater commenced with geologists and engineers providing water supplies for the British and Allied troops in the North African and associated campaigns, using percussion drilling rigs. Studies were carried out into the availability of supplies in shallow ground, typically wadi beds, and extended from Libya through to Syria.

The work was carried out principally by F. W. Shotton and G. L. Paver. Reports were published immediately post-war as a series of papers entitled 'Water Supply in the Middle East Campaigns' (e.g. Shotton 1946*a*, *b*; Addison & Shotton 1946; Paver 1947*a*, *b*). An illustration of the type of study undertaken is shown in Figure 2. It is interesting to note that in 1940 a groundwater head distribution for the Western Desert of Egypt, based upon oases and some well information, was published by the Swedish engineer Hellstrom (1940). Later work has shown no fundamental change in the understanding of the distribution.

Post-Second World War political change and British influence

Following the Second World War in 1945, from a hydrogeological perspective, and indeed many others, the future areas of interest were defined through a number of significant political events: the translation of Trans-Jordan into Jordan in 1946, the creation of Israel in 1948, the abdication of King Farouk of Egypt in 1952 and the independence of Syria in 1944, Libya in 1951 (following British and French administration since 1941), of Morocco and Tunisia in 1955–1956 and of Algeria in 1962. British 'protection' of the Arabian Gulf States and Southern Yemen ceased between 1961 and 1971.

The creation of the new independent states together with those existing pre-war meant that there were no extensive administrative areas as seen during the days of French and British rule, nevertheless the 'colonial' influences remained through imparted language so that geology and eventually hydrogeology continued to be strongly influenced by the French in Lebanon, Syria, Morocco, Tunisia and Algeria and by the British in Jordan. Elsewhere the Israelis and Egyptians embarked upon their own studies, while Saudi Arabia and Libya chose to employ international expertise. In the Arabian Gulf states much of eventual British hydrogeological participation has been based on the goodwill generated by British oil industry, banking and military support. Overall the legacy of British political involvement in Middle Eastern affairs in hydrogeological terms has proved very small when compared to that of the French.

Through the 1950s the main interest was in geology, to some extent promoted by oil exploration, with work in Jordan by the Irish geologist David Burdon (1959). In North Africa, Syria and Lebanon French and German geologists started to lay the foundations for hydrogeological studies. In 1966 the first comprehensive geological map of Saudi Arabia and

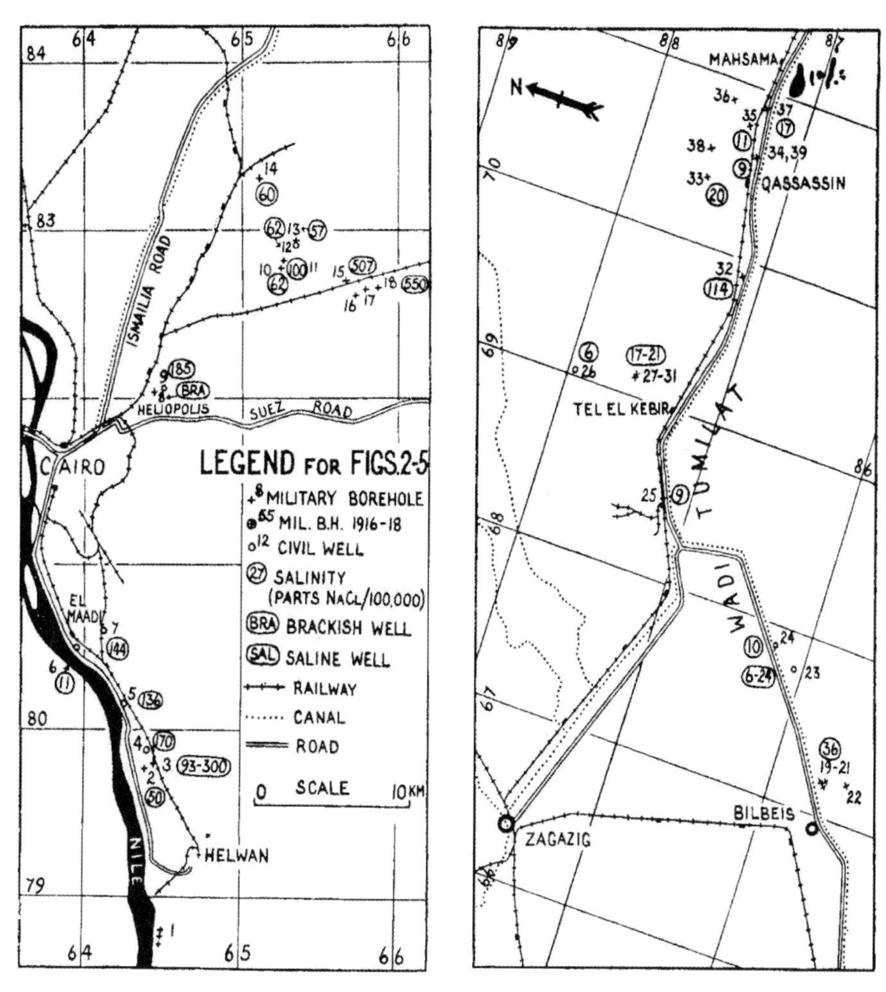

Fig. 2. Example of hydrogeological mapping during the Second World War North- African campaign (after Shotton 1946).

the Arabian Peninsula was published by the United States Geological Survey (Powers *et al.* 1966).

Despite local political upheavals the general socio-economic conditions in the region began to develop in the 1950s and early 1960s with some oil revenues and a certain amount of international financial aid. Population growth together with migration from rural to urban areas occurred. Requirements for irrigation increased and attention started to be paid to exploration for groundwater supplies. Multilateral and bilateral aid packages began to target water resources with international consulting groups recruited for specific projects. From a British perspective the political history of the region restricted future hydrogeological participation to British aid projects mainly in Jordan, competitive international tenders and individuals working for international or local government agencies.

British contributions to modern hydrogeological studies in the region

The United Nations played an important role through the Food and Agriculture Organisation (FAO) in Rome, with much of the modern hydrogeological groundwork in the Middle East at the time initiated by David Burdon (an Irish national, but a Fellow of the Geological Society of London), who worked extensively throughout Syria, Jordan and Saudi Arabia (e.g. Burdon *et al.* 1954; Burdon & Otkun 1968; Burdon 1982).

Jordan

The consulting civil engineering firms Rolfe, Raffety & Partners and Sir M. MacDonald & Partners carried

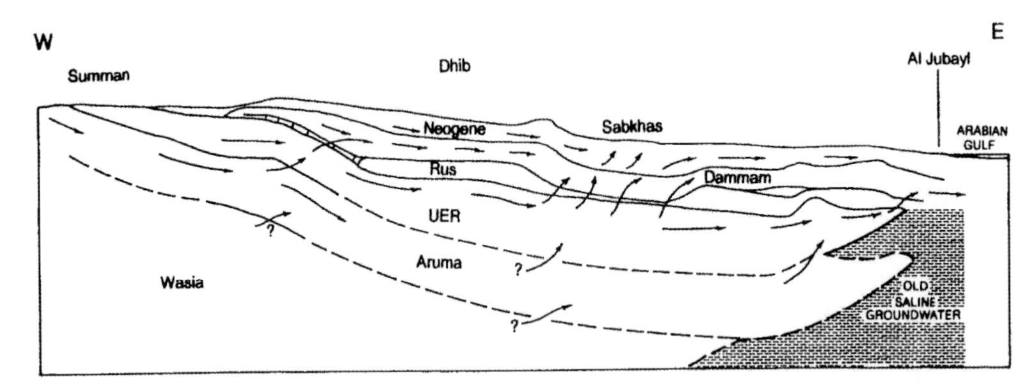

Fig. 3. Hydrogeological cross-section across eastern Saudi Arabia demonstrating the typical cross-formational flow of a large groundwater basin and the control of saline groundwaters at the basin discharge end (after Bakiewicz *et al.* 1982).

out the first major British hydrogeological projects in the region, in Jordan, through the British aid programme. Rolfe, Raffety & Partners (1964) undertook a geological and water resources assessment of the country to the west of the River Jordan, and Sir M. MacDonald & Partners (1961–65) to the east. Hunting Technical Services (HTS) carried out detailed geological mapping in the northeast and the groundwater resources were estimated through a comprehensive recharge study (Lloyd *et al.* 1966). Groundwater development in the lava fields in the north of the country was initiated by HTS (1965).

The work in eastern Jordan was followed up by a major hydrogeological study funded through the FAO with British and Jordanian personnel. The main limestone aquifers were defined (Parker 1970) and the first numerical groundwater model in the Middle East was constructed, for the Cambro-Ordovician sandstone aquifer in the south (Lloyd 1969).

Western Jordan was taken over by Israel in 1967, but British hydrogeological involvement continued in the east with studies mainly by Howard Humphreys (1977, 1978) developing a water supply for Aqaba and a national water use strategy. In 1982 the firm carried out a detailed exploration of the southern plateau to determine a water supply for the important Shadiya phosphate plant (Howard Humphreys 1982*a*) and new modeling of the Cambro-Ordovician aquifer (Howard Humphreys 1982*b*). More recently Haiste Kirkpatrick (1995) has undertaken an assessment of the Cambro-Ordovician aquifer with a view to providing a water supply for Amman through a conveyance of some 300 km.

Saudi Arabia

In Saudi Arabia groundwater abstractions on a large scale commenced in the country in the early 1940s, related to oil industry reservoir enhancement but no systematic hydrogeological appraisal was undertaken until the 1970s. Raikes & Partners, British consultants based in Rome, started establishing a climate monitoring network and some hydrogeological assessment into sebkha (saline mudflat) discharges (Pike 1970). In 1968 the government divided the country into six areas and a socio-economic study of the northern part of the country, including an initial survey of wells, was carried out by the Economist Intelligence Unit in 1973. In 1979 the British Arabian Advisory Company carried out a national review of groundwater resources (British Arabian Advisory Company and Water Resources Development Department 1979).

The most significant British contribution to the nation's hydrogeology came in the mid- to late-1970s with the assessment of the famous Um er Radhuma (UER) limestone aquifer by Groundwater Development Consultants (International) Ltd. (GDC) of Cambridge (Bakiewicz *et al.* 1982). The study demonstrated the significance of describing groundwater in terms of a system of inter-related aquifers as illustrated in Figure 3, a concept that has proved important in the regional sedimentary basins of the area. Disappointingly, the British hydrogeological involvement subsequently has been minimal and restricted to individuals working for Aramco, the national oil company, or as private consultants.

Arabian Peninsula (excluding Saudi Arabia)

In the countries fringing the Arabian Peninsula interest in groundwater increased as oil revenues became important. As in Saudi Arabia, the oil industry in Bahrain had abstracted groundwater for some considerable time, and supplies had been abstracted for municipal and irrigation requirements to such an extent that over-abstraction was a problem as early as the mid-1960s (Wright 1967). As an extension to

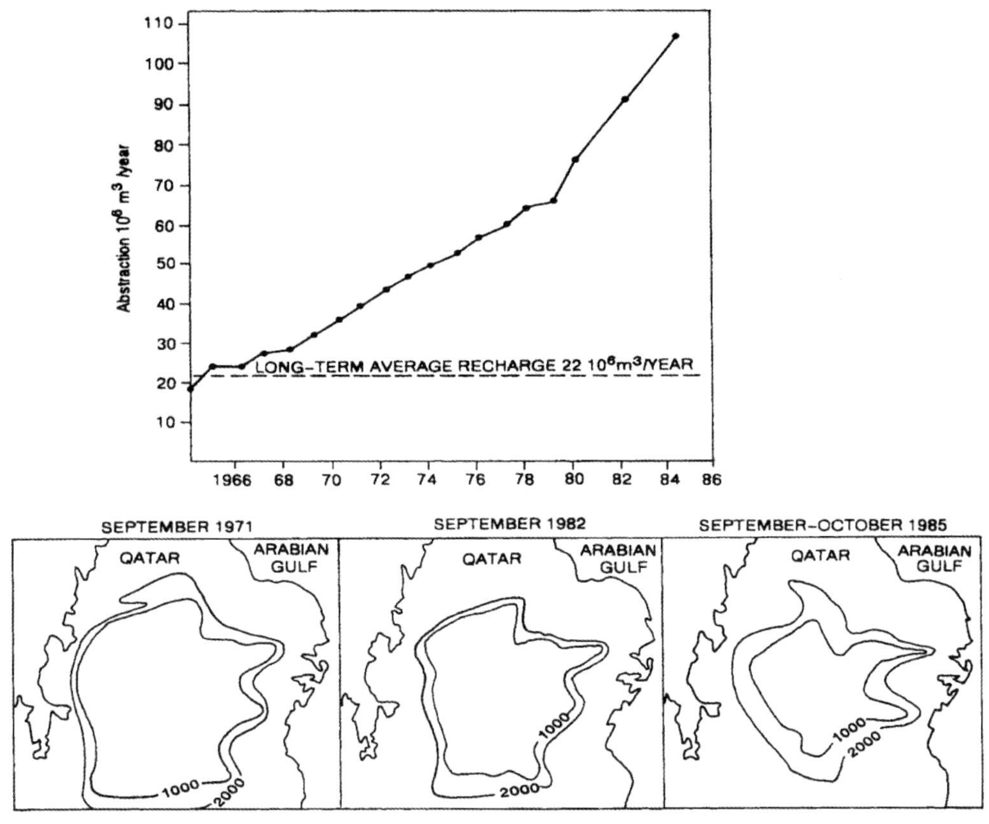

Fig. 4. Over-abstraction in Qatar in the 1970's and 80's, giving rise to diminution of the fresh-water lens (after Lloyd *et al.* 1987).

the Saudi UER study noted above, GDC (1983) recommended the abstraction of brackish groundwater (13000–15000 mg l^{-1}, total dissolved solids) for a reverse osmosis desalination plant that started operation in 1984 and is still operating effectively.

In Qatar, the initial systematic groundwater study was carried out by the FAO with British personnel involvement in 1978–1981. A master plan review for water resources was undertaken by Halcrow-Balfour in 1981. The FAO and Halcrow studies conclusively demonstrated over-abstraction, which was further illustrated by Atkinson & Eccleston (1986) and Lloyd *et al.* (1987). In acknowledgement of the declining groundwater head situation and the ingress of saline water (see Fig. 4) the Ministry of Municipal Affairs and Agriculture retained Entec in 1994 to carry out well injection testing with a view to recharging desalinated sea water.

From the early 1960s Sir William Halcrow & Partners was actively engaged in hydrogeological survey work throughout the United Arab Emirates (UEA, formally the Trucial States). Initially, the firm was involved in the development of individual well-fields supplying Sharjah, Dubai, Ajman and smaller settlements, but in 1967 was appointed Government Agent to undertake the Trucial States Water Resources Survey (Halcrow 1969), work which formed the foundation for the nation's water supply. Subsequently, while groundwater is still part of the water supply system it has been supplanted by desalinated water for much of the urban water needs.

During the early 1970s, Halcrow carried out a similar programme to that in the UAE, in the Dhofar area of Oman, with Sir Alexander Gibb & Partners completing a water resources assessment of the northern part of the country in 1976. As part of the latter project stable isotope studies were carried out by the Institute of Hydrology. With the growing importance of hydrocarbons in the country hydrogeological support studies were carried out for the oil industry (Parker 1985). Subsequently, a succession of small scale, but important projects were carried out by British consulting firms in the Oman through to the present day and the employment occurred of many British hydrogeologists in the Omani Ministry of Water Resources. Project examples are given in Sir M. MacDonald & Partners (1982), W.S. Atkins (1990), Mott Macdonald (1990, 1994) and Travers

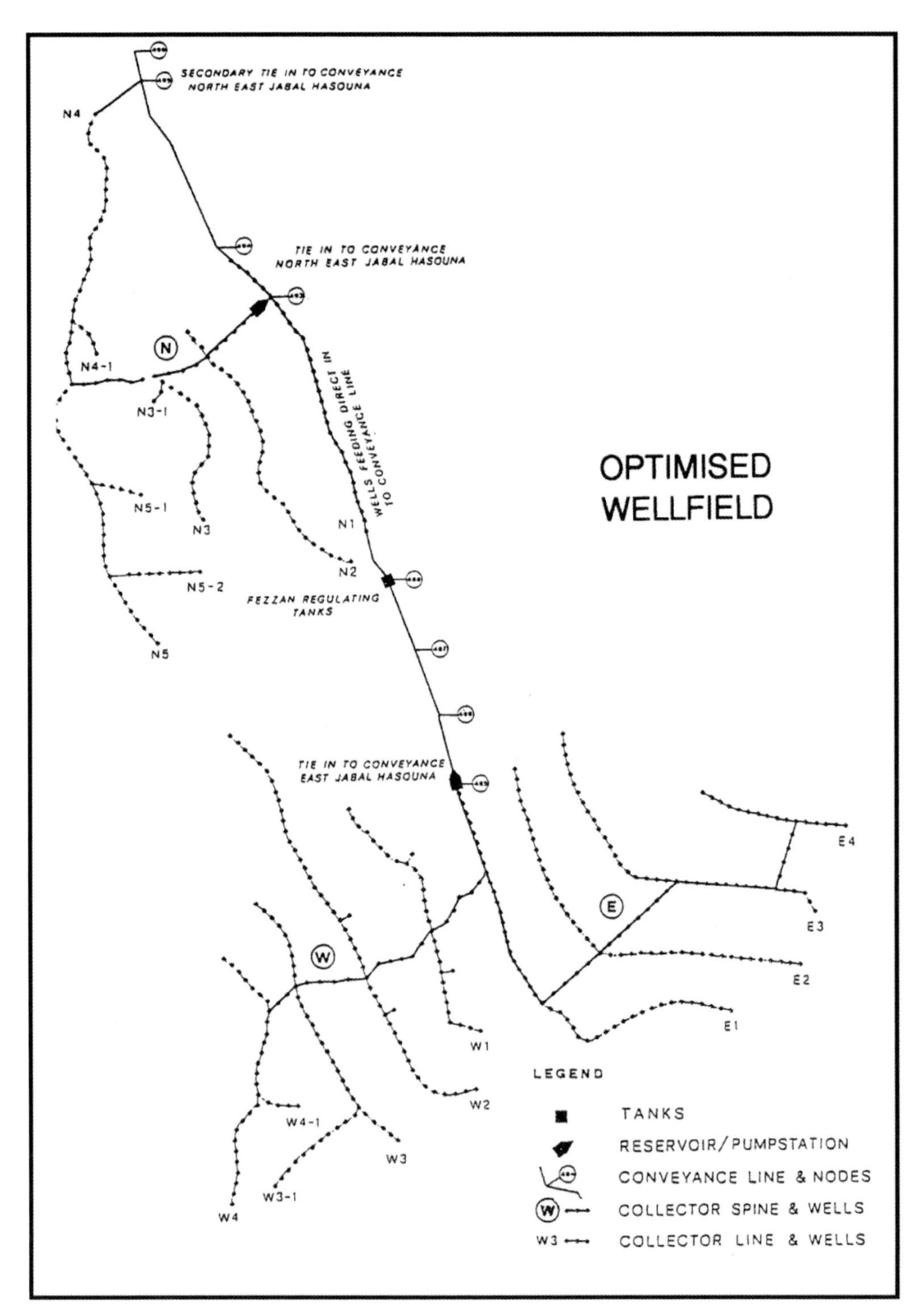

Fig. 5. The lay-out of the Libyan Great Man-Made River Phase II wellfield of 440 wells, optimized for topography and electricity supply following groundwater modelling (after Lloyd *et al.* 1997).

Morgan (1993). Many concern assessments of wadi groundwater resources and the potential for impoundment and artificial recharge. Control of saline groundwater has been a feature, as has water resources master planning (Mott MacDonald 1991) with the integration of water sources to include treated water (Dames & Moore 1992).

Since the end of the Second World War the political evolution leading to the present-day Yemen has been complex and is outside of the scope of this discussion. British hydrogeological contributions have been limited in a country with severe water resources problems. Moseley (1966, 1971) undertook some preliminary work in the south of the country with the Royal Engineers. Between 1974 and 1977 the Institute of Geological Sciences (now the British Geological Survey) carried out hydrogeological studies as part of a wider British aid project in the Intermontane Plains around Dhamar and Yerim (Chilton 1980) and the Wadi Rima area of the Tihama Coastal Plain (Morris 1979). More recent British involvement has been through some irrigation based projects, for example in the Hadhramaut (Binnie and Partners 1987) and for the Sana'a water supply (Howard Humphreys 1983). In 1988–93 staff from the Institute of Hydrology participated in a nation-wide water resources study financed by the United Nations Development Programme.

North Africa

With the Egyptians mostly undertaking their own hydrogeological studies and the French language influence in the region, as noted above, the British hydrogeological involvement in North Africa has been limited. From 1967 to 1974 the Institute of Geological Sciences (British Geological Survey) carried out studies in the Kufra and Sirte basins of eastern Libya (Wright *et al.* 1982) and in 1981 a consortium of Binnie & Partners and W.S. Atkins carried out a water resources master plan for Algiers that included a groundwater resources appraisal of the Mitidja, the large alluvial basin to the southwest of the town.

Otherwise, apart from some minor projects in Libya and the work of individuals (e.g. Lloyd 1990) no major British hydrogeological work was carried out until the North American civil engineering consultants Brown & Root were established in the UK in 1988 and the major Libyan Great Man-Made River Project was administered and staffed through the UK. British hydrogeologists were employed and together with Libyan hydrogeologists and other consulting groups (e.g. Geomath of Pisa, Italy) the groundwater resources assessment for Phases I and II of the project were undertaken (Pim & Binsariti 1994; Lloyd *et al.* 1997). The project entails the conveyance of groundwater from large wellfields in the interior desert of the country to coastal areas. The Phase II wellfield is shown on Figure 5 and was located following a groundwater modelling representation of the 864000 km^2 Hamadah-Muzuq basin. The Phase II study has been important in demonstrating the concept of unconfined head (storage) depletion (of the order of 1mm a^{-1}) providing the long-term groundwater throughput in the regional basins, rather than any 'modern' recharge input. The most recent major British hydrogeological study in Libya has been a review of flowing well potential in the Wadi Bay area, carried out by Halcrow (2002).

Conclusions

From the foregoing summary of British hydrogeological participation in the region it may be concluded that some notable contributions have been made generally to the understanding of arid zone hydrogeology and big basin hydrodynamics, and specifically to groundwater resources assessment in countries such as Jordan and Libya and some of the states fringing the Arabian Peninsula.

The British hydrogeological influence has been less than that of the French regionally, because of the 'colonial' legacy. Nevertheless, as the need for more integrated water supplies incorporating groundwater, surface waters, desalinated water and treated waters develops in many of the countries it is likely that British involvement through the large consulting companies will increase and with it greater hydrogeological participation.

References

ADDISON, H. & SHOTTON, F. W. 1946. *Water supply in the Middle East campaigns: III-Collecting galleries along the Mediterranean coast of Egypt and Cyrenaica.* Water and Water Engineering, August, 11p.

ATKINS, W. S. International. 1990. *Groundwater recharge schemes for the Ibri/Araqi area.* Report to Min. Agriculture and Fisheries, Sultanate of Oman.

ATKINSON, G. W. & ECCLESTON, B. L. 1986. *Water resources and their utilization in the Arab world: country report for Qatar.* Arab Fund for Economic and Social Development, Kuwait Symposium, 50p.

BAKIEWICZ, W., MILNE, D. M. & NOORI, M. 1982. Hydrogeology of the Um er Radhuma aquifer, Saudi Arabia. *Quarterly Journal of Engineering Geology,* **15**, 105–126.

BALL, J. 1900. *Kharga oasis: Its topography and geology.* Survey Department, Cairo.

BALL, J. & BEADNELL, H. J. L. 1903. *Bahariya oasis: Its topography and geology.* Survey Department, Cairo

BALL, J. & BEADNELL, H. J. L. 1927. Problems of the Libya desert: the artesian water supplies of the Libyan desert. *Geographical Journal,* **70**, 34–38.

BEADNELL, H. J. L. 1901*a*. *Dakhla oasis: Its topography and geology*. Survey Department, Cairo.
BEADNELL, H. J. L. 1901*b*. *Farafra oasis: Its topography and geology*. Survey Department, Cairo.
BEADNELL, H. J. L. 1909. The mutual interference of flowing wells. *Geological Magazine*, **6**, 23–26.
BEADNELL, H. J. L. 1911. The underground waters of the Kharga oasis. *Cairo Scientific Journal*, **52**, 1–8.
BINNIE AND PARTNERS & W. S. ATKINS INTERNATIONAL. 1982. *Water resources project for Greater Algiers*. Report to Ministry of Water, Government of Algeria.
BINNIE AND PARTNERS. 1987. Wadi Hadhramaut agricultural development project. Report to the Ministry of Agriculture and Agrarian Reform, Government of Algeria.
BLAKE, G. S. 1930. *The mineral resources of Palestine and Trans-Jordan*. Geological Advisor in Palestine, Publication **2**.
BLAKE, G. S. 1937. Mineral deposits of Palestine and Trans-Jordan. *Palnews-Economic Annual of Palestine*, Vol.**III**, 137–141.
BRITISH ARABIAN ADVISORY COMPANY AND WATER RESOURCES DEVELOPMENT DEPARTMENT. 1979. *National Water Plan Volume 1*, Water Resources of Saudi Arabia, 272p.
BURDON, D. J. 1959. *Handbook of the Geology of Jordan*. Benham & Co. Ltd., Colchester, 82p.
BURDON, D. J. 1982. Hydrogeological conditions in the Middle East. *Quarterly Journal of Engineering Geology*, **15**, 71–82.
BURDON, D. J. & OTKUN, G. 1968. Hydrogeological control of development in Saudi Arabia. *Proceedings of the 20th International Geological Congress*, **12**, 145–153.
BURDON, D. J., MAZLOUM, S. & SAFADI, C. 1954. Groundwater in Syria. *International Association Scientific Hydrology Conference*, Rome, 377–388.
CATON-THOMPSON, G. & GARDNER, E. W. 1934. The prehistoric geography of the Kharga oasis. *Geographical Journal*, **80**, 369–406.
CHILTON, P. J. 1980. *Hydrogeology of the Montane plains, Yemen Arab Republic*. ODA, Land Resources Division, Report **YAR-01–46**.
DAMES AND MOORE INTERNATIONAL. 1992. *Water and wastewater master plan for Salalah*. Report to Ministry of State and Governor of Dhofar, Sultanate of Oman.
ECONOMIST INTELLIGENCE UNIT. 1973. *Socio-Economic Development Plan for Northern Region of Saudi Arabia*. Report to the Ministry of Agriculture and Water, Government Saudi Arabia.
ENTEC. 1994. *Artificial recharge of groundwater in northern Qatar*. Report to the Ministry of Municipal Affairs and Agriculture, State of Qatar.
GARDNER, E. W. 1932. Some problems of the Pleistocene hydrography of the Kharga oasis. *Geological Magazine*, **69**, 386–421.
GIBB, SIR A. & PARTNERS. 1976. *Water resources survey of northern Oman*. Report to the Ministry of Electricity and Water, Sultanate of Oman.
GROUNDWATER DEVELOPMENT CONSULTANTS. 1983. *Reverse osmosis desalination: aquifer C investigation*. Report to the Ministry of Electricity and Water, State of Baharain.
HAISTE KIRKPATRICK INTERNATIONAL. 1995. *Qa Disi aquifer study, Jordan: Long-term management of aquifer resources*. Report to the Ministry of Water and Irrigation, Government Jordan.
HALCROW, SIR W. & PARTNERS. 1969. *Report on the water resources of the Trucial States*. Report to the Government of the Trucial States.
HALCROW BALFOUR. 1981. *Master water resources and agricultural development plan*. Report to the Ministry of Industry and Agriculture, Government Qatar.
HALCROW. 2002. *Wadi Bay irrigation project: Groundwater resources study*. Report to the Government of the Daewoo Engineering and Construction Ltd. Libya.
HELLSTROM, B. 1940. *The subterranean water of the Libyan desert*. Bulletin of the Royal Institute of Hydraulics, Stockholm, No. **26**.
HOWARD HUMPHREYS. 1977. *Aqaba water supply hydrogeological study*. Report to Water Supply Corp. Government of Jordan.
HOWARD HUMPHREYS. 1978. *Water use strategy, north Jordan*. Report to Central Water Authority, Government of Jordan.
HOWARD HUMPHREYS. 1982*a*. *Shidiya Phosphate Mine Groundwater Supply*. Report to Central Water Authority, Government of Jordan.
HOWARD HUMPHREYS. 1982*b*. *Modelling of the Disi Sandstone aquifer*. Report to Water Authority, Government of Jordan.
HOWARD HUMPHREYS. 1983. *Hydrogeolgy and development of the Tawilah-Mejd Zir sandstone and other sources – Sana'a water supply*. Report to National Water and Sewage Authority, Government of Yemen.
HUNTING TECHNICAL SERVICES. 1965. *Wadi Dhuleil investigation*. Report to Central Water Authority, Government of Jordan.
IONIDES, M. G. & BLAKE, G. S. 1939. *Report on the water resources of Trans-Jordan and their development: Incorporating a report on geology, soils and minerals and hydrogeological correlations*. Crown Agents, London, 372p.
LLOYD, J. W. 1969. *The hydrogeology of the southern desert of Jordan*. UNDP/FAO Publications, Technical Report, **1**, Special Fund 212, 120p.
LLOYD, J. W. 1990. Groundwater conditions and development in the Eastern Sahara. *Journal of Hydrology*, **119**, 71–87.
LLOYD, J. W., BINSARITI, A., SALEM, O., EL SUNNI, A., KWAIRI, A. S., PIZZI, G. & MOORWOOD, H. 1997. The groundwater assessment for the Western Jamahiriya System Wellfield, Libya. *Proceedings of the 30th International Geological Congress*, **22**, 258–269.
LLOYD, J. W., DRENNEN, D. & BENNELL, B. 1966. A groundwater recharge study in north-eastern Jordan. *Proceedings of the Institution of Civil Engineers*, **37**, 701–721.
LLOYD, J. W., PIKE, J. G., ECCLESTON, B. L. & CHIDLEY, T. R. E. 1987. The hydrogeology of complex lens conditions in Qatar. *Journal of Hydrology*, **89**, 239–258.
MACDONALD, SIR M. & PARTNERS. 1965. *East Bank Jordan water resources report*. Report to Central Water Authority, Government of Jordan.
MACDONALD, SIR M. & PARTNERS. 1982. *Power and urban water supply: Feasibility study of the Wadi Dayqah.*

Report to the Ministry of Electricity and Water, Sultanate of Oman.
MORRIS, B. L. 1979. *Hydrogeology of Wadi Rima, Yemen Arab Republic*. ODA Land Resources Division, Report, **YAR-01-41**.
MOSELEY, F. 1966. Exploration for water in the Aden Protectorate. *Royal Engineers Journal*, 124–142.
MOSELEY, F. 1971. Problems of water supply, development and use in Audhali, Dathina and eastern Fadhli, Southern Yemen. *IGS, Overseas Geology and Mineral Resources*, **10** (4), 309–327.
MOTT MACDONALD. 1990. *Groundwater recharge schemes for the Salalah plains*. Report to the Ministry of Water Resources, Sultanate of Oman.
MOTT MACDONALD. 1991. *National water resources master plan*. Report to the Ministry of Water Resources, Sultanate of Oman.
MOTT MACDONALD. 1994. *Detailed investigations for up to 100 ha irrigated land: Nejd region*. Report to the Ministry of Agriculture and Fisheries, Sultanate of Oman.
PARKER, D. H. 1970. *The hydrogeology of the Mesozoic-Cainozoic aquifers of the western highlands and plateau of east Jordan*. UNDP/FAO Publications, Technical Report, **2**, Special Fund 212, 100p.
PARKER, D. H. 1985. *The hydrogeology of the Cainozoic aquifers in the PDO concession area*, Sultanate of Oman. Petroleum Development Oman, Sultanate of Oman.
PAVER, G. L. 1947*a*. *Water supply in the Middle East campaigns: VII-Syria and the Lebanon*. Water and Water Engineering, February, 16p.
PAVER, G. L. 1947*b*. *Water supply in the Middle East campaigns: X-Trans-Jordan*. Water and Water Engineering, September, 8p.
PIKE, J. G. 1970. Evaporation of groundwater from coastal playas (sabkhah) in the Arabian Gulf. *Journal of Hydrology*, **11**, 79–88.
PIM, R. H. & BINSARITI, A. 1994. The Libyan Great Man-Made River Project. Paper 2: The water resource. *Proceedings of the Institution of Civil Engineers, Water and Maritime Energy*, **106**, 123–145.
POWERS, R. W., RAMIREZ, L. F., REDMOND, C. D. & ELBERG, E. L. 1966. *Geology of the Arabian Peninsula: Sedimentary geology of Saudi Arabia*. US Geological Survey Prof. Paper 560-D, 147p.
ROLFE, RAFFETY & PARTNERS. 1964. *Hydrogeological studies in the Jerico syncline*. Report to Natural Resources Authority, Government of Jordan.
SANDFORD, K. S. 1934. Geological observations on the north-western frontiers of the Anglo-Egypt Sudan and the adjoining parts of the southern Libyan desert. *Abs. Proc. Geological Survey of Egypt*, 114–115.
SHOTTON, F. W. 1946*a*. *Water supply in the Middle East campaigns: V-The desert between the Nile delta and the Suez Canal*. Water and Water Engineering, October, 12p.
SHOTTON, F. W. 1946*b*. *Water supply in the Middle East campaigns: IV-Water supplies adjacent to the Cairo-Alexandria road*. Water and Water Engineering, September, 12p.
TRAVERS MORGAN. 1993. *Detailed hydrotechnical recharge studies in the Wadi Gulaji*. Report to the Ministry of Water Resources, Sultanate of Oman.
WRIGHT, E. P. 1967. *Groundwater resources of Bahrain Island*. Institute of Geological Sciences, London, 32p.
WRIGHT, E. P., BENFIELD, A. C., EDMUNDS, W. M. & KITCHING, R. 1982. Hydrogeology of the Kufra and Sirte basins, eastern Libya. *Quarterly Journal of Engineering Geology*, **15**, 83–103.

British hydrogeologists in West Africa – an historical evaluation of their role and contribution

ROBIN HAZELL

Little Margate, Bodmin, Cornwall, PL30 4AL, UK

Abstract: The Colonial Office established and funded geological surveys in British West African colonies, from 1903 until self government in c.1960. Provision of water supplies, at first a minor component of the services provided, later often dominated departmental activities. Understanding of the nature of groundwater mirrored the state of the art elsewhere: supply kept pace with demand. Exploration of sedimentary basins led to development of major aquifers. In the 1930s innovative refinements of geophysical siting and well sinking techniques were developed. From 1980 major water borehole programs were largely supervised by British consultants, who continued to pioneer siting and construction techniques.

Britain established and governed four colonies in West Africa for a little over half a century from start to finish. (Table 1). Whatever the merits and demerits of the colonial system, the establishment of clean and adequate water supplies was a considerable achievement and an undisputed benefit. In this period the major sedimentary aquifers were explored, and the water geologists were British. This paper is an outline account of their achievements. Component data are dispersed in many publications and reports, and the bibliography covers the larger or more important geographical areas explored. For reasons of space, minor recorded work has not been listed. To save wearisome repetition, British personnel mentioned in the text are tagged 'Br.'

The four territories which comprised British West Africa are vastly disparate in population, in total area, and in the aggregate areas of major aquifers (Figs 1 & 2). Nigeria has an area and population greater than that of France and Spain combined; The Gambia is smaller than Wales. The attention given to Nigeria is not therefore disproportionate.

Table 1. *The British presence in West Africa*

Territory	Began	Ended
Nigeria	1903: Protectorates established.	1960
Gold Coast (became Ghana)	1902: Total British control	1956
Sierra Leone	1894: Protectorate; 1908: Colony	1961
The Gambia.	1894: Protectorate	1965

The exploration phase of the first half century

Government geological surveys

The geological survey departments of British West Africa came into being during the second decade of the 20th century (Table 2) and flourished until independence. The stated objectives were groundwater mapping and mineral surveys. As a mineral resource, groundwater was from the outset the responsibility of the geological surveys.

For the first few years, the departments had no onshore base. Field work was carried out in the dry season, between November and May. When the rains came, geologists returned to U.K. and enjoyed the hospitality of the Imperial Institute and Imperial College. Table 3 shows how, in the early days, geologists travelled. With few railways or motorable tracks, most work was done on foot. In 1927, Dr. Colin Raeburn and his wife landed at Lagos (Nigeria) with stores and equipment for six months' work, engaged porters and walked inland. Until 1951, though project areas could be reached by road, mapping was carried out on foot, with good results.

Dry statistics cannot do justice to the hardships endured by the pioneering geologists in the 'white man's grave'. By the 1940s, most colonial servants survived blackwater and yellow fever, but retirement was obligatory at the age of 40 to 45 years. The average life expectancy thereafter was 18 months. During the dry 'mapping' season geologists were under pressure to stay in the field. In rain forests, sedimentary formations are deeply weathered and fresh exposures occur only in the beds of youthful streams. Aquifers in jointed coal seams form resistant waterfalls and to view them it was necessary to share a muddy pool with predators and parasites. In dryer terrain, in the absence of borehole samples, sedimentary beds were often examined from a swinging bucket at the bottom of a well, often over 50 m below surface. Maps were drawn on starch-impregnated linen dosed with talc to absorb the sweat, by the light of a kerosene pressure lamp.

From: MATHER, J. D. (ed.) 2004. *200 Years of British Hydrogeology*. Geological Society, London, Special Publications, **225**. 229–237. 0305-8719/04/$15

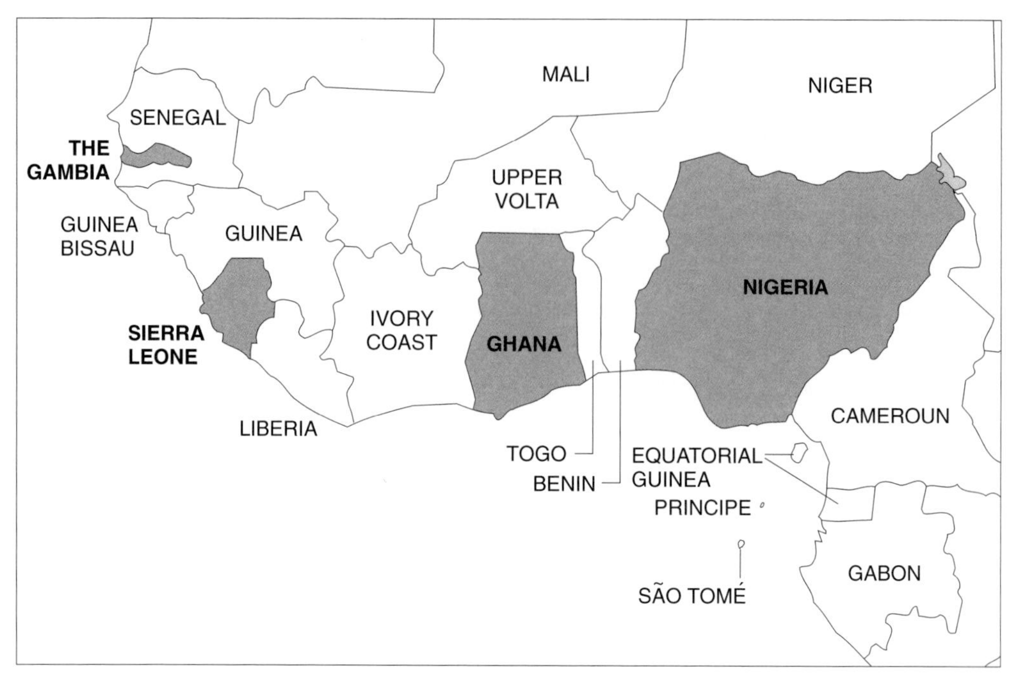

Fig. 1. Territories of British West Africa.

Field equipment comprised a hammer, head-pan, compass, notebook, water bottle and strong boots, until about 1950, when air photos and pocket stereoscopes were introduced.

The evolution of water geologists, 1896–1961

Before the advent of geological survey departments, Arthur Beeby-Thompson spent some months in the Gold Coast in 1896. Though better known for his pioneering work in oil, he also worked on water supply (Beeby-Thompson 1961). Near Accra he sited and supervised drilling for water, though only saline water was found.

Geologists recruited to the government geological survey departments were nearly all newly graduated with no specialized knowledge, each spending upwards of a year under the wing of a more experienced member of staff. The type of work they subsequently did was imposed by the Director in response to the exigencies of the time. Water geologists evolved with help from two published groundwater textbooks (Dixey 1931; Tolman 1937) and several treatises on hydraulics and hydrology, but were for the most part guided by the mistakes and achievements of those who went before (Table 4).

Many of the principles and practices developed during this phase were later incorporated in the new specialization of hydrogeology. Long before Penman devised his formulae, evaporation from open water in the north of the Gold Coast was estimated as 2.1 m a^{-1}, very close to the true value. From the earliest days, the principles of groundwater movement and storage were well understood, as was seasonal groundwater fluctuation (Geological Survey of Gold Coast Annual Report 1916). Reports, couched in beautiful English, were a joy to read. In 1931 Brynmor Jones described the Chad Basin elegantly thus: 'a porous water bearing formation overlain by an impervious formation which seals down the water'. Terms like 'aquifer' and 'aquiclude' were rarely used. Water geologists subsequently moving on to the UK establishment rose to occupy key posts.

Geophysics in the exploration phase

Resistivity was almost the only method used in groundwater work (Table 5). This robust tool was wielded by geologists with all the confidence of youth. Innovative techniques were developed to suit local conditions. The instrument used was the A.C. Megger Earth Tester, with some inadvertent applications of electric shock treatment as a by-product. This instrument was later superseded by the D.C. Geophysical Megger. The Paver, an ingenious DC box using the appropriate technology of the time was used in Sierra Leone and (in 1961) in Nigeria for

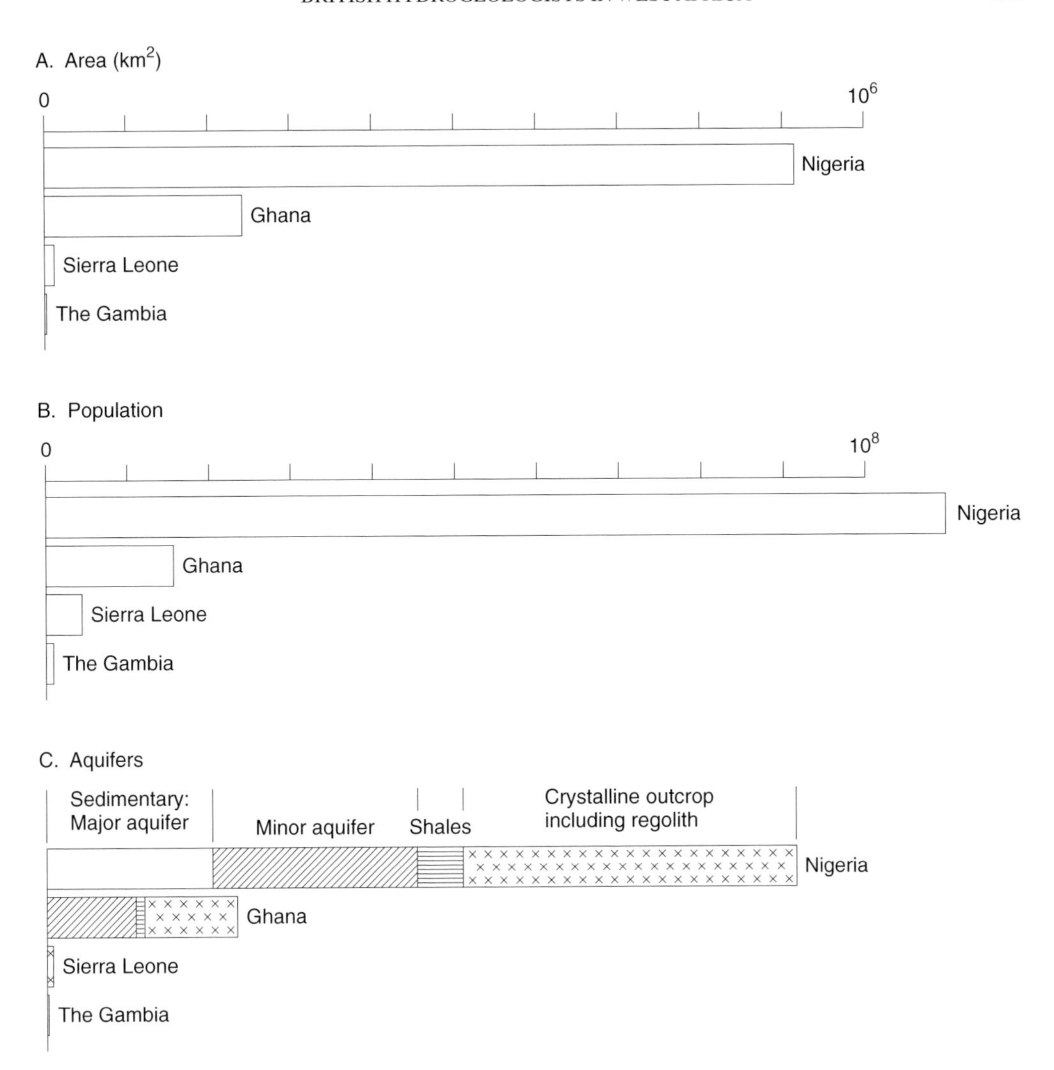

Fig. 2. Areas, populations and aquifers in the four territories.

hand-held down-hole logging. Data from Depth Probes (later called Vertical Electric Sections) were interpreted either by mathematical analysis or by curve matching. The Wenner configuration was used in Nigeria. In Ghana Cooper was more imaginative and experimented with variations in dipole spacing (Cooper 1936). A magnetic agitation technique (Manne and Sahasrabuddhe 1925) was considered useful only in shallow jointed aquifers and never tried (Raeburn & Jones 1934); with hindsight this may have been an opportunity missed in crystalline terrain.

A major gravity survey of the Nigeria sector of the Chad Basin was carried out by the Directorate of Overseas Surveys in 1960–1961 (Cratchley & Jones 1965), providing important data for use in groundwater exploration.

Exploration of important aquifers

In the exploration phase, mapping of important aquifers overlapped and complemented basic geological mapping (Table 6). Much of the impetus in exploration and development in the natural world came from local administrators who, in the course of regular travels were able to identify areas of need. They had regular contact with decision makers, who in turn had ready access to the directors of technical departments. Given the gentlest of pressures, the directors of geological survey would energize their water supply sections, often at short notice.

In order that supply could keep pace with demand, groundwater investigations often dominated departmental activities. In 1953, the entire professional

Table 2. *Establishment of geological surveys*

Territory	Formed in:	Based onshore from:
Nigeria	(Mineral Survey 1903) Geological Survey 1914	1929
Gold Coast (later Ghana)	1913	1926
Sierra Leone	1918	1926
The Gambia*	1980	1980

*The Gambia, a narrow riverine strip of sedimentary terrain, had no geological survey until long after independence, when mapping of potential oil reservoirs aroused attention.

Table 3. *Early travel by geologists in the Gold Coast*

Method of travel	1916	1925
Bicycle or on foot	79%	40%
Car or lorry	0%	29%
Train	20%	28%
Launch/canoe	1%	3%

staff of the Gold Coast geological survey (6 in number) were so engaged.

The sedimentary basins were all identified and for the most part their groundwater potential evaluated. The weathered zone in crystalline terrain was developed for water supplies, but drilling and excavation methods were not capable of penetrating the basal regolith. Consequently a full understanding of the properties of this important 'minor' aquifer was not achieved until considerably later. Table 6 shows progress in mapping up to 1961.

Water points

Prior to 1930, water points were adapted from local techniques, some of great antiquity:

(1) surface catchments into cisterns,
(2) improvements to dug 'native' wells,
(3) encasement of springs,
(4) seepage-fed ponds (tapkis).

In the Gold Coast, a local method of collecting surface water and ducting it into a cistern (biliga) was adapted and standardised (Fig. 3). The cisterns were dug through lateritic crust and belled out into shales, which needed no lining.

Between 1930 and 1933, the Geological Survey of Nigeria (Cochran 1937) experimented and perfected the 'government' 1.2 m diameter lined dug well, which became the standard for anglophone and francophone West Africa. In Nigeria alone, over 1500 'Cochrans' were constructed up to 1939 (Table 7). In 1933, Beeby-Thompson introduced drive point tube wells, known as Abyssinian wells (Beeby-Thompson 1961).

In West Africa there was no pre-existing tradition of building small dams in suitable terrain, and during the colonial period opportunities to build them in impervious lowlands were missed. Indigenous seepage-fed ponds, annually recharged, were common in northern Nigeria. In the Gold Coast the geological survey improved existing ponds, reducing evaporation cheaply by covering them with zana matting spread over wire mesh.

In 1945, the Henderson Box was devised in the Gold Coast for abstracting water free of detritus from running streams. This was the forerunner of the Cansdale Box, which closely resembles it. It is a simple box, opening downwards, lying just above the stream bed.

From 1951 in Nigeria and from 1964 in Gold Coast, programmes of rotary drilling by British contractors overtook and ultimately superseded the drilling of individual boreholes by percussion. The latter method was suited to government employees, as sudden decisions were not needed and supervision was hardly required. Moreover, until the early 1950s there was no perceived urgency. A driller could go on leave and on return could complete the borehole he left behind. Rotary mudflush drilling was adapted from oil well practice and modern screening and aquifer development were introduced.

Table 4. *Chronology of water supply*

Territory	Water supply activity first recorded	Systematic water surveys recorded	Water supply section formed
Nigeria	1933	1928	1947
Gold Coast	1896 (Beeby-Thompson)	1920 (Cooper)	1937
Sierra Leone	1919	1953	—
The Gambia	No data	1972 (Howard Humphries)	—

Table 5. *Early recorded use of resistivity*

Territory	Year	Comment
Nigeria	1932	Deep phreatic sandstones
Gold Coast/Ghana	1931	Voltaian basin
Sierra Leone		Paver
The Gambia	–	–

The development phase of the second half century

The basic exploration and classification of the major aquifers of Nigeria, Ghana and Sierra Leone had been achieved before independence. When expatriates departed, consultants and contractors largely replaced government surveys in the execution of groundwater projects. British consultants played a major part.

Prior to independence, a programme of training of indigenous geologists had been put in place, and the first graduates appeared – in Ghana in 1953, in Nigeria in 1957, and subsequently became Directors of their respective geological surveys. Acceleration of the movement towards independence overtook plans for further training, so that an inevitable shortage of indigenous qualified geologists took some years to overcome. In 1971 the Geological Survey of Nigeria had 131 scientific posts (Federal Government of Nigeria 1987) though lack of funds inhibited field work.

Water projects were no longer sustained by funds from the governments, which came to rely more on overseas aid and bank loans. Crucial policy decisions were made by the Agencies to mass-produce boreholes for villages, while dug wells became a minor component of water supply. Drilling programmes were put out to tender at international level and high-tech rigs were deployed. A rapid expansion in village water supplies was thus achieved, reaching a maximum in the water decade of the 1980s, when several thousand rural water boreholes were constructed in Nigeria. This leap was dramatic and effective, but when the big projects were finished there were no local drilling contractors to fill the vacuum.

With the leap in groundwater abstraction, sustainability for the first time became an issue. To determine 'safe yields', the Geological Survey of Nigeria, in co-operation with the United States Geological

Table 6. *Sedimentary aquifers identified and mapped by geological survey departments*

Territory	Basin or aquifer system	Age	Identified/Explored/Mapped	Year*
Nigeria	Chad Basin	Cretaceous-Quaternary-	Raeburn & Jones	1928
			First deep test	1945
			Barber & Jones	1958
			Barber	1965
	Sokoto/Iullemiden basin	U. Cretaceous-Eocene (CT)	Jones	1948
	Kerri Kerri (C.T.)	Eocene	Carter	1963
			Du Preez	1965
			Water Surveys Nigeria	1977
	Niger/Benue basin	L. Cretaceous	Carter	1963
			Du Preez	1965
	Anambra Platform	U. Cretaceous	Bain	1924
			Shell d'Arcy	1954
			Simpson	1954
	Dahomey Basin	U. Cretaceous-Eocene	Jones &Hockey	1964
			Hazell	1969
	Coastal Plain	Eocene-Holocene	Shell d'Arcy	1939
			Grove	1951
			Hazell	1960
			Jones & Hockey	1964
	Delta	Recent	Shell d'Arcy	1958
			Hazell	1959
	Alluvium	Quaternary-Recent	Water Surveys Nigeria	1984
Gold Coast	Voltaian Basin	Paleozoic	Cooper	1930
			Junner & Bates	1945
Sierra Leone	Bullom Formation	Tertiary		
The Gambia		Upper Cretaceous	Cooper	1926
			(Howard Humphreys)	(1975)

* Date of publication.

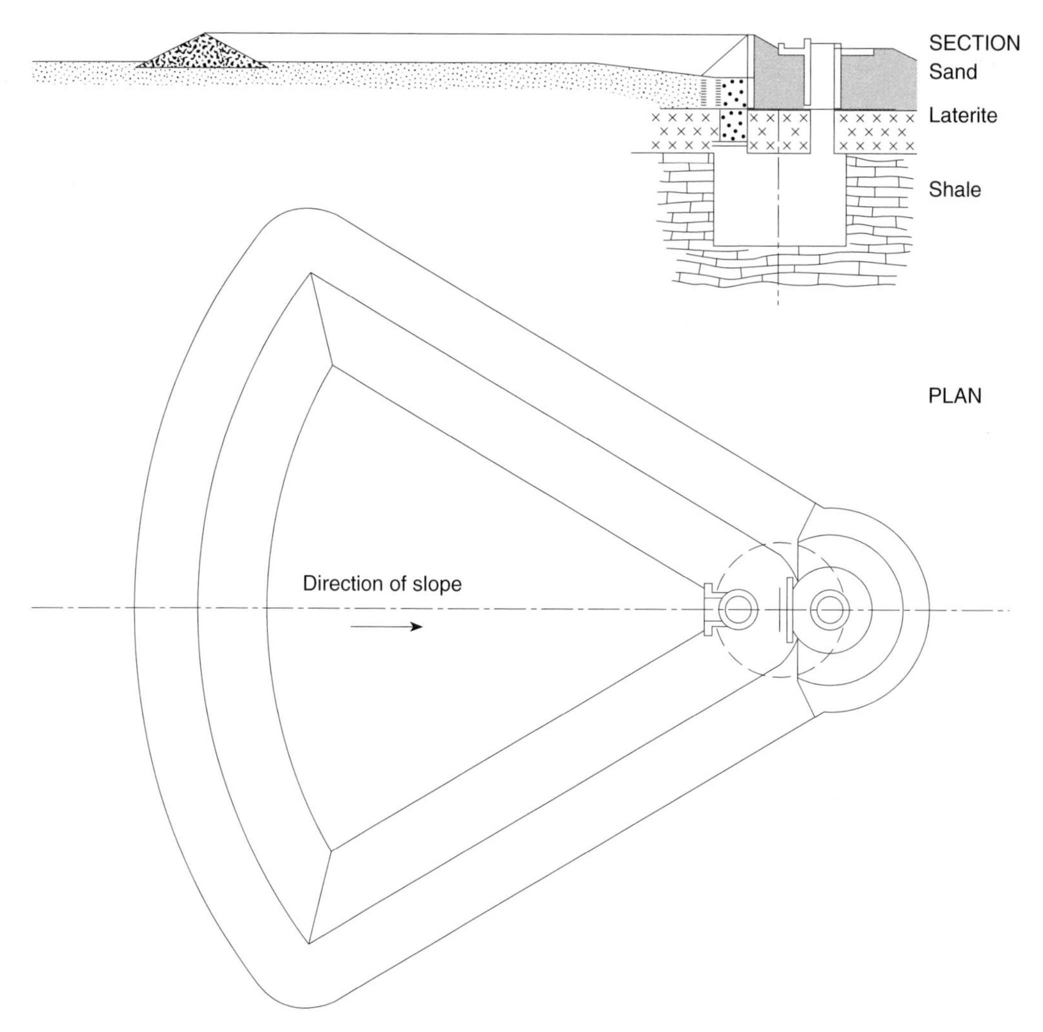

Fig. 3. Single biliga After Cooper, Annual Report GS Gold Coast 1937.

Survey, carried out investigations of the Nigerian Chad and Sokoto basins. State governments ignored the resulting recommendations; uncontrolled over-abstraction of groundwater from the Chad Formation ensued, while paradoxically the Tertiary (Continental Terminal) aquifer of Sokoto was under-used.

British hydrogeologists from 1961

1961–1970 A major investigation was carried out in Cameroun, Tchad and Nigeria by the Chad Basin Development Commission. The dominating groundwater component was headed by a British and a Swiss hydrogeologist.

In central and southern Nigeria a British hydrogeologist propounded the groundwater element of three State-wide master plans (Eastern Region, Mid-West Region, Benue Plateau State) and the Geological Survey of Nigeria provided an update on the Eastern Region (Monkhouse 1963). In Ghana, water drilling continued, at the rate of 35 (percussion) boreholes a year, until 1965. Near Lagos, Nigeria, in 1969, an artesian borehole, sited by the author, was drilled successfully to 700 m depth for the Guinness brewery and qualified for the Guinness book of records.

1971–1980 A Nigerian Federal groundwater data bank (the first of many) was assembled by a team headed by British personnel (G.P. Jones, R. Hazell).

A groundwater study for the first World Bank funded project in Nigeria, the Gombe Agricultural Development Project, was completed by British consultants. Also for the World Bank a major water

Table 7. *Dug wells in Nigeria to 1945*

Year (19–)	29	30	31	32	33	34	35	36	37
Number	33	28	75	132	176	189	125	120	98
Year (190)	38	39	40	41	42	43	44	45	
Number	89	58	35	36	4	75	94	98	

resources assessment of The Gambia river basin was carried out by consultants Howard Humphreys.

1981–1991 World Bank Agricultural Development Projects in Nigeria followed, involving siting and drilling of several thousand rural water supply boreholes. Over half of the hydrogeological input was British. Alluvial aquifers were evaluated in northern Nigeria (Hazell & Barker 1995).

In the Gambia, Howard Humphreys undertook a programme of borehole rehabilitation. In Senegal, the British Geological Survey carried out a groundwater recharge study in the West African Sahel (Edmunds 1991)

1991–2002 In Nigeria British consultants carried out a 500 borehole drilling programme in Kaduna State; groundwater mapping in Lagos State and the British Geological Survey was involved in a groundwater resource evaluation in the southern lowlands of Benue State. In Rusafiya, also in Nigeria, Dr. W. P. Ede was the British senior hydrogeologist in a rural water supply and sanitation programme (Ede 1993).

In Ghana the British Geological Survey carried out:

(1) drought alleviation studies in 1995–1997 (with the Ghana Water & Sewerage Corporation),
(2) groundwater quality studies in northern and central areas (Smedley *et al.* 1996), and;
(3) groundwater resource evaluation of the Afram Plains of central Ghana (Davies & Cobbing 2002).

Geophysics, post-1961

Resistivity continued to be the work-horse for water borehole siting in crystalline terrain. Imaging was brought into use in 1976 in crystalline terrain in Bauchi, Nigeria to locate master joints (Acworth 1981). A little later gridding, using a VLF device (the Fracture Finder by ELF) with a locally generated energy source, was successfully used on the same targets.

In 1977, with the imminent proliferation of water drilling projects, it became clear that a rapid assessment geophysical tool was needed. For this purpose, electromagnetic traversing was introduced to Northern Nigeria by British consultants. After standard mapping had identified suitable crystalline targets, a siting method was developed which combined EM with resistivity. This became the standard technique in 1981. (Beeson & Jones 1985; Reynolds 1987; Hazell *et al.* 1992). As knowledge of local conditions increased, EM traversing largely supplanted resistivity, which was reserved for use only in ambiguous situations. In the 1990s, technicians were trained to carry out EM and resistivity traverses and to interpret the data.

Water well design and drilling

Following independence, close supervision of well digging was no longer practicable, and productivity declined. It was appreciated that drilled wells would wax and dug wells would wane. Percussion drilling in crystalline terrain was too slow to be effective and rotary drilling was inadequate. In 1966, a pneumatic down-the-hole hammer was used to drill into quartzites in northern Nigeria at the author's instigation. This paved the way for serious exploration and development of regolith aquifers in West Africa. In sediments, rotary mudflush drilling continued to prove to be most effective in Nigeria; UPVC casing became popular, though stainless steel helical wire wound screen continued to dominate.

In Ghana, percussion drilling was preferred until the 1970s, when Canadian geologists supervised a large programme of rural water borehole drilling. These were fitted with hand pumps. Initially highly successful, incursion by mica and lack of maintenance put most of the boreholes out of action.

In Sierra Leone, the West German Prakla Seismos sited (using resistivity) and drilled 600 rural water boreholes in crystalline rocks.

The water supply decade (1980–1990) brought in a ten-fold increase in the rate of rural groundwater development in Nigeria. British and Canadian hydrogeologists evolved a design for boreholes and hand pumps which became the norm for rural groundwater supplies. The concept of Village Level Operation and Maintenance (VLOM) of hand pumps was developed.

Standardization achieved the greatest speed for the greatest number, but it became apparent during this period that there was a need for small 'appropriate technology' drilling rigs. The Eureka, and later the Oxfam Portarig were introduced for shallow drilling, mainly in alluvium.

The author is grateful for help from surviving staff of the Colonial Service, the facilities of the libraries of the Geological Society and the Geological Survey of Nigeria. The author shares with many a debt of gratitude to all water geologists, living and dead, who worked in West Africa in difficult conditions.

References

ACWORTH, R. I. 1981 *The evaluation of groundwater resources in the crystalline basement of Northern Nigeria*. PhD thesis, Birmingham University.

BAIN, *et al.* 1924. *The Nigerian coalfield*. Geological Survey of Nigeria Bulletin No. 6.

BARBER, W. 1965. *Pressure water in the Chad Formation of Bornu and Dikwa Emirates, north-eastern Nigeria*. GSN Bull No. 35.

BARBER, M. & JONES, D. G. 1960. *The geology and hydrology of Maiduguri, Bornu Province*. Geological Survey of Nigeria Records 1958.

BEEBY-THOMPSON, A. 1961. *Exploring for Water*. Villiers Publications, London, 101–124.

BEESON, S. & JONES, C. R. C. 1988 The combined EMT/VES geophysical method for siting Boreholes. *Groundwater*, *26*.

CARTER, J. D. 1963. *The geology of parts of Adamawa, Bauchi and Bornu Provinces in north-Eastern Nigeria*. GSN Bull. No.30.

COCHRAN, H. A. 1937. *The technique of well sinking in Nigeria*. GSN Bull. No.16.

COOPER, W. G. G. 1926. *A rapid survey of the Gambia Colony*. Gold Coast Geological Survey Bulletin.

COOPER, W. G. G. 1934. Electrical prospecting. Mineralogical Magazine, **51**, 275–279.

COOPER, W. G. G. 1936. *Water supply investigation, Northern Territories*. Gold Coast Geological Survey Annual Report.

COOPER *et al.* 1930–1931. *Reports on the Water Supply of the coastal area of the Eastern Province of The Gold Coast*. Government of Gold Coast Sessional Paper XXVII.

CRATCHLEY, C. R. & JONES, G. P. 1965. *An interpretation of the geology and gravity anomalies of the Benue Valley, Nigeria*. Overseas Geological Surveys, Geophysical division. Geophysical Paper No. **1**.

DAVIES, J. & COBBING, J. 2002. *Low permeability rocks in Sub-Saharan Africa. An assessment of the Afram Plains, Eastern Region, Ghana*. British Geological Survey Internal Report, **CR/02/137N**.

DIXEY, F. 1931. *A practical handbook of water supply*. Murby.

DU PREEZ, J. W. & BARBER, M. 1965. The distribution and chemical quality of groundwater in Northern Nigeria. *GSN. Bull. No. 36*.

EDE, D. P. 1993. Rural water supply and sanitation demonstration project (Rusafiya) *Terminal report; UNDP/WB Water Supply and Sanitation Programme*.

EDMUNDS, W. M. 1991. Groundwater recharge in the west African Sahel. *NERC News*, **April 1991**, 8–10.

GROVE, A. T. 1951. *Land use and soil conservation in parts of Onitsha & Owerri Provinces*. GSN Bull. No.21.

HAZELL, R. *et al.* 1992. *The hydrogeology of crystalline aquifers in northern Nigeria and Geophysical methods used in their exploration*. From GSL special publication No.66.

HAZELL, R. & BARKER, M. 1995. Evaluation of alluvial aquifers for small-scale irrigation in part of the southern Sahel, West Africa. *Quarterly Journal of Engineering Geology*, **28**, 75–90.

JONES, B. 1948. *The Sedimentary rocks of Sokoto*. GSN Bull. No.18.

JONES, H. A. & HOCKEY, R. D. 1964. *The geology of part of south-western Nigeria*. GSN Bull.31.

MANNE, H. & SAHASRABUDDHE, D. 1925. *Experiments with the automatic water finder in the Trap Region of western India*. Bulletin 72, Department of Agriculture, Bombay.

MONKHOUSE, R. A. 1963. *Groundwater in the Eastern region of Nigeria: A compilation of present knowledge*. Geological Survey of Nigeria, Enugu.

REYNOLDS, J. M. 1987. The role of surface geophysics in the assessment of regional groundwater potential in northern Nigeria. *In:* COLSHAW, M. B., BELL, F. G., CRIPPS, J. C. & O'HARA (eds) *Planning and Engineering Geology*. Geological Society Engineering Geology, Special Publications, 185–190.

SIMPSON, A. 1954. *The Nigerian Coalfield*. GSN Bull.No.24.

SMEDLEY, P. L., EDMUNDS, W. M. & PELIG-BA, K. B. 1996. Mobility of arsenic in groundwater in the Obuasi gold-mining area of Ghana: some implications for human health. *In:* APPLETON, J. D., FUGE, R. & MCCALL, G. J. H. (eds) *Environmental Geochemistry and Health*. Geological Society, London, Special Publications, **113**, 163–181.

TOLMAN, C. F. 1937. *Ground Water*. McGraw Hill, New York.

Additional References

BARBER, W. & CARTER, J. D. 1955. *Water Supply of Nafada, Potiskum & Damaturu sheets*. GSN. Report No. 1157.

BUCHANAN, M. S. 1955. Water Supply of Igala division.

BUCHANAN, M. S. 1956 Water Supply Ogoja Province. *GSN Report No. 5052*.

CARTER, J. D. 1953. The hydrology of southern Idoma Division. GSN Report No. 1086

CARTER, J. D. 1956. Groundwater in Western Nigeria. GSN Report No. 1185.

CARTER, J. D. & BARBER, M. 1956. The rise in the water-table in parts of Potiskum Division, Bornu Province. *GSN Records*.

DAVIES, J. & MACDONALD, A. M. 1999. Final report: The groundwater potential of the Oju/Obiarea, Eastern Nigeria. *BGS Technical Report N- WC/99/32*.

DIYAM CONSULTANTS. 1987. Kano State shallow aquifer study. *Unpublished report*.

DU PREEZ, J. W. The Water Supply Drilling Program in Orlu Division. *GSN Report No.953,*

DU PREEZ, J. W. 1947. The hydrology of Gumel Emirate. *GSN Annual Report*.

DU PREEZ, J. W. 1947. The hydrology of Katsina town. *GSN Annual Report*.

DU PREEZ, J. W. & RICHARDS, H. J. 1955. The water supply of Katsina Town. ***GSN Records***.

DU PREEZ, J. W. & RICHARDS. 1955. The hydrology of Gumel Emirate, Kano Province. *GSN Records.*

FALCONER, J. D. 1911. The Geology and geography of northern Nigeria. *Macmillan*

FEDERAL GOVERNMENT OF NIGERIA. 1987. *Minerals and Industry in Nigeria with notes on the history of the Geological Survey in Nigeria.*

GROVE, A. T. 1959. A note on the former extent of Lake Chad. *Geog. Journ.*

GEOLOGICAL SURVEY OF GOLD COAST/GHANA. *1913–1964 Annual reports.*

GEOLOGICAL SURVEY OF NIGERIA, *1930–1961. Annual reports.*

HAZELL, R. 1954. Water Supply of Nsukka Division. *GSN Report No. 5096.*

HAZELL, R. 1955. Water Supply of Kontagora and Bida Divisions. *GSN Report No. 1149.*

HAZELL, R. 1956. Hydrology of Udi Division. *GSN Report No. 5167.*

HAZELL, R. 1956. The Water Supply of Okigwe Division. *GSN Report No. 5190.*

HAZELL, R. 1957. A Water-table Survey in the Orlu Area. *GSN Report No. 5194.*

HAZELL, R. 1958. The Water Supply of Obubra Division. *GSN Report No. 5195.*

HAZELL, R. 1958. The Water Supply of Afikpo Division. *GSN Report No. 5196.*

HAZELL, R. 1958. The Water Supply of Ogoja Division. *GSN Report No. 5197.*

HAZELL, R. 1960. Water Supply of the Udi-Nsukka plateau. *GSN Report No. 1231.*

HAZELL, R. 1960. Groundwater in the Eastern Region of Nigeria. *GSN Report No. 5198.*

HAZELL, R. 1969. *Maestrichtian aquifer under Ikaja.* Internal report to Guinness International.

JONES, B. 1933. The geology and water supply of Daura and Katsina Emirates. *GSN* Annual Report.

JONES D.G. 1957. The rise in the water-table in parts of Daura & Katsina Emirates, Katsina Province. *GSN Records.*

JONES, C. R. C. & BEESON, S. 1985. The EM/VES technique for borehole siting, Kano State. In: Advances in groundwater detection and extraction. *International conference on arid zone Hydrology and water resources, Maiduguri.*

JONES, M. J. 1985. The weathered zone aquifers of the basement complex in areas of Africa. *QJGS 18.*

JUNNER & BATES. 1945. Reports on the geology and hydrology of the coastal area west of the Akwapim range. *Gold Coast Geol. Surv. Memoir No.7.*

JUNNER & HIRST. 1946. Reports on the geology and hydrology of the Voltaian basin. Gold Coast. Gold Coast Geol. Surv. Memoir No.8.

MCREARY KORESKI. 1966. Water resources inventory for the Govt. of Sierra Leone. Phase 1. *San Fransisco.*

MALCOLM MCDONALD PARTNERS. 1986. Rural water supply report *to Kano State Agricultural Development Project Development Project and Rural Development Authority.*

MONKHOUSE, R. M. 1961. Groundwater in the Coastal Plain Sands east of the Calabar River. GSN. Report No. 5201

MONKHOUSE, R. M. 1963. The water-table in the Onitsha-Orlu area. *GSN Report No. 5202.*

MINERAL SURVEY OF S. NIGERIA. 1913. *Colonial Reports Misc. No. 89*

RAEBURN, C. & DU PREEZ, J. W. 1946. Water resources, minor irrigation schemes and soil Conservation, Jos plateau. Nigeria. *Conf. Africaine des Sols, Goma, Congo Belge. 1946*

RAEBURN, C. & DU PREEZ, J. W. 1948. Water resources of part of Plateau Province. *GSN Annual Report.*

RAEBURN, C. & JONES. BRYNMOR 1934. The Chad Basin; geology and water supply. *GSN Bull. No.15.*

RICHARDS, H. J. 1955. Report on Water Supply Conditions in Awgu Division. *GSN Report No.1161.*

RICHARDS, H. J. 1964. Water Supply of Awka division. *GSN Report. No. 5164.*

RICHARDS, H. J. Report on Well Siting in Abakaliki and Afikpo Divisions. *GSN Report No.5059.*

RICHARDS, H. J. Report on Water Supply Conditions in Bende Division. *GSN Report No. 5154.*

SCHULTZ INTERNATIONAL 1976 Hadejia River basin study. Unpublished report.

THOMPSON, J. 1953. The Hydrology of Potiskum division. *GSN Report No. 1111.*

THOMPSON, J. 1956. The geology and hydrology of Gombe, Bauchi Province. *GSN Records.*

UNDP/WORLD BANK. Rural Water Supply and Sanitation Demonstration Project (Rusafiya), Terminal Report, UNDP/World Bank Water Supply and Sanitation Programme.

WATER SURVEYS NIGERIA. 1977. Unpublished report: Water resources in the Gombe Agricultural Project Area. *GADP.*

WATTS and WOOD, W. E. Hand dug wells. *Intermediate Technology Publication.*

British attempts to develop groundwater and water supply on Gibraltar 1800–1985

EDWARD P. F. ROSE[1], JOHN D. MATHER[1] & MANUEL PEREZ[2]

[1]*Department of Geology, Royal Holloway, University of London, Egham, Surrey TW20 0EX, UK (e-mail: mather@jjgeology.demon.co.uk)*
[2]*AquaGib Limited, Suite 10b, Leanse Place, 50 Town Range, Gibraltar (e-mail: mperez@lyonnaise.gi)*

Abstract: The 6 km^2 peninsula of Gibraltar is unusual hydrogeologically as, in effect, a small but high limestone island, subject to a Mediterranean climate of cool wet winters and warm dry summers. Provision of an adequate water supply for its town and garrison has been a continuing problem, particularly as the population has grown from about 3000 in the 18th century to over 30000 by the end of the 20th. The narrow peninsula is dominated by the Rock, a mass of Lower Jurassic dolomite and limestone whose main ridge has peaks over 400 m high. Early supplies of potable water were from roof and slope rainwater runoff, and from shallow wells in the Quaternary sands that cover 'shales' flanking the Rock at low levels. Intermittent hydrogeological studies through the 19th and 20th centuries, notably in association with the British Geological Survey in 1876, 1943–1952, and 1974–1985 attempted to develop inferred groundwater resources within the sandy isthmus which links the Rock to southern Spain and in the Rock itself. Problems resulted from inadequate understanding of the geology, of recharge, of the behaviour of aquifers containing saline water at depth and of the need to protect aquifers from pollution. Failure to extract adequate groundwater led to development of a separate supply of saline sanitary water to reduce demand for potable water and innovative attempts to improve slope catchment of rainwater, before near-total commitment to desalination for potable supplies in 1993.

Gibraltar, ceded to Britain by the Treaty of Utrecht in 1713 (Jackson 1987), is a peninsula that juts north-south from Spain at the western strait entering the Mediterranean Sea (Fig. 1). Little over 5 km in length and 1.6 km in maximum natural width, it occupies an area of some 6 km^2, divisible into three regions:

(1) The Isthmus. A low-lying plain, <3 m above present sea level, joins Gibraltar to the Spanish mainland (Figs 1 & 2). It extends from the sheer North Face of the Rock northward for about 800 m to the frontier fence and beyond to La Linea in Spain. Since 1942 the Gibraltar airfield and its associated buildings have been developed on this region, adjacent to a cemetery in use since 1804, largely obscuring its natural surface area.

(2) The Main Ridge. A sharply-ridged crest with peaks over 400 m above sea level extends from the North Face of the Rock southward for nearly 2.5 km (Figs 1, 2 & 3). In east-west profile the ridge is asymmetric, with steep eastern scarps contrasting with a gentler western slope that is broadly consistent with the dip of the underlying bedrock. A fortified town, mostly low on the western slope, has been developed and garrisoned since 1160 under successive Moorish, Spanish and British administrations (Rose 2001).

(3) The Southern Plateaux. South of the Main Ridge, the Rock slopes steeply down to Windmill Hill Flats, a plateau inclined gently southwards from 130 m to 90 m above sea level (Figs 1 & 3). This plateau is bordered, most clearly to the south and east, by a steep cliff-line that leads down to Europa Flats – a plateau that slopes gently south from about 40 m down to 30 m above sea level. Steep cliffs fringe this plateau at its seaward margin. Surface topography here is primarily the result of Quaternary marine erosion and tectonic uplift: the two plateaux are wave-cut platforms, and their fringing cliffs the product of shoreline processes (Rose & Hardman 2000).

Bedrock geology

The geology of Gibraltar has been studied intermittently for nearly 250 years (Rose & Rosenbaum 1992; Rose 2004), notably by Ramsay & Geikie (1876, 1878), Bailey (1952), G. B. Alexander (Rose & Cooper 1997), and Rose & Rosenbaum (1990*a*, *b*, 1991). Their differing interpretations had implications for predictions of the strata likely to be encountered in deep boreholes. According to Ramsay & Geikie (1878), Gibraltar was a mass of Jurassic limestone, overlain by 'shales' in the (western) town area. It was truncated by a normal fault at the North Face that downthrew limestone with presumed overlying Tertiary sandstone beneath the sands covering the Isthmus and bisected by a NW–SE 'Great Main

From: MATHER, J. D. (ed.) 2004. *200 Years of British Hydrogeology*. Geological Society, London, Special Publications, **225**. 239–262. 0305-8719/04/$15 © The Geological Society of London.

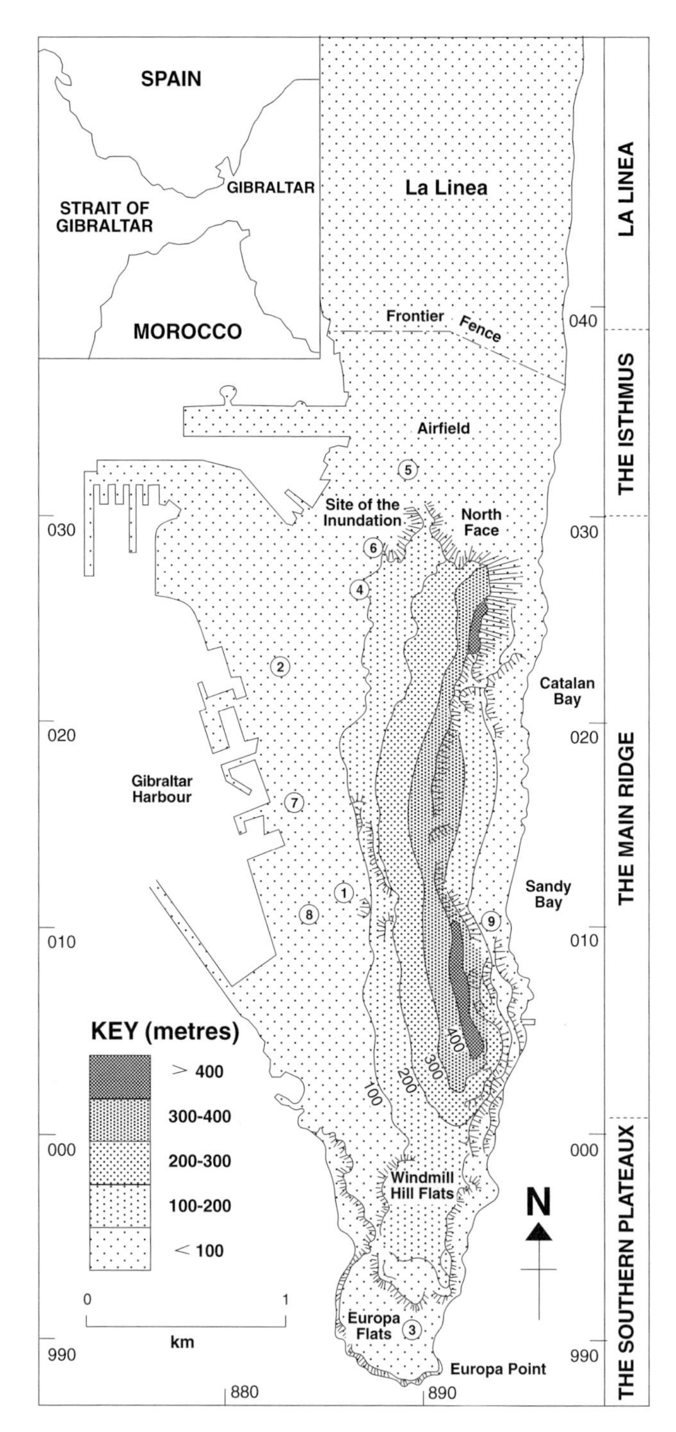

Fig. 1. Map of Gibraltar showing localities mentioned in the text, with major topographic regions indicated to the right, and regional diagram inset top left. 1. Botanic (=Alameda) Gardens; 2. John Mackintosh Square; 3. Nun's Well; 4. Moorish Castle (=Tower of Homage); 5. Site of Governor's garden; 6. Site of Orillon Spring; 7. Trafalgar Cemetery; 8/9. Ends of Admiralty Tunnel. Place names, contours at 100 m intervals above mean sea level, hachured cliff lines and UTM grid co-ordinates are based on a © Crown copyright/MOD map, Gibraltar Town Plan scale 1:5000, map series M984 edition 5–GSGS, and reproduced with the permission of the Controller of Her Majesty's Stationery Office.

Fig. 2. Aerial view from Spain SW across the Isthmus to the North Face and east coast of the Main Ridge of Gibraltar, showing part of the airfield, cemetery (mostly in shadow), and frontier fence in near foreground. © Estoril Ltd., Gibraltar: published with permission.

Fig. 3. Aerial view of Gibraltar from the SE, showing the two Southern Plateaux in the foreground, the Main Ridge behind, the harbour and part of the town area to the left (west), and parts of La Linea and the Spanish coastland in the background. From Rose & Rosenbaum (1989*a*), courtesy of the Institution of Royal Engineers.

Fault' that separated westward-dipping, uninverted limestone in the Main Ridge from eastward-dipping inverted strata in the Southern Plateaux. During the Second World War, excavation of numerous tunnels, and quarrying on the northern and eastern flanks of the Rock, enabled Sapper A. L. Greig of the Royal Engineers to observe 'shales' outside the western area. His observations provided the basis for Bailey's (1952) re-interpretation of the Rock as a klippe of overturned dolomitic limestone, underlain rather than overlain by 'shales' and thrust over Tertiary flysch sandstones (Rose & Rosenbaum 1989*a*). Soon after the War ended, Captain G. B. Alexander of the Royal Engineers generated an unpublished geological map which depicted the 'Gibraltar Limestone' as a unit both underlain and overlain by 'shales', in which only the sequence in the Main Ridge is inverted – that in the Southern Plateaux is not (Rose & Rosenbaum 1989*b*).

This last interpretation is maintained by the most recent geological map (Rosenbaum & Rose 1991*a*), and in the most recent detailed account of the bedrock (Rose 2000*b*). The Rock is currently interpreted as the much-eroded outlier of a sequence of Early Jurassic shallow-water carbonates thrust over deeper-water 'shales' (Fig. 4). A major east-west transcurrent fault somewhere beneath the Isthmus is believed (Rose & Rosenbaum 1994) to separate the Jurassic strata of the Rock from Tertiary flysch sandstones/mudstones visible north of La Linea, across the Spanish border. A major NW–SE fault zone (the Great Main Fault of Ramsay & Geikie, 1878) is inferred to separate inverted, westerly-dipping carbonates of the Main Ridge from uninverted, easterly-dipping carbonates of the Southern Plateaux.

The three regions of Gibraltar differ in terms of geology (Fig. 4) as well as geomorphology (Fig. 1):

(1) The Isthmus is formed by uncemented Quaternary sands and clays plus conglomerates or scree breccias, proved in shallow boreholes to overlie 'shales' at a depth of some 20 m in the west and uncertain bedrock over 60 m in the east (Rose & Rosenbaum 1991, figs 3.1 & 10.5).

(2) The Main Ridge is formed by the Gibraltar Limestone Formation, an overturned sequence which dips at moderate to high angles to the WSW. 'Shales' occur at low levels both east and west of the Ridge: the Catalan Bay Shale Formation dips westward beneath inverted Gibraltar Limestone at the base of cliffs along much of the Ridge's eastern and northern margins; the near-vertical Little Bay Shale Formation underlies much of the town area along the western margin of the Ridge, and crops out in the Great Main Fault zone. Neither of the 'shale' formations is well exposed, for they are generally obscured by buildings or by Quaternary sediments. Structurally, the Rock is now interpreted as the remnant of a nappe emplaced during the Betic Orogeny, presumably by the Early Miocene (Bailey 1952; Rose & Rosenbaum 1994; Rose 2000*b*). A thrust plane is inferred to lie beneath it, at shallowest depth in the town area to the west, but has not knowingly been observed in excavations.

(3) The Southern Plateaux essentially lie south of the Great Main Fault. They are eroded into uninverted Gibraltar Limestone which dips at moderate to high angles to the ESE. Shales crop out only as small patches along the west coast.

Two lithologies thus dominate bedrock: 'limestone' and 'shales'. The Gibraltar Limestone Formation is well-cemented, crystalline and generally medium- to thickly-bedded, increasingly dolomitic towards its base. It has been mapped (Rose & Rosenbaum 1990*a*; Rosenbaum & Rose 1991*a*) in terms of four members: three relatively thin and predominantly of dolomite, succeeded by a much thicker member, mostly of true limestone. Facies analysis (Bosence *et al.* 2000) indicates deposition as shallow-marine peritidal carbonates, arranged in metre-scale, shallowing-upward cycles. Dolomitization and/or cementation (Qing *et al.* 2001) have obliterated the primary porosity. The Catalan Bay Shale comprises medium-bedded grey cherty limestones 0.2–0.3 m thick alternating with thinner beds of reddish-grey fissile mudstones, structurally underlain by younger thinly-bedded cherts (Rose 2000*b*). The Little Bay Shale consists largely of purplish-red mudstones with thin 0.05–0.15 m beds of grey-green chert.

The pattern of faulting in the bedrock is complex, with three major fault sets NW–SE, NE–SW and north-south (Fig. 4). Wright *et al.* (1994) noted that if the NW–SE and NE–SW faults are conjugate shears, the north-south faults would be tensional features – an inference possibly confirmed by the north-south trend of the most prominent solution features.

Joint patterns have been analysed from discontinuity mapping and scanline surveys undertaken to monitor the stability of some of the 50 km of unlined tunnels and chambers concentrated in the 2.5 km-long Main Ridge, extending both north-south and east-west, but coverage is patchy. There is as yet no overall analysis. Bedding and joints commonly reveal three orthogonal discontinuity sets. Observations generally within the tunnels indicate that most bedding planes and joints are tight, but locally both steeply-dipping bedding planes and master joints show evidence of movement (slickensides, and narrow bands of breccia and clay gouge) and/or development of solution fissures.

Quaternary geology

Volumetrically, the most important superficial deposits flanking the Rock are thick Quaternary scree

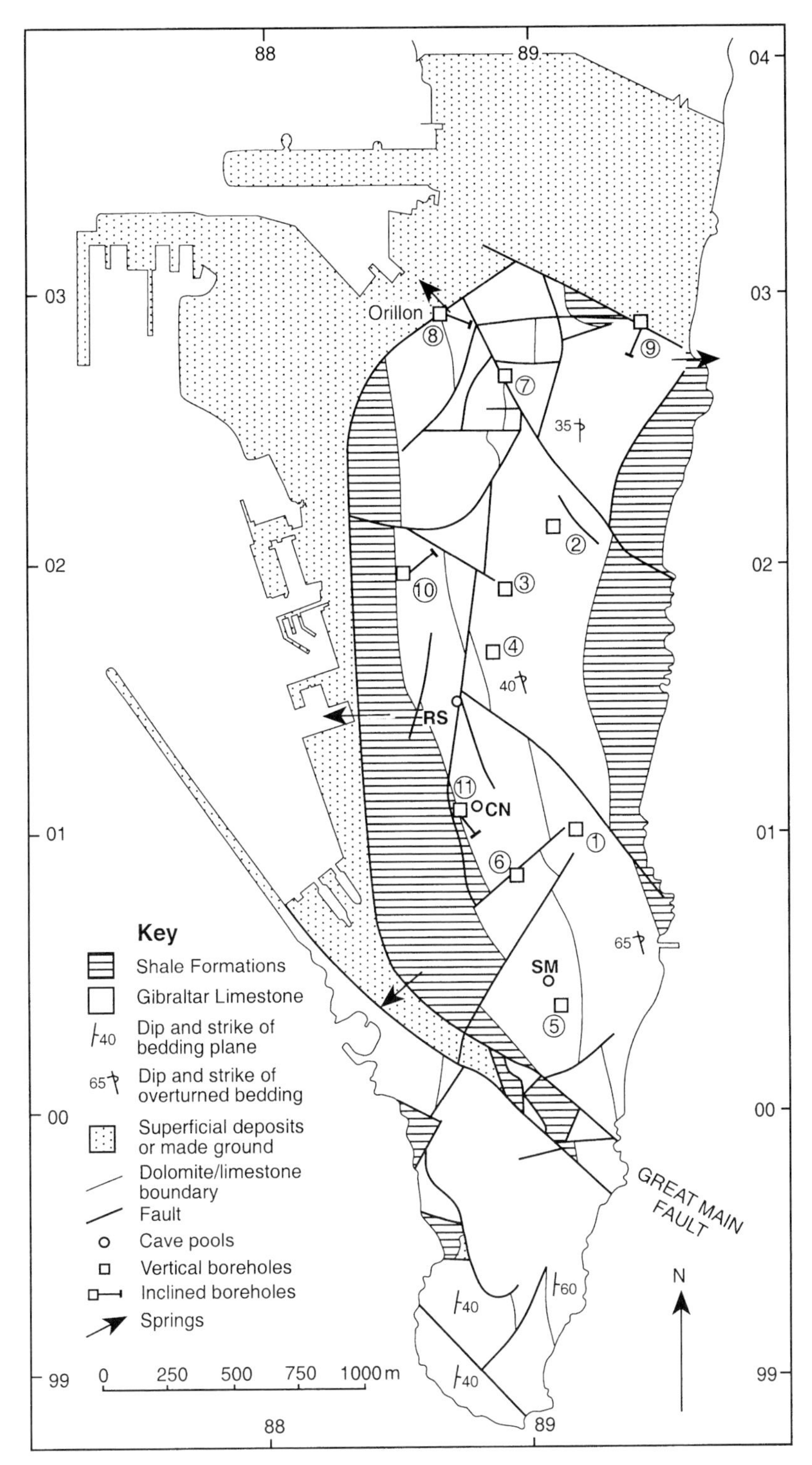

Fig. 4. Outline geological map of Gibraltar, with 1000 m UTM grid co-ordinates, showing positions of exploration boreholes drilled in 1979–1980, plus sites of low-level (RS: Ragged Staff; CN: Comcen) and high-level (SM: St Michael's) cave pools, and springs. After Wright *et al.* (1994), courtesy of The Geological Society.

breccias (Fig. 5), overlain by sands deposited under conditions fluctuating with respect to both climate and relative sea level (Rose & Rosenbaum 1990*b*; Rose & Hardman 2000).

The Isthmus Sands are largely unconsolidated sands and subordinate clays, seemingly deposited finally under marginal marine conditions as a tombolo (spit-bar) in Holocene times.

The poorly-cemented Alameda Sands (formerly known as the Red Sands) that underlie much of the town area and extend to the Botanic Gardens (Fig. 1) further south, are quartz-rich, with grains which are well rounded and well sorted, with a mode in the fine sand fraction (125–250 μm), and a poor to moderate calcite cement. Up to 16 m in thickness, these are former beach sands re-deposited by wind action. Their typical dark reddish brown colour is seemingly the result of post-depositional soil-forming processes.

The Catalan Sands are banked against the 1-km-long eastern cliffs of the Rock, apparently from near sea level to a maximum height of *c.* 200 m. These poorly-cemented, yellow-brown sands are moderately well-sorted and medium (250–500 μm) grained, the grains subrounded and highly spherical. Grains are dominantly (*c.* 80%) of quartz. The grain size, roundedness, sphericity and high degree of sorting of the sands, together with the large scale cross bedding visible in a quarried face at the toe of the slope, are all features consistent with their deposition by wind action. The sands apparently originated on a marine beach to the east of the Rock, before being blown westwards to accumulate as dunes against the Main Ridge.

Groundwater recharge

The present-day climate is mild Mediterranean, with warm dry summers alternating with cooler, wetter winters. Mean maximum daily temperatures are 24 °C for the warmest month (August) and 13 °C for the coldest month (January). Daily rainfall records have been kept for one or more low-level stations on the Rock since 1790. The frequency distribution of annual values is normal, with a mean of 838 mm, a standard deviation of 262 mm, and a range from 381 to 1956 mm (Wright *et al.* 1994). Secular plots of raw and smoothed data of departures from the mean show no long-term trends to wetter or drier conditions but do show some obvious periodicities (Wright *et al.* 1994, fig. 4). Recharge to the main Rock has been estimated, by means of soil moisture (29–39%) and chloride balance (25%), as equivalent to some 410000 m^3a^{-1} (Wright *et al.* 1994, p. S24).

Early sources of water supply

Moors from Morocco seized Gibraltar from the Visigoths in the year 711, but founded their first city there only in 1160 (Jackson 1987). Sometime afterwards they routed water seeping from the western Rock via channels and imported earthenware pipes to a cistern near the harbour for maritime use (Palao 1979). From 1462 until 1704 Gibraltar was permanently occupied by Christian Spaniards. According to Finlayson (1994), the earliest reference to a water supply for the town as such comes from the writings of Alonso Hernandez del Portillo (1624), almost mid way through this period. Portillo (as translated by Finlayson 1994, p. 60) records that by then 'the city contained many wells and fountains of very sweet and healthy water', plus cisterns for the collection of rainwater. It had also been served by an aqueduct (constructed or enhanced in 1571), which by 1624 was broken and dry. The origin of many of these works, however, could apparently be traced back to the period of Moorish occupation.

Following the capture of Gibraltar by the British and Dutch in 1704 during the War of Spanish Succession, and the award of sovereignty to Britain in 1713, a number of visitors described its contemporary water sources. Robert Poole spent 28 days on Gibraltar in 1748. He mentions the Spanish aqueduct, by then refurbished, describing its water as 'soft and well tasted' (Benady 1996, p. 68). A more detailed description is provided by Thomas James (1771), who served six years with the garrison artillery from 1755. He describes how the aqueduct brought water from the 'Red Sands' (=Alameda Sands of Fig. 5) south of the town to a fountain near the town centre – a route illustrated by Palao (1979) as beginning downslope (due west) of the present Botanic Gardens. According to James (1771, p. 345), 'this aqueduct is extremely well executed, it was begun by the conde de la Corsana, under the directions of a jesuit, taken from an aqueduct at Carthage.' It followed the line of the earlier Moorish water route, which James was able to follow from earthenware pipes then still *in situ*.

According to James 'the water is extremely good, and very much purified by its filtering through that immense body of sand, before it arrives at the aqueduct; it will keep for many years and it is reported, that formerly this water was greatly valued, insomuch that several used to come to Gibraltar from the different parts of Spain on purpose to drink it for particular disorders. I have drank of it of 15 years bottling, as clear and as pure as when it first ran in the aqueduct' (James 1771, p. 346).

The original site of the fountain is marked by the steps called 'Fountains Ramp' on the western side of John Mackintosh Square (Fig. 1). The fountain facia itself (Fig. 6), which dates from the refurbishment of

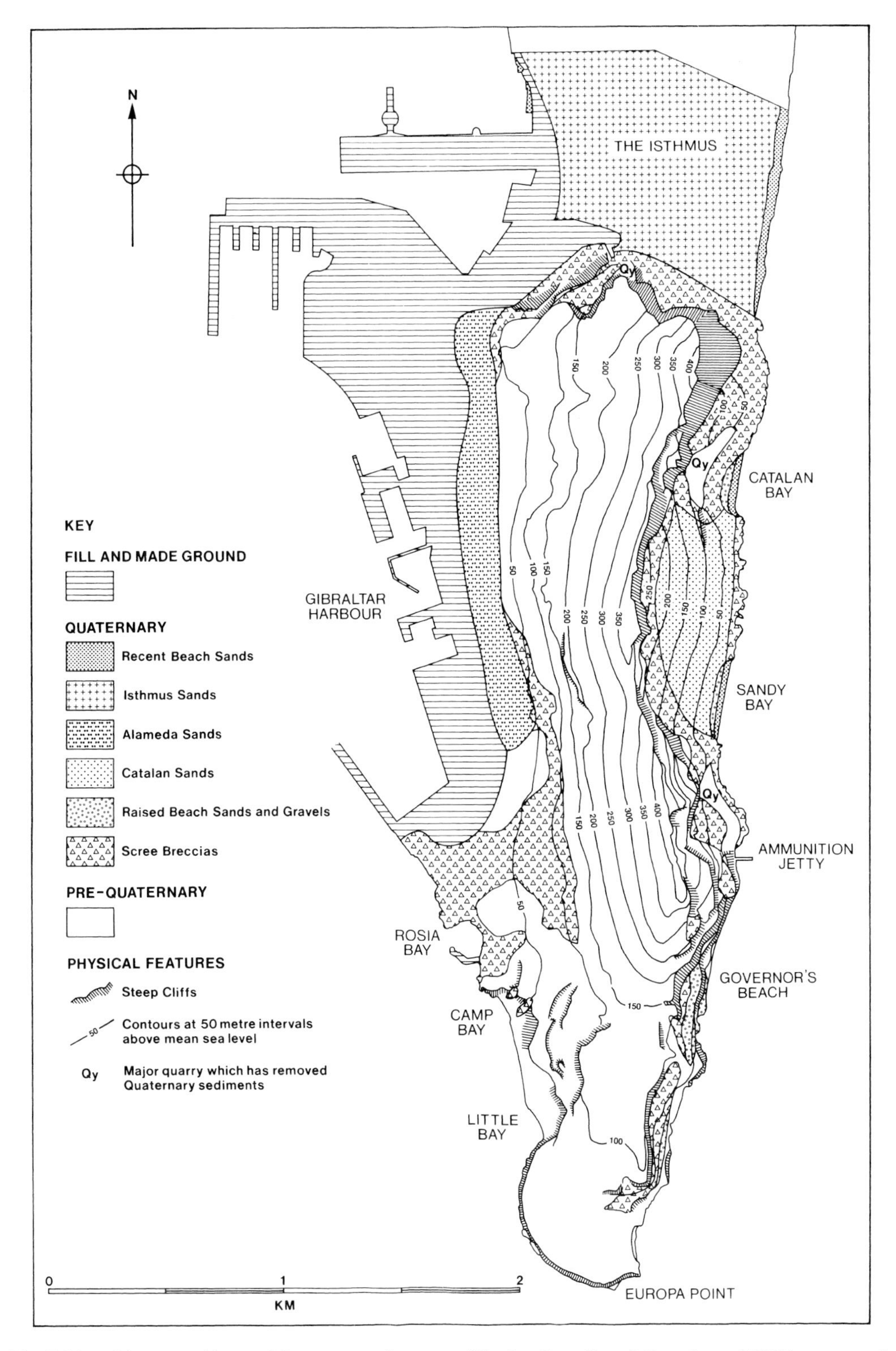

Fig. 5. Map of the most widespread Quaternary sediments on Gibraltar. From Rose & Rosenbaum (1990*b*), courtesy of the Institution of Royal Engineers.

Fig. 6. Aqueduct fountain head of 1694. The four lion-headed vents represent war, pestilence, death and peace (James 1771; Palao 1979).

Fig. 7. The *noria*: raising water from a shallow well on the Isthmus, NW of the North Face. Part of a 19th Century engraving.

the aquifer in 1694, has been re-erected nearby, some 20 m to the NW. One of the original ventilation shafts to the aqueduct is preserved *in situ* about 1 km to the south. James (1771, p. 347) provides an engraving of the fountain as it looked in 1755, and modern descriptions of the aqueduct, including details of its route and fountain, have been published by Palao (1979).

Robert Poole also visited the 'Nuns' Bathing Place' (now known as Nun's Well) (Fig. 1), on Europa Flats. This he described as 'a place in the ground, paved at the bottom with brick, about 20 feet [6 m] square, and had formerly different partitions; perhaps for different sexes, with a building over it . . . There is now but a small depth of water in the bath, which is said to be only when it falls from the heavens, and not from any spring therein' (Benady 1996, p. 78). James (1771) describes the water as infested by small leeches which frightened the garrison troops forced to drink it. The 'bath' is a cistern, probably Moorish in origin (Palao 1979), supplied during rainy periods with runoff from Windmill Hill above. Other cisterns were associated with buildings. For example, above the northern town, the keep of the largely 14th Century Moorish Castle (now known as the Tower of Homage: Fig. 1) incorporated a shaft-like reservoir into which rainwater from the roof was channelled for storage and use (Lyonnaise des Eaux 1996).

The third water-gathering device described by Robert Poole was a well in the Governor's garden (Fig. 1). This was a large diameter shallow well, a *noria*, in which water was raised by wheels turned by donkeys or oxen walking around a platform (Fig. 7). It was still by far the most important well supplying the garrison over 100 years later (Barrack and Hospital Improvement Commission 1863, p. 268).

During the 18th Century, the water supply situation seems to have been difficult but manageable. Provided numbers did not exceed about 3000 in total, the civilian population and British garrison could cope by combined use of the old Spanish aqueduct and shallow wells yielding brackish water, both in the town and on the Isthmus, supplemented by some roof catchment of rainwater. However, when numbers began to increase in the early years of the next century (Sawchuk 2001, table 3.1), problems began to emerge, particularly in years when rainfall was below average.

1800 to 1863 – the period leading to the creation of the Sanitary Commission

In December 1800, Governor O'Hara wrote to his superiors in London that because of drought conditions he had ordered his Chief Engineer to sink a well on the Isthmus, immediately below the North Face of the Rock (Sawchuk 2001). This eased the water supply problem temporarily, and another well was sunk within the garrison walls to provide a more secure source. O'Hara also proposed that greater attention should be given to the construction of water storage tanks.

This call was taken up by one of his successors,

Sir George Don, who was Governor from 1814 to 1831. James Anton (1998, p. 87), describing Gibraltar in the 1820s, recorded that 'there are several public draw-wells which serve for the consumption of the garrison and some private ones for irrigating the gardens: and Sir George Don, on assuming the lieutenant-governorship of the garrison, adopted measures to prevent a general scarcity of water, by enjoining every builder of a house to sink a tank to receive the rain falling from the roof. This, in a few years when old buildings give place to new, will afford an ample supply to the inhabitants, if the order be enforced'.

Don also sank wells on the Isthmus, and proposed the construction of tanks at the base of gullies on the face of the Rock close to the Moorish Castle to collect runoff for supplying individual houses and to clear the sewers in dry weather (Finlayson 1994). Although the yield of wells on the Isthmus was described by Don (1818) as 'immense', the water was frequently judged to be unwholesome and described as 'at all times brackish' (Sawchuk 2001, p. 90).

The botanist E. F. Kelaart, writing in 1846, had rather a different view of the Isthmus wells. According to his account (Kelaart 1846, p. 21):

> 'The wells on the neutral ground [=Isthmus], although only a few feet deep, afford abundance of fresh water, and they have not been known to fail even in the driest seasons. The source of this abundant flow of fresh water, in almost a sandy desert, and so near the sea, is a subject of some interesting speculations. The only way I can account for it is, by attributing its source to the mountains of Spain [north of La Linea] on the principle of artesian springs, or as some suppose, there may be at no considerable depth below the sands, a layer of impenetrable rock which prevents the rainwater from percolating to unfathomable depths and thereby makes the neutral ground a reservoir for water.'

Kelaart can seemingly be credited with establishing the myth that a large reservoir of artesian fresh water might lie beneath the Isthmus. Kelaart (1846) also mentions the aqueduct, which was still providing water to the fountain in the town's central square.

By 1860, water was thus supplied from the aqueduct; rainfall collected in tanks and cisterns; and public and private wells. However, according to returns quoted by Sayer (1862), out of 959 houses on Gibraltar 213 were without either a well or cistern, and out of 16 303 inhabitants 5799 were forced to buy water from street vendors. Sayer, who was the Civil Magistrate at Gibraltar, recorded (1862, p. 475) that 'To many houses tanks are attached, in which during the rainy season the water is collected, but . . . in many dwellings, especially among the poorer classes, no such convenience exists, and the poor creatures are dependent for the water they require upon the hawkers who distribute it through the city in small barrels carried on donkeys or mules'. Such water was 'conveyed from the isthmus' (Barrack and Hospital Improvement Commission 1863, p. 275).

The work of the Sanitary Commissioners and their engineer Edward Roberts.

In 1863 the problems of Gibraltar's water supply were highlighted in a government report on the sanitary condition of Britain's Mediterranean stations (Barrack and Hospital Improvement Commission 1863). Its authors, Captain Douglas Galton of the Royal Engineers and the physician Dr John Sutherland, had visited the Mediterranean region from late September through October and November 1861. Their report incorporated comments by Garrison-Quartermaster Hume made to the Governor in 1857: 'In the first place the inhabitants owe nothing to the British Government for the small supply of water they have had for 150 years . . . it would appear that from our conquest in 1704 down to within the last 15 years, nothing was done to collect water or to increase the supply, nothing whatsoever' (Barrack and Hospital Improvement Commission 1863, p. 274). The report summarized (p. 34) water supply as 'bad, deficient, and costly', drainage as 'defective' to 'very bad' and sanitation as 'most offensive and dangerous to health'. By this time the town population had grown to over 16000. Additionally, the garrison numbered some 6000–7000. Yet 'The military population [was] at all times on a water allowance just as if the garrison were in a state of siege' (p. 30) – 2.5 gallons (11.4 litres) per person per day and 'utterly incommensurate to the soldiers' wants in such a climate'.

Water from the four main categories of supply was sent to the War Department chemist for analysis. The results (Table 1) given in the report indicated (p. 31) that:

> 'The chemical constitution of these waters is extraordinary, both as regards the amount and character of their impurities, when compared with what are considered waters sufficiently pure for town use . . . The Gibraltar waters contain an extraordinary quantity of nitrates . . . The amount of organic matter is excessive, and there is also an excessive amount of chlorides . . . The well water and aqueduct water must derive their nitrates and organic matter by infiltration from a subsoil saturated with sewage . . . The chlorides in the wells indicate the presence of sea water.'

Water from the aqueduct and town wells was therefore deemed 'quite unfit for any but the most ordinary purposes' (p. 274), and these were soon closed as sources of potable supply. Although sewage was undoubtedly the pollutant of the town wells, this was not so obviously the case for the aqueduct. The aqueduct also tapped the Alameda Sands aquifer, but at its southern outcrop – an area formerly used as Gibraltar's principal burial ground.

Rainfall stored in cisterns was deemed 'the least objectionable [potable water], and ought to be chosen for extending the supply' (p. 32). The report therefore recommended construction of rainwater catchment areas on the Rock slopes above the town, one of them to supply a 1.5 million gallon (some 6750 m^3) reservoir then under construction near the Moorish Castle (Fig. 1). Additionally, it was deemed 'essential that [for sanitary purposes] a brackish or sea water supply be obtained . . . [but] Sea water, if it can possibly be avoided, should not be used for washing or watering the streets or yards' (p. 33).

Following the report's presentation to Parliament in 1863, and within months of a cholera epidemic in the summer of 1865 which claimed over 400 deaths, a new establishment of Sanitary Commissioners for Gibraltar was officially constituted by an Order in Council dated 20 November 1865 (Sawchuk 2001). Amongst their duties were the preparation of schemes for the supply of fresh water to the town and garrison and for a new system of drainage with a supply of water for flushing. Earlier in 1865 a civil engineer, Edward Roberts, had been sent to Gibraltar by the War Office to prepare plans for tanks to store runoff from the Rock (Roberts 1870). Roberts became the Engineer to the Sanitary Commissioners and over the next few years formulated schemes to improve both drainage (Roberts 1867) and water supply (Roberts 1870).

In December 1867 the Commissioners 'ordered a well to be sunk and borings to be made, on the Isthmus . . ., which were commissioned and carried on for some time, but without success, from want of proper tools and apparatus' (Roberts 1870, p. 10). However, below average rainfall in subsequent years induced the Commissioners to try again and, starting on 17 May 1869, a well (No 1) was sunk about midway between the eastern and western beaches, some 886 yards (810 m) north of the North Face of the Rock (Roberts 1869; Alton 1870) (Fig. 8). A large supply of good quality water was encountered at 13 feet (4 m) below the surface, after passing through strata consisting of a mixture of sand with boulders and shells. In order to ascertain whether the water was merely inflow from the surface, another well (No 2) was sunk about 80 yards (73 m) to the east. So much water inflow occurred at 13 ft that further deepening was abandoned. Pumping trials carried out with a pumping level at 11 ft (3.4 m) below ground surface yielded 500 to 600 gallons per minute for 72 hours, i.e. 3.3–3.9 Mld^{-1}. Recovery was rapid and it was concluded that there was an abundant supply of water.

Table 1. *Analyses of water samples from Gibraltar*

	Isthmus well	Aqueduct	Rainwater cistern	Town well
Ca	110	204	15	233
Mg	24	63	151	62
Na	42	159	20	447
HCO_3	326	602	754	222
SO_4	46	127	16	154
NO_3	68	174	9	623
Cl	65	249	31	700
Organic matter	58	142	72	110

Columns represent the four main categories yielding supposedly potable water in 1861. Data in mgl^{-1}, converted from grains per Imperial Gallon as recorded by the Barrack and Hospital Improvement Commission (1863). Na and K in the original analyses are recalculated as Na. Additionally, each analysis recorded a trace of ammonia.

Following the pumping trials, attempts were made to deepen No 1 well by boring through the fine white quartz sand full of water (quicksand) found at the base of the well. Clay was reached at 30 ft (*c*. 9 m) below ground surface and penetrated to 3 ft (*c*. 1 m). It was judged imprudent to go much deeper (Alton 1870).

The Commissioners obtained four separate analyses of the water which, although rather hard, was almost entirely free from organic matter. Water from the wells had a salinity of around 6.5 grains per Imperial Gallon (93 mgl^{-1} NaCl) and was considered to be 'good and potable' (Roberts 1870, p. 17). The success of these wells persuaded the Sanitary Commissioners that they could provide a permanent fresh water supply and a rising main was constructed to reservoirs above the town. From these piped water was made available to those house owners who would pay for it. Pipes were also laid to what is now John Mackintosh Square and a fountain installed. This was inaugurated on 8 December 1869 by the Governor's wife, Lady Airey (Benady 1993). Unfortunately, neither the scheme nor the fountain were successful. Water from the Isthmus wells became brackish very quickly after pumping commenced and concern was later voiced that it might be susceptible to pollution, e.g. from the cemetery (Sawchuk 2001). It was therefore used mainly for sanitary purposes. Thus developed Gibraltar's unusual dual water supply and distribution systems: one for potable water and the other for sanitary water. The latter eventually developed into the present-day salt water system whereby seawater is pumped, stored and distributed to all households for flushing toilets, fire-fighting, street cleansing and other purposes where potable water is not essential (Rose 2000*a*, fig. 9.9). The 'Airey Fountain' soon became dry and dusty and was replaced in 1879 by a

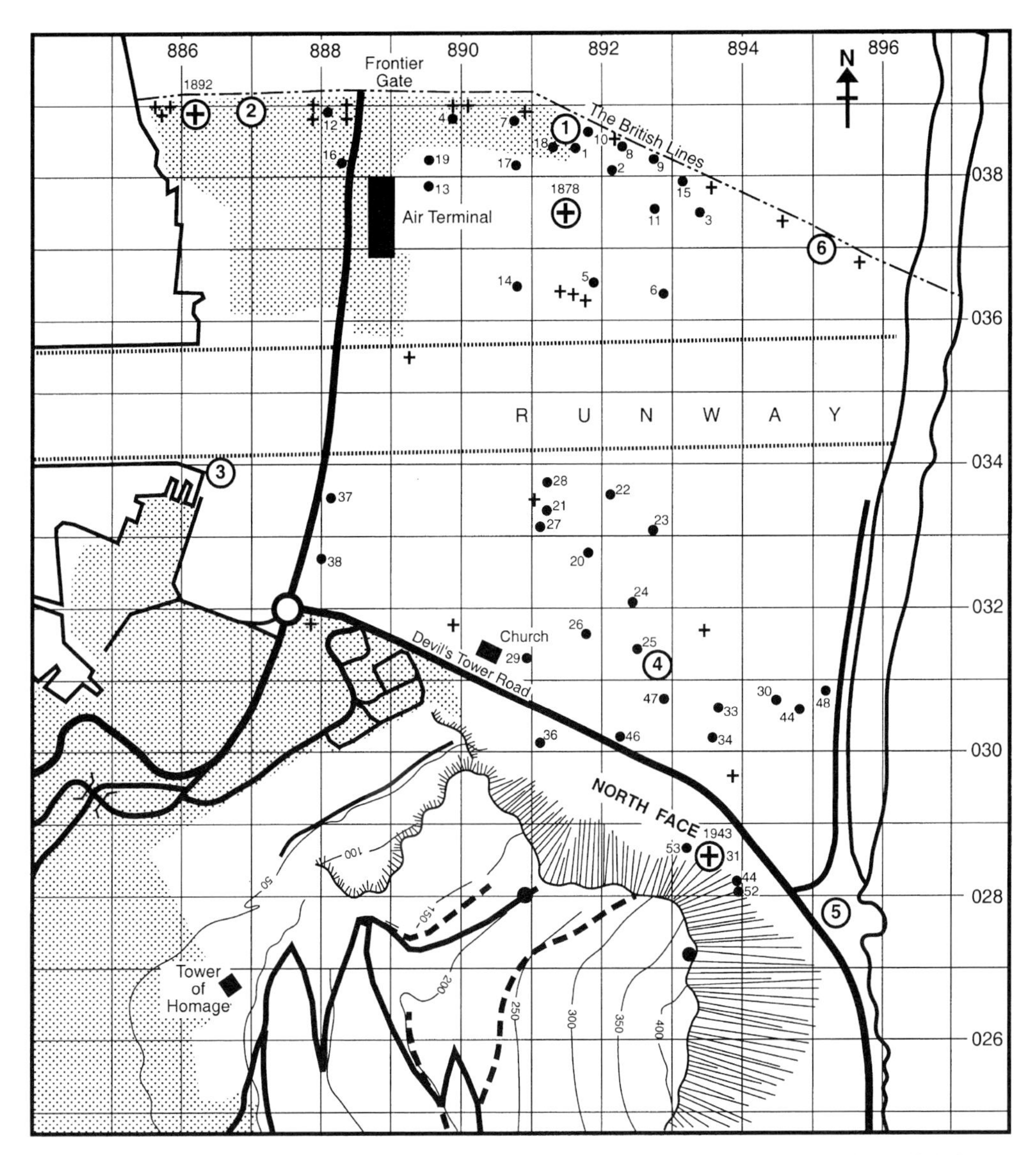

Fig. 8. Map of the Isthmus, showing wells and boreholes sited between 1869 and 1985. ⊕ deep boreholes of 1878, 1892 & 1943; + drive wells emplaced by Beeby Thompson in 1933; • wells and boreholes as catalogued by Murray (1975), numbers 1 to 19 being waterworks 'North Front' wells, 20 to 29 irrigation shafts in the cemetery, 31 the Devil's Dyke bore of 1943, plus miscellaneous wells and boreholes (numbers seemingly missing from the sequence refer to boreholes elsewhere on the peninsula); ①–⑥ boreholes drilled by the British Geological Survey in 1985. Topography, contours, grid lines at 50 m intervals, main roads and urban areas (stippled) are based on a © Crown copyright/MOD map, as for Fig. 1, and reproduced with permission.

more ornate affair installed to commemorate the stay of His Royal Highness the Duke of Connaught in Gibraltar, as described and illustrated by Benady (1993).

The Sanitary Commissioners were also charged with improving the drainage of Gibraltar. This involved the provision of water to flush the sewers that would be built. Roberts (1867) noted that water in the Inundation (Fig. 1), a lagoon created in the SW of the Isthmus early in the 18th century to bar access to the town except by a narrow causeway, was brackish rather than saline. He thought that it was fresh water made brackish by contact with the sea and suggested that it should be used for flushing instead of seawater. He also thought that it might be possible to separate the fresh water before it mixed with seawater, particularly water from the largest spring, near the Orillon Battery, at the western base of the North Face (Figs 1 & 4). This was attempted by the Royal Engineers, but without success (Roberts 1870).

The first hydrogeological survey – 1876

At the request of the Colonial Office, Andrew Ramsay and James Geikie, officers of the Geological Survey of Great Britain and Ireland, visited Gibraltar in the autumn of 1876 for the purpose of inspecting and reporting upon the water supply to the town and garrison (Ramsay & Geikie 1878; Geikie 1895). Such a project had been proposed in 1865, following completion of the first detailed trigonometrical survey of the Rock and at the same time as the formation of the Sanitary Commissioners, but had not been implemented then because of the outbreak of cholera (Rose 2004). The visit was subsequently postponed many times, but was effected at last in 1876, by which time Andrew Ramsay had succeeded Roderick Murchison as Director General of the Geological Survey.

Ramsay and Geikie sailed from England on 14 September 1876, reaching Gibraltar five days later, and departed on 25 October, returning to England on 30 October. The final report (Ramsay 1877), dated 16 January 1877, began with a general account of the geology of the Rock. It was accompanied by a series of maps and sections, coloured geologically, copies of which are currently preserved in the library of the British Geological Survey at Keyworth, Nottingham.

Ramsay noted that the Jurassic limestone was traversed by numerous fissures and caverns which served as channels during rainy seasons, allowing water to flow rapidly away. He considered that if the water table in the limestone was at even a moderate height above sea level, strong springs would be expected at or somewhere above sea level on the eastern side of the Rock and at the limestone/shale junction on the western side. Such springs were rare (Fig. 4), and when they did occur, they were brackish (Roberts 1870). Thus the evidence suggested that the limestone was highly permeable and that no head of fresh water could develop. Ramsay, therefore, concluded that he had 'no confidence that any available supply of water could be found by sinking a deep artesian well in any part of the limestone rock' (Ramsay 1877, p. 6).

The other possible source of potable water reviewed by Ramsay was further development of the sands on the Isthmus. The wells sunk by the Sanitary Commissioners yielded water of variable salinity. In some years (e.g. 1869) the water was 'good and potable' but at other times (e.g. 1875) it was 'quite unfit for domestic use' (Ramsay 1877). The increased salinity was caused by increasing concentrations of sodium and magnesium chlorides. These, Ramsay considered, were the result either of sea spray driven on to the land during gales, or of groundwater evaporation during the hot summer months. He clearly felt that the increase in salinity was related to processes close to the surface rather than intrusion of poor quality water from depth.

A clay horizon had been discovered at around 9 m below the surface during the drilling of No 1 well in 1869. Ramsay considered that although seawater might sometimes percolate through the sands above this clay, it did not follow that it should percolate through the sands beneath it. Therefore, it was important to investigate both the quantity and quality of water from these lower sands. He also felt that it was important to ascertain the properties of the bedrock that underlay these sands. The rocks exposed to the north of La Linea on the adjacent Spanish mainland are chiefly sandstones and he considered it possible that these could form the bedrock and, because of the elevation of these local Spanish hills, that they could provide a source of artesian water from beneath the Isthmus.

His report recommended that potential water supply both from the lower sands and the underlying bedrock should be thoroughly investigated. The report is one of the first hydrogeological reports produced by the Geological Survey and certainly its first for a region overseas. Following his visit to Gibraltar, Ramsay was a member of the Organising Committee of the National Water Supply Congresses convened by the Society of Arts in 1878 and 1879, and he spoke briefly at the Second Congress (Ramsay 1879). It is clear from his remarks that he had considerable experience of advising on water supply issues and this was probably the reason for his nomination to lead the Gibraltar study.

On the basis of the work carried out by Ramsay and Geikie, a borehole to a depth of 73.5 feet (22.4 m) was drilled between 29 November 1878 and 22 February 1879 (Figs 8 & 9). A detailed log (Tulloch 1890) shows the upper sands to be around 30 feet (9 m) in thickness, below which is 4 feet (1.2 m) of clay and 2.25 feet (0.7 m) of coarse sands (the lower sands), after which the borehole entered limestone either as compact beds or boulders mixed with sand. In other words, of the two potential groundwater sources identified, the lower sands were too thin to yield significant volumes of water and the underlying bedrock was limestone rather than sandstone.

Rather perversely, Geikie's biography (Newbigin & Flett 1917, p. 88) records that in 1879 Ramsey wrote to him 'in jubilant spirits because their joint work at Gibraltar had proved correct, although

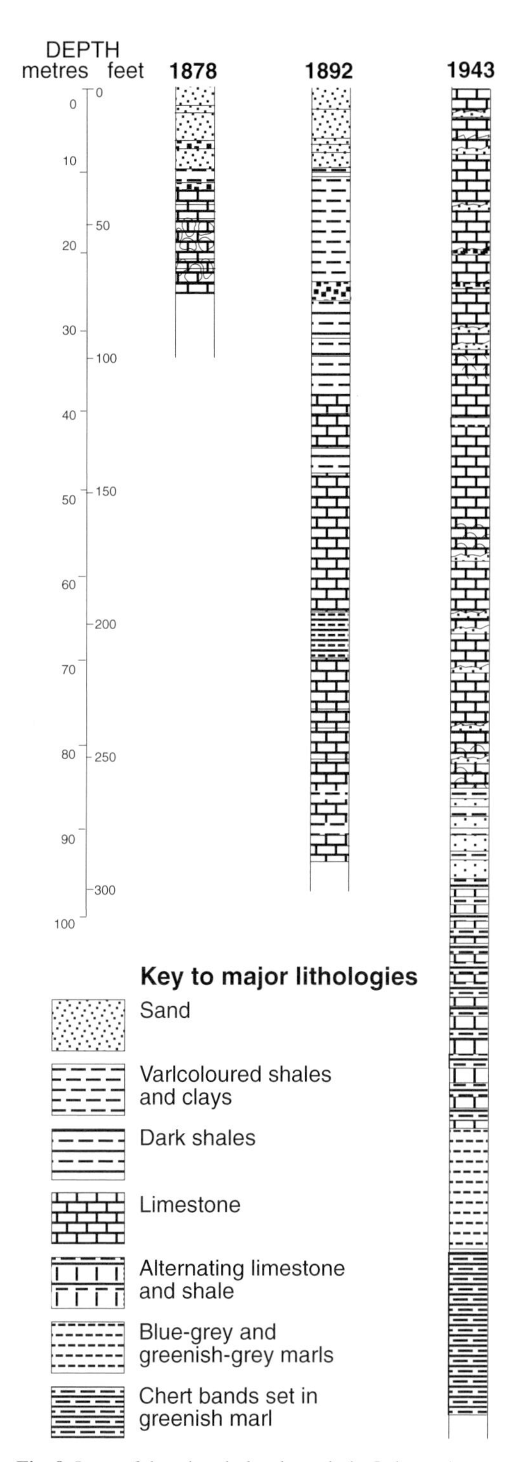

Fig. 9. Logs of deep boreholes through the Isthmus (see Fig. 8): 1878 – Ramsay bore; 1892 – Roberts Well bore; 1943 – Devil's Dyke bore. See Rose & Rosenbaum (1991) for details of lithology and stratigraphy.

certain borings had seemed at first to cast doubt upon some of their conclusions.' However, there seems no doubt that their conclusions relating to the hydrogeology of the Isthmus were fundamentally in error.

The report of Hector Tulloch (1890) and its aftermath

Despite the efforts of the Sanitary Commissioners problems remained and, in 1890, Major Hector Tulloch of the Royal Engineers, seconded as Chief Engineering Inspector to the Local Government Board, was sent out from London to report on both water supply and sewerage (Tulloch 1890). Tulloch reviewed three options for increasing potable water supplies: the Isthmus, the interior of the Rock, and surface catchments. Although he did not agree with Ramsay and Geikie that the sandstones of the adjacent Spanish mainland would be found beneath the Isthmus, he felt that the borehole put down in 1878 did not prove this conclusively as it penetrated only 3 ft 4 inches (1 m) of massive limestone, which could have been a large boulder. He therefore proposed a deeper borehole, close to the sea on the western side of the Isthmus, to prove once and for all if these sandstones existed.

Ramsay (1877) had dismissed the Rock itself as a possible source of water supply. However, Tulloch again did not agree with him. He considered that precipitation falling on the Rock, which clearly did not form rivers and streams, must pass into its heart and ultimately down to the level of the sea, below which it could not pass 'in consequence of the greater specific gravity of the salt water. It probably lies on the surface of the latter like oil on water' (Tulloch 1890, p. 23). Because of the presumably impermeable shales on the western side of the Rock which would prevent this fresh water moving to the west, Tulloch considered that it moved north-south along fissures in the limestone and accounted for the freshwater discharges seen near the Inundation (Figs 1 & 4). Although no discharges were observed on the eastern seaboard, Tulloch found evidence in the orientation of caves which led him to believe that fresh water was discharging along the southern part of the east coast. He suggested that this flow of fresh water could be tapped not by sinking wells but by driving 'an adit right into the heart of the Rock in an east and west direction so as to intercept the water flowing north and south' (Tulloch 1890, p. 26).

Tulloch proposed that the adit should start in the west at the Trafalgar Cemetery (outside the south gate of the town) and be driven eastward to the coast between Catalan Bay and Sandy Bay (Fig. 1). This idea had been mooted previously by a civil engineer, John Dixon and a shaft and short gallery constructed at the Trafalgar Cemetery. Dixon's venture was a failure, for salt water was pumped almost immediately. However, Tulloch considered that this was because pumping had begun 'before getting into the heart of the mountain' and that the water entering the gallery came not from the interior of the mountain but from the sea (Tulloch 1890, p. 26).

The adit proposed by Tulloch followed the line of a sewer, not constructed but proposed some years previously by the resident Colonial Engineer, Captain J. Buckle, late of the Royal Engineers, to an outfall on the eastern side of the peninsula. Tulloch envisaged that the two schemes might proceed side by side and stated that 'Practically, my proposal is this: drive a tunnel through the hill as if for a sewer. After it has been driven 300 yards (274 m), search for water. If water is found, continue the tunnel as a water conduit and intercepting channel. If water is not found, continue the tunnel through the mountain as a sewer' (Tulloch 1890, p. 42). If potable water was found and the tunnel was used for water supply then he proposed an alternative sewage outfall at the extreme south of the Gibraltar peninsula: Europa Point.

The third option considered by Tulloch was to increase the volume of water collected from surface runoff. The Sanitary Commissioners already had one reservoir at the Moorish Castle which received rainfall collected from a few acres [hectares] of ground lying above it. The limestone slope in this area had been rendered impermeable by sealing fissures and covering the surface with concrete. Although Tulloch considered that some further use of such catchments could be made, he concluded that 'this would be a herculean task, and the cost would be out of all proportion to the results obtained' (Tulloch 1890, p. 29).

When Tulloch's report was received in Gibraltar it generated some controversy. For example, Isaac Henry, writing in August 1892, commented that 'Major Tulloch's project of an adit through the Rock, is nothing short of a wild experiment which nothing can justify' (Henry 1892, p. 23). Henry also thought little of the idea of drilling a deep borehole on the Isthmus, but the proposed borehole was in fact drilled between March and September 1892 by Mr Thom, a British contractor (Buckle 1892). The borehole (Figs 8 & 9) reached a depth of 302 feet 6 inches (92.2 m), with a final diameter of 6 inches (152 mm). After penetrating 31 feet (9.4 m) of shelly sands, some 3 feet 6 inches (1 m) of clay was intersected before the borehole entered 'shales'. Hard limestone was reached at 119 feet (36.3 m), where it contained sea water in large quantities. Buckle (1892, p. 2) concluded that 'This important experimental borehole . . . has finally eliminated a much debated and even hopeful source of water supply for Gibraltar from the sandstone of the Queen of Spain's Chair [the hill north of La Linea where a Queen of Spain had supposedly watched her troops lay siege to Gibraltar] passing as formerly supposed under our boundary.'

The idea of the horizontal gallery through the Rock was finally abandoned when Lord Ripon (1893), then Secretary of State for the Colonies, informed the Officer Administering the Government of Gibraltar that he had 'decided that the search for water in the Rock should be abandoned'. His objections were that delays would occur in carrying out improvements to the drainage system and that the water if found would probably be brackish. The scheme favoured was to increase the use of surface catchments, a route considered too expensive by both Roberts and Tulloch.

Development of rainwater catchments and underground reservoirs

Following closure of many of the town's shallow wells because of pollution (Copland 1896), from 1896 plans for the extension of the use of water catchments were drawn up by William Wallace Copland, who was now the engineer to the Sanitary Commissioners. Between 1898 and 1900 four reservoirs were excavated within the Rock on the western side of the Main Ridge (Fig. 10, reservoirs 1 to 4), each being 200 feet long, 50 feet high, with an average width of 20 feet (about $60 \times 15 \times 6$ m) and total capacity of five million gallons (nearly 23000 m^3) (Gonzalez 1966). The excavations were lined with bricks and rendered impervious. These reservoirs were fed with water collected from 14 acres (about 55000 m^3) of rock outcrop on the NW upper slopes of the Rock. The slopes were cleared of surface vegetation and fissures sealed with grout. The water thus collected was conveyed through open channels and pipes into the reservoirs.

From 1903 onwards increased efforts were made to collect rainfall from surfaced catchment areas, the main Government catchments eventually occupying an area of 243000 m^2. The most spectacular were constructed on the east side of Gibraltar, on the Catalan Sand slope (Fig. 11). The first 40000 m^2 were constructed in 1903 (Gonzalez 1966). Rockfall boulders embedded in the slope were blasted away, the surface was trimmed as evenly as possible, and a channel and footpath were constructed at the lower perimeter of the collecting area. Timber piles were driven through the sand and a timber framing was nailed to them, covered with corrugated iron sheets. Rainwater falling on this catchment area flowed down into the channel at the foot of the catchments and thence via a tunnel into the reservoirs within the Rock. The catchment area increased by a further 56000 m^2 between 1911 and 1914 (and eventually by another 40000 m^2 below the existing catchments

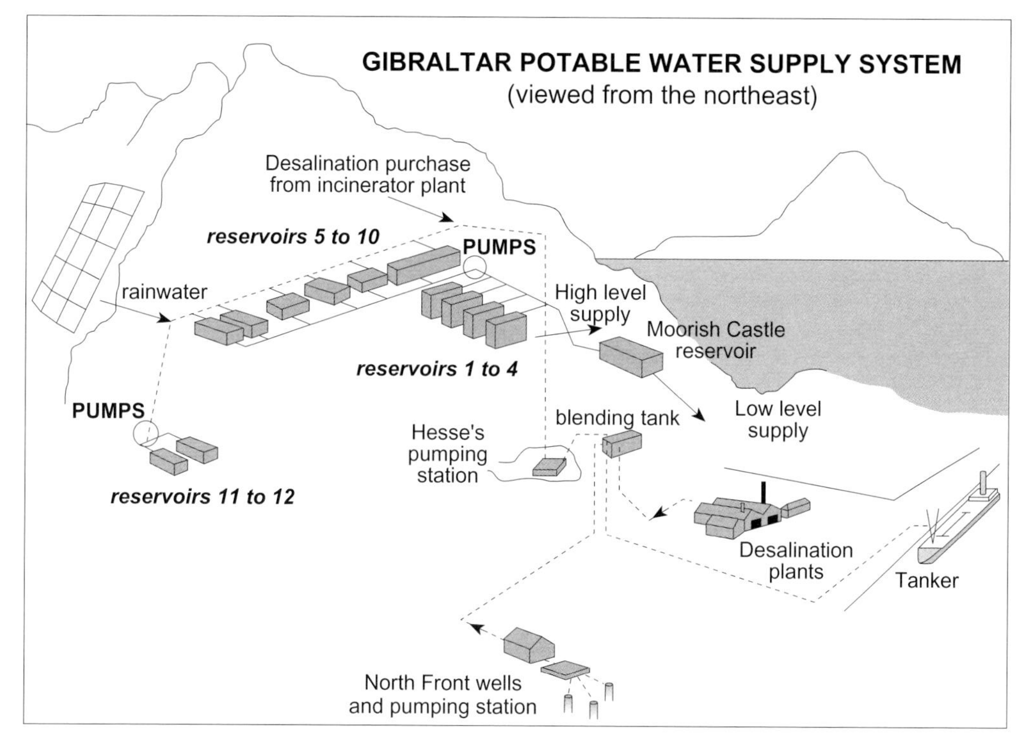

Fig. 10. Cross-section diagram of collection, storage, and supply of domestic potable water as at 1992, showing main underground reservoirs and associated tunnel systems. From Rose (1998) courtesy of The Geological Society, London, based on a diagram provided by AquaGib Ltd.

between 1958 and 1961). In addition to these municipal catchments, some 175 000 m^2 of prepared rock-face or surfaced slope, together with appropriate storage, were dedicated to naval or military use.

The four reservoirs completed by 1900 were increased by a further reservoir (Fig. 10, reservoir 5), of two million gallons (some 9000 m^3) capacity, constructed between 1911 and 1914. Five more reservoirs were excavated between 1928 and 1945 (Fig. 10, reservoirs 6 to 10), with a total capacity of five million gallons (nearly 23 000 m^3). Two more reservoirs (Fig. 10, reservoirs 11 to 12), each of one million gallons (about 4600 m^3) capacity, were constructed between 1958 and 1961. Including the original Moorish Castle reservoir (Fig. 10), the eventual waterworks system thus had a total of 13 reservoirs with overall capacity of 16 million gallons (over 73 000 m^3). From the town system, annual potable water consumption is currently over one million m^3.

The years immediately preceding the Second World War

Despite this increasing use of water catchments, water supply problems continued. Further wells were sunk on the Isthmus and seven supply wells were in existence by 1932 (Pearce 1934). Some of these wells were deepened by drilling boreholes in the base of their shafts. Some were used to supply sanitary water, whilst others contributed to the potable water supplies. During years of low rainfall, the catchments did not capture enough rain and the salinity of well waters increased, resulting in great difficulties in supplying the water required. The year 1931 was particularly difficult, and well water levels were so low that at times pump suctions were partly exposed. In addition, sand was being pumped from the wells, damaging the pumps (Pearce 1934).

Various solutions were proposed. In July 1932 a water diviner, Henri Tremolet, claimed that 'from experiments he had been able to make at a distance, he had discovered that there was a good source of drinking water brought down from the last buttresses of the Spanish Chain of Mountains which was lost at sea at a certain depth and that this water passed under the east portion of the Rock of Gibraltar in a fault which is not very deep' (Pearce 1934, p. 2). Tremolet offered to come to Gibraltar to carry out the necessary investigations with his divining rod, but the City Council would not consider the payment of fees or expenses unless satisfactory results were

Fig. 11. The main east coast water catchments above Sandy Bay viewed from the south prior to decommissioning in 1993 (see Figs 2 & 10). Sheets of corrugated iron fastened to wooden battens cover the smoothed upper surface of a 200 m-high Quaternary scree breccia and sand slope. Hotel developments at the toe of the slope indicate scale. From Rose & Rosenbaum (1990*b*), courtesy of the Institution of Royal Engineers.

obtained. Clearly, he was not certain enough of his ground to agree to such conditions for his visit did not take place.

In February 1933, the *Compagnie Hydraulique Afrique* offered to bore for water but the City Council resolved to ask for a British expert to visit the Colony. As a consequence, in August 1933 Arthur Beeby Thompson, an experienced consultant, arrived in Gibraltar and spent two months carrying out a series of investigations (Beeby Thompson 1934, 1969). With the aid of Norton tube well equipment loaned by the Royal Engineers, 25 1.25 inch [32 mm] tubes were driven to explore the groundwater potential of the unconsolidated Quaternary sediments, 22 on the Isthmus (Beeby Thompson 1934) (Fig. 8). These were driven by him personally to depths of up to 40 ft (*c.* 12 m) and he records that sometimes a whole hour would be spent driving the tubing a single foot (Beeby Thompson 1934). He also inspected galleries and tunnels within the Rock itself and natural caverns discovered by the Admiralty during tunnel excavations.

On the Isthmus, 14 wells were driven close to the frontier fence, and a wide variety of sediment types was encountered. Beeby Thompson noted that the static levels of water in sands above and below the clay layer at 30 feet (*c.* 9 m) depth, first discovered in 1869, were different, indicating that the clay isolated water in the lower sands from that in the upper sands. He further realized that water from the lower sands was of poorer quality, having a high salinity. He suggested that a large body of water with a low salinity awaited extraction from the upper sands of the Isthmus. This, he inferred, could be extracted through driven tubes or drilled wells suitably distributed over the area, especially along the frontier fence. However, drawdowns and the depth of screens would need to be controlled, to exclude the bacteriologically-contaminated 'top waters' (Beeby Thompson 1934).

Beeby Thompson was also impressed by the presence of large caverns within the heart of the rock, particularly the Ragged Staff and St Michael's cave systems (Fig. 4). The Ragged Staff caves, discovered by the Admiralty near sea level when tunnelling near the harbour in 1903, were associated with a complicated system of fissures. They had a standing water level 2.5 feet (0.76 m) above datum and contained water that was up to 30 feet (*c.* 9 m) deep. He proposed that pumping tests should be carried out in the Ragged Staff caves to measure the capacity of the cave system, accompanied by periodic analyses of salinity to show whether or not there was an influx of seawater as pumping progressed. As an addendum, he suggested that geophysical methods be employed to locate hidden rock fissures and caverns (Beeby Thompson 1934).

The City Council rejected the idea of a geophysical survey (Pearce 1934), but agreed to implement trials of driven tube wells into the upper aquifer of the Isthmus. A nest of five tube wells was driven and coupled together. A second site was selected, but when good yields were obtained it was immediately converted to an additional well (No 8) (Fig. 8). At the end of 1934, Pearce (1934) recorded that water from this well was then being used generally for potable supplies. Unfortunately, the success of this well meant that Beeby Thompson's scheme for obtaining water from a network of widely distributed, interconnected driven wells on the Isthmus was never adequately explored.

The Second World War and the period to 1950

During the Second World War, tunnelling activity on Gibraltar reached a peak (Rosenbaum & Rose 1991*b*). A.L. Greig, who had been trained in geology as part of a course in Oil Technology at Imperial College, London, was based on the Rock at this time, initially only as a military driver (Rose & Rosenbaum 1989*a*). However, his value as a geologist was recognized by the Tunnelling Companies then operating, and he was transferred to the Royal Engineers. When E. B. Bailey, Director of the Geological Survey of Great Britain, stopped briefly at Gibraltar, in transit to and from Malta to provide hydrogeological advice, Greig's local knowledge enabled him to rapidly re-appraise the geology of the Rock and provide advice on water supply (Bailey 1943*a*, *b*).

Bailey (1943*a*) concluded that the Rock itself could be expected to yield fresh water from fissures in the limestone and suggested that 'The most profitable plan of exploration would be to follow downwards in a shaft some fissure which during rainy weather admits a considerable flow of water to one of the already existing tunnels. In pumping such a fissure below sea level there would always be a chance of drawing in the sea. Risk of sea penetration is, however, much reduced in this case by the occurrence of underlying and flanking shales'. He also accepted that 'Spanish Tertiaries' could lie close below the surface of the Isthmus, and suggested an inclined borehole from the northern outcrop of shales below the North Face of the Rock to tap their groundwater content. Bailey was a very charismatic individual and clearly imparted the idea of 'unlimited supplies of fresh water from Spain in Artesian Wells' (Clifton 1943) to Fortress Headquarters in Gibraltar. However, his recommendations on water supply effectively ignored most previous groundwater work on the peninsula.

The recommended borehole (subsequently known as the Devil's Dyke Borehole) (Figs 8 & 9) was begun in March 1943 and abandoned in May at a depth of 515 feet (*c.* 157 m) in cherts, when the drilling speed decreased markedly and the consumption of diamonds rose significantly (Mason-MacFarlane 1943). The borehole, drilled at an inclination of 70 degrees to the south, was logged by Greig. He concluded, in a note accompanying the borehole log, that the 'relatively undisturbed condition of the strata in the last 20 feet [6 m] of the core do not appear such as one would associate with big overthrusts and the possibility of encountering Spanish Tertiaries at depth under the Rock seems therefore much reduced as a result of this borehole' (quoted by Rose & Rosenbaum 1991, p. 55).

Greig himself put forward ideas on possible sources of water supply (Greig 1943). He inferred that the two extremities of the Main Ridge presented the best hopes of tapping groundwater. In his view, the best conditions for water accumulation were furnished by a strongly disturbed area in which the limestone was open enough to collect and store inflows from a wide catchment area.

Further comments on water supply are contained in an undated and unsigned nine-page typescript (Anon. *c.*1946) which from its style and content was probably generated by Lieutenant (later Captain) G. B. Alexander of the Royal Engineers – who worked on Gibraltar between 1945 and 1948 (Rose & Rosenbaum 1989*b*; Rose & Cooper 1997). This favours two possible schemes, said to be essentially similar to those put forward by E. B. Bailey in January 1944 (in a report now seemingly untraceable). The first scheme involved the construction of a gallery, slightly to the SW of the Orillon spring, driven to just below sea level in the face of the Rock, with a series of horizontal diamond drill holes directed laterally from this gallery parallel to the cliff face at about 1 foot (0.3 m) above mean sea level. The idea was to tap the northern flow of water within the Rock first postulated by Tulloch (1890). The second scheme involved drilling boreholes in the Admiralty Tunnel east-west through the centre of the Rock (Fig. 1), to measure water table elevation accurately. If this stood at not less than 4 feet (1.2 m) above sea level it was proposed that a shaft should be sunk to sea level and horizontal holes drilled with a diamond drill for a distance of 200 feet (*c.* 60 m) east and west of the shaft about 1 foot above sea level. Both schemes involved skimming water from just below the water table with a large area of collection to reduce drawdown and the upconing of the fresh water/salt water interface. It was recognized that any fresh water would be floating on underlying saline water and that careful abstraction was needed.

The continuing search for supplies 1950 to 1975

Despite the numerous reports, there was little action to follow, and water supply problems continued to affect Gibraltar. Additional wells were sunk on the Isthmus, the last one being No 19 in 1950 (Gonzalez 1966) (Fig. 8). As a last resort, water was imported by tanker from Southampton, Rotterdam or Morocco. Where possible, a new unused oil tanker on its maiden voyage to the Middle East was used (Anon. 1949; Gonzalez 1966; Doody 1981). Water from the condensers of steam machinery, used to lift seawater to various reservoirs for the supply of sanitary water, had provided some 0.9 million gallons (4000 m^3) of distilled water per annum since the 1890s (Lyonnaise des Eaux 1996). In 1953 the volume of distilled water was increased by the installation of two compression distillation units of American manufacture, and in 1964 a 320 m^3d^{-1} MSF (Multi-Stage Flash) plant using waste heat from the Kings Bastion Generating Station was installed (Gonzalez 1966). This was followed by the commissioning in 1969 of a 1200 m^3d^{-1} MSF plant sited near Eastern Beach on the southeast coast of the Isthmus and in 1973 by a 1360 m^3d^{-1} Vertical Tube Evaporator plant sited on reclaimed land adjacent to the North Mole of the harbour.

A spectacular but transient local feature of Gibraltar is the Levanter cloud that results when moisture-laden east winds strike the precipitous eastern cliffs and are deflected upwards, to condense over the town and harbour to the west. A pilot scheme to collect water from these winds by condensation on metallic meshes was activated for eight months from mid-1957 (Hurst 1959) and further experiments were carried out in 1967–1968 (Canessa 1968). However, nothing came of these studies. Professional water diviners visited Gibraltar in 1952 and 1967 and one of these, a Mr G. K. Rogers, sited three wells (Manning 1967). His recommendations were not followed up. Beeby Thompson (1969, p. 370) records that on a visit to Gibraltar in 1952 he was informed that 'The authorities had been induced to test the claims of diviners, geophysicists, rainmakers employing a seeding with silver iodine and those with electronic devices, with no success'.

Attempts were made to obtain water from two of the three natural caverns that contained it. In 1954–1955, Lower St Michael's Cave (Fig. 12), with a pool surface level of around 850 feet (260 m) above sea level, was pumped out, but the pool refilled only slowly and not at all in dry weather (Manning 1967). Some water was pumped from the Ragged Staff Pool, with pool surface close to sea level, but was used purely as make-up water to 'add taste' to distilled water (Cruse 1975). The hydraulics of the Ragged Staff Pool were investigated by following the dilution of Iodine-131 tracer rapidly mixed with water in the pool at low tide (Smith & Clark 1963). The volume of the pool was determined as 3075 m^3 and the fresh water throughput at 26 m^3hr^{-1}, a flow figure (0.6 Mld^{-1}) quite large compared with the possible recharge.

Fig. 12. Lower St Michael's Cave: the lake. From Rose & Rosenbaum (1990*b*), courtesy of the Institution of Royal Engineers.

Attempts were also made to tap specialist expertise available to the armed forces through the part-time (Territorial Army) officers within the Engineer Specialist Pool administered by Central Volunteer Headquarters Royal Engineers in the UK (Rose & Hughes 1993). Major Peter Manning (in civilian life a geologist with the Northern Ireland office of the Institute of Geological Sciences) visited Gibraltar in 1967 to determine the likelihood of augmenting the water supply by use of a Royal Engineers well drilling team (Manning 1967). He inferred that there was no scope for further development on the Isthmus but that water within fault zones in the limestone with their associated brecciation might act as conduits and yield significant volumes. He selected three sites but cautioned that 'At the moment the above mentioned boreholes rank, though better than other sites, as little more than wildcatting'.

Lieutenant-Colonel Kenneth Cruse (in civilian

life a well drilling contractor) visited and reported in 1975. One of his conclusions was that 'It would be most useful to drill a deep borehole close to the frontier and as far as possible from an existing well. This would settle once and for all the question of whether or not the Spanish Tertiaries underlie the Isthmus at a reasonable depth with regard to possible water abstraction' (Cruse 1975, p. 16). Thus after well over 100 years, the dream of a major source of artesian water derived from the Spanish mainland lived on, despite all the failed attempts to prove it in the past.

Cruse noted that the salinities of groundwater in general were rising and thought that this was almost certainly due to the use of seawater for street washing and flushing the sewers. He recommended four sites for drilling into the Rock, and also suggested further work on experimental pumping of the Ragged Staff Caverns and another low-level pool, known as the Troubled Waters Pool (= Comcen Pool of Fig. 4), off the Admiralty Tunnel. No action appears to have been taken on the recommendations of either Manning or Cruse.

The work of the Institute of Geological Sciences 1974–1985

Between 1974 and 1985 hydrogeological studies were carried out intermittently by the Institute of Geological Sciences (from 1965 successor to the Geological Survey of Great Britain, and from January 1984 again re-named, as the British Geological Survey). These studies were conducted in association with the Gibraltar Public Works Department and geologists of the Engineer Specialist Pool. At last, a properly planned and executed exploration programme was carried out. The work concentrated initially on evaluating the groundwater resources associated with the limestone of the Rock itself, but later considered storing surface runoff from the airport tarmac, and/or using it for artificial recharge of the Isthmus sands (Wright *et al.* 1994).

An initial visit by K. H. Murray took place in August/September 1974. A survey of all existing wells and boreholes was carried out, fissures and suspected faults were mapped and three boreholes were drilled to prove the extent of any freshwater lenses within the limestone (Murray 1975). Murray thought that the potential for further development on the Isthmus was limited, but two boreholes into the limestone beneath the North Face drilled to depths of 16.5 and 12.3 m respectively were encouraging, although water quality was poor. His third borehole, drilled from a tunnel in the geographical centre of the Rock to a depth of around 3 m below OD, found at least 0.8 m of oil floating at the water table. Investigation of the source of this oil led to the nearby Troubled Waters Pool, otherwise known as Comcen Pool (Fig. 4), in a natural cavern breached by tunnelling. Here fuel oil was known to have leaked from a broken pipeline nearby in the past. Murray considered that, in view of the considerable volume of rainfall infiltrating into the Rock, it was worth undertaking further investigations and suggested these should take the form of deep boreholes at sites previously suggested by Greig (1943) and Manning (1967).

Test pumping of the Troubled Waters Pool in 1978, under extremely difficult conditions, demonstrated a hydraulic connection between it and the Ragged Staff Pool at similarly low level in the Rock, some 400 m to the north. However, this was thought to be indirect as contaminants from Troubled Waters were not detected at Ragged Staff. The rate of recharge proved to be very low and it was concluded that further investigations at Troubled Waters were not justified (Murray 1979).

Planning for a drilling programme commenced in August 1977 and drilling was eventually implemented between November 1979 and July 1980 (Murray & Wright 1980; Wright 1981). All the boreholes were sited to intersect fracture zones (Fig. 4). Seven vertical holes were drilled to between 7 and 11 m below the water table. Five of these were either poorly productive or rapid up-coning of brackish water resulted when they were pumped. Six further holes were drilled at shallow inclinations (5–8 degrees below the horizontal), being designed to develop the thin freshwater lens more efficiently. Of these sites, two (at the Orillon and North Face) were the most successful, and produced fresh to slightly brackish water for quite lengthy periods when they were test pumped in 1980–1983 (Wright 1981; Shedlock 1982; Wright & Shedlock 1983; Wright *et al.* 1994). The thickness of potable water observed in the boreholes at the time of initial completion ranged from 2 to 15 m with a broad transition zone below. It was concluded that such a thin lens would be extremely difficult to develop and that a flexible abstraction regime would be needed to cope with the lengthy periods of above or below average rainfall. Pollution was also a cause for concern and it was emphasized that while fuel oil and seawater were stored within or transferred by pipeline through the Rock there was always a chance of accidental leakage (Wright *et al.* 1994).

Some 270000 m^3a^{-1} of rainfall was estimated to run to waste from the airport tarmac. This water had been rejected as a potable source in the past because of possible hydrocarbon pollution and the cost of storage reservoirs. Wright (1981) noted that the problem of storage might be solved by using the natural aquifers, either the limestone or the Isthmus sands. The upper aquifer on the Isthmus, originally defined by Beeby Thompson (1934), was inferred to have insufficient surplus storage available and the suggestion was to displace brackish water in the lower aquifer by fresher water from the airport tarmac.

Following several seasons of monitoring the quality of the runoff, it was established (Wright & Shedlock 1983) that its hydrocarbon values lay within World Health Organisation limits of potability, as did trace metal values except perhaps those for lead and iron, and that runoff could therefore be suitable for artificial recharge without pre-treatment. A drilling programme was consequently initiated in January 1985 (Shedlock & Wright 1985; Shedlock 1985). Boreholes were drilled at six sites (Fig. 8) between January and April at depths of up to 65 m. Basement shales were encountered at depths of some 20 m in boreholes on the western side of the Isthmus but much greater thicknesses of relatively coarse material, perhaps suitable for artificial recharge, were encountered in the centre and to the east where these shales were not reached even at depths of 65 m. The final available report (Shedlock 1985) mentions a series of pumping tests to be carried out in the boreholes to evaluate the most productive horizons. However, there is no record of this work having been carried out. The need for such studies was overtaken by decisions to pursue other solutions to the problems of supplying potable water, to a resident population now exceeding 30000.

Reliance on desalination – post 1984

In 1984, the existing relatively inefficient distillation plants came to the end of their operational life and two new MSF plants were commissioned. These were each rated at 1350 m^3d^{-1} and were designed to use waste heat from the Waterport Electricity Generating Station but have always relied on oil fired boilers. In 1991, the complex was expanded to include two reverse osmosis plants, each rated at 240 m^3d^{-1}. Moreover, in 1990 an agreement was reached for the construction of a new refuse incinerator which would produce steam from waste heat and provide an additional 650000 m^3a^{-1} of potable water – a venture which was only partially successful, with water production ceasing in 2001.

In 1991, the water supply service of Gibraltar, long in public ownership, was partly privatized. A joint venture company was formed between the Government of Gibraltar and the French company *Lyonnaise des Eaux*. This company, Lyonnaise des Eaux (Gibraltar), recently renamed AquaGib, is currently responsible for Gibraltarian water supplies under a licence agreement with the Government of Gibraltar. The company increased desalination capacity by 2000 m^3d^{-1} by commissioning two new reverse osmosis plants in 2001. Reliance is now placed almost exclusively on desalination to provide a potable water supply, but groundwater is still extracted from the Isthmus wells and used to give taste to the distilled water. The dual system which provides separate distribution networks for potable and sanitary water still persists, the sanitary system being provided by sea water pumped from intakes in the SW and NW of Gibraltar. Problems of corrosion and encrustation of the old cast iron sea water mains are being eliminated by renewing these mains using high density polyethylene pipes (Lyonnaise des Eaux 1996). Some four million m^3 of sanitary water are pumped annually, about four times the total of potable water.

Discussion and conclusions

The systematic, as distinct from *ad hoc*, development of groundwater resources on Gibraltar began with studies initiated by the newly-formed Sanitary Commission in 1865 and carried forward by their engineer Edward Roberts. However, even before he provided the first piped water supply for the city, the source of fresh water beneath the sands of the Isthmus had become a subject of debate and the idea that it might originate from artesian springs, whose flow derived from the mountains of Spain, had been put forward (Kelaart 1846).

Andrew Ramsay and James Geikie of what is now the British Geological Survey were, in 1876, the first of a long line of consultants to visit the Rock specifically to provide water-supply advice. Ramsay (1877) thought that the brackish nature of water derived from the Isthmus wells was the result of sea spray or evaporation rather than the up-coning of saline water from depth. It has been noted by Davis (1978) that Charles Darwin (1845), in the *Voyage of the Beagle*, had described the flotation of fresh water on denser seawater. This was one of the most widely read books of its time and must have been known to both Ramsay and Geikie. However, they clearly did not make any connection between the conditions described by Darwin and those on Gibraltar. On Ramsay's recommendation, a borehole was drilled on the Isthmus to search for supposedly downthrown 'Spanish Tertiaries', but at a final depth of only 73.5 ft (22.4 m) it proved inconclusive.

The next such visitor, in 1890, was Hector Tulloch, an engineer. Tulloch disagreed with Ramsay in most respects. Through comparisons with the situation on the island of Malta, he recognized that on Gibraltar fresh water probably floated on the surface of underlying salt water and that, providing the surface of the fresh water was not lowered too much, water good enough for potable use could be obtained. His idea of a horizontal adit through the heart of the Rock to intercept water flowing north and south was scientifically sound but local politics and cost probably account for the failure to implement it. Following his visit, the second borehole to look for the 'Spanish Tertiaries' was drilled, but again failed to find them. It would seem reasonable to suppose that this 302 ft (92 m) borehole finally eliminated these rocks as a possible source of potable water, but this was not always to be the view of visiting specialists.

Arthur Beeby Thompson, who came to Gibraltar in 1933, was an experienced water engineer and recognized the relationship which existed between fresh water and underlying saline water. His delineation of both an upper and a lower aquifer on the Isthmus, and his ideas for skimming fresh water from the upper aquifer, were based on sound hydrogeological principles, but the need to provide increased potable water as a matter of urgency meant that his longer-term ideas were never followed through.

Brief visits in 1943 by the Geological Survey of Great Britain's Director, E. B. Bailey, failed to advance the exploitation of groundwater on Gibraltar. Rather, they had a negative impact. He suggested a renewed attempt to drill for the elusive 'Spanish Tertiaries' and recommended following fissures down through the rock and pumping them despite the fact that vertical holes had been shown to yield brackish water very quickly in the past. The 515 ft (157 m) inclined borehole, drilled on his recommendation, again failed to find either 'Spanish Tertiaries' or artesian water. Perhaps Bailey took advice from colleagues when he returned to the UK as two schemes attributed to him, outlined in a manuscript (Anon. *c*.1946) thought to have been written by G. B. Alexander, propose skimming water from just below the water table using galleries and horizontal boreholes.

Both P. I. Manning and P. K. Cruse, visiting in 1967 and 1975 respectively, recommended drilling sites in the limestone. However, these were for vertical boreholes to be emplaced by a Royal Engineers well drilling team. Thus they took no account of groundwater conditions and the need to skim water from the surface of freshwater lenses. Subsequent studies by the British Geological Survey (known as the Institute of Geological Sciences between 1965 and 1984) looked for the first time at the thickness of freshwater lenses but ignored previous experience when they tried to exploit the thin lenses through vertical boreholes. Later, inclined boreholes were more successful but the thinness of the lenses and pollution from both seawater used for road washing and leakage of fuel oil stored within the Rock made exploitation extremely difficult.

None of the visits and studies by visiting geologists and engineers for well over a century resulted in any practical improvement to the development of groundwater sources on Gibraltar. The typescript attributed to Alexander makes reference to this and gives two reasons. First, a failure on the part of the geologists and engineers to define the questions proposed boreholes were designed to solve. Second, an unwillingness on the part of the authorities to incur expense for purely exploratory drilling. Another reason is that visiting experts were not usually called in until a potentially serious water shortage had developed and there was no time to implement recommendations which did not result in an immediate improvement in the potable supply.

Despite the precarious nature of water supplies on Gibraltar it would appear that groundwater protection has never been a major issue. Shallow wells in the town area became polluted by sewage. The first major groundwater supply, provided by an aqueduct from the Alameda Sands, was lost when the aquifer was developed as a burial ground. The sands on the Isthmus risk pollution from the airport, whose development has also reduced direct recharge to this aquifer. Half the catchment of the Isthmus wells is across the border in Spain – an area now largely built over, whose development is outside Gibraltar's control. Moreover, groundwater within the Gibraltar Limestone has been contaminated by leakage of fuel oil stored on or within the Rock. It is also susceptible to contamination by seawater used for washing roads and flushing sewers, and by leakage from the salt water mains.

In view of the muddled approach to groundwater exploration taken by the British in Gibraltar it is perhaps comforting to recognize that groundwater could never have been a total solution to the water supply problems. The current population uses over one million m^3 of potable water annually (say 3 Mld^{-1}). If the recharge rate calculated by Wright *et al.* (1994) for the main Rock is applied to the whole of Gibraltar, this equates to a recharge of a little over 3 Mld^{-1}. However, in developing thin freshwater lenses such as occur on Gibraltar allowances need to be made for the impact of drought periods and also the increase in the thickness of the transition zone between freshwater and underlying saline water brought about by pumping (Mather 1975). Taking these into account and considering the thin lenses, which were defined by Wright *et al.* (1994), it is likely that much less than 2.0 Mld^{-1} is recoverable on a sustainable basis. Thus whatever approach had been taken to develop groundwater resources alternative supplies would have needed to be sought elsewhere.

We thank copyright owners as indicated in figure captions for permission to reproduce illustrations; former members of the British Geological Survey (notably K. H. Murray, S. L. Shedlock and the late E. P. Wright) for helpful discussion; G. McKenna (Chief Librarian and Archivist, British Geological Survey), T. J. Finlayson (Gibraltar Government Archives) and L. Swift (Gibraltar Garrison Library) for access to archive documents; J. McGrow of the Royal Holloway Geology Department for drawing Figures 1, 4, 8 & 9; and two referees for constructive comment on the manuscript.

References

ALTON, G. 1870. *Remarks on the water supply for Gibraltar*. Gibraltar Garrison Library, Gibraltar.

ANON. *c*.1946. *Notes on water supply.* CEG (Chief Engineer Gibraltar) 576. [British Geological Survey archives, Wallingford].

ANON. 1949. Portsmouth water for Gibraltar. *Journal of the Institution of Water Engineers*, **3**, 514.

ANTON, J. 1998. Description of Gibraltar in the 1820s. *Gibraltar Heritage Journal*, **5**, 77–102.

BAILEY, E. B. 1943*a*. Gibraltar geology. Preliminary notes covering visit 5–7 February 1943. *In*: HOBDEN, S. C. L. & ROSE, E. P. F. 1973. *An engineering geology reconnaissance of the Sandy Bay – Catalan Bay area of Gibraltar*. Unpublished report, Central Volunteer Headquarters Royal Engineers, Camberley, Appendix 3 of Annex C, 9 p. [British Geological Survey library, Keyworth].

BAILEY, E. B. 1943*b*. Second report covering visit 2–3 April 1943. *In*: HOBDEN, S. C. L. & ROSE, E. P. F. 1973. *An engineering geology reconnaissance of the Sandy Bay – Catalan Bay area of Gibraltar*. Unpublished report, Central Volunteer Headquarters Royal Engineers, Camberley, Appendix 4 of Annex C, 6 p. [British Geological Survey library, Keyworth].

BAILEY, E. B. 1952. Notes on Gibraltar and the Northern Rif. *Quarterly Journal of the Geological Society, London*, **108**, 157–175.

BARRACK AND HOSPITAL IMPROVEMENT COMMISSION. 1863. *Report of the Barrack and Hospital Improvement Commission on the sanitary condition and improvement of the Mediterranean stations*. Eyre & Spottiswoode, for HMSO, London.

BEEBY THOMPSON, A. 1934. *Report on water supply of Gibraltar*. Gibraltar Garrison Library, Gibraltar.

BEEBY THOMPSON, A. 1969. *Exploring for water*. Villiers Publications, London.

BENADY, T. 1993. Gibraltar's main square. *Journal of the Friends of Gibraltar Heritage Society*, **1**, 11–17.

BENADY, T. 1996. Gibraltar in 1748. Described by Robert Poole. *Gibraltar Heritage Journal*, **3**, 61–90.

BOSENCE, D. W. J., WOOD, J. L., ROSE, E. P. F. & QING, H. 2000. Low- and high-frequency sea-level changes control peritidal carbonate cycles, facies and dolomitization in the Rock of Gibraltar (Early Jurassic, Iberian Peninsula). *Journal of the Geological Society, London*, **157**, 61–74.

BUCKLE, J. 1892. *Letter to the Chairman of the Sanitary Commissioners*. (Undated, but seemingly late 1892.) [Gibraltar Government Archives].

CANESSA, E. A. J. 1968. *Extraction of water from Levanter cloud. Preliminary notes.* Unpublished report, Gibraltar Government Public Works Department. [British Geological Survey archives, Wallingford].

CLIFTON, N. M. 1943. *Letter from the Chief Engineer's Office, FHQ Gibraltar, to Brigadier H. E. Heathrow, War Office London, dated 30 April 1943.* [British Geological Survey archives, Wallingford].

COPLAND, W. W. 1896. *Letter to the Chairman of the Sanitary Commissioners, 17 March 1896.* [Gibraltar Government Archives].

CRUSE, P. K. 1975. *The abstraction of further underground water in Gibraltar.* Unpublished report, Central Volunteer Headquarters Royal Engineers, Camberley. [British Geological Survey archives, Wallingford].

DARWIN, C. 1845. *Journal of researches into the natural history and geology of the countries visited during the voyage of H.M.S. "Beagle" round the world, under command of Capt. FitzRoy, R.N.* John Murray, London.

DAVIS, S. N. 1978. Flotation of fresh water on sea water, an historical note. *Groundwater*, **16**, 444–445.

DON, SIR G. 1818. *Letter to Secretary of State the Earl Bathurst, 10 September 1818.* [Gibraltar Government Archives].

DOODY, M. C. 1981. Gibraltar's water supply. *Journal of the Institution of Water Engineers and Scientists*, **35**, 151–154.

FINLAYSON, T. J. 1994. The history of Gibraltar's water supply. *Gibraltar Heritage Journal,* **2**, 60–72.

GEIKIE, A. 1895. *Memoir of Sir Andrew Crombie Ramsay.* Macmillan, London.

GONZALEZ, F. J. 1966. The water supply in Gibraltar. *Aqua: the Quarterly Bulletin of the International Water Supply Association*, **2**, 58–67.

GREIG, A. L. 1943. *An account of the geology of Gibraltar.* Unpublished report, Royal Engineers, Gibraltar. [British Geological Survey archives, Wallingford].

HENRY, W. I. 1892. *Remarks on Major Tulloch's scheme of water supply and sewerage for Gibraltar.* Published privately for the author, Gibraltar. [Gibraltar Government Archives].

HURST, G. W. 1959. Collection of water from cloud at Gibraltar. *Journal of the Institution of Water Engineers*, **13**, 341–352.

JACKSON, W. G. F. 1987. *The Rock of the Gibraltarians.* Associated University Presses, Cranbury New Jersey.

JAMES, T. 1771. *The history of the Herculean Straits, now called the Straits of Gibraltar; including those parts of Spain and Barbary that lie contiguous thereto.* Rivington, London, 2 vols.

KELAART, E. F. 1846. *Flora Calpensis, contributions to the botany and topography of Gibraltar and its neighbourhood.* John Van Vorst, London.

LYONNAISE DES EAUX. 1996. *Gibraltar water supply.* Publicity material, Lyonnaise des Eaux (Gibraltar), Gibraltar.

MANNING, P. I. 1967. *Report on the feasibility of drilling for groundwater in Gibraltar.* Unpublished report, Central Volunteer Headquarters Royal Engineers. [British Geological Survey archives, Wallingford].

MASON-MACFARLANE, F. M. 1943. *Letter from the Governor and Commander-in-Chief, Gibraltar, to Director of Fortifications and Works, War Office London, dated 14 May 1943.* [British Geological Survey archives, Wallingford].

MATHER, J. D. 1975. Development of the groundwater resources of small limestone islands. *Quarterly Journal of Engineering Geology*, **8**, 141–150.

MURRAY, K. H. 1975. *Report on research into the possibility of finding potable groundwater in Gibraltar.* Institute of Geological Sciences: unpublished report **WD/75/1**.

MURRAY, K. H. 1979. *Report on test pumping of Comcen Pool ("Troubled Waters"), Gibraltar.* Institute of Geological Sciences: unpublished report **WD/OS/79/5**.

MURRAY, K. H. & WRIGHT, E. P. 1980. *Report on the 1978–80 Gibraltar groundwater investigation*. Institute of Geological Sciences: unpublished report **WD/OS/80/14**.

NEWBIGIN, M. I. & FLETT, J. S. 1917. *James Geikie. The man and the geologist*. Oliver & Boyd, London.

PALAO, G. 1979. *Gibraltar: "our heritage"*. Ferma, Gibraltar.

PEARCE, W. H. 1934. *Report on water supply from wells at North Front*. Unpublished report, City Engineer, Gibraltar. [Gibraltar Government Archives].

PORTILLO, A. H. DEL 1624. *Historia de Gibraltar*. [Reprinted in 1994 by the Centro Asociado de la U.N.E.D., Algeciras, with introduction and notes (in Spanish) by Antonio Torremocha Silva.]

QING, H., BOSENCE, D. W. J. & ROSE, E. P. F. 2001. Dolomitization by penesaline sea water in Early Jurassic peritidal platform carbonates, Gibraltar, western Mediterranean. *Sedimentology*, **48**, 153–163.

RAMSAY, A. C. 1877. *Report on the question of the supply of fresh water to the town and garrison of Gibraltar.* Geological Survey of the United Kingdom, London.

RAMSAY, A. C. 1879. In discussion of essays submitted on suggestions for dividing England and Wales into watershed districts. *Journal of the Society of Arts*, **27**, 610.

RAMSAY, A. C. & GEIKIE, J. 1876. *Geological map of Gibraltar*. Scale 1:2500. Eight sheets mounted as one sheet, hand coloured, on base map of plan of the Fortress and Peninsula of Gibraltar surveyed 1865 by Lt Charles Warren RE (plus 3 sheets comprising 7 geological sections). Geological Survey of the United Kingdom, London.

RAMSAY, A. C. & GEIKIE, J. 1878. On the geology of Gibraltar. *Quarterly Journal of the Geological Society, London*, **34**, 505–541.

RIPON, EARL. 1893. *Letter from Secretary of State for the Colonies to the Officer Administering the Government of Gibraltar. No. 110 of 6 September 1893*. [Gibraltar Government Archives].

ROBERTS, E. 1867. *Report on the drainage of Gibraltar, accompanying plans and estimates for a new system of drainage and a supply of water for flushing. Prepared by direction of the Rt Honble the Secretary of State for War, for the consideration of the Sanitary Commissioners of Gibraltar*. Sanitary Commissioners, Gibraltar.

ROBERTS, E. 1869. *Log book*. Unpublished Manuscript. [AquaGib Limited].

ROBERTS, E. 1870. *Report on a proposed scheme for a supply of fresh water to the town and garrison of Gibraltar*. Gibraltar Garrison Library Press, Gibraltar.

ROSE, E. P. F. 1998. Environmental geology of Gibraltar: living with limited resources. *In*: BENNETT, M. R. & DOYLE, P. (eds) *Issues in environmental geology: a British perspective*. Geological Society, London, 81–121.

ROSE, E. P. F. 2000*a*. Geology and the fortress of Gibraltar. *In*: ROSE, E. P. F. & NATHANAIL, C. P. (eds) *Geology and warfare: examples of the influence of terrain and of geologists on military operations*. Geological Society, London, 236–274.

ROSE, E. P. F. 2000*b*. The pre-Quaternary geological evolution of Gibraltar. *In*: FINLAYSON, J. C., FINLAYSON, G. & FA, D. A. (eds) *Gibraltar during the Quaternary: the southernmost part of Europe in the last two million years*. Gibraltar Government Heritage Publications, Gibraltar, 1–29.

ROSE, E. P. F. 2001. Military engineering on the Rock of Gibraltar and its geoenvironmental legacy. *In*: EHLEN, J. & HARMON, R. S. (eds) *The environmental legacy of military operations*. Reviews in Engineering Geology **14**, Geological Society of America, Boulder, Colorado, 95–121.

ROSE, E. P. F. 2004. *Founders of Gibraltarian geology*. Gibraltar Government Heritage Publications, Gibraltar, in press.

ROSE, E. P. F. & COOPER, J. A. 1997. G. B. Alexander's studies on the Jurassic of Gibraltar and the Carboniferous of England: the end of a mystery? *Geological Curator*, **6**, 247–254.

ROSE, E. P. F. & HARDMAN, E. C. 2000. Quaternary geology of Gibraltar. *In*: FINLAYSON, J. C., FINLAYSON, G. & FA, D. A. (eds) *Gibraltar during the Quaternary: the southernmost part of Europe in the last two million years*. Gibraltar Government Heritage Publications, Gibraltar, 39–84.

ROSE, E. P. F. & HUGHES, N. F. 1993. Sapper geology: part 3. Engineer Specialist Pool geologists. *Royal Engineers Journal*, **107**, 306–316.

ROSE, E. P. F. & ROSENBAUM, M. S. 1989*a*. Royal Engineer geologists and the geology of Gibraltar. Part I – Tunnelling through the Rock. *Royal Engineers Journal*, **103**, 142–151.

ROSE, E. P. F. & ROSENBAUM, M. S. 1989*b*. Royal Engineer geologists and the geology of Gibraltar. Part II – The age and geological history of the Rock. *Royal Engineers Journal*, **103**, 248–259.

ROSE, E. P. F. & ROSENBAUM, M. S. 1990*a*. Royal Engineer geologists and the geology of Gibraltar. Part III – Recent research on the limestone and shale bedrock. *Royal Engineers Journal*, **104**, 61–76.

ROSE, E. P. F. & ROSENBAUM, M. S. 1990*b*. Royal Engineer geologists and the geology of Gibraltar. Part IV – Quaternary "ice age" geology. *Royal Engineers Journal*, **104**, 128–144.

ROSE, E. P. F. & ROSENBAUM, M. S. 1991. *A field guide to the geology of Gibraltar.* The Gibraltar Museum, Gibraltar.

ROSE, E. P. F. & ROSENBAUM, M. S. 1992. Geology of Gibraltar: School of Military Survey Miscellaneous Map 45 (published 1991) and its historical background. *Royal Engineers Journal*, **106**, 168–173.

ROSE, E. P. F. & ROSENBAUM, M. S. 1994. The Rock of Gibraltar and its Neogene tectonics. *Paleontologia i Evolució*, **24–25**, 411–421.

ROSENBAUM, M. S. & ROSE, E. P. F. 1991*a*. Geology of Gibraltar. Single sheet 870 x 615 mm : Side 1, cross-sections and solid (bedrock) geology map 1:10000, Quaternary geology, geomorphology, and engineering use of geological features maps 1:20000; Side 2, illustrated geology (combined bedrock/Quaternary geology) map 1:10000, plus 17 coloured photographs/figures and explanatory text. *School of Military Survey Miscellaneous Map 45.*

ROSENBAUM, M. S. & ROSE, E. P. F. 1991*b*. *The tunnels of Gibraltar*. The Gibraltar Museum, Gibraltar.

SAWCHUK, L. A. 2001. *Deadly visitations in dark time: a*

social history of Gibraltar. Gibraltar Government Heritage Publications, Gibraltar.

SAYER, F. 1862. *The history of Gibraltar and of its political relation to events in Europe from the commencement of the Moorish dynasty in Spain to the last Morocco war, etc.* Saunders, Otley & Co., London.

SHEDLOCK, S. L. 1982. *Pump testing of the North Face borehole and data logging, Gibraltar.* Institute of Geological Sciences: unpublished report **WD/OS/82/10**.

SHEDLOCK, S. L. 1985. *Deep drilling programme in the Isthmus of Gibraltar.* British Geological Survey: unpublished report **WD/OS/85/21**.

SHEDLOCK, S. L. & WRIGHT, E. P. 1985. *Drilling programme in the Isthmus of Gibraltar: 1985. Interim report.* British Geological Survey: unpublished report **WD/OS/85/12**.

SMITH, D. B. & CLARK, W. E. 1963. An investigation of the hydraulics of an underground pool with iodine-131. *In*: INTERNATIONAL ATOMIC ENERGY AGENCY. *Radioisotopes in hydrology. Proceedings of the symposium on the application of radioisotopes in hydrology held by the IAEA in Tokyo, 5–9 March 1963.* IAEA, Vienna, 77–88.

TULLOCH, H. 1890. *Report on the water supply and sewerage of Gibraltar.* Waterlow & Sons, London.

WRIGHT, E. P. 1981. *Gibraltar water supply: groundwater studies – review and progress report.* Institute of Geological Sciences: unpublished report **WD/OS/81/23**.

WRIGHT, E. P. & SHEDLOCK, S. L. 1983. *Gibraltar: groundwater resources study progress report and summary review.* Institute of Geological Sciences: unpublished report **WD/OS/83/4**.

WRIGHT, E. P., ROSE, E. P. F. & PEREZ, M. 1994. Hydrogeological studies on the Rock of Gibraltar. *Quarterly Journal of Engineering Geology,* **27**, S15–S29.

The first use of geophysics in borehole siting in hardrock areas of Africa

RON BARKER

School of Geography, Earth and Environmental Sciences, University of Birmingham, Edgbaston, Birmingham, B15 2TT. UK (e-mail: R.D.Barker@bham.ac.uk)

Abstract: The first resistivity soundings applied to borehole siting in Basement areas of Africa were probably measured by Dr. Sydney Shaw in 1933 in Southern Rhodesia, now Zimbabwe. Using a Megger Earth Tester and Wenner array, soundings were measured at a selection of both dry and successful boreholes, and compared. Re-examination of Shaw's soundings using modern technology shows that the sounding curves are of surprisingly high quality even on today's standards and indicate a strong relationship between interpreted depth to unfractured granite and drilling results. His survey helped the adoption in Africa of resistivity methods as standard practice in the location of drilling sites for water supply.

The current debate concerning the optimum geophysical technique to use in borehole siting in Basement areas might give the impression that the application of geophysical techniques to such problems is a relatively recent phenomenon. This would be a wrong assumption, for what were probably the first geophysical surveys for water-supply borehole siting were measured near Bulawayo, Southern Rhodesia (now Zimbabwe) more than 70 years ago.

In 1933 Sydney Herbert Shaw (1903–1991), a young assistant lecturer in the Department of Geology at the University of Birmingham, was interested in the potential applications of the new geophysical techniques which were then being developed. It was only a few years previously that Frank Wenner (1874–1915) and Conrad Schlumberger (1874–1954) had developed the basic techniques for measuring earth resistivity (Wenner 1915; Van Nostrand & Cook 1966), and instruments such as the Megger Earth Tester (Drysdale & Jolley 1924; Low *et al.* 1932) and Gish-Rooney double-commutating device were now available. Shaw had just returned from working as assistant to mining geologist and geophysicist Arthur Broughton-Edge (1845–1953), as part of the Imperial Geophysical Experimental Survey in Australia (Broughton-Edge & Laby 1931), and while there had become team leader of the electrical section. Shaw had previously attended lectures in geophysics, some of which had been given by Broughton-Edge, at the Royal School of Mines, and with his field experience in Australia was amply qualified to follow his developing geophysical interests at Birmingham. Thus he was well placed to seize the opportunity offered to him by the Irrigation Division of the Department of Agriculture and Lands of Southern Rhodesia to test resistivity techniques in borehole siting in Africa (Shaw 1934*a*, 1934*b*).

African field survey

The area selected was part of the Nata Reserve, in the Bulalima-Mangwe district, about 100 km WNW from Bulawayo (Fig. 1). Here the country is almost flat and is underlain by granite basement which outcrops in the south and dips below the Kalahari and Karoo sediments to the north. The Karoo beds consist mainly of sandstone overlain by basalts in places. Outcrops are infrequent and give little clue to the nature of the underlying geology.

24 boreholes had been drilled, mainly into the granite overlain by sediments, although some had been drilled into basalts. Not all the boreholes had produced acceptably high yields of water and a subsequent study of the drilling logs suggested that a necessary condition for success was the presence of a sufficient cover of sediments. However, with no surface evidence to indicate the areas of thick sediment overburden, the selection of drilling sites was an unresolved problem. Perhaps geophysical techniques could help.

Shaw set up camp 60 km north of Plumtree (Anon. 1933) and decided first to measure resistivity soundings at existing boreholes and compare the results obtained at dry boreholes with measurements at high-yielding sites. Further tests were then made to locate suitable sites where new boreholes could be drilled.

Sounding measurements and interpretation

Two types of instrument were used for the sounding measurements, a Megger Earth Tester (Fig. 2), which applied alternating current supplied from a hand-rotated generator in the instrument itself and gave a resistance reading directly in ohms and

From: MATHER, J. D. (ed.) 2004. *200 Years of British Hydrogeology*. Geological Society, London, Special Publications, **225**. 263–269. 0305-8719/04/$15

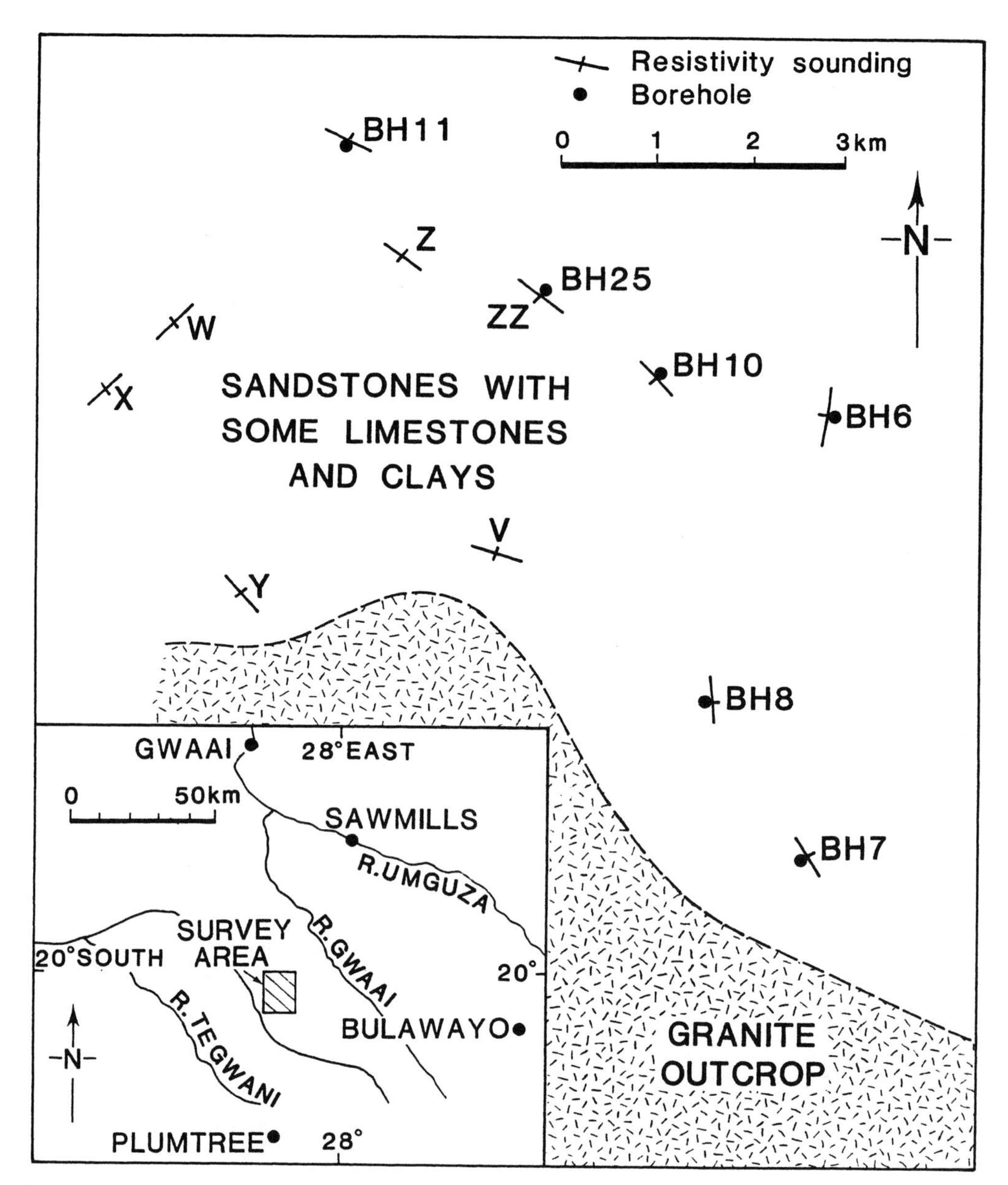

Fig. 1. Location of resistivity soundings and boreholes.

a Broughton-Edge potentiometer, which employed direct current and required special non-polarising electrodes. The results from the two instruments agreed to within 4%, although the Broughton-Edge potentiometer was able to measure much lower resistances and operate in areas of high contact resistance. Shaw recognized that a good time to carry out a survey was immediately after a shower of rain when the surface had been saturated and good contact could be made with the ground. The Wenner electrode arrangement was employed for the sounding measurements.

In 1933, good interpretation procedures were unavailable and sounding curves were plotted on linear graph paper with simple, inaccurate and empirical interpretations being carried out. If, however, Shaw's curves are plotted in a currently conventional double-logarithmic format (Fig. 3), typical granite basement soundings are revealed. This type of plot also indicates clearly the occasional

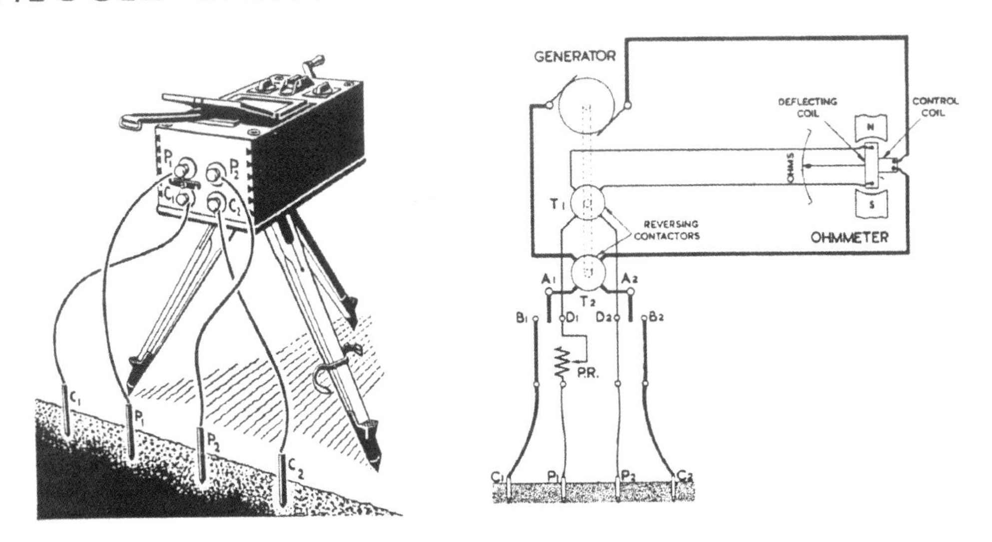

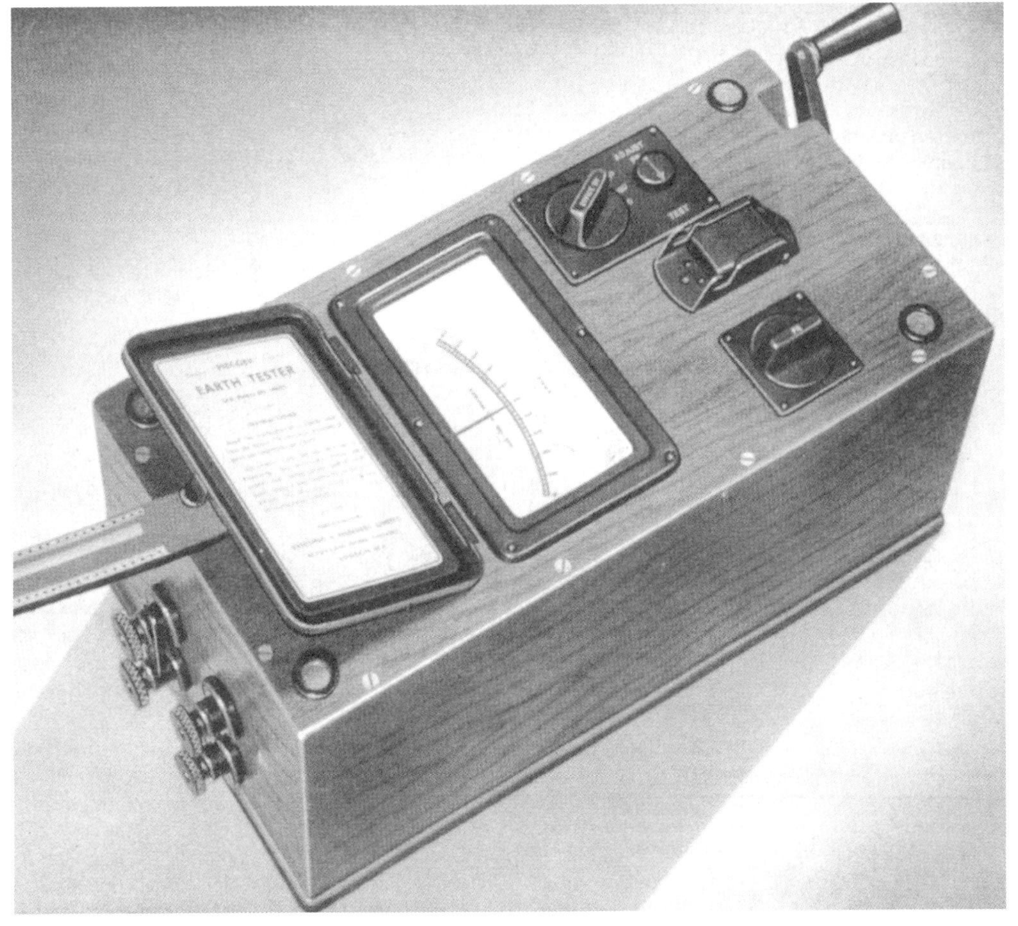

Fig. 2. The Megger Earth Tester as used by Shaw in Southern Rhodesia (from manufacturer's literature: Anon. 1950).

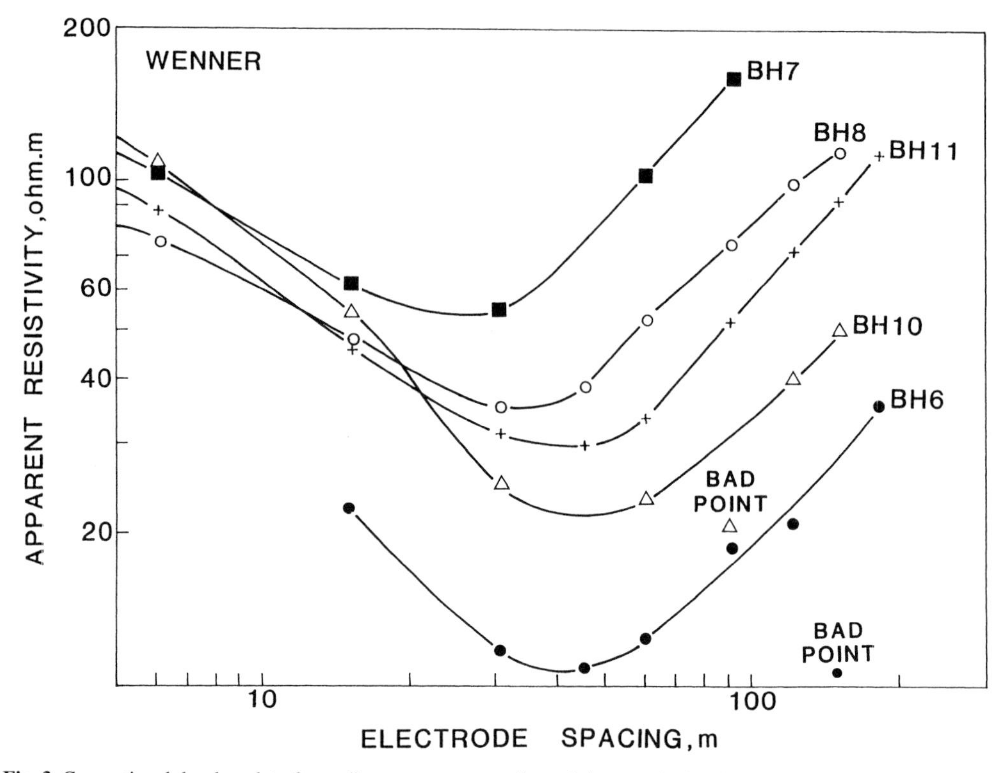

Fig. 3. Conventional, log-log plot of sounding curves measured at existing granite boreholes in Nata Reserve, Southern Rhodesia (data from Shaw 1934*a*).

poor measurement point, as Shaw sometimes did have problems with high contact resistances. On a linear plot erroneous points are not quite so obvious and were generally interpreted by Shaw as horizontal subsurface layers.

The soundings shown in Figure 3 were measured in the granite area over the boreholes listed in Table 1. The sounding curves show the three-layer form with which we are now so familiar: i.e. a high-resistivity surface layer of dry sand overlying a low-resistivity layer representing moist or saturated sand, or sandstone which overlies unfractured high-resistivity granite. Shaw noticed some differences between soundings measured over the dry boreholes BH7, BH8 and BH11 and those measured over the high yielding boreholes BH6 and BH10. Firstly, all the dry boreholes reached granite bedrock at a much shallower depth than the successful boreholes. Secondly, the resistivity of the sandstone at the successful boreholes was much lower.

On the basis of these preliminary results, Shaw went on to use resistivity sounding to test new sites. Several soundings (V, W, X and Y) (Fig. 4) were measured near an area of granite outcrop where all boreholes drilled had until then proved unsuccessful. The soundings confirmed thin, probably high-resistivity sandstone overlying granite and on the experience of the earlier work, were thought to indicate unsuitable conditions for drilling. Soundings Z and ZZ, measured along the road between boreholes BH10 and BH11, appeared to indicate good prospects and a borehole, BH25, was sunk on sounding ZZ, which is interpreted as 75 m of sandstone overlying unfractured granite. This borehole produced a good yield of 0.8 ls^{-1}. In fact the thickness of saturated sandstone is greater at Z, but this was misinterpreted by Shaw as being an inferior site and was anyway close to the dry borehole BH11.

Discussion and conclusions

It is seen that Shaw's deductions have been substantially verified although, of course, the sample of soundings and boreholes is quite small.

The accuracy of the original resistivity interpretations was relatively poor because the only techniques available to Shaw were empirical; good analytical interpretive checks had still to be developed. If modern computer techniques are used to interpret Shaw's curves, a good correlation between the base of the low-resistivity layer and the depth to

Table 1. Summary of borehole and sounding data

Borehole	BH6	BH7	BH8	BH10	BH11	BH25
Elevation (m)	1253	1271	1261	1247	1224	1242
Thickness of soil (m)	4.3	0	1.8	1.2	0	0
Thickness of sandstone (m)	>64	29*	40	66	48	62**
Yield (l/s)	0.5	-	-	0.4	-	0.8
Sounding Interpretation						
Soil thickness (m)	8	6	7	8	10	15
Soil resistivity (Ωm)	60	150	70	140	70	300
Sandstone thickness (m)	72	34	43	62	50	60
Sandstone resistivity (Ωm)	10	40	30	15	25	20

* includes limestone and weathered granite
** includes weathered granite

granite is observed (Fig. 5). A slight bias towards greater values on the computed depths suggests that the resistivity technique responds to the depth to unfractured granite, whereas the driller probably recorded the shallower depth to more fractured granite.

Most of the curves are affected by equivalence, an interpretational problem (Kunetz 1966) about which Shaw and his contemporaries were ignorant. Equivalence implies that although individual layer thicknesses and resistivities might not be unambiguously evaluated from a sounding curve, the product of thickness and conductivity (longitudinal conductance) can be uniquely determined. Subsequent workers have sometimes found a correlation between the longitudinal conductance of the aquifer and the

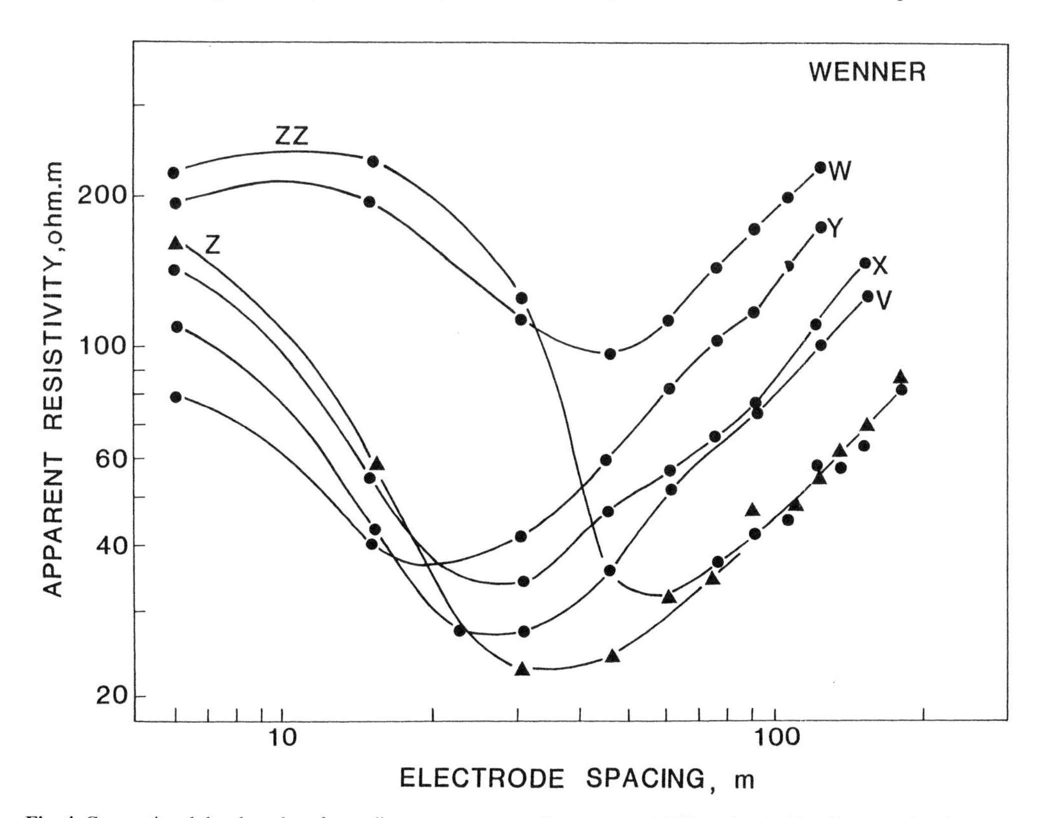

Fig. 4. Conventional, log-log plot of sounding curves measured at proposed drilling sites in Nata Reserve, Southern Rhodesia (data from Shaw 1934*a*).

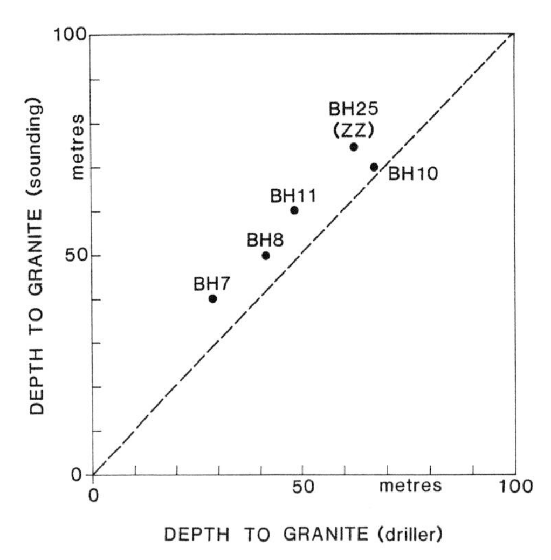

Fig. 5. Relationship between depth to granite determined from resistivity sounding interpretation and that confirmed by drilling.

borehole yield (Worthington 1977), although Shaw's work is not sufficient to test this possibility here.

There is no doubt that the success of Shaw's work was a result of the reasonably high quality of the sounding curves which he was able to measure. Most of the curves were smooth and can be closely reproduced by computer modelling of horizontally-layered structures. Where irregularities, possibly lateral resistivity effects, or bad points were recorded, Shaw fell into the trap of interpreting these also as horizontal layers. Fortunately, such occurrences were few and did not prevent Shaw developing a soundly-based borehole-siting methodology.

Although the area investigated by Shaw is not typical of *in situ* weathered granite hardrock, it does provide an analogous hydrogeological situation. During the years since Shaw's pioneering work, the philosophy behind borehole siting in hardrock areas has remained largely unaltered. Although many tens of thousands of resistivity soundings have been measured in borehole-siting projects in Africa, the technique has developed only very slowly, sadly perhaps because of a lack of understanding of the important spurious effects of near-surface lateral-resistivity variations (Barker 1979). These 'electrode effects' often cause irregularities in sounding curves which, if not identified for what they are, may be interpreted as subsurface geological layers. This is a major problem in developing countries, where empirical techniques (Sanker Narayan & Ramanujachary 1967; Ballukraya *et al.* 1983; Muralidharan 1996) are still widely employed in sounding interpretation. This has often led to low success rates in drilling and sometimes to the complete abandonment of geophysical surveys. However, the measurement of good-quality data (as demonstrated by Shaw) and their careful interpretation using modern computer techniques provides a powerful tool for the groundwater exploration in hardrock areas.

Shaw's eight-week visit to Southern Rhodesia very much helped the adoption in Africa of resistivity methods as standard practice in the location of drilling sites for water supply. A number of water departments in the African Colonies began adopting electrical methods (Way 1938, 1942; Cooper 1950) and the war campaigns in Africa led to the establishment of specific geophysical units for water finding (War Office 1945). The Megger used by Shaw was subsequently purchased by the Irrigation Department of Southern Rhodesia, where it was still in use 14 years later (Anon. 1947). Shaw's interests extended to geophysical borehole logging, and possibly the first boreholes logged geophysically in the UK were those on the University of Birmingham campus (Shaw 1937). Shaw went on to become Director of the Directorate of Overseas Geological Surveys and maintained an interest in geophysical surveying for water supply. He was awarded the OBE in 1958 and the CMG in 1963 (Hepworth 1991).

Shaw (1963) provides a useful review of geophysical surveys carried out for water supply investigations in various parts of the world. Since then, electrical sounding together with electromagnetic surveying has become the mainstay of siting technology (Barker *et al.* 1992), although in the last few years electrical-imaging techniques have also become very important (Kellett *et al.* 2000).

I should like to thank the late Dr. S. Shaw for providing me several years ago with some background material to his work in Africa. Although this is believed to be the first use of resistivity sounding in borehole siting in Africa, I should be pleased to hear of earlier work or indeed of any groundwater geophysical surveys conducted in Africa before 1940.

References

ANON. 1933. New techniques for water surveys. *Bulawayo Chronicle*, **July 15th 1933**.

ANON. 1947. Successful search for water by geophysical methods. *Rhodesia Herald*, **11th March**.

ANON. 1950. *Megger Earth Testers for geophysical prospecting*. Publication **250a**, Evershed and Vignoles Ltd, London.

BALLUKRAYA, P. N., SAKTHIVADIVEL, R. & BARATAN, R. 1983. Breaks in resistivity sounding curves as indicators of hard rock aquifers. *Nordic Hydrology*, **14**, 33–40.

BARKER, R. D. 1979. Signal contribution sections and their use in resistivity studies. *Geophysical Journal of the Royal Astronomical Society*, **59**, 123–129.

BARKER, R. D., WHITE, C.C. & HOUSTON, J. F. T. 1992.

Borehole siting in an African accelerated drought relief project. *In*: WRIGHT, E. P. & BURGESS, W. G. (eds) *Hydrogeology of Crystalline Basement Aquifers in Africa*. Geological Society, London, Special Publications **66**, 183–201.

BROUGHTON-EDGE, A. B. and LABY, T. H. (eds) 1931. *The Principles and Practice of Geophysical Prospecting*. Report of the Imperial Geophysical Experimental Survey, Cambridge University Press, 372p.

COOPER, W. G. G. 1950. *Electrical aids in finding water*. Nyasaland Geological Survey Bulletin, No. **7**.

DRYSDALE, C. V. & JOLLEY, A. C. 1924. *Electrical Measuring Instruments*, **2**, 385–388, Benn, London.

HEPWORTH, J. V. 1991 Sydney Herbert Shaw (1903–1991). *Annual Report of the Geological Society of London*, 29–30.

KELLETT, R. L., ANSCOMBE, J. R., BAUMAN, P. D., HANKIN, P. & ENGELBRECHT, L. 2000. Geophysical mapping of groundwater potential in a rural water supply project: Malawi, Africa. *Proceedings of the 6th Meeting of Environmental and Engineering Geophysics, Bochum, Germany*.

KUNETZ, G. 1966. *Principles of Direct Current Resistivity Prospecting*. Geoexploration Monograph **1**, Gebrüder Borntraeger, Berlin.

LOW, B., KELLY, S. F. & CREAGMILE, W. B. 1932. Applying the Megger Ground Tester in electrical exploration. *A.I.M.E. Geophysical Prospecting*, **97**, 114–125.

MURALIDHARAN, D. 1996. A semi-quantitative approach to detect aquifers in hard rocks from apparent resistivity data. *Journal of the Geological Society of India*, **47**, 237–242.

SANKER NARAYAN, P. V. & RAMANUJACHARY, K. R. 1967. An inverse slope method of determining absolute resistivities. *Geophysics*, **32**, 1036–1040.

SHAW, S. H. 1934*a*. Geophysical prospecting – a study of the resistivity method in connection with the investigation of underground-water supplied in the Nata Reserve, Southern Rhodesia, *Transactions of the Institution of Mining and Metallurgy, London*, **44**, 1–27.

SHAW, S. H. 1934*b*. Discussion on 'Geophysical prospecting – a study of the resistivity method in connection with the investigation of underground-water supplied in the Nata Reserve, Southern Rhodesia' *Proceedings of the Institution of Mining and Metallurgy*, 1–18.

SHAW, S. H. 1937. Geo-electrical measurements in drill holes. *Mining Magazine*, **56**, 17–23 (with discussion 201–208).

SHAW, S. H. 1963. Some aspects of geophysical surveying for ground-water. *Journal of the Institution of Water Engineers*, **17** (3), 23–36.

VAN NOSTRAND, R. G. & COOK, K. L. 1966. *Interpretation of Resistivity Data*. U.S. Geological Survey Professional Paper **499**.

WAR OFFICE 1945. *The Location of Underground Water by Geological and Geophysical Methods*. Military Engineering, **6**, Water Supply Supplement 1.

WAY, H. J. R. 1938. Some results of the application of the resistivity method to the water problems of Uganda. *Bulletin of the Geological Survey of Uganda*, **3**, 161–179.

WAY, H. J. R. 1942. An analysis of the results of prospecting for water in Uganda by the resistivity method. *Transactions of the Institution of Mining and Metallurgy, London*, **51**, 285.

WENNER, F. 1915. *A method of measuring earth resistivity*. U.S. Bureau of Standards Bulletin, **12**, 469–478.

WORTHINGTON, P. F. 1977. Geophysical investigations of groundwater resources in the Kalahari Basin. *Geophysics*, **42**, 838–849.

The development of groundwater in the UK between 1935 and 1965 – the role of the Geological Survey of Great Britain

R. A. DOWNING

20, Springfield Park, Twyford, Reading RG10 9JH, UK

Abstract: After the drought of 1933–1934 the Geological Survey became responsible, under the Inland Water Survey, for collecting and collating data on groundwater. In 1935 a Water Unit was formed for this purpose. Following the Water Act of 1945, the Survey advised the Government on aspects of the Act relating to groundwater. The Act led to the introduction of quantitative hydrogeology in England and Wales. The groundwater resources of the main aquifers were assessed, well hydraulic theory was applied to British aquifers, and geophysical techniques and new instrumentation introduced.

1921 was the driest year over England since 1788. The rainfall was less than 60% of average over an area to the SE of a line from the Wash to Torquay and less than 50% in east Kent. Just over 10 years later a second major drought, extending from November 1932 until November 1934, afflicted southern and central areas of England and Wales. This drought of 1933–1934 finally and conclusively demonstrated the need for a survey of the nation's water resources, a requirement that had been discussed for over 50 years and indeed had been raised at the 1878 meeting of the British Association for the Advancement of Science. Following the 1932 annual meeting, the Association appointed a committee to look into the prospects of an Inland Water Survey of Britain and the immediately succeeding 1933–1934 drought was a timely event that emphasised the desirability of such a study. At this time, groundwater resources were developed in a piecemeal fashion as the need arose; national coordination did not exist.

A joint deputation of the British Association and the Institution of Civil Engineers put the case for a national study to the Minister of Health in 1934. The outcome was that an Inland Water Survey Committee was appointed in 1935 by the Minister of Health for England and Wales and the Secretary of State for Scotland with the objective of creating a mechanism for the collection and correlation of records dealing with rivers and underground waters. The Geological Survey of Great Britain was made responsible for the recording and assessment of underground water, the preliminary work being entrusted, in 1935, to F. H. Edmunds. The Survey decided to collaborate fully with the Inland Water Survey in May 1936, surveys of groundwater becoming part of its accepted duties. The Director of the Survey, Bernard Smith, made Edmunds responsible for the work on a permanent basis, assisted from October 1936 by J. Lee, thereby creating 'a very modest Water Unit' (Bailey 1952), although both Edmunds and Lee remained members of the Southern District Field Unit. When the legendary Sir Edward Bailey became Director in 1937, he recognized groundwater as an issue of national importance and he formalized the arrangement, establishing a separate Water Unit. The staff was strengthened when T. R. M. Lawrie joined in December 1937, to be followed by J. R. Earp in 1938.

The accumulation and collation of surface water data by the Inland Water Survey were less than successful primarily because of a lack of adequate financial support. In contrast the Geological Survey, supported by the Department of Scientific and Industrial Research, successfully carried out its task and continued to do so until World War II intervened in 1939. This success was probably due, at least in part, to the tradition in the Survey for collecting all relevant geological data, including a multiplicity of well records. The needs of the Inland Water Survey allowed the collection of water data to be put on a systematic footing (Bailey 1952, p. 161).

The value of the Inland Water Survey to the Geological Survey was the recognition of the latter's position regarding groundwater. This led, in addition to the establishment of a Water Unit, to closer liaison with officials in the Ministry of Health which, in turn, eventually led to fruitful discussions and proposals about water legislation.

Bernard Smith, delivered the Cantor Lectures before the Royal Society of Arts in 1935, addressing geological aspects of underground water supplies (Smith 1935), a topical subject in view of the previous two years. His comprehensive review presented the understanding of the Survey at the time, derived from field surveys and the special surveys that led to the water supply memoirs. The occurrence of groundwater in British rocks, including the properties that influenced yields, was described, illustrated by many examples from across the country. Smith recognized that the collection and coordination of data had been the prime objective of the Survey but,

From: MATHER, J. D. (ed.) 2004. *200 Years of British Hydrogeology*. Geological Society, London, Special Publications, **225**. 271–282. 0305-8719/04/$15 © The Geological Society of London.

although the principles of the occurrence of groundwater and its links with geology were well understood, there had been little or no attempt at systematic correlation of data on a regional basis. This was probably because the prime purpose of the Survey was making geological maps and much of the detailed knowledge of groundwater came from answering hydrogeological enquiries centred on specific localities. Smith hoped this restricted attitude would change and in this respect he was very prescient.

A Central Advisory Water Committee was established in 1937 to advise the Minister of Health for England and Wales on general water policy including the conservation and allocation of water resources. Eventually, in 1944, a command paper was published by the Ministries of Health and of Agriculture and Fisheries entitled 'A National Water Policy'. The Survey was consulted in its preparation and staff contributed an appendix on 'The influence of geological factors on water supply in Great Britain'.

The only Geological Survey publication dealing with groundwater that post-dated the water supply memoirs initiated by Geikie and preceded World War II, was 'The Water Supply of the County of London from Underground Sources' published in 1938 and written by Stevenson Buchan, who at the time was a member of the Southern England Field Unit and not the Water Unit. The Survey saw this as the last of the County Water Supply Memoirs (Bailey 1944), but it was also anticipated that it would be the first of a new style of water memoirs; regrettably this proved to be wide of the mark. However, in view of the imminence of war in 1939, this publication was particularly opportune as it provided up-to-date details of the location and capacity of each important borehole and pump in London and was invaluable during the bombing assault on London, being used by Buchan as the basis for a plan for emergency water supply to the city from private wells.

With the outbreak of war in 1939, the attention of the Survey with respect to groundwater focussed on the need to provide water supplies for new airfields, military camps and factories. Many staff were diverted to this work. One outcome was a considerable increase in data relating to the location of wells and boreholes, details of water levels and yields and the occurrence of groundwater generally. Information considered essential to the war effort was published in a series known as Wartime Pamphlets; about half were concerned with groundwater. The pamphlets dealing with groundwater were based on the quarter-inch-to-the-mile Geological Survey maps following an agreement between the Survey and the Inland Water Survey Committee (Bailey 1944). One of the first publications based on these proposals was Wartime Pamphlet No 4 which dealt with the Water Supply from Underground Sources of the Oxford-Northampton District and it included a general discussion of the hydrogeology (Woodland 1940). Work had actually begun on this region in 1938 when it was intended that it would be one of the first of a new series of regional water supply memoirs.

Of all the pamphlets concerned with ground water, the most outstanding was that of the Cambridge-Ipswich District by Austin Woodland. This was Wartime Pamphlet No 20 published in 1946. Part X, General Discussion, was a masterly review of the hydrogeology of East Anglia south of an east-west line through Lowestoft and extending almost to the Thames Estuary. Details of the geology, groundwater levels and chemical quality of the groundwater were summarized in a series of maps. A map of the yield potential of the Chalk was included based on a linear relationship of yield to borehole diameter. Although such a relationship is incorrect, the small diameters of boreholes in East Anglia reduced the error and the maps revealed the close relationship between high yields from the Chalk and valley sites. Although this was probably already recognized by water engineers, it had not appeared in papers published up to this time and Woodland was the first to express it clearly and graphically in map form. Using the maps in this pamphlet, an estimate could be made very rapidly for any prospective site of the likely yield of water, its quality and the depth and diameter of the borehole necessary. Woodland also estimated the groundwater resources based on data from infiltration gauges. The content of the pamphlet set the style for the reports that were to be prepared in the 1950s. It was said that Part X of Pamphlet No 20 was published exactly as Woodland submitted it – not a word had to be changed.

Another pamphlet, by William Anderson, described the distribution of groundwater resources in NE England, referring to the high yields from the fissured and brecciated, in places cavernous, Magnesian Limestone. The occurrence of very saline waters in the Coal Measures was also reviewed (Anderson 1945*a*, *b*).

The extent of the wartime programme was summarised by Bailey (1952, p. 246) who referred in particular to the roles of Edmunds, Woodland, Lee and Buchan. Edmunds had an encyclopaedic knowledge of the practical aspects of the link between geology and groundwater in England. During the war he published (Edmunds 1941) a comprehensive review of the underground water supply in England and Wales, a panoramic overview which covered not only a description of the aquifers but also a discussion of the hydrological cycle, the relationship of geology to groundwater flow and a review of hydrogeological terms. Plate I was a hydrogeological map. The paper included much interesting detail about local hydrogeology, for example a description of an unusual occurrence for Britain. On the night of September 8–9 1937 a small earthquake in Sussex fractured the impermeable seal of a perched water table in a permeable zone of the Weald Clay at Warnham and deprived several houses of a previously reliable water supply. The paper (and the earlier one by Smith (1935)) is valuable in an historical context as the contents, together with the contributions to the discussion by senior water engineers of the time, provide a 'snapshot' of the outlook and approach to hydrogeology in the period before the Water Act of 1945 heralded a more quantitative approach to the subject. The paper was actually based on the subject matter of a handbook on the 'Water Supply of England and Wales from Underground Sources' which was being prepared by the Geological Survey at the time of the outbreak of war but, as a consequence of the war, was never published.

The Water Act, 1945

The Water Act marked the beginning, in the UK, of quantitative hydrogeology. Geology became but one factor in hydrogeological ideology, not the principal factor. The water balance of aquifers and the hydraulics of aquifers, linking rock properties to groundwater flow and yields, thereby providing more realistic interpretations of cause and effect, became the new primary objectives. The Water Act, the first piece of major legislation affecting water supply for over 100 years, was part of the social revolution of the immediate post-war years – it was part of a new idealistic social policy put into practice by the Attlee government. The White Paper, published in 1944 as a prelude to the Act, was entitled 'A National Water Policy'. It underlined the availability of ample water for all needs, the problem being not one of total resources but of organization and distribution. Sectional interests were to be subordinate to the national interest. 60 years on this ideal has still to be achieved.

The Water Act affected the Survey, and in particular the Water Unit, in several ways. Firstly, the Ministry of Health (and from 1951 the Ministry of Housing and Local Government (MHLG)) was responsible both for the provision of adequate public water supplies and the conservation of water resources in England and Wales. The Ministry examined and had authority to approve schemes submitted by Local Authorities, Statutory Water Companies and Industry. Under Section 14 of the Act, areas were defined in which groundwater was overdeveloped or where this was anticipated. Organisations or individuals wanting to abstract water in such areas required a licence to do so from the Ministry. The Water Unit advised the Ministry on important licence applications. All new large-yielding wells, and existing wells for which an increased yield was sought, were test-pumped under the supervision of the Water Unit. The tests lasted 14 or 21 days and were, if possible, timed for the autumn when groundwater levels were low. The timing gave water engineers more confidence in the long-term reliability of the yield and the length of the test afforded sufficient time for significant interference effects in nearby wells (generally within a radius of 2 to 3 miles (3 to 5 km) of the pumped well) to become apparent. On the basis of the tests, the Unit recommended the yield and conditions to be stipulated in the licence and identified any nearby wells that would be significantly affected by the abstraction so that remedial work could be put in hand.

Prior to the Water Act, it was not uncommon for a well to lose its supply because of the interference effect of a neighbouring, deeper, higher yielding well without there being any liability for damages.

Secondly, following regulations issued in 1947 under Section 6 of the Act, returns of annual abstraction from all major wells had to be made to the Survey at the end of October each year. Returns were required from concerns abstracting over 50000 gallons (227 m^3) per day except in Dorset, Gloucester, Hampshire, Lincolnshire, Norfolk, Suffolk and Wiltshire, where the figure was 20000 gallons (90 m^3) per day. The National Coal Board also made returns of groundwater abstracted for the purpose of mine-drainage. The first returns were made in October 1948 when over 4000 were received specifying total annual abstraction, the abstraction rate and rest and pumping water levels. These data provided, for the first time, an indication of the use of groundwater on a national basis. The total abstraction was 1549 Mm^3 with 608 Mm^3 from the Chalk and 347 Mm^3 from the Bunter Sandstone.

The third major impact of the Water Act was that under Section 7 details of new wells and boreholes more than 50 feet (15 m) deep had to be sent to the Survey by well-drillers.

In 1945 Stevenson Buchan became responsible for the Water Unit which was renamed the Water Department. He was promoted to District Geologist in 1952. The Survey had accelerated the collection of data about wells and boreholes and aquifers during the war and this continued after the war when Buchan initiated further countrywide data-collection surveys to obtain basic information: the location of wells and their siting on 6-inch maps, the recording of details of their construction and depth, strata penetrated, the estimation of the ground level, the measurement of water levels and the recording of any other relevant information. These basic data were collected in the field by a group of Experimental Officers, in particular T. K. Tate, Rebecca

Moseley, Beta Campbell and Cicely Willis. The well-record collection at this time (late 1940s) comprised some 90000 wells and boreholes. Maintaining and expanding, as well as validating and cataloguing, this data-base continued to be an important task and between 1945 and 1965 more than 40 Experimental Officers were concerned with the work at various times. Most were young ladies who became known as the 'Water babies'. The Water Department was itself generally referred to in the Survey as 'The Water Cart'.

The Water Department was housed in the upper of the two galleries that formed the Bridge connecting the Geological and Science Museums in South Kensington The office was in an open plan style with a separate room partitioned off for Buchan. The permanent geological staff was increased in the late 1940s by the transfer of Jack Ineson and, in 1950, David Gray from the South East Field Unit. In addition members of the field staff were seconded to the Department for periods of time or to carry out specific investigations. This was a continuation of a policy advocated by Bailey that field geologists should have experience of groundwater problems.

Even in the late 1940s the problems that were to be of concern for many years into the future were apparent, including saline intrusion in north Kent and NE Lincolnshire, disposal of mine drainage from the Kent Coalfield into the Chalk and the need for an observation-well network to record seasonal variations in groundwater levels. With regard to the last requirement, a pioneer project, developed between the Survey and Norfolk County Council, began in April 1950 with the measurement on a monthly basis of water levels in 60 wells distributed throughout the county. Artificial recharge of aquifers was also a subject under consideration with papers published by Buchan (1955, 1964) who also advised the Metropolitan Water Board on their pioneering experiments in the Lee Valley.

The important role of advising the Ministry of Housing and Local Government on national policy generally and the distribution and availability of resources and important licence applications in particular, required a much more detailed knowledge of regional hydrogeology than had hitherto existed. Surveys were started to assess the groundwater resources of the principal aquifers, the distribution of undeveloped resources and the definition of regions of over-development. In the early to mid-1950s George Bisson worked in the Lee Valley, Harry Wilson in north Kent, David Gray in east Yorkshire, David Land in NE Lincolnshire, Nottinghamshire and the Birmingham-Lichfield region, Wyndham Evans in south Lincolnshire, Dick Downing in northern East Anglia and north Kent, John Price in Lincolnshire and John Day in Teeside.

The assessment of the groundwater resources required reliable information about infiltration into aquifers – the difference between rainfall and evaporation for permeable outcrops. The estimation of infiltration was the Holy Grail. Over the years many attempts had been made to calculate it (see Rodda *et al.* 1976) but not uncommonly a value of 10 inches (250 mm) per year was simply assumed for areas in eastern and southern England.

The areal variation of rainfall was well established, the difficulty was measuring evaporation. When Penman of the Rothamsted Experimental Station published his classic work on calculating evaporation in 1948, it was possible to assess groundwater resources with much greater accuracy, particularly where permeable rocks such as the Chalk, Triassic sandstones and Jurassic limestones cropped out. The method developed by Penman (and extended later, in 1967, by Grindley of the Meteorological Office) became the standard method for assessing the regional distribution of infiltration. Penman (1948) derived a relationship between evaporation from open water and evaporation from bare soil and grass. He concluded that evaporation from freshly wetted bare soil is 90% of that from open water. Later he introduced the concept of soil moisture deficit (Penman 1950*a*, *b*). This involved assigning a root constant to each type of vegetation, usually assumed to be between 75 and 200 mm. Subsequent studies by Headworth (1970) and at the Institute of Hydrology (Gardner *et al.* 1991) indicated that the early assumptions were too high and that estimates of recharge derived from meteorological parameters can give values that are too low (Rushton & Ward 1979). Nevertheless, at the time (the 1950s and 1960s) Penman's work was a big step forward for which he was elected a Fellow of the Royal Society.

In the post-war years, statutory water undertakings and water companies were expanding their water mains networks to take in outlying areas and villages not previously connected to a mains supply. It was a period of rejuvenation and construction – an improvement of basic facilities for many. The use of groundwater for irrigation was also increasing rapidly. Exploration for new groundwater resources was a necessary outcome and in the 1950s the Survey was associated with many major groundwater development schemes, particularly in the Chalk and Triassic sandstones. Groundwater was required for new towns, such as Stevenage, that were being developed and also for the industrial expansion that was introduced after the war. The Survey's staff tested many new wells on behalf of the Ministry of Housing and Local Government. In 1952 and 1953 for example, 12 and 15 major pumping tests were carried out.

Ineson began to investigate the application to British aquifers of the methods proposed in the USA by Theis in 1935 and extended by Jacob, for the analysis of non-steady state flow of groundwater. The Theis method and its derivatives represented a paradigm shift – a permanent change in the way groundwater development was approached. It was an enduring event the impact of which was delayed in Europe because of World War II. The involvement of the Survey in pumping tests all over England provided Ineson with the basic data he required. Despite the fact that fracture flow is significant in British aquifers, he found that the techniques could indeed be applied (Ineson 1952, 1953). In coming to his conclusions he analysed over 300 tests.

The fact that non-equilibrium methods could be applied to British aquifers was a considerable step forward in the entire approach to hydrogeological studies in the UK. The previous empirical style for well-test analysis gave way to a more quantitative viewpoint. It represented a complete change of outlook. Ineson's role in applying well hydraulic theory to British aquifers was crucial to the Geological Survey's interpretation of pumping tests and providing advice to the Ministry of Housing and Local Government on borehole yields and licence conditions at a time when the use of groundwater was increasing to meet demand as rural areas were connected to water mains.

In 1951, David Gray began applying geophysical techniques to groundwater problems, with particular reference to electrical methods. The first equipment employed was an electrical resistivity set, powered by a hand-cranked generator, which had been used by the Army during World War II to locate water supply boreholes during the desert campaigns in North Africa and the Middle East. It was presented to the Geological Survey by the Royal Engineers. Initially surface resistivity surveys were used to locate structures in the Chalk favourable for well sinking. However, the most fruitful application in those early days proved to be the use of downhole resistivity logging to supplement data obtained during drilling and for which purpose the same equipment was modified. The electrode configuration was devised to cope with borehole diameters ranging from 6 to 48 inches (150 to 1220 mm). The lead electrodes were cast in-house. A decade was to pass before capital became available to modernize the equipment.

Most boreholes in the Chalk were drilled by percussion methods and stratigraphical correlation between boreholes was not usually possible. The initial objective was to see if individual horizons in the Chalk – Totternhoe Stone, Melbourn Rock, Chalk Rock and Top Rock – maintained their distinct physical properties, particularly porosity, over wide areas allowing correlation by electrical resistivity methods. If so, there was a need to know whether the technique was applicable to large diameter water-filled boreholes. Investigations began in 1952 in Hertfordshire and it was soon evident that bands of high and low resistivity existed in the Chalk. These 'marker bands' were seen to have stratigraphical significance, initially in Hertfordshire and eventually throughout the Chalk's outcrop and subcrop (Gray 1958). The stratigraphical significance of the marker bands was confirmed when the Survey drilled the Fetcham Mill borehole near Leatherhead in 1960–1961 (Gray 1965). Ultimately, geophysical logs were used to recognize stratigraphical and lithological units in the Chalk over increasingly extensive areas (Murray 1986; Mortimore & Pomerol 1987).

The continued development of instrumentation led to downhole temperature, electrical conductivity, flow metering and CCTV systems. These instruments enable studies to be made of the flow characteristics of aquifers, particularly levels of water entry and exit in fissured formations. Conductivity logs were used to examine the problem of saline intrusion in Grimsby, north Kent and along the south coast (Ineson & Gray 1963; Gray 1964*a*; Tate *et al.* 1970). The suite of logs suitable for groundwater studies increased to include in addition to resistivity logs, natural gamma radiation, neutron, density of the formation and acoustic velocity of the formation.

The expansion of the role of the Survey in monitoring and conserving groundwater entailed the use of automatic water level recorders, the first of which was installed in a well at the National Gallery in Trafalgar Square, London. In addition to Norfolk County Council, other local authorities were persuaded to make regular measurements of groundwater levels and the Survey installed continuous automatic recorders at key locations. By 1956 regular, generally monthly, measurements were being made by the local authorities for Bridlington, Norfolk, Brighton, Bognor, Peterborough and Worthing, as well as Grimsby Water Board and several industrial concerns.

From early days the analysis of water samples for both chemical and bacterial quality had been collected by the Survey. After the 1945 Act, the Survey commissioned analyses of samples for specific purposes but generally continued to rely on analyses from public water authorities and industry to increase their data base. In the late 1940s and in the 1950s full chemical analyses of the major constituents in groundwater were not generally available. Instead partial chemical analyses were more usual comprising total dissolved solids (TDS), temporary (or carbonate) and permanent (or non-carbonate) hardnesses, chloride and not uncommonly nitrate, ammonia and iron. These were sufficient to interpret the predominant chemical changes in groundwater on a regional scale. Most of the regional reports included maps of both types of hardness, chloride content and occasionally TDS. The purpose was to provide a means of assessing the chemical quality of a groundwater in an aquifer prior to drilling. The gradual change in the composition of groundwater as it flowed down gradient below confining clays had been known for many years. The processes which changed hard calcium-bicarbonate water, found in aquifers at outcrop, into a soft sodium-bicarbonate water below cover were described by W. W. Fisher in a series of papers, between 1901 and 1904, dealing with waters in the Chalk (Fisher 1901), Lower Greensand (Fisher 1902) and Jurassic limestones (Fisher 1904). Thresh (1912) and Whitaker & Thresh (1916) suggested the softening of groundwater in the Chalk and Thanet Formation below the London Clay was caused by base exchange (ion exchange).

During a regional study of the groundwater resources of the Chalk in north Kent in 1956, a good down-gradient sequence of analyses of the principal ions in groundwater in the Chalk became available for the Hoo Peninsula and the Isle of Grain. They suggested possible more comprehensive interpretations of the sequence. A series of zones were identified each representing a type of water. This preliminary work was extended to cover the London Basin by interpreting a limited number of analyses of the principal ions in conjunction with many more partial analyses and was published eventually by Ineson & Downing (1963). Nitrate and sulphate reduction were seen as processes contributing to the changes. Bacteria were suspected of being involved. Nitrate-reducing bacteria were identified by the government's Chemical Research Laboratory (which existed at that time alongside the National Physical Laboratory at Teddington) in water samples collected from wells in the Chalk and Lincolnshire Limestone, but sulphate-reducing bacteria proved more elusive. The microbiologists at the Chemical Research Laboratory believed that they could remain adsorbed on the rock matrix. The role of these bacteria in groundwater processes was overlooked for many years, inorganic reactions being more generally accepted. The balance was not restored until Chapelle began publishing on groundwater microbiology (for example Chapelle 1993).

Saline intrusion was an increasing problem at several localities where the Chalk formed the coastal boundary with the sea particularly near Grimsby, Hull, Ipswich, north Kent and Brighton, as well as in the Triassic sandstones on Merseyside. Studies of the extent of the saline intrusion were made by means of repeated conductivity surveys, supported by chemical analyses of samples, often obtained with depth samplers, for total dissolved solids and chloride. A summary of the extensive programme in and around Grimsby, which started in 1950, was published by Gray (1964*b*). The initial survey of the chemical quality of the water below the Grimsby area was made by David Gray using a bicycle to visit and sample wells, boreholes and water collection tanks (often hazardously placed high in the structural framework of warehouses) in the docks and surrounding industrial centre. In those early days transport for fieldwork was very limited. The Department only had one or two old saloon cars of doubtful reliability, cast-offs from Whitehall government departments. Eventually Land Rovers provided a very acceptable addition to the small transport pool.

Brighton's Water Department was in need of additional water resources in the mid-1950s. Problems with supplies were being encountered because of increasing salinity in boreholes near the coast particularly at the site at Balsdean. Because of this the Department pursued a drilling programme in an east-west valley near Falmer, some distance from the coast. Buchan, in consultation with Needham Green and Sydney Warren, the Brighton Water Engineer and Chemist respectively, proposed that boreholes near the coast should be pumped in winter and thereby intercept the strong flow of groundwater to the sea at this time, and inland boreholes should be used preferentially in the summer, reducing abstraction from those nearer the coast according to the chloride content of the water. This seasonal use of boreholes proved to be very successful over the succeeding years (Headworth & Fox 1986).

A further continuous problem was the contamination of groundwater in the Chalk by the disposal of saline mine-drainage from the collieries of the Kent Coalfield. Recognized in the 1930s, it was given wider attention by Buchan (1962) but many years were to elapse before action to rectify the problem was initiated (Headworth *et al.* 1980). Under the Water Act 1945, groundwater pollution was only controlled to the extent that Section 21 made it an offence to pollute or to be likely to pollute groundwater that was or could be used for public supply. Obtaining proof of such pollution was not easy and thereby the legislation was to a large extent ineffective. But in the 1940s and 1950s pollution of groundwater was not seen as a serious problem, more a matter for local interest and action. Nevertheless, concern was being expressed in some quarters for there was evidence that waste disposal practices and some other forms of pollution were affecting ground water quality. The World Health Organisation invited Buchan (together with A. Key 1956) to review the extent of groundwater pollution in Europe, and Ineson (1964) examined the problems that were becoming evident as a consequence of the increasing use of petroleum products.

During the late 1950s and early 1960s the number of geologists in the Department increased significantly and studies of the hydrogeology and assessments of the groundwater resources of the two main aquifers continued and expanded. Investigations were made of the Chalk of the South Downs and catchments in Hampshire – the Test, Itchen and Meon. The Triassic sandstones of the Needwood and Stafford basins, the vales of York and Mowbray and Devon and West Somerset were also studied. The programme extended eventually beyond these aquifers to include the Magnesian Limestone in Durham, the Coal Measures of the East Midlands and South Wales, the Carboniferous Limestone of North and South Wales, the Upper Greensand of SW England, the Jurassic limestones of the Cotswold Hills and the Pleistocene Crag of East Anglia. Geologists in the Survey's Edinburgh office reviewed the distribution of groundwater in Scotland (Earp & Eden 1961). During the course of the work in the South Downs, a water balance of the catchment feeding the Bedhampton Springs, near Havant, gave support to the method of assessing infiltration into permeable rocks from estimates of rainfall and actual evaporation (Day 1964). Ineson also continued his work on quantitative aspects of groundwater flow in the Chalk (Ineson 1959*a*, *b*; 1962*a*).

The use of automatic water-level recorders to collect continuous records for an increasing number of wells revealed that fluctuations of groundwater levels due to earthquakes could be detected in confined aquifers. The Chilean earthquakes of May and June 1960 affected groundwater levels over a wide area in England. In 1962 fluctuations of levels in the Chalk, Lower Greensand and Jurassic limestones were also recognized as a result of changes in atmospheric pressure caused by the USSR's tests of nuclear weapons in the atmosphere in October 1961 and August 1962 (Ineson 1960, 1962*b*).

The Water Department began using neutron probes in the early 1960s to investigate soil moisture and moisture movement in the unsaturated zone, thereby marking the beginning of programmes to examine the role of the unsaturated zone in the recharge and pollution of groundwater. Work also commenced on electrical analogue models of groundwater flow.

The Water Department operated an enquiry service, free of charge, dealing with both major and minor enquiries from government departments, water companies, industrial concerns and the general public, especially farmers. The advice to farmers was provided as a service to the Ministry of Agriculture Fisheries and Food and was much appreciated by the farming community, even featuring in the long-running radio serial about country life, *The Archers*. Some 1200 to 1300 enquiries a year were answered in the early 1960s.

Some enquiries led to more extensive investigations. An example was a review of the mineral and thermal waters of England and Wales initiated following an enquiry from the medical world about the composition of these waters. The results of the study were subsequently incorporated in a later review of such waters published in 1969 by Edmunds *et al.* During the course of the initial review it became apparent that the principal source of basic information about saline groundwaters in Britain was held by the British Petroleum Co Ltd. Many chemical analyses, linked to particular stratigraphical horizons, were available for waters associated with the occurrence of oil in the Carboniferous rocks of the East Midlands. Through N. L. Falcon, BPs Chief Geologist at the time, R. A. Downing was generously given access to the many files. These data were collated and eventually summarised by Downing (1967) and Downing & Howitt (1969).

The hydrogeological programme occasionally led to new stratigraphical information or a new interpretation of the stratigraphy. Examination of material from a borehole drilled for water supply at Melton in Yorkshire, just to the north of the Humber, provided the first evidence that calcareous Corallian Beds occur in SE Yorkshire, south of the Market Weighton axis (Gray 1955). In the 1950s the accepted sequence for the Crag deposits of East Anglia was a tripartite division with the Norwich Crag at the base overlain by the Chillesford Clay and then the Weybourne Crag. Examination of the well and borehole records in the Survey's data base during a survey of the aquifers of northern East Anglia in the mid-1950s revealed that the Chillesford Clay did not apparently occur as a continuous clay deposit but most likely represented a clay facies developed near the top of the Norwich Crag at two main horizons (Downing 1959), a view similar to that expressed by the Survey's geologist, H. B. Woodward, as long ago as 1881.

Several staff of the Water Department (Buchan, Ineson, Gray, Price, Tuson and Downing) presented papers at a symposium on the Groundwater Supply of East Anglia which was part of the 1961 Meeting of the British Association in Norwich. In the days before the general advent of major scientific meetings, this was probably the first in the UK devoted to the regional occurrence of groundwater resources.

As the work load and the number of staff increased, the accommodation on the Bridge became increasingly overcrowded. Staff began to be housed on the Third Gallery of the Museum – referred to as the Reserve and Study Gallery, which was not open to the general public. When the Atomic Energy Division vacated the Sixth Floor of the offices at the rear of the Museum in 1956, the Water Department eagerly moved in. This gave much needed office space and also, for the first time, rooms that could be used as laboratories. In 1960 Jack Ineson became head of the Department when Buchan was promoted to be in charge of the Special Services Division.

The Central Advisory Water Committee published a report in 1959 on the 'Growing Demand for Water'. It recommended that hydrological surveys should be made beginning in areas where the expected surplus of water supply was lowest. These surveys required the comprehensive examination by river basins of rainfall, runoff, public and private sources of water supply, effluent discharges, reuse of water and potential sites for water storage. The studies were made by the Surface Water Survey of the Ministry of Housing and Local Government and the Water Department of the Geological Survey was responsible for groundwater aspects. As the programme developed, the field staff based in the Survey's office in Leeds became responsible for surveys in areas covered by that office.

The object of the studies was to make available comprehensive information about water resources and the demands on them. Steps to remedy any deficiency in supply remained the responsibility of the water undertakers.

These studies required estimation of groundwater resources in the aquifers of major river basins, often where they were overlain by Pleistocene deposits. Two surveys covered the Essex Rivers and Stour, and the East Anglian Rivers (Ministry of Housing and Local Government 1961, 1963), areas where Pleistocene cover is extensive. In these regions groundwater discharge deduced from the base-flow component of total river flow was used to assess the infiltration through boulder clays. The uneven distribution of the groundwater discharge to the river system led to the suggestion that groundwater could be over-developed during the summer at sites sufficiently distant from the river system so that groundwater flow into the rivers was not affected until the following winter. In a similar manner dry-weather flows in rivers could be supplemented by discharging groundwater, from sites remote from the rivers, into the rivers. These views were summarised in papers by Ineson & Downing (1964, 1965).

The hydrogeological survey of the Great Ouse Basin (Ministry of Housing and Local Government 1960) drew attention to the large reserve of groundwater in the Chalk of this basin. As a consequence the Ministry commissioned Binnie and Partners to make proposals for the development of the resources of the basin (Binnie and Partners 1965), proposals that led in the late 1960s to the Great Ouse Groundwater Pilot Scheme.

By the late 1950s and early 1960s there were difficulties in applying the licensing procedure for new groundwater sources. It proved to be virtually impossible to demonstrate whether or not a new well was detrimentally affecting river flows. The consequences of abstracting groundwater were disputed at public enquiries and it was not possible to present convincing data relating to the relationship between groundwater abstraction and river flow. Licences were issued with time limits attached, so that the consequences of the abstraction could be evaluated before the licence was confirmed. However, establishing the consequences was often impossible as the groundwater abstraction in such difficult cases represented only a small proportion of the river flow.

The summer of 1959 was very dry and the lack of rainfall had a significant effect on rivers in SE England that were dependent on groundwater discharge, even though the previous winter had had an average rainfall. The close relationship between groundwater discharge and dry-weather-flows was becoming increasingly obvious to a wider audience of engineers and administrators in the government. The results of the hydrological surveys were also emphasising the advantage of developing surface and groundwater in conjunction within entire major river basins. These views were to lead to the fundamental changes in outlook as expressed in the Water Resources Act of 1963.

When the extent of the changes that would result from the Water Resources Act became apparent, the Yorkshire Ouse and Trent river boards each commissioned the Water Department to prepare reports on the geology and groundwater resources of the river basins under their jurisdiction. These reports, which were for areas that had not been included in the hydrological surveys, were eventually published (Gray *et al.* 1969; Downing *et al.* 1970).

In the run-up to the implementation of the Water Resources Act and the formation of the Water Resources Board, the future Director and Secretary of the Board visited the Water Department ostensibly to explain the role of the new organisation but also to persuade staff to transfer to it. This was a difficult period, one of great uncertainty with respect to the long-term future of both the Water Department and the new Board. The Survey's Directorate and many of the Water Department's staff were in favour of secondment of some staff to the new organisation but this was not acceptable (although it subsequently became evident that staff from the Ministry of Housing and Local Government were, in fact, seconded to the Board). Eventually, in 1965, Ineson agreed to transfer to the Water Resources Board and he was accompanied by six other members of the Department. They took with them the role of evaluating and developing groundwater resources on a national basis.

The Water Department, which became the Hydrogeological Department in 1965 under David Gray, had a remit to carry out hydrogeological research. The split of the Department was forced upon the Survey. From a national point of view, as seen at the time, it was unfortunate and undesirable to divide the one small group working on hydrogeology when the work-load was about to expand exponentially. The Survey Board actually envisaged that the Water Department would advise and assist the Water Resources Board and the new river authorities as well as the Ministry of Housing and Local Government. However, this did not materialize and, with hindsight, it was inevitable that, with the increasing interest in groundwater, component organisations in the revamped water industry would require their own specialist staff. This marked the beginning of the expansion of the role of hydrogeologists in the water industry. The changes did allow a relatively rapid and successful expansion of the Hydrogeological Department into areas of research previously excluded, especially experimental and theoretical geochemistry, studies of the physical properties of aquifers and eventually pollution control and the application of information technology to various hydrogeological problems.

The role of the Water Department: an assessment

On reflection, the Water Department's role between 1945 and 1965 was one of significant achievement. At that time staff were graded as either Geologists or Experimental Officers. There were never more than 10 Geologists supported by 10 to 15 Experimental Officers in the Department, with most of the EOs largely engaged on maintaining, updating and cataloguing the data archive and assisting with pumping tests. This small group was in reality managing the technical aspects of groundwater necessary for the successful implementation of the Water Act of 1945. They were applying hydrogeology to the practical needs of society and the work programmes were intended to solve specific problems arising from groundwater legislation. It must be said that they did not really appreciate this; they were not conscious of living through this view of their work. They were much more closely associated with the ethos of the Geological Survey itself, which was to advance knowledge of geology (and in their eyes hydrogeology) in the UK – research was the ideal and took pride of place in the work effort. However, in retrospect, their principal role can be seen to have been concerned with regional and local groundwater management and the establishment of a reliable national hydrogeological baseline. Eight of the staff were paid for by the Ministry of Housing and Local Government, an indication of the extent of the Department's contribution to the running of the Water Act. Surprisingly, providing this duty does not seem to have been appreciated, or even understood, by the Geological Survey Board for it is never referred to in the annual Summary of Progress of the Geological Survey. Even the Water Resources Act of 1963 and its drastic effect on the Water Department received only minimal attention, simply recording the transfer of seven members of staff to the new organisation.

The main advances that can be credited to the Water Department are:

(1) The hydrogeology of the main aquifers was examined in some detail, on a regional basis, for the first time. The groundwater resources were assessed, the abstraction documented and the potential reserves defined.
(2) Groundwater abstraction data were collected and collated as required under Section 6 of the Water Act. This was initially organized by Ineson with, in later years the assistance of a clerk seconded from the Ministry of Housing and Local Government for about two months each year, and then, finally, with a permanent clerk on the Survey's own strength. Section 6 of the Water Act was repealed by the Water Resources Act, 1963. The last returns under Section 6 were requested on a voluntary basis in an attempt to maintain continuity of the long-term record. This proved to be a forlorn hope as action by the new river authorities, under Section 114 of the Water Resources Act, was slow to materialize and annual abstractions from the main aquifers on a national basis were no longer collated, and still are not.
(3) Aquifer properties were determined at many sites and the regional pattern of transmissivity values in the Chalk became more widely recognized.
(4) Suitable instrumentation for measuring water levels and other parameters was developed and geophysical techniques were shown to have wide application in the study of groundwater flow.
(5) As a result of regional studies, the areal variations in groundwater chemistry were recognized.
(6) Through cooperation with local authorities, regular measurements of groundwater levels were made in many areas. By the mid-1960s regular manual observations of groundwater levels (at least at monthly intervals) were made at some 760 sites, mainly on the Chalk. These were supplemented with data from automatic recorders at 113 sites of which the Survey was responsible for 96 installations, again mainly on the Chalk. As the data accumulated analyses of the seasonal fluctuations of the levels were used to assess the changes in groundwater storage particularly in the Chalk. Details of groundwater levels in England and Wales during 1963 were published by the Survey (Lovelock 1967); the publication was in effect the first Groundwater Yearbook. The measurements of groundwater levels by the local authorities were not sent to the Survey after the end of 1965 when collating them became the

responsibility of the Water Resources Board. Ineson (1965) described the principles underlying the design of groundwater networks. These principles were essentially adopted in the subsequent government-funded Hydrological Networks established by the river authorities under the Water Resources Act.

(7) By 1965 a hydrogeological map of NE Lincolnshire was nearing completion, which marked the beginning of a series of maps produced by the Survey covering almost all the main aquifer units of the UK.

All this was accomplished with an extremely small capital budget.

Between 1945 and 1965 the use of groundwater in England and Wales increased to over 2000 Mm^3a^{-1}. During this period of expansion, the Survey's advice to the Ministry of Housing and Local Government was carried out with a complete absence of bureaucracy. If a statutory water undertaking wished to increase its groundwater sources it invariably approached the Survey's staff to discuss a favourable location for the site, the likely yield and the anticipated quality of the water. It would then apply to the Ministry for a licence and following advice from the Survey the Ministry may have requested a 14 or 21-day pumping test to be supervised by the Survey's staff. On the basis of the test the Survey identified any interference with other users and advised on an appropriate long-term yield and any conditions that should be attached to the licence. If there were many objections to the proposal a public inquiry would normally be necessary when a Survey geologist would commonly act as an assessor and assist the inspector appointed to take the inquiry.

The staff of the Water Department had close informal links with many engineers in charge of local water authorities and companies that depended on groundwater. There was a less close link at a personal level with the Ministry's staff which was regrettable and somewhat unsatisfactory but not a serious drawback.

Undoubtedly the main downside of the entire period was a failure to publish the results of the work of the Department, particularly the regional studies. Many internal reports, complete with maps of rest water levels, aquifer depths and thicknesses and chemical properties of the water, were prepared. These reports, now available on open file at the British Geological Survey, are of historical value as they portray groundwater conditions in England in the 1950s. There were no official publications from the Department until 1964 by which time Ineson was in charge of the Water Department. An opportunity to begin educating a wider spectrum about the occurrence and distribution of groundwater was lost and this was undoubtedly unfortunate from a national viewpoint especially when seen in the light of what was expected and needed from the water industry to implement the Water Resources Act.

One bright spot was the resumption by the Water Department of the publication of summaries of the data held in the Well Record Collection. This was a continuation of the policy that really began with the Water Memoirs introduced by Geikie and which was continued in the Wartime Pamphlet Series. Publication of the new Well Catalogue Series for England began in 1964 and by the end of 1965 24 had been issued. Each covered a one-inch-to-one-mile geological map. Details of the geological sequence, construction of the well, rest water levels, pumping water levels and pumping rates, as well as the basic chemical composition of the water were given, if available, for each site. The catalogues included brief summaries of the hydrogeology of the area.

Publication of regional groundwater studies only began in 1966 with David Land's work on the Triassic sandstones of Nottinghamshire and the Birmingham-Lichfield District followed in 1967 by Jack Ineson's study of the South Wales Coalfield.

The staff of the Water Department were all traditionally educated geologists; they had to enlighten themselves in the ways of hydrogeology from each others experiences and from the literature, especially that of the US Geological Survey's publications – the Water Supply Papers, the Professional Papers and the papers by O. E. Meinzer (1923*a*, *b*), the visionary geologist in charge of the USGS's Groundwater Division. An up-to-date textbook on groundwater did not become available until D. K. Todd published 'Groundwater Hydrology' in 1959. However, before that, in 1957, Ineson completed his PhD thesis at the University of London. This was a comprehensive review of British hydrogeology as known at the time, as well as the successful application of Theis's work to British aquifers – a *tour de force* (Ineson 1957). Furthermore, beginning in 1958, Ineson gave a course of lectures on hydrogeology at the Chelsea College of Technology (now part of the University of London). These were continued annually until his death in 1970. Glyn Jones, who had joined the Water Department in 1960, resigned at the end of 1964 to become Lecturer in Hydrogeology at University College London, under Professor S. E. Hollingworth, himself an ex-Survey geologist, thereby establishing the first full-time course in hydrogeology at a British university.

It is evident that the Water Department set the scene for the increasingly important role hydrogeology assumed in water resources development in the UK. It provided the framework necessary for the effective management of groundwater resources on a national basis – a general recipe as to how to proceed.

D. A. Gray kindly read an early draft of this paper. He made many constructive comments and pointed out important omissions, which have now been incorporated.

References

ANDERSON, W. 1945*a*. *Water supply from underground sources of north-east England*. Wartime Pamphlet No **19**, Part III, Geological Survey of Great Britain.

ANDERSON, W. 1945*b*. On the chloride waters of Great Britain. *Geological Magazine,* **82**, 267–273.

BAILEY, E. B. 1944. Geological Survey in relation to underground water. *Journal of the British Waterworks Association,* **XXVI**, 121–125.

BAILEY, E. B. 1952. *Geological Survey of Great Britain.* T Murby and Co., London.

BINNIE AND PARTNERS, 1965. *Water resources of the Great Ouse Basin.* Ministry of Housing and Local Government, London.

BUCHAN, S. 1938. *The water supply of the County of London from underground sources.* Memoir of the Geological Survey of Great Britain

BUCHAN, S. 1955. Artificial replenishment of aquifers. *Journal of the Institution of Water Engineers,* **9**, 111–163.

BUCHAN, S. 1962. Disposal of drainage water into the Chalk from coal mines in Kent. *Proceedings of the Society for Water Treatment and Examination.*, **11**, 101–105.

BUCHAN, S. 1964. The problem of groundwater recharge: artificial recharge as a source of water. *Journal of the Institution of Water Engineers*, **18**, 239–246.

BUCHAN, S. & KEY, A. 1956. Pollution of groundwater in Europe. *Bulletin of the World Health Organisation*, **14**, 949–1006.

CHAPELLE, F. H. 1993. *Groundwater microbiology and geochemistry.* Wiley and Sons, Chichester.

DAY, J. B. W. 1964. *Infiltration into a groundwater catchment and the derivation of evaporation.* Research Report No **2**, Water Supply Papers of the Geological Survey of Great Britain.

DOWNING, R. A. 1959. A note on the Crag in Norfolk. *Geological Magazine*, **96**, 81 –86.

DOWNING, R. A. 1967. The geochemistry of groundwaters in the Carboniferous Limestone in Derbyshire and the East Midlands. *Bulletin of the Geological Survey of Great Britain No 27*, 289–307.

DOWNING, R. A. & HOWITT, F. 1969. Saline groundwaters in the Carboniferous rocks of the English East Midlands in relation to the geology. *Quarterly Journal of Engineering Geology*, **1**, 241–249.

DOWNING, R. A., LAND, D. H., ALLENDER, R., LOVELOCK, P. E. R. & BRIDGE, L. R. 1970. *The hydrogeology of the Trent River Basin.* Hydrogeological Report No **5**, Institute of Geological Sciences, London.

EARP, J. R. & EDEN, R. A. 1961. Amounts and distribution of underground waters in Scotland. *Water and Water Engineering*, **65**, 255–259.

EDMUNDS, F. H. 1941. Outlines of underground water supply in England and Wales. *Transactions of the Institution of Water Engineers*, **XLVI**, 15–104.

EDMUNDS, W. M., TAYLOR, B. J. & DOWNING, R. A. 1969. Mineral and thermal waters of the United Kingdom. *International Geological Congress, Report of the 23rd Session Czechoslavakia* 1968, **18**, 139–158.

FISHER, W. W. 1901. On alkaline waters from the Chalk. *The Analyst*, **26**, 202–213.

FISHER, W. W. 1902. Alkaline waters from the Lower Greensand. *The Analyst*, **27**, 212–219.

FISHER, W. W. 1904. On the salinity of waters from the Oolites. *The Analyst*, **29**, 29–44.

GARDENER, C. M. K., BELL, J. P., COOPER, J. D., DARLING, W. G. & REEVE, C. E. 1991. Groundwater recharge and water movement in the unsaturated zone. *In*: DOWNING, R. A. & WILKINSON, W. B. (eds) *Applied Groundwater Hydrology.* Clarendon Press, Oxford, 54–76.

GRAY, D. A. 1955. The occurrence of the Corallian Limestone in east Yorkshire, south of Market Weighton. *Proceedings of the Yorkshire Geological Society*, **30**, 25–34.

GRAY, D. A. 1958. Electrical resistivity marker bands in the Lower and Middle Chalk of the London Basin. *Bulletin of the Geological Survey of Great Britain*, **15**, 85–95.

GRAY, D. A. 1964*a*. Instrumentation in groundwater studies. *Water and Water Engineering*, **68**, 185–188.

GRAY, D. A. 1964*b*. *Groundwater conditions of the Chalk of the Grimsby area, Lincolnshire*. Research Report No **1**, Water Supply Papers of the Geological Survey of Great Britain.

GRAY D. A. 1965. The stratigraphical significance of electrical resistivity marker bands in the Cretaceous strata of the Leatherhead (Fetcham Mill) Borehole, Surrey. *Bulletin of the Geological Survey of Great Britain*, **23**, 65–114.

GRAY, D. A., ALLENDER, R. & LOVELOCK, P. E. R. 1969. *The groundwater hydrology of the Yorkshire Ouse River Basin.* Hydrogeological Report No **4**, Institute of Geological Sciences, London.

HEADWORTH, H. G. 1970. The selection of root constants for the calculation of actual evaporation and infiltration for Chalk catchments. *Journal of the Institution of Water Engineers and Scientists*, **24**, 431–446.

HEADWORTH, H. G. & FOX, G. B. 1986. The South Downs Chalk aquifer: its development and management. *Journal of the Institution of Water Engineers and Scientists*, **40**, 345–361.

HEADWORTH, H. G., PURI, S. & RAMPLING, B. H. 1980. Contamination of a Chalk aquifer by mine drainage at Tilmanstone, East Kent. *Quarterly Journal of Engineering Geology*, **13**, 105–117.

INESON, J. 1952. Notes on the theory and formulae associated with pumping tests for the determination of formation constants. *Journal of the Institution of Water Engineers*, **6**, 443–458.

INESON, J. 1953. Some observations on pumping tests carried out on Chalk wells. *Journal of the Institution of Water Engineers*, **7**, 215–225.

INESON, J. 1957 *The movement of groundwater as influenced by geological factors and its significance.* PhD thesis, University of London.

INESON, J. 1959*a*. Yield-depression curves of discharging wells with particular reference to Chalk wells and their relationship to variations in transmissibility. *Journal of the Institution of Water Engineers*, **13**, 119–163.

INESON, J. 1959*b*. The relation between the yield of a discharging well at equilibrium and its diameter with particular reference to a Chalk well. *Proceedings of the Institution of Civil Engineers*, **13**, 299–316.

INESON, J. 1960. The effect of the Chilean earthquakes of May and June 1960 on groundwater levels in England. *Proceedings of the Geological Society of London*, **No 1583**, 10.

INESON, J. 1962*a*. A hydrogeological study of the permeability of the Chalk. *Journal of the Institution of Water Engineers*, **16**, 449–463.

INESON, J. 1962*b*. Fluctuations of groundwater levels due to atmospheric pressure changes from nuclear explosions. *Nature*, **195**, 1092–1093.

INESON, J. 1964. Pollution of water and soil by miscellaneous petroleum products. *Journal of the British Waterworks Association*, **46**, 307–326.

INESON, J. 1965. Groundwater principles of network design. *International Symposium on Hydrometeorological Networks, Quebec*, World Meteorological Organisation and Association of Scientific Hydrology.

INESON, J. 1967. *Groundwater conditions in the Coal Measures of the South Wales Coalfield*. Hydrogeological Report **No 3**, Geological Survey of Great Britain.

INESON, J. & DOWNING, R. A. 1963. Changes in the chemistry of groundwaters in the Chalk passing beneath argillaceous strata. *Bulletin of the Geological Survey of Great Britain*, **20**, 176–192.

INESON, J. & DOWNING, R. A. 1964. The groundwater component of river discharge and its relationship to hydrogeology. *Journal of the Institution of Water Engineers*, **18**, 519–541.

INESON, J. & DOWNING, R. A. 1965. Some hydrogeological factors in permeable catchment studies. *Journal of the Institution of Water Engineers*, **19**, 59–80.

INESON, J. & GRAY, D. A. 1963. Electrical investigations of borehole fluids. *Journal of Hydrology*, **1**, 204–218.

LAND, D. H. 1966. *Hydrogeology of the Bunter Sandstone in Nottinghamshire*. Hydrogeological Report **No 1**, Geological Survey of Great Britain.

LAND, D. H. 1966. *The hydrogeology of the Triassic sandstones in the Birmingham-Lichfield district*. Hydrogeology Report **No 2**, Geological Survey of Great Britain.

LOVELOCK, P. E. R. 1967. *Groundwater levels in England during 1963*. Research Report **No 3**, Water Supply Papers of the Geological Survey of Great Britain.

MEINZER, O. E. 1923*a*. *Outline of groundwater hydrology with definitions*. United States Geological Survey Water Supply Paper, **No 494**.

MEINZER, O. E. 1923*b*. *The occurrence of groundwater in the United States*. United States Geological Survey Water Supply Paper, **No 489**.

MINISTRY OF HOUSING AND LOCAL GOVERNMENT, 1960. *River Great Ouse Basin Hydrological Survey*. HMSO, London.

MINISTRY OF HOUSING AND LOCAL GOVERNMENT. 1961. *Essex Rivers and Stour Hydrological Survey*. HMSO, London.

MINISTRY OF HOUSING AND LOCAL GOVERNMENT. 1963. *East Anglian Rivers Hydrological Survey*. HMSO, London.

MORTIMORE, R. N. & POMEROL, B. 1987. Correlation of the Upper Cretaceous White Chalk in the Anglo-Paris Basin. *Proceedings of the Geologists Association*, **98**, 97–143.

MURRAY, K. H. 1986. *Correlation of electrical resistivity marker bands in the Cenomanian and Turonian Chalk from the London Basin to East Yorkshire*. British Geological Survey Report **17**, No 8, HMSO, London.

PENMAN, H. L. 1948. Natural evaporation from open water, bare soil and grass. *Proceedings of the Royal Society*, **193**, 120–146.

PENMAN, H. L. 1950*a*. Evaporation over the British Isles. *Quarterly Journal of the Royal Meteorological Society*, **96**, 372–383.

PENMAN, H. L. 1950*b*. The water balance of the Stour catchment area. *Journal of the Institution of Water Engineers*, **4**, 457–469.

RODDA, J. C., DOWNING, R. A. & LAW, F. M. 1976. *Systematic Hydrology*. Newnes – Butterworths, London.

RUSHTON, K. R. & WARD, C. J. 1979. The estimation of groundwater recharge. *Journal of Hydrology*, **41**, 345–361.

SMITH, B. 1935. *Geological aspects of underground water supplies*. The Cantor Lectures, The Royal Society of Arts, London.

TATE, T. K., ROBERTSON, A. S. & GRAY, D. A. 1970. The hydrogeological investigation of fissure flow by borehole logging techniques. *Quarterly Journal of Engineering Geology*, **2**, 195–215.

THRESH, J. C. 1912. The alkaline waters of the London Basin. *Chemical News*, **105**, 25–27 & 37–44.

TODD, D. K. 1959. *Groundwater Hydrology*. Wiley and Sons, Chichester.

WHITAKER, W. & THRESH, J. C. 1916. *The water supply of Essex from underground sources*. Memoir of the Geological Survey of Great Britain.

WOODLAND, A. W. 1940. *Water supply from underground sources of the Oxford-Northampton District*. Wartime Pamphlet **No 4**, Geological Survey of Great Britain.

WOODLAND, A. W. 1946. *Water supply from underground sources of the Cambridge-Ipswich District*. Wartime Pamphlet **No 20**, Part X, General Discussion. Geological Survey of Great Britain.

WOODWARD, H. B. 1891. *The geology of the country around Norwich*. Memoir of the Geological Survey of Great Britain.

Jack Ineson (1917–1970)
The instigator of quantitative hydrogeology in Britain

R. A. DOWNING[1] & D. A. GRAY[2]

[1]*20, Springfield Park, Twyford, Reading, Berkshire, RG10 9JH, UK*
[2]*46, Bonnersfield Lane, Harrow, Middlesex, HA1 2LE, UK (e-mail hammerer@beeb.net)*

Abstract: Jack Ineson will always be associated with introducing quantitative methods to British hydrogeology. A geologist with a sound knowledge of mathematics and statistics, unusual for the time, he seized the opportunity in 1948 to apply to British aquifers the burgeoning theory of well hydraulics initiated by Theis. Ineson's career was mainly spent with the Geological Survey of Great Britain, now the British Geological Survey, but in the period 1965–1970 as Chief Geologist of the Water Resources Board. It was, however, a relatively short career with the start postponed by the Second World War and tragically truncated in June 1970 as a direct consequence of his experiences in the war.

Jack Ineson was born in Otley in 1917. He had many of the characteristics of a Yorkshireman; although not blunt by any means, he was direct and straightforward, did not mince words and was generally unable to tolerate fools gladly. He had a sense of humour which did not extend to frivolity.

He went to the University of Durham in 1936 with an Entrance Scholarship, intending to read theology but came under the inspiring influence of Arthur Holmes and made what must have been a dramatic switch to geology, graduating with honours in that subject in 1939. During his time at Durham he played rugby for the University and was Senior Man at University College – a position reserved for a student of undoubted all-round ability and with leadership qualities.

Originally destined to be a petrologist, he began research for a PhD under Holmes but the war intervened and a few months after graduating he enlisted in the army and was commissioned in the Royal Northumberland Fusiliers. Posted to Malaya, his unit was captured by the Japanese at the fall of Singapore in February 1942. He spent over three years in captivity working on the notorious Burma-Siam railway and it was largely due to his initial good physique and his strength of character that he was able to withstand the experience. But his health was permanently affected by the privation, malnutrition, disease and severe hardship. Whilst in captivity he taught himself Russian with the help of a fellow prisoner – an achievement in itself in the circumstances. He rarely mentioned this period other than occasionally to describe some humorous incident; indeed, he disapproved of it becoming known. When he was repatriated he weighed only 38 kg (6 stones) and throughout his life he suffered recurring bouts of malaria and was soon exhausted by physical exertion. But he had the constant support of his wife, Mary, who undertook many of the physical tasks associated with domestic life, giving him much freedom to pursue his interests.

Fig. 1. Jack Ineson 1917–1970

After repatriation Ineson joined the Geological Survey of Great Britain, in 1946, at South Kensington, and immediately began mapping the Cretaceous in Kent, subsequently contributing to the Maidstone memoir. But after 18 months as a field geologist he transferred to the Water Department under Stevenson Buchan, to undertake work more suited to his approach to research. At that time the Survey was being called upon to implement those sections of the 1945 Water Act concerned with groundwater. The resources had to be assessed for the first time, a data

From: MATHER, J. D. (ed.) 2004. *200 Years of British Hydrogeology*. Geological Society, London, Special Publications, **225**. 283–293. 0305-8719/04/$15

base created and an administrative system established to obtain returns of groundwater use. Ineson threw himself into these tasks wholeheartedly, in particular organizing, coordinating and maintaining the records of groundwater use, which provided for the first time quantitative information about the development of aquifers in England and Wales.

The implication for groundwater hydraulics of the realization by Theis that the flow of water through rocks is analogous to the conduction of heat in solids was not appreciated in Europe until the late 1940s because of poor communications during the war. It was in 1948 that Jack Ineson was asked to investigate the application of the new methods of analysis to the fissured aquifers typical of the UK. The basic purpose of the work was the urgent need to design pumping tests that would provide reliable data from which to forecast long-term 'safe yields' for the many wells being drilled to meet the increasing post-war demand for groundwater for industrial and public supplies. Previously the basis for development was merely conventional wisdom, widely believed but with little basis in fact. Ineson was at the time applying new theoretical principles to the solution of a longstanding problem and his contribution was both crucial and pivotal. Through the Geological Survey's involvement in testing wells throughout eastern and southern England, he had much data to analyse and this work brought him into contact with most of the country's water engineers who were concerned with the expanding use of groundwater. His advice was in demand continuously. In a paper published in 1958 he described the procedure for testing new wells adopted by the Survey.

The early results of his researches were incorporated in his 1957 PhD thesis, which was virtually a textbook on British hydrogeology, a *tour de force*. The principal conclusions were published in a series of papers in the Proceedings of the Institution of Civil Engineers and the Journal of the Institution of Water Engineers. He demonstrated that the non-equilibrium methods could indeed be applied to fissured aquifers such as the Chalk and he recognized the dual porosity nature of that formation. An examination of the relation between well yields and well diameter for Chalk wells led to the recognition of the importance of 'two flow phases' linked to movement through the fissure system and seepage from the mass of the Chalk. By analysing pumping tests, but also by applying statistical methods to limited data about well yields, and the relationship of yields to well diameters, he determined the transmissivity in the vicinity of many pumped wells and was thereby able to map relative values for the permeability of the aquifer in England. The link between permeability and chalk valleys was established quantitatively – something previously based largely on conventional wisdom. His wide-ranging interests extended to studies of groundwater chemistry as well as early papers on the risks to the quality of groundwater from pollution by oil products. He also made studies of the regional hydrogeology of the English Midlands and South Wales. Recurrent themes in his papers were the importance of an interpretation of the three-dimensional geological pattern related to lithology, sedimentary sequences, structure and erosional features and the need for close cooperation between groundwater hydrologists and water engineers.

In 1960 he recognized the effect of the major earthquake in Chile on groundwater levels in the UK. Subsequently he detected and analysed fluctuations caused by atmospheric pressure changes produced by nuclear tests in the atmosphere. Tipped off by the geophysicists at the Blacknest Seismological Station that the Russians had set off a nuclear device, he would rush, with one or two colleagues, to a well in South Kensington to set up a recorder with appropriate scales, picking up the fluctuations some 20–30 minutes after receiving the telephone call.

Jack Ineson became Head of the Water Department in the Geological Survey in 1960 and in the run-up to the Water Resources Act,1963, attention began to be focussed on the need to develop groundwater without producing a significant impact on river flows. Ineson was closely involved in assessing the groundwater component of river flow and the development of resources in the context of basin management.

In 1965 he was appointed Chief Geologist of the newly-formed Water Resources Board (WRB) and immediately became concerned with implementing the Water Resources Act, 1963, and ensuring that groundwater received proper recognition in policy decisions. This was a period of great optimism. Logical development of river basins, with the conjunctive use of groundwater and surface water, seemed possible at last, hydrometric schemes were established, groundwater studies of major aquifers initiated. These investigations, started by Ineson, contributed to the three major water resource studies of the WRB and ultimately to the National Plan. He summarized the views of the WRB regarding the development of groundwater resources in England and Wales in an important paper published by the Institution of Water Engineers in 1970. This stressed the significance of the combined use of surface water and groundwater within the overall objective of effective river basin management. In this and another paper co-authored in 1967 by N.A.F. Rowntree, emphasis was placed on the need for the optimum use of storage in aquifers not only for water supply development but also for the requirements of the environment and the river system in the widest sense.

Jack Ineson served on and chaired many committees both nationally and internationally. Despite his heavy administrative duties, he found time to continue writing scientific papers. In all he was an author of almost 50 publications and at the time of his death eight were in the press. One was the major book 'Ground-water Studies' published by UNESCO in 1972. He contributed significantly to the text but he was also one of four editors, with one American and two Russians; his contribution to editorial discussions was simplified by his ability to speak and read Russian.

He was an outstanding hydrogeologist. His main contributions to the science were, firstly, his introduction and emphasis on quantitative methods which had a profound influence on the approach to British groundwater engineering in the 1950s and 1960s. Secondly, his contribution to resources planning and development, ensuring that groundwater had a recognized role and made an appropriate input at regional and national levels at a time when many geologists and engineers just did not understand groundwater. Thirdly, he passed on his knowledge for 12 years from 1958 by delivering a pioneering course of evening lectures on hydrogeology at the (then) Chelsea College of Science and Technology in the University of London. In these lectures he clarified the mysteries of groundwater flow for a wide audience of geologists and engineers. They contained many practical examples taken from his experiences in England and Wales.

Jack Ineson worked hard but had many other interests, not least watching motor racing and following rugby. He was an accomplished pianist and organist and could speak fluent French and German as well as Russian; he translated texts in all three languages. He always had time to offer advice. Give him a text and he would return it in a few days neatly corrected and annotated with relevant points. He was cautious and careful in expressing his views – not a man to publish and be damned. Points were argued at length from all possible angles and his classical background meant that it was not unusual to spend an afternoon discussing and arguing about the style and structure of a paragraph. While often hard to convince he was always fair in his final assessment. He had firmly held principles and a professional integrity and was well respected and liked by his colleagues. His wide-ranging scientific knowledge gave him the versatility to collaborate readily with other disciplines and to provide momentum during the rapid expansion of hydrogeology in the UK in the decade of the 1960s. His approach was always in line with Pasteur's view that there is no such thing as Applied Science only the application of Pure Science.

Time alters perspective with startling rapidity and it is perhaps difficult now to appreciate Ineson's contribution from a bit higher up the hill. But things that seem obvious with hindsight were not so at the time. Thoughts are shaped by others and during his life he undoubtedly had the stature to lay the foundations for a modern approach to groundwater in the UK. His research and his presence were influential at a decisive time.

In 1961 the Geological Society honoured him with the award of the Murchison Fund and in 1970 the Institution of Water Engineers followed suit with the Whitaker Medal – both for fundamental contributions to groundwater studies. He would have been delighted to see how hydrogeology has developed since his death, although no doubt dismayed by the failure to apply successive legislation to prevent the deterioration of groundwater quality. He would have been honoured to receive the respect of a discriminating, critical group of younger scientists through the establishment of the annual Ineson Lecture in his memory. But it is a sad reflection on how Britain treated and regarded its ex-servicemen in that he was employed for 20 years as a 'temporary civil servant' without pension rights because his health had been undermined by his experiences as a prisoner of war.

Those who knew Jack Ineson recognized that he was a rather unique individual. He belonged to the tradition and generation which regarded public service as a duty willingly met. Few who came into contact with him realized or appreciated the underlying health problems that he always had to overcome. But overcome them he did very successfully, until his untimely death from a heart attack on June 3rd 1970, whilst travelling home from work.

Principal publications

1952. Notes on the theory and formulae associated with pumping tests for the determination of formation constants. *Journal of the Institution of Water Engineers*, **6**, 443–448.

1953. Some observations on pumping tests carried out on Chalk wells. *Journal of the Institution of Water Engineers*, **7**, 215–225.

1956. *Darcy's law and the evaluation of "permeability"*. International Association of Hydrogeologists, Publication No. **41**, 165.

1957. *The movement of groundwater as influenced by geological factors and its significance*. PhD Thesis, University of London.

1958. Pumping-test procedure. *Proceedings of the Symposium on Groundwater, 1955*, Central Board of Geophysics, Calcutta, Publication **No.4**, 335–342.

1959. Yield-depression curves of discharging wells with particular reference to Chalk wells and their relationship to variations in transmissibility. *Journal of the Institution of Water Engineers*, **13**, 119–163.

1959. The relation between the yield of a discharging well at equilibrium and its diameter with particular reference to a Chalk well. *Proceedings of the Institution of Civil Engineers*, **13**, 299–316.

1960. The effect of the Chilean earthquakes of May and June 1960 on groundwater levels in England. *Proceedings of the Geological Society*, London, No. **1583**, 10.

1962. Fluctuations of groundwater levels due to atmospheric pressure changes from nuclear explosions. *Nature*, **195**, 1082–1083.

1962. A hydrogeological study of the permeability of the Chalk. *Journal of the Institution of Water Engineers*, **16**, 449–463.

1963. (With D. A. GRAY). Electrical investigations of borehole fluids. *Journal of Hydrology*, **1**, 204–218.

1963. (With R. A. DOWNING). Changes in the chemistry of groundwaters in the Chalk passing beneath argillaceous strata. *Bulletin of the Geological Survey of Great Britain*. No. **20**, 176–192.

1964. Pollution of water and soil by miscellaneous petroleum products. *Journal of the British Waterworks Association*, **46**, 307–326.

1964. (With R. A. DOWNING). The groundwater component of river discharge and its relationship to hydrogeology. *Journal of the Institution of Water Engineers*, **18**, 519–541.

1965. (With R. A. DOWNING). Some hydrogeological factors in permeable catchment studies. *Journal of the Institution of Water Engineers*, **19**, 59–80.

1965. Aquifer elasticity. *Proceedings of the Geological Society, London*, No. **1621**, 41–42.

1965. Groundwater principles of network design. *In*: *International Symposium on Hydrometeorological Networks, Quebec*. World Meteorological Organisation and Association of Scientific Hydrology.

1967. *Groundwater conditions in the Coal Measures of the South Wales Coalfield*. Water Supply Papers, Hydrogeological Report No. **3**, Geological Survey of Great Britain.

1967. (With N. A. F. ROWNTREE). Conservation projects and planning. *Journal of the Institution of Water Engineers*, **21**, 275–290.

1970. The significance of oil pollution in the water resources field. *In*: *Seminar on Water Pollution by Oil*. Aviemore, Scotland, Institution of Water Pollution Control, 143–151.

1970. Development of groundwater resources in England and Wales. *Journal of the Institution of Water Engineers*, **24**, 155–177.

1972. (With R. H. BROWN, A. A. KONOPLYANTSEV & U. S. KOVALEVSKY, eds). *Ground-water Studies: an International Guide for Research and Practice*. UNESCO, Paris.

Stevenson Buchan (1907–1996): field geologist, hydrogeologist and administrator

D. A. GRAY[1] & J. D. MATHER[2]

[1]46, Bonnersfield Lane, Harrow, Middlesex HA1 2LE, UK
[2]Department of Geology, Royal Holloway University of London, Egham, Surrey, TW20 0EX, UK

Abstract: Stevenson Buchan, a Scot educated at Aberdeen University, joined the Geological Survey of Great Britain in 1931. Assigned to the Southern District Unit he became involved in hydrogeology when he revised the 6-inch to the mile maps of Greater London. This led to the publication of a memoir on the water supply of the County of London in 1938. During the war he was involved in the search for water and coal. Appointed Head of a new Water Department when hostilities ceased he was given responsibility for overseeing the statutory obligations placed on the Survey by recent Water Acts. He developed around him a group of able colleagues creating the first groundwater research group of any size in the UK. Promoted to Assistant Director in 1960, he was very active internationally and in his role as an enabler and administrator played an important role in the development of hydrogeology in the UK.

Stevenson Buchan (1907–1996) was born on the 4th March 1907 at 32, Queen Street, Peterhead. He was the only child of James Buchan (1864–1943), a fish trader and his second wife Christian Ewen Stevenson (1870–1933). Stevenson, or 'Steve' as he was always known to his friends and colleagues, was educated at the Central School in Peterhead and at Peterhead Academy before going up to Aberdeen University in 1925. There he graduated with a first class honours degree in geology in 1929 and remained at Aberdeen as a demonstrator in the Geology Department whilst at the same time undertaking research for a doctorate. In his undergraduate days he was a Prizeman and Medallist and made time for a 'blue' in cross country running and was President of both the Geological and Scientific Societies.

His research work was concentrated on the Peterhead Granite and on the dykes which intruded it. His first paper described some of these dykes and was read at a meeting of the Edinburgh Geological Society on 21st January 1931 (Buchan 1932). He examined 25 occurrences of dykes along the coastal section south of Peterhead, looking at their petrography and age relationships. However the main thrust of his research was on the petrology of the Granite itself which he divided into two separate masses, defined as the Cairngall and Peterhead Granites. This work was never published in full but he gave a paper on the subject to the British Association meeting in Aberdeen in September 1934 some months after the award of his doctorate (Buchan 1934).

Field mapping with the Geological Survey (1931–1939)

Much of his thesis must have been written in his own time as, in 1931 after two years as a demonstrator at Aberdeen, he was one of only two new appointments to the geological staff of the Geological Survey of Great Britain (GSGB). Based initially at the Museum of Practical Geology in Jermyn Street, he was assigned to the Southern District Unit under its District Geologist, Henry Dewey. His first field season found him mapping Lower Greensand strata on the Sevenoaks sheet (287) where he was trained in the Westerham district by Henry George Dines.

In 1932 he returned to the Sevenoaks sheet for a short period but then joined a group working in Dorset on the Bridport sheet (327) where he mapped Jurassic and Cretaceous rocks. He also began office work on the preparation of 6-inch to the mile maps of the London area. In 1933 publication of these 6-inch maps was sanctioned and Buchan continued to be heavily involved. By 1937 a total of 42 sheets had been published on all of which his contribution is recognized by the acknowledgement 'Modifications and additional notes by S. Buchan'. In 1933 he also continued his mapping around Sevenoaks where he recognized the importance of the Atherfield Clay from the point of view of water supply. The Clay acted as an impermeable base to the overlying Hythe Beds throwing out strong springs. In addition for a period in 1933 he was loaned to the Shetland Islands Unit where he mapped the Island of Foula as part of the primary survey.

He again spent part of 1934 in Shetland where he mapped igneous and metamorphic rocks in the Walls Peninsula and on the island of Papa Stour. 1934 saw the completion of the Shetland mapping but the Western Shetland sheet on which he had worked was not published until 1971. However, in 1934, his main activity was surveying the culverted streams and rivers of Greater London and their associated alluvium to provide revision data for the 6-inch to the mile maps.

From: MATHER, J. D. (ed.) 2004. *200 Years of British Hydrogeology*. Geological Society, London, Special Publications, **225**. 287–293. 0305-8719/04/$15

Fig. 1. Dr Stevenson Buchan CBE and his wife Barbara with the Director of the Institute of Geological Sciences, Sir Kingsley Dunham, at the time of his retirement in 1971.

1935 and 1936 again saw him mapping on the Sevenoaks sheet and, whilst mapping the Chartwell Estate he encountered Winston Churchill building his swimming pool and was invited to lunch with the family. 1935 was an important year for Buchan as it was the beginning of his career as a hydrogeologist. Since the publication of 'Records of London Wells' (Barrow & Wills 1913) over 1000 new records and a wealth of detail relating to chemical analyses of water, water levels and yields had been received. It was decided not to produce a second edition which would have been far too large but to produce a totally new volume on a more restricted area to include only the County of London.With his experience of the 6-inch to the mile maps of the London area Buchan was an ideal choice to prepare this new memoir and he was entrusted with the work. It was in 1935 also that F. H. Edmunds began to devote time to research for the Inland Water Survey of Great Britain leading to the eventual creation of a separate Water Unit. However, the London memoir remained under the direction of the Southern District Unit.

On 27th January 1937 Buchan married Barbara Hadfield (born 1914) from Droylsden, Manchester, whom he had met on a visit with a friend who was engaged to her sister. They settled in Banstead in Surrey where he was to make his home for the next 35 years and where his two children, Anne and Stuart, were born.

During 1937 considerable progress was made with the Sevenoaks and Bridport memoirs, the 6-inch standards were prepared and proofs of the one-inch to the mile maps were recorded as well in hand. The Sevenoaks map was eventually published in 1950 and the memoir in 1969 with Buchan as one of the authors (Dines *et al.* 1969). The Bridport map had been engraved prior to war breaking out in 1939 and was published in 1940 but the memoir was delayed until 1958 (Wilson *et al.* 1958). Buchan's only contribution to the latter was a section on water supply. The summers of 1937 and 1938 saw his final mapping work in Southern England on the Isle of Sheppey and the area around Gillingham where he contributed to the Chatham (272) and Faversham (273) sheets. Buchan had by then become a valued member of staff as he was entrusted with the training of J. R. Earp who joined in the autumn of 1938. At this time Buchan was credited with working out a number of criteria by which the various beds of the Eocene could be distinguished in the field. The Faversham map was published in 1953 but the memoir not until 1981 (Holmes 1981) some years after Buchan had retired. The Chatham map appeared in 1951 but no memoir to record Buchan's contribution has ever been published.

During 1937 Buchan completed the memoir on the water supply of the County of London and it was published the following year (Buchan 1938*a*). 27 water supply memoirs had been published previously, the first in 1899 (Whitaker & Reid,1899) and the London memoir was to be the last in the traditional format. Records of 1080 wells were recorded, the falling water levels in the Lower London Tertiaries and Chalk were illustrated by successive plots of the piezometric surface, and saline intrusion was recognized in the vicinity of the Eastern Docks. The memoir also highlighted the number of abandoned wells which existed creating a potential danger to public water supplies through the entry of contaminated water via deteriorating lining tubes. He discussed these problems at meetings of the British Association in Cambridge in August 1938 (Buchan 1938*b*), the Institution of Water Engineers (Buchan 1939*a*) and the Geologists' Association (Buchan 1939*b*). At about this time he also published a short paper on the reinterpretation of some clay outliers around Tunbridge Wells as Grinstead Clay rather than Weald Clay (Buchan 1938*c*) and contributed to a paper with Survey colleagues on the mapping of head deposits (Dines *et al.* 1940).

The war years

During the war years Buchan was involved in the search for water and coal (Bailey 1952). His first task was to compile a list of London wells that could be used for emergency water supplies. Subsequently, until the middle of 1942, his efforts were directed towards locating groundwater supplies for military camps and airfields as well as industrial sites. He was the lead author on Wartime Pamphlet No. 10, 'The Water Supply of South-East England from Underground Sources', 9 parts of which were issued during 1940 and 1941 (Buchan *et al.* 1940–1941) and on Wartime Pamphlet No. 15, 'The Water Supply from Underground Sources of the Reading-Southampton District', 8 parts of which appeared during 1941 and 1942 (Buchan *et al.* 1941–1942). These Wartime Pamphlets were essentially revisions of the old water supply memoirs which were by then well out of date.

His contribution received praise from his Director who wrote 'Among the many outside the Water Unit, who cooperated in the good work, Buchan deserves special mention' (Bailey 1952 p. 247). His contribution, and particularly his London water supply memoir, was also recognized by the Geological Society with the award of the Lyell Geological Fund in 1944. At a joint meeting between the Geological Society and the Institution of Water Engineers held on the 19th April 1944, Buchan outlined the work carried out in SE England stressing its value for water resources studies and town and country planning (Buchan 1944). According to Wilson (1985) Buchan was also involved with W. B. R. King, who was Chief Geological Advisor to Supreme Headquarters, Allied Expeditionary Force, in preparations for the Normandy invasion and its aftermath.

For much of his time from mid-1942 Buchan was based in Chesterfield where he investigated the outcrops of coal seams in the Notts-Derby-Yorks Coalfield in order to locate sites where coal could be extracted rapidly and cheaply using opencast methods. After the war, on 7th November 1945, he gave a lecture on this work to the Geological Society (Buchan1946) but otherwise it does not seem to have been published.

The Water Department: 1945–1960

Stevenson Buchan's pre-war and wartime water supply experience was of fundamental significance in the further development of his career. Initially it led to him making a major contribution to the GSGB input to the Water Acts of 1945 and 1946, which for the first time recognized that the country's groundwater resources should be co-ordinated at a national level. During the preceding century the Survey had accumulated thousands of well records, voluntarily supplied by well-sinkers. Under the new Acts, which were implemented from 1948, these now had to be notified formally for addition to the well record collections. Additionally, major groundwater abstractors were required to provide annual returns of their abstractions and of the water levels in their wells. Overdeveloped areas of the major aquifers, characterized by falling water levels, had to be defined and a licensing system to control the problem for both new and existing wells was introduced.

Against that background it is interesting to speculate how groundwater work would have developed within GSGB and Britain had Buchan's 1945 request to accept a lectureship at Cambridge University been granted. However, his was a reserved occupation under wartime regulations and the Director advised that 'In the conditions now existing it would be well-nigh impossible to provide a substitute capable of doing his work'. So the hydrogeological phase of his career began and, following the promotion of F. H. Edmunds to lead the Southern District Unit, he was appointed as the Head of a new Water Department, with responsibility to oversee the statutory obligations the recent Acts placed upon GSGB. A Scot with a winning smile, he was well suited to the task. By the end of 1946 he had eight support staff and had been promoted to

Principal Geologist. From 1948 he was joined by numerous colleagues, some on short-term secondment to the Department, others permanent staff. He had a great ability to generate enthusiasm in others even for trivial tasks. His receipt of the ensuing reports was equally enthusiastic, although they rarely re-emerged for publication! He was promoted to the rank of District Geologist in 1952 and in 1953 the value of his work was recognized by the Institution of Water Engineers and Scientists by the award of the Whitaker Medal.

Buchan personally supervised the conversion of the previously somewhat haphazard record collection into a well-founded data base. He instituted regional surveys, mainly by young women geologists who located both historic and new wells, sited them on 6-inch plans, estimated their Ordnance Datum level, measured the water levels and recorded other available information, such as water quality. These 'Water Babies' then registered the records and from 1964 onwards published synopses of the information in a Well Catalogue Series. He also persuaded the Norfolk County Council to initiate monthly measurements of the water levels in a network of some 60 Chalk wells and boreholes. These projects were the forerunners of the present-day databases and networks of observation wells which provide the factual base on which the Environment Agency and privatized water companies control the development and operation of the major aquifers throughout the country. Significantly at the time they enabled Buchan's new colleagues joining the Department to undertake regional and local studies and to quantify their groundwater resource assessments. Their work formed the basis of advice to government, local authorities, industry and the public. Regrettably it was not until 1966 that publication of these regional reports began, with a study of the Bunter Sandstone of Nottinghamshire (Land 1966).

Buchan himself spent a lot of his time in these early days publicising the work of his Department. He lectured to groups and organisations such as the Geologists' Association, the Institution of Sanitary Engineers, the British Waterworks Association and the Institution of Water Engineers emphasizing the information which was being collected and its importance to water resources and planning (e.g. Buchan 1951, 1953). He also wrote short articles in specialist and trade journals such as the Lea Valley Growers Association Newsletter, Farmer and Stockbreeder, Sprinkler Irrigation Manual and the National Farmers Union Watercress Branch News Sheet.

In 1948 he was involved in the organisation of the 18th Session of the International Geological Congress (IGC) held in London. Between the 9th and 24th of August he led an excursion looking at the Hydrogeology of England and Wales. All the important aquifers were traversed, pumping stations were visited in the two most prolific aquifers, the Chalk and the Triassic sandstones, and geological aspects of some of the major impounding works were studied (Buchan 1948). It was on this excursion that initial discussions began which eventually led to the formation of the International Association of Hydrogeologists. A provisional committee was formed at the next IGC meeting in Algiers in 1952 and a formal constitution was agreed in Mexico City in 1956 (Day 1992). Buchan himself became the sixth President from 1972 to 1977.

The permanent staff members were encouraged to undertake research in subjects of their choice. First in the field and pre-eminent was Jack Ineson who began to apply to pumping tests in the dual-porosity Chalk aquifer the analytical methods developed in the USA by Theis and Jacob, before himself developing a yield/drawdown analytical tool (Ineson 1959). David Gray applied geophysical techniques, particularly down-hole logging methods, which became significant in the interpretation of the non-steady state flow problems typical of the Chalk (Gray 1958). Dick Downing began to analyse the river/groundwater interface and the chemical reactions which took place in groundwater as it moved through varying strata (Ineson & Downing 1963), John Day and colleagues began to synthesize many well records in compiling hydrogeological maps. (Day *et al.* 1967).

Buchan's own interests lay in natural groundwater quality, the occurrence of pollution and the application of artificial recharge to overdeveloped aquifers. Jointly with Dr. A. Key of the Public Health Laboratories he was requested by the WHO to report on the pollution of groundwater in Europe (Buchan & Key 1956). Following a review of experiences in a range of European countries, which was based on replies to a questionnaire sent out by WHO, the authors discussed a range of issues concerning groundwater pollution. They noted for example that:

> '. . . the use of agricultural fertilizers is growing. Although it is the intention of the farmer that these nutrients should find their way into the crops and be harvested, some of the constituents are undoubtedly washed into the subsoil and hence contaminate groundwater . . . Much of the sulfate of ammonium sulfate probably enters the subsoil as calcium sulfate and adds permanent hardness to the water . . . Part of the nitrogen may also drain away, in which form it will largely remain' (Buchan & Key 1956 p. 989).

They also noted the general absence of groundwater pollution in Europe and emphasized the importance of natural processes in groundwater protection. The discussion at the end of this paper is still worthwhile reading, even for todays groundwater professionals.

Buchan was keen on collaborating closely with other professionals involved in the use of groundwater supplies. He joined engineers of the Metropolitan Water Board in investigating the possible use of artificial recharge to offset the de-watering of the Lower London Tertiaries and the Chalk in the Lee Valley, and later with others in the Bunter Sandstone of Nottinghamshire (Buchan 1955, 1964). His 1955 paper summarized experience of artificial recharge in various countries and then looked at the scope for such recharge in England. He concluded that, if water of suitable quality could be found, artificial recharge could be used to reduce the areas of saline infiltration along coasts and estuaries, to restore Chalk water levels in London and levels in the Triassic sandstones of the Midlands and to displace connate water in East Anglia and Lincolnshire.

He was one of the first scientists to look at the evolution of natural groundwater compositions and examined the sequential changes which took place as water moved through the soil and the unsaturated and saturated zones (Buchan 1958). For this paper, in 1958, he was awarded the Gans Medal of the Society for Water Treatment and Examination. He was also involved in the design of a scheme to dewater an extensive area of Magnesian Limestone in the Durham Coalfield in order to ensure reasonably dry conditions in the underlying Coal Measures (Armstrong *et al.* 1959). He was elected a Professional Associate of the Institution of Water Engineers and an Honorary Fellow of the Institute of Public Health Engineers.

Member of the Directorate: 1960–1971.

The final phase of Buchan's career began in 1960 when he was appointed Assistant Director of the GSGB responsible for five Special Service departments, including the Geological Museum and work in Northern Ireland. At the time he was absent in the USA undertaking a lecture tour as Visiting Scientist under the auspices of the American Geological Institute. In 1965 the GSGB merged with the Overseas Geological Surveys to become the Institute of Geological Sciences (IGS), within the newly created Natural Environment Research Council (NERC) and Steve was heavily involved in the administrative procedures. Two years later he was appointed Chief Geologist and in 1968 promoted to Deputy Director, responsible for all work in the UK.

Heavily involved internationally he travelled widely and presided at numerous international conferences. As well as the International Association of Hydrogeologists, he was President of the Groundwater Commission of the International Association of Hydrology and chaired the Standing Committee on Hydrogeological Maps, acting as a Co-Editor of the European Hydrogeological Map. In the UK he was appointed the British Delegate to the UNESCO Co-ordinating Committee for the International Hydrological Decade, and for several years chaired the Royal Society's Hydrology Sub-Committee. On behalf of the latter he compiled two bibliographies on British hydrology which were published by the Royal Society (Buchan 1961*a*, 1971).

Whilst Assistant Director he continued to pursue his own research interests and to publicise hydrogeological work within GSGB and then IGS. In 1962 he drew attention to the problems which had developed in the Chalk aquifer in Kent as a result of the disposal of drainage water from the Tilmanstone Colliery into a system of ditches on the Chalk outcrop (Buchan 1962). He continued his interest in artificial recharge (Buchan 1961*b*, 1963*a*, 1964) advocating regional management of large units of the country in order to facilitate flexible use of both surface and groundwaters (Buchan1963*a*). He also published a series of papers on more general aspects of hydrogeology including the relationship between geology and water supplies (Buchan 1961*c*, 1963*b*), and groundwater balances (Buchan 1965, 1968*a*)

He always retained his Scottish interests and in 1962 was elected a Fellow of the Royal Society of Edinburgh. In 1968 he took on the role of President of Section C (Geology) of the British Association for the Advancement of Science when it held its Annual Meeting in Dundee. This was 34 years after he had first addressed members of the Association in Aberdeen in 1934. He took as the theme of his Presidential address 'Geology in Action' and covered a whole range of applied issues ranging from water and fuels to ores, building stones and industrial minerals. In his conclusion he deplored the fact that university training was not equipping enough young geologists for the modern world and that there were not enough specialists available to meet the challenges of our modern society (Buchan 1968*b*).

Retirement

Steve Buchan retired from IGS in 1971 in which year he was awarded the CBE for his services to hydrogeology and to Government science. He retired to Rockland St. Mary in Norfolk where he perfected his technique of producing a range of powerful liqueurs! However, he continued his involvement with the International Association of Hydrogeologists and in fact his term as President (1972–1977) occurred after his retirement. In July 1977 he was Chairman of the Organising Committee of the 13th Congress of the Association at Birmingham, the first such congress to be held in the

UK. The objective of the meeting was to discuss the results of applied investigations and research undertaken to overcome both natural and artificial obstacles to the optimal development of groundwater. Buchan was also instrumental in the formation of the Hydrogeological Group of the Geological Society and was its first Chairman from 1974 to 1976.

He died on 24th July 1996 and was cremated at Earlham, Norwich. Obituaries were published in The Daily Telegraph (29th September 1996), the Annual Report of the Geological Society of London for 1996, the International Association of Hydrogeologists 'News and Information' Issue C 12 1996 and the Yearbook of the Royal Society of Edinburgh in 1998.

Discussion

Stevenson Buchan played a pivotal role in the history of British hydrogeology. Although it is unlikely that he ever met those pioneering Victorian hydrogeologists William Whitaker and Joseph Lucas, his life overlapped theirs by some 20 years. He represented a link between an age when data were simply collected and applied directly to water resources assessments and one in which hydrogeological processes were investigated and modelled. His own career, up to the end of World War II, involved data collection (e.g. Buchan 1938*a*, 1940–1941, 1941–1942). After the war, although he became more involved in research (e.g. Buchan 1958), his own contributions were not directly innovative. Of more significance is that he developed around him a group of able colleagues who were encouraged to develop their own research interests and so created the first groundwater research group of any size in the UK.

He was a man of energy and resource, who devoted himself wholeheartedly to the task in hand, moving smoothly through the ranks of the Geological Survey from junior Geologist to Deputy Director. There seems no doubt that the 1939–1945 war materially influenced his career and directed him into hydrogeology rather than field mapping or university teaching. The war years gave him an opportunity to demonstrate that he could shoulder responsibility and when the war ended he was rewarded with a responsible post within a developing and expanding organisation.

He pursued a policy with respect to publication which many of his younger colleagues found frustrating because it did not give the dissemination of results the priority which many of them thought was deserved. In the post war years of the late 1940s and 1950s, at a time of expanding groundwater use and when there was a real need for explanatory memoirs on regional hydrogeology, the Survey did not issue any official publications on groundwater. Thus an opportunity was missed to raise the profiles of both groundwater as a potential resource and that of the Water Department itself.

As a founder member of the International Association of Hydrogeologists, he was a well known international figure and had a fund of anecdotes about leading hydrogeologists of his day. He was conscious that sound groundwater development needed to involve both geologists and engineers and he contributed enthusiastically to joint meetings. On behalf of the Survey Stevenson Buchan played a significant role in the drafting of the 1945 Water Act and later in the formation of IGS when the GSGB merged with the Overseas Geological Surveys.

On his retirement Buchan wrote to one of us that 'I am satisfied that I managed to achieve what I set out to do – lay the solid foundations for the Geological Survey to help people to use their water resources fully and properly'. Overall his contribution to the development of hydrogeology in the UK was one of major significance, particularly in his role as an enabler and administrator during the years from 1945 to 1971.

References

ARMSTRONG, G., KIDD. R. R. & BUCHAN, S. 1959. Dewatering scheme in the South Durham Coalfield. *Transactions of the Institution of Mining Engineers*, **119**, 141–152.

BAILEY, E. B. 1952. *Geological Survey of Great Britain.* Thomas Murby, London.

BARROW, G. & WILLS, L. J. 1913. *Records of London Wells.* Memoir of the Geological Survey. HMSO, London.

BUCHAN, S. 1932. On some dykes in East Aberdeenshire. *Transactions of the Edinburgh Geological Society*, **12**, 323–329.

BUCHAN, S. 1934. The petrology of the Peterhead and Cairngall granites. *Report of the Annual Meeting, British Association for the Advancement of Science, Aberdeen 5–12 Sept 1934,* British Association, London, **303**.

BUCHAN, S. 1938*a*. *Water Supply of the County of London from Underground Sources.* Memoirs of the Geological Survey of Great Britain, HMSO, London.

BUCHAN, S. 1938*b*. Pollution and exhaustion of London's underground water supply. *Report of the Annual Meeting, British Association for the Advancement of Science, Cambridge 17–24 August 1938,* British Association, London, **418**.

BUCHAN, S. 1938*c*. Notes on some outliers of Grinstead Clay around Tunbridge Wells, Kent. *Proceedings of the Geologists' Association*, **49**, 407–409.

BUCHAN, S. 1939*a*. The Underground Water Supply of the County of London. *Transactions of the Institution of Water Engineers*, **43**, 129–153.

BUCHAN, S. 1939*b*. Report of a Lecture on London's Underground Water Supply. *Proceedings of the Geologists' Association*, **50**, 147–148.

BUCHAN, S. 1944. Contribution to discussion on "Sources of Water in relation to Town and Country Planning". *Proceedings of the Geological Society, Quarterly Journal of the Geological Society of London,* **100**, lxxxvii.

BUCHAN, S. 1946. Report of a lecture on "The Geology of Opencast Coal in Great Britain". *Proceedings of the Geological Society, Quarterly Journal of the Geological Society of London,* **102**, iii.

BUCHAN, S. 1948. Hydrogeology of England and Wales, *International Geological Congress 18th Session – Great Britain, 1948, Guide to Excursion A.18, August 9th to August 24th, 1948.* 41p.

BUCHAN, S. 1951. The work of the Water Department of the Geological Survey. *Journal of the Institution of Sanitary Engineers,* **50**, 5–14.

BUCHAN, S. 1953. Estimation of yield (underground). *Journal of the Institution of Water Engineers,* **7**, 205–214.

BUCHAN, S. 1955. Artificial replenishment of aquifers. *Journal of the Institution of Water Engineers,* **9**, 111–163.

BUCHAN, S. 1958. Variation in mineral content of some ground waters. *Proceedings of the Society for Water Treatment and Examination,* **7**, 11–29.

BUCHAN, S. 1961*a*. *Selected Bibliography of Hydrology (with abstracts) United Kingdom for the years 1955–1959 inclusive.* British National Committee for Geodesy and Geophysics, Hydrology Subcommittee, Royal Society, London.

BUCHAN, S. 1961*b*. Natural purification of water during artificial replenishment of the ground. *Journal of the Society of Health,* **81**, 294–299.

BUCHAN, S. 1961*c*. Water supplies and geology. *Science Survey (London),* **2**, 84–93.

BUCHAN, S. 1962. Disposal of drainage water from coal mines into the Chalk in Kent. *Proceedings of the Society for Water Treatment and Examination,* **11**, 101–105.

BUCHAN, S. 1963*a*. Conservation by integrated use of surface and ground waters. *In*: *Conservation of Water Resouces in the United Kingdom.* Proceedings of the Symposium organised by, and held at, the Institution of Civil Engineers on 30 and 31 October and 1 November,1962, Institution of Civil Engineers, London, 181–185.

BUCHAN, S. 1963*b*. Geology in relation to ground-water. *Journal of the Institution of Water Engineers,* **17**, 153–164.

BUCHAN, S. 1964. The problem of ground-water recharge: artificial recharge as a source of water. *Journal of the Institution of Water Engineers,* **18**, 239–246.

BUCHAN, S. 1965. Balancing the world's water budget. *Discovery,* **26**, 10–14.

BUCHAN, S. 1968*a*. Estimation of groundwater balances in England. *Bulletin of the International Association of Scientific Hydrology,* **3**, 126–134.

BUCHAN, S. 1968*b*. Geology in action. *The Advancement of Science,* **25**, 213–226.

BUCHAN, S. 1971. *United Kingdom Hydrology Bibliography (Annotated) 1960–1964.* Hydrology Subcommittee of the British National Committee for Geodesy and Geophysics, Royal Society, London.

BUCHAN, S. & KEY, A. 1956. Pollution of ground water in Europe. *Bulletin of the World Health Organisation,* **14**, 949–1006.

BUCHAN, S., ROBBIE, J. A., BUTLER, A. J. & HOLMES, S. C. A. 1941–1942. *Water Supply from Underground Sources of Reading-Southampton District* (1/4 inch geological sheets 19 and 23 eastern halves). Geological Survey Wartime Pamphlet, **15**. [Parts 2 & 7 issued 1941; parts 1, 3, 4, 5, 6 & 8 issued 1942.]

BUCHAN, S., ROBBIE, J. A., HOLMES, S. C. A. & EARP, J. R. 1940–1941. *Water Supply of SE England from Underground Sources* (1/4 inch sheets 20 and 24). Geological Survey Wartime Pamphlet, **10**. [Parts 1, 2, 3, 5, 6 & 7 issued 1940; parts 4, 8 & 9 issued 1941.]

DAY, J. B. W. & OTHERS. 1967. *Hydrogeological Map of North and East Lincolnshire.* Institute of Geological Sciences, London.

DAY, J. B. W. 1992. A brief account of the International Association of Hydrogeologists. *Applied Hydrogeology,* **0**, 47–50.

DINES, H.G., HOLLINGWORTH, S.E., EDWARDS W., BUCHAN, S. & WELCH, W.B.A. 1940. The mapping of head deposits. *Geological Magazine,* **77**, 198–226.

DINES, H. G., BUCHAN, S., HOLMES, S. C. A. & BRISTOW, C. R. 1969. *Geology of the Country around Sevenoaks and Tonbridge (Explanation of one-inch Geological Sheet 287, New Series).* Memoirs of the Geological Survey of Great Britain, HMSO, London.

GRAY, D. A. 1958. Electrical resistivity marker bands in the Lower and Middle Chalk of the London Basin. *Bulletin of the Geological Survey of Great Britain,* **15**, 85–95.

HOLMES, S. C. A, 1981. *Geology of the country around Faversham Memoir for 1:50000 geological sheet 273.* Memoirs of the Geological Survey of Great Britain, HMSO, London.

INESON, J. 1959. Yield-depression curves of discharging wells with particular reference to Chalk wells and their relationship to variations in transmissibility. *Journal of the Institution of Water Engineers,* **13**, 119–163.

INESON, J. & DOWNING, R. A. 1963. Changes in the chemistry of groundwater of the Chalk passing beneath argillaceous strata. *Bulletin of the Geological Survey of Great Britain,* **20**, 176–192.

LAND, D. H. 1966. *Hydrogeology of the Bunter Sandstone in Nottinghamshire.* Hydrogeological Report No **1**, Geological Survey of Great Britain.

WHITAKER, W. & REID, C. 1899. *The water supply of Sussex from underground sources.* Memoirs of the Geological Survey, HMSO, London.

WILSON, H. E. 1985. *Down to Earth. One hundred and fifty years of the British Geological Survey.* Scottish Academic Press, Edinburgh.

WILSON, V., WELCH, F. B. A., ROBBIE, J. A. & GREEN, G. W. 1958. *Geology of the Country around Bridport and Yeovil (Explanation of sheets 327 and 312).* Memoirs of the Geological Survey of Great Britain, HMSO, London.

Groundwater studies in the Institute of Geological Sciences between 1965 and 1977

D.A. GRAY.

46, Bonnersfield Lane, Harrow, Middlesex HA1 2LE, UK

Abstract: Until 1965 the Water Department of the Institute of Geological Sciences had two principal functions – data collection and a national groundwater advisory service, but a limited research role. A change in the legislation then led to the loss of the advisory service but a major increase in the research role, backed by the support of successive Directors. After two years in which outstanding commitments were met and six new staff recruited, a portfolio of research projects was introduced. These were principally applied projects in the UK and overseas, but with a continuing background of fundamental studies. The most successful projects involved hydrogeochemistry; mechanisms of matrix flow in the Chalk; diffuse pollution from agriculture; point source pollution from landfills; groundwater/surface water interactions; automation of well records for publication; hydrogeological maps and the application of hydrogeology to civil engineering projects. The development of advisory work overseas was also important. The presence of the Hydrogeological Department in the Institute of Geological Sciences and its introduction of innovative techniques and ideas had a significant influence on the development of the hydrogeological community in the UK.

This review documents the early development of groundwater studies in the Institute of Geological Sciences (IGS) formed in 1965 by the merger of the Geological Survey of Great Britain (GSGB) and the Overseas Geological Surveys (OGS). The foundations were laid in 1935 when F.W. Edmunds headed a Water Unit within the GSGB with a twofold function: primarily data collection and analysis, and secondarily a national groundwater advisory service. However, in the post-war years, first under Stevenson Buchan to 1960 and then under Jack Ineson to 1965, research directed to water resources was introduced.

Initially Ineson adapted to the UK's fissured aquifers the well hydraulics theories developed by Theis (1935): subsequently other research developed. The 1963 Water Resources Act was implemented in 1965 and led the Water Department (WD) to change its role with research becoming the dominant component and with the advisory service directed to overseas territories and maintained for Scotland and Northern Ireland. The present review covers the period between the change in 1965 and 1977 when a major reorganization of IGS coincided with the first phase of the transfer of the WD out of London to Wallingford. Its purpose is to place the development of the Department's programmes within that historical framework rather than to review the detailed science.

The report of the Chairman of the Geological Survey Board for 1963 records that 'With the passing of the Water Resources Act 1963, we envisage that the Water Department will be called upon in future to advise and assist in geological matters related to underground water . . .'. That view, taken two years before the implementation of the Act, could not have been further from reality. At an early stage the Director of the new Water Resources Board (WRB) decided to recruit hydrogeologically qualified staff to work internally and arranged for the transfer in June 1965 of seven staff from the WD, including the then Head, Jack Ineson.

The replacement in late 1966 of the staff transferred to the WRB enabled the new Head of the WD to enlarge the research portfolio. This took place slowly because four of the six new recruits had little or no previous hydrogeological experience and had to learn 'on the job'. By 1967 the major revisions of the programme had been achieved and the Department's title was changed to the Hydrogeological Department (HD) under its first Chief Hydrogeologist, David Gray. The entire staff were never accommodated in a single area but were disadvantageously dispersed in pre-fabricated offices on the roof of the Geological Museum in Exhibition Road, London, SW7 and in a variety of spare offices elsewhere in the building. From 1974 some staff of a newly formed sub-section of the HD, entitled the Environmental Pollution Unit (ENPU), moved to Harwell, Oxfordshire where they shared offices and worked with the Hazardous Materials Service (HMS) of the UK Atomic Energy Authority (UKAEA). In the summer of 1977 the first stage of the dispersal of the HD out of London took place when 15 staff transferred to the Maclean Building at Wallingford, Oxfordshire already occupied by the Institute of Hydrology. As part of the simultaneous reorganization of IGS the HD, from 1974 under the leadership of John Day, was administered by the new Geophysics and Hydrogeology Division, led by Gray who had been appointed an Assistant Director also in 1974. In view of expanded work programmes two inter-disciplinary Task Forces were created within the Division (Woodland 1978):

From: MATHER, J. D. (ed.) 2004. *200 Years of British Hydrogeology*. Geological Society, London, Special Publications, **225**. 295–318. 0305-8719/04/$15

these were able to call on expertise from any relevant Unit of the IGS. The Geothermal Task Force was based initially in the Applied Geophysics Unit housed at Keyworth, Nottinghamshire, but was later transferred to the HD and headed by Dick Downing, whilst the Radioactive Waste Disposal Task Force was led by John Mather within ENPU at Harwell. From 1966 Edmund Wright supervised overseas activities and in 1974 was formally appointed Advisor to ODA on Overseas Hydrogeological and Geothermal matters.

As the 1966 recruits and most of their colleagues have now retired from their official roles, sufficient time has elapsed to warrant a review of the programmes of work in which they were involved. In that historical context the original names of the countries and projects are retained, as are the original units of measurements and the stratigraphical terms. The scale of the Department's work is reflected in the 334 internal reports compiled over the period. The British Geological Survey holds copies of them but space prevents reference to them and only peer-reviewed and IGS publications are quoted as references.

The legislative framework

The evolution of the groundwater programme between 1965 and 1977 was influenced by no fewer than four Acts of Parliament and one change in the funding mechanism (Wilkinson & Gray 1981). The first was the implementation in 1965 of the provisions of the Water Resources Act 1963. However, the statutory requirement under the 1945 Water Act for well sinkers to notify the IGS of their intention to sink wells of more than 50 feet in depth was not amended and the data collection programme continued for the period of this review and beyond.

The Science and Technology Act 1965 established the Natural Environment Research Council (NERC) as from June 1965 and it assumed responsibility for the IGS and for funding its Science Budget. That funding function was amended in 1972 when the Government accepted the Rothschild recommendations leading to the transfer of the costs of Government commissioned research from the NERC to the relevant Departments of State. In the case of the HD some two thirds of the funding eventually lay with the Overseas Development Ministry (ODM; Overseas Development Administration (ODA) from 1970) and the departments of the Environment (DOE) and Energy (DEn), supplemented in some programmes by funding from the European Economic Community (EEC) and commercial contracts.

The 1973 Water Act also influenced the HD's operation. It replaced the WRB with the Central Water Planning Unit (CWPU) within the DOE and established ten Regional Water Authorities. The Control of Pollution Act, 1974 was the final legislative procedure to affect the HD's programme; the newly constituted Waste Disposal Authorities were required to consult the IGS on groundwater pollution risks associated with discharge of effluents to the sub-surface.

Yet further changes to the water industry were foreshadowed in the 1977 'White Paper' entitled 'The Water Industry in England and Wales: The Next Step'. Its publication coincided with the first stage of the transfer of the Department to Wallingford, and with the internal re-organization in IGS referred to earlier. These coincidences act as an appropriate point at which to end this history.

Changes in the strategic role

Before 1965 The Water Department of the GSGB had provided '. . . the framework necessary for the effective management of groundwater on a national basis . . .' (Downing 2004). Implementation of the 1963 Water Resources Act eliminated that role, but did not inhibit the creation of a framework for the effective management of hydrogeological research in the IGS. That took time to achieve, and could not have happened without the Directorate's fundamental belief that such a role was valid for the country's principal geological establishment and should enjoy sustained support. Initially that support came from Sir James Stubblefield, then Director, followed shortly by Sir Kingsley Dunham, with Stevenson Buchan as Deputy to each and a strong protagonist of groundwater research.

The chronological account of the research programme that follows necessarily has to ignore the human aspects. These ranged from the steep learning curves required of the new recruits, through the acquisition by all staff of the then new IT disciplines, to the development of new field and laboratory techniques, commonly adapted to meet overseas conditions.

The legislative and administrative changes also affected the author as he abandoned his role of researcher to become an administrative 'enabler'. The somewhat daunting brief agreed with the Directorate was for him to convert the Department from its former roles into an organization able to undertake fundamental and applied research into the role of groundwater throughout the hydrological cycle and in its behaviour as a geological agent. The need was clear in that international literature had already appeared which covered academic and applied studies in fields within which the UK had not been active.

It was accepted that the necessary changes could be brought about neither rapidly, nor without new staff. There should be 'young' recruits with the potential to undertake research, as well as an experienced, senior officer able to work immediately on overseas hydrogeological problems and to supervise others in such work. Ideally the former group should have been able to develop their own research topics and techniques – as under Buchan a decade earlier. Yet the funding mechanism was so different that such a relaxed regime was inhibited and relatively rapid results and publications were paramount. Against that background seven new areas of investigation were seen as having a measure of priority. In no order of priority, these were 1) hydrogeochemistry and associated physicochemical reactions within aquifers; 2) groundwater pollution at local and regional levels; 3) the physical properties of aquifers and the associated sedimentological and structural control on groundwater movement; 4) analogue and digital computer modelling of groundwater flow; 5) infiltration rates in the unsaturated zone; 6) the groundwater/surface water interface; and 7) the hydrogeological aspects of civil engineering.

All of those fields of study were covered to a greater or lesser extent, whilst the advisory roles for overseas territories, Scotland and Northern Ireland were developed fully. The programmes were initially deliberately organized with individuals or small teams exploring separate topics, but as integration of the several activities became practicable there were clear scientific benefits. These were achieved only after a few years, by when the 'new' staff had gained practical experience with projects both in the UK and overseas.

The strategic role changes reflected those that the science of hydrogeology as a whole was then undergoing. They were caused partly by the recognition that within an environmental context there were issues to be considered other than simply the abstraction of groundwater, and partly by the ability to address those issues through the development and application of techniques unknown before the mid- to late-1960s. To the extent that the Department recognized these new, over-riding considerations at an early stage its role was pivotal within the UK. There were three phases to the programme of work. First, a period of two years in which outstanding commitments were completed, then a period between 1967 and 1970 during which new fields of research were initiated and finally from 1971 the introduction of a forward looking programme once the Department's research capability had been established.

Completion of outstanding commitments: 1965–1966

Most of the early programming related to the completion of previous studies and ensuring the continuity of data collection. Data collation and distribution became of major significance, particularly for the 'Water Babies' – the women geologists who at that time checked the validity of the records in the field and published the resulting Well Catalogues. The WRB had to be provided with copies of the approximately 85000 well records collected by the GSGB since 1845, as well as the associated one inch and 6-inches to the Mile site maps and plans. The WRB in turn provided copies to the 27 newly constituted River Authorities and 2 River Conservancies. The Board were also provided with the annual returns of groundwater abstraction required under the Water Act 1945; these had been collated by Ineson from their inception and an analysis of them for the period 1948–1963 was published (Water Resources Board 1966).

The data collation aspects emphasized the value of a Hydrogeological Map programme, in that each map summarized the stratigraphical and hydrogeological information derived from the innumerable well records in the areas covered. Under the supervision of Day, the first in the series, begun in 1964 under Ineson and relating to North and East Lincolnshire, was nearing completion whilst that for the Dartford (271) Sheet was started. The collection of water level data was continued and the number of observation wells for which regular measurements were received rose to 760. An analysis of the 1963 measurements had been started in 1964, but staff shortage delayed its completion (Lovelock 1967): this became the first Groundwater Yearbook, subsequently published annually by the WRB (Downing 2004).

The newly constituted Yorkshire Ouse and Hull River Authority and Trent River Authority agreed to the issue of regional reports commissioned by their predecessor River Boards (Gray *et al.* 1969; Downing *et al.* 1970). The results were also published of an earlier resource assessment of the Hertfordshire Colne (Ineson 1965*a*), of a paper on groundwater networks (Ineson 1965*b*), as well as work on permeable catchments (Ineson & Downing 1965), on the global water balance (Buchan 1965), on the stratigraphical significance of widespread electrical resistivity marker bands in the Chalk (Gray 1965) and on the Hydrogeology of the Triassic Sandstones of the central Midlands (Land 1966).

The water resource advisory service was transferred to the WRB from 1st April 1966. About then the water industry declared its intention to metricate all working units. Clearly manual conversion of the numerical stratigraphical and hydrogeological data in the 85000 well records would not be practicable and computerization would be essential. Staff involved with the compilation of Well Catalogues, of which four were published in 1966, undertook the necessary program preparation. Initially the work was undertaken in association with ICL Ltd. but it was subsequently transferred to The Rutherford

Atlas Computer Laboratory. The Hydrogeological Map Programme was expanded and to avoid publication delays the series was printed by Messrs. Cook, Hammond and Kell of London and issued by the IGS.

A fresh field of research started by Philip Lovelock was the first stage in the creation of a Physical Properties Laboratory. Initially only measurements of Effective Porosity and Inter-granular Permeability could be made, but the range of measurements and scope of the research were subsequently expanded considerably. The first measurements were on core samples from the Carboniferous rocks of the Spilmersford Borehole, East Lothian.

The move towards the development of new research programmes began in the latter part of 1966 when six new Scientific staff joined the Department. Edmund Wright, experienced in overseas work, was recruited to lead the overseas activities anticipated following the merger with the OGS. Ronald Kitching, a soil physicist, was recruited to resume the programmes of analogue and digital modelling of groundwater flow abandoned when the staff involved transferred to the WRB. He also headed an extensive Infiltration Research Project concerned with the rates of recharge of a range of aquifers. Coming from non-hydrogeological backgrounds the four other recruits were initially directed to the literature covering the five main research fields in which they would be working – hydrogeochemistry, groundwater/surface water relations, groundwater pollution, groundwater resource assessment and hydrogeology related to civil engineering.

The development of a research site on the Lower Greensand at Liss, Hampshire, as part of the Infiltration Project, was the only large new activity designed to be started during this period. However, an exception arose as a result of the occurrence of the Aberfan Colliery Disaster on 21st October 1966. IGS was immediately required to prepare evidence for the Treasury Solicitor to submit to the subsequent Aberfan Tribunal. Dr. Austen Woodland, Assistant Director for Northern England and Wales, was the geological witness but, because the cause of the disaster lay with the discharge of groundwater to the surface beneath colliery Tip No.7, the Head of the WD provided written evidence on an analysis of the available data (Gray 1969*a*). Initially these were obtained from other organizations but were subsequently augmented by on-site tracer studies (Fig. 1). These substantiated the evidence submitted earlier in that they confirmed that rapid groundwater flow occurred through the fissured zones of the Brithdir Sandstone beneath Tip No. 7, which had been placed in tension by mine subsidence. Conversely, in areas under compression tracer movement was not identified (Mather *et al.* 1969).

Development of a research portfolio 1967–1970

With outstanding commitments met and new staff in place, Departmental effort was directed towards developing the research potential. Initially all funding was for either Science Budget or Government commissioned research activities, but from 1967 onwards commercially commissioned projects became increasingly important. An early task was the completion of the publication of previous studies, such as that into the Coal Measures of South Wales (Ineson 1967). Three new laboratories were developed during this period. The first for Groundwater Modelling came into use in 1969 but those for Hydrogeochemistry and Physical Properties research were not completed until 1970. Much of the Department's earlier work had been undertaken within the Humber Basin and new research was developed selectively within a generalized Humber Basin Project.

Database studies

The Hydrogeological Map Programme saw the publication of the first sheet covering North East Lincolnshire (Day *et al.* 1967), followed by that for the Dartford (271) Sheet (Day *et al.* 1968). A 'Model Hydrogeological Map' was drafted for a Standing Committee of the International Association for Scientific Hydrology (IASH) chaired by Stevenson Buchan. It was later developed into the 'International Legend for Hydrogeological Maps' – published jointly by the IASH, the International Association of Hydrogeologists (IAH), the IGS and UNESCO (UNESCO 1970). The two sheets of the Hydrogeological Map covering the outcrops of the Kentish Chalk and Lower Greensand (Benfield *et al.* 1970; Day *et al.* 1970) were also issued.

Work continued under Betty Harvey on the computerization of the Well Catalogue Series and the subsequent creation of an automated national database for the Well Record Collection. In a pilot study the core of the system was an 'Aquifer list' to which 11 other lists were connected, permitting multi-level retrieval. A second input program converted the numerical data from Imperial to Metric units and the output took the form of a restyled Well Catalogue, subsequently termed a Metric Well Inventory.

Regional and resource studies

To introduce the new staff to regional studies two projects were started within the Humber Basin. The groundwater hydrology and resources of the Chalk of East Yorkshire was the first (Foster 1974; Foster &

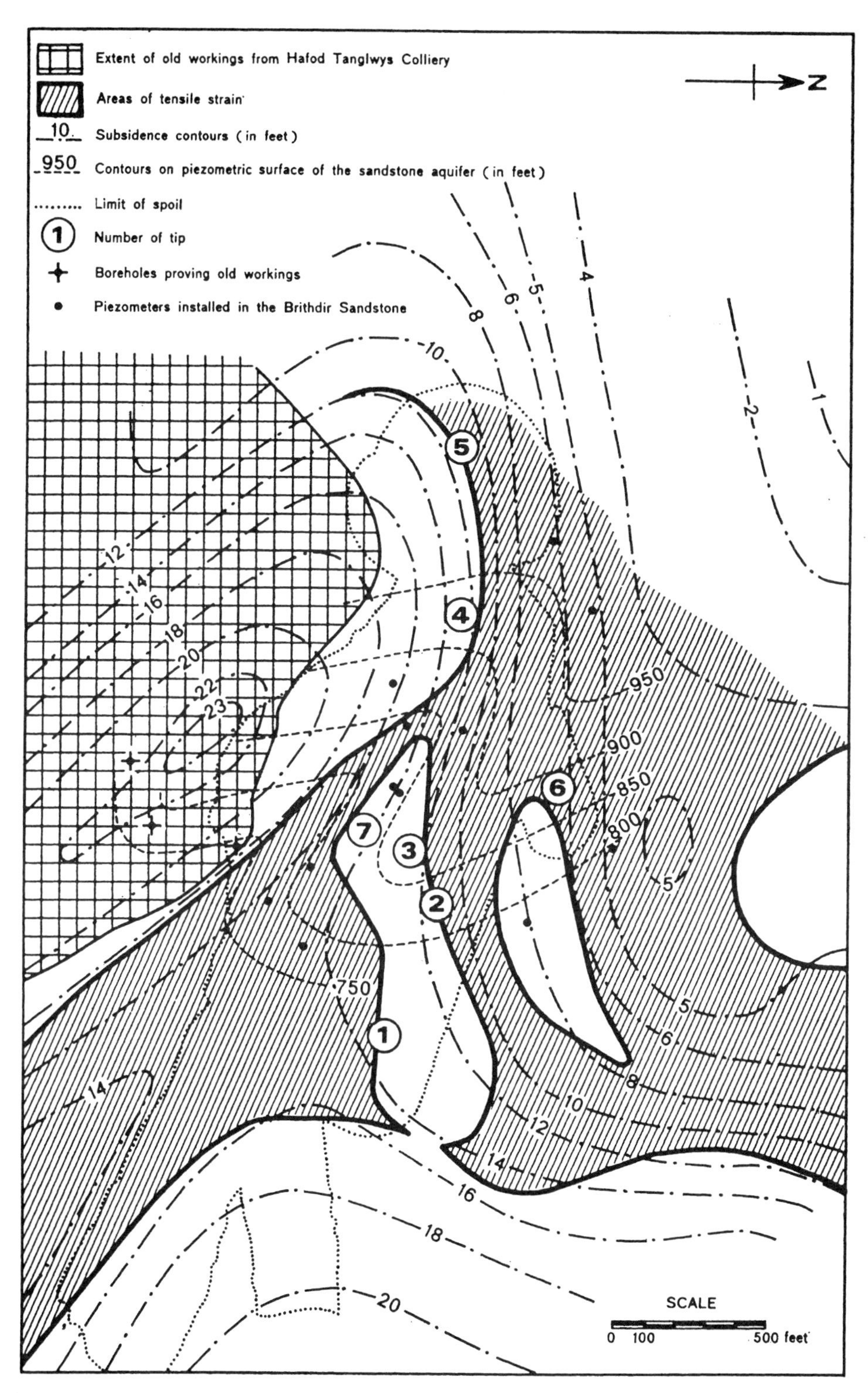

Fig. 1. A map of the Aberfan colliery tip site showing mining subsidence and associated features (from Mather *et al.* 1969, fig.2).

Milton 1974). Groundwater movement in the Chalk of this region being dominated by fissure flow, a possible relationship between the Specific Capacity of wells and joint orientation and density was investigated. Three major joint sets, regarded as shear features, were identified in 17 quarries and were differentiated from minor joints, tentatively seen as tensional in origin. Maximum and minimum water level data at over 100 sites were related to groundwater discharges at 60 sites. Extensive analysis was applied to the results of an autumn pumping test at Etton Pumping Station and was augmented by the results of an earlier borehole-logging flow study. Water samples were collected for chemical and thermo-nuclear tritium analysis along two transects. A smaller study was also undertaken into the groundwater hydrology of the Jurassic strata between the Market Weighton Axis and the Humber.

Regional hydrogeochemical research, which had been interrupted by the staff transfers to the WRB, was resumed under Michael Edmunds. The first study augmented earlier work on the mineral and thermal waters of the UK (Edmunds *et al.* 1969). An investigation was also undertaken into the occurrence of trace elements in the Carboniferous Limestone of the Derbyshire Dome. Preliminary results showed that the thermal component was enriched in six elements and impoverished in two (Edmunds 1971). The acquisition of atomic absorption and UV/visible spectrometers in the new Hydrogeochemical Laboratory enabled work to proceed on a new project concerning the Lincolnshire Limestone (Fig. 2). Trace elements in the oxygenated zone showed little variation. However, down-dip from a newly-recognized Eh front, systematic change could be related to several factors, including Eh, pH, the presence of sulphide and the progressive movement of the fresh water induced by pumping (Edmunds 1973; Edmunds & Walton 1983). In another project around Yate north of Bristol, high levels of strontium were measured in the groundwaters in both the Keuper Marl and the Coal Measures.

An unusual resource problem was investigated using borehole-logging techniques: thermometric and flow-profiling methods permitted the identification of separate levels of entry of fresh and saline waters into three boreholes in Triassic sandstones at the Woodfield Pumping Station, East Shropshire. A 'scavenger' pump was installed to remove the saline component and this led to a reduction of the chloride content of the fresh supply from 348 mgl^{-1} to 206mgl^{-1} when pumping at 1 Mgd. from No. 3 Borehole (Tate & Robertson 1971).

Advisory work in relation to groundwater resources developed in Northern Ireland and Scotland, with input from the respective offices. In Northern Ireland work on the fluvio-glacial gravels of the Enler valley was supplemented by an investigation into the underlying Permian, which yielded some 20 000 gph. Investigations were also started into the groundwater prospects of the Chalk and Hibernian Greensand in the Larne area, as well as in the alluvial deposits of the River Braid at Ballymena. Work was also undertaken at Bolea, Co. Londonderry and near Limavady. In Scotland pumping tests were supervised on wells drilled into the Permian sandstones of the Dumfries Basin.

Applied research studies

Then, as now, the surface water/groundwater interface was subject to much research. The Thames Conservancy needed to resolve this in terms of the major Thames augmentation scheme then under consideration. At the Conservancy's invitation in 1967 the Department initiated a continuing programme of borehole logging in the Lambourn and Winterbourne valleys to identify the levels of entry of water into both riverside and river-remote wells in the Chalk. Thermometric and flow profiling at times of high and low groundwater level, during rest and pumping conditions, proved effective (Tate & Gray 1971).

As part of the Infiltration Project a drainage lysimeter was designed and constructed at Styrrup, Nottinghamshire, as part of the Humber Basin Project. It consisted of a block of Triassic Lower Mottled Sandstone with a surface area of 100 m^2 and isolated by vertical impermeable membranes sealed into Upper Permian Marl at a depth of 5 m. Infiltration was measured by maintaining the water-level inside the block at the same level as that outside, as well as by neutron-scattering observations and monitoring the rate of infiltration of radioactive tracers irrigated onto the surface (Kitching & Bridge 1974).

The completion of the Physical Properties Laboratory enabled new equipment to be installed. A temperature-controlled centrifuge provided the capacity to determine the Centrifuge Moisture Equivalent – effectively the Specific Yield of the sample, as in a subsequent paper (Price 1977). Equipment was also installed to permit the measurement under sterile conditions of the permeability of coarse-grained, highly permeable sandstone samples. Rapid progress was made in a study of regional variations in the physical properties of the principal arenaceous aquifers in England and Wales. The sampling of rock outcrops was completed in each of the zones into which the countries had been divided and samples were also obtained from 34 widespread boreholes. The work on the Permo-Triassic aquifers was completed and the results published later (Lovelock 1977). 600 plugs of rock cut from the cores of 15 offshore oil

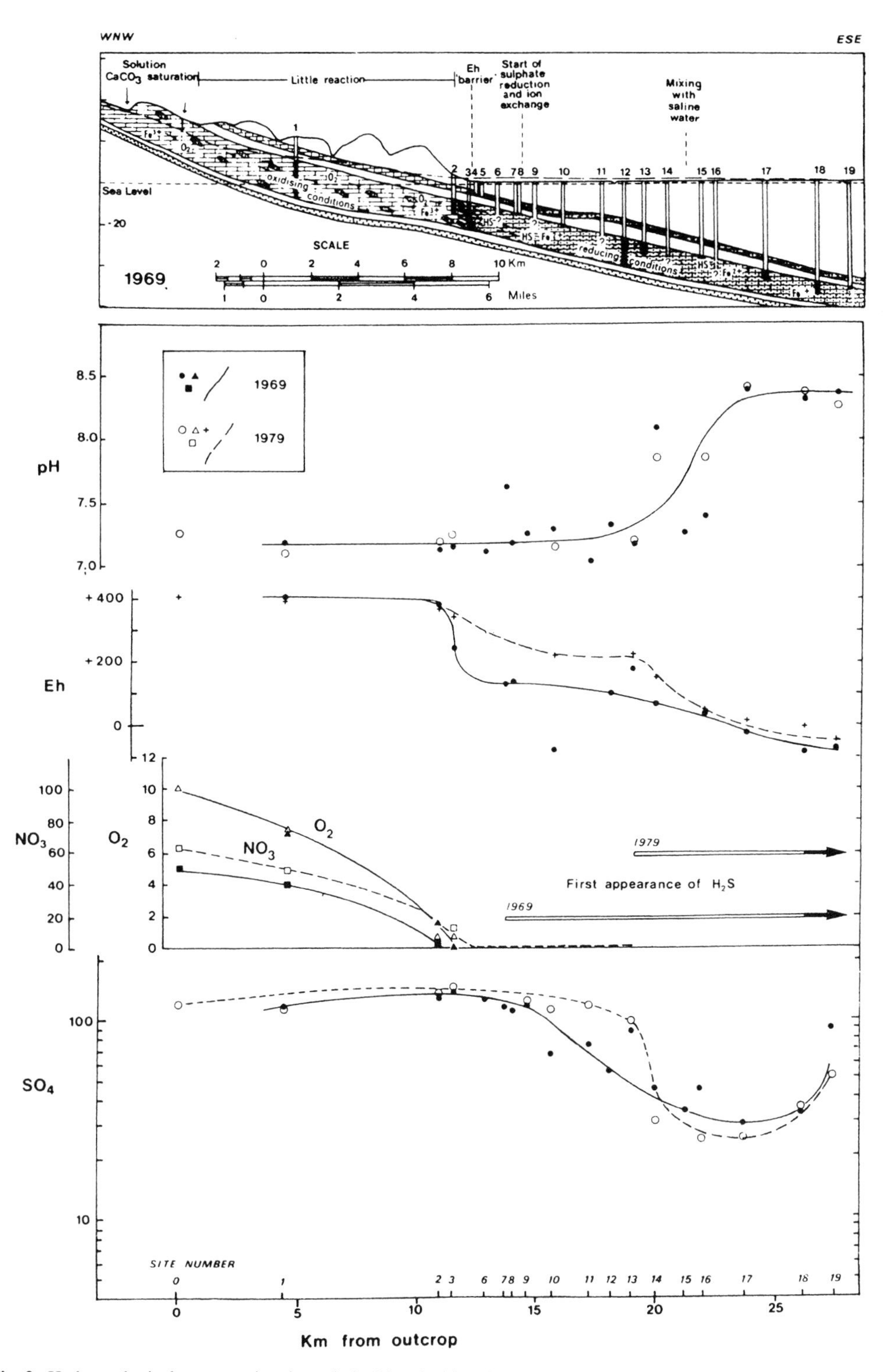

Fig. 2. Hydrogeological cross-section through the Lincolnshire Limestone from east of Grantham to Spalding. The hydrogeochemical trends (pH, Eh, NO_3^-, SO^{2-}, H_2S) are shown for both 1969 and 1979 (from Edmunds & Walton 1983, fig.1).

and gas wells were tested in 1970 for the Petroleum Division of the Department of Trade and Industry: this marked the beginning of a continuing, extensive programme.

A new series within the Department's Water Supply Papers entitled 'Technical Communications' was introduced. The first reviewed the literature on the application of radioactive isotopes in groundwater research (Mather 1970), whilst the second concerned laboratory measurement of soil and rock permeability (Lovelock 1970).

Commercially commissioned research

In 1968 the Greater London Council (GLC) commissioned a major investigation as to whether the proposed flood protection system for central London should take the form of a fixed barrage with shipping locks or a mobile barrier. The former would lead to the elimination of tidal effects on the groundwater in the Flood Plain deposits upstream of the barrier and to their permanent saturation to a higher level than had been the case, whereas a mobile barrier would not change the regime except for local effects during periods of barrier closure. The Department was commissioned to investigate the advantages and disadvantages of the two systems by determining the hydrogeological conditions in the alluvial deposits and forecasting the effects both on them and on the building foundations, basements, tunnels, cut-and-cover tube-lines and other drainage systems in the Flood Plain deposits.

As a first stage in this lengthy investigation the distribution of the fine- and coarse-grained materials in the deposits was determined by an analysis of the many records of trial boreholes located on the riverine deposits from upstream at Kew downstream to Thamesmead (Mather *et al.* 1971). More detailed work in central London involved drilling a line of 12 boreholes extending from Charing Cross southwards to beyond the Festival Hall. Water level recorders were installed in these and a programme of hydrochemical sampling was also mounted. The results of these investigations were summarized and proposals made for further work (Gray 1969*b*). As part of this extended programme, 39 observation boreholes were distributed widely over the Flood Plain at locations where the hydrogeological conditions were thought to be either variable or particularly relevant to the natural and artificial drainage systems. Water-level recorders or piezometers were installed in them to determine the time-based variations in the groundwater regime. Analysis showed that this regime was controlled to a significant extent by the many artificial features (Fig. 3); these included the Victorian main drains, the drainage systems of the District and Circle lines which act as asymmetric, curvilinear sinks, and the extremely complex and variable make-up of the river walls (Gray & Foster 1972).

The first overseas commissioned project related to the occurrence of 'fresh' groundwater within the desert petroleum provinces of Cyrenaica in Libya. British Petroleum in association with Nelson Bunker Hunt, commissioned a report for presentation to the (then Royal) Libyan Government. It arose because Sir Peter Kent, then Chief Geologist of BP, and his colleagues in both BP and Bunker Hunt, had recognized the potential significance for Libya of the widespread occurrence of such water at sites of petroleum exploration wells throughout their Cyrenaican concession areas. Following submission of the initial report, the Department was invited to extend the survey to cover 25 000 km^2. An analogue model of the anisotropic, post-Eocene alluvial deposits in the initial area was constructed and then developed further with new data gathered at many well sites by Departmental staff.

Groundwater with a total dissolved solid content of less than 1500 mgl^{-1} was found over extensive areas. The distribution of the fresher water could be related partly to variations in the present hydraulic gradient, even though ^{14}C dating indicated that the last significant recharge took place more than 29 000 years BP (Fig. 4). This second stage report was presented to the Government in Tripoli in June 1969 and the interest it generated led to a visit to the IGS in July by the then Libyan Minister for Agriculture. The implications for possible large-scale resource development and well-field construction were outlined to him. Emphasis was placed on the fact that the resources were 'fossil' – they would not be renewed under the present or foreseeable climatic regimes, so that the water levels would decline as development took place. A supplement to the regional report recorded the occurrence of nitrogenous deposits in the post-Eocene deposits and their effect upon the quality of the water. The results of these several surveys were published (Wright & Edmunds 1971).

An early commissioned project in the UK related to the search by CERN (Geneva) for a site for a proposed 300 GeV proton-synchroton. IGS staff joined colleagues from the Building Research Establishment to characterize the geology and geotechnical aspects of the proposed British site on the Chalk at Mundford, Norfolk (Anon. 1967): data on the hydrogeology were included.

Government funded overseas activities

Most overseas work during this period was funded by the ODM as part of the British Technical Aid Programme. The first study by Wright concerned the

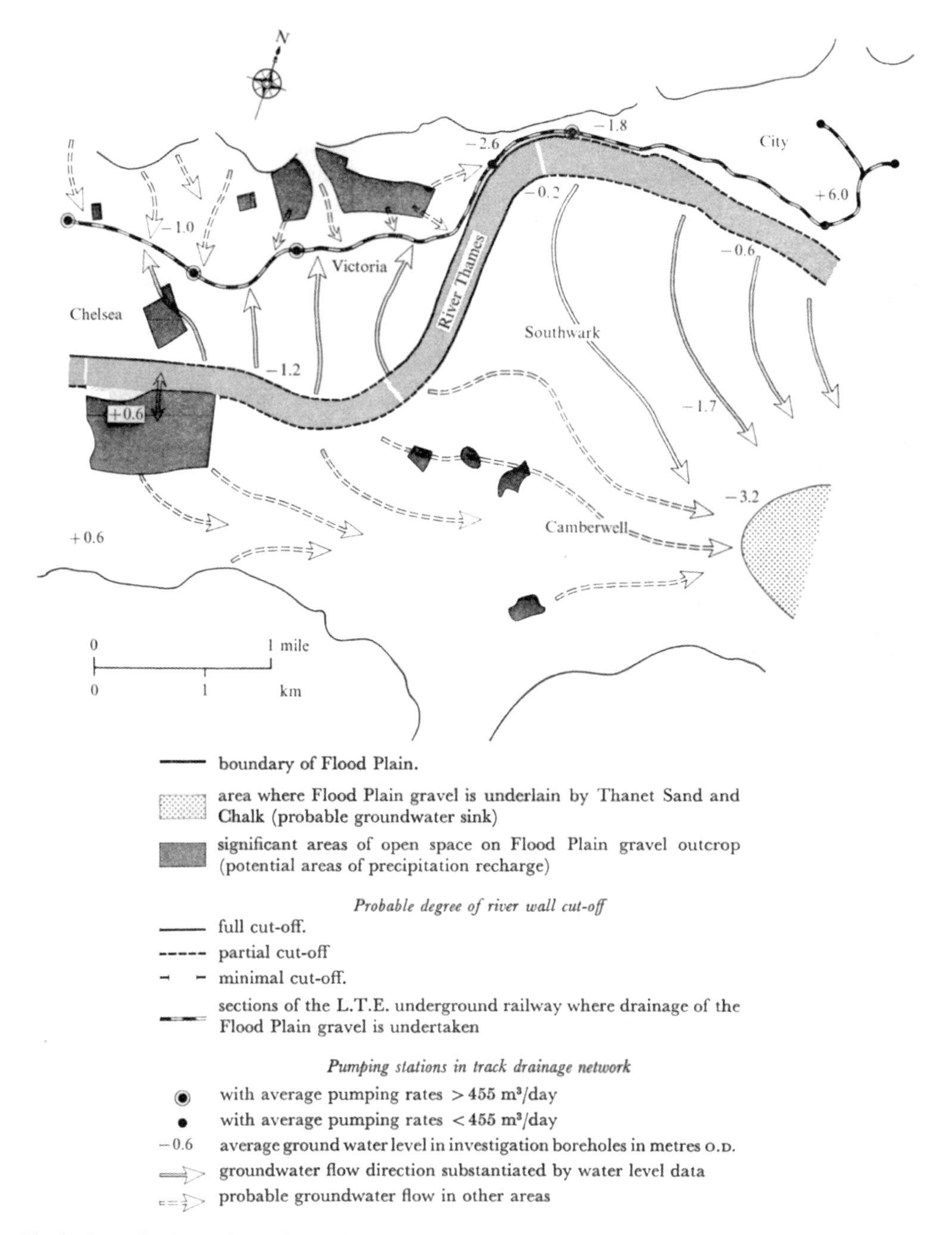

Fig. 3. Generalized groundwater flow regime in the Flood Plain deposits of central London (from Gray & Foster 1972, fig.6).

groundwater resources of Bahrain. He recommended a series of activities which extended over several years and resulted in the establishment of a resource authority. Other smaller studies related to the availability of water resources for coastal areas in north Morocco and towns in British Honduras.

In August 1968 a severe earthquake occurred in NE Iran around Dasht-I-Biaz, with reports of the destruction of many traditional water supply tunnels – termed 'quanats'. An offer of assistance in developing new groundwater supplies in the region was made to the Iranian Government. Its acceptance

Fig. 4. Petrified (?Pleistocene) reed bed and lake deposits, north of Jebel Dalma, Libyan desert, indicating that the climate was once less arid (from IGS Annual Report 1968, Plate VII. Photo I.F.Mercer. IPR/36-16C British Geological Survey. ©NERC. All rights reserved.)

resulted in Day and Wright, together with a Departmental Land Rover, being flown to Iran by the RAF. They worked for three months in the Nimbalok and Gonabad regions of Khorossan Province and reported on the abstraction potential of the two plains (Fig. 5).

The first component of a lengthy series of investigations into the groundwater resources of small islands developed when the Land Resources Division (LRD) of the Directorate of Overseas Surveys requested hydrogeological assistance in a land-use survey for the Bahamian Government. Mather was seconded for the work which began on Abaco (Fig. 6) and Eleuthera islands, where major fresh-water lenses were identified (Little *et al.* 1977). In the Tertiary limestones of Grand Cayman and Cayman Brac comparable lenses some 50 ft. and 30 ft. thick respectively were located (Mather 1972). There as elsewhere, the principal issue was the rate of abstraction achievable without inducing saline intrusion (Mather 1975). Similarly, in Miocene limestones on Anguilla, investigations were made into the distribution of fresh and saline groundwaters.

John Lloyd, an experienced hydrogeologist who had worked extensively in the Middle East, joined the Department in 1970 and was seconded for two years to the Chilean Copper Corporation to provide hydrogeological advice on a wide range of mine-associated activities. Following that secondment he joined Birmingham University to lecture on hydrogeology.

The continuing programme: 1971–1977

During this period changes occurred in the HD's staffing, funding and accommodation. Staff numbers increased from 32 in 1971 to 63 in 1977 and the senior staff changes detailed earlier took place in 1974. The scale of groundwater investigations in both Northern Ireland and Scotland increased to the extent that individual staff members were appointed to funded posts in each office in 1974 and 1977 respectively. For much of the period the DOE was financing half of the Department's activities, whilst the ODA directly supported up to 14 staff for overseas work, which involved geothermal studies from 1974. Additional funding came from the DEn and the EEC for a geothermal programme in the UK, and from the EEC for extensive pollution control studies. Commercially commissioned contracts were maintained at a high

Fig. 5. Villagers watching a pumping test on a new borehole in the Nimbalok Plain (photo D.A.Gray).

level. A major increase in groundwater pollution studies led to the creation of the Environmental Pollution Unit (ENPU) within the HD, although based at Harwell.

Database studies

Contributions from Belgium, France and the Republic of Ireland for the Hydrogeological Map of Europe were incorporated in Sheet B4 (London) (Buchan & Day 1976): work proceeded in Scotland and Northern Ireland on the more northerly Sheet B3. The Hydrogeological Map of northern East Anglia was published (Mosely *et al.* 1976) and a national 'Ten Mile' Hydrogeological Map of England and Wales was presented at the Birmingham Conference of the International Association of Hydrogeologists (Day & Shepard-Thorn 1977).

The implementation of the Water Act 1973 caused changes in the distribution of new well records in that copies were sent to the ten new Regional Water Authorities, as well as to the Central Water Planning Unit that replaced the WRB. Preparation of Metric Well Inventories continued using the IT system developed previously and their publication began (Harvey 1973; Harvey *et al.* 1973); by the end of the period seven had been issued.

Fig. 6. Dept sampling a vertical sided sinkhole near Dundas Town, Abaco, Bahamas (from IGS Annual Report 1971, Plate 10. Photo J.D. Mather).

Regional and resource studies

Work continued on the regional study of the Chalk of East Yorkshire where the use of groundwater for river augmentation was under investigation. The layered hydraulic characteristics of the formation had been demonstrated by relating the piezometric levels in the aquifer to the groundwater recession curves of streams fed by it (Foster 1974). This layered condition was of major significance for any large-scale developments (Foster & Crease 1975; Foster & Milton 1976). The groundwater resources around Hull had been developed extensively over many years and comparison of observations made in 1973 with others from1951 and 1967 indicated *inter alia* the state of saline intrusion into the aquifer from the Humber (Foster *et al.* 1976a).

An assessment was made of the hydraulic behaviour of the Triassic, Penrith Sandstone at the Cliburn Pumping Station, Westmorland. The results of core-analysis, pumping tests and borehole logging indicated that near horizontal fissures dominated the transmissivity, whereas the inter-granular matrix controlled the storage coefficient. Despite complex flow conditions two hydraulic boundaries were identified (Lovelock *et al.* 1975). The paper attracted the award of the Whitaker Medal of the Institution of Water Engineers.

A long-term study of the hydrogeochemistry of the principal arenaceous aquifers of the UK was started in the Bunter Sandstone around Wolverhampton (Edmunds & Morgan-Jones 1976). Later, in the Worksop-Retford region an oxidation/reduction boundary was identified some 6 kms east of the Retford outcrop. Following an earlier study of the interstitial waters extracted from a cored borehole at Harpford, Devon, a new survey of the groundwaters of South Devon was funded by the DOE. Particular attention was given to the relationship between the occurrence of iron and the trace elements (Walton 1982). Another regional hydrogeochemical study indicated that the brines from various sectors of the Durham coalfield had differing origins (Edmunds 1975).

In Scotland a reconnaissance study of the groundwater potential in Fife and Kinross included a drilling programme with associated borehole-logging and pumping tests (Foster *et al.* 1976*b*). A report was also prepared on the resources of central Ayrshire: recommendations were made for investigations of the Coal Measures and Permian sandstones of the Maughlin Basin and for the gravels of the Darvil-Newmills area. As part of a multi-unit study, investigations were undertaken into the hydraulic conditions affecting the Old Red Sandstone beneath the site for possible oil-storage caverns at Hunterstone, Ayrshire. Despite complex flow conditions it was clear that fissure-flow accounted for some 90% of groundwater movement. Laboratory determinations of the physical properties of core samples, the results of pumping tests and borehole-logging were incorporated in a comprehensive review of the proposals (Institute of Geological Sciences 1976). An assessment was made of the groundwater potential on Shapinsay, Graemay and Eday on behalf of the Orkney Islands Council. In Northern Ireland drilling and testing of the resources of the Bunter Sandstone around Englishtown and Lisburn were undertaken and the hydrogeology of the valley gravels of Ulster was reviewed (Price & Foster 1974).

Geothermal studies

At the request of the DEn and the Energy Technical Support Unit (ETSU), the IGS Director, Sir Kingsley Dunham, compiled an internal IGS report on the geological aspects of the geothermal energy of the UK. Following the associated recommendations, a contract was let to the Institute, jointly by ETSU and the EEC, for a long-term, multi-unit assessment of the geothermal potential of the UK. Two aspects of such energy were to be assessed. The low enthalpy potential of groundwaters in the deeper sedimentary basins represented one source. The second was the 'Hot Dry Rock' potential which, in the UK, would be restricted to the geothermal heat within granitic intrusions, particularly those having an abnormally high content of radioactive minerals. From 1977 the studies were undertaken by staff from various IGS units working in a Geothermal Task Force. Initially based in the Applied Geophysics Unit, it was subsequently transferred to the HD under Downing's leadership.

The first project began in 1977 and collated the extensive pre-existing geothermal, geological, hydrogeological, geophysical and hydrochemical data in a catalogue (Burley & Edmunds 1978). A major component of the data was provided by other agencies active in this field of research, in particular the geology departments of Oxford University, Imperial College, the Open University and the Camborne School of Mines. Their research continued, some under sub-contract to IGS.

Field studies began with assessments of the data relating to the thermal waters of the Bath/Bristol district and in the mines in the Cornubian granites. When deep exploration boreholes were being drilled for any purpose the opportunity was taken of making heat flow and associated observations at various depths within them. For example, such investigations were undertaken in the Wessex Basin where a borehole funded by the DEn was being drilled at Winterbourne Kingston, Dorset. Drill-stem tests on permeable formations in the borehole, supplemented by geophysical logs and measurements on core material, indicated that the bulk transmissivity of the Bunter Sandstone in the borehole would be around 17 m^2d^{-1}, whilst interstitial fluid samples implied that much of the groundwater in the sandstone would be a hypersaline brine (Edmunds *et al.* 1982). Within the Hot Dry Rock programme the field work included a CCTV inspection of a borehole drilled for the Camborne School of Mines at their Troon research site in Cornwall. Following the detonation of an explosive charge in the borehole, aimed at artificially fracturing the granite at a specific depth, a classical explosion pattern was recorded. Subsequent to the present period the Task Force undertook many surveys throughout the UK and extensive publications were issued (Downing & Gray 1986; British Geological Survey 1988; Barker *et al.* 2000).

Applied research studies

In the continuing investigation of the surface/groundwater interface, joint use of hydraulic analysis, hydrochemical sampling and borehole-logging was employed in a riverside Chalk well at Taplow, Buckinghamshire. Analyses were made for 12 chemical constituents of water sampled from the river, the alluvial deposits and the Chalk. Of these potassium, strontium and phosphate were identified as significant in estimating the volume of induced recharge and in predicting that volume under given hydraulic conditions. Three borehole-logging methods were also relevant, namely observations of the temperature, conductivity and oxygen content of the water column. The programme was continued over some years to determine the significance of seasonal variations (Edmunds *et al.* 1976). At a comparable site near Dorney, Buckinghamshire, trace elements confirmed that about 30% of the water was river derived.

Research was undertaken into the relationships between the physical properties of the Chalk and the composition of the interstitial pore water. Edmunds & Bath (1976) described the centrifuge technique used to extract the water. Plugs of chalk cut from a cored borehole into the Upper and Middle Chalk at Boxford in the Lambourn valley were centrifuged and the interstitial water collected. These pore waters were up to ten times more mineralized than the water in the well column and chemical differences between the pore waters from the different stratigraphical levels could be related partly to the flow characteristics in a nearby well (Edmunds *et al.* 1973). Similar studies were mounted using material from a borehole at Faircross, Hampshire. Interstitial waters from the Upper, Middle and Lower Chalk helped to elucidate the freshwater diagenesis and evolution of the groundwater mineralization. The physical properties of the upper 50 m of the Upper Chalk differed from those of the lower 50 m: the upper section had a relatively constant porosity of 40–45% and a permeability of 3–5 mD, whilst the lower section showed greater variability.

The Infiltration Project was expanded in 1973 by the construction of two large lysimeters in the Lower Chalk overlying Gault Clay at Reach, Cambridgeshire: one was seeded with grass, the second with cereal. Later, a new type was installed at the Fleam Dyke Pumping station near Cambridge: a 5 m cube of Chalk was isolated using linked steel piling. As part of the same Project the use of lysimeters in a wide range of hydrogeological applications was explored at a Symposium held in IGS in 1976 and attended by some 100 participants (Kitching & Day 1980). Analyses were published of the results of lysimeter monitoring of infiltration over a period of some years into the Lower Greensand (Kitching 1974) and into the Bunter Sandstone (Kitching *et al.* 1977). The relevance to rates of infiltration was well illustrated by the severely diminished recharge recorded during the drought year of 1976. A broader assessment of the effects of that drought on groundwater resources was published (Day & Rodda 1978).

An unusual project began with the design and construction of a sand-tank model of a new concept in well construction. It consisted of a central well intersected at depth by several radially-oriented, angled boreholes, subsequently dubbed 'rakers'. The objective was to improve the yield/drawdown characteristics of the central well by effectively increasing its diameter. It was anticipated that in a fissured formation the improvement could be greater than the threefold improvement seen in the intergranular condition in the sand model. In 1974 an opportunity arose for a field trial of the concept. In collaboration with Norwich Corporation Water Department three 'rakers' were drilled to intersect a central borehole into the Chalk at Lyng, near Norwich, and greatly increased the yield (Day *et al.* 1978).

A project in Scotland involved the examination of the groundwater components of streams draining metamorphic and igneous catchments. The hydrograph of the Ullt Uaine stream in Argyllshire, draining unweathered mica-schist, indicated a negligible groundwater component. By contrast, a not inconsiderable component could be seen in the hydrograph of the Dulnain stream in Inverness-shire draining fissured, crystalline igneous rocks. Moreover, low-flow hydrographs from tributary streams indicated that the contributions from different drift deposits could be differentiated from that derived from the solid geology.

The newly developing groundwater age-dating systems were applied to several projects. The use of ^{14}C in Libya has been referred to above. In the UK a paper on the use of thermo-nuclear tritium as a tracer in a variety of hydrogeological applications was published in association with the Wantage Research Laboratory (Mather & Smith 1973). Another was concerned with the use of both systems in assessing the recharge to the Lower Greensand of the London Basin (Mather *et al.* 1973).

Commercially commissioned research

On the basis of evidence it had commissioned, the Greater London Council decided that the Thames Barrier would be formed by rotating gates and not a fixed barrage. Accordingly the scope of the HD's programme was modified to resolve the effects of such a structure on the hydrogeology of the alluvial deposits adjoining its proposed site, as well as both up- and downstream of it. Work on the regime adjoining the site included studies of the mode of propagation, transfer and dissipation of the groundwater tide. A study was mounted into the possible change of that regime in the event of a tidal surge of any defined height in the river. A mathematical model simulated the conditions downstream of the barrier site and was again employed to predict the groundwater fluctuations resulting from such surge tides. In the light of these and other extensive flood protection studies in connection with the Barrier project, a Royal Society Symposium entitled 'Problems associated with the subsidence of south-eastern England' was organized by Sir Kingsley Dunham and Gray. 16 papers from 14 organizations were presented, including one from the HD (Gray & Foster 1972).

Work continued in Libya on a new contract for the Libyan Arab Republic. In the Jalo district the post-Eocene deposits were divided into three, an upper post-Middle Miocene group of poorly consolidated deposits, underlain by Lower and Middle Miocene limestones, dolomites, sandstones and clays resting on Oligocene sediments. This stratigraphical work permitted the selection of areas for groundwater development and a drilling programme began in December 1971. The next year up to 11 staff were involved in the extended Cyrenaican project. Analysis of records from extensive drilling programmes covering the area and associated with pumping tests on some wells, indicated that there were two principal regional aquifers containing water of a quality suitable for irrigation and, over much of the region, for municipal purposes. Data from the region included new gamma-ray borehole logs. On the basis of these logs the earlier stratigraphical analysis of the post-Middle Miocene formations was revised, particularly with respect to the occurrence of clays and their effects upon the hydrology, as indicated in the mathematical model of the regime. Delays to some drilling in the second phase of the project, which extended from the Sirte basin into the Kufra basin, deferred the development of the modelling; only post-Eocene formations had been tested and the potential of the Nubian in the south still required evaluation. The region was subsequently fully investigated (Edmunds & Wright 1979; Edmunds 1980; Wright *et al.* 1982).

Civil engineering projects undertaken included a study of the drainage from the 'cut-and-cover' road tunnel on the A1(M) at Hatfield, Hertfordshire and an examination of the groundwater conditions around Brunel's historic railway tunnel under the Thames at Rotherhithe. Investigations were also mounted on behalf of the National Coal Board into the drainage problems likely to be encountered during the construction of both vertical and angled mine shafts for the then proposed coalmine in the Selby area of East Yorkshire. The work included laboratory determinations of the hydraulic properties of the local Bunter Sandstone and Magnesian Limestone, as well as pumping tests and geophysical logging of boreholes at Wistow.

Pollution control studies

Pollution of groundwater had been the subject of intermittent investigation since the inception of the Department, characterized by a classic study of such pollution in Europe (Buchan & Key 1956). However, from 1972 it became a central topic of funded research at both a local 'point pollution' level, as in contamination derived from landfill leachates, and at a regional 'diffuse pollution' scale, in which leachates from agricultural fertilisers were then dominant. From 1975 a third field of research – the disposal of radioactive wastes – developed and was seen as another aspect of pollution control. Most of the studies on both 'point pollution' and radioactive waste disposal were undertaken by the staff of ENPU, led by Mather, whilst that for 'diffuse pollution' was handled by the staff involved in regional studies, principally guided by Stephen Foster.

Studies at the local level had been triggered by public disquiet at 'fly-tipping' of cyanide wastes. The DOE commissioned the IGS to examine the geology and groundwater pollution risk at several sites in the Midlands and North at which this had occurred: further action was recommended at some locations. Following these incidents the IGS was commissioned to review the risk of groundwater contamination at all landfill waste disposal sites, either active or disused, known to local authorities in England and Wales. The objective was to assemble all relevant information in a standard format. The mechanism adopted to generate the ensuing database began with a questionnaire completed by the

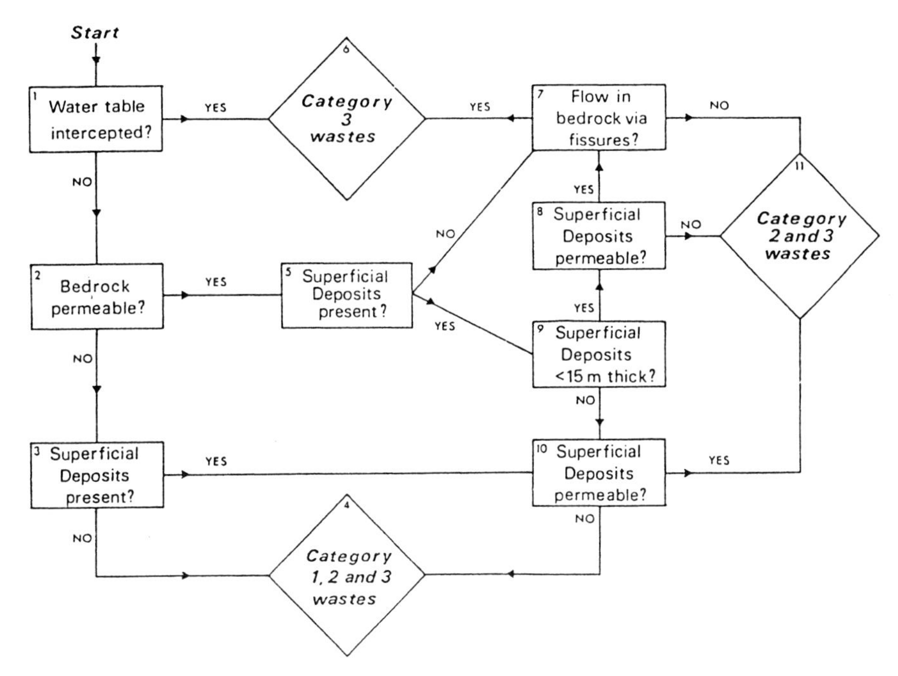

Fig. 7. Flow diagram to assist in the selection of landfill wastes: Category 1 – Hazardous wastes, Category 2 – Domestic and related wastes, Category 3 – Inert wastes (from Gray *et al.* 1974, fig.2).

Local Authorities. The IGS field staff identified the geology of each site and members of ENPU assessed the pollution risk. These data were then passed to the relevant River Authority for their comments. Finally the returns were co-ordinated by ENPU. Of the 1321 questionnaires registered during 1972, 1050 had been processed by the year end and the trend indicated that some three to four percent could represent a hazard to local groundwater resources. The survey finished in 1974, by when 3054 disposal sites, together with 268 spoil heaps and lagoons of the National Coal Board, were entered into the database. Of those totals only 51 sites (less than 2%) were finally thought likely to be a serious risk to groundwater resources.

In 1973 the DOE let two new associated contracts jointly to ENPU and the HMS at Harwell. The first was to evaluate the use of thermo-nuclear tritium in assessing the groundwater pollution hazard at not less than 25 landfill sites. This was shown to be a relatively inexpensive but successful technique. The objectives of the second contract were to determine which toxic wastes, if any, could be safely disposed of in landfills and under what conditions, and to define general criteria for the selection of sites for both domestic and industrial wastes. Over a period of some three years the joint ENPU/HMS team investigated conditions at some seven landfill sites. Preliminary results were presented of the multi-authority reviews of the pollution hazard arising at landfill sites and criteria were proposed (Fig. 7) for the selection of landfill sites based on the types of waste to be deposited (Gray *et al.* 1974). The relevant legislation was also reviewed (Guiver & Gray 1974).

DOE continued to fund other landfill research, including work on the *in situ* attenuation and control of landfill leachates (Mather 1977; Mather & Bromley 1977). The construction of the boreholes required for this purpose was also detailed (Harrison 1976). Four large, wholly-enclosed lysimeters were constructed at the junction of the Lower Greensand and the Kimmeridge Clay at Uffington, Berkshire. Their purpose was the investigation of rates of movement and of *in situ* reactions of various irrigated pollutants within the strata. Access to the interiors of the lysimeters was via a central access trench (Fig. 8) from which suction probes extracted fluids from the unsaturated zone, with full drainage emerging at the sand/clay boundary (Black *et al.* 1977). A comprehensive summary of all of the DOE's Landfill Research Programmes with several organizations was published (Department of the Environment 1978).

Concern was being expressed, by the water industry in particular, at the increasing evidence of 'diffuse pollution', epitomized then by rising levels of nitrate in groundwater supplies in the major aquifers, but later by other contaminants. Rising nitrate levels had been widely reported previously, internationally and nationally, and were so widespread as to suggest that

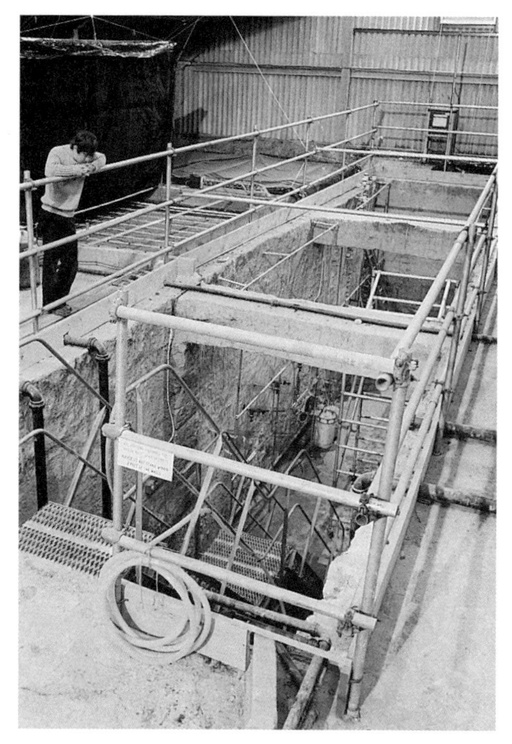

Fig. 8. Landfill Research Programme research site on the Lower Greensand where flow of pollutants through unsaturated sandstone was being investigated. Recording monitoring equipment in the access trench (right), general view of lysimeter showing access trench and storage tanks (left) (from IGS Annual report 1975, Plate 13.IPR/36-16C British Geological Survey ©NERC. All rights reserved).

the contamination could be regional. The application of nitrogenous fertilisers for intensive agriculture appeared to be the cause. In the UK, for example, several pumping stations in the cereal growing areas of East Yorkshire were affected (Figs 9a, b). Most water samples prior to 1970 had nitrate concentrations between 2.5 and 4.5 mgl^{-1} NO_3-N whereas from 1972 some concentrations were in excess of the WHO recommended limit of 11.3 mgl^{-1} NO_3-N. These rises appeared particularly disturbing in that the absence of thermo-nuclear tritium in most samples implied that the concentrations pre-dated 1953, when nitrate applications were at a far lower level than from the late 1950s onwards. The implication that there might be even higher concentrations in the water infiltrating through the unsaturated zone was emphasized in an important paper by Foster & Crease (1974). They recognized the complexity of the flow regime in that zone in the Chalk so that the tritium results were open to interpretation. The authors were awarded the Whitaker Medal of the Institution of Water Engineers for their paper. Further investigations of the tritium anomaly demonstrated the major significance of diffusion in formations such as the dual porosity Chalk (Foster 1975). These results were reinforced in a subsequent paper concerned with pore size measurements in the Chalk (Price *et al.* 1976), as well as in later publications (Foster & Smith-Carington 1980; Barker & Foster 1981).

A contract from the DOE led to a geographical and stratigraphical expansion of this work on agricultural leachates at new research sites on the Lincolnshire Limestone and the Chalk. Three conclusions were reached from the several assessments of such pollution to Chalk groundwaters. Firstly, the evidence showed that the leachates led to higher levels of NO_3, SO_4 and Ca in the pore spaces in the unsaturated zone; secondly, the average rate of leaching was estimated to fall between 45 and 75 kg of nitrogen per hectare per year since 1964; thirdly, following the 1976 drought most leaching occurred in October/November 1976. The vulnerability of groundwater resources in British aquifers to pollution by such leachates was reviewed by Foster (1976). Evidence on the HD's work on this topic was given in 1977 to the Royal Commission on Environmental Pollution, under the Chairmanship of Prof. H.L. Kornberg, and was later summarized (Foster & Young 1980).

(a)

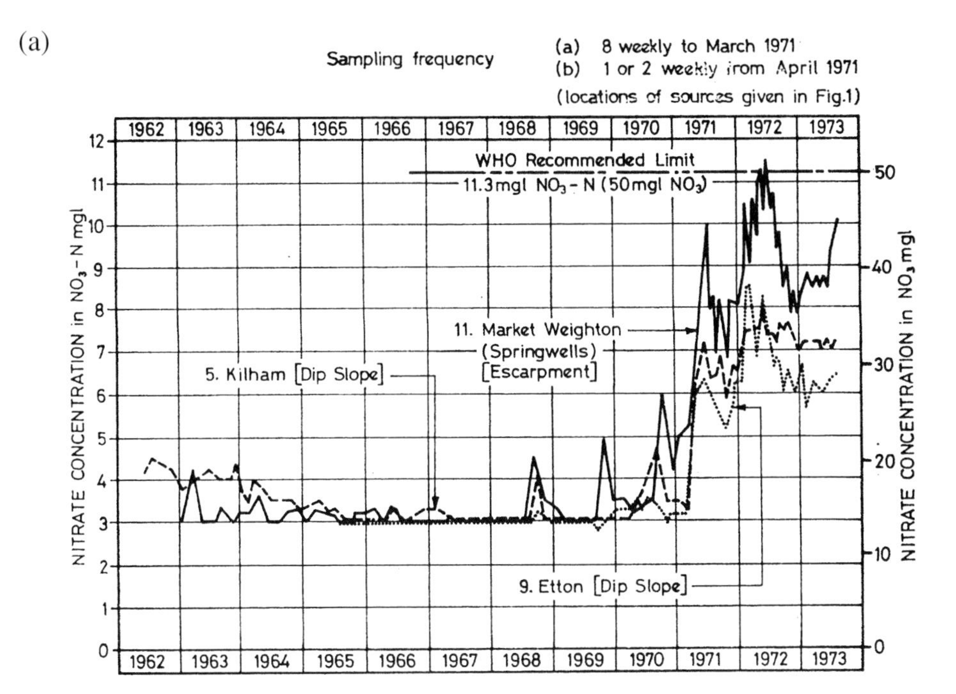

(b)

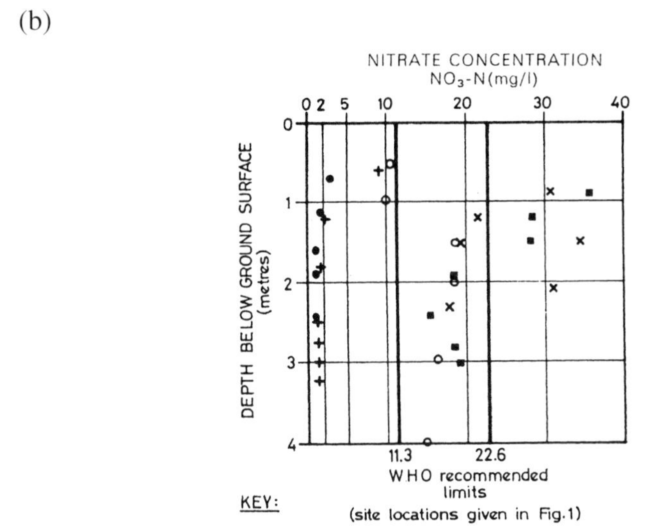

KEY:

A: ● Bridlington: natural unfertilised grassland for over 50 years; soil 1.0m. thick; sampled 5th. February 1973

B: ■ Burton Fleming: arable field regularly used for cereals receivig about 100kg N/ha/annum; soil 0.2m. thick; sampled 8th. January 1973

C: × North Newbald: arable field used regularly for cereals; soil 0.15m. thick; sampled 10th. November 1972

D: + Market Weighton: unfertilised grassland for 7 years; previously arable field receiving about 40kgN/ha/annum; soil 0.4m. thick; sampled 19th. March 1973

E: ○ Octon: arable field used for cereals receiving 85kgN/ha/annum; soil 0.15mn. thick; sampled 25th. April 1973

Fig. 9(a). Trends of nitrate concentration in selected sources (from Foster & Crease 1974, fig. 2). **(b)**. Nitrate concentrations of Chalk pore water in the unsaturated zone (from Foster & Crease 1974, fig. 4).

In 1975 the Royal Commission on Environmental Pollution, then Chaired by Sir Brian Flowers, sought evidence from the IGS on the concept of geological disposal of radioactive wastes. Written and verbal contributions to the Commission by Gray were supplemented by verbal evidence from Sir Kingsley Dunham, then recently retired, his successor as Director Dr. Austen Woodland and Mr Dennis Ostle (Head, Radioactive and Metalliferous Minerals Unit). Members of the Commission were accompanied by Gray when they visited the Asse Salt Mine in West Germany, where low and intermediate-level wastes were being placed in a salt diapir. Associated interest from the DOE, the EEC and the UKAEA resulted in research contracts for the IGS, centred on ENPU at Harwell and headed by Mather. In 1977 the Radioactive Waste Disposal Task Force was formed under his supervision.

The first contract was from the UKAEA and led to an internal IGS Working Party established to define criteria relevant to geological disposal in the UK (Gray *et al.* 1976). A contract with the EEC related to the UK component (Greenwood *et al.* 1976) of a catalogue of geological formations suitable for the disposal of high-level and/or alpha-emitting wastes throughout the member states (Anon. 1979). A third contract, let jointly by the UKAEA and the EEC, concerned the disposal of high and intermediate-level wastes in crystalline rocks in the UK. The Loch Doon Granite in Galloway was the first intrusion selected for study. Preliminary fieldwork was undertaken but access problems prevented its completion. Desk studies started under another contract relating to the disposal of high-level wastes to argillaceous and evaporitic formations. In different aspects of radioactive waste disposal, the staff of ENPU were commissioned by British Nuclear Fuels Ltd. to assist in the investigation of sub-surface conditions at the Company's Windscale site (Holmes & Hall 1980), at the low-level disposal trenches at Drigg (Williams *et al.* 1985), and at the Dounreay Nuclear Establishment.

Government commissioned overseas activities

Staff worked in 21 territories on ODA-funded Technical Aid projects. Much of the advisory work involved one or more short-term visits, but longer term secondments were made for several projects, such as those in India and the Yemen Arab Republic. Due to space limitations only selected activities are referred to and for most of these the results are recorded only in internal reports.

In Cyprus a gravel aquifer at Akrotiri was prone to both seasonal over-development for irrigation purposes and to periodic flooding. The formation constants for the gravel were derived from a series of pumping tests and water-level monitoring and the results incorporated in a digital model, which was subsequently calibrated (Kitching 1975). The success of this study led the Food and Agriculture Organization of the UN to commission a model of another gravel aquifer in Cyprus, near Morphou. The first overseas component of the Infiltration Project was also developed in Cyprus and aimed at determining the rate of groundwater recharge in a semi-arid zone. Six small areas were selected for lysimeters in a sandy aquifer in the Mesaoria, whilst a 10 m$\times$7 m lysimeter was installed near Paralimni with the help of the Royal Engineers. A drilling programme at Akrotiri formed part of the project and provided four geochemical profiles in the unsaturated zone. A broader hydrogeochemical project, also aimed at determining recharge in semi-arid zones, was developed later in other territories (Edmunds & Walton 1980).

Projects involving the groundwater resources of small islands continued. In the Caribbean fresh water lenses were identified in the coral islands of the Turks and Caicos, and Barbuda, as well as in alluvial materials found in two groups of volcanic islands – the British Virgin Islands and the Grenadines. In the Pacific similar investigations were mounted into the resources of some atolls of the Gilbert and Ellice Islands. The supplies in use on Tarawa were heavily polluted and became saline during droughts; only on the islets of Bonriki and Buota were the fresh-water lenses sufficiently thick to supply sustainable supplies – even there limited to some 40000 gpd (Mather 1978). Elsewhere, in the Seychelles Archipelago some fresh water aquifers were located in unconsolidated reef debris on the coastal fringes of the islands.

Resource studies at a variety of scales were mounted in several territories. In Nepal the groundwater irrigation potential of the floodplain deposits and the terrace gravels of rivers in the Nawal Parasi district were investigated and the resources of the floodplains found to be significant. Sites were suggested for alternative groundwater supplies for the domestic supply of the British Military Cantonment at Dharan. Limited exploration was recommended for a cattle-watering project in Ethiopia, with a warning that the prospects in the region were poor. The potential was evaluated of Costa Rica's most important aquifer – the Colima Lavas – under development for the supply of San Jose. Two activities were mounted in Jordan: data were collected relative to an aquifer in the Wadi Duhliel, for which the recharge area is in Syria, whilst in the south of the country a review was undertaken of a mathematical model of the Qa' Disi aquifer developed by Messrs Raikes and Partners. A computerized field data system was provided to the Geological Survey of Indonesia and was installed for a project in north Sumatra. In Zambia a preliminary study was undertaken in association with Colquhoun and Partners, of the Kakontwe Limestone as a source

Fig. 10. Installation of prototype 'minipuit' (from IGS Annual Report 1973, Plate 11. Photo British Materials Engineering Limited. IPR/36-16C British Geological Survey. ©NERC. All rights reserved).

of supply for Ndola; a model was prepared and subsequently calibrated.

Longer-term activities to which staff were seconded included the Bhopal project in India. This was undertaken jointly with counterparts from the Indian Central Groundwater Board and initiated an assessment of the methodologies of groundwater development in the Deccan Traps. Extensive geophysical logging of exploration boreholes featured in the programme (Buckley *et al.* 1978: summarized by Buckley & Oliver 1990). Staff seconded to the Geological Survey in Botswana worked on resource developments and subsequently on a national development programme proposed previously by Wright. An LRD team working in the Yemen Arab Republic included staff seconded from the HD who undertook groundwater surveys in the Montane Plains and the Wadi Rima region.

An ODA team, including Day as a member, investigated the prospects for cattle rearing and transport in Mauritania, Chad and Mali for which groundwater supplies would be critical. A similar later visit to Eastern Niger led to suggestions for a new form of well construction in such arid regions with shallow water tables. Prefabricated, fibreglass, sectional well casings and screens were envisaged which would be easy to transport and install. An ODA contract led to British Composite Engineering Ltd. funding the design and construction of prototype 'minipuits' and a successful trial installation was mounted in the UK (Fig. 10). Operational units were subsequently installed in Chad and the Gambia.

From 1974 overseas geothermal studies were mounted under Wright's supervision. Preliminary studies in the Caribbean region were undertaken in the islands of St. Lucia and Montserrat and in Central America in Panama. However, the only development during the present period was on St. Lucia where the geothermal potential of the Soufriere area was subject initially to geophysical and hydrogeochemical reconnaissance studies. The high-temperature system in a fractured, hydrothermally altered dacite was shown to be vapour-dominated. Stable isotope and tritium analyses were used to investigate the water/steam interaction with the wall rock and to infer a turn-round time of at least 20 years (Bath 1977). Three of the five exploratory boreholes sunk in the vicinity of Sulphur Springs produced steam (Fig. 11). Although a production of 1.5 megawatts (electric) might be possible, a high gas content necessitated a further testing programme directed towards a greater understanding of the system as a whole (Williamson 1979).

An overview

The 1965 brief had been to convert the Department from its former role to a body able to undertake fundamental and applied research in any field of hydrogeology. The extent to which that was achieved can be assessed by examining the programmes completed. Amongst the most successful were the:

(1) theoretical and practical advances in groundwater hydrogeochemistry, including isotopic applications,
(2) resolution of many uncertainties in groundwater flow mechanisms, especially within the matrix of the Chalk,
(3) recognition of the 'diffuse pollution' arising from agricultural leachates, and the need for their control,
(4) investigation of the many problems associated with the control of landfill leachates and other sources of 'point pollution',
(5) resolution of a range of induced recharge problems and associated issues of the groundwater/surface water interface,
(6) major resource study in Cyrenaica, Libya,
(7) creation of an automated database for the well record collection and its metrication,
(8) hydrogeological map programme,

Fig. 11. Discharge of wet steam from Borehole 4 during the St Lucia geothermal project (from IGS Annual Report 1975, Front cover. Photo D.K. Buckley. IPR/36-16C British Geological Survey. ©NERC. All rights reserved).

(9) resolution of many civil engineering problems, such as those of the Thames barrier, and;
(10) extensive development of advisory activities in many overseas territories in Africa, Asia, Central and South America, as well as in numerous small islands in the Caribbean and Pacific.

Associated with the major change in role, were the administratively demanding changes in the funding mechanisms. Essentially a conversion from a Science Budget organization to one in which the principal finances were derived from Government and EEC funds, as well as from commercially commissioned research in the UK and overseas.

The successful completion of most of these programmes commonly involved the integration of several techniques. For example, in the previously unknown Cyrenaican groundwater province, the stratigraphical and sedimentological data, derived partly from drilling records and partly from geophysical logs, had to be merged with pumping test results and associated hydrochemical information to permit the generation of analogue and digital groundwater models. Now routine, such large-scale comprehensive studies, involving a wide variety of largely new specialisms, were then just developing. Of major assistance in such studies were the newly created laboratories and the introduction of IT systems. Initially, not the least of the issues involved was that of ensuring that these various specialist topics were integrated fully. The scientific legacy created by these several teams of specialist staff and their ability to provide an integrated input to virtually any hydrogeological project, enabled the Department to claim that within this review period the terms of the original 1965 brief were being met.

One of the consequences of the several legislative changes from 1965 onwards was a major increase in the number of publicly funded individuals concerned full-time with hydrogeological issues. Prior to 1965 the numbers involved were less than 20, excluding those employed by consulting companies on public works. By 1980 that total had increased dramatically to an estimated minimum of 178 (Wilkinson & Gray 1981). Of those 65 were in IGS, 30 in the Water Research Centre, 60 in the Water Authorities, 8 in the Institute of Hydrology and 15 in the Universities. Whether or not this dispersal of the limited, hydrogeological expertise was either desirable or efficient is a matter for debate. However, it arose through the developing de-centralization of the water industry, culminating later in its privatization. At each stage of the process the authorities decided that to meet their new responsibilities they required their own specialist staff and these were becoming available through the introduction of several university hydrogeological courses. By the mid-1970s it became clear that a UK discussion forum was becoming desirable and the staff of the HD were prominent in the formation in 1974 of the Hydrogeological Group of the Geological Society of London. Stevenson Buchan was the first Chairman, Mather the first Secretary and the inaugural lecture was delivered by Gray (1975). Two years later the British National Chapter of the International Association of Hydrogeologists was formed and its Committee, again chaired by Buchan and with John Lloyd as its local Secretary, organized a Congress in 1977 at Birmingham.

As a further indication of the Department's role within the hydrological community, it's senior staff served on the committees and sub-committees of many national and international organizations. In a national context the principal involvements lay within the Royal Society, DOE, DEn, ODA, NERC, the Scottish Development Department, the Greater London Council, the Geological Society, the Institution of Civil Engineers and several of the Water Authorities. Internationally the organizations included the EEC, IAH, IASH, the International Hydrological Decade, the International Association of Geochemistry and Cosmochemistry, the International Atomic Energy Agency, the OECD's Nuclear Energy Agency and the International Energy Agency. Senior staff were also called upon to act as Geological Assessors at Public Inquiries held in connection with major national hydrogeological proposals. Again such varied commitments have continued long past the period reviewed in this paper.

The author gratefully acknowledges the helpful comments on an early draft by R. Downing and J. Mather. He also thanks the referees for their constructive criticisms and J. Mather for his subsequent suggestions. Permission to reproduce copyright figures has been granted in the following terms; Figure 1. *Journal of Hydrology*, **9**, Mather, Gray & Jenkins, The use of tracers to investigate the relationship between mining subsidence and groundwater occurrence at Aberfan, South Wales, 136–154, Copyright (1969), fig. 2 with permission from Elsevier: Figure 2. *Journal of Hydrology*, **61**, Edmunds & Walton, The Lincolnshire Limestone – Hydrogeological Evolution over a Ten-year period, 201–211, Copyright (1983), Fig. 1 with permission from Elsevier: Figure 3. *Philosphical Transactions of the Royal Society*, **A**, **272**, Gray & Foster, Urban influences upon Groundwater conditions in Thames Flood Plain deposits of Central London, 245–277, (1972) fig. 6: Figures 4, 6, 8, 10 & 11 by permission of the British Geological Survey. ©NERC. All rights reserved. IPR/36–16C: Figure 5 by permission of the author: Figures 9a & b. Journal Institution of Water Engineers, **28**, Foster & Crease, Nitrate pollution of Chalk Groundwater in East Yorkshire – a Hydrogeological Appraisal, 178–194 (1974), figs 2 & 4 by permission of the Chartered Institution of Water and Environmental Management.

References

ANON. 1967. *Sites for the proposed CERN 300 GEV proton Synchroton*. Report on the geological and geotechnical investigations on the Mundford site in West Norfolk, England, CERN, Geneva, **Cern/664/Rev. 11**.

ANON. 1979. *European Catalogue of Geological Formations having Favourable Characteristics for the Disposal of Solidified High-level and/or Long-lived Radioactive Wastes. Vol.1. Introduction and Summary*. European Economic Commission, Brussels.

BARKER, J. A. & FOSTER, S. S. D. 1981.A diffusion exchange model for solute movement in fissured porous rock. *Quarterly Journal of Engineering Geology*, **14**, 17–24.

BARKER, J. A., DOWNING, R. A., GRAY, D. A., FINDLEY, J., KELLAWAY, G. A., PARKER, R. H. & ROLLIN, K. E. 2000. Hydrogeothermal studies in the United Kingdom. *Quarterly Journal of Engineering Geology and Hydrogeology*, **33**, 41–58.

BATH, A.H. 1977. Chemical interaction in a geothermal system in St. Lucia, West Indies. *In: Proceedings Second International Symposium on Water Rock Interaction (WRI-2)*. Strasbourg, France III, 170–179.

BENFIELD, A. C., BRUCE, B. A. & DAY, J. B. W. 1970. *Hydrogeological Map of the Chalk and Lower Greensand of Kent. Sheet 2. Lower Greensand*. Institute of Geological Sciences, London.

BLACK J. H., BOREHAM, D., BROMLEY, J., CAMPBELL, D. J. V., MATHER, J. D. & PARKER, A. 1977. Construction and instrumentation of lysimeters to study pollutant movement through unsaturated sand. *In*: *Proceedings of the Groundwater Quality – Measurement, Prediction and Protection Conference, 1976*. Water Research Centre, Medmenham, 327–340.

BRITISH GEOLOGICAL SURVEY. 1988. *Geothermal Energy in the United Kingdom: review of the British Geological Survey's Programme 1984–1987*. British Geological Survey. Keyworth.

BUCHAN, S. 1965. Balancing the World's water budget. *Discovery*, **26**. 10–14.

BUCHAN, S. & KEY. A. 1956. Pollution of groundwater in Europe. *Bulletin World Health Organization* **14**, 949–1006.

BUCHAN, S. & DAY, J. B. W. 1976. *International Hydrogeological Map of Europe. B4 London*. Bundesanstalt fur Geowissenschaften und Rohstoffe & UNESCO, Hannover and Paris.

BUCKLEY, D.K. & OLIVER, D. 1990. Geophysical logging of water exploration boreholes in the Deccan Traps, Central India. *In*: HURST, A., LOVELL, M.A. & MORTON, A.C. (eds) *Geological Applications of Wireline Logs*. Geological Society, London, Special Publications, **48**, 153–161.

BUCKLEY, D.K., SINGH, B.K. & DEV BURMAN, G.K. 1978. Borehole logging and Deccan Trap hydrogeology in the Betwa Basin, MP. *In: Proceedings of the Groundwater Conference, Bhunbaneshwar, India*.

BURLEY, A.J. & EDMUNDS, W.M. 1978. *Catalogue of geo-*

thermal data for the land area of the United Kingdom. Department of Energy, London.

DAY, J.B.W. & RODDA, J.C. 1978. The effects of the 1975–76 drought on groundwater and aquifers. *In*: *Proceedings. Royal Society Conference on scientific effects of the 1975–76 drought*. Royal Society, London, 55–68.

DAY, J.B.W. & SHEPHARD-THORN, E.R. 1977. The 10-mile Hydrogeological Map of England and Wales. *In*: *Memoirs, Congress of the International Association of Hydrogeologists, Birmingham*, **XIII**, B1–B13.

DAY, J.B.W., MOSELEY, R., ROBERTSON, A.S. & MERCER. I.F. 1967. *Hydrogeological Map of North and East Lincolnshire*. Institute of Geological Sciences. London.

DAY, J.B.W., MERCER, I.F., BRUCE, B.A. & WILLIAMS, I. 1968. *Hydrogeological Map of the Dartford (Kent) District*. Institute of Geological Sciences. London.

DAY, J.B.W., MOSELEY, R., BRUCE, B.A. & MERCER, I.F. 1970. *Hydrogeological Map of the Chalk and Lower Greensand of Kent*. Sheet 1. Chalk. Institute of Geological Sciences. London.

DAY, J.B.W., ROWE, K., PRICE, M. & TATE, T.K. 1978. *The use of inclined boreholes for well development*. Journal of the Institution of Water Engineers and Scientists, **32**, 329–340.

DEPARTMENT OF THE ENVIRONMENT. 1978. *Co-operative Programme of Research on the Behaviour of Hazardous Wastes in Landfill Sites. Final Report of the Policy Review Committee*. HMSO, London.

DOWNING, R. A. 2004. The development of groundwater in the UK between 1935 and 1965 – the role of geological Survey of Great Britain. *In*: MATHER, J.D. (ed.) *200 years of British Hydrogeology*. Geological Society, London, Special Publications, **225**, 271–282.

DOWNING, R. A. & GRAY, D.A. (eds) 1986. *Geothermal Energy – The Potential in the United Kingdom*. HMSO, London.

DOWNING, R.A., LAND, D.H., ALLENDER, R., LOVELOCK, P.E.R. & BRIDGE, L.R. 1970. *The hydrology of the Trent River Basin*. Water Supply Papers, Hydrogeological Report, No. **5**, Institute of Geological Sciences. London.

EDMUNDS, W.M. 1971. *Hydrogeochemistry of groundwaters in the Derbyshire Dome with special reference to trace constituents*. Institute of Geological Sciences Report **71/7**, HMSO, London.

EDMUNDS, W.M. 1973. Trace element variations across an oxidation- reduction barrier in a limestone aquifer. *In*: *Proceedings of a Symposium on Hydrogeochemistry, Tokyo, 1970*. Clarke Co., Washington, 500–526.

EDMUNDS, W.M. 1975. Geochemistry of brines in the Coal measures of North-east England. *Transactions of the Institution of Mining and Metallurgy*, **84**, 339–352.

EDMUNDS, W. M. 1980. The hydrogeochemical characteristics of groundwaters in the Sirte Basin, using strontium and other elements. *In*: *The Geology of Libya*. University of Tripoli, Tripoli, **2**, 703–714.

EDMUNDS, W. M. & BATH, A. H. 1976. Centrifuge extraction and chemical analysis of interstitial waters. *Environmental Science and Technology*, **10**, 467–472.

EDMUNDS, W. M. & MORGAN-JONES, M. 1976. Geochemistry of groundwaters in the British Triassic Sandstones. I. The Wolverhampton-East Shropshire area. *Quarterly Journal of Engineering Geology*, **9**, 73–101.

EDMUNDS, W. M. & WALTON, N. R. G. 1980. A geochemical and isotopic approach to recharge evaluation in semi-arid zones, past and present. *In*: *Arid Zone Hydrology, investigations with isotope techniques*. International Atomic Energy Agency, Vienna, 47–68.

EDMUNDS, W. M. & WALTON, N. R. G. 1983. The Lincolnshire Limestone – Hydrogeological Evolution over a Ten-year Period. *Journal of Hydrology*, **61**, 201–211.

EDMUNDS, W. M. & WRIGHT, E. P. 1979. Groundwater recharge and palaeoclimate in the Sirte and Kufra basins, Libya. *Journal of Hydrology*, **40**, 215–241.

EDMUNDS, W. M., BATH, A. H. & MILES, D. L. 1982. Pore fluid geochemistry of the Bridport Sands and Sherwood Sandstone, Winterbourne Kingston Borehole. *In*: *The Winterbourne Kingston Borehole, Dorset, England*. Institute of Geological Sciences Report **81/3**. HMSO, London, 149–164.

EDMUNDS, W. M., LOVELOCK, P. E. R. & GRAY, D. A. 1973. Interstitial water chemistry and aquifer properties in the Upper and Middle Chalk of Berkshire, England. *Journal of Hydrology*, **19**, 21–31.

EDMUNDS, W. M., OWEN, M. & TATE, T. K. 1976. *Estimation of induced recharge of river water into Chalk boreholes at Taplow, Berkshire, using hydraulic analysis, geophysical logging and geochemical methods*. Institute of Geological Sciences Report **76/5**, HMSO, London.

EDMUNDS, W. M., TAYLOR, B. J. & DOWNING, R. A. 1969. Mineral and thermal waters of the United Kingdom. *Mineral and Thermal Waters of the World. A – Europe*. *In*: International Geological Congress, Prague, **18**, 139–158.

FOSTER, S. S. D. 1974. Groundwater storage – riverflow relations in a Chalk aquifer. *Journal of Hydrology*, **23**, 299–311.

FOSTER, S. S. D. 1975. The Chalk groundwater tritium anomaly – a possible explanation. *Journal of Hydrology*, **25**, 159–163.

FOSTER, S. S. D. 1976. The vulnerability of British groundwater resources to pollution by agricultural leachates. *Ministry of Agriculture, Fisheries and Food, Technical Bulletin*, No. **32**, 68–91.

FOSTER, S. S. D. & CREASE, R. I. 1974. Nitrate pollution of Chalk groundwater in East Yorkshire – a hydrogeological appraisal. *Journal of the Institution of Water Engineers*, **28**, 178–194.

FOSTER, S. S. D. & CREASE, R. I. 1975. Hydraulic behaviour of the Chalk aquifer in the East Yorkshire Wolds. *Proceedings of the Institution of Civil Engineers, II*, **50**, 181–188.

FOSTER, S. S. D. & MILTON, V. A. 1974. The permeability and storage of an unconfined Chalk aquifer. *Hydrological Science Bulletin*, **19**, 485–500.

FOSTER, S. S. D. & MILTON, V. A. 1976. *Hydrological basis for the large-scale development of groundwater storage capacity in the East Yorkshire Chalk*. Institute of Geological Sciences Report **76/3**, HMSO, London.

FOSTER, S. S. D. & SMITH-CARINGTON, A. K. 1980. The interpretation of tritium in the Chalk unsaturated zone. *Journal of Hydrology*, **46**, 343–364.

FOSTER, S. S. D. & YOUNG, C. P. 1980. Groundwater contamination due to agricultural land-use practices in the United Kingdom. *Studies and Reports in Hydrology*. UNESCO, Paris. **36**, 268–282.

FOSTER, S. S. D., PARRY, E. L. & CHILTON, P. J. 1976*a*. *Groundwater resource development and saline water intrusion in the Chalk aquifer of North Humberside*. Institute of Geological Sciences Report **76/4**, HMSO, London.

FOSTER, S. S. D., PARRY, E. L. & CHILTON, P. J. 1976*b*. *Groundwater storage in Fife and Kinross – its potential as a regional resource*. Institute of Geological Sciences Report **76/9**, HMSO, London.

GRAY, D. A. 1965. The stratigraphical significance of electrical resisitivity marker bands in the Cretaceous strata of the Leatherhead (Fetcham Mill) Borehole, Surrey. *Geological Survey of Great Britain Bulletin*, **22**, 65–115.

GRAY, D. A. 1969*a*. Groundwater conditions in the Pennant Sandstone. 139–145. *In*: WOODLAND, A.W. (ed.), Geological Report on the Aberfan Tip Disaster of October 21st, 1966. Technical Reports submitted to the Aberfan Tribunal, HMSO, Welsh Office.

GRAY, D. A. 1969*b*. Appendix 4. *In*: *Thames flood prevention: First Report of Studies*. Greater London Council, 173–197.

GRAY, D. A. 1975. The scope of hydrogeology. *Quarterly Journal of Engineering Geology*, **8**, 175–191.

GRAY, D. A. & FOSTER. S. S. D. 1972. Urban influences upon groundwater conditions in Thames Flood Plain deposits of Central London. *Philosophical Transactions of the Royal Society. A*. **272**, 245–277.

GRAY, D. A., ALLENDER, R. & LOVELOCK, P. E. R. 1969. *The groundwater hydrology of the Yorkshire Ouse River Basin*. Water Supply Papers, Hydrogeological Report, No. **4**, Institute of Geological Sciences, London.

GRAY, D. A., MATHER, J, D. & HARRISON, I. B. 1974. Review of groundwater pollution from waste disposal sites in England and Wales, with provisional guidelines for future site selection. *Quarterly Journal of Engineering Geology*, **7**, 181–196.

GRAY, D. A., GREENWOOD, P. B., BISSON, G., CRATCHLEY, C. R., HARRISON, R. K., MATHER, J. D., OSTLE, D., POOLE, E. G., TAYLOR, B. J. & WILLMORE, P. L. 1976. *Disposal of highly-active, solid radioactive wastes into geological formations – relevant geological criteria for the United Kingdom*. Institute of Geological Sciences Report **76/12**,HMSO, London.

GREENWOOD, P. B., BISSON, G., POOLE, E. G. & TAYLOR, B. J. 1979. *Catalogue of Geological Formations having Favourable Characteristics for the Disposal of Solidified High-level and/or Long-lived Radioactive Wastes. Vol. 9. United Kingdom*. EEC, Brussels.

GUIVER, K. & GRAY, D. A. 1974. Legislation controlling Waste Disposal. *Public Health Engineer*, **No. 8**, 45–50.

HARRISON, I. B. 1976. Construction of investigatory boreholes at landfill sites. *Surveyor*, **147**, 22–24.

HARVEY, B. I. 1973. *Records of wells in the area around Norwich. Well Inventory Series [Metric units]* Sheet 161. Institute of Geological Sciences, HMSO, London.

HARVEY, B. I., LANGSTON, M. J., HUGHES, M. D. A. & CRIPPS, A. C. 1973. *Records of wells in the area around Bury St, Edmunds. Well Inventory Series [Metric units]* Sheet 189. *Institute of Geological Sciences*, HMSO, London.

HOLMES, D. C. & HALL, D. H. 1980. *The 1977–1979 geological and hydrogeological investigations at the Windscale Works, Sellafield, Cumbria*. Institute of Geological Sciences Report **80/12**. HMSO, London.

INSTITUTE OF GEOLOGICAL SCIENCES. 1976. *An investigation into the geological feasibility of storing crude oil in unlined caverns in the Hunterston Peninsula, Ayrshire. Parts 1, 2 & 3*. Institute of Geological Sciences, London.

INESON, J. 1965*a*. The Water Resources of the River Colne Catchment (Hertfordshire). *In:* HAWKESWORTH, J. M. W. *et al.* (ed.) *The North West Home Counties*. Regional Advisory Water Committee, HMSO, London.

INESON, J. 1965*b*. Groundwater principles of network design. *In*: *International Symposium on Hydrometeorological Networks. Quebec*. World Meteorological Office and International Association of Scientific Hydrology.

INESON, J. 1967. *Groundwater conditions in the Coal Measures of South Wales*. Water Supply Papers, Hydrogeological Report No. **3**, Institute of geological Sciences, London.

INESON, J. & DOWNING, R. A. 1965. Some hydrogeological factors in permeable catchment studies. *Journal of the Institution of Water Engineers*, **19**, 59–80.

KITCHING, R. 1974. *Infiltration studies on the Lower Greensand outcrop near Liss, Hampshire*. Institute of Geological Sciences Report, **74/10**, HMSO, London.

KITCHING, R. 1975. *A mathematical model of the Akrotiri Plio-Pleistocene gravel aquifer, Cyprus*. Institute of Geological Sciences Report **75/2**, HMSO, London.

KITCHING, R. & BRIDGE, L. R. 1974. Lysimeter installations in sandstone at Styrrup, Nottinghamshire. *Journal of Hydrology*, **23**, 219–232.

KITCHING, R. & DAY, J. B. W. (eds) 1980. *Two-day meeting on lysimeters*. Institute of Geological Sciences Report **79/6**, HMSO, London.

KITCHING, R., RUSHTON, K. R. & WILKINSON, W. B. 1975. Groundwater yield estimation from models. *In*: MONRO, M. (ed.) *Engineering Hydrology Today*. The Institution of Civil Engineers, London.

KITCHING, R. SHEARER, T. R. & SHEDLOCK, S. L. 1977. Recharge to Bunter Sandstone determined by lysimeters. *Journal of Hydrology*, **33**, 217–232.

LAND, D. H. 1966. *Hydrogeology of the Triassic Sandstones of the Birmingham – Lichfield district*. Water Supply Papers, Hydrogeological Report No. **2**, Institute of Geological Sciences. London.

LITTLE, B. G., BUCKLEY, D. K., CANT, R., HENRY, P. W. T., JEFFERIS, A., MATHER, J. D., STARK, J. & YOUNG, R. N. 1977. *Land resources of the Bahamas: a Summary*. Land Resources Study **27**, Land Resources Division, Ministry of Overseas Development, Surrey, 133 p.

LOVELOCK, P. E. R. 1967. *Groundwater levels in England and Wales during 1963*. Water Supply Papers, Research Report No. **3**, Institute of Geological Sciences, London.

LOVELOCK, P. E. R. 1970. *Laboratory measurment of soil and rock permeability*. Water Supply Papers, Technical Communication No. **2**, Institute of Geological Sciences, London.

LOVELOCK, P. E. R. 1977. Aquifer properties of Permo-

Triassic sandstones in the United Kingdom. *Bulletin. Geological Survey of Great Britain*, No. **56**, 1–49.

LOVELOCK, P. E. R. PRICE, M. & TATE, T. K. 1975. Groundwater conditions in the Penrith Sandstone at Cliburn, Westmorland. *Journal of the Institution of Water Engineers and Scientists*, **29**, 157–174.

MATHER, J. D. 1970. *The use of radioisotopes in groundwater studies*. Water Supply Papers, Technical Communication No. **1**, Institute of Geological Sciences, London.

MATHER, J. D. 1972. The geology of Grand Cayman and its control over the development of lenses of potable groundwater.. *In*: *Transactions, Sixth Caribbean Geological Conference, Isla de Margarita, Venezuela, 1971*. 154–157.

MATHER, J. D. 1975. Development of the groundwater resources of small limestone islands. Quarterly *Journal of Engineering Geology*, **8**, 141–150.

MATHER, J. D. 1977. *Attenuation and control of landfill leachates*. 79th Annual Conference of the Institute of Solid Waste Management, Torbay, 14p.

MATHER, J. D. 1978. Saline intrusion and groundwater development on a Pacific atoll. *In*: *Proceedings 5th Salt water Intrusion meeting, Medmenham, England, 4–6 April 1977*, 126–135.

MATHER, J.D. & BROMLEY, J. 1977. Research into leachate generation and attenuation at landfill sites. *In*: *Papers of the Land Reclamation Conference. 1976*. Thurrock Borough Council, Grays, Essex, 377–400.

MATHER, J. D. & SMITH, D. B. 1973. Thermonuclear tritium – its use as a tracer in local hydrogeological investigations. *Journal of the Institution of Water Engineers*, **27**, 187–196.

MATHER, J. D., GRAY, D. A., ALLAN, R. A. & SMITH, D. B. 1973. Groundwater recharge in the Lower Greensand of the London Basin – results of tritium and carbon-14 determinations. *Quarterly Journal of Engineering Geology*, **6**, 141–152.

MATHER, J. D., GRAY, D. A. & HOUSTON, J. F. T. 1971. *Distribution of the flood plain deposits of the Thames between Chiswick and Beckton*. Water Supply Papers, Research Report, No. **4**, Institute of Geological Sciences, London.

MATHER, J. D., GRAY, D. A. & JENKINS, D. G. 1969. The use of tracers to investigate the relationship between mining subsidence and groundwater occurrence at Aberfan, South Wales. *Journal of Hydrology*, **9**, 136–154.

MOSELEY, R., WOODWARD, C. M., DAY, J. B. W. & LANGSTON, M. J. 1976. *Hydrogeological Map of Northern East Anglia*. Institute of Geological Sciences, London.

PRICE, M. 1977. Specific yield determinations from a consolidated sandstone aquifer. *Journal of Hydrology*, **33**, 147–156.

PRICE, M. & FOSTER, S. S. D. 1974. Water supplies from Ulster valley gravels. *Proceedings of the Institution of Civil Engineers, Part 2*, **57**, 451–466.

PRICE, M., BIRD, M. J. & FOSTER, S. S. D. 1976. Chalk pore-size measurements and their significance. *Water Services*, **80**, 596–600.

TATE, T. K. & ROBERTSON, A. S. 1971. *Investigations into high salinity groundwater at the Woodfield Pumping Station, Wellington, Shropshire*. Water Supply Papers, Research Report No. **6**. Institute of Geological Sciences, London.

TATE, T. K. & GRAY, D. A. 1971. *Borehole logging investigations in the Chalk of the Lambourn and Winterbourne valleys of Berkshire*. Water Supply Papers, Research Report No. **5**, Institute of Geological Sciences, London.

THEIS, C. V. 1935. The relation between the lowering of the piezometric surface and the rate and duration of discharge of a well using groundwater storage. *Transactions of the American Geophysical Union*, **16**, 519–524.

UNESCO. 1970. *International Legend for Hydrogeological Maps*. UNESCO, Paris: International Association of Scientific Hydrology, Gentbrugge, Belgium: International Association of Hydrogeologists, Paris: Institute of Geological Sciences, London. [English, French, Spanish, Russian].

WALTON, N.R. 1982. *A detailed hydrogeochemical study of the groundwaters from the Triassic sandstones of south-west England*. Institute of Geological Sciences Report **81/5**, HMSO, London.

WILLIAMS, G. M., STUART, A. & HOLMES, D. C. 1985. *Investigations of the geology of the low-level radioactive waste burial site at Drigg, Cumbria*. British Geological Survey Report, **17**, No. 3, HMSO, London.

WILLIAMSON, K. H. 1979. A model for the Sulphur Springs Geothermal Field, St. Lucia. *Geothermics*, **8**, 75–83.

WILKINSON, B. R. & GRAY, D. A. 1981. The organization and objectives of hydrogeological research in the United Kingdom. *In*: *A Survey of British Hydrogeology*. Royal Society, London, 7–22.

WOODLAND, A. 1978. Report of the Director, Annual Report for 1977. Institute of Geological Sciences, London

WRIGHT, E.P. & EDMUNDS, W.M. 1971. Hydrogeological studies in central Cyrenaica, Libya. *In*: *Proceedings. Symposium on the Geology of Libya*. University of Tripoli, Tripoli, 459–481.

WRIGHT, E. P., BENFIELD, A. C., EDMUNDS, W. M. & KITCHING, R. 1982. Hydrogeology of the Kufra and Sirte basins, eastern Libya. *Quarterly Journal of Engineering Geology*, **15**, 83–103.

WATER RESOURCES BOARD. 1966. *Annual Report for 1966, Appendix D*. HMSO, London.

Norman Savage Boulton (1899–1984): civil engineer and groundwater hydrologist

R. A. DOWNING[1], W. EASTWOOD[2] & K. R. RUSHTON[3]

[1]*20, Springfield Park, Twyford, Reading, RG10 9JH, UK*
[2]*45, Whirlow Park Road, Sheffield, S11 9NN, UK*
[3]*106, Moreton Road, Buckingham, MK18 1PW, UK*

Abstract: Norman S. Boulton (Fig. 1) was a civil engineer who achieved international recognition for his work on groundwater hydraulics. He recognized that in unconfined aquifers water is released from storage by drainage under gravity from the pore-spaces in the cone of depression as it expands. This 'delayed yield' gave a characteristic S-shape to the log-log, time-drawdown graph of water levels in an observation well near a pumping well. Boulton developed a mathematical solution that reproduced the three segments of the curve. Most of his career was spent in academia mainly at the University of Sheffield where he was Professor of Civil Engineering between 1955 and 1964. His work embraced studies of structural engineering and soil mechanics as well as groundwater flow.

Norman S. Boulton (Fig. 1) will be remembered by the groundwater fraternity, both in the UK and internationally, for significant contributions in the 1950s–1970s to the theory of groundwater hydraulics and its practical applications. But this was merely one aspect of his professional work for he also made fundamental contributions in the fields of structural engineering and soil mechanics from the early 1930s.

Norman Boulton was born on 8th May 1899 at Penarth in South Wales, the son of William S. Boulton who became Professor of Geology in the University of Birmingham. He was educated at Penarth County School and King Edward's School, Birmingham between 1909 and 1918. In the later stages of World War I, he was commissioned in the Royal Garrison Artillery.

After the war Boulton entered the University of Birmingham to read for a degree in civil engineering, graduating in 1922 as a B.Sc. with first class honours and gaining a Bowen Research Scholarship on the basis of the results of the examination. He followed this with post-graduate research on hydraulics, receiving an M.Sc. in 1923. He was awarded the Dudley Docker Research Scholarship and continued research in hydraulics for a further year.

He cut his professional teeth in 1924 with the Public Works Department of the City of Birmingham where he gained valuable wide-ranging, practical experience of the drainage and sewerage problems that face a large local authority. However, his inclination was towards academic life and in 1929 he was appointed Lecturer in Civil Engineering at Armstrong College, Newcastle, later King's College, then part of the University of Durham.

His long association with the University of Sheffield began in 1936 when he was appointed Senior Lecturer in charge of the Department of Civil Engineering and in 1955 he assumed the newly-created Chair of Civil Engineering, retiring in 1964. During his time at Sheffield Boulton pioneered the teaching of soil mechanics in the UK as a specific subject. His Department was one of the first in the country to do so and under his guidance it soon became one of the largest.

Fig. 1. Professor Norman S. Boulton

From: MATHER, J. D. (ed.) 2004. *200 Years of British Hydrogeology*. Geological Society, London, Special Publications, **225**. 319–322. 0305-8719/04/$15 © The Geological Society of London.

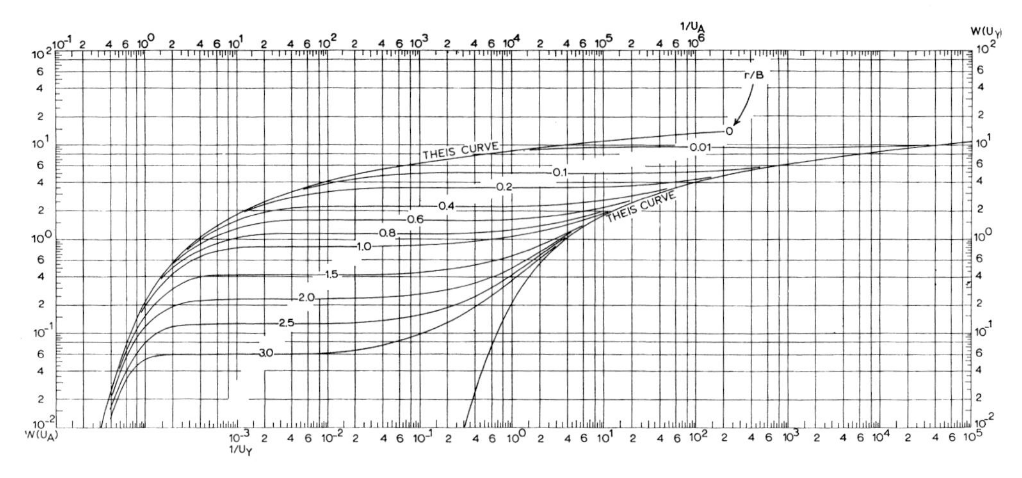

Fig. 2. Family of Boulton-type curves for analysing pumping test data in unconfined aquifers.

Boulton applied his outstanding mathematical and intellectual abilities to a wide range of physical problems. He did important work in the fields of soil mechanics and structural engineering as well as groundwater hydrology. His earliest publications dealt with solutions to the constitutive equations governing thermal and plastic stresses in structural members, a subject to which he continued to contribute into the 1960s. Boulton's interest in groundwater hydrology was also evident early in his career. This possibly developed because of his previous work on hydrology as well as the role of sub-surface water in soil mechanics. His father's interest in hydrogeology was no doubt an influence. His involvement in the subject increased as a consultant dealing with practical problems in the groundwater field, possibly in some cases with his friend W. G. Fearnsides, who was Professor of Geology at Sheffield.

Boulton's fundamental work on the theory of groundwater flow dates from the early 1930s when he examined the time-variant flow to a pumped well in a confined aquifer and the effect of the pumping on the piezometric surface. His solution to this problem of non-steady flow in confined aquifers anticipated by several years the solution published by Theis (1935). Many ideas in science ahead of their time make no impact and this was the fate of Boulton's manuscript for when he submitted it for publication it was not accepted, a great disappointment to him personally and very unfortunate for the status of hydrogeology in Britain. One of the fundamental equations of well hydraulics could have been named after him instead of Theis. In the 1930s to early 1950s Boulton was unfortunately ploughing a lone furrow in the UK with respect to his research on well hydraulics.

His most significant contributions to groundwater hydrology began with the formulation and solution of transient flow in unconfined aquifers. He recognized that unconfined aquifers respond to pumping from a well in a different manner from confined aquifers leading to his concept that the water table has a delayed response to pumping which he referred to as 'delayed yield'. The germ of this concept came from his background in soil mechanics, observing water slowly draining from soil samples in laboratory experiments. In unconfined aquifers water levels in wells near the pumped well decline at a slower rate than that described by the Theis equation. Log-log, time-drawdown curves consequently show a typical S-shape with three recognizable distinct segments. During the first segment, which covers only a very short period after the start of pumping, an unconfined aquifer reacts in the same manner as a confined aquifer with water released instantaneously from storage by the compaction of the aquifer and by the expansion of the water. However, the effects of gravity drainage are reflected in the second segment with vertical flow components recognized in the vicinity of the pumped well. The slope of the time-drawdown curve decreases relative to the Theis curve because water drains into the well from pore spaces above the cone of depression as a result of the dewatering of the aquifer as the water table falls. In the third segment the time-drawdown curve once again tends to conform to a Theis-type curve but with the storage now reflecting the specific yield and the storativity no longer increasing with time (Fig. 2).

Boulton developed a semi-empirical mathematical solution that reproduced the three segments of the curve, work that necessitated a vast amount of time-consuming calculations with a mechanical, hand calculator. He introduced a more rigorous basis to the analysis of data from pumping tests on unconfined aquifers.

His mathematical description of the drainage process turned out to be identical to that introduced by Barenblatt and his co-workers (1960) in relation to pumping tests in fractured porous rocks. These models represent the flow between the porous rock matrix and the fractures as a pseudo-steady-state interporosity flux that is proportional to head difference: this gives the same exponentially declining behaviour as envisaged by Boulton. Consequently, the pseudo-steady-state type curves widely used for analysing pumping tests in fissured aquifers, such as those published by Bourdet & Gringarten (1980), are essentially identical to Boulton's original time-drawdown curves for an unconfined aquifer displaying delayed yield. An improved approach to modelling both interporosity flow in fractured rock and inter-layer flow in multi-layered aquifers, sometimes referred to as the transient interporosity model, can in most practical situations be approximated by Boulton's model, at least in relation to the late-time data.

Alternative approaches to treating unconfined aquifers have been developed, particularly by S.P. Neuman (1972; 1975 and 1979) and Boulton's colleague, T.D. Streltsova (1972*a*, 1972*b*; 1973). These methods now tend to be used more than Boulton's, partly as a result of software availability, but a common mistake is to regard these methods as superseding Boulton's model rather than alternatives to it. While his model represents the delayed drainage explicitly, other approaches produce delayed yield only as an artefact, resulting from a finite diffusive response time over some distance to the water table.

Boulton's contributions to groundwater hydrology extended well beyond his formal retirement from the University of Sheffield. He developed his ideas in cooperation with other workers but particularly with T. D. Streltsova. She believed that the sigmoid shape of the time-drawdown curve in unconfined aquifers was caused by vertical flow components in the aquifer whereas Boulton was convinced of the physical reality of delayed drainage. Eventually they agreed that both phenomena occurred and following this reconciliation they cooperated with further analytical solutions with Boulton carrying out the mathematical analysis and Streltsova identifying the problems. In 1976 they considered the effect of pumping an unconfined, anisotropic aquifer from a partially penetrating, large-diameter well and developed a well-function that described the first segment of the S-curve. They turned their attention in 1977 to fissured aquifers and considered transient flow from matrix blocks into fractures and also transient flow to a pumped well in a two-layered, water-bearing formation.

Boulton and Streltsova received the Horton Award in 1975 from the American Geophysical Union for the best paper published in Water Resources Research in that year. This paper presented new equations for calculating formation constants that took into consideration a wide range of factors that affected pumping test data including the compressibility and anisotropy of the aquifer, partial penetration of the pumping well depth at which drawdown is measured in an observation well and the existence of a low-permeability layer above the water table. Both confined and unconfined conditions were considered.

Boulton was always careful to explain the range of validity of his solutions and the limitations of his work. The danger of trying to force groundwater to mathematics was recognized. His papers were very intensive and compact with few superfluous words. Careful concentrated effort was necessary to understand the message in its entirety

The 1950s–1970s was a fertile time for studies of transient groundwater flow following the paradigm shift introduced by C.V. Theis, supplemented subsequently by C.E. Jacob. Norman Boulton metaphorically rubbed shoulders with the handful of men, including M. S. Hantush, H.H. Cooper, J.G. Ferris, I.S. Papadopulos and L.K.Wenzel, who widened the scope and application of well hydraulic theory to meet specific geological and engineering conditions. Much of Boulton's work on groundwater hydraulics also had application in petroleum engineering. As well as his papers, Boulton contributed to the discussion of many papers presented at the Institution of Civil Engineers over the years, discussions ranging from structural engineering and the durability of materials to specialized aspects of hydrology and groundwater hydrology, reflecting a wide breadth of interest and knowledge. He was a Fellow of the Institution and very active in the local affairs of the organisation; on two occasions he was Chairman of the Yorkshire Association.

Norman Boulton was a tall, dignified man of unfailing kindness and courtesy who carried himself well and looked you straight in the eye. He was always impeccably dressed, managing to maintain this throughout World War II despite the restrictions of clothes rationing that existed then. He was a patient man who spoke carefully and deliberately, with no liking for small-talk. He always maintained the highest standards in his work. He was extremely shy and modest, reticent in fact, and often this made him reluctant to express his views. His shyness and desire to avoid the limelight would extend to passing on consulting work that he had started to colleagues rather than become involved in personal relationships. He certainly did not believe in self-promotion. Music was one of his great pleasures, particularly the music of Elgar and its associations with Worcestershire and the Malvern Hills. He was an accomplished violinist, playing as a young man in the City of Birmingham Orchestra. His love of music reflected a sensitivity which he did not readily display. He was also a keen swimmer, taking a daily swim in the university's pool until just before his death.

When he retired, in 1964, Boulton was honoured by the University of Sheffield with the title Emeritus Professor of Civil and Structural Engineering and in 1966 the University of Birmingham awarded him a D.Sc. (in civil engineering) for his original contributions to engineering science, which rightly had brought him international acclaim. Norman Boulton's interest in well hydraulics, the specialized aspect of the subject he first studied as a post-graduate research student, continued until shortly before his death on 10th August 1984 at the age of 85.

Principal publications relating to groundwater hydrology.

1942. The steady flow of groundwater to a pumped well in the vicinity of a river. *Philosophical Magazine, Series* 7, **33**, 34–50.
1951. The flow pattern near a gravity well in a uniform water-bearing medium. *Journal of the Institution of Civil Engineers*, **36**, 534–550.
1954. The drawdown of the water table under non-steady conditions near a pumped well in an unconfined formation. *Proceedings of the Institution Civil Engineers*, **3**, 564–579.
1955. Unsteady radial flow to a pumped well allowing for delayed yield from storage. *In: Proceedings General Assembly, Rome, Publication* 37, *International Association of Scientific Hydrology.* 472–477.
1958. (with G.S. Dhillon). A field method for measuring the permeability of sandstone cores. *In: Proceedings General Assembly, Toronto, Volume II, International Association of Scientific Hydrology.* 183–192.
1963. Analysis of data for non-equilibrium pumping tests allowing for delayed yield from storage. *Proceedings of the Institution of Civil Engineers*, **26**, 469–482, and 1964. Discussion. *Proceedings of the Institution of Civil Engineers*, **28**, 603–610.
1965. The discharge of a well in an extensive unconfined aquifer with a constant pumping level. *Journal of Hydrology*, **3**, 124–130.
1970. Analysis of data from pumping tests in unconfined anisotropic aquifers. *Journal of Hydrology*, **10**, 369–378.
1971. (with J.M.A. Pontin). An extended theory of delayed yield from storage applied to pumping tests in unconfined anisotropic aquifers. *Journal of Hydrology*, **14**, 53–65.
1973. The influence of delayed drainage on data from pumping tests in unconfined aquifers. *Journal of Hydrology*, **19**, 157–169.
1975. (with T. D. Streltsova). New equations for determining the formation constants of an aquifer from pumping test data. *Water Resources Research*, **11**, 148–153.
1976. (with T.D. Streltsova). The drawdown near an abstraction well of large diameter under non-steady conditions in an unconfined aquifer. *Journal of Hydrology*, **30**, 29–46.
1977. (with T.D. Streltsova). Unsteady flow to a pumped well in a fissured water-bearing formation. *Journal of Hydrology*, **35**, 257–269.
1977. (with T.D. Streltsova). Unsteady flow to a pumped well in a two-layered water-bearing formation. *Journal of Hydrology*, **35**, 245–256.
1978. (with T.D. Streltsova). Unsteady flow to a pumped well in a fissured aquifer with a free surface level maintained constant. *Water Resources Research*, **14**, 527–532.
1978. (with T.D. Streltsova). Unsteady flow to a pumped well in an unconfined fissured aquifer. *Journal of Hydrology*, **37**, 349–363.

This paper has been compiled with the assistance of the University of Sheffield and the Institution of Civil Engineers. The authors acknowledge constructive comments and additions to the text by Professor J.A. Barker.

References

Barenblatt, G. E., Zheltov, I. P. and Kochina, I. N. 1960. Basic concepts in the theory of seepage of homogeneous liquids in fissured rocks. *Journal of Applied Mathematics and Mechanics*, **24**, 1286–1303.
Bourdet, D. & Gringarten, A. C. 1980. *Determination of fissure volume and block size in fractured reservoirs by type-curve analysis*. Paper 9293, Society of Petroleum Engineers at 1980 Annual Fall Technical Conference and Exhibition, Dallas.
Neuman, S. P. 1972. Theory of flow in unconfined aquifers considering delayed response of the water table. *Water Resources Research*, **8**, 1031–1045.
Neuman, S. P. 1975. Analysis of pumping test data from anisotropic, unconfined aquifers considering delayed gravity response. *Water Resources Research*, **11**, 329–342.
Neuman, S. P. 1979. Perspective on "Delayed yield". *Water Resources Research,* **15**, 899–908.
Streltsova, T. D. 1972*a*. Unconfined aquifer and slow drainage. *Journal of Hydrology*, **16**, 117–124.
Streltsova, T. D. 1972*b*. Unsteady radial flow in an unconfined aquifer. *Water Resources Research*, **8**, 1059–1066.
Streltsova, T. D. 1973. On the leakage assumption applied to equations of groundwater flow. *Journal of Hydrology*, **20**, 237–253.
Theis, C. V. 1935. The relation between the lowering of the piezometric surface and the rate and duration of discharge of a well using groundwater storage. *Transactions of the American Geophysical Union*, **16**, 519–524.

Groundwater in a national water strategy, 1964–1979

R. A. DOWNING

20, Springfield Park, Twyford, Reading RG10 9JH, UK

Abstract: The Water Resources Board was formed in 1964, an outcome of the Water Resources Act of 1963. Its remit was to advise the Government and the new river authorities on 'the proper use of water resources in England and Wales'. It made three major regional studies of water resources and, in 1973, advocated a national water strategy. The Water Resources Board was disbanded in 1974 following the reorganization of the water industry under the Water Act of 1973 which created the regional water authorities. In the 1970s, a decline in the rate of population growth together with an economic recession reduced the demand for water and the Board's proposals were not fully implemented. After 1974 the Central Water Planning Unit continued the Water Resources Board's role until it too was disbanded in 1979.

The Water Resources Act of 1963 made consideration of the role of groundwater in a national strategy for water possible for the first time. This became part of the main objective of the Water Resources Board (WRB) which was formed on 1st July 1964, a product of the Act. Simultaneously, 29 river authorities were created (Fig. 1) and became responsible, *inter alia,* for the conservation and use of water resources in England and Wales. The remit of the WRB was to advise the Government and the river authorities on the proper use of water resources, including measures for augmenting and redistributing resources by transferring water as necessary, from one area to another. It became possible to integrate the development of both surface water and groundwater within a river basin. This reorganization partly stemmed from the drought of 1959, which was a signal event, an irrefutable occurrence of nature that altered attitudes. Following the drought, the Central Advisory Water Committee (CAWC) recommended to the Government, in 1962, that a national body should be formed to pursue a national policy for water and coordinate the work on water resources in the various parts of England and Wales.

The WRB was a rather unique organization for a government agency. It operated more or less in the manner of a private company. The eight members of the board met monthly under the chairmanship of Sir William Goode, virtually a full-time, hands-on, no-nonsense chairman, a barrister who completed a career in the Colonial Civil Service as Governor of North Borneo. Each member of the board had responsibility for a specific aspect of the organization's programme. Although coming under the Ministry of Housing and Local Government (MHLG), the WRB acted virtually as an independent body, issuing reports and holding press conferences as the need arose. This attitude stemmed from both the composition of the board and the fact that many of the senior staff were recruited either from firms of consulting engineers or universities and were not steeped in the Civil Service ethos.

The Director was Norman Rowntree, a consulting civil engineer who had worked in the water industry throughout his career. The Secretary was N. H. Calvert who was seconded from the MHLG where he had been largely responsible for drafting the Water Resources Act. A man with a clear incisive mind and an education in geography, Calvert was totally at ease with the purpose and objectives of the Act, which made him the ideal person to guide the administrative and legal aspects of the complex series of work programmes that the WRB initiated to meet its objectives. Other key figures were B. Rydz, Head of Planning, V. K. Collinge, Head of Research, Dr J. Ineson, Chief Geologist and A. G. Boulton, responsible for surface water.

Groundwater was part of Ineson's responsibilities. He had joined the WRB from the Water Department of the Geological Survey, together with Mrs M. C. Davies, A. G. P. Debney, R. A. Downing, M. Hoyle, H. J. Richards and L. M. J. Standon-Batt and who together formed the initial nucleus of the Geology Division. Over the next few years the Division's strength increased, notable additions being A. B. Birtles, N. R. Brereton, K. J. Edworthy M. Fleet, E. M. Gray, J. B. Joseph, N. K. Lambert, M. Lees, R. A. Monkhouse, J. A. Naylor, D. A. Nutbrown, D. B. Oakes, E. Parry, J. M. A. Pontin, M. J. Reeves, R. L. H. Satchell, A. C. Skinner, C. D. N. Tubb, W. B. Wilkinson, B. P. J. Williams, C. E. Wright and C. P. Young. The Senior Engineers at the WRB included L. E. Taylor who had been responsible for groundwater supplies at the Metropolitan Water Board and had practical experience of artificial recharge (Taylor 1964).

In the 1960s there was widespread opposition to the construction of surface reservoirs, the opposition being led by both local and national pressure groups. They achieved considerable success with the rejection by the government of several inland storage

From: MATHER, J. D. (ed.) 2004. *200 Years of British Hydrogeology*. Geological Society, London, Special Publications, **225**. 323–338. 0305-8719/04/$15 © The Geological Society of London.

schemes and the delay of decisions on others. This climate of opinion proved to be favourable for those advocating greater use of groundwater, as society became increasingly sensitive to the loss of good quality agricultural land in particular, and alleged adverse effects on the environment in general.

The early years (1964–1971)

One of the first tasks of the WRB, in cooperation with the new river authorities, the statutory water undertakings and the water companies, was to assess the future demand for water and the resources to meet it. Three regional studies were made beginning with South East England, then Northern England and finally Wales and the Midlands (Fig. 1).

The first report, 'Water supplies in South East England', was published in 1966. Its conclusions, in so far as they related to groundwater, emphasized the importance of assessing the feasibility of augmenting the dry weather flows of the rivers Thames and Great Ouse by pumping groundwater into them, as well as studying the practicality of artificial recharge of aquifers, and developing the groundwater resources 'in and around Peterborough', that is the resources of the Lincolnshire Limestone.

Interest in augmenting river flows with groundwater had been generated in the early 1960s as a result of the hydrological surveys made by the MHLG. These surveys, to which the Geological Survey contributed with respect to groundwater, led to the conclusion that the uneven annual flow of rivers, and indeed groundwater discharge, could be reduced, with benefits to both dry weather river flows and water supplies relying on river intakes, by pumping groundwater into the rivers during the summer and the autumn from wells remote from the rivers (Ineson & Downing 1964). At that time the Thames Conservancy was considering solving London's water supply problems by pumping groundwater from the Chalk of the Berkshire Downs into tributaries of the Thames, an idea proposed initially by Guthrie Allsebrook, a well-driller of Reading, in a letter to the Ministry of Health as early as 1922. He repeated his advice to the Ministry and to the Thames Conservancy in 1934 and yet again in 1949 suggesting his idea as an alternative to the proposed Enborne Valley Reservoir Scheme in west Berkshire (Hardcastle 1978, pers. comm. 2004).

The Hydrological Survey of the Great Ouse Basin (Anon. 1960) revealed appreciable reserves of groundwater in the Chalk and in view of the anticipated large increase in the population of SE England (Anon. 1964) the MHLG commissioned Binnie and Partners (1965) 'to study and report upon the exploitable reserves of water in the Great Ouse Basin with the view to the preparation of a comprehensive programme of coordinated development of the groundwater and surface water reserves'. As a consequence exploitation of groundwater in the Chalk of the Thames and Great Ouse basins for river regulation became key components of the WRB's plan for SE England.

At the time knowledge of the principles of groundwater flow and the place of groundwater in the hydrological cycle was rather limited, even among water engineers. The effect of groundwater abstraction on river flows was largely a matter of opinion and generally not based on rigorous observational evidence. A view commonly held was that 'deep groundwater' could be exploited without affecting spring discharges which were deemed to have a shallow origin, a view not shared by the WRB. Ineson was largely responsible for insisting that pilot schemes were necessary both to test the feasibility of the proposals and provide reassurance for the general public likely to be affected. This led to the Lambourn and Great Ouse pilot schemes which investigated, *inter alia*, the effect on river flows of pumping from wells adjacent to and more remote from the perennial rivers. At a time when groundwater modelling was just becoming possible, pilot schemes were essential to demonstrate unequivocally the effect of abstracting groundwater on river flows and provide sufficient evidence to convince an inspector at a public inquiry.

The Thames Conservancy, with their consulting engineers Herbert Lapworth and Partners (later Rofe, Kennard and Lapworth), was responsible for the Lambourn Scheme while the Great Ouse Scheme, in the Thet Valley, was a combined study by the Great Ouse River Authority, Binnie and Partners and the WRB. Geological conditions in the two catchments are very different. The Lambourn Valley is in a typical chalk downland setting with the Chalk cropping out in the valley floor, while in the Great Ouse area the valley floors are generally covered by alluvium and boulder clay which tend to seal the rivers from the underlying Chalk. The two pilot schemes were major groundwater studies extending over several years and were unique for that time.

Groundwater levels in the Chalk and overlying Tertiary sands in the London Basin had been steadily declining with increasing abstraction since the early 19th Century, some 150 years. With respect to the WRB's overall terms of reference, such a large volume of dewatered aquifer, 150 Mm^3 (Water Resources Board 1972), represented a very significant potential storage volume and the feasibility of recharging the aquifers had clearly to be investigated. At the height of the 'Cold War' another important reason for an investigation was that groundwater in the confined Chalk aquifer below London would be a vital source of water in the event of a nuclear attack.

The Metropolitan Water Board had been recharg-

Fig. 1. River authorities as defined by the Water Resources Act 1963. The areas of the Water Resources Board's three regional studies are also shown.

ing the Chalk in the Lee Valley since 1953 (Boniface 1959) but although the annual decline in the groundwater level had been reduced, only 40% of the water recharged was recovered as it flowed down the hydraulic gradient away from the recharge wells, a discouraging result for the Metropolitan Water Board. However, under the Water Resources Act, Thames Conservancy and the Lee Conservancy Catchment Board became responsible for the confined aquifer and effective management of the entire basin was a practical proposition. Consequently, the WRB began a study to identify areas in the basin suitable for artificial recharge. The relationship between groundwater levels and abstraction over some 150 years was examined to provide basic data for use in an electrical analogue model of the basin. Four areas were proposed as suitable for recharge, the most favourable being the Lower Lee Valley where the Metropolitan Water Board was already practising the technique (Water Resources Board 1972 and 1972–1974).

The Lincolnshire Limestone was (and still is) a major source of water in Lincolnshire. An assessment by the WRB of the water balance of the aquifer revealed that licences issued for groundwater abstraction in south Lincolnshire exceeded the estimated infiltration and, bearing in mind the need to maintain the flow of rivers dependent on spring flow from the limestone, the aquifer was potentially overdeveloped (Downing & Williams 1969). The report proposed that the aquifer could be used in conjunction with the rivers Welland and Nene and that consideration should be given to recharging the aquifer artificially from the two rivers. But these were proposals too far. The WRB was at the time intent on promoting the construction of a reservoir at Empingham (subsequently completed as Rutland Water) and was facing widespread opposition to all reservoirs. An alternative proposal to the reservoir was far from welcome. After internal debate the WRB determined to support the construction of Empingham. This was a wise decision, as a conjunctive use scheme involving groundwater, with all the uncertainties surrounding success at that time, was a little too premature. The requirement was for a guaranteed source of water for the region which the reservoir could reliably supply.

Essex was a region where a serious water deficit was anticipated and the situation in the Stour Valley was typical of the county. By the 1960s the water resources of the valley were already extensively exploited and a comprehensive study of the hydrology was initiated in 1966 by the WRB, in conjunction with the Essex River Authority, the Institute of Geological Sciences (now the British Geological Survey) and the Meteorological Office. The study, completed in 1969, investigated the complex groundwater system in the Chalk and overlying Tertiary, Pleistocene and Recent deposits and led to proposals for the future combined development of groundwater and surface water. Although an interim report was issued in 1969 pressure of other commitments meant that it was only in 1977 that a final report was made available by the Central Water Planning Unit (Central Water Planning Unit 1977*a*).

In 1966 the WRB and the river authorities began to investigate methods of providing additional water resources in the north of England, an area not so well endowed with respect to aquifers of regional significance, and in fact virtually limited to the sandstones of the Permo-Triassic west and east of the Pennines. The terrain favoured surface reservoirs and initial groundwater studies were restricted to the Magnesian Limestone in the Tees Valley, the Chalk of east Yorkshire, and the Permo-Triassic sandstones of west Cumberland, the Clwyd Valley and the Fylde, investigations intended to provide for local needs rather than regional requirements (Water Resources Board 1970*a*). Historically, the most interesting was the work in the Fylde, an extension of the pioneering developments of Frank Law who, well ahead of his time, had conceived a scheme, that later developed into the Lancashire Conjunctive Use Scheme, for the integrated use of surface reservoirs, river intakes and groundwater (Law 1965). Advantage was taken of the juxtaposition of upland reservoirs in the Bowland Forest of Lancashire and the Triassic sandstones in the lowlands of the Fylde to the west. The objective was to take water from the source best able to satisfy demand at any give time, using the reservoirs as a base load, supplementing this with river intakes as necessary and, finally, in dry periods, introducing groundwater (Oakes & Skinner 1975). Of the total yield of 260000 m^3/d, groundwater could supply almost 190000 m^3/d in dry periods.

The third regional study, Wales and the Midlands, began in 1968. It necessarily focused on the resources of the Severn and Trent basins and the principal groundwater component was the Permo-Triassic sandstones. The River Trent was a potential resource of major significance but its use was inhibited as it was severely polluted (Water Resources Board 1971). To tackle this problem the Trent Research Programme was conceived to examine the validity of the wide range of possible solutions available; the estimated cost (in 1968) was £0.5 million. The main hydrogeological component was the possibility of artificially recharging the Permo-Triassic sandstones of Nottinghamshire from the River Trent and to see to what extent infiltration through the unsaturated zone would improve the quality of the contaminated water in the river. This task was allocated to the WRB's staff, but, concurrently, the Trent River Authority reviewed the potential of the river gravels adjacent to the Trent as a source of groundwater and also if replenished by artificial recharge. The gravels soon proved to be an inadequate aquifer and attention focused on the sandstones. After preliminary studies a recharge basin of 900 m^2 was constructed in 1970 at Edwinstone, near Mansfield, about 1½ km from the River Maun, a tributary of the Trent with a similar undesirable quality to that of the main river below Nottingham. After settlement and aeration, water from the Maun was recharged into the sandstones over a two-year period and changes in the quality were monitored during flow through the sandstones. About 500 m^3/d were recharged, initially without, and ultimately with, a filter on the floor of the basin.

A parallel study began in 1970 in Clipstone Forest, near Mansfield, to assess the practicality of recharging the sandstone through wells. Fully-treated mains water was used to recharge a well 900 mm in diameter and 60 m deep.

An expanding programme (1971–1974)

In February 1971 the WRB created a National Planning Division under R. G. Sharp with responsibility for defining the future water strategy for England and Wales, a strategy that was ultimately published in a report generally referred to as 'The National Plan' (Water Resources Board 1973). Simultaneously the groundwater, surface water and research programmes were amalgamated in a new Resources Division under H. J. Richards, for Jack Ineson had died suddenly in 1970 at the height of his career. His death was a severe blow to the WRB at a crucial time when the coordination of much hydrogeological work that he had initiated was beginning to produce results.

As the staff of the river authorities with hydrogeological experience increased by the early 1970s, the WRB's groundwater staff increasingly assumed more of a consultancy role providing additional technical contributions to projects, such as geophysical logging services, designing and operating numerical models, and in specialist roles as for example organizing aerial thermal infrared imagery (Davies 1973) and studies of isotopes in groundwater using the skills of the Atomic Energy Authority (AERE) at Harwell. Nevertheless, the WRB's geological staff remained actively involved in many schemes, in particular those concerned with artificial recharge and, of course, the regional programmes required to identify a national water strategy. The groundwater resources of the river gravels in the Middle Thames Valley were evaluated to assess their potential use for water supply (Naylor 1974) and the wider issues of using groundwater in the Great Ouse Basin were examined by Wright (1974).

By 1971 the Lambourn and Great Ouse Pilot Schemes had been completed, both indicating that groundwater in the Chalk could indeed be used to augment river flows, although the Lambourn Scheme also showed unequivocally that wells alongside rivers flowing over outcrop chalk could not be used for this purpose because the water simply recirculated back into the aquifer (Thames Conservancy 1972; Institution Civil Engineers 1978; Great Ouse River Authority 1972; Backshall *et al.* 1972). A pumping test in 1969 in the Lambourn Valley yielded 71 000 m^3/d from nine wells over 4 months, five of the wells were situated alongside the river. The net gain in river flow was 43%, the relatively low figure being due to the riverside wells. In the Great Ouse Pilot Area 18 wells were pumped for 8 months in 1970 at a total rate of 60000 m^3/d. In the final stages of the test the wells provided the entire flow of the river, with leakage through the riverbed amounting to less than 10% of the abstraction rate; the net gain was about 70%.

On the basis of the pilot scheme Thames Conservancy sought approval for Stage 1 of its proposed overall groundwater scheme, which was designed to increase the flow of the Thames in dry periods by 115000 m^3/d. A Public Inquiry was held in 1972 and approval was given. A Parliamentary Order was issued in 1977 for the first stage of the Great Ouse Scheme which was expected to increase the flow of the Great Ouse by almost 200000 m^3/d. Hesitant steps were also being taken at that time towards the eventual use of the Lincolnshire Limestone in conjunction with the Empingham Reservoir.

The WRB published four reports on artificial recharge in the London Basin (Water Resources Board 1972–1974) dealing with the hydrogeology, the use of an electrical analogue model, engineering and economic aspects and the results of pilot recharge works. The quality problems caused by high iron and sulphate concentrations in the groundwater in the aquifer after recharge raised their heads at this time but were not regarded as insuperable and this did indeed prove to be so. On the basis of the pilot studies an extension of the Metropolitan Water Board's scheme in the Lee Valley was recommended and it is to the credit of O. T. Addyman, the Engineer to the Lee Conservancy Catchment Board, that he accepted the proposals and began to implement the scheme, the success of which could not then be guaranteed. By the end of 1973 sufficient progress had been made for submission of an application to the Department of the Environment for an operational scheme comprising 13 wells to recharge 55000 m^3/d for some 150 days with the aim of providing a reliable yield in a dry year of 70000 m^3/d.

The experiments to artificially recharge contaminated river water into the Permo-Triassic sandstones at Edwinstowe indicated that an average recharge rate of 0.36 m/d could be achieved with a filter on the sandstone. The filter extended the time before clay and silt in the water and microbial growths reduced the infiltration rate to unacceptable levels; cleaning the filter then became necessary. The recharge process improved the water quality considerably. The well recharge experiment at Clipstone Forest suggested that rates of 5000 m^3/d should be possible for an extensive period using a fully efficient well and treated water (Satchell & Edworthy 1972; Edworthy 1978).

By the early 1970s, in addition to the projects started in the 1960s, the river authorities had initiated many groundwater schemes. The early schemes to augment rivers with groundwater were based on the Chalk, a fissured aquifer with a high transmissivity and a low storage coefficient, not ideal properties for the purpose. Water engineers at the time were slow to appreciate that the Permo-Triassic sandstones with a lower, but still high, transmissivity and a much higher storage coefficient offered a more suitable medium. But the potential of a regional resource of groundwater existing in the Triassic sandstones of Shropshire was recognized in 1970 and when in October of that year the Secretary of State for Wales, after a Public Inquiry, refused an application by the Severn River Authority to investigate the Dulas reservoir site, in Wales, the potential of the sandstones assumed greater significance and a pilot study began to examine the feasibility of using them to regulate the Severn.

The practicality of developing the Chalk for river augmentation under the confined conditions that existed over much of East Anglia was examined with a pilot scheme in the Waveney Valley (Downing *et al.* 1981) and similar field programmes were started in the Upper Wylye Valley (Avon and Dorset River Authority 1978), the Vale of York (Reeves *et al.* 1974), and the Itchen Valley (Southern Water Authority 1979). Recharge of the Lower Greensand at Hardham, in Sussex, by means of basins had begun in 1968 and ultimately demonstrated that recharge rates of 1.5 m/d were feasible (Southern Water Authority 1979).

For many years development of the Chalk of the South Downs had been bedevilled by the risk of saline intrusion. Detailed analysis of this problem was obviously a precursor to further development of the Chalk's resources. With this in mind boreholes were drilled close to the sea in Brighton to examine the detailed form of the saline ingress (Monkhouse & Fleet 1975). They were an early stage in the Sussex River Authority's study of the South Downs. New production wells were drilled to evaluate the various methods of further development. These studies focused on the Brighton and Worthing Chalk Blocks between the rivers Arun and Ouse.

The essence of the many schemes carried out between 1964 and 1974 has been distilled by Rodda *et al.* (1976), Downing *et al.* (1981), Headworth *et al.* (1983) and Downing (1986). Overviews of the philosophy for the regional development of groundwater resources were given by Ineson & Rowntree (1967), Ineson (1970) and Downing *et al.* (1974).

Concern had been expressed since the 1930s about the practice of discharging saline minedrainage water from collieries in east Kent into the Chalk, a practice that sterilized a significant groundwater resource in an area already considered to need additional water supplies. Objections from the local water companies to the colliery owners and, from 1947, to the National Coal Board had fallen on deaf ears. In the early 1970s the Director of the WRB personally approached the Chairman of the NCB, Lord Robens, and the practice stopped in 1973 when a pipeline was laid from Tilmanstone Colliery to Betteshanger Colliery which did discharge its minewater to the sea. As a consequence the WRB, the Kent River Authority and the Thanet Water Board began to investigate the distribution of the saline water in the aquifer by drilling observation boreholes and using both downhole and surface geophysical techniques, with the object of assessing whether at least some of the saline water could be removed by pumping to waste.

The discovery of a canister marked cyanide in a landfill in the Midlands reinforced the concern that existed about the possible effects of waste disposal generally on groundwater quality. An immediate outcome was the Deposit of Poisonous Waste Act in 1972 (subsequently overtaken by the Control of Pollution Act of 1974). The WRB, in cooperation with the Water Pollution Research Laboratory produced a code of practice for the disposal of wastes in landfills.

At this time several water companies were expressing concern to the Department of the Environment and the WRB about the rising nitrate levels in groundwater; the cause was attributed to the use of fertilizers by agriculture. On the basis of studies of tritium in the unsaturated zone of the Chalk (see below), it seemed possible that nitrate was accumulating in the unsaturated zones of the main aquifers and a programme to investigate this by dry-core drilling of the Chalk was put in hand by the Water Research Centre soon after it was established in 1974.

Data collection and research

Under Sections 15 and 18 of the Water Resources Act, the river authorities were required to prepare and submit for the WRB's approval, hydrometric schemes for the measurement of hydrological parameters, and, under Section 14, make a survey of the water resources and water demand in their areas and formulate proposals for conserving, redistributing or augmenting their water resources. The river authorities, which had been created from river boards, were suffering from considerable work overload in meeting the plethora of requirements of the Water Resources Act. Although their staffs included civil engineers, inherited from the ancestral river boards, the virtual absence of trained hydrogeologists was a serious handicap. Because of this the Geology Division at the WRB was heavily involved in the 1960s in advising the river authorities on groundwater aspects of hydrometric schemes, was well as the approach to Section 14 surveys and methods of carrying out and interpreting pumping tests on new and existing wells. One recurring difficulty was 'licences of right'. These related to the statutary rights of water companies prior to the Water Resources Act, to abstract groundwater, generally in excess of their current requirements; the rights had been granted in anticipation of future needs and in many cases were excessive. These entitlements were a weakness in the Water Resources Act but were not unforeseen and the rights had been allowed to stand to placate opposition to the legislation. The number of licences of right issued for groundwater was 32479, allowing an annual abstraction of 4000 Mm^3 (Taylor 1975); the total groundwater abstraction in 1963 was only 2478 Mm^3.

The hydrometric schemes were concerned with the measurement of rainfall, evaporation, river flow and river quality as well as the provision of a network of observation boreholes to monitor changes in groundwater storage and quality. The schemes were supported by central government funds amounting to 50% of the total cost. Between 1964 and 1974 the capital expenditure approved for groundwater networks was almost £0.6 million, at 1973 values, a substantial sum for the time. The WRB advised on the density of networks and by 1974 the density in the major aquifers was one borehole per 27 km^2 with a continuously monitored borehole every 175 km^2; the national network then comprised almost 2000 boreholes.

The hydrometric schemes have provided far-reaching benefits for the planning and management of water resources. Good management practice depends upon reliable long-term hydrological data and it was recognized that the cost of obtaining data represented only a small percentage of the value of the resource. The data collected from observation borehole networks was published by the WRB in 'Groundwater Year books', the first, covering the three years 1964 to 1966, was issued in 1970 and included details of 540 selected boreholes (Water Resources Board 1970*b*)

The extensive data from the many pumping tests in the Great Ouse Pilot Scheme and other schemes with which the WRB was involved in the Chalk and Triassic sandstones were analysed by J. M. A. Pontin. This work led to cooperation with N. S. Boulton of the University of Sheffield and an extension of the latter's theory of delayed yield from storage in unconfined aquifers (Boulton & Pontin 1971).

There were obvious limitations to the application of the theories of well hydraulics based on ideal conditions, to the interpretation of tests on groups of wells intended for river augmentation. The way forward was to use values of aquifer properties derived from pumping tests, as basic data in the design of groundwater models; the first model made at the WRB was a digital model of the Lambourn Pilot Area (see below).

As field investigations gathered pace in the late 1960s both the WRB and the river authorities required information that could only be provided by geophysical borehole logging. Although the Geological Survey had developed such equipment in the UK, it was needed for its primary purpose – research – and a day-to-day service as required by the WRB could obviously not be reliably provided. Accordingly, the WRB formed a Geophysical Unit, under L.M. J. Standon-Batt, for work on the various schemes being promoted by the WRB and the river authorities.

The thermonuclear tests in the atmosphere by the USA and USSR in the 1950s and 1960s released many radioactive fission products into the atmosphere; one of these was tritium, the radioactive isotope of hydrogen, which increased in rainfall many-fold as it was incorporated in the water molecule. The UK Atomic Energy Authority set up an Isotope Analysis Laboratory at the Wantage Research Laboratory, under D. B. Smith and R. L. Otlet, to examine whether or not tritium and other fission constituents could be used in hydrological studies and the WRB became associated with some of the resulting experiments. Examination of water from deep boreholes in the Chalk showed no evidence of tritium. This led to the analysis of the base flow of the chalk-fed River Lambourn for tritium. The very low values that were found again confounded expectations. The question was: where was the tritium that must have infiltrated the Chalk? It was to prove to be a defining moment. Smith, being a physicist, was unencumbered by preconceived ideas about the behaviour of the Chalk as a typical fissured limestone and considered, from the obvious appearance of a piece of porous chalk, that the tritium was in the matrix of the aquifer in the unsaturated zone. Dry-core drilling of the aquifer revealed that this was indeed so, the tritium moving slowly down at about $1 ma^{-1}$. The results of that work (Smith *et al.* 1970) demonstrated the important role that the matrix plays in both the recharge of groundwater in the Chalk and how solutes move through the aquifer, a fact not previously fully appreciated. Thereafter, tritium analysis played a part in the interpretation of most of the WRB's groundwater projects.

Models of aquifers

Modelling groundwater flow in aquifers became a reality in the early 1960s (e.g. Skibitzke 1963) and models began to revolutionize the interpretation of hydrogeological phenomena particularly the relationship between cause and effect including the links between groundwater abstraction and the resultant change in river flows. The first models were electrical analogues using resistance-capacitance networks.

Fig. 2. Electrical analogue model of the London Basin.

The WRB made two large models, the first was of the Chalk and Tertiary sands in the London Basin and the second of the Chalk in the Great Ouse Pilot Area, which was subsequently extended to incorporate the Stage I development proposal. After the WRB published its report on the groundwater resources of the Lincolnshire Limestone (Downing & Williams 1969), K. R. Rushton of the University of Birmingham, began his long association with the aquifer by making an electrical analogue model of the southern part of the limestone and thereby assisting the WRB to interpret the optimum method of using the water resources of the aquifer.

The model of the London Basin represented an area of 3600 km^2 comprising a mesh of 900 nodes (Fig. 2). It was a steady-state (resistor network) model because of the nature of the data available, covering a long time-span of 150 years of historical records. A steady-state model was also made of the Dagenham area and a non-steady-state model of a cross-section of the confined basin. These models were used to estimate the likely response of the aquifer to artificial recharge. The model of the Chalk in the Great Ouse catchment was a non-steady state model of the Little Ouse and Thet sub-catchments representing an area of some 650 km^2. It was used to evaluate Stage I of the scheme for the regulation of the river system with groundwater from the Chalk. A number of individuals were involved in making these various analogue models but in particular C. D. N. Tubb was heavily committed and he must be credited with successfully calibrating them and producing the results for which they were intended.

Electrical analogue models are time-consuming to build and calibrate and by 1969 attention had begun to turn towards mathematical solutions using computers although at that time operational costs and computer storage precluded their use for large models.

The WRB's first mathematical model was made in 1969 by D. B. Oakes and J. M. A. Pontin of the Chalk aquifer and river system in the Lambourn Valley. It was successfully calibrated against both natural conditions and the data from the pumping tests of the Chalk wells in the Lambourn Pilot Area. Subsequently it was used to assess the consequences of developing the groundwater in the Chalk by a number of possible methods. The work was eventually published in 1976 (Oakes & Pontin 1976). By 1972 mathematical models were being designed by the WRB to examine the operation of the Lancashire Conjunctive Use Scheme (Oakes & Skinner 1975), groundwater development of the Triassic sandstones in Shropshire and the Vale of York (Reeves *et al.* 1974), as well as the optimum use of the resources of the Chalk of the South Downs between the rivers Arun and Ouse (Nutbrown *et al.* 1975). The abstraction of groundwater from confined aquifers for river regulation was examined (Birtles & Wilkinson 1975) and a one-dimensional model was used by Oakes & Wilkinson (1972) to identify the factors controlling the time-delay between abstraction from wells at different distances from a river and the effect on the flow of the river. Mathematical models had become a standard tool for assessing different methods of developing an aquifer and determining the consequences of development outside the range of meteorological conditions that could be tested in field experiments. This work on mathematical models of aquifers was largely in the hands of A.B. Birtles, D. A. Nutbrown and D. B. Oakes.

The WRB also used models to simulate the regional development of water resources comprising groups of individual sources, both surface and groundwater, to assess the optimum use and reliable yields from such systems. One such model was made by Armstrong & Clarke (1972) to study regional plans for SE England involving various combinations of surface water and groundwater storages. Jamieson & Sexton (1972) developed a similar model incorporating the rivers Welland, Nene and Great Ouse with the Lincolnshire Limestone, the Chalk of the Great Ouse Basin and storage in the Empingham Reservoir.

The WRB, in producing its 'National Plan', used models to derive the cost of different development programmes and to simulate the construction and operation of the various programmes and thereby obtain detailed estimates of costs (For the estimated costs of groundwater schemes see Water Resources Board 1973, vol 2, Tables A5 & A6).

Modelling techniques clearly transformed the understanding of hydrological and hydrogeological phenomena. They cast a shaft of light on the inherent uncertainties and imprecisions of the field experiments which were invariably carried out with less than ideal and variable weather conditions.

'The National Plan'

By 1972 the WRB was focused on collating the results of the three regional studies, as well as the many parallel investigations of groundwater sources, potential reservoir sites and storage sites in estuaries. The examination of the potential for, and desirability of, transferring water between river basins and the comparison of alternative strategies and costs for resource development to meet the demand for water at various times to the year 2001 were now very relevant. For this purpose England and Wales was considered as a whole, the basic philosophy being that they comprise a relatively small area with wide variations in the availability of water and the demand for it from NW to SE. A further factor was that the potential for upland storage is greater in the north and west and the principal aquifers are in the Midlands and SE England. Only if the two countries were coordinated as a single entity could the best use be made of both the surface and groundwater resources. Therefore, the WRB recommended an integrated strategy of development that included the maximum use of groundwater and groundwater storage (but used intermittently for river regulation or in combination with surface sources), a small number of large inland reservoirs (because of the widespread opposition to such projects) and the inter-regional transfer of water by means of the river systems (Fig. 3). The underlying basis for the use of groundwater in the proposals was summarized by the WRB (1973, vol 2, Appendices G & H).

By 2001 use of water for public supply was anticipated to be 28 million m^3/d, with direct net industrial

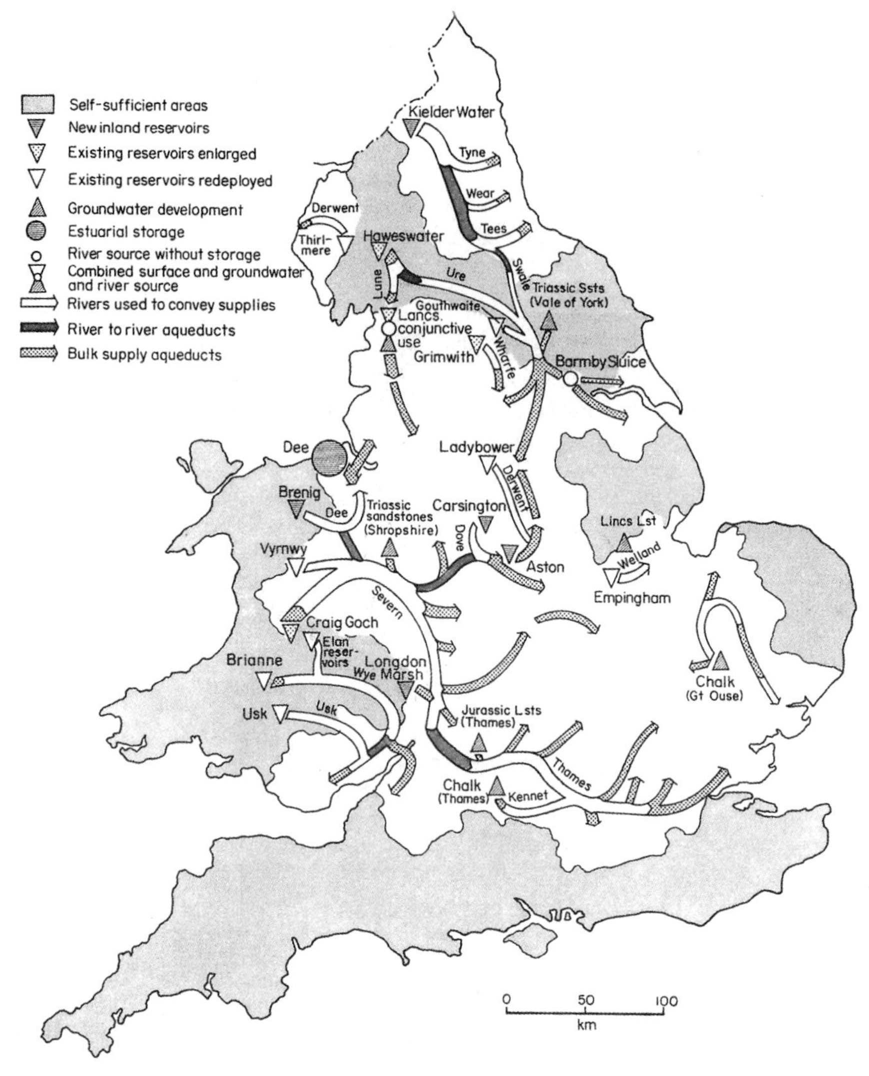

Fig. 3. Strategy of water resources development in England and Wales in AD 2000 as preferred by the Water Resources Board.

and agricultural use a further 2 million m^3/d; some 12 million m^3/d of new resources were thought to be needed to meet the resulting deficiency in yield requiring new storage capacity of 2000 million m^3 of which 25% would be in aquifers. The groundwater schemes that were expected to be in use were: augmentation of river flows with groundwater from the Chalk in the Thames and Great Ouse basins and from the Triassic sandstones in the Severn Basin and the Vale of York; conjunctive use of groundwater in the Triassic sandstones in the Fylde with surface resources and in the Trent Basin with rivers from the Peak District; probably artificial recharge of the Chalk in the London Basin and possibly of the Triassic sandstones in the Middle Trent Basin (see Water Resources Board 1973, Vol 2, Tables A5 & A6). The WRB emphasized the need to prevent any further contamination of aquifers, sound advice but, as events proved, not given sufficient attention.

Yet more reorganization

The key component of the WRB's thinking for the future development of water resources depended upon maintaining the quality of the rivers. It was soon evident that weaknesses in the Water Resources Act of 1963 and hence in the Board's remit, were

Fig. 4. Regional Water Authorities as defined by the Water Act 1973.

that it did not have any responsibility for water quality, nor have powers to construct and operate works other than for research purposes. In 1969 the Labour Government reconvened the Central Advisory Water Committee (CAWC) to consider once again the organization of the water industry and much of the WRB's life was conducted against this background.

The WRB, in its evidence to the CAWC, made a case for multipurpose river basin management by 'river basin authorities' with a strong national water authority at the centre which had clear executive powers and a Water Research Centre responsible for research on both clean and dirty water. In the Water Act of 1973 the Conservative Government accepted the need for multipurpose regional water authorities (Fig. 4) but opted for a weak National Water Council (with virtually no executive powers) although it did centralize research in a new Water Research Centre. The changes were compatible with the government's reorganization of local government, which took place at the same time, devolving power to the regions rather than the centre. *En passant,* it is perhaps interesting to observe that river basin management by multipurpose authorities was proposed by Edwin Chadwick as long ago as 1842 (Chadwick 1842; Evans 1993) some 130 years before the 1973 Act introduced this rather obvious practical proposal.

As a result of the changes inaugurated by the 1973 Act, the WRB ceased to exist on 1st April 1974, ten years after it was created. Some of the staff were dispersed to two new units of the Department of the Environment – the Central Water Planning Unit and the Water Data Unit – and to the new Water Research Centre, which also incorporated the Water Pollution Research Laboratory and the Water Research Association. Many others joined the new regional water authorities.

Retrospect

The publication of the WRB's 'National Plan' (Water Resources Board 1973) coincided, unfortunately, with a fourfold increase in the price of oil, which precipitated a world energy crisis and led to changes that affected every corner of the world's economy not excluding the water industry of England and Wales. Previously water resources had been developed to meet peak demands under drought conditions following the Victorian ideal that the maintenance of good public health depended upon a cheap, pure, plentiful water supply. Now the cost of providing water began to be examined in relation to benefits. This coincided with a fall in anticipated demand, both domestic and industrial, because of a decline in the rate of population growth and the economic recession. In 2001 the population of England and Wales was 52 million, 3 million less than the figure anticipated in 1972. As already noted, in 1973 abstraction for public supply had been forecast to double by 2001 to 28 million m^3/d (based on the projected growth rates and an estimate of per capita consumption), whereas by the end of the century it was only 15 million m^3/d, and, with industrial and agricultural abstraction, the total was less than 20 million m^3/d compared with 30 million m^3/d forecast in 1973; so there was less need for additional water supplies. But perhaps more importantly in this context, the new autonomous regional water authorities, sensing a change in the balance of power, began to look towards self-sufficiency rather than inter-regional development. Given this outlook, the changed economic situation and the growing uncertainties over population growth, a national plan for water development began gradually to lose its attraction. The only groundwater schemes that were developed as originally envisaged were the regulation of the Severn from the Triassic sandstones in Shropshire, the Lancashire Conjunctive Use Scheme and the artificial recharge of the London Basin; the Thames and Great Ouse river augmentation schemes using groundwater (which had been expected to yield 455 and 330×10^3 m^3/d respectively) were only partially developed. However, in the ensuing years many smaller river augmentation schemes based on groundwater were commissioned (Owen *et al.* 1991). 'Demand management' became the new sound-bite with attention eventually focusing on the enormous volume of water leaking from water mains, some of which recharged aquifers unintentionally, and also on the use of meters to curtail domestic consumption. There was a sea change in attitudes to water supply – a transformation of previous concepts.

So in retrospect what did the WRB achieve? What was its legacy? Essentially, a monumental change in attitudes to water resources development was engendered, not least the use of groundwater for river regulation, its use in conjunction with surface resources, and the artificial recharge of aquifers (Taylor 1975, 1987). Major groundwater experiments examined the principles of such proposals in the main aquifers, covering the principal geological situations at outcrop and under confined conditions; each cost of the order of £0.3–0.5 million. These proposals were a break from the traditional abstraction of groundwater continuously for direct supply; the objective changed to the effective use of the large storage capacity of aquifers. Perhaps more importantly, the WRB applied the concept that river basins are the appropriate basic unit for water resources management, a concept that became a practical reality as a result of the Water Resources Act.

The WRB was a leading proponent of mathematical models not only for groundwater systems but also for analysing the regional resources systems of which they formed part. These models increased the understanding of aquifer behaviour manyfold particularly the cause and effect link between abstraction from wells at different locations and spring flow.

The ten years that the WRB existed was an extraordinary decade for British hydrology and hydrogeology. British scientists featured prominently on the international stage for the WRB's philosophy on comprehensive basin management and the need for pragmatic research was much admired. Furthermore, the Board generated an interest in water affairs among the general public, featuring frequently in daily news items as it continually projected views and ideas on water development.

In its enthusiasm for developing groundwater within integrated water resource systems, the WRB was well aware of the need to maintain adequate rivers flows and provide for all the users of rivers (Ineson & Rowntree 1967). This was the basis for river regulation and in fact defining 'minimum acceptable flows' (MAF) was part of its remit under the Water Resources Act, although the concept proved difficult to implement mainly because a realistic, practical MAF was very difficult to define. The 'sustainable development' of water resources within an environmental concept was an accepted fact of life long before the term became fashionable in the 1990s. Ecological studies of rivers and the effect of discharging groundwater into them, especially Chalk-fed rivers, were important parts of the WRB's programmes for all the groundwater pilot schemes. The comprehensive approach was intended to serve all users of water including recreational and amenity interests as well as the paramount basic need for water supply. The objective was to manage water resources development while preserving the aquatic environment.

For the first time accurate figures became available for the quantities of water licensed and abstracted in England and Wales, essential data about the demand for water. However, with respect to groundwater, the collection of abstraction data for individual aquifers by the Geological Survey under Section 6 of the Water Act ceased and the regional water authorities were slow to instigate a new system under the Water Resources Act of 1963. In fact national data of abstraction from individual aquifers are still not available.

The government's decision to abolish the WRB was received with disbelief for it was held in high esteem in many quarters (e.g. Evans 1993). It had clearly met its original objectives in the form of three comprehensive regional studies, coordinated into a national plan, as well as many supporting programmes of significance. Some would say the WRB had an ideal life-cycle. Formed for a particular purpose – defining a national water strategy for England and Wales, completed with a bold solution, and then disbanded; the perfect sequence although rarely attained, particularly by government agencies. The WRB pioneered a new progressive attitude to water development in England and Wales, particularly the use of groundwater storage as a means of groundwater development and in recognition of this its Director, Norman Rowntree, received a knighthood.

The Central Water Planning Unit

The Central Water Planning Unit (CWPU) was literally a unit of the Department of the Environment, a product of the dismantlement of the WRB. It was required to provide planning advice to the government, the new National Water Council and the regional water authorities on the augmentation of water resources, on water quality, and on pollution prevention; the Director was O. Gibb, a civil engineer who had been with the WRB throughout its life. The programme of the Unit was supervised by a Steering Committee. Initially the professional staff comprised 45 engineers and scientists but with time this declined to only about 25; about 25% were concerned with groundwater programmes.

The Unit's policy with respect to groundwater development was to collaborate with the regional water authorities on, and provide financial support for, schemes relevant to a national strategy for the use of water. This included river augmentation studies in Shropshire and in the Waveney and Itchen valleys. Identifying the optimum method of developing the resources in the Chalk of the South Downs to avoid saline intrusion was assisted with the use of mathematical models to examine, for example, appropriate pumping regimes (Nutbrown 1976*a*, *b*). These additional modelling studies confirmed the view that the maximum use of the groundwater resource could be assisted by developing the South Downs as a single unit.

The Unit also pursued a policy of encouraging prototype artificial recharge schemes in the main aquifers. Investigation of the Chalk in the Lee Valley was already underway and to complement this schemes were supported to examine both lagoon and well recharge of the Lower Greensand at Hardham (Southern Water Authority 1979; O'Shea 1984) and of the Triassic sandstones in the Worfe Catchment of the Severn Basin (Jones 1983). The results of the entire artificial recharge programme in the 1960s and 1970s were reviewed by Edworthy & Downing (1979).

The Unit also pursued its own programme of groundwater resources studies and research. At the request of the North West Water Authority, the groundwater resources of the Permo-Triassic sandstones in the Vale of Eden were assessed (Monkhouse & Reeves 1977). Considerable volumes of water were being pumped from the Coal Measures for mine drainage. The volume abstracted from each coalfield and analyses of its quality were collected but the poor quality precluded any consideration of its use for water supply (Rae 1978).

A deep borehole was drilled through the Chalk into the Lower Greensand at Sompting, in Sussex, to measure the aquifer properties of the confined Lower Greensand in an area where it is, as yet, unexploited. The groundwater proved to be of good quality and with a potential yield in excess of 2500 m^3/d from a borehole 300 mm in diameter (Young & Monkhouse 1980).

Mathematical models continued to be used to interpret a wide range of groundwater problems (e.g. Birtles & Nutbrown 1976). Reference has been made to their use in the South Downs programme. A model was also designed for the Thames Water Authority of the Lambourn and Kennet valleys (Morel 1980) and, at the request of the Yorkshire Water Authority, the existing model of the Triassic sandstones in the Vale of York was used to design well-fields and devise pumping schedules to minimize capital and running costs of a river augmentation scheme.

The rising concentration of nitrate in groundwater was seen as a serious problem in the 1970s. The Unit analysed all available data to predict probable future concentrations in water supplies and to identify groundwater sources at risk (Central Water Planning Unit 1978), but the field studies to investigate the nature of the problem and its extent were in the hands of the Water Research Centre and the Institute of Geological Sciences (now the British Geological Survey).

Radioactive and stable isotopes (carbon-14, deuterium and oxygen-18) were used in cooperation with the UKAEA, to study the flow of groundwater through the Chalk, the Lincolnshire Limestone and the Lower Greensand on a regional scale and where they are confined, and to estimate their 'ages' and hence assess the storage times of groundwater in the aquifers. (Smith *et al.* 1976; Downing *et al.* 1977, 1979; Evans *et al.* 1979). The overall conclusion of these studies was that the deep groundwaters are complex mixes from different sources and include components of very old waters recharged during the Pleistocene period. The work helped to confirm the important role of the matrix in the flow process and of diffusion between fissures and the matrix in the migration of solutes in groundwater (Smith *et al.* 1970; Foster 1975; Oakes 1977).

The public inquiry into the Severn-Trent Water Authority's Carsington Reservoir Scheme was reopened in 1976. The Unit took the rather unusual step for a unit in the Department of the Environment, of proposing an alternative to the water authority's scheme. This involved the conjunctive use of groundwater storage in the Triassic sandstones of Nottinghamshire with water in the River Derwent. The proposal envisaged that the flow of the Derwent would be adequate for 80% of the time and groundwater would be used for the remaining 20% but it was not accepted by the Inspector taking the Inquiry and the Carsington Reservoir was approved. This 'safe decision' was disappointing as an opportunity for an imaginative major development of a ready-made groundwater reservoir, one of the largest in England, was lost. After 1974 enthusiasm for groundwater as a regional resource gradually abated amongst many water engineers. It was still perceived to be a riskier venture than a surface reservoir and reflected the uncertainty that continued to surround hydrogeological principles at the time. A further factor was the absence of a strong driving force such as had existed with the WRB.

One of the last groundwater projects undertaken by the Unit was the coordination of an assessment of the groundwater resources of the UK for the European Economic Community (now the European Union) as part of a programme involving all the member states. The report on the UK, illustrated by 11 maps at a scale of 1:500000, was published by the European Community in 1982 and the report only was also issued by the CWPU (Monkhouse & Richards 1979). This remains the only published record of the groundwater resources of the UK, or indeed of England and Wales.

During the Unit's relatively short life of five years, 15 reports concerned with groundwater were published (Central Water Planning Unit 1979) as well as many papers by members of the staff, despite operating in a climate of continuous uncertainty as regards meeting its original remit and purpose. By 1975 the revised forecasts of the future growth of the population, together with the economic recession, were leading to a reduced demand for additional water supplies. As already noted, this led the regional water authorities, after reappraising their existing resources, to realize that their size allowed them to adopt a policy of self-sufficiency. The fact that they managed to maintain water supplies during the severe drought in 1976 (Central Water Planning Unit 1977*b*; Day & Rodda 1978), even though there was considerable disruption in some areas, gave them confidence to pursue this policy which marked a change of perspective.

One of the primary reasons for creating regional water authorities had been to maintain and improve the quality of rivers so they would be suitable for water supply if regulated by reservoirs and aquifers. Ironically, the authorities inherited the existing sewerage and sewage treatment systems of local government at a time of financial stringency, with the water authorities responsible for monitoring and reporting on their own waste-water discharges. In the analogy of the time, the gamekeeper had, in effect, also become the poacher. The regional water authorities were perceived to have failed in one of their principal tasks, as they were now seen as the major polluters of rivers and river water quality declined in the late 1970s. Later large investments in waste-water treatment turned this around.

The government had also realized that after 1974 the organization at the centre was flawed. As far as the CWPU was concerned it was difficult to review, modify or influence the national strategic element of water planning and operations without statutory authority. The Labour Government, therefore, instigated another review of the water industry. A central feature of the resulting White Paper in 1977 was a proposal to establish a National Water Authority which would replace the weak National Water Council and incorporate the CWPU. However, following the general election in 1979, the Conservative Government announced it was not going to proceed along this course and the CWPU was disbanded. It was to be a further ten years before the need for a strong central regulatory authority was to be accepted.

With hindsight, the period 1964–1979 can best be seen in two phases: the first covered the life-span of the WRB which was completed with a proposal for a national water strategy; the second covered the first five years of the regional water authorities when an economic recession and revised forecasts of demand for water reduced the need to develop new resources as quickly as had been envisaged by the WRB. Furthermore, these authorities soon found that they had to give priority to investment in the 'dirty water' side of their business and, from that point of view, it was fortunate that the time frame for water resources development could be extended. It is unequivocal, however, that the period 1964–1979 saw a tremendous advance in the science and practice of hydrogeology and the development of groundwater resources in England and Wales, an advance in which the WRB and CWPU played no small part. The WRB demonstrated the merit of a national water strategy – the ideal expressed in the Water Act of 1945 – but an ideal still to be realized.

The author is indebted to L. E. Taylor who kindly reviewed and improved the manuscript.

References

ANON. 1960. *Hydrological survey of the Great Ouse Basin.* Ministry of Housing and Local Government, London.

ANON. 1964. *The South East Study.* HMSO, London.

ARMSTRONG, R. B. & CLARKE, K. E. 1972. Water resource planning in south east England. *Journal of the Institution of Water Engineers*, **26**, 11–47.

AVON AND DORSET RIVER AUTHORITY. 1973. *Upper Wylye Investigation.* Poole.

BACKSHALL, W. F., DOWNING, R. A. & LAW, F. M. 1972. Great Ouse Groundwater Study. *Water and Water Engineering*, **76**, 215–223.

BINNIE AND PARTNERS. 1965. *Report on the water resources of the Great Ouse Basin.* Ministry of Housing and Local Government, London.

BIRTLES, A. B. & NUTBROWN, D. A. 1976. The use of groundwater modelling techniques in water resource planning. *Water Services*, **80**, 533–538.

BIRTLES, A. B. & WILKINSON, W. B. 1975. Mathematical simulation of groundwater abstraction from confined aquifers for river regulation. *Water Resources Research*, **11**, 571–580.

BONIFACE, E. S. 1959. Some experiments in artificial recharge in the Lower Lee Valley. *Proceedings of the Institution of Civil Engineers*, **14**, 325–338.

BOULTON, N. S. & PONTIN, J. M. A. 1971. An extended theory of delayed yield from storage applied to pumping tests in unconfined anisotropic aquifers. *Journal of Hydrology*, **14**, 53–65.

CENTRAL WATER PLANNING UNIT. 1977*a*. *The hydrology of the Stour Valley, Essex, with particular reference to the groundwater resources.* Reading.

CENTRAL WATER PLANNING UNIT. 1977*b*. *The 1975–76 drought: a hydrological review.* Reading.

CENTRAL WATER PLANNING UNIT. 1978. *Nitrate and water resources with particular reference to groundwater.* Reading.

CENTRAL WATER PLANNING UNIT. 1979. *Annual Report for 1978/79.* Reading.

CHADWICK, E. 1842. *The sanitary conditions of the labouring classes.*

COMMISSION OF THE EUROPEAN COMMUNITIES. 1982. *Groundwater resources of the European Community.* Th. Scafor GmbH, Hannover.

DAVIES, M. C. 1973. A thermal infrared linescan survey along the Sussex coast. *Water and Water Engineering*, **77**, 392–396.

DAY, J. B. W. & RODDA, J. C. 1978. The effects of the 1975–76 drought on groundwaters and aquifers. *Proceedings of the Royal Society, London,* **A368**, 55–68.

DOWNING, R. A. 1986. Development of groundwater. *In*: Brandon, T. W. (ed.) *Groundwater: occurrence, development and protection.* Institution of Water Engineers and Scientists, London, 485–542.

DOWNING, R. A. & WILLIAMS, B. P. J. 1969. *The groundwater hydrology of the Lincolnshire Limestone with special reference to the groundwater resources.* Water Resources Board, Reading.

DOWNING, R. A., ASHFORD, P. L., HEADWORTH, H. G., OWEN, M. & SKINNER, A. C. 1981. The use of groundwater for river regulation. *In*: *A survey of British hydrogeology*. The Royal Society, London, 153–171.

DOWNING, R. A., OAKES, D. B., WILKINSON, W. B. & WRIGHT, C. E. 1974. Regional development of groundwater resources in combination with surface water. *Journal of Hydrology*, **22**, 155–177.

DOWNING, R. A., PEARSON, F. J. & SMITH, D. B. 1979. The flow mechanism in the Chalk based on radioisotope analyses of groundwater in the London Basin. *Journal of Hydrology,* **40**, 67–83.

DOWNING, R. A., SMITH, D. B., PEARSON, F. J., MONKHOUSE, R. A. & OTLET, R. L. 1977. The age of groundwater in the Lincolnshire Limestone, England, and its relevance to the flow mechanism *Journal of Hydrology,* **33**, 201–216.

EDWORTHY, K. J. 1978. *Artificial recharge through a borehole.* Technical Report, **86**, Water Research Centre, Medmenham.

EDWORTHY, K. J. & DOWNING, R. A. 1979. Artificial recharge and its relevance in Britain. *Journal of the Institution of Water Engineers and Scientists,* **33**, 151–172.

EVANS, G. V., OTLET, R. L., DOWNING, R. A., MONKHOUSE, R. A. & RAE, G. 1979. Some problems in the interpretation of isotope measurements in United Kingdom aquifers. *In*: *Isotope Hydrology 1978, Vol II.* International Atomic Energy Authority, Vienna, 679–708.

EVANS, H. R. 1993. The structure and management of the British Water Industry, 1945–91. In: *Water and Environmental Management Yearbook.* Institution of Water and Environmental Management, London, 64–71.

FOSTER, S. S. D. 1975. The Chalk groundwater tritium anomaly – a possible explanation. *Journal of Hydrology*, **25**, 159–165.

GREAT OUSE RIVER AUTHORITY. 1972. *Great Ouse Groundwater Pilot Scheme, Final Report.* Cambridge.

HARDCASTLE, B. J. 1978. From concept to commissioning. *In*: *Thames Groundwater Scheme. Institution of Civil Engineers, London,* 1–27.

HEADWORTH, H. G., OWEN, M. & SKINNER, A. C. 1983. River augmentation schemes using groundwater. *British Geologist*, **9**, 50–54.

INESON, J. 1970. Development of groundwater resources in England and Wales. *Journal of the Institution of Water Engineers,* **24**, 155–177.

INESON, J. & DOWNING, R. A. 1964. The groundwater component of river discharge and its relationship to hydrogeology. *Journal of the Institution of Water Engineers,* **18**, 519–541.

INESON, J. & ROWNTREE, N. A. F. 1967. Water resources in the UK: conservation projects and planning. *Journal of the Institution of Water Engineers*, **21**, 275–290.

INSTITUTION OF CIVIL ENGINEERS. 1978. *Thames Groundwater Scheme.* London.

JAMIESON, D. G. & SEXTON, J. R. 1972. The hydrological evaluation of regional water resource systems in the UK. *In*: *International Symposium on water resource planning.* Mexico.

JONES, H. H. 1983. Investigations for artificial recharge of the Triassic sandstones aquifer near Stourbridge, UK. *Journal of the Institution of Water Engineers and Scientists,* **37**, 9–27.

LAW, F. 1965. Integrated use of diverse sources. *Journal of the Institution of Water Engineers,* **19**, 413–457.

MONKHOUSE, R. A. & FLEET, M. 1975. A geophysical investigation of saline water in the Chalk of the south coast of England. *Quarterly Journal of Engineering Geology,* **8**, 291–302.

MONKHOUSE, R. A. & REEVES, M. J. 1977. *A preliminary appraisal of the groundwater resources of the Vale of Eden, Cumbria.* Central Water Planning Unit, Reading.

MONKHOUSE, R. A. & RICHARDS, H. J. 1979. *Groundwater resources of the United Kingdom.* Centreal Water Planning Unit, Reading.

MOREL, E. H. 1980. The use of a numerical model in the management of the Chalk aquifer in the Upper Thames Basin. *Quarterly Journal of Engineering Geology,* **13**, 153–165.

NAYLOR, J. A. 1974. *The groundwater resources of the river gravels of the Middle Thames Valley.* Water Resources Board, Reading.

NUTBROWN, D. A. 1976*a*. *Optimum development of combined resources.* Central Water Planning Unit, Reading.

NUTBROWN, D. A. 1976*b*. A model study of the effects of artificial recharge. *Journal of Hydrology,* **31**, 57–65.

NUTBROWN, D. A., DOWNING, R. A. & MONKHOUSE, R. A. 1975. The use of a digital model in the management of the Chalk aquifer in the South Downs, England. *Journal of Hydrology,* **27**, 127–142.

OAKES D. B. 1977. The movement of water and solutes through the unsaturated zone of the Chalk of the United Kingdom. *In*: *Theoretical and Applied Hydrology.* Proceedings of 3rd International Hydrological Symposium, Colorado State University, Fort Collins, 447–459.

OAKES, D. B. & PONTIN, J. M. A. 1976. *Mathematical model of a chalk aquifer.* Technical Report 24, Water Research Centre, Medmenham.

OAKES, D. B. & SKINNER, A. C. 1975. *The Lancashire Conjunctive Use Scheme groundwater model.* Technical Report 12, Water Research Centre, Medmenham.

OAKES, D. B. & WILKINSON, W. B. 1972. *Modelling of groundwater and surface water systems: theoretical relationships between groundwater abstraction and base flow*. Water Resources Board, Reading.

O'SHEA, M. J. 1984. Borehole recharge of the Folkestone Beds at Hardham, Sussex, 1980–81. *Journal of the Institution of Water Engineers and Scientists*, **38**, 9–25.

OWEN, M., HEADWORTH, H. G. & MORGAN-JONES, M. 1991. Groundwater in basin management. *In*: DOWNING, R. A. & WILKINSON, W. B. (eds) *Applied Groundwater Hydrology*. Clarendon Press, Oxford, 16–34.

RAE, G. 1978. *Mine drainage from coalfields in England and Wales: its distribution and relationship to water resources*. Central Water Planning Unit, Reading.

REEVES, M. J., BIRTLES, A. B., COURCHEE, R. & ALDRICK, R. J. 1974. *Groundwater resources of the Vale of York*. Water Resources Board, Reading.

RODDA, J. C., DOWNING, R. A. & LAW, F. M. 1976. *Systematic Hydrology*. Newnes-Butterworths, London.

SATCHELL, R. L. H. & EDWORTHY K. J. 1972. *Artificial recharge: Bunter Sandstone*. Water Resources Board, Reading.

SKIBITZKE, H E. 1963. The use of analogue computers for studies in groundwater hydrology. *Journal of the Institution of Water Engineers*, **17**, 216–230.

SMITH, D. B., DOWNING, R. A., MONKHOUSE, R. A., OTLET, R. L. & PEARSON, F. J. 1976. The age of groundwater in the Chalk of the London Basin. *Water Resources Research*, **12**, 392–404.

SMITH, D. B., WEARN, P. L., RICHARDS, H. J. & ROWE, P. C. 1970. Water movement in the unsaturated zone of high and low permeability strata using natural tritium. *In*: *Isotope Hydrology 1970*. International Atomic Energy Authority, Vienna, 73–87.

SOUTHERN WATER AUTHORITY. 1978. *The Candover Pilot Scheme, Final Report*. Worthing.

SOUTHERN WATER AUTHORITY. 1979. *Lagoon recharge experiments at Church Farm, Hardham, Sussex*. Worthing.

TAYLOR, L. E. 1964. The problem of groundwater recharge with special reference to the London Basin. *Journal of the Institution of Water Engineers*, **18**, 247–254.

TAYLOR, L. E. 1975. A decade of water resources planning in England and Wales. *In*: *Proceedings of 2nd World Congress, International Water Resources Association*. New Delhi, Volume II, 361–371.

TAYLOR, L. E. 1987. The planning and development of water resources in England and Wales, 1965–1985. *In*: *Proceedings of International Symposium on Water for the future*. Rome, 355–365.

THAMES CONSERVANCY. 1972. *Report on the Lambourn Valley Pilot Scheme (1967–1969)*. Reading.

WATER RESOURCES BOARD. 1966. *Water supplies in South East England*. Reading.

WATER RESOURCES BOARD. 1970*a*. *Water Resources in the North*. Reading.

WATER RESOURCES BOARD. 1970*b*. *Groundwater Year Book 1964–1966*. HMSO, London.

WATER RESOURCES BOARD. 1971. *Water resources in Wales and the Midlands*. Reading.

WATER RESOURCES BOARD. 1972. *The hydrogeology of the London Basin*. Reading.

WATER RESOURCES BOARD. 1972–1974. Artificial recharge of the London Basin, Vols I–IV. Reading.

WATER RESOURCES BOARD. 1973. *Water resources in England and Wales, Volumes 1 and 2*. HMSO, London.

WRIGHT, C. E. 1974. *Combined use of surface and groundwater in the Ely Ouse and Nar catchments*. Water Resources Board, Reading.

YOUNG, B. & MONKHOUSE, R. A. 1980. The geology and hydrogeology of the Lower Greensand of the Sompting Borehole, West Sussex. *Proceedings of the Geologists Association, London*, **91**, 307–313.

Recollections of a golden age: the groundwater schemes of Southern Water 1970–1990

H. G. HEADWORTH[1]

Calle Aljibe, 3, 04118 San José, Níjar, almería Spain

Abstract: The creation of the river authorities and Water Resources Board in 1965 and the Water Authorities, Water Research Centre and Central Water Planning Unit in 1974 led to an explosion of groundwater investigation and development in England and Wales. In Southern Water's region, from the Hastings Beds of the Wealden Series to the Recent beach gravels at Dungeness, a dozen or so schemes were carried out to investigate and develop aquifers and manage their groundwater resources. Six schemes are described here, including artificial recharge in Sussex, groundwater augmentation in Hampshire and the assessment of saline contamination from minewater disposal in East Kent.

The period from the end of the 1960s to the middle 1980s was a golden age for hydrogeology in the UK. The Water Resources Act 1963 triggered this with the creation of the river authorities and the Water Resources Board (WRB) in 1965. This Act gave the impetus for region-wide resources studies and investigations financed through the new abstraction licensing system. By the late 1960s several major schemes were under way in England and Wales and these were given an added boost with the creation of the water authorities, Central Water Planning Unit (CWPU) and Water Research Centre (WRc) in 1974. The integration of the oft-opposing interests of water supply and river conservation into the autonomous and self-financing water authorities led to an explosion of hydrogeological innovation and groundwater development (Downing & Headworth 1989). This chapter chronicles six schemes of particular hydrological interest undertaken in the area of Southern Water, by its predecessor river authorities between 1965 and 1974 and then in the water authority (Headworth 1994). Some schemes were continued by Southern Water PLC after privatization in 1989. The six schemes presented are (Fig. 1):

(1) South Downs aquifer management, Sussex 1968–1985.
(2) Itchen river augmentation, Hampshire 1970–1994.
(3) Hardham artificial recharge, Sussex 1973–1981.
(4) Tilmanstone minewater contamination, Kent 1973–1981.
(5) Isle of Wight Lower Greensand aquifer development 1975–1990.
(6) Denge Beach gravels, Kent 1980–1984.

Nearly all of these schemes were managed by dedicated working groups which reported periodically to steering committees bolstered by specialist advisory staff from the Water Research Centre and the then Central Water Planning Unit. Most of them came under the author's overall control through their respective working groups. These investigations drew on engineering staff from Southern's multifunctional divisions as well as its scientific staff. Their contributions are recognized gratefully. Certainly, without the interest and active involvement of the Divisional Engineers few of these schemes would have got under way.

Southern's numerous annual and internal reports for these schemes, many now available, are not listed in the references, which concentrate on published and conference papers.

South Downs aquifer management

The South Downs Groundwater Project started in 1971 following the submission of a scheme to the Water Resources Board by Geoff Fox in Sussex River Authority under Section 18 of the Water Resources Act. The South Downs comprises five discrete coastal chalk 'blocks' separated by tidal rivers (Fig. 2). The initial work centred on the Brighton and Worthing blocks, but later it extended to the Chichester block to the west and the smaller Eastbourne and Seaford blocks to the east. Most of what follows describes the work related to the two principal central blocks.

The contribution and involvement in all aspects of this scheme over many years from modelling to aerial infra-red scans by many staff in the WRB,

[1] Howard Headworth was Hydrogeologist with Hampshire River Authority from 1965 and Principal Hydrogeologist with Southern Water Authority from 1974. From 1988 he was Managing Director of Southern Science Ltd, a wholly-owned subsidiary of Southern Water PLC. He now lives in southern Spain.

From: MATHER, J. D. (ed.) 2004. *200 Years of British Hydrogeology*. Geological Society, London, Special Publications, **225**. 339–362. 0305-8719/04/$15 © The Geological Society of London.

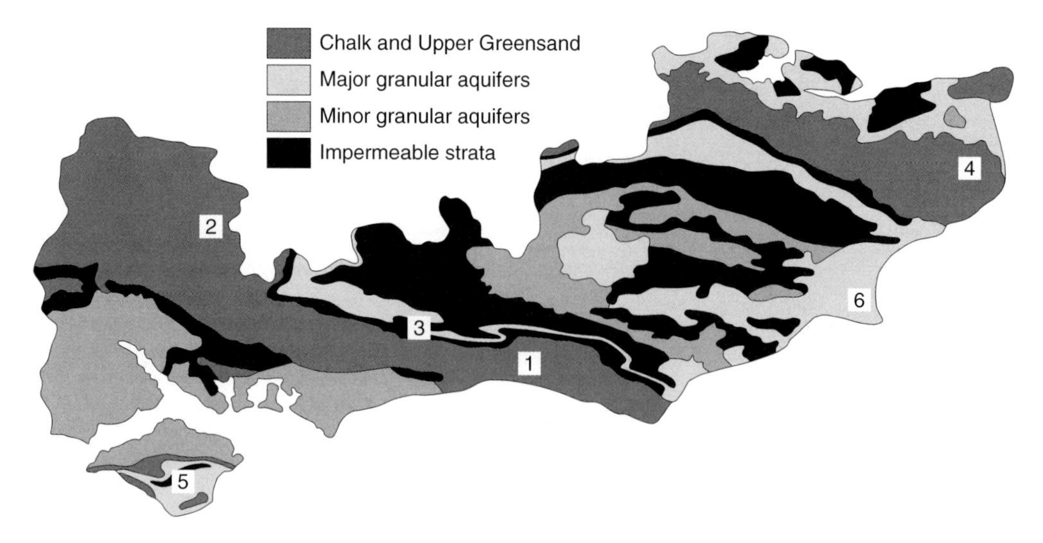

Fig. 1. Simplified geology for Southern Water's region and the location of the schemes referred to in paper.

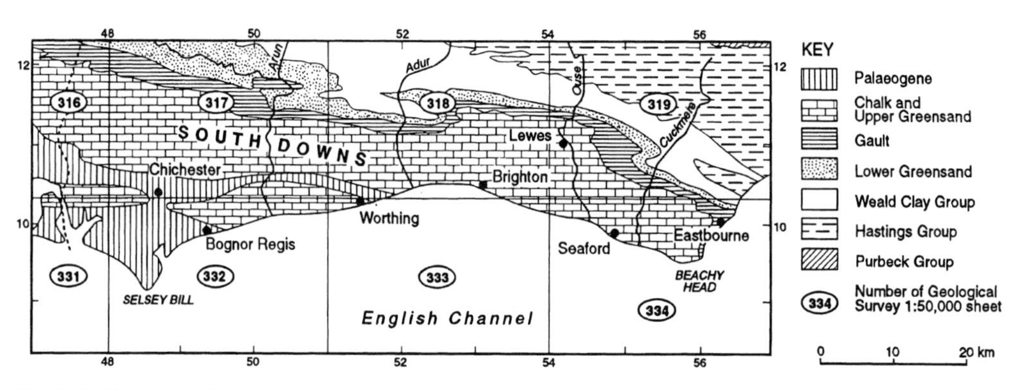

Fig. 2. Solid geology of the five Chalk Blocks of the South Downs.

CWPU, WRc and the British Geological Survey (BGS) is gratefully acknowledged. In no other Southern Water scheme was their involvement as great as in the South Downs study (Headworth & Fox 1986; Miles 1992).

Brighton Corporation abstraction policy

In the late 1960s the 'South Downs policy', in its water resources context, was synonymous with the Brighton Corporation 'storage and leakage' abstraction policy (Mustchin 1974), although the investigation as such was aimed at maximising groundwater usage and obtaining a greater understanding of the behaviour of the coastal aquifers. The Brighton abstraction policy started in 1957 after problems were encountered with certain coastal sources, notably Balsdean (the positions of pumping stations are shown on Fig. 3), due to their being pumped with undue enthusiasm by their local controllers. With water levels then in two crucial mid-Downs sources close to their sea-level adits (0 mOD), some action was clearly needed. The policy emerged of permitting Brighton's large-yielding central sources (Mile Oak, Patcham and Falmer) to recover their levels in the winter by reducing their abstractions while, at the same time, concentrating pumping as much as possible from the coastal sources (Fig. 4). Output from the coastal sites would be maintained into the Spring and early Summer months until their chloride concentrations started to rise. Then their output would be reduced and pumping switched to the primary sites inland. These coastal and inland sources were aptly termed the 'leakage and storage' stations respectively; the

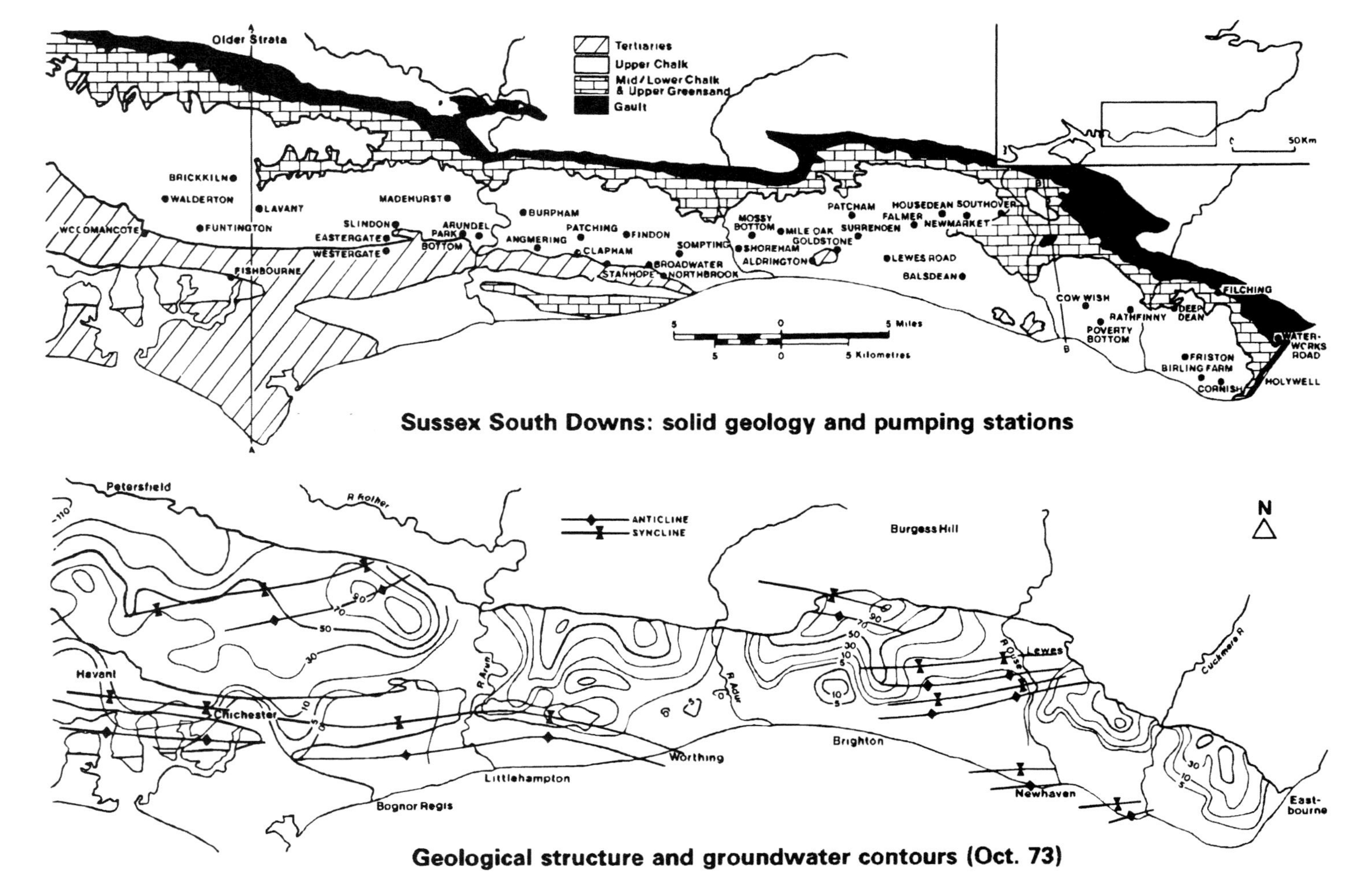

Fig. 3. Location of public water supply pumping stations (upper) and geological structure and groundwater contours (October 1973) (lower).

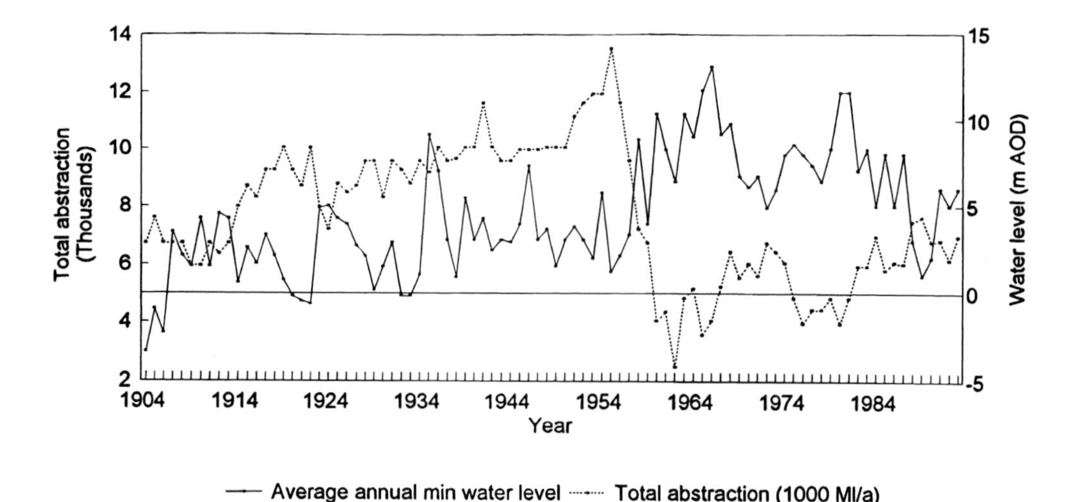

Fig. 4. Annual average minimum water levels for Mile Oak, Patcham and Falmer and total annual abstractions (1904–1990).

former intercepting winter outflows from the Chalk and the latter abstracting groundwater from the centre of the Chalk block.

Success of policy

This 'storage and leakage' abstraction policy has stood the test of time and is still in force, although the regime in recent years has been as much geared to operational efficiency as conserving groundwater storage. Nevertheless, as much through careful stewardship of the sources as through policy switches in abstraction between them, the policy has proved successful. Two important factors have aided its success. Firstly the very close control of service reservoir levels, pumping levels and daily pumping regimes, and secondly the number of sources added since 1972 which has allowed the effects of the abstractions to be spread over more of the aquifer.

The results of mathematical modelling (see below), showed that for steady-state conditions, groundwater levels, when averaged over the whole of the 193 km^2 Brighton block area, are 0.6 m higher than they were in 1953, despite the 33% increase in total output since then. For the 62 km^2 central zone of the aquifer, where most of the active aquifer storage exists, the model showed that under steady-state conditions the present average groundwater levels are 1.9 m higher than those which occurred under 1953 conditions. Moreover, if the present-day output had been pumped in 1953 with the abstraction regime then in force, the improvement in levels since then is shown to be 4.3 m. This figure is considered to represent the truest measure of the benefit of the Brighton abstraction policy.

Worthing block

The 'storage and leakage' abstraction policy has not been applied as such in the Worthing block for three reasons. Firstly, the Worthing block is narrower so that aquifer storage is less readily mobilized; secondly, for most of its length the coastal frontage of the Worthing block is protected by Tertiary deposits; and lastly, there are fewer abstraction sources in the higher part of the block to facilitate switching sources seasonally.

Chloride concentrations at the leakage stations

In the Brighton block the leakage stations are at Shoreham, Balsdean and, to an extent, at Goldstone. In the Worthing block, Sompting in the east and Burpham in the west are the leakage stations, while the source at Arundel might be considered one at the eastern end of the Chichester block, although not exhibiting salinity. Each of the remainder exhibit quite distinct salinity features (Warren 1962). Some, such as Balsdean, Shoreham and Burpham are fast and 'flashy', others, such as Goldstone and Sompting, are slow and gradual. This is typical of coastal sources in fissured aquifers.

Sites tested as part of South Downs investigation

The South Downs Investigation centred on locating, testing and developing new sources of supply so that the groundwater resources of the 'blocks' could be developed to meet the growing water demand of the coastal seaside towns. During the course of the 12 years of the investigation some 20 boreholes were drilled and tested and 11 were subsequently developed and integrated into the supply networks of the various water undertakings (Fig. 3).

The methodology adopted generally was to sink three 250 mm diameter trial boreholes at each test site. These were spaced in a way most suited to the site conditions. They were drilled to about 50 m below the water level with sufficient lining to permit acidization. If bailer yields were promising, acidization followed and the boreholes were tested for several hours with the water discharged overground nearby. Subsequently, boreholes showing sufficient promise were re-drilled to large diameter (610–760 mm), and the others retained for observation. If none of the test boreholes showed promise, then one was retained as a long-term observation borehole and the others filled in. Generally, this method of testing proved successful.

In summary, in the Seaford Block, the most easterly of the five Chalk blocks, Rathfinny was developed as a new source and adopted by the then Eastbourne Waterworks Company. One at Mount Caburn was not proceeded with because of its difficult location. In the Brighton Block, new sources at Withdean, Housedean Farm and Mossy Bottom were developed, while sites at Balmer Down, Houndean Farm, Benfield Valley, and Ladies Mile were unsuccessful. In the Worthing Block, sources at Angmering, Clapham and Warningcamp were tested successfully and developed, but others at Myrtle Grove, Annington Farm and Lychpole Farm yielded negligible quantities. Lastly, in the Chichester Block to the west, a new source was developed by Portsmouth Water Company at Brickkiln Farm as an 'outstation' to their Chichester source, while new sources were successfully tested for Southern Water at Madehurst and Tortington. One at Park Bottom, Arundel was deemed to be too environmentally sensitive.

Around 1981 all the potential dry valley sites in the main blocks were reviewed and it was concluded that apart from some sites for which, in any case, access was a problem, there were no more dry valleys worth exploring.

Eastbourne/Seaford chalk blocks

Here the main issue centred around Eastbourne Waterworks Company's new Rathfinny source which lay within its statutory supply area, but close to two Mid Sussex Water Company sources. It would have been impossible for Southern Water to licence this source because of the clear derogation of these which would have occurred. Consequently, Eastbourne transferred Rathfinny to Mid Sussex so that it could be operated collectively with their own sources, thus promoting sound aquifer management.

Chichester Block

Most of the Chichester Block fell within the supply area of Portsmouth Water Company which was well endowed with high-yielding licensed sources in the Chalk. It is noteworthy to record the fabulous springs at Havant and Bedhampton which have a combined output of over 100 Ml/d. This water arises at the southern margin of the Tertiary syncline but it is derived from a substantial outcrop on the Chalk Downs to the north. John Day of BGS considered that groundwater flowed beneath the Tertiary strata through an upward arch in the synclinal axis, focussing groundwater flow to these springs (Day 1964).

Infra-red aerial survey

In 1973 WRB organized an aerial infra-red linescan survey along the Sussex coast to see if shoreline springs could be detected (Davies 1973; Brereton & Downing 1975). The survey was run in February between Bognor Regis and Eastbourne but while it was moderately successful, it did not lead to any great surprises.

Downhole geophysics

The South Downs project was the focus of a great deal of downhole geophysical work done by WRB, WRc and later Southern Science. Numerous 200–230 m observation boreholes were drilled along the Brighton frontage. The logging showed, as expected, that there is no conventional Ghyben-Herzberg wedge front between fresh and saline water. Moreover, each borehole had unique salinity features. Some showed a simple increase of salinity with depth, increasing and diminishing on the tide. Others had complex salinity profiles with the saline front moving in rapidly on the tide along major fissures and dissipating equally rapidly on the falling tide (Monkhouse & Fleet 1975).

Mathematical models

Several mathematical models have been constructed of the various blocks, the first by Dan Nutbrown at WRB in 1971–1973 (Nutbrown 1975; Nutbrown *et al.* 1975). Later, Robert Bibby at WRc recalibrated it using extended and improved recharge data. The idea was to use the models through the summer months to forewarn operators of impending problems. However, this concept was flawed because the model could not accommodate aquifer salinities and, moreover, aquifer fissuring is so unpredictable that once pumping levels fall below those previously encountered one is in the realm of the unknown and borehole sources can fail at any time. Subsequently, Terry Keating took over the Brighton Block model for Southern Water and the figures above relating to historic storage values are based on his work.

In the 1980s a conventional finite-difference model of the Chichester Block was successfully constructed by Jim Thomson in Southern Water. This model represented reasonably the hydrogeological conditions in this block. In particular, it confirmed the easterly deflection of southward-flowing groundwater by a low impermeable zone and thereby accounted for the substantial flows emanating around Arundel.

Later work

Some later work was carried out by Southern Science on the South Downs project. This concentrated on optimizing the existing licensed quantities, with some new sources being developed and test pumped for lengthy periods. A review of all the groundwater abstraction licences has been undertaken since and recommendations made concerning variations to the licences of a number of sources to facilitate their most effective operation.

In the 1990s the British Geological Survey, as part of their National Groundwater Survey and in collaboration with others, undertook a major review of the South Downs chalk aquifer (Jones & Robins 1999).

Itchen River Augmentation

Background

The Itchen River Augmentation Scheme had its origins in a desk study which the author carried out in 1970 following the First Hampshire Periodic Survey in that year. The Itchen catchment is distinctly unusual. Above its main river gauging station near Eastleigh, to the north of Southampton, it has a surface area of 360 km^2 but its groundwater catchment determined from biannual well surveys is 484 km^2 (Figs. 5 & 6). Moreover, this varies seasonally from 477 to 490 km^2 being greatest in the autumn (Headworth 1967). Three tributaries feed the upper catchment to Easton above Winchester, namely the Candover, Alre and Cheriton Stream. The central of the three, the River Alre is the largest, while a considerable proportion of groundwater bypasses the Candover and Cheriton Streams to discharge along the main River Itchen above Easton. Between the confluence of these three tributaries below Alresford and Easton, some 6 kms downstream, the river increases in flow by 75%.

Alresford is a major cress-growing area and artesian boreholes located at the 12 cress farms provide a significant proportion of the flows. The 1970 desk study postulated river augmentation boreholes in each of these three tributary catchments, as well as to the north and south of the main river above Easton to intercept the bypassed flow.

In January 1972 Hampshire River Authority (HRA) applied to the WRB for approval for a pilot scheme in the Candover catchment under Section 18 of the Water Resources Act 1963. Later, an application to authorize the planned river discharge was sought under the then recently-issued Water Resources Act 1971 which Parliament had rushed through to plug a gap in the 1963 Act. HRA was the only organization to seek to limit its liability using this short Act for the potential consequences of a river discharge (mainly flooding). However, this did not prevent a one-day local public inquiry in January 1973. The discharge order was subsequently obtained.

Stage I: The Candover pilot scheme

The early 1970s was the 'age of the pilot scheme' for river augmentation schemes using groundwater (Thames, Great Ouse, Shropshire, Wylye) (Downing *et al.* 1981; Headworth *et al.* 1983). With the idea of using upland boreholes to augment the summer flows of chalk streams still unproven, it was considered essential that the efficacy of the Itchen scheme was demonstrated by means of a pilot scheme. The Candover valley was chosen although the risk was recognized that the effects of pumping could spread into adjacent catchments. Much time and effort was spent meeting farmers, landowners and householders to explain what the scheme entailed. Public meetings were held and a lot of goodwill was generated. Local residents and farmers were content to let the National Farmers Union (NFU) and Country Landowners Association (CLA) act for them and a joint agreement was made with both organizations providing protection to the various local interests (Headworth & Giles 1977).

Watercress interests

Over half the watercress grown in the UK comes from Hampshire and the largest concentration of cress farms is in and around Alresford (Headworth 1978). The cress growers were understandably very nervous about the effects which the Candover

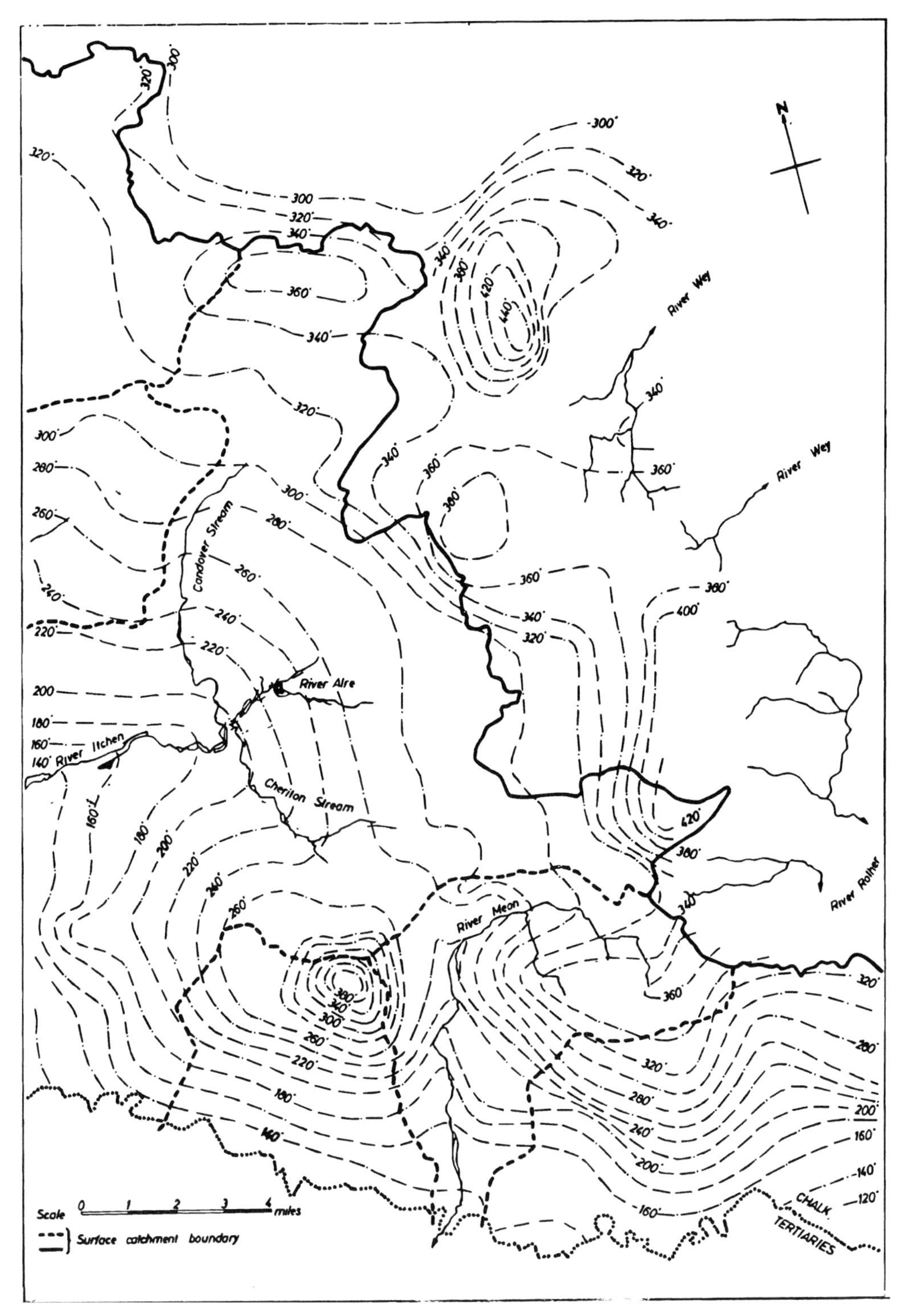

Fig. 5. Groundwater contours (autumn 1967) for the upper Itchen Catchment.

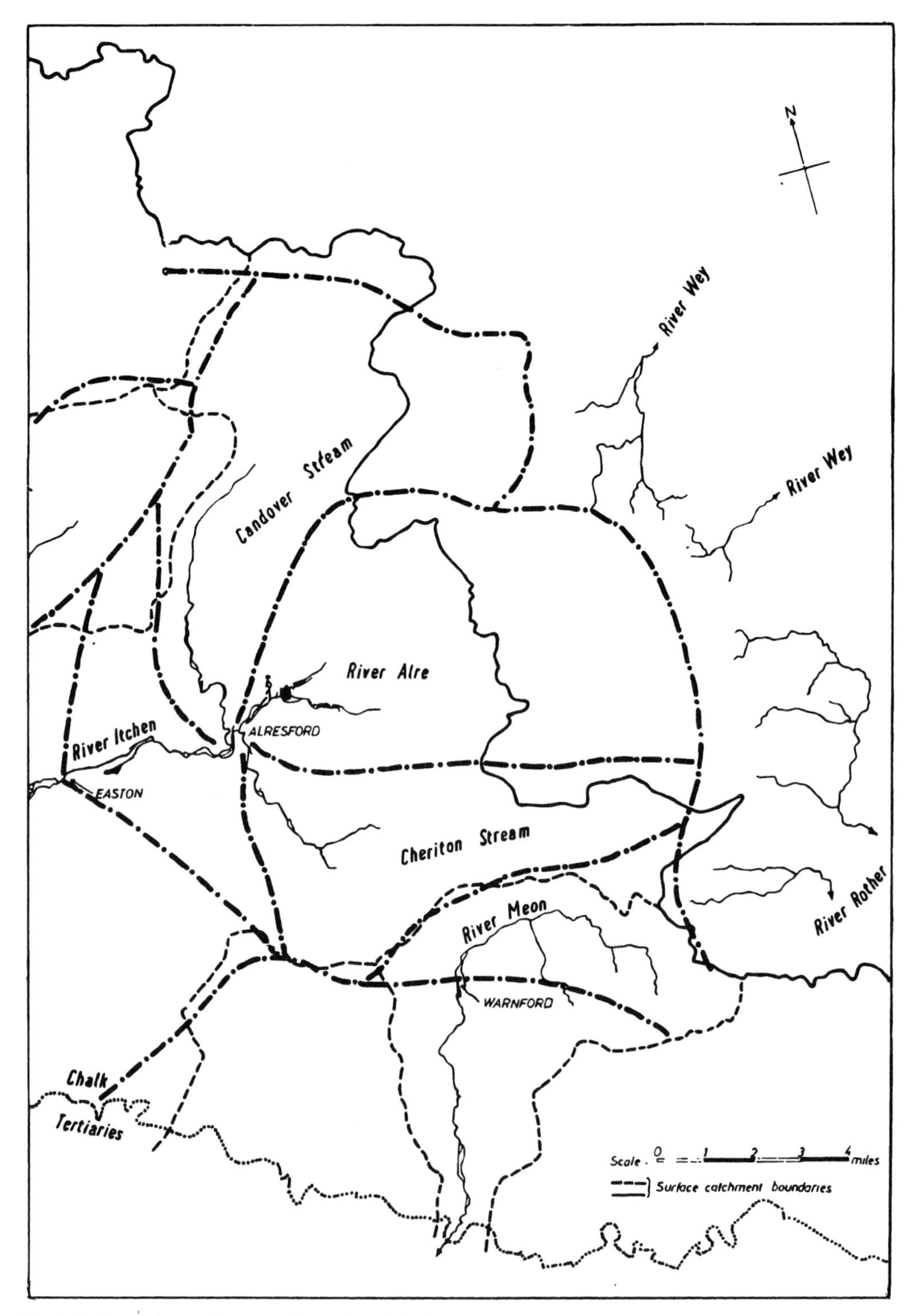

Fig. 6. Surface and groundwater catchment boundaries for the upper Itchen tributaries (autumn 1967). The thicker alternatively dashed and dotted lines show groundwater catchment boundaries.

scheme might have on their spring-fed beds and their artesian boreholes. The latter, not being pumped, were seen to be at risk from the effects of the Candover scheme pilot boreholes. Consequently, regular meetings with all the farmers were initiated in 1972 and maintained throughout the scheme. Generously, the cress growers were prepared to let the pilot scheme proceed unhindered and not seek to block it as they might have done. From 1972 the flows from some 60 artesian boreholes were measured fortnightly at 12 cress farms.

Initial drilling and testing

The pilot scheme comprised the sinking and testing of six full-size boreholes, two each at three different sites. They were located around the village of Preston Candover, some 7 kms up the valley from discharge points at Northington. The latter were a short distance below the perennial head of the river. Numerous observation boreholes were drilled at the outset in and around the valley using a small percussion rig purchased by HRA. Drilling of the six pilot production holes started in 1974 under the direction of Derek Giles, ably supported by his team of Mike Packman, Terry Keating and later Valerie Lowings. In order not to be seen to be pre-judging the success of the scheme, 13 kms of temporary, ductile-iron pipeline were laid over-ground from the three borehole sites to the two discharge points above and below the village of Northington.

Scientific measurements, testing and modelling

The drilling, testing and pipeline construction were part of a comprehensive scientific programme of measurement designed to assess the effects on the river by the discharges. In 1972 innovative and lengthy field investigations were designed and undertaken by biologist Peter Soulsby to assess the impact on river macrophytes and invertebrates. Equally detailed fisheries studies were carried out by Robin Templeton and later Don Paterson. Mike Beard measured potential changes in river and groundwater chemistry. Some first-class science was done in all these fields.

With six months continuous testing being planned, it would have been impossible to project natural river flow and natural groundwater level recessions over a period of six months to assess the effects of pumping. Consequently, a parallel set of readings were essential in order that true comparisons could be made. The Cheriton Stream catchment to the south of Alresford was used for this purpose. Comprehensive measurements were instigated in 1972 for all the resources and scientific studies.

The six 250 mm diameter pilot boreholes were tested individually in 1974. Large diameter production boreholes were drilled alongside them in the winter of 1974–1975 and in the summer of 1975 repeat tests were conducted. The full test proper started in early May 1976. This proved an ideal year to test the scheme fully since groundwater levels, river flows and river water levels were at a record low with a return period of 1 in 150 years. Not only did the discharges double river flows but, importantly, the discharge of cool groundwater in the hot summer months saved much of the fish life from severe distress. The six months testing was phased out so that all pumps were switched off at the end of November. Following very heavy rainfall in the autumn and winter of 1976–1977 the extensive 14 km-wide cone of depression filled up rapidly and river flows were back to normal by the following spring.

It proved too difficult to construct a conventional two-dimensional finite-difference model so Keating (1982) used a simpler lumped parameter model to simulate a thin highly-permeable layer above the main aquifer which ‘creamed off’ much of the water from the system (Headworth *et al.* 1982). This model proved successful and gave the clue to the very unusual conditions which existed in the upper part of the catchment where a thin layer (less than 5 m) with very high T and high S values existed.

Main results

The six production boreholes were tested with their combined pumping rate declining from 31 to 26 Ml/d over the trial period. 5.1 Mm^3 were abstracted in total. This represented 25% of the average annual flow of the Candover Stream and 42% of the total flow in 1973, the previous driest years. The cumulative net gain in stream flow in the Candover Stream was 80% of the volume pumped. This volume represented 60 mm of rainfall over the 65 km^2 Candover catchment and 25 mm over the whole area of the cone of depression which developed. The cone was almost circular and was some 14 kms across (Fig. 7). The catchment was found to be unusual with the presence of a thin, 5 m thick, highly permeable zone below the water table which had a storativity of 5% to 7%. Over a greater depth, transmissivity and storativity averaged 4300 m^2/d and 1.3% respectively. Across the whole cone of depression, the average storativity was 1.1%. This increased from 0.8% to 3% during the period of the test as the cone developed progressively into the peripheral higher-S zones at or close to the water table. So extensive was the cone of depression that it extended into adjacent catchments and other rivers were affected. Taking

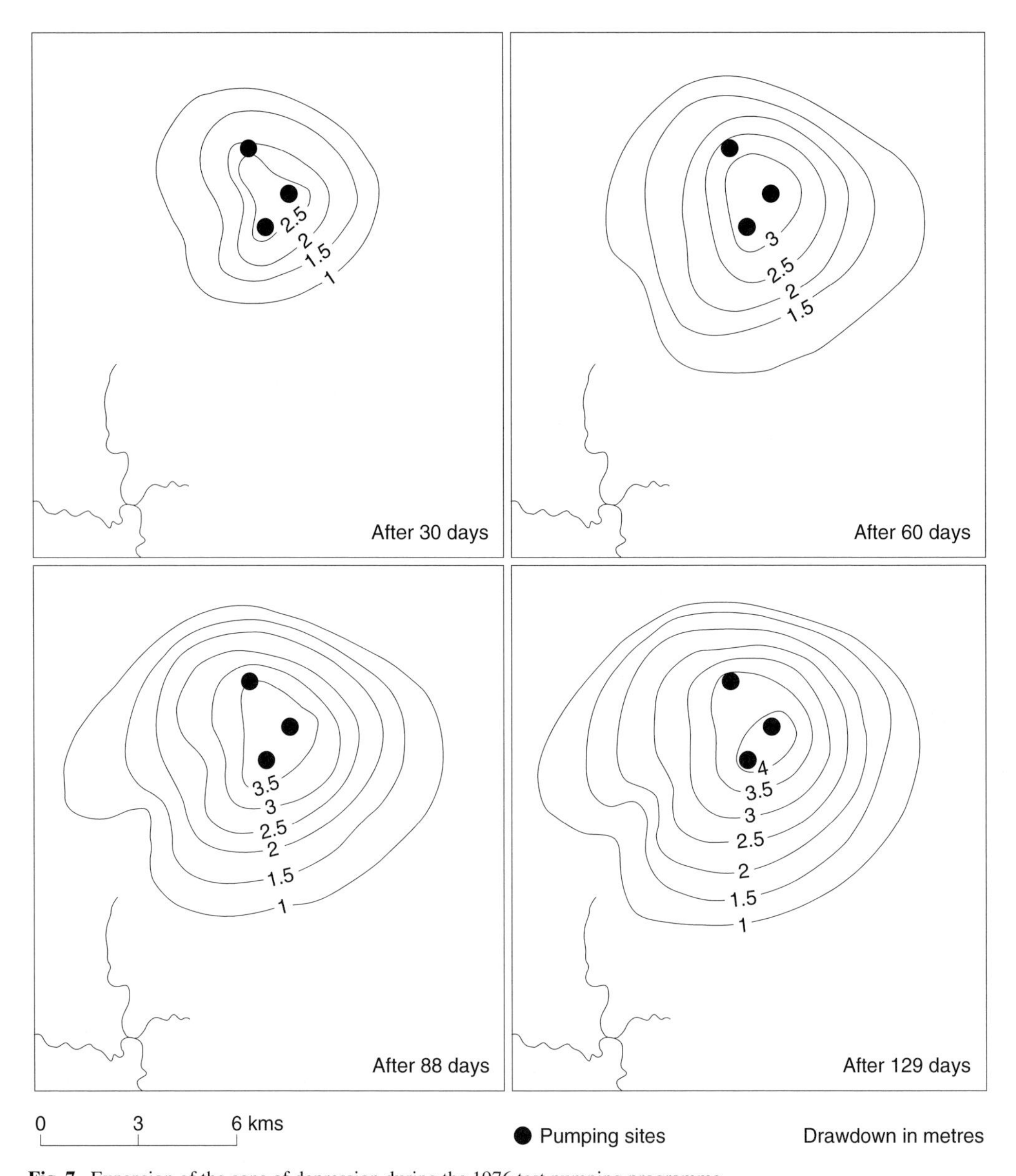

Fig. 7. Expansion of the cone of depression during the 1976 test pumping programme.

the depletion of these rivers into account the overall cumulative net gain for the Candover Scheme was 70% of the total pumped volume (Fig. 8).

It is pleasing to conclude that so successful were the consultations, discussions, public meetings and publicity information during the life of the scheme that it was authorized without a public inquiry being held. The final construction phase of the scheme was undertaken by Divisional Engineers between 1981 and 1984 and it was used for the first time in November 1985 and later in 1989–1990. Soon afterwards the then National Rivers Authority took over responsibility for the scheme.

Stage II: the Alre scheme

This had its origins in a desk study undertaken by Valerie Lowings in November 1980. The Candover scheme had already shown the vulnerability of cress farms to large-scale pumping in the upper reaches of these catchments. It was clear that the only conceiv-

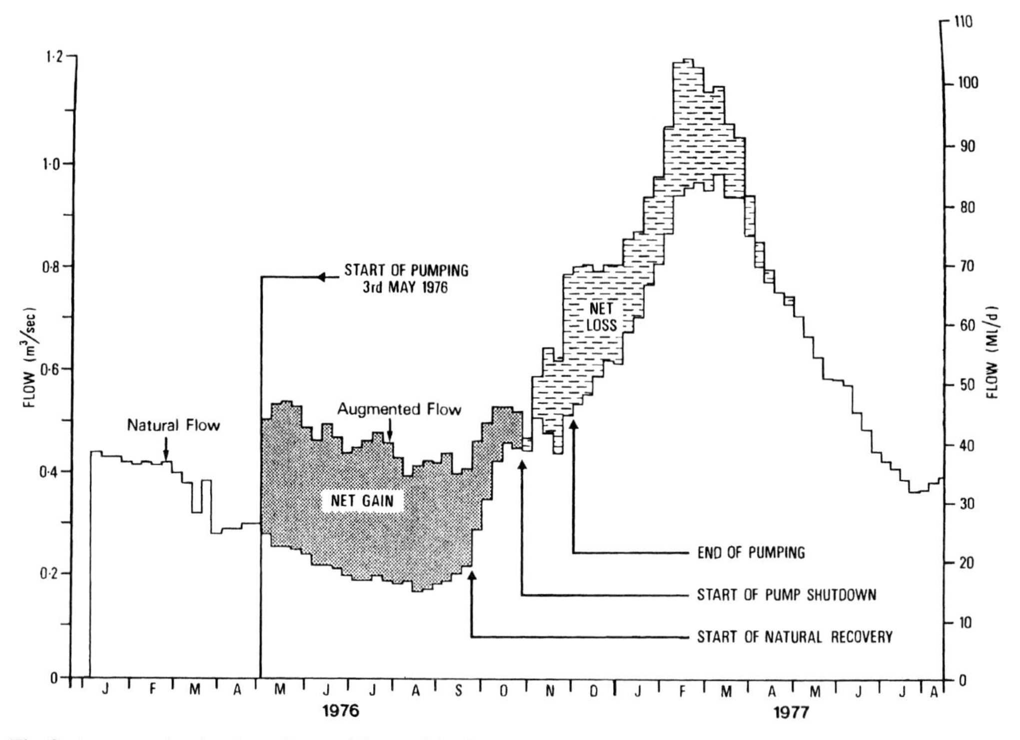

Fig. 8. Augmented and estimated natural flows of the Candover Stream during the 1976 test pumping programme.

able way of developing groundwater for augmentation in the Alre catchment was to pipe the pumped water directly to the cress farms themselves so that they could have direct benefit of the abstractions. Water temperature is a key element in the successful growing of cress in the winter. Consequently, a buried pipeline, which would help conserve the temperature of the pumped water, had to be envisaged from the start. Thus the dye was cast for two of the controlling elements of this scheme (Lowings & Midgley 1989).

As part of the author's original concept in 1970 it was envisaged that boreholes sited on the flanks of the Itchen valley might provide a short but useful boost to flows at the end of a summer of augmentation. However, while sites were actively sought and tested on the north side of the river, yields were poor and this element of the scheme was abandoned.

As for the Candover scheme, an extensive network of observation boreholes and gauging stations was established at the outset. In all, 16 pilot boreholes were drilled and tested but only four were developed into production boreholes. Each had a yield exceeding 14 Ml/d. The 'up-catchment' boreholes were amongst those with the poorest yields. These four production boreholes were successfully tested individually in 1984 for up to two weeks each (Giles & Lowings 1989).

CCTV inspection and the testing showed that karstic solution features in the Chalk resulted in transmissivities that were generally much higher than those in the Candover catchment while storativities were lower. These conditions are not ideal for obtaining high net gains.

An asbestos cement pipeline was laid and buried at the outset. Two discharge points were selected at the cressbeds at Bishops Sutton and Drayton, on the southern and northern tributaries of the Alre respectively, and water from two boreholes was discharged to each via separate pipelines.

Full testing and results

Operational testing started in May 1989 and, because of the severity of the drought that year, pumping had to continue at full rate until mid-December. One of the four production boreholes suffered a dramatic reduction in yield from its initial 14 Ml/d at the start to less than 1 Ml/d by the end of the summer. This was a result of the dewatering of the large karst-like fissure supplying the majority of the borehole yield. However, the high yields of the remaining three boreholes were sustained throughout the year. In total, 10.7 Mm^2 of water was

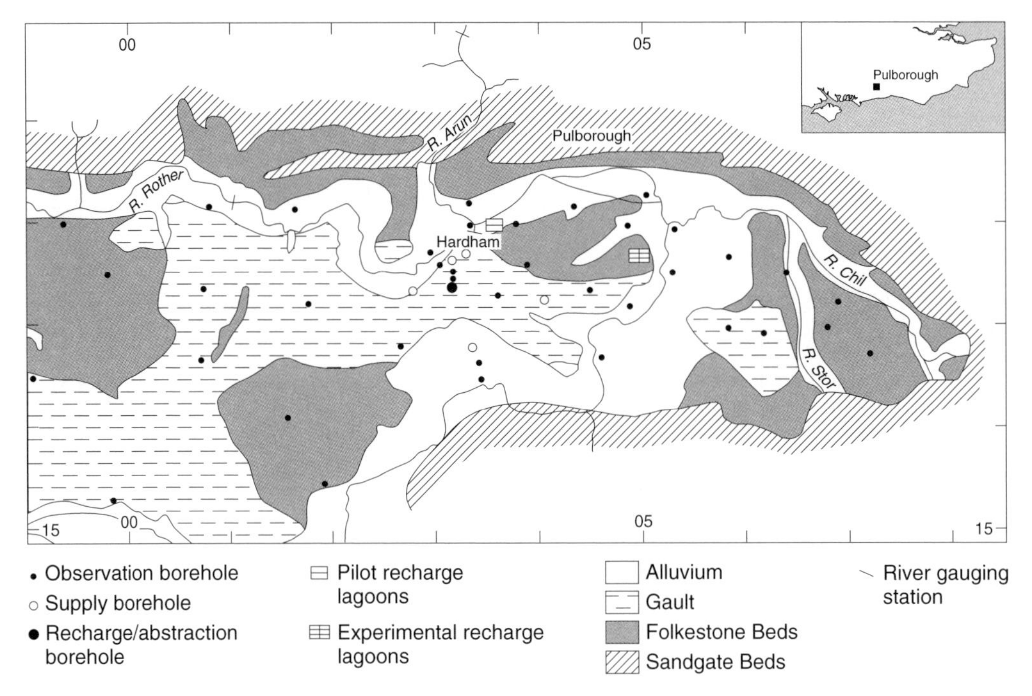

Fig. 9. Location map and geology of the Hardham Basin, Sussex.

pumped, more than twice the volume pumped during the 1976 testing of the Candover Scheme.

Compensation discharges to the cressbeds in the Alre catchment were provided by valve-controlled off-takes from the main pipeline. Unfortunately, because of unforeseen problems with mains power failures, it was deemed safer for the cressgrowers to rely on their own boreholes, pumps and standby generators.

The cone of depression from the pumping spread over a very large area into the Candover, Cheriton and Meon catchments, as well as the Alre catchment (Barker 1990). This necessitated remedial works on 21 private sources, including the lowering of existing pumps, the deepening of boreholes, drilling others and connections to mains supplies.

As with the Candover Scheme, a high profile was given to maintaining good public relations, particularly with the local parish councils, landowners, cress-growers and licence holders. Again, this paid dividends, as the scheme was ultimately promoted and licensed without the need for a public inquiry.

Under true operational conditions in a severe drought, it is intended to operate the Candover Scheme first (due to its higher net gains) and to bring the Alre Scheme on line if needed at a later date (Giles *et al.* 1988). It is not envisaged that the large volumes of water pumped in 1989 from the Alre Scheme would ever be required. On this basis, an application was made by the Environmental Agency to abstract 5.1 Mm^3 of water per year, about 50% of that abstracted from the Alre Scheme in 1989. This was granted in April 1994. So nearly 25 years after the first desk study was undertaken the whole Itchen River Augmentation Scheme became operational.

Hardham recharge

The Hardham Artificial Recharge Investigation was started by Sussex River Authority in 1968 with Geoff Fox maintaining his involvement throughout. The lagoon recharge experiments were conducted by Marcus Tague supported by David Izatt. Mike O'Shea oversaw the latter stages of the investigation which involved borehole recharge. WRc made an enormous contribution through Ken Edworthy, Jeremy Joseph, Sue Cullen and Robert Bibby.

The Geology of the Hardham Basin

Geologically, the Hardham Basin is unusual. It lies along the axis of the Wiggonholt syncline but at Hardham the Lower Greensand strata are folded downwards to form an oval-shaped basin 9 kms long and 3 kms wide (Fig. 9). The basin is endowed with two sand aquifers, the Folkestone Beds and Hythe

Beds, which are separated by the clayey Sandgate Beds. The principal aquifer, the Folkestone Beds averages about 50 m in thickness. The basin is filled with Gault Clay up to 50 m thick in the centre. The river valleys of the Arun and Rother meander across the northern and eastern margins of the basin and their valley deposits mask the older sediments with ribbons of clayey alluvium. Valley gravels are also present to confuse the simple picture of the solid geology.

Recharge at Hardham

Several factors led to the instigation of artificial recharge at Hardham. North West Sussex Water Board started river abstraction with treatment in 1954 and commenced groundwater abstraction from both aquifers in 1964, but the Hythe Beds were soon abandoned due to the high iron content of its groundwater. By 1971 groundwater comprised 65% of the total abstraction at Hardham. Groundwater levels had began to drop quickly and cracks appeared in structures sited on the soft alluvial clays. In an attempt to arrest both of these effects, two large lagoons were dug into the reworked sands and gravels which overlie the Folkestone Beds near the treatment works. Extensive recharge trials were conducted between 1967 and 1970 using treated river water. These showed that rates of recharge of 2.5 Ml/d could be sustained as could a vertical infiltration rate of 1.4 m/d (Fox 1969; Ellson 1973).

By 1972 the Water Resources Board was introducing new hydrogeological ideas into the UK and encouraged the resumption of recharge trials at Hardham but, through the leverage of financial support, sought a 'purer' site for the work where recharge could be carried out directly into undisturbed Folkestone Beds strata. The site chosen was alongside the river Arun not far to the south.

Much of the Folkestone Beds outcrop is covered by clayey alluvium which severely limits direct natural recharge. Consequently, the average precipitation recharge to the basin is only 9.1 Ml/d and this is very small in relation to its storage capacity. When the basin is full it overflows over its western lip and contributes to river flows. However, when the groundwater levels are drawn down these outflows are prevented and the average annual recharge to the basin is increased to 15 Ml/d by induced recharge from the rivers. This limited natural recharge to the basin explains why groundwater levels fell so quickly in 1971.

Experimental lagoon recharge

Three narrow, trapezoidal rectangular lagoons, each 160 m long, were constructed in 1971–1972 close to the river Arun. Extensive recharge trials were undertaken between 1972 and 1974 using one, two or three lagoons and water pumped from the river (Izatt *et al.* 1979). No pre-treatment was undertaken. The results showed that:

(1) individual rates of recharge varied between 2.6 and 5.9 Ml/d for each lagoon with a maximum of 6.8 Ml/d when they were used together.
(2) rates of recharge declined quickly because of clogging and if untreated river water was to be used then regular bed cleaning would be required.
(3) no more than two lagoons are worth recharging at the same time. One is best, and;
(4) pre-settlement of the river water is advantageous while some chlorination would be beneficial.

In essence, lagoon recharge is basically low-tech and easy and would readily sustain aquifer storage. However, because of problems with access, the lagoons were back-filled in 1978, the experimental work having finished.

Artificial recharge using boreholes

With the completion of the lagoon trials, attention turned to the feasibility of borehole artificial recharge into the Folkestone Beds using winter surplus, fully-treated water (Monkhouse & Phillips 1978). With the cleaning of boreholes being far more difficult than the floor of a clogged lagoon it was clear that very high quality water with a turbidity of no more than 1 or 2 FTU was necessary for successful and prolonged borehole recharge.

A full-size, abstraction/recharge borehole was constructed in 1976 and lengthy recharge experiments undertaken, led by Mike O'Shea. Great care was taken with this borehole which was drilled by reverse circulation methods to 63 m below ground level through 15 m of Gault Clay. The final diameter was 1050 mm and it was finished with stainless steel wire-wound screen. 10 m of blank lining was installed at the bottom to house the abstraction pump and recharge valve outlet. Around the annular space, filled by graded sand pack, six 60 mm diam. backwash tubes were installed to depths between the recharge water level and the pumping water level. This was so as to permit clean water to be injected into the pack and thereby wash any particulate matter from the pack and hopefully from the sides of the borehole. Four 'tremie' tubes were inserted in the top of the sand pack so that it could be topped-up.

Recharge experiments were conducted in 1980 and 1981 (O'Shea 1984). Rates of almost 4.3 Ml/d were achieved which were comparable to abstraction rates. Several small-diameter observation boreholes were drilled close to the recharge borehole and a novel method adopted for using them. Their bases were sealed and they were filled with water. This water column was monitored for temperature since a static water column reflects temperature profiles in the aquifer into which they lie. These demonstrated that the recharged water did not mix with the native groundwater, but displaced it over a wide front, fingering outwards along particularly permeable layers. The main lessons to be learnt from the borehole experiments were:

(1) recharge can be successfully achieved in the Folkestone Beds,
(2) careful design is essential although a constant-rate recharge outlet valve would be adequate for operational recharge and this would facilitate the use of a smaller diameter borehole,
(3) back-wash tubes and expensive stainless-steel wire-wound screen are unnecessary,
(4) very high quality water is necessary for artificial recharge and some pre-chlorination is necessary if it is not of potable-standard, and;
(5) with potable-standard water then only short period of pumping to waste are needed to 'clean' the borehole between phases of recharge.

Aquifer model

A two-dimensional, finite-difference model of the Hardham Basin was constructed by Robert Bibby of WRc using a variable grid size. It provided for the varying thickness of the Folkestone Beds so that with a uniform value for permeability, modelled transmissivity varied from 500–1000 m^2/d. A uniform value of storativity of 20% was adopted. The model allowed for induced leakage at times of river flooding when the Rother and Arun were out of their banks. In summary, the model found that full basin storage was 158 Mm^3. A good calibration was obtained with the model using an overall aquifer permeability of around 15 m/d and transmissivity between 500 and 1000 m^2/d.

The problems with consolidation

Quite soon after groundwater abstraction commenced in 1964, problems were encountered by the water board with the box culvert which housed the water mains across the Arun river valley. Cracks in this structure arose from the consolidation of the soft peaty alluvial clays in the valley due to abstraction from the Folkestone Beds beneath. Problems with the culvert continued for years and indeed the settlement effects were much more widespread.

To guard against allegations of ground settlement in the future. Professor Tom Hanna of Sheffield University was retained to advise on the installation of boreholes to monitor any long-term settlement of the clays in the Hardham area. Some boreholes possessed in-situ transducers to measure pore water pressure changes in the clay. Others comprised telescopic tubes resting on the bottom of the borehole with magnetic targets attached to them at various depths lodged in the clays in the sides of the hole. These magnetic targets move upwards following any settlement of the clays. All were measured periodically by portable field equipment.

Tilmanstone minewater contamination

The contamination of the Chalk aquifer in East Kent from the prolonged discharge of saline mine water drainage provides an illuminating if not salutary example of bureaucratic indifference and pigheadedness in the face of sound technical warnings.

In 1973 Kent River Authority commenced an investigation into the extent of the problem in conjunction with the WRB and this was continued by Southern Water with the WRC in the years thereafter. Those principally involved were Shami Puri, Graham Warren, Brian Rampling, Robin Brereton, Robert Bibby and David Oakes.

Herbert Lapworth (1930) had expressed early concern about the potential long-term problem to Chalk water resources from the disposal of mine water at Snowdown and Tilmanstone collieries (Figs 10 & 11). Stevenson Buchan (1962) repeated these warnings in 1962 although by 1948 the local water undertakings had began already to show concern about the pollution plume developing down-gradient of Tilmanstone mine (Unitt 1959). No real progress was made until they constructed and tested a supply borehole in 1968 in order to produced indisputable evidence of aquifer contamination. Even then no action was taken and the Coal Board continued to dispose of mine drainage into soakage ditches near the mine. By 1974 a pipeline had been laid from Tilmanstone to Betteshanger's mine drain and after 64 years the discharge to the Chalk finally ceased (Rampling 1974).

Origins of the contamination

Three coal mines operated in the Kent coalfield at Snowdown, Betteshanger and Tilmanstone. Betteshanger, the one closest to the sea, disposed of its minewater safely to the sea via a pipeline. Snowdown, located 5 kms to the west of Tilmanstone, discharged its drainage onto the Chalk (Foster Brown 1922; Plumptre 1959). This ceased in 1935 although a

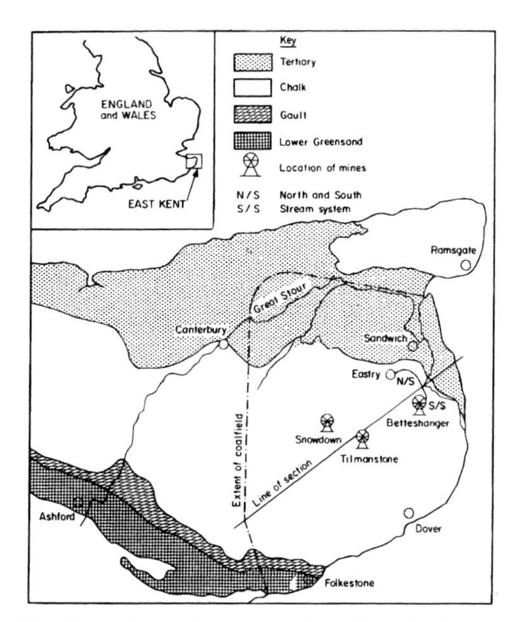

Fig. 10. Outline geological map of east Kent showing the extent of the concealed coalfield and the location of the former mines.

considerable pollution plume then remained. At Tilmanstone, where mine water was discharged into nearby lagoons, later analyses showed that between 1907 and 1974 some 187×10^6 m^3 of water had been soaked into the Chalk containing some 318000 metric tonnes of chloride. Until 1952 the quantities abstracted were very small with the chloride content less than 500 mg/l. With the opening of a deeper seam, the chloride rose to 1500 mg/l, increasing further to 3000 mg/l by 1965 when the discharge reached 11 Ml/d. If action had been taken earlier, even as late as 1960, the problem would have been significantly less.

Regular sampling and analysis and river gauging by Kent River Authority (KRA) of the North and South Streams which drained the pollution plume showed that by the middle 1980s some 47000 tonnes of chloride had dissipated from the aquifer, that is only 15% of that discharged from the mine.

Studies and the investigation

In 1948 the water undertaking commenced sampling 40 wells in the area in order to try to quantify the extent of the chloride pollution around Tilmanstone. Taking the 200 mg/l isochlor as the limit, the pollution plume was some 27 km^2 in area and fanned out north-eastwards along two broad arms down-gradient of the mine (Figs 12 & 13). Yet in 1972 little was known about the true area and depth of the pollution

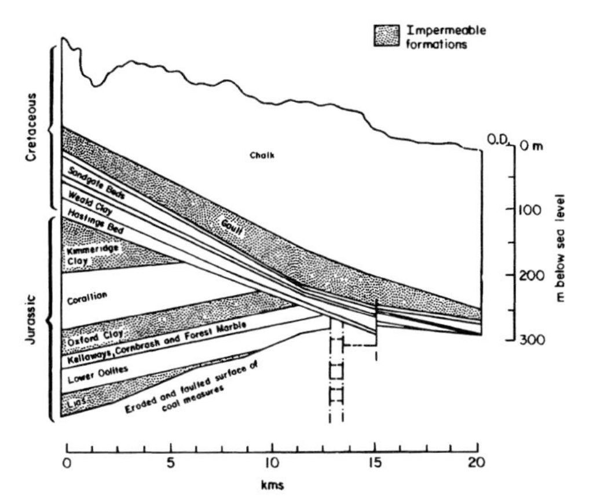

Fig. 11. Geological section across Kent coalfield.

plume or how the chloride pollution was distributed at depth in pores and fissures. Nor was it known how long it would take to recover naturally or through rehabilitation (the natural concentration of chloride in the Chalk in this area is around 50 mg/l, being influenced by salt-bearing rain).

In 1973, WRB commissioned a surface resistivity geophysical survey by Huntings Surveys which defined the pollution plume (Brereton 1974). Taking a resistivity of 30 ohm-m as defining the margin of the salinity plume, an area of 10 km^2 was deduced, as well as the presence of two levels of pollution in the aquifer (Brereton 1976). However, it was considered that the well sampling probably provided a better overall figure for the surface extent of the pollution plume.

Several cored GRP-lined boreholes were drilled at three sites and to different depths along the axis of the main dry valley running down from Tilmanstone. Sampling and analysis, down-hole logging and short test pumping showed that the pollution had extended to a great depth in the aquifer. The results tied in well with the surface geophysical survey. Stratification of the pollution existed within the Chalk but in general the extent of the pollution diminished with depth.

As mentioned above, a large-diameter borehole was sunk at Eastry in 1968 to demonstrate once-and-for-all the existence of the pollution plume. It was located down-gradient of Tilmanstone and along the main axis of the plume. It had chlorides of around 1300 mg/l. In 1977 it was connected to Betteshanger mine by pipeline so that water pumped from it could be piped to the sea. 15 months continuous pumping took place in 1977 and 1978 at rates around 5 Ml/d. No overall reduction in chloride concentration occurred, with it declining from 1350 mg/l to 1200 mg/l during the summer months, but rising again to 1500 mg/l in the winter and at other times during wet periods. The

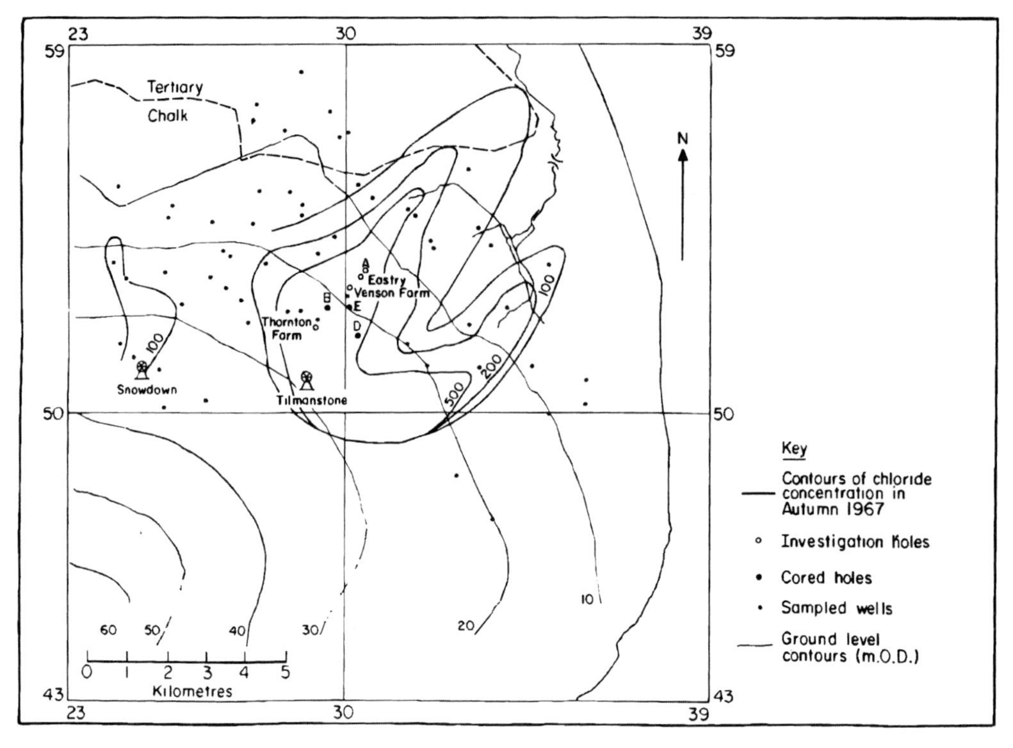

Fig. 12. Isochlores of contamination plumes from Tilmanstone and Snowdown mines for 1967.

results showed clearly that aquifer rehabilitation would be an extremely long and expensive process (Headworth *et al.* 1980).

Snowdown mine

An illuminating parallel to Tilmanstone occurred at Snowdown mine and it proved of vital importance to the study. There, by 1935 a total of 40000 tonnes of chloride had been discharged into the Chalk. A major public groundwater source with extensive adits existed at Wingham only 4 kms down-gradient. The natural chloride concentration at Wingham was 32 mg/l. This had risen to 110 mg/l in 1930 and 200 mg/l in 1934 when the abstraction rate had risen to 8.4 Ml/d. After failing to get the mine owner to abate the pollution, the water undertaking laid a pipeline in 1935 to convey the minewater from Snowdown into the Little Stour, 10 kms to the north. Chlorides continued to rise reaching a maximum of 330 mg/l in 1936, thereafter declining to 70 mg/l in the early 1950s and only 45 mg/l in the 1980s. The slow decline in chloride from a relatively modest level of aquifer contamination had considerable implications for the future rehabilitation of the aquifer at Tilmanstone.

Mathematical models

Models proved to be the key to understanding the Tilmanstone pollution (Bibby 1979, 1981). Two models were constructed. Firstly, a hydrodynamic, finite-element model was constructed using computed recharge data, river flow data and known boundary conditions. Once this was calibrated it was linked to a dispersion model which had parameters for aquifer dimensions, chalk porosity, fissure sizes, chalk block sizes, permeability and storativity, plus the historic volumes of water and contaminant discharged. The historic case of the Snowdown mine drainage was indispensable in calibrating the model over an extended period.

The results explained the failure to achieve any aquifer clearance during the 1977–1978 testing at Eastry. The model showed that some 87% of the chloride discharged onto the Chalk remained in the system and that this would reduce to 30% by 2008. It was thus evident that aquifer rehabilitation would be an extremely long and expensive process. Significantly, the model showed that it would be futile to attempt to rehabilitate the aquifer artificially. This was because groundwater pumped from Eastry would discharge naturally into the North and South streams within a year or two and, therefore, pumping

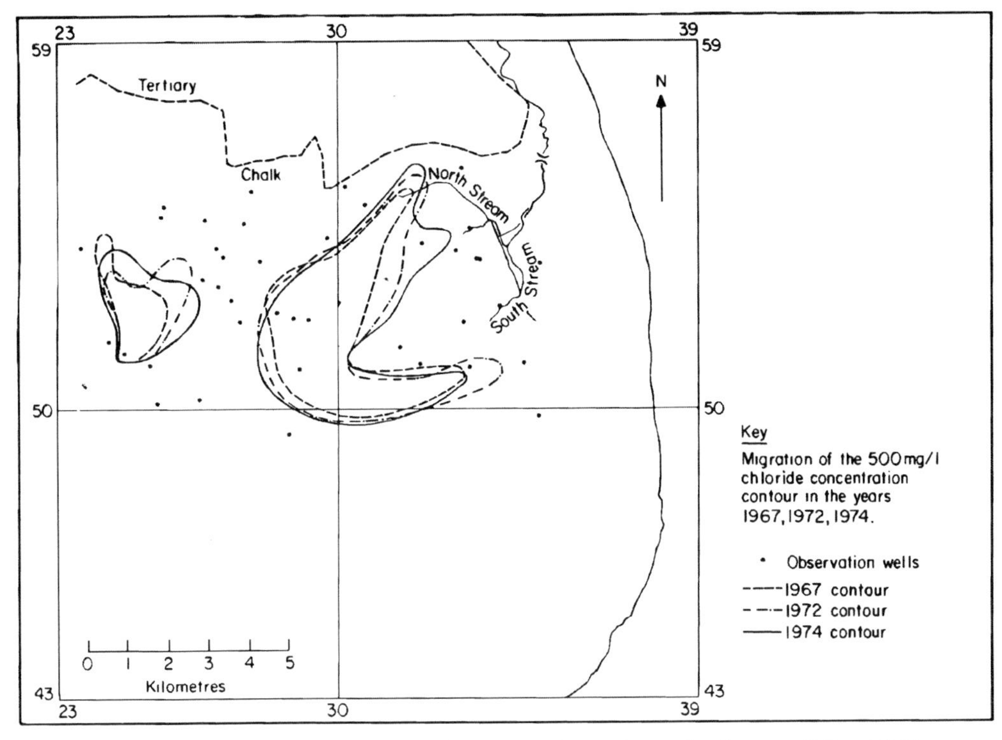

Fig. 13. Migration of contamination plumes from Tilmanstone and Snowdown mines between 1967 and 1974.

would only advance the clearance process by a small amount.

Lower Greensand river augmentation, Isle of Wight

The Isle of Wight groundwater scheme evolved into a river augmentation scheme using groundwater contained in Lower Greensand strata (Fig. 14). It was conceived and executed by Mike Packman. Obtaining substantial yields from the difficult, fine-grained Lower Greensand aquifer of the Isle of Wight was a notable achievement (Packman 1985).

Being a popular holiday centre, the demand for water on the Island nearly doubles during the peak period. The severe drought of 1976 spurred Southern Water to investigate the marginal aquifer of the Lower Greensand whose grain-size is significantly finer than on the mainland and which comprises predominantly glauconitic or limonitic silty fine-medium sands with silts and clays. It also exhibits extreme vertical and lateral variations in lithology. Existing borehole yields were poor and they often suffered from sand ingress and/or encrustation due to the ferruginous nature of the groundwater. The Lower Greensand strata cover a relatively small area and show wide variations in conditions from unconfined to totally confined. In its thicker parts it is multi-layered and leaky. Because of this it was decided that any development should utilize the storage potential of the confined aquifer. The scheme devised involved augmenting the River Eastern Yar with groundwater at times of low flow to permit increased river abstraction and treatment at Sandown.

An investigation was undertaken between 1977 and 1981. Following the sinking of fully-cored boreholes, two trial production boreholes were sited towards the northern and southern margins of the area with the greatest groundwater development potential. Promising results were obtained, with sand-free yields of over 3 Ml/d. Following this, six further production boreholes were drilled between 1984 and 1987; two were used for direct supply (Fig. 14). All of them were constructed by the reverse-circulation rotary method using an inhibited polymer drilling mud as the flushing medium and lined with stainless steel wire-wound screen and plain casing. Borehole depths varied from 93 to 180 m with finished diameters of up to 333 mm. A combination of poly-phosphate dispersants, high pressure jetting and air-lift clearance pumping was found to be the most effective method for yield development with

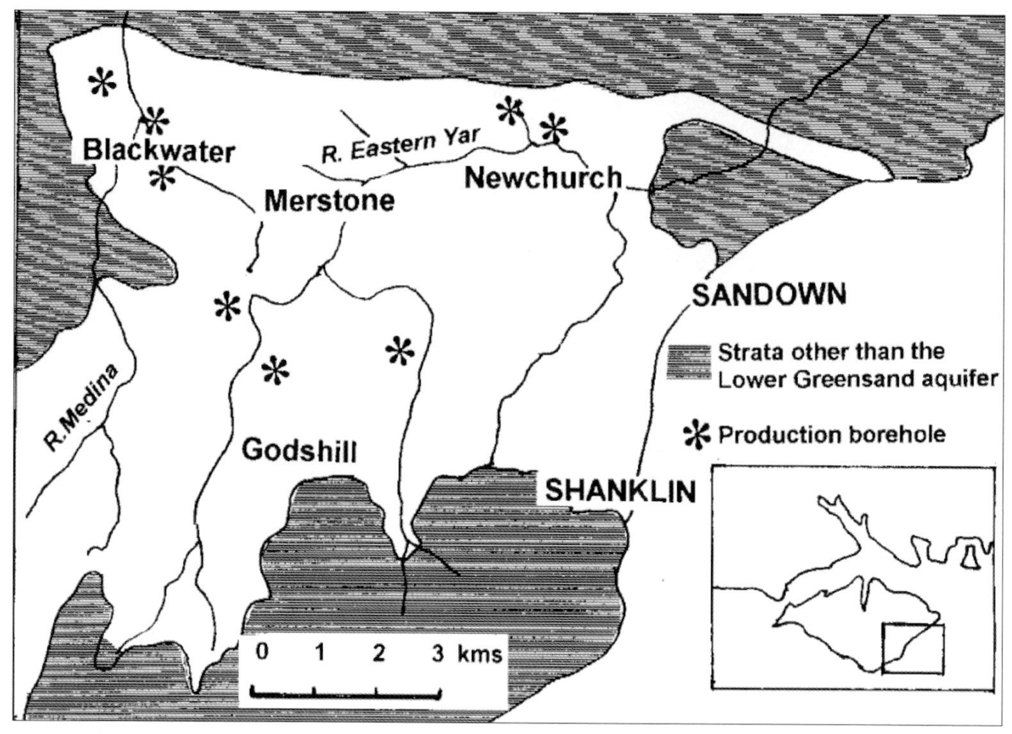

Fig. 14. Lower Greensand outcrop of the Isle of Wight and the location of the groundwater scheme's production boreholes.

Wellmaster flexible rising mains selected to reduce the build up of iron encrustation.

Three river augmentation boreholes (Stage 1A) were located in the upper part of the Eastern Yar catchment so that groundwater could be discharged directly into the river for re-abstraction at Sandown. The three Stage 1B boreholes were situated in the lower part of the Medina catchment and the pumped water transferred by pipeline to the River Eastern Yar. The boreholes were test pumped in 1985 with yields varying from less than 2 Ml/d to over 3.5 Ml/d. The 1989 test of all six augmentation boreholes lasted six months and achieved a maximum combined rate of some 16 Ml/d which declined by only 10% at the end. The piezometric surface for the Lower Greensand aquifer was defined successfully which enabled the extent and configuration of the cone of depression to be accurately determined.

No significant derogation of any of the Protected Right sources in the area was detected, nor were any of the environmentally sensitive areas, identified by the Nature Conservancy Council, affected adversely by the scheme. However one of the boreholes did significantly deplete flows in a tributary of the River Medina and therefore the scheme allowed for a compensation discharge to be made to the stream. The regression model, using data from a control catchment, showed that the cumulative net gain fell gradually from 100% to 70% during the test. Following an environmental study by the NRA, the latter issued an abstraction licence for the scheme in 1991, to be reviewed after seven years.

Denge aquifer management

The Denge Hydrogeological Study was undertaken in the glare of opposing interests following inconclusive public enquires into further aggregate extraction from Denge Beach. It is pleasing to record that a good deal of trust and co-operation developed between the parties during the study leading to an agreed final report. The study provided a detailed and thorough understanding of an unusual aquifer and involved different techniques and ideas from any other carried out in Southern Water region.

Denge Beach

Denge Beach is unique in Britain comprising 14 km^2 of beach shingle, of which the SE tip forms Dungeness. Inland is Romney Marsh (Fig. 15). The

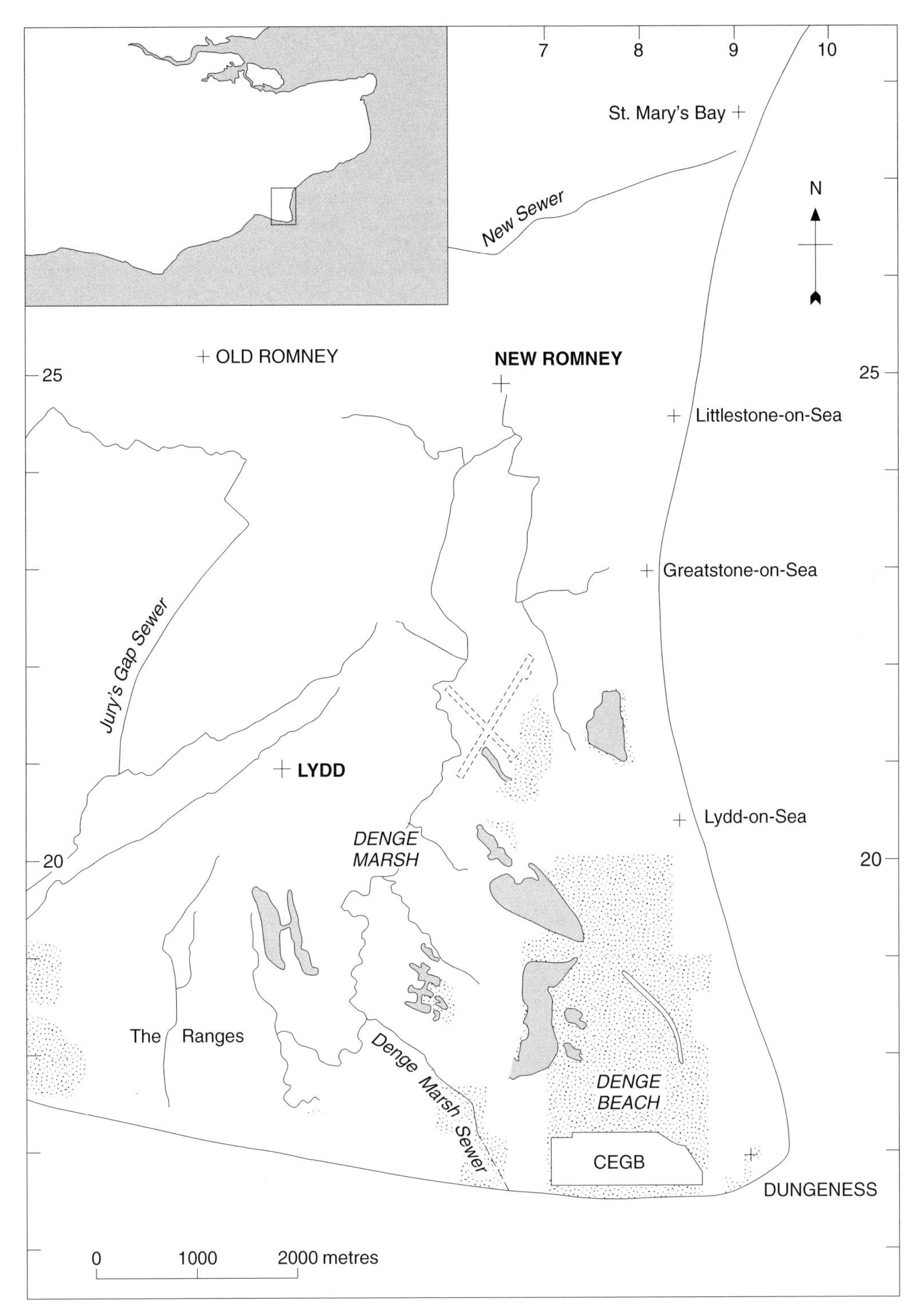

Fig. 15. Location map of the Dungeness peninsula in east Kent.

beach shingle meets a quarter of the sand and gravel needs of SE England. Their freedom from topsoil and ease of working makes them extremely valuable to the aggregate operators. However, the beach gravels also provided an important source of water for Folkestone and District Water Company (FDWC) which, at the time of the study, abstracted 2400 Ml/a from 19 shallow wells. Yet the company held licences for more than double this amount and it was the unobtainable magnitude of these quantities which contributed to the problem.

The gravel deposits increase in thickness southwards from 4 m to 14 m and sit on 22 to 25 m of alluvial sand which overlie the sandstones of the Hastings Beds (Lewis & Balchin 1932; Scott 1963). The shingle deposits exhibit distinctive ridges which represent discrete stages of growth of the beach and provide shelter for plants, animals and birds. The uniqueness and wildness of the area, its importance as a landing site for migrating birds and, not least, its importance to the water company and aggregate companies, provided a perfect setting for a classic encounter.

Origins of study

The origins of the study lay in the conflicts of interests between the three gravel extractors at Denge and FDWC (Kent County Council 1982). During the middle and late 1970s there were a series of public inquiries over applications by two aggregate companies to extend their gravel operations at Denge. Their consultants argued that the creation of large lakes by the extraction of gravel substantially increased the amount of water stored in the aquifer and this was beneficial to the water company (Fig. 16). The water company and Southern Water, on the other hand, argued that the creation of large open water areas increased evaporative loss and diminished groundwater resources. Both aggregate companies appealed against the refusals by KCC to grant them planning permissions for further gravel winning. In his decision letter of 1979 on the appeal by Amy Roadstones (ARC) the Secretary of State concluded that there was a need for further hydrological studies of the Denge Beach aquifer. The many parties involved in the appeal accepted this and agreed to fund jointly a two-year study. ARC was represented technically by Aspinwall and Company. The study was carried out by Southern Water and led by the author. Members of the study group were Terry Keating, Simon Blackley, Paul Shaw (Southern Water), Paul Powell (FDWC), Cliff Thurlow (KCC), Campbell Latchford (ARC), Rod Aspinwall, Geoff Bonney and John Miles (Aspinwalls).

Drilling and aquifer properties

More than 80 150 mm-diameter observation boreholes, 300 mm test holes and small-diameter piezometers were installed at Denge during the study. Six lines of holes were sunk perpendicular to the shoreline to establish the incidence of saline intrusion and these provided a remarkable insight into the nature of the intrusion during storm events. FDWC's numerous pumping sources comprised large-diameter concrete rings sunk into the gravel. Test pumping data from them were difficult to interpret and therefore two 300 mm test holes were drilled and tested.

The numerous test pumpings of the gravels showed that in the north, where the saturated thickens is only 2.5 m to 3.5 m, permeabilities are 800–1000 m/d, while in the centre, aquifer thicknesses are 5 m to 7 m and permeabilities are around 300 m/d. Permeability increases again to 500 m/d in the south where saturated thicknesses are 10 m. Storativities from the test pumpings were in the range from 6 to 20%, with an arithmetic mean of 14.5%.

Groundwater levels

The pattern of groundwater flow at Denge Beach is straight-forward. A groundwater divide runs southwards close to the seaboard diminishing in elevation from 2.5 m OD. Inland, groundwater flows south and southwestwards towards the main gravel lakes. On the western extremity the Denge Marsh Sewer acts as a drain and constant-head boundary.

Least-squares and double-mass statistical techniques on data from tubewells showed that water levels had declined since 1960 to the tune of 0.2 m to 0.5 m. This was mainly due to the effects of pumping which had tripled during the period. Around the gravel pits groundwater levels were lowered by 0.6 m to 0.7 m. This confirmed a property of open, water-filled, gravel pits which flatten groundwater gradients by acting as a 'sink' to surrounding groundwater flows. Evaporation exacerbates this effect (Oteri 1983).

Aquifer salinities and levels

Resistivity and electrical conductivity methods were used to determine the thickness of the gravels and their groundwater salinities along the eastern and southern seaboards (Barker *et al.* 1980). These surveys showed that around the whole of the southern and eastern seaboards there is a saline zone which diminished in concentration inland. Over most of Denge, the chloride concentrations in the

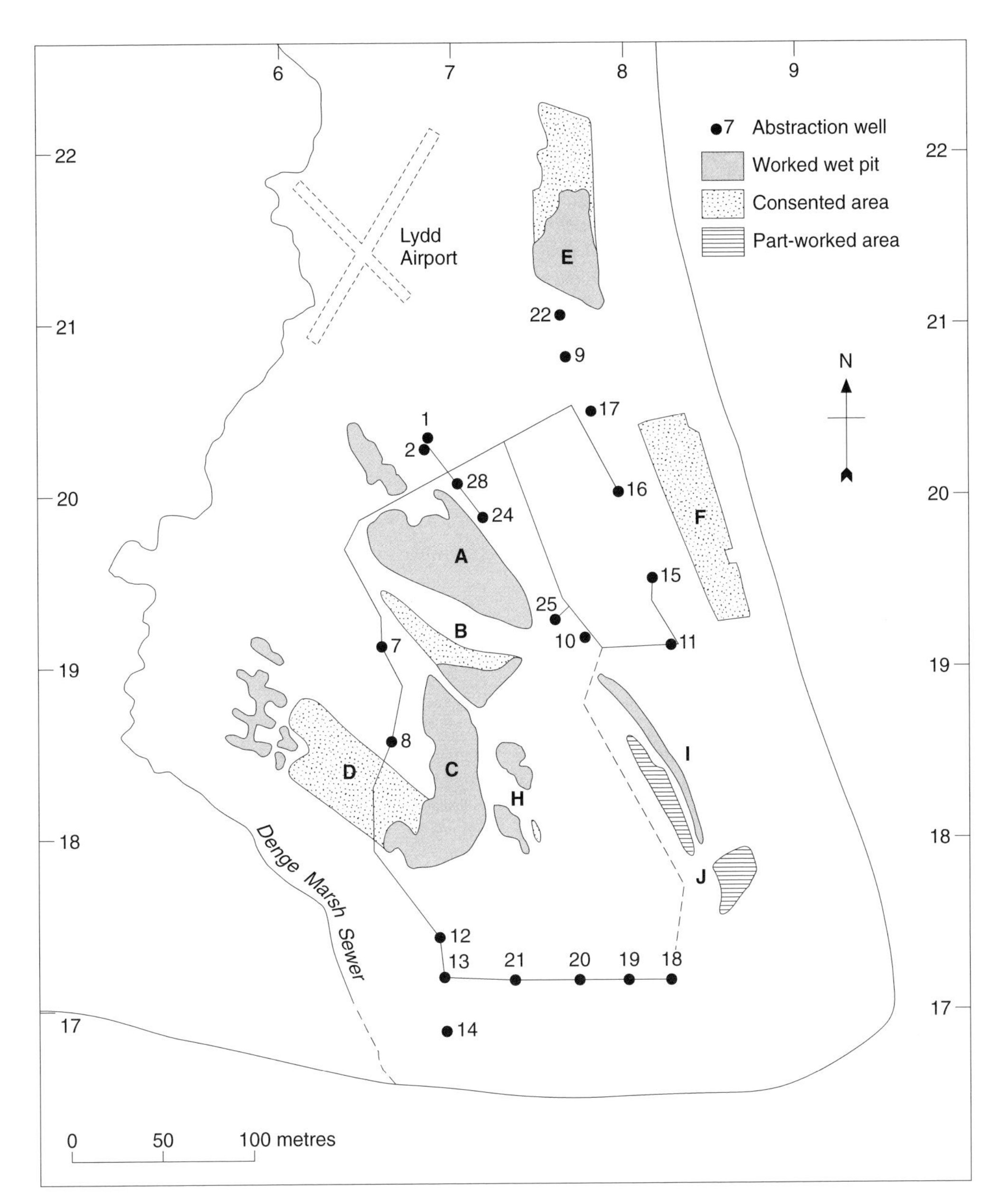

Fig. 16. Denge beach with location of the gravel lakes and public water supply sources.

shingle are between 20 mg/l and 50 mg/l and these extend to within 500 m of the east coast and almost to the beach on the south coast. Within this 500 m coastal zone along the east coast, the chlorides rise to 2000 mg/l and to over 10000 mg/l along the beach margin.

A mass balance salinity model for Denge Beach as a whole showed that the effects of regular pulses of salinity entering the aquifer along this critical length of eastern seaboard would, over a period of time, raise the average concentrations of chloride in the aquifer. Consequently, a decline in the elevation of this coastal groundwater divide would pose a very serious threat to the Denge aquifer since it would be 'overtopped' more frequently during storm surges. This became a key controlling element in assessing how the aquifer should be managed and exploited in the future.

Coastal aquifer responses

Water level recorders placed on the lines of observation boreholes placed at right-angles to the shoreline showed that along the northern section of the east coast the groundwater table has an autumn elevation of 2.7 m OD with a groundwater divide separating seaward and inland groundwater flow some 500 m from the coast at this point. This elevation proved critical. Over the two and a half years of records there were four occasions when high tides caused a saline pulse to move inland beyond the edge of the storm beach. However, around the southern end of the Ness groundwater outflows are substantial and prevent any saline incursion into the aquifer.

Aquifer model

A crucial part of the study was the construction of a fine-element model comprising some 130 elements (Keating 1983, 1984). Its main findings were:

(1) evaporation caused by the gravel pits equalled around 20% of the annual abstraction by the water company,
(2) the storage in the gravels above the underlying sands amounted to 17 Mm^3 of which only 6.4 Mm^3 lay above mean sea level. This latter figure equalled the average annual recharge of Denge Beach and was 2.5 times FDWC's abstraction,
(3) over Denge as a whole, groundwater levels were 0.26 m lower than those which would have occurred if no gravel extraction or water abstraction had taken place. Of this, 0.18 m (69%) was attributable to public water supply since 1961 and the remainder (31%) to gravel extraction,
(4) of the seven postulated additional or extended gravel pits at Denge, having a total area of 26% of the then worked areas, each would cause an average lowering of the water table up to 0.1 m, depending on their size, and a total of 0.24 m, and;
(5) if the water company wished to increase its abstraction, the new sources could only be located at the southern part of Denge Beach. Abstraction near the east coast must cease or be avoided in the summer months.

Postscript

Shortly after the Denge Study report was issued (Southern Water *et al.* 1984), a major sea surge overwhelmed the southern shore of Denge Beach near the nuclear power station. The storm beach, which was fed constantly by Southern Water, was breached and sea water invaded the western flank of Denge Beach via the Denge Marsh Sewer. Over-land flow of the sea water occurred over the exposed shingle for several kilometres inland. Two of FDWC's principal sources, some 3 kms inland, became contaminated with high salinity of several thousand milligrams per litre, although the re-establishment of the normal groundwater gradient saw this gradually abate and the sources eventually brought back into supply.

This major event almost turned on its head the principal findings of the Denge study which had envisaged the overtopping of the groundwater divide along the *eastern* seaboard as posing the biggest threat to the water resources of Denge. Nevertheless, it helped to underline the frailty of the Denge aquifer to such natural events and showed that the loss of the water resource is more a question of when than if.

Conclusion

With the valuable assistance of outside agencies, Southern Water and its predecessor bodies conducted many wide-ranging groundwater and aquifer studies in its area in the years roughly between1970 to 1990. Most of these were undertaken 'in the public's eye' or arose from conflicts of interests. If lessons can be learnt from this work it is that openness, honesty, thorough local consultation and publicity go a long way to overcome genuine anxieties and secure success.

It is fair to say the shear scale, variety and technical innovation of work done across England and Wales as a whole, made this a golden era for British hydrogeology.

References

BARKER, M. I. 1990. *The Chalk structure of the groundwater catchment of the River Alre, Hampshire, and its relationship to well yields.* MSc thesis, University College, London.

BARKER, R. D., SOMERTON, I. W. & GRIFFITHS, D. H. 1980. *Geophysical survey over the Denge gravels, Dungeness, Kent.* Report GEORUN 25, December, University of Birmingham Applied Geophysical Unit.

BIBBY, R. 1979. *A numerical model of contamination by mine drainage water of the Chalk aquifer, Tilmanstone, Kent.* Water Research Centre, Report **ILR 1005**.

BIBBY, R. 1981. Mass transport of solutes in dual-porosity media. *Water Research Association,* **17**(4), 1075–1081.

BRERETON, N. R. 1974. *A resistivity survey of the Tilmanstone area in East Kent.* Water Research Centre, Report **ILR 369**.

BRERETON, N. R. 1976. *The depth distribution of mine drainage water in the Chalk aquifer at Tilmanstone, Kent.* Water Research Centre, Report **ILR 580**.

BRERETON, N. R. & DOWNING, R. A. 1975. *Satellite imagery of the UK with special reference to water resources development and management*. Water Services, June.

BRITISH GEOLOGICAL SURVEY. 1997. *Chalk Aquifer Study: Hydrogeology of the Chalk of the South Downs*. National Environmental Research Council.

BUCHAN, S. 1962. Disposal of mine drainage water from coal mines into the Chalk of Kent. *Proceedings of the Society of Water Treatment and Examination*, **11**, 101–105.

DAVIES, M. C. 1973. A thermal infra-red linescan survey along the Sussex coast. *Water and Water Engineering*, **77**, 392.

DAY, J. 1964. *Infiltration into a groundwater catchment and the derivation of evaporation*. Research report No **2**. Geological Survey and Museum,

DOWNING, R. A. & HEADWORTH, H. G. 1989. Keynote address: Hydrogeology of the Chalk in the UK: the evolution of our understanding. *In*: *Proceedings of the International Chalk Symposium*. Brighton Polytechnic, 4–7 September 1989.

DOWNING, R. A., ASHFORD, P. L., HEADWORTH, H. G. & OWEN, M. 1981. The use of groundwater for river augmentation. *In*: *A Survey of British Hydrogeology*. The Royal Society, 152–172.

ELLSON, T. R. 1973. Artificial recharge investigations of Folkestone Sands – Hardham, Sussex. *Journal of the Institution of Water Engineers*, **27**, 163.

FOSTER BROWN, E. O. 1922. Underground water in the Kent Coalfield and their incidence in mining development. *Proceedings of the Institute of Civil Engineers*, **215**, 291–302.

FOX, G. B. 1969. *An artificial recharge experiment at Pulborough, Sussex*. Association of River Authorities Year Book.

GILES, D. M. & LOWINGS, V. A. 1989. Variation in the character of the Chalk aquifer in East Hampshire. *In*: *Proceedings of the International Chalk Symposium*, Brighton Polytechnic, 4–7 September 1989.

GILES, D. M., LOWINGS, V. A. & MIDGLEY, P. 1988. Regulation of the River Itchen by seasonal groundwater abstraction. *In*: *Fourth International Symposium on Regulated Rivers*. Loughborough University.

HEADWORTH, H. G. 1967. *Problems of groundwater catchments*. Association of River Authorities Year Book.

HEADWORTH, H. G. 1978. Hydrogeological characteristics of artesian boreholes in the Chalk of Hampshire, *Quarterly Journal of Engineering Geology*, **11**, 139–144.

HEADWORTH, H. G. 1994. *Recollections of a Golden Age: The groundwater schemes of Southern Water 1970–90*. Internal Report, Southern Water PLC.

HEADWORTH, H. G. & FOX, G. B. 1986. The South Downs Chalk aquifer: its development and management. *Journal of the Institution of Water Engineers and Scientists*, **40**, 345–361.

HEADWORTH, H. G. & GILES, D. M. 1977. The Candover Scheme (as part of the Itchen Groundwater Regulation Scheme). *In*: *Symposium on Water Output from Multiple Sources, Manchester*. Institute of Civil Engineers Hydrological Group.

HEADWORTH, H.G., KEATING, T. & PACKMAN, M. J. 1982. Evidence of a shallow highly-permeable zone in the Chalk of Hampshire. *Journal of Hydrology*. **55**, 93–112.

HEADWORTH, H. G., OWEN, M. & SKINNER, A. C. 1983. River augmentation schemes using groundwater. *British Geologist*, **9**(2), 50–54.

HEADWORTH, H. G., PURI, S. & RAMPLING, B. H. 1980. Contamination of a Chalk aquifer by mine drainage at Tilmanstone, East Kent, UK. *Quarterly Journal of Engineering Geology*, **13**, 105–117.

IZATT D., FOX, G. B. & TAGUE, M. 1979. Lagoon recharge of the Folkestone Beds at Hardham, Sussex. *Journal of the Institution of Water Engineers and Scientists*, **33**, 217.

JONES, H. K. & ROBINS, N.S. (eds) 1999. *The Chalk Aquifer of the South Downs*. Hydrogeological Report Series of the British Geological Survey.

KEATING, T. 1982. A lumped parameter model of a Chalk-aquifer system in Hampshire, UK. *Groundwater*, **20**, 430–436.

KEATING, T. 1983. Landsat imagery for assessing changes in lake areas at Dungeness on the Kent coast. *Journal of the Institution of Water Engineers and Scientists*, **37**, 290–294.

KEATING, T. 1984. Recharge into a shingle beach. *Journal of Hydrology*, **72**, 187–194.

KENT COUNTY COUNCIL. 1982. *Dungeness Countryside Plan*.

LAPWORTH, H. 1930. *Hydrogeological Survey*. Joint Committee East Kent Water Supplies, Austens, Canterbury, UK.

LEWIS, W. V. & BALCHIN, W. G. V. 1932. The formation of Dungeness foreland. *Geographical Journal*, 258–285.

LOWINGS, V. A. & MIDGLEY, P. 1989. Stages of construction and operation of a scheme to augment river flows using seasonal groundwater abstraction. Presentation to regional meeting of the *British Hydrological Society*.

MILES, R. F. 1992. *Maintaining supplies in adverse conditions*. Institution of Water Engineers and Scientists, Brighton conference.

MONKHOUSE, R. A. & FLEET, M. 1975. A geophysical investigation of saline water in the Chalk of the south coast of England. *Quarterly Journal Of Engineering Geology*, **8**, 291.

MONKHOUSE, R. A. & PHILLIPS, S. 1978. *The design, construction and maintenance recharge wells*. Technical Note, **25**, Central Water Planning Unit.

MUSTCHIN, C. J. 1974. *Brighton's water supplies from the Chalk 1834–1956. A history and description of the heading system*. Brighton Corporation Water Department.

NUTBROWN, D. A. 1975. Identification of parameters in a linear equation of groundwater flow. *Water Resources Research*, **2**(4), 581.

NUTBROWN, D. A., DOWNING, R. A. & MONKHOUSE. R. A. 1975. The use of a digital model in the management of the Chalk aquifer in the South Downs, England. *Journal of Hydrology*, **27**, 127.

O'SHEA, M. 1984. Borehole recharge of the Folkestone Beds at Hardham, Sussex, 1980–81. *Institution of Water Engineers and Scientists*, **38**, 9–24.

OTERI, AKOMENO ULAYA-EGBE. 1983. Delineation of saline intrusion in the Dungeness shingle aquifer using surface geophysics. *Quarterly Journal of Engineering*

Geology, **16**, 43–71.

PACKMAN, M. J. 1985. Phased investigation and development of a marginal aquifer. *The Johnson Journal*, **1**, 4–11.

PLUMPTRE, J. H. 1959. Underground waters of the Kent Coalfield. *Transactions of the Institute of Mining Engineers*, **119**, 155–169.

RAMPLING, B. H. 1974. *Some alternatives to Chalk borehole supplies*. Institution of Water Engineers, South East Section.

SCOTT, J. 1963. The shingle succession at Dungeness. *Journal of Ecology*, **53**, 21–31.

SOUTHERN WATER PLC, FOLKESTONE AND DISTRICT WATER COMPANY, ARC (SOUTH EASTERN) LTD & KENT COUNTY COUNCIL. 1984. *The Denge Hydrogeological Study*. Joint Report, 121 p.

UNITT, J. L. 1959. *Water supply in north-east Kent and some engineering problems. Institution of Public Health Engineers.*

WARREN, S. C. 1962. Some notes on an investigation into seawater infiltration. *Proceedings of the Society of Water Treatment and Examination*, **2**, 343.

Developments in UK hydrogeology since 1974

F.C. BRASSINGTON

Rick Brassington Consultant Hydrogeologist, 12 Culcheth Hall Drive, Culcheth, Warrington, WA3 4PS UK

Abstract: The last quarter of the 20th century has seen the most rapid development of hydrogeology in terms of both the depth of understanding of hydrogeological processes and of the hydrogeology of Britain, with an associated rapid growth in the number of people employed as professional hydrogeologists in the UK. The four main influences that brought about these developments are changes in the structure of the UK water industry and environmental regulators, particularly in England and Wales; influences of EC directives on UK environmental regulation; a growing public awareness of environmental issues and the pressure they applied on successive governments; and developments in computing power, software development and electronic instrumentation. The paper examines the fields in which hydrogeologists have worked during the last three decades and concludes that it has been the richest period for hydrogeological achievement in the history of the science.

The development of hydrogeology in the UK over the last quarter of the 20th century, as both a science and a profession, is closely linked to the evolution of the British water supply industry in England and Wales. The early arrangement of many single function public sector organizations developed into multi-functional water authorities, which a decade and a half later became private sector companies with environmental regulation carried out by government bodies. This evolution was driven by political considerations, new legislation and more recently European Directives. Such changes influenced both routine hydrogeological work and the research and development programmes that have built the modern science and profession.

This story contains numerous strands of scientific endeavour, many being interlinked, with some having their roots several years or even decades before the water industry changes of 1974. The evolution of hydrogeological science and practice took place against a background of wider changes in society and technological advances, which both provided significant influences. During the last half of the century, a growing public awareness of environmental issues and the ability of public opinion to influence government resulted in the creation of pressure groups, environmental campaigns and new political parties that gradually influenced new laws at European and UK government level. These in turn have led to new lines of research especially into contaminant transport and the close relationship between surface water and groundwater systems. Technological developments, largely in terms of computing power, programming and the development of electronic instruments for field measurements and data capture have also influenced the way that the science has progressed. Other developments have been in vocational training, especially at post-graduate level and the growth of specialist research centres in government organizations and universities. This period also saw increasing support for professional development from many applied geologists, a movement strongly supported by most professional hydrogeologists.

Many British hydrogeologists spent a significant part of their careers working overseas and later brought the experience gained to their work in the UK. They were employed by the British Geological Survey (BGS), local Geological Survey organizations and government departments, as well as UK-based consultants working for local governments or on UNESCO, World Bank or other aid-agency funded projects. The countries involved have tended to be members of the Commonwealth and other states where there have been long-established links with the UK. A summary of the type of projects undertaken by UK hydrogeologists during the 1980s has been provided by Simpson (1991).

Laying the foundations – the period before 1974

The Water Act of 1945 was the first legislation in England and Wales to define a national water resources management policy (Evans 1993). It introduced a system to control groundwater abstractions in areas of high demand by licences given by central government. Responsibilities were given to the Geological Survey to take an active role in the government's control of groundwater resources resulting in its Water Department becoming the dominant influence on the development of hydrogeological practice and scientific development in Britain at that time (Downing 1993, 2004*a*). Since 1945, the organization now known as the British Geological Survey (BGS) has undergone name changes from

From: MATHER, J. D. (ed.) 2004. *200 Years of British Hydrogeology*. Geological Society, London, Special Publications, **225**. 363–385. 0305-8719/04/$15 © The Geological Society of London.

the Geological Survey of Great Britain to the Institute of Geological Sciences and then to the present British Geological Survey. To save confusion it is referred to as the BGS throughout.

During the decades before 1974, public water supplies were provided by local authorities often combining resources as 'water boards'. A number of the early groundwater resources studies in the UK were carried out by the water engineers who managed these organizations, usually with a pragmatic approach based on observations and operational experience. Lyon (1949) for example, described aspects of the hydrogeology of the Coventry district concentrating on the yield of public supply wells and differences in groundwater chemistry. A few years later, Lyon, (by then the Manager of the West Cheshire Water Board) encouraged Hibbert (1956) in a study of the hydrogeology of the Wirral Peninsula that used new hydrogeological techniques being developed both in the UK and the USA.

Lapworth (1948) studied the recharge of the Chalk aquifer in Kent using percolation gauges, water table fluctuations, spring flows and rainfall records to quantify recharge into the aquifer. Over the same period, Penman (1948, 1950) provided a method for estimating evapo-transpiration that allowed groundwater resources to be assessed more accurately based on the new concepts of soil moisture deficits related to different plant types. Penman's study of the Stour catchment (Penman 1950) and other long-term catchment studies by others confirmed the general applicability of Penman's method (Day 1964). The method takes into account land use, the nature of the soil and a root constant assigned to each vegetation type representing the plant's water needs. Initially, root constants between 75 and 200 mm were used but Headworth (1970) showed that 25–50 mm is more appropriate for the thin chalk soils of Hampshire. Other workers showed that the recharge mechanism is more complex than that envisaged by Penman although modern techniques to estimate recharge are based on Penman's work. Downing *et al.* (1978) showed that in dry periods recharge can bypass the effect of the soil moisture condition; and Rushton & Ward (1979) discussed the complexity of recharge and showed that an underestimate can occur when estimates are based solely on meteorological parameters. Penman's method was extended by Grindley (1967) and is the basis of the Meteorological Offices Rainfall and Evaporation Calculation System (MORECS) described by Thomson *et al.* (1981).

The close relationship between groundwater and river flows was recognized by Buchan (1953) and Ineson & Downing (1964, 1965) who estimated groundwater resources from river base-flows using methods developed from those first proposed by Barnes (1939). Later workers such as Wright (1968) showed that the groundwater recession depends on numerous catchment characteristics such as total area, topographical slopes and annual run-off although catchment geology is usually the most significant factor. This approach has since been used widely as a method of quantifying groundwater resources (Wright 1975).

A major contribution to British hydrogeology was made by Jack Ineson working mainly at the BGS's Water Department (Downing & Gray 2004). He became Head of the BGS's Water Department in 1960 and moved to the newly formed Water Resources Board as Chief Geologist in 1965. He was the principal author of the chapter on hydrogeology (Ineson *et al.* 1969) in the Institution of Water Engineers' handbook on water engineering practice (Skeat 1969), which led directly to the later publication of the Institution of Water Engineers and Scientists' groundwater handbook (Brandon 1986). In a series of papers, (Ineson 1952, 1953, 1956) he demonstrated that the Theis method of pumping test analysis (Theis 1935) can be applied to the Chalk despite the aquifer's evident heterogeneous nature and also recognized the significance of the dual-porosity nature of the Chalk describing it as having 'two flow phases' causing the 'delayed yield' effect. Boulton developed special type curves for analysing pumping test data from unconfined aquifers and was the first to provide such curves to take account of delayed yield (Boulton 1951, 1963; Boulton & Streltsova 1975). This work established principles enabling other workers to develop more sophisticated methods of pumping test analysis and modelling groundwater flow to wells (Downing *et al.* 2004).

Ineson's interest in the study of pumping tests extended to the definition of water well hydraulics using observations from tests carried out on a large number of Chalk wells. Ineson (1959*a*) related yield to well diameter to enable the yield of a large diameter well to be estimated from tests on a small-diameter trial borehole and showed that using the Dupuit equation (Dupuit 1863) under-estimates the effect of increases in diameter on the yield of wells in fissured aquifers such as the Chalk. Ineson (1959*b*) derived a series of yield-depression (specific capacity) type-curves for wells in both Chalk and sandstone aquifers demonstrating that individual wells have an upper limit to their pumping rate and providing practical assistance to yield estimates from pumping test data. Ineson died in 1970 aged 53 years. His major contribution to British hydrogeology over almost three decades is commemorated by the annual Ineson Lecture organized jointly by the Geological Society's Hydrogeological Group and the British Chapter of the International Association of Hydrogeologists.

The Water Resources Act 1963 created 29 river authorities in England and Wales out of the 32 pre-existing river boards and established the Water

Resources Board (WRB) at national level to co-ordinate water resources planning and provide technical assistance to the river authorities (Evans 1993). Both the WRB and the new river authorities employed small numbers of hydrogeologists. The Essex River Board was the first part of the UK water industry outside central government to employ a hydrogeologist. This was in anticipation of new Water Resources Act responsibilities when it became the Essex River Authority. By the late 1960s most of the river authorities with significant groundwater resources had appointed geologists or engineers with water well experience, yet even at this time there were less than 20 hydrogeologists working in the British water industry. Numbers slowly increased so that by the end of the river authority era in 1974, it is estimated that there were approaching 100 hydrogeologists working in the water industry with possibly a similar number in central government organizations and academia.

The river authorities' responsibilities for groundwater management were the assessment of new abstraction licences, evaluating groundwater resources in many cases by detailed field investigations and to design and construct a network of observation boreholes. The nine-year river authority period saw the start of many schemes, many of which were inherited and completed by the river authorities' successors, the water authorities and, in a few cases, by the National Rivers Authority (NRA) that was formed in 1989. One scheme (the Shropshire Groundwater Scheme) is still being developed by the NRA's successor, the Environment Agency, in 2003.

Water industry restructuring

The 1974 re-organization of the water industry was a consequence of the local government restructuring throughout England and Wales (Evans 1993). The Water Act 1973 established ten regional water authorities to replace the river authorities, water boards and sewage treatment authorities (part of the local councils) from April 1974. Figure 1 shows the river authority areas and Figure 2 those of the water authorities. The water authorities operated over areas based on surface water catchments so that the whole of the water cycle in that area was under the control of a single statutory body (There were however, a significant number of small deviations from the catchment boundaries covering the water supply function. These reflected the network of water distribution pipes inherited from the earlier water boards). The water authorities established strong water resources sections with an in-house capability of water resources management. It was a period of increasing demand for hydrogeologists for regulatory work. In addition, the new bodies continued the research projects that had been originally established under the WRB's influence. The 1970 and 1980 decades saw the greatest activity in British hydrogeology in terms of major research projects carried out to quantify groundwater resources and investigate groundwater quality.

The Water Act 1973 disbanded the WRB and amalgamated its research capability with the water industry's research organization, the Water Research Association (the research organization funded by the water boards) to form the Water Research Centre (WRC later known as WRc), at that time largely funded by the water authorities. The data gathering part of the former WRB became the Central Water Data Unit with a role of advising government on water resources issues, later to be subsumed into the Department of the Environment. WRC quickly assumed an influential role in the development of hydrogeological activities, although the emergence of strong in-house capabilities in the water authorities meant that it never entirely achieved the same influence as had the WRB.

A fundamental restructuring of the water industry in England and Wales took place in 1989 when the ten water authorities were privatized to become water supply and sewage treatment utility companies. The water authorities' environmental regulatory duties were transferred to a new national body, the NRA. Further changes in environmental regulation in the UK took place in April 1996 with the formation of the Environment Agency in England and Wales to replace the NRA, Her Majesty's Inspectorate of Pollution (HMIP) and the waste regulation authorities (based in the county councils).

Privatization resulted in the hydrogeological teams in each water authority being split, with most staff moving to the NRA and a small number moving to the private water company. The six years of the NRA's existence saw an increase in the number of hydrogeologists working in environmental regulation, although many were recent graduates seconded from consultancy companies. This trend has resulted in a major part of the Environment Agency's hydrogeological work being undertaken by consultants on a contract basis. The NRA set up a small national team to co-ordinate the groundwater protection policy. With the formation of the Environment Agency this subsequently became established as the National Centre for Groundwater and Contaminated Land to lead research in these fields and provide specialist advice to the regional staff as required. Since privatization, many water companies have reduced their in-house capability in hydrogeology, preferring to rely largely on external consultants. Notable exceptions are Thames Water and the Vivendi Water Partnership that supplies water to the NW of London, around Ipswich and in east Kent. Both these

Fig. 1. The administration areas of the 29 river authorities, also showing pre-1974 county boundaries.

companies have a high reliance on groundwater sources.

Water resources studies

The Water Resources Act 1963 required the river authorities to assess the available water resources within their area and the present and future demands. By the end of their existence, most had published reports on water resources, with the studies made to prepare these reports giving rise to many more detailed groundwater studies.

These assessments showed that in some cases, groundwater development had already exceeded the available resources and was having unacceptable

Fig. 2. The administrative areas of the ten regional water authorities.

environmental consequences. For example, the River Worfe catchment in Shropshire is underlain by the Sherwood Sandstone aquifer. A series of public water supply wells had depleted the baseflow to the extent that during the summer months the flow at the confluence with the Severn was less than the total sewage discharge into the Worfe catchment (Severn River Authority 1974).

The report on water resources made by the Mersey and Weaver River Authority (1969) showed that the extent of groundwater abstraction in the Sherwood Sandstone aquifer around Liverpool and in the Trafford Park area of Manchester had caused groundwater levels to fall significantly, reaching below sea level in some places. The extent of the depletion in groundwater levels was sufficient to cause groundwater quality deterioration from saline intrusion in coastal areas and, further in land, the up-coning of deep-seated saline groundwater bodies (Anon. 1981). As a consequence, the River Authority established a groundwater management policy designed to reduce the quantities abstracted in the over-pumped areas and to limit the new abstraction in areas where resources were becoming hard pressed (Mersey and Weaver River Authority 1973).

A major responsibility of the Water Resources Board was to formulate a national water resources policy for England and Wales (Ineson & Rowntree 1967), accomplished through three major regional studies in South East England, Northern England and Wales and the Midlands (Downing 2004*b*). The results of these studies were then integrated into a national plan (Water Resources Board 1973). At the time that the Water Resources Act 1963 legislation was being framed, an assessment of the future water demand based on the population growth trends indicated, by the 1951 and 1961 censuses, future water shortages in England and Wales. These population trends and the associated water demand forecasts were subsequently shown to be considerable overestimates.

The WRB encouraged a number of the river authorities to undertake specific water resources investigations, many in relation to the use of groundwater. These projects included the investigation of the feasibility for regulating the flow of rivers using groundwater abstracted from aquifers within the catchment (Downing *et al.* 1981; Headworth *et al.* 1983; Owen *et al.* 1991). The principal of river augmentation involves water being pumped from wells into a river at times when its flow falls below that required by the demands on the river. These demands may be for water supply or the maintenance of adequate flows for irrigation, navigation, fisheries, wildlife, general amenity or the dilution of effluents. As the period of groundwater pumping continues, two potential impacts on the river develop: part of the groundwater flow which would have naturally discharged to the river as a component of base-flow is diverted to the pumping well; and water is induced to flow from the river by extended pumping that causes a reversal of the water table gradient. As pumping continues therefore, the actual increase in river flow resulting from the discharge of well water becomes progressively less. The amount by which the natural flow is augmented by the pumped discharge is termed net gain and is expressed as a percentage of the water discharged into the river.

The net gain depends on the transmissivity and storativity of the aquifer, the permeability of the riverbed and the position of the wells in relation to the river. For aquifers with a high transmissivity such as the Chalk and other limestones, wells need to be sited a considerable distance from the river to produce sufficient net gain unless the riverbed is effectively impermeable. In less transmissive aquifers or where the riverbed is impermeable, the wells can be sited close to the watercourse.

The main areas for the early studies were in the catchments of the rivers Thames, Great Ouse, Severn, Itchen and Waverney, each with different geological conditions. The Thames and Itchen schemes are concerned with abstraction from the Chalk at outcrop contrasting with the Great Ouse scheme where the Chalk is partially covered by boulder clay and there are relatively impermeable river beds. The Shropshire Groundwater Scheme to regulate the River Severn is concerned with Triassic

Sandstones that are overlain by glacial materials, which vary both in lithology and thickness. The Waverney scheme involved the confined Chalk aquifer.

There were more than 24 such water resource studies in England and Wales by the end of the 1980's (Owen *et al.* 1991). In some cases projects were continued after privatization by the water companies. For example, a long series of research projects carried out by the Southern Water Authority that included resource assessment, river augmentation and artificial recharge experiments were completed by its successor, Southern Water (Headworth 1994, 2004). The flow of the River Severn is regulated both by releases from the Llyn Clywedog reservoir in Mid-Wales and groundwater abstracted from a series of boreholes in the Sherwood Sandstone aquifer in Shropshire. This scheme, known as the Shropshire Groundwater Scheme was initiated by the Severn River Authority and WRB in 1970 and, after a series of investigations completed by Severn Trent Water Authority, is now owned and operated by the Environment Agency. The Shropshire Groundwater Scheme has been implemented on a phased basis with the last phase still not fully commissioned in 2004.

At the first meeting of the new NRA Board in 1988, it was decided to initiate the Alleviation of Low Flows Scheme (ALF), a programme to restore the flow of rivers impacted by principally groundwater abstraction. This problem was most serious in the Chalk catchments of southern and eastern England and resulted from groundwater developments in rural areas that had taken place over the previous 30 years or so, largely for public water supplies (Brassington 1992). The NRA proposed a number of schemes to restore flows including reductions in the quantities pumped either in total or on a seasonal basis, lining permeable beds of rivers and the construction of new wells to supplement river flows during summer periods. In many respects these investigations were similar to those initiated by the WRB more than 20 years earlier and described above.

The relationship between groundwater abstraction and river flow continues to attract significant attention. Rushton (2002) showed that reducing groundwater abstraction is only likely to have a small benefit in terms of improved river flows and then only to close-proximity rivers. Acreman *et al.* (2000) also question the extent that groundwater abstraction causes surface water and wetland degradation and stress the need for objective site-specific investigations. Kirk & Herbert (2002) describe an analytical method of determining the extent of an impact on river flows that is intended for general use within the Environment Agency abstraction licensing teams.

The Natural Environment Research Council (NERC) has initiated a five-year (1999–2006) Lowland Catchment Research programme (LOCAR) to examine water resources issues in a broad context (National Environmental Research Council, 2003). Three catchments have been selected, two underlain by the Chalk aquifer and one by the Sherwood Sandstone. The project will examine surface and groundwater supplies, changes in water quality and their impacts on fisheries and wetlands. It comprises an integrated approach to address the problems being caused by possibly drier summers and wetter winters, changes in farming, urban expansion, new industrial sites and road building in terms of altered water flows and water chemistry, as well as by increasing competition between rural, urban and ecological water demands. The study is supported by the Department of Environment, Food and Rural Affairs (DEFRA), the Environment Agency, English Nature and other nature conservation organizations, as well as bodies representing farming and business interests. It is probably the largest single government funded academic hydrogeology-based research programme at the start of the 21st century.

Artificial groundwater recharge

Artificial groundwater recharge was one of the options for water supply development actively considered by the WRB who undertook their own research programme on the Sherwood Sandstone in the area between Nottingham and Edwinstowe (Satchell & Edworthy, 1972). Extensive investigations were carried out into the use of infiltration basins and spray irrigation to test the viability of recharging the aquifer using polluted river water. Part of the objective was to discover the extent to which the recharge process would provide treatment to the polluted river water. Other experiments in the same area tested the use of recharge boreholes where the recharge water was obtained from boreholes penetrating the same aquifer.

A practical experiment of recharging water down a supply borehole for part of the year was carried over an extensive period starting in the 1880s by Thames Water and its predecessors in the Lee Valley. Encouraged by the success of these simple experiments further research was carried out during the mid-1970s and an operational scheme brought online in December 1977 (Hawnt *et al.* 1981; Flavin & Joseph 1983). Thames Water has extended the use of artificial recharge in public water supplies by its Enfield-Haringey Scheme (O'Shea *et al.* 1995). This scheme was commissioned in the mid-1990s and Thames Water is currently investigating the potential for artificial recharge at two sites in south London.

Extensive studies into artificial recharge were also carried out at Hardham in West Sussex involving the recharge of the Folkestone Beds aquifer with water from the River Arun using recharge lagoons. These trials proved successful and were followed by a programme of borehole recharge experiments showing that these techniques can also be successful (Edworthy *et al.* 1981; Headworth 1994).

In addition to supplementing fresh groundwater resources, artificial recharge techniques have been developed over the last decade or so to store potable water in aquifers that naturally contain brackish groundwaters or even brines (Pyne 1995). This method, called Aquifer Storage and Retrieval (ASR), has attracted the attention of several English water companies that have water shortage problems. The method allows water to be injected at times when surface water resources are high and the injected water then recovered when other supplies are low. The environmental impacts are potentially low as the part of the aquifer system receiving the injected water is not part of the fresh water circulation. The key issues are the proportion of the injected water that can be recovered and whether the volumes recovered decrease with time. Potential problems include the precipitation of oxides, particularly iron with the introduction of oxygenated water into an anoxic environment, and a consequential loss of potential storage and decrease in hydraulic conductivity. A number of investigations have been carried out in the UK with that for Wessex Water (Eastwood & Stansfield 2001) being one of the first to be reported.

Rising groundwater levels

In the UK, as with most industrialized countries, a shift from heavy industry resulted in a reduced water demand with significantly less water being abstracted from factory boreholes. Groundwater abstraction had been a feature of many old industrial centres and had drawn groundwater levels down, maintaining them at a significant depth below their natural levels over many decades. The reduced abstraction caused groundwater levels to rise beneath a number of city centres, notably London, situated on the Chalk aquifer, and Birmingham, Liverpool, Manchester and Nottingham on the Sherwood Sandstone aquifer (Wilkinson & Brassington 1991). A survey of UK case histories (Brassington 1990), identified 18 examples where groundwater levels had risen mainly as a consequence of reduced abstraction by industry although changes in dewatering for quarrying and coal mining have had similar effects.

The first major study concerned the rising groundwater levels in the Chalk aquifer beneath London and was completed in 1989 (Simpson *et al.* 1989). The report pointed out that there was no public body with the responsibility to monitor and formulate a strategy to combat the problem. The government responded by making the National Rivers Authority responsible for monitoring rising groundwater and suggesting remedial measures. Lucas & Robinson (1995) describe a groundwater modelling exercise to predict the level changes across the London area in both the Chalk and Tertiary aquifers in order to assist the NRA and Thames Water to develop groundwater sources that would have an advantageous effect in limiting the rise in critical areas.

The rising groundwater levels in Birmingham were investigated using a computer model by Knipe *et al.* (1993) who found that the total recharge had remained fairly constant at around 40 Ml/d throughout the modelled period 1870–1989. During this period increasing urbanization reduced the effective rainfall from 39 to 31.2 Ml/d although this was offset by water mains leakage that increased from an estimated 2.3 Ml/d in 1900 to 8.5 Ml/d in 1989. The observed rise in water levels indicates that there is a large groundwater resource that can be abstracted without a serious impact on the aquifer. Benefits of re-establishing significant groundwater abstraction would be the control of the rising water levels and a reduction in the flooding impacts. A scheme has since been established to abstract groundwater in the south and east of the city centre avoiding areas with polluted groundwater and the potential of inducing recharge from the Rive Tame. Severn-Trent Water discharges the abstracted groundwater into the River Tame and abstracts a similar volume from the River Trent for public supplies.

Significant rises in groundwater levels have taken place in the Sherwood Sandstone aquifer in the city centre area of Liverpool (Brassington & Rushton 1987). After the government gave the NRA the duty to measure and report on rising groundwater levels the NRA North West Region commissioned a PhD project (Ion 1996) to investigate the changes in groundwater levels that had occurred in both Liverpool and the Trafford Park area in Manchester. In both cases the rises were a direct consequence of reduced industrial abstraction and have caused problems of flooded basements in Liverpool, with some concern over the integrity of submerged structures and the potential mobilization of pollutants from contaminated ground in Manchester. The higher water table elevations in Liverpool have caused the permanent flooding of a railway tunnel (Fig. 3). A number of options to manage this problem were discussed by Gallagher & Brassington (1994) and the one chosen is the construction of a series of dewatering boreholes close to the tunnel location. The first part of the scheme was implemented during 1999 with additional work commencing during 2003. Unlike the Birmingham area the water abstracted is

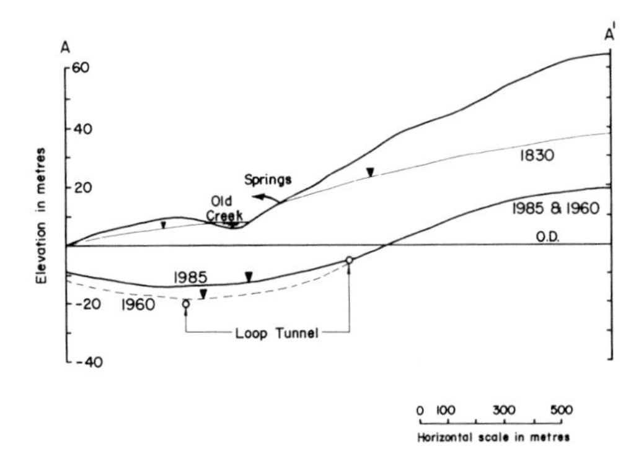

Fig. 3. Changes in groundwater levels over 150-year period in central Liverpool were caused by growing groundwater abstractions followed by a gradual cessation. During this period railway tunnels were constructed above the water table that are now suffering severe flooding problems (Brassington & Rushton 1987).

not used for local water supplies and is discharged to waste, largely because of poor water quality.

The closure of many deep coalmines has resulted in a general rebound in groundwater levels in the complex multiple aquifer systems formed by Coal Measures sandstones. Groundwater flow and storage within these aquifers has been substantially modified by the interconnections provided by shafts, roadways, adits and workings. Fracturing caused by subsidence has also made changes to the hydraulic characteristics of these aquifers and in some other aquifers that overlie the Coal Measures strata. Younger (1993) provided an early warning of some of the problems that are likely to be caused by the water table rebound effect after the cessation of deep mine dewatering. The problematic impacts are likely to include flooding of underground structures, increased flows in sewers, and the introduction of poor quality, iron-rich groundwater both into other aquifers and surface water systems,

Whitworth (2002) predicted the rates of groundwater level recovery that would be expected in many coalfields across the UK. He assumed that the additional storage provided by the voids created by mining would have a significant influence on the rates of rise. Long-term monitoring however, has shown that the rates of recovery are very similar to those observed in aquifers following a period of pumping from an abstraction borehole.

Groundwater chemistry and contamination

By the start of the 1970s, there was growing government concern over the pollution of groundwater-based public water supplies especially from the use of agricultural chemicals and landfill sites. Studies of groundwater quality became a dominant influence on hydrogeological work from the mid-1970s onward. Many base-line studies were carried out on the hydrogeochemistry of major unpolluted aquifers, with much of the earlier work undertaken by the BGS. Ineson & Downing (1963) reported on the changes in groundwater chemistry that take place down-dip in the Chalk aquifer with increasing concentration of the dissolved constituents. Later examples of aquifers studied include the Carboniferous Limestone (Edmunds 1971), the Sherwood Sandstone in east Shropshire (Edmunds & Morgan-Jones 1976) and the east Midlands (Edmunds *et al.* 1982), the Jurassic Limestones of Gloucestershire (Morgan-Jones & Eggboro 1981), the Lower Greensand (Morgan-Jones 1985), the Suffolk Chalk (Heathcote & Lloyd 1984) and many others.

Interest on the potential use of radioactive isotopes in groundwater studies developed in the late 1960s (Mather 1968) with most attention given to the possible use of thermo-nuclear tritium as a tracer. Tritium (H^3) as an isotope of hydrogen is able to form part of the water molecule and therefore is an ideal tracer. Studies of the age of recharge water percolating through the vadose zone were carried out using tritium (Smith *et al.* 1970; Smith & Richards 1971; Mather & Smith 1973).

Core samples of the aquifer were obtained and the pore water was removed from the rock sample using high-speed centrifuge techniques (Edmunds & Bath 1976) and then the tritium content measured at the Atomic Energy Authority's Harwell Laboratory. The pore water recovery technique was used in many other studies of groundwater contamination and the variation in groundwater chemistry within an aquifer.

A significant contribution to the understanding of hydrogeochemical processes was made by John Andrews, particularly in the role of dissolved noble gases and radioactive elements (Andrews 1991). Studies of these elements have been used to assess groundwater migration and age, the palaeoclimate at the time of recharge and a number of geochemical processes that affect groundwater chemical evolution.

Saline groundwater

The occurrence and distribution of saline water in major aquifers were studied from the mid-1970s (Lloyd 1981). Saline groundwater is found in most of the aquifers in the UK with a potential impact on the development of freshwater reserves. Saline groundwaters can be divided into three classifications, very old deep-seated waters underlying fresh groundwater, saline waters occurring in the confined parts of fresh groundwater aquifers or recent seawater intrusion.

The Anglian Water Authority and the University

of Birmingham jointly investigated the distribution of saline groundwater in the Chalk aquifer around Grimsby, Lincolnshire (Anon. 1978). A complex saline groundwater system was discovered with deep-seated, very saline groundwaters occurring near the coast. A high concentration of groundwater abstraction in the Grimsby area was producing a lowering of groundwater heads in the coastal area causing both seawater intrusion and up-coning of the deep-seated saline groundwater body. The investigation into the saline groundwater in the Permo-Triassic sandstone aquifer in north Cheshire and south Lancashire by the North West Water Authority, again in conjunction with the University of Birmingham, showed that sea water intrusion occurred in coastal areas but that a deep-seated saline groundwater body existed at depth throughout much of the aquifer and that this very saline water body had a greater potential to impact on water supplies (Anon. 1981). The nature of the interface between the brine and the fresh groundwater was defined by Brassington *et al.* (1992) using data from focussed electric geophysical logs in coal exploration boreholes that fully penetrated the aquifer (Fig. 4). The interface takes the form of a zone of decreasing salinity varying between 30 m and 250 m in thickness. The depth to the top of this zone also varies. It is thought that the boundary represents a diffusion zone with its depth and thickness controlled by the fresh groundwater circulation.

Operational systems and groundwater management protocols have been developed for sources threatened by saline groundwater. The use of a scavenger pump in the Woodfield borehole near Telford, Shropshire was described by Tate & Robertson (1971). This borehole penetrates the Sherwood Sandstone aquifer to 120 m depth. It was found that water with a usable quality could be abstracted by setting the main pump at 20 m depth and a scavenging pump at 90 m. A system of using coastal boreholes in conjunction with others further inland in the Chalk aquifer of the Brighton area first started in 1957 enabled maximum abstraction without causing saline intrusion (Headworth & Fox 1985 1986). During summer months the coastal boreholes are pumped until the sodium chloride content of the pumped water rises. Abstraction is then shifted to the inland sources. Groundwater levels then recover in the coastal area thereby preventing the inland movement of seawater.

Waste disposal

The Deposit of Poisonous Wastes Act 1972 was a response to the illegal disposal of hazardous industrial waste – often reputed to be cyanide – buried surreptitiously in domestic waste landfills or even tipped into sewers and water courses. It was replaced by the Control of Pollution Act 1974 which established the Waste Disposal Authorities (WDAs) (county councils in England and district councils in Wales) and also gave powers to the new regional water authorities to influence solid waste disposal and the disposal of liquid waste into landfills. The regulatory function of the WDAs passed to the Environment Agency in 1996.

The government commissioned the BGS to undertake a multi-authority review of landfills in 1973 (Gray *et al.* 1974). The work evaluated the threat of water pollution from 2494 landfill sites in England and Wales that contained hazardous industrial wastes. Desk studies suggested that only 51 of these sites represented a serious pollution risk. The study had highlighted a need for a greater understanding of the pollution threats from solid waste disposal and the Department of the Environment initiated a three-year programme involving a joint team from the BGS and the Atomic Energy Authority's Harwell Laboratories, and a second joint team from WRB and the Water Pollution Research Laboratory. The research programme concentrated on the generation of leachate within landfills, its movement and attenuation, and leachate management within a landfill environment resulting in an influential report that generally became known as the 'Brown Book' (Department of the Environment 1978). This research programme has continued since that time and its management was taken over by the Environment Agency from the Department of the Environment in 1996.

The need to manage and treat leachate has led to it forming an important part of landfill-related research. As part of the initial programme, a number of small-scale experiments were carried out both on a laboratory scale and using the sedimentation tanks at a disused sewage works. The results of these experiments are comparable with and complementary to leachate generation studies carried out in other countries (Mather & Young 1981). The on-going programme examined both the minimization of leachate generation and treatment methods. A good summary of this work was provided by Robinson (1996) and has been used to produce recent guidance by the Environment Agency (2000*a*).

The need to prevent leachate from discharging into groundwater systems led to an increasing reliance on engineered containment. Research into the use of natural materials and artificial membranes has continued since the late 1970s (Williams *et al.* 1991) with a recent example of testing the effect of desiccation on natural clays used to line a landfill described by Philip *et al.* (2002).

The significance of landfill gas generation was realized in the early 1980s with migration away from landfill sites presenting health and safety problems to the public. Raybould & Anderson (1987) reported

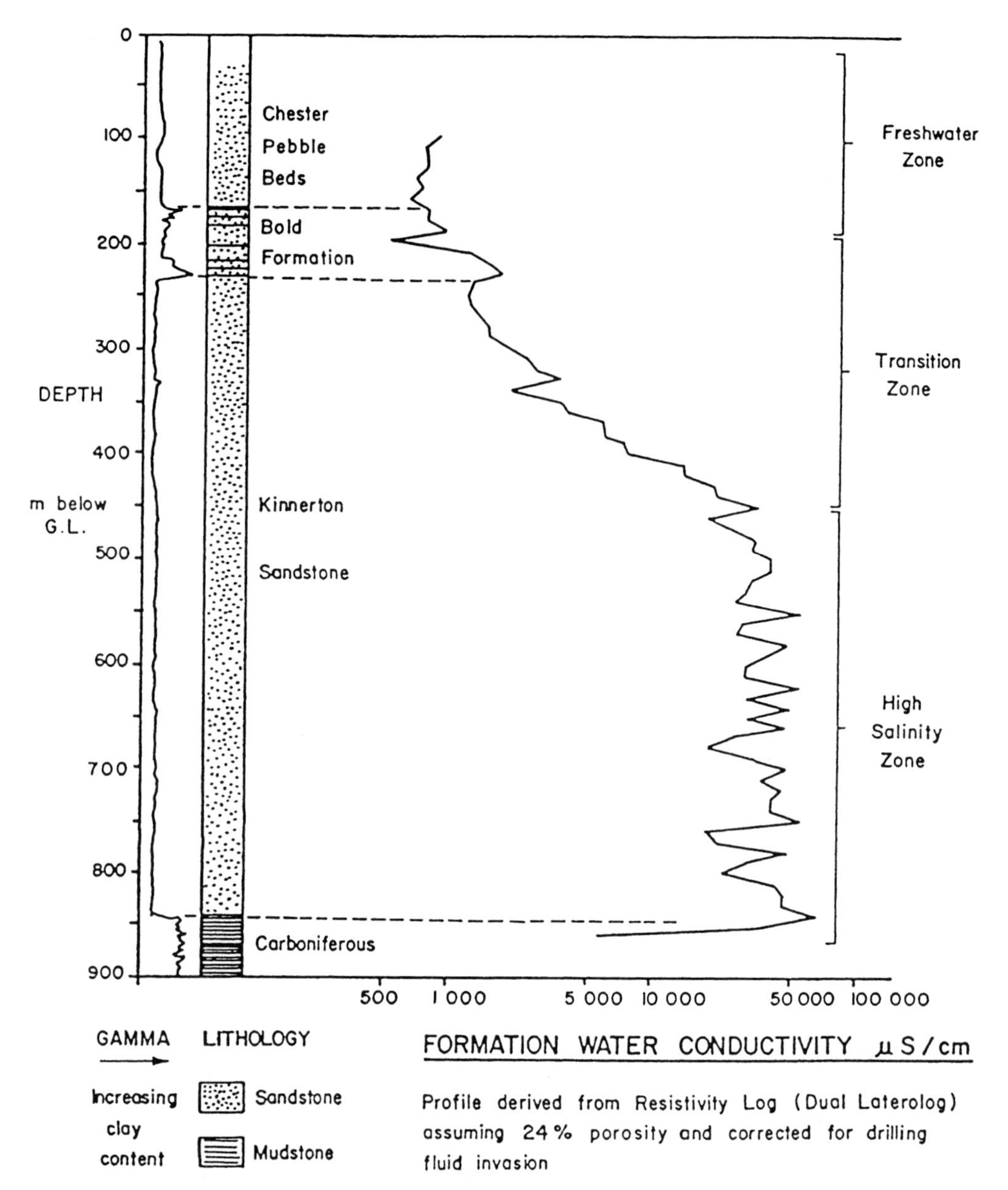

Fig. 4. A profile of the groundwater conductivity throughout the full thickness of the Permo-Triassic Sandstone aquifer on the western limb of the Cheshire Basin was constructed from a focussed electric log run in a coal exploration borehole. The upper part of the profile has freshwater conductivities around 800 µSiemen/cm whereas the deep-seated groundwater has values approaching 50000 µSiemen/cm. The transition zone between these values is some 250 m thick (Brassington *et al.* 1992).

on an incident in St Helens, Merseyside in June 1984 where landfill gas was discovered in houses more than 200 m from a landfill. Investigations into the cause of an explosion that destroyed a bungalow (fortunately without loss of life) in at Loscoe, Derbyshire on 24th March 1986 (Fig. 5) showed the need for sophisticated gas management to minimize the safety hazards (Williams & Aitenkenhead 1991). Continuing research on these aspects has led to the modern practice of control and in many instances for landfill gas being used to generate electricity fed into the national grid.

Investigations concentrating on the effect of the unsaturated zone were carried out during the 1970s

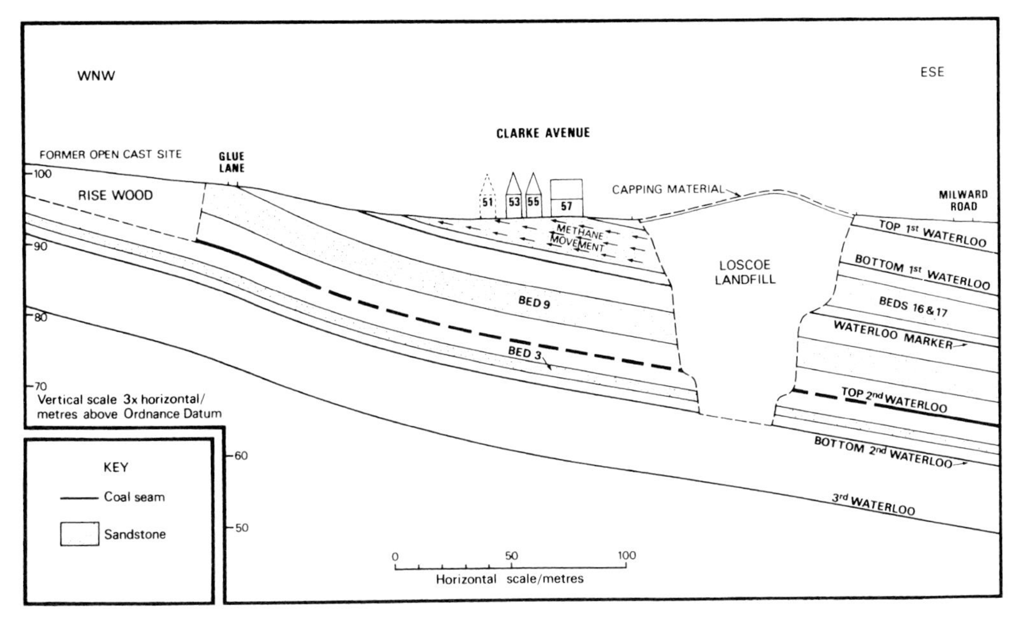

Fig. 5. A landfill that infilled old workings for brick clay, sandstone and coal at Loscoe in Derbyshire lies close to houses. Migrating landfill gas along sandstone beds caused a build-up beneath houses in Clarke Avenue that resulted in an explosion on 24th March 1986 destroying No 51 albeit without loss of life (Williams & Aikenhead 1991).

because of the potential for physical, chemical and biochemical processes to attenuate leachate (Department of the Environment 1978; Mather & Young 1981). The work that began in 1973, included laboratory and lysimeter-based experiments together with field studies using existing landfills. The presence of an unsaturated zone beneath the landfill site was found to play an important role in the attenuation of leachates because of the time taken for leachate to percolate from the base of the landfill to the water table providing an opportunity for chemical and biochemical processes to remove some of the chemicals within the leachate.

These studies resulted in optimistic statements on the effectiveness of the unsaturated zone that were beginning to be questioned by the late 1970s with a number of detailed long-term studies. About this time, the Severn-Trent Water Authority initiated detailed investigations into a number of landfills located on the Sherwood Sandstone in Nottinghamshire (Harris & Parry 1982; Harris & Lowe 1984). An important feature of these investigations has been sampling the unsaturated zone to produce a profile of the leachate components above the water table. Boreholes were drilled to recover continuous core samples from which the pore waters (including leachate) were removed by high-speed centrifuge techniques. Profiles were obtained in 1978, 1981, 1985, 1987 and 1991 in order to assess changes in the vertical distribution of the leachate components with time and have been reported in a number of publications with summaries in Williams *et al.* (1991) and Lewin (1992). Data from this work showed that attenuation of organic components was minimal because the lack of buffering minerals in the aquifer rock is not conducive to bacterial activity. The research has continued as part of the Department of the Environment landfill research programme since 1984 and was taken into the Environment Agency's R&D programme from 1996.

Several landfills in former sand and gravel quarries overlying the Upper Chalk, for example, at Ingham, Suffolk, have also been used for long-term study of the effects on the Chalk aquifer. As in the Sherwood Sandstone sequential vertical profiling was carried out and it was found that the high buffering capacity of the Chalk is conducive to microbial metabolism resulting in the degradation of organic compounds with time (Williams *et al.* 1991). The Environment Agency is continuing an active research programme in this area for example, with a recent report on a site at Thriplow near Cambridge (Environment Agency 2000*b*).

The foot and mouth outbreak of 2001 required a large number of disposal sites on individual farms for burning dead animals and at a smaller number of large special sites for the burial of carcasses. The hydrogeological assessment of the impact of these sites was co-ordinated by the Environment Agency using its own staff and consultants working with officials from the Ministry of Agriculture Fisheries and Food and the army units drafted in to manage the sites. The long-term impacts of these sites are still undergoing assessment and it is likely to be another year or more before formal reports are available.

Nitrate in groundwater

In the early 1970s public water supplies from groundwater sources in some areas suffered from a gradual increase in nitrate concentrations to levels that gave rise to concerns over public health. A study of groundwater in the Chalk aquifer in Yorkshire, (Foster & Crease 1974) showed the rising nitrate concentrations to be related to agricultural practices, particularly the increased use of nitrate-based fertilizer in cereal production. These authors expressed serious apprehension regarding the long-term trends found in groundwater sources because of the possibility of very slow downward percolation of pollutants through the unsaturated zone to the water table. This implied that the full effects of the major changes in arable agriculture since the early 1950s might not be perceived for many years to come.

The increasing nitrate concentration in groundwater-based public water sources was particularly evident in eastern and southern England where extensive arable cultivation is the dominant land-use. Early investigations by both BGS and WRC concentrated on the unsaturated zone so that future predictions of nitrate in groundwater could be made. Drilling programmes were undertaken, with emphasis on areas of intensive arable farming, to obtain porewater samples from rock cores recovered from boreholes in all the major aquifers. The porewaters were analysed for nitrates and other determinands to establish quality profiles through the unsaturated zone. Pore water profiles of thermo-nuclear tritium were used to help understand the rate of movement of solutes through the saturated zone, as illustrated in Figure 6. When compared to solute migration it was found that de-nitrification can take place in the unsaturated zone (Foster 2000).

A large body of data was collected which showed that nitrate losses from arable land were increasing, probably as a result of increased applications of organic fertilizer and that the concentration of nitrate in many groundwater sources was expected to rise for many years (Foster & Young 1981). Several water authorities also carried out their own research programmes in order to have local data on which to plan any necessary action to replace sources with nitrate concentrations too high for their continued use.

Permanent grassland proved not to be a significant source of nitrate except after ploughing when the accumulation of organic nitrogen in the soil is mineralized and made susceptible to leaching. Although general conclusions were drawn from this work, problems remained on the more detailed interpretation of nitrate profiles resulting from uncertainties regarding the mechanism of solute transport in the unsaturated zone, particularly in the Chalk aquifer. The research of the 1970s also involved the use of a number of research catchments where the occurrence and behaviour of nitrate in the saturated aquifer was studied. Although the data was limited, these studies showed that there is considerable variation in nitrate concentration, both spatially and with depth, in aquifers that are subjected to diffuse pollution (Foster & Young 1981). During the 1980s, nitrate pollution remained the most important aspect of groundwater contamination from agricultural practices (Parker *et al.* 1991). Studies of the unsaturated zone have continued and provided better understanding of solute flow mechanisms and given more precise information for predictions of nitrate concentrations. The techniques and results obtained from the nitrate research programme have also been used to assist in more recent groundwater contamination studies.

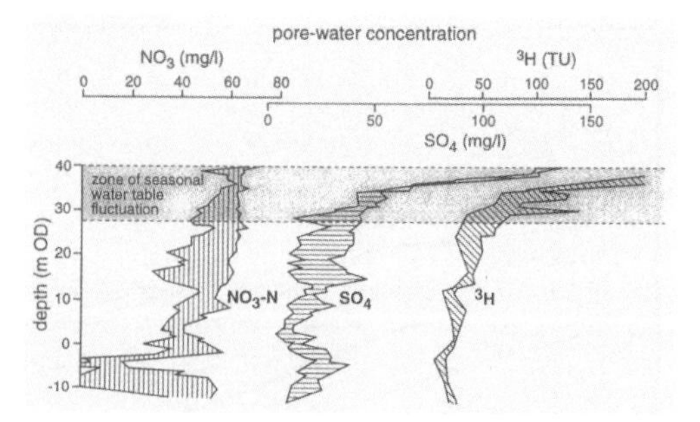

Fig. 6. Profiles of saturated zone pore-water chemistry in the Norfolk Chalk showing marked vertical stratification and the use of tritium as a tracer (Foster 2000).

A detailed history over the past 25 years of hydrogeological investigations to define the processes that control pollution from nitrate sources has been given by Foster (2000). The problem of high nitrate concentrations in water supplies is still ongoing with each water company recently being required by the Drinking Water Inspectorate to formulate proposals to ensure that nitrate concentrations in potable supplies fall below the statuary maximum concentration.

Contamination from other agricultural activities

In addition to the use of chemical fertilizers, modern agricultural practice uses a variety of chemicals for controlling unwanted plant growth and insects. Most of these chemicals used as herbicides and insecticides are persistent in the very long-term and will not breakdown in the natural environment. Their long-term use and the widespread application method have ensured that much of the groundwater resource in rural areas contains low concentrations of the chemicals used as pesticides. A typical example was described by Gomme *et al.* (1992) where pesticides were found to be widespread in a Chalk catchment near Cambridge. Research has been undertaken in this field for more than two decades and is summarized by Parker *et al.* (1991), Foster (2000) and Mather *et al.* (1998).

Agricultural practices also give rise to potential groundwater contamination by micro-organisms that infect livestock. *Escherichia coli* and other faecal coliforms are derived from sewage disposal as well as animal manures and even wild animals. The fact that groundwater may be contaminated by the protozoa cysts of *Cryptosporidium parvum* and *Giardia lamblia* is now well established (Bridgman *et al.* 1995). Cryptosporidium gives rise to the greatest concern, as it is an endemic infection amongst intensively reared livestock and the cysts go into stasis and may survive for long periods in a groundwater environment. A number of water supply contamination incidents have occurred involving public water sources. Private water supplies give rise to great concern, as the majority are untreated (Clapham & Franklin 1998). Foster (2000) describes recent work that has been undertaken to assess the potential hazards from these sources including a methodology for groundwater hazard assessment.

Sewage treatment

A government and EC funded research programme was undertaken during the late 1970s and early 1980s into the impact on groundwater quality caused by the spreading of untreated and partially treated sewage effluents over outcrops of the major aquifers (Baxter *et al.* 1984). At a site near Stourbridge in the West Midlands, Spears (1987) reported that porewaters below a site used for sewage effluent spreading were enriched in some metallic species. There was strong evidence that metals were held at shallow depth in the rock matrix by filtration and possibly by reaction with oxyhydroxides. A finding that the metallic concentration in porewaters decreased with depth was attributed to dilution by the groundwater water body.

Urban pollution

The development of industry and large urban centres has progressed hand-in-hand since the 18th century. A combination of concentrating activities in relatively small areas combined with large-scale use of potentially contaminating materials has inevitable led to significant groundwater pollution. Lloyd *et al.* (1991) discuss the potential sources of groundwater contamination sources in urban areas ranging from leaking sewers, to spillages of industrial chemicals and waste disposal. Mather *et al.* (1998) provide a number of recent case histories of this type of groundwater pollution.

Research into the occurrence of groundwater pollution in urban areas only gained momentum during the late 1980s although specific examples had been identified earlier. Solvents are used in a great number of industrial processes e.g. dry-cleaning and as de-greasing agents. These materials are generally significantly more dense than water and have a relatively low solubility, which means that rapid and deep penetration of the immiscible phase into aquifers can be anticipated (Schwille 1981). The maximum recommended concentrations in drinking water are very low, which implies that even a small spill of only a few litres volume could potentially contaminate many million litres of groundwater.

Rivett *et al.* (1990) reported on the widespread pollution of British groundwaters by chlorinated solvents using extensive contamination of the Sherwood Sandstone in Birmingham as the main example. A detailed investigation into the chlorinated hydrocarbon contamination in the Coventry area is summarized by Lerner *et al.* (1993). Perhaps the groundwater contamination incident that received the greatest publicity is the pollution of a source owned by the Cambridge Water Company by chlorinated solvents that originated from a factory operated by Eastern Counties Leather PLC (Misstear *et al.* 1998). The case was heard in the High Court, the Court of Appeal and the House of Lords and has many implications including on the collection of hydrogeological evidence.

Both leakage and spillage from fuel oil use and storage forms another major source of groundwater pollution at filling stations, airports and even from domestic heating fuel tanks. Hydrogeologists have been involved in clean-up operations from such spillages over much of the period. Several case histories are given in Mather *et al.* (1998) on methods used to investigate and clean-up spills.

Nuclear waste

The search for a solution for nuclear waste disposal in the UK has occupied hydrogeologists and others for almost 30 years. The first structured research programme into the disposal of high-level nuclear waste was initiated in 1975 and abandoned in 1981 (Mather 1997). Early nuclear waste disposal was either sea dumping or burial in trenches at Drigg in Cumbria or Dounreay in Scotland. In 1975, the BGS was commissioned to advise on areas with suitable geology for the disposal of high-level wastes. Initial attention was concentrated on a wide range of crystalline rocks. By late 1976 however, a programme to consider the potential of argillaceous rocks and evaporites for such disposal was started. Other programmes were carried out on identifying sites for the disposal of low- and intermediate-level wastes but were abandoned in 1981 due in part, to the difficulty of obtaining planning permission to enable exploratory drilling to be completed.

The Nuclear Industry Radioactive Waste Executive (which was reconstituted as UK Nirex Ltd in 1985) was formed in 1982 to mange Britain's low-level and intermediate-level nuclear wastes with the responsibility for high-level wastes remaining with British Nuclear Fuels Limited (BNFL). Initial Nirex studies were for a shallow repository for radioactive wastes. This programme was abandoned in 1987 again largely due to difficulties caused by public opposition. The option then pursued was for a deep multi-purpose repository to take both low- and intermediate-level wastes. Between 1987 and 1989 Nirex narrowed the search for a potential site to 12 locations, which were then reduced to two using a method of multiple attribute decision analysis.

In 1989 Nirex announced proposals to investigate the suitability of the Dounreay and Sellafield sites for the construction of a deep repository for the disposal of low- and intermediate-level radioactive wastes. Sellafield on the west Cumbria coast has been the site of nuclear activity since the construction of the experimental nuclear generation station at Calder Hall in the early 1950s and is now the site of British Nuclear Fuel's reprocessing plant. Dounreay on the north coast of Caithness has been the site used for fast-breeder reactor experiments. In each case the known geology showed basement rocks at depths of 500 m–1000 m. The initial studies suggested that there is little to distinguish the two sites on the basis of their geology and in July 1991 UK Nirex announced that it was to concentrate further investigations at Sellafield as the nuclear reprocessing plant at this site generates some 60% of the wastes destined for disposal and disposal at this site reduces risks of further transport (Chapman & McEwen 1991; Chaplow 1996).

The geological investigations at Sellafield have included cores from 20 deep boreholes totalling 20000 m depth supplemented by extensive seismic surveys and airborne geophysical surveys. Extensive hydrogeological testing has been undertaken to enable a conceptual model to be developed and the groundwater system to be modelled using numerical methods. It is considered by some that this has been the most extensive, detailed and costly hydrogeological investigation to be carried out in Britain. The investigations show that the Borrowdale Volcanic Group forms the bedrock with the upper surface some 400–600 m below ground level. Tests show that groundwater flow occurs through fractures but further work is needed before the flow can be characterized in sufficient detail for a decision to be made on the construction of a repository at Sellafield (Nirex 1993; Chaplow 1996). UK Nirex applied for planning permission to construct an underground testing laboratory but this was refused in 1997 bringing the programme of investigations to an abrupt end. Since then the government has not decided on an alternative disposal strategy although a consultative document was published in late 2001 (DEFRA 2001).

The strategy for deep disposal is fraught with political difficulty and the issues have been aired recently by Lord Oxburgh in his presidential address to the Geological Society (Oxburgh 2002). Whatever the long-term outcome, the hydrogeological investigations into nuclear waste disposal provide a good example of how government decisions have had a fundamental influence on hydrogeological projects and the professional careers of many individual hydrogeologists.

Groundwater protection

The level of concern regarding groundwater contamination from the agricultural use of chemical fertilizers and from waste disposal and other industrial activities encouraged several water authorities to develop their own groundwater protection policies. Dowse & Selby (1975) reported on a range of industry-related contamination sources in part of the Trent River basin and recognized the threat to existing public supply groundwater sources. Following these concerns, the Severn-Trent Water Authority became the first to publish and implement a groundwater protection policy in 1976. The policy document defined protection zones around groundwater sources in which certain potentially contaminating developments would be opposed by the water authority (Selby & Skinner 1979). In 1978 the Southern Water Authority published its Policy Guidelines, which were adopted as a policy in 1985 (Southern Water Authority 1985). The Severn-Trent approach specified fixed 1 km diameter protection zones around major sources of supply and also specified additional protection to the rest of their area by defining major aquifers, minor aquifers and non-aquifers. The Southern Water policy developed this approach by defining different-sized protection zones based on the nature of the aquifer and the size of the abstraction and varying in size from 0.5–2.5 km.

In 1992 the National River Authority published its Groundwater Protection Policy (Environment Agency 1998) that built on these ideas with protection zones defined for all public water supply sources and major sources used in food manufacture. These zones are based on a standard assessment of the travel times for water to flow through the groundwater system (Harris 1998). The methodology is predominantly based on computer models of groundwater flow except in the more complex cases or where the available data are severely limited, when the assessment is based on a conceptual model and simplified calculations. The catchment for each groundwater source is divided into an inner zone (50 day travel time), an outer zone (400 day travel time) and the total borehole catchment. Source Protection Zones (SPZs) have been defined for more than 2000 sources with maps showing their location available on the Environment Agency's website. A series of policy statements are set out in the Groundwater Protection Policy document that defines potentially risky activities (such as landfill) that the Agency will oppose within each zone or allow subject to certain conditions. This approach means that potential developers and planners have a clear understanding of the Agency's attitude before any new proposals are sent to them thereby reducing the number of unsuccessful proposals made. Each proposal is considered by the Agency on an individual basis within the limitations of the policy.

In addition to considerations of the groundwater source, the policy also seeks to reduce the risks of groundwater contamination by defining aquifers in different vulnerability categories. A simple classification is used of Major, Minor and Non-Aquifers (a misleading term as these rocks usually contain groundwater and often support very small groundwater supplies). A series of policy statements have been drawn up for each aquifer type setting out activities that the Agency would not wish to see in a similar way to the source protection zones. To aid them in this work the Environment Agency have had a series of some 53 vulnerability maps drawn up based on soil types and covering England and Wales.

The Groundwater Protection Policy is considered by many to be the first land-use planning tool to cross the technical/planning boundary. In 1995, the Chartered Institution of Water and Environmental Management (CIWEM) presented their Centenary Award to the NRA team that developed the policy in recognition of its significance to groundwater protection and in setting standards on the international scene.

The Groundwater Protection Policy was adopted by Environment Agency in 1996 and the same principles have been used by the Scottish Environment Protection Agency (SEPA 1998; Fox 2000).

Geothermal energy

Increasing oil prices during the mid-1970s led to the investigation of a number of alternative sources of energy including the geothermal potential in the UK. The UK research programme has been summarized by Downing *et al.* (1991) and Barker *et al.* (2000) who provide a comprehensive list of the relevant reports. BGS was commissioned to assess the UK potential by initiating a three-fold investigation that included a study of the special variation of heat flow, an assessment of the potential for low enthalpy, direct-heat applications of hot groundwaters, and the investigation of high enthalpy resources associated with granites that act as Hot Dry Rock (HDR) reservoirs using the circulation of injected water. Parts of the programme were sub-contracted to Oxford University, Imperial College of Science and Technology, the Open University and Bath University. The Camborne School of Mines undertook a separate project to research the rock mechanics of the HDR process based on the Carnmenellis Granite in Cornwall.

The HDR project in Cornwall involved three boreholes drilled to a depth of 2 km with the successful development of fractures through which water was circulated although there are serious technical difficulties to be overcome before the technique can be exploited. Future work is expected to be on a collaborative basis involving French, Japanese and British teams.

The recent investigations have identified the Sherwood Sandstone aquifer as having the greatest potential for low enthalpy geothermal energy production. The identified resource is almost 70×10^{18}J, equivalent to over 2500 million tonnes of coal. Practical considerations will limit the ability for these resources to be exploited and it is estimated that it is unlikely that more than 50 schemes will be developed in the UK. A scheme at Southampton provides district heating with an energy output of 2000–12000 kilowatt.

Although low enthalpy resources exist in the Upper Palaeozoic aquifers development is frustrated by the difficulty in predicting the extensive fracture systems at depths necessary to give suitably high temperatures. Some of the natural discharges from these systems have been exploited as spa waters at Bath, Buxton and Matlock.

Scotland and Northern Ireland

Public water supplies in both Scotland and Northern Ireland are largely based on surface water sources in contrast to SE England. As a result, groundwater in these two countries was given little attention before the mid-1970s (Robins *et al.* 2004). Groundwater is used extensively for private supplies in rural areas with great reliance on springs and increasingly on boreholes.

Some of the first reports were published by BGS

on the groundwater resources of the Ulster valley gravels (Price & Foster 1974) and groundwater in the Devonian and Carboniferous rocks of Fife and Kinross (Foster *et al.* 1976). In implementing the Control of Pollution Act 1974, only the Forth Rivers Purification Board appointed a hydrogeologist working on problems of discharges from abandoned coal mines (Henton 1979), and landfill in the Forth area. Occasionally she was also asked to work on sites in other parts of Scotland on behalf of other rivers purification boards. BGS appointed a hydrogeologist in its Belfast office in 1974 and in its Edinburgh office in 1977. A hydrogeological map of Scotland was published in 1984 and a report on the country's hydrogeology four years later (Robins 1990). The hydrogeological map of Northern Ireland was published in 1994 and its companion report two years later (Robins 1996).

Hydrogeological work is largely associated with impact assessments for new developments, waste disposal and contamination issues and is largely carried out by consultants. Groundwater matters are a research interest at the Macaulay Institute, the Scottish Agricultural College and at Queen's University, Belfast. Recent projects in Scotland and Northern Ireland have been reported by Robins & Misstear (2000).

The water industry in Scotland has not been privatized along the English model and since April 2002 is the responsibility of one multi-functional water authority (Scottish Water). Responsibilities for environmental regulation lie with the Scottish Environment Protection Agency (SEPA), formed in 1996. Water resources are not managed through a licensing system in Scotland although SEPA plans to introduce controls in areas of high demand. In Northern Ireland responsibilities for water supply and environmental regulation lie with the Department of the Environment for Northern Ireland.

The impact of electronics and computing

As with most scientific endeavour over this period, the rapid progress in electronic engineering to develop the personal computer and electronic equipment to measure and store field data, coupled with the associated advances in computing have enabled great leaps in improved understanding of hydrogeological processes. Jones & Brassington (1991) discuss the development of these trends during the 1980s that have continued to the present day. Field instruments are now smaller, more reliable, and can measure an ever-increasing range of parameters. The large growth in computing power available from desktop machines also means that increasingly large data sets can be examined and greater value obtained from them in terms of better understanding. In addition, sophisticated groundwater models can be used in a Windows environment on desktop computers and are becoming a tool available for routine work rather than the special investigations where they were employed in the past.

Early groundwater modelling techniques used can be grouped into four types: porous media methods using sand tanks; miscellaneous analogue models using viscous fluids and membranes; electrical analogue models based on the similarity of Ohm's law and Darcy's law; and digital computer models (Prickett 1975, 1979). The first three categories of modelling involved lengthy activities to construct a purpose made model to represent a particular system so that the model development could take several years to complete. Electrical analogue models were the more usual to be employed and these required a network of electronic components to be built which effectively produced a computer dedicated to a particular aquifer (Rushton & Herbert 1966). The evolution of digital models from the analogue technique involved a phase where computers were used to control the analogue electronic network (Rushton & Redshaw 1979). The development of increasingly more powerful computers has meant that digital computer models eventually became the standard approach in most hydrogeological assessments. The USGS computer code MODFLOW (McDonald & Harbaugh 1984) is considered to be the most widely used groundwater modelling software in the world. It is now available in a Windows format with front-end and back-end programmes that allows it to be used for a wide range of applications on a desktop computer. In addition, there is a wide range of other computer codes to simulate the migration of solutes in contamination studies as well as modelling the complex interactions of groundwater and surface water systems.

Foster (2000) discusses the potential benefits of technological innovation in agricultural practice that may have the effect of reducing diffuse groundwater pollution. These include microcomputer-controlled sensors to manage the application of agro-chemicals, the design of new pesticides that are less mobile in groundwater systems and more controversially, the development of GM crops that are more effective in nutrient uptake and require less harmful pesticides.

Basin studies

A comprehensive understanding of all the geological processes involved in the evolution of individual sedimentary basins provides an important background on availability of mineral resources including fossil fields and groundwater to aid their economic exploitation (Galloway & Hobday 1996). Over the past few years a number of comprehensive studies have been completed by BGS examining the development of sedimentary basis in terms of the geology and natural resources. Downing & Penn (1992) examined the influence of groundwater circulation in the movement of hydrocarbon resources in the Wessex Basin and identified periods of compaction-dominated flow during the Jurassic and late Cretaceous in contrast with gravity-induced flow at other times. They showed that oil migration from Lias and Oxford Clay source rocks took place at these times.

As part of a broader study (Plant & Jones 1991) and Barker *et al.* (1999) discuss the role of fluid flow in Palaeozoic rocks of northern England during Carboniferous and Permian times in the mineralization that produced the ore field of the North Pennines and Cumbria.

A detailed assessment of the Cheshire Basin (Plant *et al.* 1999) also included the role of groundwater and brine movement as part of the basin evolution and the development of the modern freshwater system that is used extensively for water supply.

The results have recently been published of an EU-funded research programme into the evolution of groundwater since the late Pleistocene (Edmunds & Milne 2001). The PALAEAUX project was carried out by scientists from eight EU states and Switzerland as part of the Environment and Climate Programme (1994–1998). This programme demonstrated that fresh groundwater reserves that were recharged during lowered sea-level conditions associated with glacial periods are found to greater depths in some areas of Europe than was previously realized. Such groundwater resources are often of very high quality and are therefore of strategic importance for the water supplies in coastal areas where population fluctuations and the related demands for water are often seasonal. In view of their excellent quality the use of such waters warrants careful management and husbanding.

Training, professional development and the working environment

An increasing need for trained hydrogeologists has seen the growth of postgraduate courses in a number of universities. The first taught MSc in hydrogeology was initiated at the University College in the University of London (UCL) in 1965 under the leadership of Glyn Jones, a hydrogeologist who had previously worked for the BGS. A second MSc course was established at Birmingham University in 1971. Since then other MSc courses have been established in groundwater engineering at the University of Newcastle upon Tyne in 1984, and in hydrogeology at the Universities of East Anglia (UEA) in 1992 and at Reading in the same year. Undergraduate teaching in hydrogeology is available at these universities and a number of others, notably the University of Bristol where hydrogeology has formed part of the undergraduate options for more than 20 years. University-based research centres were established associated with these MSc courses with a notable exception being the research team concentrating on groundwater contamination established at Bradford and since moved to Sheffield.

Recently changes in the allocation of grants has seen the closure of the MSc courses at UCL and UEA and the course at Reading will close at the end of the 2002/2003 academic year. The consequential reduction in trained hydrogeologists has already created recruitment problems particularly within the Environment Agency.

During the period covered by this paper many British professional geologists supported the creation of an organization to regulate the geological profession in the UK. Although the Geological Society played a fundamental role at the inception of this process, the main development was made by those who actively supported the Association for the Promotion of an Institution of Professional Geologists and its successor, the Institution of Geologists (Brassington 2002). A significant number of those active in these latter organizations were hydrogeologists. The Institution of Geologists reunited with the Society in 1991 and since then the Geological Society has developed its role as the profession's regulator.

The Hydrogeological Group was formed in 1974 as a specialist group of the Geological Society and has been an important focus for the exchange of information between British hydrogeologists since then. Regular scientific meetings have engendered a regular flow of publications both in the Society's journal and as Special Publications (see reference list for examples). The current membership of the Hydrogeological Group is about 1050. The British Chapter of the International Association of Hydrogeologists (IAH) has enjoyed support from British hydrogeologists with a current membership of about 250. The British Chapter has organized a number of international meetings held in the UK and joins with the Hydrogeological Group in sponsoring the annual Ineson Lecture.

The *Transactions* (re-named *Journal* in 1946) of the Institution of Water Engineers (and its successor the Institution of Water Engineers and Scientists (IWES) formed in 1974), and to a lesser extent the *Proceedings of the Institution of Civil Engineers,* were the first journals for the regular (but not frequent) publication of hydrogeological papers in Britain. IWES amalgamated with the Institution of Public Health Engineers and the Institute for Water Pollution Control in 1987 to form the Institution of Water and Environmental Management (renamed the Chartered Institution of Water and Environmental Management in 1995). The *Journal of the Chartered Institution of Water and Environmental Management* continues to contain a number of groundwater related topics as does the *Proceedings of the Institution of Civil Engineers*.

The Geological Society began publication of the *Quarterly Journal of Engineering Geology* in 1967, which included a significant number of hydrogeological papers from its first issue. The number of hydrogeological papers increased making it necessary to publish four special supplements in 1994, 1995 (two supplements) and 1996 entirely devoted to hydrogeology papers. These were in addition to those published in the usual volumes, reflecting the large number submitted for publication. From 2000 the name of the journal was changed to the *Quarterly Journal of Engineering Geology and Hydrogeology* to better reflect its scope.

Other British journals that publish the occasional groundwater related papers are the *Proceedings of the Geologists' Association*, the *Transactions of the Institution of Mining and Metallurgy* and industry-related Journals such as *Waterline*, the journal of the Industrial Water Society. British hydrogeologists also have papers published in the *Journal of Hydrogeology* (formally called *Applied Hydrogeology*) published by the International Association of Hydrogeologists and a number of other foreign international journals such as the *Journal of Hydrology* and *Ground Water*, which is published by the National Water Well Association in the USA.

The period covered by this account has seen the working environment for hydrogeologists evolve from almost exclusively the public sector to one now dominated by consultancy companies of varying sizes. A drive to reduce employment costs in the new water companies and government bodies alike resulted in reduced staff numbers, particularly by redundancies of the older (and hence more experienced) hydrogeologists. In the main, these staff moved to consultancy organizations and continued to provide hydrogeological advice to their former employers. In many cases this loss of specialist experience has been replaced by the use of consultants working on specific projects, as advisers on term contracts or by seconded consultancy staff performing routine tasks.

Hydrogeological work has become very diverse over the period since 1974, a time when water resource evaluation was given the most attention. Hydrogeologists are still employed to help maintain water supplies by developing new sources for water companies, for factory supplies and for private individuals and to assist in improving source operation. The decline in groundwater quality over the past 15 years is continuing to demand hydrogeological input to resolve specific problems. This type of work typifies the trend which has developed over the past three decades for an increasingly more detailed understanding of aquifer behaviour and groundwater flow systems with perhaps the research programme associated with nuclear waste disposal and the LOCAR programme for integrated water resource management being the ultimate examples.

Postscript

The last three decades has seen an exponential growth in the number of hydrogeologists and the fields in which they work, with the total volume of their combined scientific endeavour probably exceeding all that preceded it. Almost all of the routine and research effort has been driven directly or indirectly by political decisions at UK government and EU level, and by the changing needs of business. As the demand for trained hydrogeologists grew it was met by new postgraduate courses based at established research centres. Recent changes in government funding have caused courses to close and a consequential skills shortage that is seen by many to have a significant environmental penalty. Along with most hydrogeological processes, the timescale for the impact of these consequences is likely to be long and therefore difficult to perceive by non-specialists. The development of the Geological Society as the regulator of the geological profession provides a potential focus to advise government and others on the future national need for trained and experienced hydrogeologists and to ensure that these are met.

Personal Note

Like the holocene, the period covered by this account is on-going. Most of the main players are still alive with the majority continuing to pursue active careers. It has been said that a historical period cannot be understood properly until the people who lived through it are dead. It will therefore, remain to future generations to fully evaluate the significance of the scientific advances made by British hydrogeologists during the latter decades of the 20th century. It is hoped that this paper will provide a starting point for that future appraisal.

This brief account cannot do full justice to the large volume of work carried out and inevitably, some contributions have been described only briefly or even omitted, which I regret. I am grateful to many friends and colleagues who have generously provided information, in particular R. Downing, H. Headworth and N. Robins and to L. Clark and J. Mather who provided valuable suggestions. Any opinions expressed are my own and not necessarily those of the Geological Society or its members.

References

ACREMAN, M. C., ADAMS, B., BIRCHALL, P. & CONNORTON, B. 2000. Does groundwater abstraction cause degradation of rivers and wetlands? *Journal of the Chartered Institution of Water Engineers and Scientists*, **14**, 200–206.

ANDREWS, J. N. 1991 Noble gases and radioelements in groundwaters. *In*: DOWNING R.A. & WILKINSON, W.B. (eds) *Applied Groundwater Hydrology.* Clarendon Press, Oxford, 243–265.

ANON. 1978. *South Humberbank Salinity Research Project.* Final Report by the University of Birmingham to the Anglian Water Authority.

ANON. 1981. *Saline Groundwater Investigation: Phase I – the Lower Mersey Basin.* Final Report by the University of Birmingham to the North West Water Authority.

ARDENT, C. R. & GRIFFIN, D. J. H. (eds) 1981. *A Survey of British Hydrogeology 1980.* The Royal Society, London.

BARKER, J. A., DOWNING, R. A., GRAY, D. A., FINDLAY, J., KELLAWAY, G. A., PARKER, R. H. & ROLLIN, K. E. 2000. Hydrogeothermal studies in the United Kingdom. *Quarterly Journal of Engineering Geology and Hydrogeology*, **33**, 41–58

BARKER, J. A., DOWNING, R. A., HOLLIDAY, D. W. & KITCHING, R. 1999. Hydrogeology. *In*: PLANT, J. A. & JONES, D. G. (eds) *Development of regional exploration criteria for buried carbonate–hosted mineral deposits: a multidisciplinary study in Northern England.* Technical Report **WP/91/1**, British Geological Survey, Keyworth, Nottingham, 119–126.

BARNES, B. S. 1939. The structure of the discharge-recession curve. *Transactions of the American Geophysical Union.* **20**, 721–725.

BAXTER, K. M & CLARK, L. 1984. *Effluent Recharge: The Effects of Effluent Recharge on Groundwater Quality.* Water Research Centre Technical Report **TR199**.

BOULTON, N. S. 1951. The flow pattern near a gravity well in a uniform water-bearing medium. *Journal Institution of Civil Engineers*, London, **36**, 534–550.

BOULTON, N. S. 1963. Analysis of data from non-equilibrium pumping tests allowing for delayed yield from storage. *Proceedings of Institution of Civil Engineers*, London, **26**, 469–482.

BOULTON, N. S. & STRELTSOVA, T. D. 1975. New equations for determining the formation constants of an aquifer from pumping test data. *Water resources Research,* **11**, 148–153.

BRANDON, T. W. (ed.) 1986. *Groundwater: Occurrence, Development and Protection.* Institution of Water Engineers and Scientists, London.

BRASSINGTON, F. C. 1990. A review of rising groundwater levels in the United Kingdom. *Proceedings of the Institution of Civil Engineers, Part 1*, **88**, 1037–1057.

BRASSINGTON, F. C. 1992. Present and future use of groundwater resources. *Modern Geology*, **16**, 91–100.

BRASSINGTON, F. C. 2002. *A brief history of the Institution of Geologists.* Geological Society, London, web site: http://www.geolsoc.org.uk/template.cfm?name=IG.

BRASSINGTON F. C. & RUSHTON, K. R. 1987. A rising water table in central Liverpool. *Quarterly Journal of Engineering Geology*, **20**, pp 151–158.

BRASSINGTON, F. C., LUCEY, P. A. & PEACOCK, A. J. 1992. The use of down-hole focused electric logs to investigate saline groundwaters. *Quarterly Journal of Engineering Geology*, **25**, 343–350.

BRIDGMAN, S. A., ROBERTSON, R. M. P., SYED, Q., SPEED, N., ANDREWS, N. & HUNTER P. R. 1995. Outbreak of Cyrptosporidiosis associated with a disinfected groundwater supply. *Epidemic & Infection*, **115**, 555–566.

BUCHAN, S. 1953. Estimation of yield (underground). *Journal Institution of Water Engineers*, **7**, 205–214.

CLAPHAM, D. & FRANKLIN, N. 1998. Cryptosporidium and Giardia Lamblia in Private Water Supplies. *Environmental Health News*, **106**(6).

CHAPLOW, R. 1996. The geology and hydrogeology of Sellafield: an overview. *Quarterly Journal of Engineering Geology*, **29**, S1–S12.

CHAPMAN, N. A. & MCEWEN, T. J. 1991. Deep disposal of nuclear waste in Britain. *In*: DOWNING, R. A. & WILKINSON, W. B. (eds) *Applied Groundwater Hydrology*. Clarendon Press, Oxford, 177–198.

DAY, J. W. B. 1964. *Infiltration into a groundwater catchment and the derivation of evaporation.* Research Report No **2**. Geological Survey of Great Britain.

DEPARTMENT OF THE ENVIRONMENT. 1978. *Co-operative programme of research on the behaviour of hazardous wastes in landfill sites.* Final Report of the Policy Review Committee. HMSO, London.

DEPARTMENT FOR THE ENVIRONMENT, FOOD AND RURAL AFFAIRS. 2001. *Managing Radioactive Waste Safely.* London.

DOWNING, R. A. 1993. Groundwater resources, their development and management in the UK: an historical perspective. *Quarterly Journal of Engineering Geology*, **26**, 335–358.

DOWNING, R. A. 2004*a*. The development of groundwater in the UK between 1935 and 1965 – the role of the Geological Survey. *In*: MATHER, J. D. (ed.) *200 Years of British Hydrogeology*. Geological Society, London, Special Publications, **225**, 271–282.

DOWNING, R. A. 2004*b*. Groundwater in a national water strategy, 1964–1979. *In*: MATHER, J. D. (ed.) *200 Years of British Hydrogeology*. Geological Society, London, Special Publications, **225**, 323–338.

DOWNING, R. A. & GRAY, D. A. 2004. Jack Ineson (1917–1970). The instigator of quantitative hydrogeology in Britain. *In*: MATHER, J.D. (ed.) *200 Years of British Hydrogeology*. Geological Society, London, Special Publications, **225**, 283–286.

DOWNING, R. A. & PENN, I. E. 1992. *Groundwater flow during the development of the Wessex Basin and its bearing on hydrocarbon and mineral resources.* Research Report **SD/91/1**, British Geological Survey, Keyworth, Nottingham.

DOWNING, R. A. & WILKINSON, W. B. (eds) 1991. *Applied Groundwater Hydrology – A British Perspective.* Clarendon Press, Oxford.

DOWNING, R. A., ASHFORD, P. L., HEADWORTH, H. G., OWEN, M. & SKINNER, A. C. 1981. The use of groundwater for river augmentation. *In*: ARDENT, C. R. & GRIFFIN, D. J. H. (eds) *A Survey of British Hydrogeology 1980.* The Royal Society, London, 153–171.

DOWNING, R. A., EASTWOOD, W. & RUSHTON, K. R. 2004.

Norman Savage Bolton (1899–1984): civil engineer and groundwater hydrologist. *In*: MATHER, J.D. (ed.) *200 Years of British Hydrogeology*. Geological Society, London, Special Publications, **225**, 000–000.

DOWNING, R. A., PARKER, R. H. & GRAY, D. A. 1991. Geothermal energy in the United Kingdom. *In*: DOWNING, R. A. & WILKINSON, W. B. (eds) *Applied Groundwater Hydrology*, Clarendon Press, Oxford, 283–301.

DOWNING, R. A., SMITH, D. & WARREN, S. C. 1978. Seasonal variations of tritium and other constituents in groundwater in the Chalk near Brighton, England. *Journal of the Institution of Water Engineers and Scientists*, **32**, 123–136.

DOWSE, L. H. & SELBY, K. H. 1975. Groundwater pollution control in an industrialised part of the Trent basin. *Water Pollution Control*, **74**, 526–543.

DUPUIT, J. 1863. *Études théoriques et pratiques sur la movement des eaux dans le canaux découverts et à travers les terrains perméables*, 2nd edition. Dunod, Paris.

EASTWOOD, J. C. & STANSFIELD, P. J. 2001. Key success factors in an ASR scheme. *Quarterly Journal of Engineering Geology and Hydrogeology*, **34**, 399–409.

EDMUNDS, W. M. 1971. *Hydrogeochemistry of groundwaters in the Derbyshire Dome with special reference to trace constituents*. Institute of Geological Sciences, Report **71/7**.

EDMUNDS, W. M. & BATH, A. H. 1976. Centrifuge extraction and chemical analysis of interstitial waters. *Environmental Science Technology*, **10**, 467–472.

EDMUNDS, W. M. & MILNE, C. J. 2001. *Palaeowaters in Coastal Europe: evolution of groundwater since the late Pleistocene*. Geological Society, London, Special Publications, **189**, 332 pp.

EDMUNDS, W. M. & MORGAN-JONES, M. 1976. Geochemistry of groundwaters in British Triassic sandstones: the Wolverhampton- east Shropshire area. *Quarterly Journal of Engineering Geology*, **9**, 73–101.

EDMUNDS, W. M., BATH, A. H. & MILES, D. L. 1982. Hydrochemical evolution of the East Midlands Triassic sandstone aquifer, England. *Geochimica Cosmochimica Acta*, **46**, 2069–2081.

EDWORTHY, K. J., HEADWORTH, H. G. & HAWNT, R. J. E. 1981. Application of artificial recharge techniques in the United Kingdom. *In*: ARDENT, C. R. & GRIFFIN, D. J. H. (eds) *A Survey of British Hydrogeology 1980.* The Royal Society, London, 141–152.

ENVIRONMENT AGENCY. 1998. *Policy and Practice for the Protection of Groundwater*. HMSO, London.

ENVIRONMENT AGENCY, 2002*a*. *Guidance on the monitoring of landfill leachate, groundwater and surface water*. Technical Report **P1–347/TR**.

ENVIRONMENT AGENCY, 2002*b*. *Effect of old landfill on groundwater quality – Phase 2 – investigation of the Thriplow landfill 1996–1997*. Technical Report **P201**.

EVANS, H. R. 1993. The structure and management of the British water industry 1945–91. *In*: *Institution of Water and Environmental Management Yearbook 1993*. Institution of Water and Environmental Management, 64–71.

FLAVIN, R. J. & JOSEPH, J. B. 1983. The hydrogeology of the Lee Valley and some effects of artificial recharge. *Quarterly Journal of Engineering Geology*, **16**, 65–82.

FOSTER, S. S. D. 2000. Assessing and controlling the impacts of agriculture on groundwater – from barley barons to beef bans. *Quarterly Journal of Engineering Geology and Hydrogeology*, **33**, 267–280.

FOSTER, S. S. D. & CREASE, R. I. 1974. Nitrate pollution of groundwater in East Yorkshire – a hydrogeological appraisal. *Journal Institution of Water Engineers*, **28**, 178–194.

FOSTER, S. S. D. & YOUNG, C. P. 1981. Effects of agricultural land-use on groundwater quality with special reference to nitrate. *In*: ARDENT, C. R. & GRIFFIN, D. J. H. (eds) *A Survey of British Hydrogeology 1980.* The Royal Society, London, 47–60.

FOSTER, S. S. D., STIRLING, W. G. N. & PATERSON, I. B. 1976. *Groundwater storage in Fife and Kinross – its potential as a regional resource*. Institute of Geological Sciences, Report No **76/9**.

FOX, I. A. 2000. Groundwater protection in Scotland. *In*: ROBINS, N. S. & MISSTEAR, B. D. R. (eds) *Groundwater in the Celtic Regions: Studies in Hard Rock and Quaternary Hydrogeology*. Geological Society, London, Special Publication, **182**, 67–70.

GALLAGHER, N. J. & BRASSINGTON, F. C. 1994. Merseyside groundwater – a regional overview and specific example. *In*: WILKINSON, W. B. (ed.) *Groundwater Problems in Urban Areas*. Institution of Civil Engineers, London, 310–329.

GALLOWAY, W. E. & HOBDAY, D. K. 1996. *Terrigenous Clastic Depositional Systems – Applications to Fossil Fuel and Groundwater Resources*. Springer-Verlag, Berlin.

GOMME, J., SHURVELL, S. HENNINGS, S. M. & CLARK, L. 1992. Hydrology of pesticides in a Chalk catchment: groundwaters. *Journal of the Institution of Water & Environmental Management*, **6**, 172–178.

GRAY, D. A., MATHER, J. D. & HARRISON, I. B. 1974. Review of groundwater pollution from wastes disposal sites in England and Wales with provisional guidelines for future site selection. *Quarterly Journal of Engineering Geology*, **7**, 181–196.

GRINDLEY, J. 1967. The estimation of soil moisture deficits. *Meteorological Magazine*, **76**, 97–108.

HARRIS, R. C. 1998. Protection of groundwater quality in the UK: present controls and future issues. *In*: MATHER, J. D., BANKS, D., DUMPLETON, S. & FERMOUR, M. (eds) *Groundwater Contaminants and their Migration*. Geological Society, London, Special Publication, **128**, 3–14.

HARRIS, R. C. & LOWE, D. R. 1984. Changes in the organic fraction of leachate from two domestic refuse sites on the Sherwood Sandstone, Nottinghamshire. *Quarterly Journal of Engineering Geology*, **17**, 57–69.

HARRIS, R. C. & PARRY, E. L. 1982. Investigations into domestic refuse leachate attenuation in the unsaturated zone of Triassic sandstones. *In*: *Effects of Waste Disposal on Groundwater and Surface Waters*. International Association of Hydrological Sciences, Publication No **139**, 147–155.

HAWNT, R. J. E., JOSEPH, J. B. & FLAVIN, R. J. 1981. Experience with borehole recharge in the Lee Valley. *Journal Institution of Water Engineers*, **35**, 437–451.

HEADWORTH, H. G. 1970. The selection of root constants for the calculation of actual evaporation and infiltration for Chalk catchments. *Journal Institution of*

Water Engineers and Scientists, **24**, 569–574.

HEADWORTH, H.G. 1994. *The Groundwater Schemes of Southern Water 1970–1990*. Southern Water, Worthing.

HEADWORTH, H. G. 2004. Recollections of a Golden Age: The Groundwater Schemes of Southern Water 1970–90. *In*: MATHER, J.D. (ed.) *200 Years of British Hydrogeology*. Geological Society, London, Special Publications, **225**, 339–362.

HEADWORTH, H. G. & FOX, G. B. 1985. The use and management of the coastal Chalk aquifer of the South Downs, Sussex, UK. In: *Hydrogeology in the Service of Man*. Memoirs of the 18th Congress of the International Association of the international Association of Hydrogeologists, Cambridge.

HEADWORTH, H. G. & FOX, G. B. 1986. The South Downs Chalk aquifer: its development and management. *Journal Institution of Water Engineers and Scientists*, **40**, 345–361.

HEADWORTH, H. G., OWEN, M. & SKINNER, A. C. 1983. River augmentation schemes using groundwater. *British Geologist*, **9**, 50–54.

HEATHCOTE, J. A. & LLOYD, J. W. 1984. Groundwater chemistry in southeast Suffolk, (UK) and its relationship to quaternary geology. *Journal of the Hydrology*, **75**, 143–159.

HENTON, M. P. 1979. Abandoned coalfields, problems of pollution. *Surveyor*, **153** (4538), 9–11.

HIBBERT, E. S. 1956. The hydrogeology of the Wirral Peninsula. *Journal of the Institution of Water Engineers*, **10**, 441–469.

HISCOCK, K. M., RIVETT, M. O. & DAVISON, R. M. (eds) 2002 *Sustainable Groundwater Development*. Geological Society, London, Special Publications, **193**.

INESON, J. 1952. Notes on the theory and formulae associated with pumping tests for the determination of formation constants. *Journal Institution of Water Engineers*, **6**, 433–458.

INESON, J. 1953. Some observations on pumping tests carried out on Chalk wells. *Journal Institution of Water Engineers*, **7**, 215–272.

INESON, J. 1956. Darcy's Law and the evaluation of permeability. *International Association of Hydrological Sciences Symposia Darcy, Dijon*, Pub No **41**, 165.

INESON, J. 1959*a*. The relation between the yield of a discharging well at equilibrium and its diameter, with particular reference to a Chalk well. *Proceedings Institution of Civil Engineers*, **13**, 299–3116.

INESON, J. 1959*b*. Yield-depression curves of discharging wells, with particular reference to Chalk wells, and their relationship to variations in transmissibility. *Journal Institution of Water Engineers*, **2** 119–163.

INESON, J. & DOWNING, R. A. 1963. Changes in the chemistry of groundwaters in the Chalk passing beneath argillaceous strata. *Bulletin of the Geological Survey of Great Britain*, **20**, 176–192.

INESON, J. & DOWNING, R. A. 1964. The groundwater component of river discharge and its relation to hydrology. *Journal Institution of Water Engineers*, **18**, 519–541.

INESON, J. & DOWNING, R. A. 1965. Some hydrogeological factors in permeable catchment studies. *Journal Institution of Water Engineers*, **19**, 59–80.

INESON, J. & ROWNTREE, N. A. F. 1967, Water resources in the UK: conservation projects and planning. *Journal Institution of Water Engineers*, **21**, 275–290.

INESON, J., DOWNING, R. A., GRAY, D. A. & JONES, G. P. 1969. Hydrogeology. *In*: SKEAT, W. O. (ed.) *Manual of British Water Engineering Practice, 4th Edition: Vol. II: Engineering Practice*. Institution of Water Engineers, London, 87–142.

ION, N. J. 1996. *The causes and effects of rising groundwater levels in Merseyside and Manchester.* Unpublished PhD thesis, University of Liverpool.

JONES, G. P. & BRASSINGTON, F. C. 1991. Data collection, storage, retrieval and interpretation. *In*: DOWNING, R. A. & WILKINSON, W. B. (eds) *Applied Groundwater Hydrology*. Clarendon Press, Oxford, 96–113.

KIRK, S. & HERBERT, A. W. 2002. Assessing the impact of groundwater abstractions on river flows. *In*: HISCOCK, K. M., RIVETT, M. O. & DAVISON, R. M. (eds) *Sustainable Groundwater Development*. Geological Society, London, Special Publication, **193**, 211–233.

KNIPE, C. V., LLOYD, J. W., LERNER, D.N. & GRESWELL, R. B. 1993. *Rising groundwater levels in the Birmingham and The engineering implications*. Construction Industry Research and Information Association Special Publication **69**.

LAPWORTH, C. F. 1948. Percolation in the Chalk, *Journal Institution of Water Engineers*, **2**, 97–120.

LERNER, D. N., GOSRK, E., BOURG, A. C. M., BISHOP, P. K., BURSTON, M. W., MOUVET, C., DEGRANGES, P. & JAKOBSEN, R. 1993. Postscript: summary of the Coventry Groundwater Investigation and implications for the future. *Journal of Hydrology*, **149**, 257–272.

LEWIN, K. 1992. The fate of leachate and landfill gas in the Sherwood Sandstone. *In*: *Conference of Groundwater Pollution*. Café Royal, London.

LLOYD, J. W. 1981. Saline groundwater associated with fresh groundwater reserves in the United Kingdom. *In*: ARDENT, C. R. & GRIFFIN, D. J. H. (eds) *A Survey of British Hydrogeology 1980*. The Royal Society, London, 73–84.

LLOYD, J. W., WILLIAMS, G. M., FOSTER, S. S. D, ASHLEY, R. P. & LAWRENCE, A. R. 1991. Rural and agricultural pollution of groundwater. *In*: DOWNING R. A. & WILKINSON, W. B. (eds) *Applied Groundwater Hydrology*. Clarendon Press, Oxford, 149 –163.

LUCAS, H. C. & ROBINSON, V. K. 1995. Modelling the rising groundwater levels in the Chalk aquifer of the London Basin. *Quarterly Journal of Engineering Geology*, **28**, S51–S62.

LYON, A. L. 1949. The hydrogeology of the Coventry District. *Journal Institution of Water Engineers*, **3**, 209–260.

MATHER, J. D. 1968, *A literature survey of the use of radioisotopes in groundwater studies*. Water Supply Paper, **No 1**, Institute of Geological Sciences, London.

MATHER, J. D. 1997. The history of research into radioactive waste disposal in the UK and the selection of a site for detailed investigation. *Environmental Policy & Practice*, **6**, 167–177.

MATHER, J. D. (ed.) 2004. *200 Years of British Hydrogeology*. Geological Society, London, Special Publication, **225**, 000p.

MATHER, J. D. & SMITH D. B. 1973. Thermonuclear tritium – its use as a tracer in local hydrogeological investigations. *Journal Institution of Water Engineers*, **27**, 187–196.

MATHER, J. D. & YOUNG, C. P. 1981. Recent research into groundwater pollution by landfills. *In*: ARDENT, C. R. & GRIFFIN, D. J. H. (eds) *A Survey of British Hydrogeology 1980*. The Royal Society, London, 3–46.

MATHER, J. D., BANKS, D., DUMPLETON, S. & FERMOUR, M. 1998. *Groundwater Contaminants and their Migration*. Geological Society, London, Special Publication, **128**, 368p.

MCDONALD. M. C. & HARBAUGH, A. W. 1984. *A modular three-dimensional finite-difference ground-water flow model*. United States Geological Survey, Washington DC, USA.

MERSEY & WEAVER RIVER AUTHORITY. 1969. *First Periodical Survey of Water Resources – Water Resources Act 1963, Section 14*. Great Sankey, Warrington.

MERSEY & WEAVER RIVER AUTHORITY. 1973. *Statement of Policy for Ground-Water Management in accordance with Proposal 13 of the First Periodical Survey*. Great Sankey, Warrington.

MISSTEAR, B. D. R., ASHLEY, P. W. & LAWRENCE, A. R. 1998. Groundwater pollution by chlorinated solvents: the landmark Cambridge Water Company case, UK. *In*: MATHER, J. D., BANKS, D., DUMPLETON, S. & FERMOUR, M. (eds) *Groundwater Contaminants and their Migration*. Geological Society, London, Special Publication, **128**, 201–216.

MORGAN-JONES, M. 1985. The hydrogeochemistry of the Lower Greensand aquifers south of London, England. *Quarterly Journal of Engineering Geology*, **18**, 443–458.

MORGAN-JONES, M. & EGGBORO, M. D. 1981. The hydrogeochemistry of the Jurassic Limestones in Gloucestershire, England. *Quarterly Journal of Engineering Geology*, **14**, 25– 39.

NATURAL ENVIRONMENTAL RESEARCH COUNCIL. 2003. *LOCAR website. http://www.nerc.ac.uk/funding/thematics/locar/*

NIREX. 1993. *The Geology and Hydrogeology of the Sellafield Area*. Interim Assessment, United Kingdom Nirex Limited, Harwell.

O'SHEA, M. J., BAXTER, K. M. & CHARALAMBOUS, A. N. 1995. The hydrogeology of the Enfield-Haringey artificial recharge scheme, north London. *Quarterly Journal of Engineering Geology*, **28**, S115–S129.

OWEN, M., HEADWORTH, H. G. & MORGAN-JONES, M. 1991. Groundwater in basin management. *In*: DOWNING R. A. & WILKINSON, W. B. (eds). *Applied Groundwater Hydrology*. Clarendon Press, Oxford, 16–34.

OXBURGH, E. R. 2002. Making a meal out of nuclear waste. *Geoscientist*, **12**, 4–13.

PARKER, J. M., YOUNG, C. P. & CHILTON, P. J. 1991. Rural and agricultural pollution of groundwater. *In*: DOWNING, R. A. & WILKINSON, W. B. (eds) *Applied Groundwater Hydrology*. Clarendon Press, Oxford, 149 –163.

PENMAN, H. L. 1948. Natural evaporation from open water, bare soil and grass. *Proceedings of the Royal Society, A*, **193**, 120–146.

PENMAN, H. L. 1950. The water balance of the Stour Catchment. *Journal Institution of Water Engineers*, **4**, 457–469.

PHILIP, L. K., SHIMELL, H., HEWITT, P. J. & ELLARD, H. T. 2002. A field-based test cell examining clay desiccation in landfill liners. *Quarterly Journal of Engineering Geology and Hydrogeology*, **35**, 345–354.

PLANT, J. A. & JONES, D. G. (eds) 1991. *Development of regional exploration criteria for buried carbonate-hosted mineral deposits: a multidisciplinary study in Northern England*. Technical Report **WP/91/1**, British Geological Survey, Keyworth, Nottingham.

PLANT, J. A., JONES, D. G & HASLAM, H. W. 1999. *The Cheshire Basin – Basin evolution, fluid movement and mineral resources in a Permo-Triassic rift setting*. British Geological Survey, Keyworth, Nottingham.

PRICE, M. & FOSTER, S. S. D. 1974. WATER SUPPLY FROM ULSTER valley gravels. *Proceedings of the Institution of Civil Engineers*, **Part 2**, **57**, 451–466.

PRICKETT, T. A. 1975. Modelling techniques for groundwater evaluation. *In*: CHOW, V. T. (ed.) *Advances in Hydroscience*. Academic Press, 1–143.

PRICKETT, T. A. 1979. Ground-water computer models – state of the art. *Ground Water*, **17**, 167–173.

PYNE, R. D. G. 1995. *Groundwater Recharge and Wells – A Guide to Aquifer Storage Recovery*. Lewis Publishers, Boca Raton.

RAYBOULD, J. G. & ANDERSON, D. J. 1987. Migration of landfill gas and its control by grouting – a case history. *Quarterly Journal of Engineering Geology*, **20**, 75–83.

RIVET, M. O., LERNER, D. N. & LLOYD, J. W. 1990. Chlorinated solvents in UK aquifers. *Journal of the Institution of Water & Environmental Management*, **4**, 242–250.

ROBINS, N. S. 1990. *Hydrogeology of Scotland*. British Geological Survey, HMSO, London.

ROBINS, N. S. 1996. *Hydrogeology of Northern Ireland*. British Geological Survey, HMSO, London.

ROBINS, N. S., BENNETT, J. R. P. & CULLEN, K. T. 2004. Groundwater versus surface water in Scotland and Ireland-the formative years. *In*: MATHER, J. D. (ed.) *200 Years of British Hydrogeology*. Geological Society, London, special publications, **225**, 000–000

ROBINS, N. S. & MISSTEAR, B. D. R. (eds) 2000. *Groundwater in the Celtic Regions: Studies in Hard Rock and Quaternary Hydrogeology*. Geological Society, London, Special Publication, **182**, 273.

ROBINSON, H. D. 1996. *A review of the Composition of Leachates from Domestic Wastes in Landfill Sites*. Report No. **CWM 072/95**, Waste Technical Division, the Environment Agency.

RUSHTON, K. R. 2002. Will reductions in groundwater abstractions improve low river flows? *In*: HISCOCK, K. M., RIVETT, M. O. & DAVISON, R. M. (eds) *Sustainable Groundwater Development*. Geological Society, London, Special Publication, **193**, 199–210.

RUSHTON, K. R. & HERBERT, R. 1966. Groundwater flow studies by resistance network. *Geotechnique*, **16**, 264–267.

RUSHTON, K. R. & REDSHAW, S. C. 1979. *Seepage and Groundwater Flow*. John Wiley & Sons, Chichester.

RUSHTON, K. R. & WARD, C. J. 1979. The estimation of groundwater recharge. *Journal of Hydrology*, **41**, 345–361.

SATCHELL, R. L. H. & EDWORTHY, K. J. 1972. Recharge of the Nottingham Bunter Sandstone. *In*: *Symposium on Advanced Techniques in River Basin Management:*

the Trent Model Research Programme. Institution of Water Engineers, London, 75–102.

SCHWILLE, F. 1981. Groundwater pollution in porous media by fluids immiscible with water. *Studies in Environmental Science*, **17**, 451–463.

SELBY, K. H. & SKINNER, A. C. 1979. Aquifer protection in the Severn-Trent Region: policy and practice. *Water Pollution Control*, **78**, 254–269.

SEPA. 1998. *Policy and Practice for the protection of groundwater in Scotland*. Policy No. **19**, Scottish Environment Protection Agency.

SEVERN RIVER AUTHORITY. 1974. *First Periodical Survey of Water Resources – Water Resources Act 1963, Section 14*. Malvern, Worcestershire.

SIMPSON, B., BLOWER, T., CRAIG, R. N. & WILKINSON, W. B. 1989. *The engineering implications of rising groundwater levels in the deep aquifer beneath London*. Construction Industry Research and Information Association, Special Publication **69**.

SIMPSON, R. W. 1991. The international scene – the involvement of British hydrogeologists. *In*: DOWNING R. A. & WILKINSON, W. B. (eds) *Applied Groundwater Hydrology*, Clarendon Press, Oxford, 314–331.

SKEAT, W. O. (ed.) 1969. *Manual of British Water Engineering Practice, 4th edition: Vol. II: Engineering Practice*. Institution of Water Engineers, London.

SMITH, D. B. & RICHARDS, H. J. 1971 Selected Environmental studies using radioactive isotopes. *In*: *Proceedings 4th UN International Conference*. Geneva.

SMITH, D. B., WEARN, P. L., RICHARDS, H. J. & ROWE, P. C. 1970. Water movement in the unsaturated zone of high and low permeability strata using natural tritium. *In*: *Isotope Hydrology*. IAEA, Vienna, 73–87.

SOUTHERN WATER AUTHORITY. 1985. *Aquifer Protection Policy*. Worthing, West Sussex.

SPEARS, D. A. 1987. An investigation of metal enrichment in Triassic Sandstones and porewaters below an effluent spreading site, West Midlands, England. *Quarterly Journal of Engineering Geology*, **20**, 117–129.

TATE, T. K. & ROBERTSON, A. S. 1971. *Investigations into highly saline groundwater at the Woodfield Pumping Station, Wellington, Shropshire*. Institute of Geological Sciences, Research Report No **6**.

THEIS, C. V. 1935. The relation between the lowering of the piezometric surface and the rate and duration of discharge of a well using ground-water storage. *Transactions of the American Geophysical Union*, **16**, 519–524.

THOMSON, N., BARRIE, I. A. & AYLES, M. 1981. *The Meteorological Office Rainfall and Evaporation Calculation System: MORECS*. Hydrological Memorandum No **454**, Meteorological Office, Bracknell, UK.

WATER RESOURCES BOARD. 1973. *Water Resources in England and Wales*. HMSO, London.

WHITWORTH, K. R. 2002. The monitoring and modelling of mine water recovery in UK coalfields. *In*: YOUNGER, P. L. & ROBINS, N. S. (eds) *Mine Water Hydrogeology and Geochemistry*. Geological Society, London, Special Publications, **198**, 61–132.

WILLIAMS, G. M. & AITKENHEAD, N. 1991. Lessons from Loscoe- the uncontrolled migration of landfill gas. *Quarterly Journal of Engineering Geology*, **24**, 191–207.

WILLIAMS, G. M., YOUNG, C. P.& ROBINSON, H. D. 1991. *Landfill disposal of wastes*. *In*: DOWNING, R. A. & WILKINSON, W. B. (eds) *Applied Groundwater Hydrology*. Clarendon Press, Oxford, 114–133.

WILKINSON, W. B. & BRASSINGTON, F. C. 1991. Rising groundwater levels – an international problem. *In*: DOWNING, R. A. & WILKINSON, W. B. (eds). *Applied Groundwater Hydrology*. Clarendon Press, Oxford, 35–53.

WRIGHT, C. E. 1968. *The influence of geology upon river flows with particular reference to the Lothians area*. Unpublished MSc thesis, Loughborough University of Technology.

WRIGHT, C. E. 1975. The assessment of regional groundwater schemes by river flow regression equations. *Journal of Hydrology*, **26**, 209–215.

YOUNGER, P. L. 1993. Possible environmental impacts of the closure of two collieries in County Durham. *Journal of the Institution of Water and Environmental Management*, **7**, 521–531.

Index

Numbers in *italic* indicate figures, numbers in **bold** indicate tables